Lecture Notes in Artificial Intelligence 8724

Subseries of Lecture Notes in Computer Science

T0224979

Toon Calders Floriana Esposito
Eyke Hüllermeier Rosa Meo (Eds.)

Machine Learning and Knowledge Discovery in Databases

European Conference, ECML PKDD 2014
Nancy, France, September 15-19, 2014
Proceedings, Part I

 Springer

Volume Editors

Toon Calders
Université Libre de Bruxelles, Faculty of Applied Sciences
Department of Computer and Decision Engineering
Av. F. Roosevelt, CP 165/15, 1050 Brussels, Belgium
E-mail: toon.calders@ulb.ac.be

Floriana Esposito
Università degli Studi "Aldo Moro", Dipartimento di Informatica
via Orabona 4, 70125 Bari, Italy
E-mail: floriana.esposito@uniba.it

Eyke Hüllermeier
Universität Paderborn, Department of Computer Science
Warburger Str. 100, 33098 Paderborn, Germany
E-mail: eyke@upb.de

Rosa Meo
Università degli Studi di Torino, Dipartimento di Informatica
Corso Svizzera 185, 10149 Torino, Italy
E-mail: meo@di.unito.it

ISSN 0302-9743   e-ISSN 1611-3349
ISBN 978-3-662-44847-2   e-ISBN 978-3-662-44848-9
DOI 10.1007/978-3-662-44848-9
Springer Heidelberg New York Dordrecht London

Library of Congress Control Number: 2014948041

LNCS Sublibrary: SL 7 – Artificial Intelligence

Typesetting: Camera-ready by author, data conversion by Scientific Publishing Services, Chennai, India

Printed on acid-free paper

Springer is part of Springer Science+Business Media (www.springer.com)

Preface

The European Conferences on Machine Learning (ECML) and on Principles and Practice of Knowledge Discovery in Data Bases (PKDD) have been organized jointly since 2001, after some years of mutual independence. Going one step further, the two conferences were merged into a single one in 2008, and these are the proceedings of the 2014 edition of ECML/PKDD. Today, this conference is a world-wide leading scientific event. It aims at further exploiting the synergies between the two scientific fields, focusing on the development and employment of methods and tools capable of solving real-life problems.

ECML PKDD 2014 was held in Nancy, France, during September 15–19, co-located with ILP 2014, the premier international forum on logic-based and relational learning. The two conferences were organized by Inria Nancy Grand Est with support from LORIA, a joint research unit of CNRS, Inria, and Université de Lorraine.

Continuing the tradition, ECML/PKDD 2014 combined an extensive technical program with a demo track and an industrial track. Recently, the so-called Nectar track was added, focusing on the latest high-quality interdisciplinary research results in all areas related to machine learning and knowledge discovery in databases. Moreover, the conference program included a discovery challenge, a variety of workshops, and many tutorials.

The main technical program included five plenary talks by invited speakers, namely, Charu Aggarwal, Francis Bach, Lise Getoor, Tie-Yan Liu, and Raymond Ng, while four invited speakers contributed to the industrial track: George Hébrail (EDF Lab), Alexandre Cotarmanac'h (Twenga), Arthur Von Eschen (Activision Publishing Inc.) and Mike Bodkin (Evotec Ltd.).

The discovery challenge focused on "Neural Connectomics and on Predictive Web Analytics" this year. Fifteen workshops were held, providing an opportunity to discuss current topics in a small and interactive atmosphere: Dynamic Networks and Knowledge Discovery, Interactions Between Data Mining and Natural Language Processing, Mining Ubiquitous and Social Environments, Statistically Sound Data Mining, Machine Learning for Urban Sensor Data, Multi-Target Prediction, Representation Learning, Neural Connectomics: From Imaging to Connectivity, Data Analytics for Renewable Energy Integration, Linked Data for Knowledge Discovery, New Frontiers in Mining Complex Patterns, Experimental Economics and Machine Learning, Learning with Multiple Views: Applications to Computer Vision and Multimedia Mining, Generalization and Reuse of Machine Learning Models over Multiple Contexts, and Predictive Web Analytics.

Nine tutorials were included in the conference program, providing a comprehensive introduction to core techniques and areas of interest for the scientific community: Medical Mining for Clinical Knowledge Discovery, Patterns in Noisy and Multidimensional Relations and Graphs, The Pervasiveness of

Machine Learning in Omics Science, Conformal Predictions for Reliable Machine Learning, The Lunch Is Never Free: How Information Theory, MDL, and Statistics are Connected, Information Theoretic Methods in Data Mining, Machine Learning with Analogical Proportions, Preference Learning Problems, and Deep Learning.

The main track received 481 paper submissions, of which 115 were accepted. Such a high volume of scientific work required a tremendous effort by the area chairs, Program Committee members, and many additional reviewers. We managed to collect three highly qualified independent reviews per paper and one additional overall input from one of the area chairs. Papers were evaluated on the basis of their relevance to the conference, their scientific contribution, rigor and correctness, the quality of presentation and reproducibility of experiments. As a separate organization, the demo track received 24 and the Nectar track 23 paper submissions.

For the second time, the conference used a double submission model: next to the regular conference track, papers submitted to the Springer journals *Machine Learning* (MACH) and *Data Mining and Knowledge Discovery* (DAMI) were considered for presentation in the conference. These papers were submitted to the ECML/PKDD 2014 special issue of the respective journals, and underwent the normal editorial process of these journals. Those papers accepted for the of these journals were assigned a presentation slot at the ECML/PKDD 2014 conference. A total of 107 original manuscripts were submitted to the journal track, 15 were accepted in DAMI or MACH and were scheduled for presentation at the conference. Overall, this resulted in a number of 588 submissions, of which 130 were selected for presentation at the conference, making an overall acceptance rate of about 22%.

These proceedings of the ECML/PKDD 2014 conference contain the full papers of the contributions presented in the main technical track, abstracts of the invited talks and short papers describing the demonstrations, and the Nectar papers. First of all, we would like to express our gratitude to the general chairs of the conference, Amedeo Napoli and Chedy Raïssi, as well as to all members of the Organizing Committee, for managing this event in a very competent and professional way. In particular, we thank the demo, workshop, industrial, and Nectar track chairs. Special thanks go to the proceedings chairs, Élisa Fromont, Stefano Ferilli and Pascal Poncelet, for the hard work of putting these proceedings together. We thank the tutorial chairs, the Discovery Challenge organizers and all the people involved in the conference, who worked hard for its success. Last but not least, we would like to sincerely thank the authors for submitting their work to the conference and the reviewers and area chairs for their tremendous effort in guaranteeing the quality of the reviewing process, thereby improving the quality of these proceedings.

July 2014   Toon Calders
                                                                Floriana Esposito
                                                                Eyke Hüllermeier
                                                                Rosa Meo

Organization

ECML/PKDD 2014 Organization

Conference Co-chairs

Amedeo Napoli   Inria Nancy Grand Est/LORIA, France
Chedy Raïssi   Inria Nancy Grand Est/LORIA, France

Program Co-chairs

Toon Calders   Université Libre de Bruxelles, Belgium
Floriana Esposito   University of Bari, Italy
Eyke Hüllermeier   University of Paderborn, Germany
Rosa Meo   University of Turin, Italy

Local Organization Co-chairs

Anne-Lise Charbonnier   Inria Nancy Grand Est, France
Louisa Touioui   Inria Nancy Grand Est, France

Awards Committee Chairs

Johannes Fürnkranz   Technical University of Darmstadt, Germany
Katharina Morik   University of Dortmund, Germany

Workshop Chairs

Bettina Berendt   KU Leuven, Belgium
Patrick Gallinari   LIP6 Paris, France

Tutorial Chairs

Céline Rouveirol   University of Paris-Nord, France
Céline Robardet   University of Lyon, France

Demonstration Chairs

Ricard Gavaldà   UPC Barcelona, Spain
Myra Spiliopoulou   University of Magdeburg, Germany

Publicity Chairs

Stefano Ferilli   University of Bari, Italy
Pauli Miettinen   Max-Planck-Institut, Germany

Panel Chairs

Jose Balcazar   UPC Barcelona, Spain
Sergei O. Kuznetsov   HSE Moscow, Russia

Industrial Chairs

Michael Berthold   University of Konstanz, Germany
Marc Boullé   Orange Labs, France

PhD Chairs

Bruno Crémilleux   University of Caen, France
Radim Belohlavek   University of Olomouc, Czech Republic

Nectar Track Chairs

Evimaria Terzi   Boston University, USA
Pierre Geurts   University of Liège, Belgium

Sponsorship Chairs

Francesco Bonchi   Yahoo ! Research Barcelona, Spain
Jilles Vreeken   Saarland University/Max-Planck-Institut,
                                      Germany

Proceedings Chairs

Pascal Poncelet   University of Montpellier, France
Élisa Fromont   University of Saint Etienne, France
Stefano Ferilli   University of Bari, Italy

EMCL PKDD Steering Committee

Fosca Giannotti   University of Pisa, Italy
Michèle Sebag   Université Paris Sud, France
Francesco Bonchi   Yahoo! Research Barcelona, Spain
Hendrik Blockeel   KU Leuven, Belgium and Leiden University,
                                      The Netherlands

Katharina Morik   University of Dortmund, Germany
Tobias Scheffer   University of Potsdam, Germany
Arno Siebes   Utrecht University, The Netherlands
Dimitrios Gunopulos   University of Athens, Greece
Michalis Vazirgiannis   École Polytechnique, France
Donato Malerba   University of Bari, Italy
Peter Flach   University of Bristol, UK
Tijl De Bie   University of Bristol, UK
Nello Cristianini   University of Bristol, UK
Filip Železný   Czech Technical University in Prague,
                           Czech Republic
Siegfried Nijssen   LIACS, Leiden University, The Netherlands
Kristian Kersting   Technical University of Dortmund, Germany

Area Chairs

Hendrik Blockeel   KU Leuven, Belgium
Henrik Boström   Stockholm University, Sweden
Ian Davidson   University of California, Davis, USA
Luc De Raedt   KU Leuven, Belgium
Janez Demšar   University of Ljubljana, Slovenia
Alan Fern   Oregon State University, USA
Peter Flach   University of Bristol, UK
Johannes Fürnkranz   TU Darmstadt, Germany
Thomas Gärtner   University of Bonn and Fraunhofer IAIS,
                           Germany
João Gama   University of Porto, Portugal
Aristides Gionis   Aalto University, Finland
Bart Goethals   University of Antwerp, Belgium
Andreas Hotho   University of Würzburg, Germany
Manfred Jaeger   Aalborg University, Denmark
Thorsten Joachims   Cornell University, USA
Kristian Kersting   Technical University of Dortmund, Germany
Stefan Kramer   University of Mainz, Germany
Donato Malerba   University of Bari, Italy
Stan Matwin   Dalhousie University, Canada
Pauli Miettinen   Max-Planck-Institut, Germany
Dunja Mladenić   Jozef Stefan Institute, Slovenia
Marie-Francine Moens   KU Leuven, Belgium
Bernhard Pfahringer   University of Waikato, New Zealand
Thomas Seidl   RWTH Aachen University, Germany
Arno Siebes   Utrecht University, The Netherlands
Myra Spiliopoulou   Magdeburg University, Germany
Jean-Philippe Vert   Mines ParisTech, France
Jilles Vreeken   Max-Planck-Institut and Saarland University,
                           Germany

Marco Wiering   University of Groningen, The Netherlands
Stefan Wrobel   University of Bonn & Fraunhofer IAIS,
                   Germany

Program Committee

Foto Afrati
Leman Akoglu
Mehmet Sabih Aksoy
Mohammad Al Hasan
Omar Alonso
Aijun An
Aris Anagnostopoulos
Annalisa Appice
Marta Arias
Hiroki Arimura
Ira Assent
Martin Atzmüller
Chloe-Agathe Azencott
Antonio Bahamonde
James Bailey
Elena Baralis
Daniel Barbara'
Christian Bauckhage
Roberto Bayardo
Aurelien Bellet
Radim Belohlavek
Andras Benczur
Klaus Berberich
Bettina Berendt
Michele Berlingerio
Indrajit Bhattacharya
Marenglen Biba
Albert Bifet
Enrico Blanzieri
Konstantinos Blekas
Francesco Bonchi
Gianluca Bontempi
Christian Borgelt
Marco Botta
Jean-François Boulicaut
Marc Boullé
Kendrick Boyd
Pavel Brazdil
Ulf Brefeld
Björn Bringmann

Wray Buntine
Robert Busa-Fekete
Toon Calders
Rui Camacho
Longbing Cao
Andre Carvalho
Francisco Casacuberta
Michelangelo Ceci
Loic Cerf
Tania Cerquitelli
Sharma Chakravarthy
Keith Chan
Duen Horng Chau
Sanjay Chawla
Keke Chen
Ling Chen
Weiwei Cheng
Silvia Chiusano
Vassilis Christophides
Frans Coenen
Fabrizio Costa
Bruno Cremilleux
Tom Croonenborghs
Boris Cule
Tomaz Curk
James Cussens
Maria Damiani
Jesse Davis
Martine De Cock
Jeroen De Knijf
Colin de la Higuera
Gerard de Melo
Juan del Coz
Krzysztof Dembczyński
François Denis
Anne Denton
Mohamed Dermouche
Christian Desrosiers
Luigi Di Caro
Jana Diesner

Wei Ding
Ying Ding
Stephan Doerfel
Janardhan Rao Doppa
Chris Drummond
Devdatt Dubhashi
Ines Dutra
Sašo Džeroski
Tapio Elomaa
Roberto Esposito
Ines Faerber
Hadi Fanaee-Tork
Nicola Fanizzi
Elaine Faria
Fabio Fassetti
Hakan
    Ferhatosmanoglou
Stefano Ferilli
Carlos Ferreira
Cèsar Ferri
Jose Fonollosa
Eibe Frank
Antonino Freno
Élisa Fromont
Fabio Fumarola
Patrick Gallinari
Jing Gao
Byron Gao
Roman Garnett
Paolo Garza
Eric Gaussier
Floris Geerts
Pierre Geurts
Rayid Ghani
Fosca Giannotti
Aris Gkoulalas-Divanis
Vibhav Gogate
Marco Gori
Michael Granitzer
Oded Green

Tias Guns
Maria Halkidi
Jiawei Han
Daniel Hernandez
   Lobato
José Hernández-Orallo
Frank Hoeppner
Jaakko Hollmén
Geoff Holmes
Arjen Hommersom
Vasant Honavar
Xiaohua Hu
Minlie Huang
Eyke Hüllermeier
Dino Ienco
Robert Jäschke
Frederik Janssen
Nathalie Japkowicz
Szymon Jaroszewicz
Ulf Johansson
Alipio Jorge
Kshitij Judah
Tobias Jung
Hachem Kadri
Theodore Kalàmboukis
Alexandros Kalousis
Pallika Kanani
U Kang
Panagiotis Karras
Andreas Karwath
Hisashi Kashima
Ioannis Katakis
John Keane
Latifur Khan
Levente Kocsis
Yun Sing Koh
Alek Kolcz
Igor Kononenko
Irena Koprinska
Nitish Korula
Petr Kosina
Walter Kosters
Georg Krempl
Konstantin Kutzkov
Sergei Kuznetsov

Nicolas Lachiche
Pedro Larranaga
Silvio Lattanzi
Niklas Lavesson
Nada Lavrač
Gregor Leban
Sangkyun Lee
Wang Lee
Carson Leung
Jiuyong Li
Lei Li
Tao Li
Rui Li
Ping Li
Juanzi Li
Lei Li
Edo Liberty
Jefrey Lijffijt
shou-de Lin
Jessica Lin
Hsuan-Tien Lin
Francesca Lisi
Yan Liu
Huan Liu
Corrado Loglisci
Eneldo Loza Mencia
Chang-Tien Lu
Panagis Magdalinos
Giuseppe Manco
Yannis Manolopoulos
Enrique Martinez
Dimitrios Mavroeidis
Mike Mayo
Wannes Meert
Gabor Melli
Ernestina Menasalvas
Roser Morante
João Moreira
Emmanuel Müller
Mohamed Nadif
Mirco Nanni
Alex Nanopoulos
Balakrishnan
   Narayanaswamy
Sriraam Natarajan

Benjamin Nguyen
Thomas Niebler
Thomas Nielsen
Siegfried Nijssen
Xia Ning
Richard Nock
Niklas Noren
Kjetil Nørvåg
Eirini Ntoutsi
Andreas Nürnberger
Salvatore Orlando
Gerhard Paass
George Paliouras
Spiros Papadimitriou
Apostolos Papadopoulos
Panagiotis Papapetrou
Stelios Paparizos
Ioannis Partalas
Andrea Passerini
Vladimir Pavlovic
Mykola Pechenizkiy
Dino Pedreschi
Nikos Pelekis
Jing Peng
Ruggero Pensa
Fabio Pinelli
Marc Plantevit
Pascal Poncelet
George Potamias
Aditya Prakash
Doina Precup
Kai Puolamaki
Buyue Qian
Chedy Raïssi
Liva Ralaivola
Karthik Raman
Jan Ramon
Huzefa Rangwala
Zbigniew Raś
Chotirat
   Ratanamahatana
Jan Rauch
Soumya Ray
Steffen Rendle
Achim Rettinger

Fabrizio Riguzzi
Céline Robardet
Marko Robnik Sikonja
Pedro Rodrigues
Juan Rodriguez
Irene Rodriguez-Lujan
Fabrice Rossi
Juho Rousu
Céline Rouveirol
Stefan Rüping
Salvatore Ruggieri
Yvan Saeys
Alan Said
Lorenza Saitta
Ansaf Salleb-Aouissi
Scott Sanner
Vítor Santos Costa
Raul Santos-Rodriguez
Sam Sarjant
Claudio Sartori
Taisuke Sato
Lars Schmidt-Thieme
Christoph Schommer
Matthias Schubert
Giovanni Semeraro
Junming Shao
Junming Shao
Pannaga Shivaswamy
Andrzej Skowron
Kevin Small
Padhraic Smyth
Carlos Soares
Yangqiu Song
Mauro Sozio
Alessandro Sperduti
Eirini Spyropoulou
Jerzy Stefanowski
Jean Steyaert
Daniela Stojanova
Markus Strohmaier
Mahito Sugiyama

Johan Suykens
Einoshin Suzuki
Panagiotis Symeonidis
Sandor Szedmak
Andrea Tagarelli
Domenico Talia
Pang Tan
Letizia Tanca
Dacheng Tao
Nikolaj Tatti
Maguelonne Teisseire
Evimaria Terzi
Martin Theobald
Jilei Tian
Ljupco Todorovski
Luis Torgo
Vicenç Torra
Ivor Tsang
Panagiotis Tsaparas
Vincent Tseng
Grigorios Tsoumakas
Theodoros Tzouramanis
Antti Ukkonen
Takeaki Uno
Athina Vakali
Giorgio Valentini
Guy Van den Broeck
Peter van der Putten
Matthijs van Leeuwen
Maarten van Someren
Joaquin Vanschoren
Iraklis Varlamis
Michalis Vazirgiannis
Julien Velcin
Shankar Vembu
Sicco Verwer
Vassilios Verykios
Herna Viktor
Christel Vrain
Willem Waegeman
Byron Wallace

Fei Wang
Jianyong Wang
Xiang Wang
Yang Wang
Takashi Washio
Geoff Webb
Jörg Wicker
Hui Xiong
Jieping Ye
Jeffrey Yu
Philip Yu
Chun-Nam Yu
Jure Zabkar
Bianca Zadrozny
Gerson Zaverucha
Demetris Zeinalipour
Filip Železný
Bernard Zenko
Min-Ling Zhang
Nan Zhang
Zhongfei Zhang
Junping Zhang
Lei Zhang
Changshui Zhang
Kai Zhang
Kun Zhang
Shichao Zhang
Ying Zhao
Elena Zheleva
Zhi-Hua Zhou
Bin Zhou
Xingquan Zhu
Xiaofeng Zhu
Kenny Zhu
Djamel Zighed
Arthur Zimek
Albrecht Zimmermann
Indre Zliobaite
Blaz Zupan

Demo Track Program Committee

Martin Atzmueller
Bettina Berendt
Albert Bifet
Antoine Bordes
Christian Borgelt
Ulf Brefeld
Blaz Fortuna

Jaakko Hollmén
Andreas Hotho
Mark Last
Vincent Lemaire
Ernestina Menasalvas
Kjetil Nørvåg
Themis Palpanas

Mykola Pechenizkiy
Bernhard Pfahringer
Pedro Rodrigues
Jerzy Stefanowski
Grigorios Tsoumakas
Alice Zheng

Nectar Track Program Committee

Donato Malerba
Dora Erdos
Yiannis Koutis

George Karypis
Louis Wehenkel
Leman Akoglu

Rosa Meo
Myra Spiliopoulou
Toon Calders

Additional Reviewers

Argimiro Arratia
Rossella Cancelliere
Antonio Corral
Joana Côrte-Real
Giso Dal
Giacomo Domeniconi
Roberto Esposito
Pedro Ferreira
Asmelash Teka Hadgu
Isaac Jones
Dimitris Kalles
Yoshitaka Kameya
Eamonn Keogh
Kristian Kersting
Rohan Khade
Shamanth Kumar
Hongfei Li

Elad Liebman
Babak Loni
Emmanouil Magkos
Adolfo Martínez-Usó
Dimitrios Mavroeidis
Steffen Michels
Pasquale Minervini
Fatemeh Mirrashed
Fred Morstatter
Tsuyoshi Murata
Jinseok Nam
Rasaq Otunba
Roberto Pasolini
Tommaso Pirini
Maria-Jose
    Ramirez-Quintana
Irma Ravkic

Kiumars Soltani
Ricardo Sousa
Eleftherios
    Spyromitros-Xioufis
Jiliang Tang
Eleftherios Tiakas
Andrei Tolstikov email
Tiago Vinhoza
Xing Wang
Lorenz Weizsäcker
Sean Wilner
Christian Wirth
Lin Wu
Jinfeng Yi
Cangzhou Yuan
Jing Zhang

Sponsors

Gold Sponsor
Winton   http://www.wintoncapital.com

Silver Sponsors
Deloitte   http://www.deloitte.com
Xerox Research Centre Europe   http://www.xrce.xerox.com

Bronze Sponsors
EDF   http://www.edf.com
Orange   http://www.orange.com
Technicolor   http://www.technicolor.com
Yahoo! Labs   http://labs.yahoo.com

Additional Supporters
Harmonic Pharma   http://www.harmonicpharma.com
Deloitte   http://www.deloitte.com

Lanyard
Knime   http://www.knime.org

Prize
Deloitte   http://www.deloitte.com
Data Mining and Knowledge
    Discovery   http://link.springer.com/journal/10618
Machine Learning   http://link.springer.com/journal/10994

Organizing Institutions
Inria   http://www.inria.fr
CNRS   http://www.cnrs.fr
LORIA   http://www.loria.fr

Invited Talks Abstracts

Scalable Collective Reasoning Using Probabilistic Soft Logic

Lise Getoor

University of California, Santa Cruz
Santa Cruz, CA, USA
getoor@cs.umd.edu

Abstract. One of the challenges in big data analytics is to efficiently learn and reason collectively about extremely large, heterogeneous, incomplete, noisy interlinked data. Collective reasoning requires the ability to exploit both the logical and relational structure in the data and the probabilistic dependencies. In this talk I will overview our recent work on probabilistic soft logic (PSL), a framework for collective, probabilistic reasoning in relational domains. PSL is able to reason holistically about both entity attributes and relationships among the entities. The underlying mathematical framework, which we refer to as a hinge-loss Markov random field, supports extremely efficient, exact inference. This family of graphical models captures logic-like dependencies with convex hinge-loss potentials. I will survey applications of PSL to diverse problems ranging from information extraction to computational social science. Our recent results show that by building on state-of-the-art optimization methods in a distributed implementation, we can solve large-scale problems with millions of random variables orders of magnitude faster than existing approaches.

Bio. In 1995, Lise Getoor decided to return to school to get her PhD in Computer Science at Stanford University. She received a National Physical Sciences Consortium fellowship, which in addition to supporting her for six years, supported a summer internship at Xerox PARC, where she worked with Markus Fromherz and his group. Daphne Koller was her PhD advisor; in addition, she worked closely with Nir Friedman, and many other members of the DAGS group, including Avi Pfeffer, Mehran Sahami, Ben Taskar, Carlos Guestrin, Uri Lerner, Ron Parr, Eran Segal, Simon Tong.

In 2001, Lise Getoor joined the Computer Science Department at the University of Maryland, College Park.

Network Analysis in the Big Data Age: Mining Graph and Social Streams

Charu Aggarwal

IBM T.J. Watson Research Center, New York
Yorktown, NY, USA
charu@us.ibm.com

Abstract. The advent of large interaction-based communication and social networks has led to challenging streaming scenarios in graph and social stream analysis. The graphs that result from such interactions are large, transient, and very often cannot even be stored on disk. In such cases, even simple frequency-based aggregation operations become challenging, whereas traditional mining operations are far more complex. When the graph cannot be explicitly stored on disk, mining algorithms must work with a limited knowledge of the network structure. Social streams add yet another layer of complexity, wherein the streaming content associated with the nodes and edges needs to be incorporated into the mining process. A significant gap exists between the problems that need to be solved, and the techniques that are available for streaming graph analysis. In spite of these challenges, recent years have seen some advances in which carefully chosen synopses of the graph and social streams are leveraged for approximate analysis. This talk will focus on several recent advances in this direction.

Bio. Charu Aggarwal is a Research Scientist at the IBM T. J. Watson Research Center in Yorktown Heights, New York. He completed his B.S. from IIT Kanpur in 1993 and his Ph.D. from Massachusetts Institute of Technology in 1996. His research interest during his Ph.D. years was in combinatorial optimization (network flow algorithms), and his thesis advisor was Professor James B. Orlin. He has since worked in the field of data mining, with particular interests in data streams, privacy, uncertain data and social network analysis. He has published over 200 papers in refereed venues, and has applied for or been granted over 80 patents. Because of the commercial value of the above-mentioned patents, he has received several invention achievement awards and has thrice been designated a Master Inventor at IBM. He is a recipient of an IBM Corporate Award (2003) for his work on bio-terrorist threat detection in data streams, a recipient of the IBM Outstanding Innovation Award (2008) for his scientific contributions to privacy technology, and a recipient of an IBM Research Division Award (2008) for his scientific contributions to data stream research. He has served on the program committees of most major database/data mining conferences, and served as program vice-chairs of the SIAM Conference on Data Mining, 2007, the IEEE ICDM Conference, 2007, the WWW Conference 2009, and the IEEE ICDM Conference, 2009. He served as an associate editor of the IEEE Transactions on Knowledge and Data Engineering Journal from 2004 to 2008. He is an associate editor of the ACM TKDD Journal, an action editor of the Data Mining and Knowledge Discovery Journal, an associate editor of the ACM SIGKDD Explorations, and an associate editor of the Knowledge and Information Systems Journal. He is a fellow of the ACM (2013) and the IEEE (2010) for contributions to knowledge discovery and data mining techniques.

Big Data for Personalized Medicine: A Case Study of Biomarker Discovery

Raymond Ng

University of British Columbia
Vancouver, B.C., Canada
mg@cs.ubc.ca

Abstract. Personalized medicine has been hailed as one of the main frontiers for medical research in this century. In the first half of the talk, we will give an overview on our projects that use gene expression, proteomics, DNA and clinical features for biomarker discovery. In the second half of the talk, we will describe some of the challenges involved in biomarker discovery. One of the challenges is the lack of quality assessment tools for data generated by ever-evolving genomics platforms. We will conclude the talk by giving an overview of some of the techniques we have developed on data cleansing and pre-processing.

Bio. Dr. Raymond Ng is a professor in Computer Science at the University of British Columbia. His main research area for the past two decades is on data mining, with a specific focus on health informatics and text mining. He has published over 180 peer-reviewed publications on data clustering, outlier detection, OLAP processing, health informatics and text mining. He is the recipient of two best paper awards from 2001 ACM SIGKDD conference, which is the premier data mining conference worldwide, and the 2005 ACM SIGMOD conference, which is one of the top database conferences worldwide. He was one of the program co-chairs of the 2009 International conference on Data Engineering, and one of the program co-chairs of the 2002 ACM SIGKDD conference. He was also one of the general co-chairs of the 2008 ACM SIGMOD conference. For the past decade, Dr. Ng has co-led several large scale genomic projects, funded by Genome Canada, Genome BC and industrial collaborators. The total amount of funding of those projects well exceeded $40 million Canadian dollars. He now holds the Chief Informatics Officer position of the PROOF Centre of Excellence, which focuses on biomarker development for end-stage organ failures.

Machine Learning for Search Ranking and Ad Auction

Tie-Yan Liu

Microsoft Research Asia
Beijing, P.R. China
tyliu@microsoft.com

Abstract. In the era of information explosion, search has become an important tool for people to retrieve useful information. Every day, billions of search queries are submitted to commercial search engines. In response to a query, search engines return a list of relevant documents according to a ranking model. In addition, they also return some ads to users, and extract revenue by running an auction among advertisers if users click on these ads. This "search + ads" paradigm has become a key business model in today's Internet industry, and has incubated a few hundred-billion-dollar companies. Recently, machine learning has been widely adopted in search and advertising, mainly due to the availability of huge amount of interaction data between users, advertisers, and search engines. In this talk, we discuss how to use machine learning to build effective ranking models (which we call learning to rank) and to optimize auction mechanisms. (i) The difficulty of learning to rank lies in the interdependency between documents in the ranked list. To tackle it, we propose the so-called listwise ranking algorithms, whose loss functions are defined on the permutations of documents, instead of individual documents or document pairs. We prove the effectiveness of these algorithms by analyzing their generalization ability and statistical consistency, based on the assumption of a two-layer probabilistic sampling procedure for queries and documents, and the characterization of the relationship between their loss functions and the evaluation measures used by search engines (e.g., NDCG and MAP). (ii) The difficulty of learning the optimal auction mechanism lies in that advertisers' behavior data are strategically generated in response to the auction mechanism, but not randomly sampled in an i.i.d. manner. To tackle this challenge, we propose a game-theoretic learning method, which first models the strategic behaviors of advertisers, and then optimizes the auction mechanism by assuming the advertisers to respond to new auction mechanisms according to the learned behavior model. We prove the effectiveness of the proposed method by analyzing the generalization bounds for both behavior learning and auction mechanism learning based on a novel Markov framework.

Bio. Tie-Yan Liu is a senior researcher and research manager at Microsoft Research. His research interests include machine learning (learning to rank, online learning, statistical learning theory, and deep learning), algorithmic game theory, and computational economics. He is well known for his work on learning to rank for information retrieval. He has authored the first book in this area, and published tens of highly-cited papers on both algorithms and theorems of learning to rank. He has also published extensively on other related topics. In particular, his paper won the best student paper award of SIGIR (2008), and the most cited paper award of the Journal of Visual Communication and Image Representation (2004-2006); his group won the research break-through award of Microsoft Research Asia (2012). Tie-Yan is very active in serving the research community. He is a program committee co-chair of ACML (2015), WINE (2014), AIRS (2013), and RIAO (2010), a local co-chair of ICML 2014, a tutorial co-chair of WWW 2014, a demo/exhibit co-chair of KDD (2012), and an area/track chair of many conferences including ACML (2014), SIGIR (2008-2011), AIRS (2009-2011), and WWW (2011). He is an associate editor of ACM Transactions on Information System (TOIS), an editorial board member of Information Retrieval Journal and Foundations and Trends in Information Retrieval. He has given keynote speeches at CCML (2013), CCIR (2011), and PCM (2010), and tutorials at SIGIR (2008, 2010, 2012), WWW (2008, 2009, 2011), and KDD (2012). He is a senior member of the IEEE and the ACM.

Beyond Stochastic Gradient Descent for Large-Scale Machine Learning

Francis Bach

INRIA, Paris
Laboratoire d'Informatique de l'Ecole Normale Superieure
Paris, France
francis.bach@inria.fr

Abstract. Many machine learning and signal processing problems are traditionally cast as convex optimization problems. A common difficulty in solving these problems is the size of the data, where there are many observations ("large n") and each of these is large ("large p"). In this setting, online algorithms such as stochastic gradient descent which pass over the data only once, are usually preferred over batch algorithms, which require multiple passes over the data. In this talk, I will show how the smoothness of loss functions may be used to design novel algorithms with improved behavior, both in theory and practice: in the ideal infinite-data setting, an efficient novel Newton-based stochastic approximation algorithm leads to a convergence rate of $O(1/n)$ without strong convexity assumptions, while in the practical finite-data setting, an appropriate combination of batch and online algorithms leads to unexpected behaviors, such as a linear convergence rate for strongly convex problems, with an iteration cost similar to stochastic gradient descent.
(joint work with Nicolas Le Roux, Eric Moulines and Mark Schmidt)

Bio. Francis Bach is a researcher at INRIA, leading since 2011 the SIERRA project-team, which is part of the Computer Science Laboratory at Ecole Normale Superieure. He completed his Ph.D. in Computer Science at U.C. Berkeley, working with Professor Michael Jordan, and spent two years in the Mathematical Morphology group at Ecole des Mines de Paris, then he joined the WILLOW project-team at INRIA/Ecole Normale Superieure from 2007 to 2010. Francis Bach is interested in statistical machine learning, and especially in graphical models, sparse methods, kernel-based learning, convex optimization vision and signal processing.

Industrial Invited Talks Abstracts

Making Smart Metering Smarter by Applying Data Analytics

Georges Hébrail

EDF Lab
CLAMART, France
georges.hebrail@edf.fr

Abstract. New data is being collected from electric smart meters which are deployed in many countries. Electric power meters measure and transmit to a central information system electric power consumption from every individual household or enterprise. The sampling rate may vary from 10 minutes to 24 hours and the latency to reach the central information system may vary from a few minutes to 24h. This generates a large amount of - possibly streaming - data if we consider customers from an entire country (ex. 35 millions in France). This data is collected firstly for billing purposes but can be processed with data analytics tools with several other goals. The first part of the talk will recall the structure of electric power smart metering data and review the different applications which are considered today for applying data analytics to such data. In a second part of the talk, we will focus on a specific problem: spatio-temporal estimation of aggregated electric power consumption from in complete metering data.

Bio. Georges Hébrail is a senior researcher at EDF Lab, the research centre of Electricité de France, one of the world's leading electric utility. His background is in Business Intelligence covering many aspects from data storage and querying to data analytics. From 2002 to 2010, he was a professor of computer science at Telecom ParisTech, teaching and doing research in the field of information systems and business intelligence, with a focus on time series management, stream processing and mining. His current research interest is on distributed and privacy-preserving data mining on electric power related data.

Ads That Matter

Alexandre Cotarmanac'h

VP Platform & Distribution
Twenga
alexandre.cotarmanach@twenga.com

Abstract. The advent of realtime bidding and online ad-exchanges has created a new and fast-growing competitive marketplace. In this new setting, media-buyers can make fine-grained decisions for each of the impressions being auctioned taking into account information from the context, the user and his/her past behavior. This new landscape is particularly interesting for online e-commerce players where user actions can also be measured online and thus allow for a complete measure of return on ad-spend.

Despite those benefits, new challenges need to be addressed such as:
- the design of a real-time bidding architecture handling high volumes of queries at low latencies,
- the exploration of a sparse and volatile high-dimensional space,
- as well as several statistical modeling problems (e.g. pricing, offer and creative selection).

In this talk, I will present an approach to realtime media buying for online e-commerce from our experience working in the field. I will review the aforementioned challenges and discuss open problems for serving ads that matter.

Bio. Alexandre Cotarmanac'h is Vice-President Distribution & Platform for Twenga.

Twenga is a services and solutions provider generating high value-added leads to online merchants that was founded in 2006.

Originally hired to help launch Twenga's second generation search engine and to manage the optimization of revenue, he launched in 2011 the affinitAD line of business and Twenga's publisher network. Thanks to the advanced contextual analysis which allows for targeting the right audience according to their desire to buy e-commerce goods whilst keeping in line with the content offered, affinitAD brings Twenga's e-commerce expertise to web publishers. Alexandre also oversees Twenga's merchant programme and strives to offer Twenga's merchants new services and solutions to improve their acquisition of customers.

With over 14 years of experience, Alexandre has held a succession of increasingly responsible positions focusing on advertising and web development. Prior to joining Twenga, he was responsible for the development of Search and Advertising at Orange. Alexandre graduated from Ecole polytechnique.

Machine Learning and Data Mining in Call of Duty

Arthur Von Eschen

Activision Publishing Inc.
Santa Monica, CA, USA
Arthur.VonEschen@activision.com

Abstract. Data science is relatively new to the video game industry, but it has quickly emerged as one of the main resources for ensuring game quality. At Activision, we leverage data science to analyze the behavior of our games and our players to improve in-game algorithms and the player experience. We use machine learning and data mining techniques to influence creative decisions and help inform the game design process. We also build analytic services that support the game in real-time; one example is a cheating detection system which is very similar to fraud detection systems used for credit cards and insurance. This talk will focus on our data science work for Call of Duty, one of the bestselling video games in the world.

Bio. Arthur Von Eschen is Senior Director of Game Analytics at Activision. He and his team are responsible for analytics work that supports video game design on franchises such as Call of Duty and Skylanders. In addition to holding a PhD in Operations Research, Arthur has over 15 years of experience in analytics consulting and R&D with the U.S. Fortune 500. His work has spanned across industries such as banking, financial services, insurance, retail, CPG and now interactive entertainment (video games). Prior to Activision he worked at Fair Isaac Corporation (FICO). Before FICO he ran his own analytics consulting firm for six years.

Algorithms, Evolution and Network-Based Approaches in Molecular Discovery

Mike Bodkin

Evotec Ltd.
Oxfordshire, UK
Mike.Bodkin@evotec.com

Abstract. Drug research generates huge quantities of data around targets, compounds and their effects. Network modelling can be used to describe such relationships with the aim to couple our understanding of disease networks with the changes in small molecule properties. This talk will build off of the data that is routinely captured in drug discovery and describe the methods and tools that we have developed for compound design using predictive modelling, evolutionary algorithms and network-based mining.

Bio. Mike did his PhD in protein de-novo design for Nobel laureate sir James Black before taking up a fellowship in computational drug design at Cambridge University. He moved to AstraZeneca as a computational chemist before joining Eli Lilly in 2000. As head of the computational drug discovery group at Lilly since 2003 he recently jumped ship to Evotec to work as the VP for computational chemistry and cheminformatics. His research aims are to continue to develop new algorithms and software in the fields of drug discovery and systems informatics and to deliver and apply current and novel methods as tools for use in drug research.

Table of Contents – Part I

Main Track Contributions

Table of Contents – Part II

Main Track Contributions

Table of Contents – Part III

Main Track Contributions

Demo Track Contributions

Nectar Track Contributions

Classifying a Stream of Infinite Concepts:
A Bayesian Non-parametric Approach

Seyyed Abbas Hosseini, Hamid R. Rabiee, Hassan Hafez,
and Ali Soltani-Farani

Sharif University of Technology, Tehran, Iran
{a_hosseini,hafez,a_soltani}@ce.sharif.edu
rabiee@sharif.edu

Abstract. Classifying streams of data, for instance financial transactions or emails, is an essential element in applications such as online advertising and spam or fraud detection. The data stream is often large or even unbounded; furthermore, the stream is in many instances non-stationary. Therefore, an adaptive approach is required that can manage concept drift in an online fashion. This paper presents a probabilistic non-parametric generative model for stream classification that can handle concept drift efficiently and adjust its complexity over time. Unlike recent methods, the proposed model handles concept drift by adapting data-concept association without unnecessary i.i.d. assumption among the data of a batch. This allows the model to efficiently classify data using fewer and simpler base classifiers. Moreover, an online algorithm for making inference on the proposed non-conjugate time-dependent non-parametric model is proposed. Extensive experimental results on several stream datasets demonstrate the effectiveness of the proposed model.

Keywords: Stream classification, Concept drift, Bayesian non-parametric, Online inference.

1 Introduction

The emergence of applications such as spam detection [29] and online advertising [1, 23] coupled with the dramatic growth of user-generated content [7, 35] has attracted more and more attention to stream classification. The data stream in such applications is large or even unbounded; moreover, the system is often required to respond in an online manner. Due to these constraints, a common scenario is usually used in stream classification: At each instant, a batch of data arrives at the system. The system is required to process the data and predict their labels before the next batch comes in. It is assumed that after prediction, the true labels of the data are revealed to the system. Also due to limited additional memory the system can only access one previous batch of data and their labels. For example, in online advertising, at each instant, a large number of requests arrive and the system is required to predict for each ad, the probability that

T. Calders et al. (Eds.): ECML PKDD 2014, Part I, LNCS 8724, pp. 1–16, 2014.

it will be clicked by each user. After a short time, the result is revealed to the system and the system can use it to adapt the model parameters.

One of the main challenges of stream classification is that often the process that generates the data is non-stationary. This phenomenon, known as concept drift, poses different challenges to the classification problem. For example, in a stationary classification task, one can model the underlying distribution of data and improve estimates of model parameters as more data become available; but this is not the case in a non-stationary environment. If we can not model the change of the underlying distribution of data, more data may even reduce the model's efficiency. Formally, concept drift between time t_1 and t_2 occurs when the posterior probability of an instance changes, that is [19]:

$$\exists x : p_{t_1}(y|x) \neq p_{t_2}(y|x) \tag{1}$$

When modeling change in the underlying distribution of data, a common assumption is that the data is generated by different sources and the underlying distribution of each source, which is called its concept, is constant over time [19]. If the classification algorithm can find the correct source of each data item, then the problem reduces to an online classification task with stationary distribution, because each concept can be modeled separately. While the main focus in classification literature is on stationary problems, recent methods have been introduced for classification in non-stationary environments [19]. However, these algorithms are often restricted to simple scenarios such as finite concepts, slow concept drift, or non-cyclical environment [16]. Furthermore, usually heuristic rules are applied to update the models and classifiers, which may cause overfitting.

Existing stream classification methods belong to one of two main categories. Uni-model methods use only one classifier to classify incoming data and hence need a forgetting mechanism to mitigate the effect of data that are not relevant to the current concept. These methods use two main approaches to handle concept drift: sample selection and sample weighting [30]. Sample selection methods keep a sliding window over the incoming data and only consider the most recent data that are relevant to the current concept. One of the challenges in these methods is determining the size of the window, since a very large window may cause non-relevant data to be included in the model and a small window may decrease the efficiency of the model by preventing the model from using all relevant data. Sample weighting methods weigh samples so that more recent data have more impact on the classifier parameters [13, 38]. In contrast to uni-model methods, ensemble methods keep a pool of classifiers and classify data either by choosing an appropriate classifier from the pool (model selection) or combining the answers of the classifiers (model combination) to find the correct label [31]. Inspired by the ability of ensemble methods to model different concepts, these models have been used in stream classification with encouraging results [16, 27, 29, 34]. The main problem is that many of these models update the pool of classifiers heuristically and hence may overfit to the data. Moreover, a common assumption among all existing ensemble methods is that all data of a batch are i.i.d. samples of a distribution that are generated from the same source and hence have the same concept. This assumption may cause several problems. For example, since the data of a batch are not necessarily from the same source or may even belong to conflicting concepts, we may not be able to classify them with high accuracy even using complex base classifiers. Moreover, since the diversity of batches of data can be very high, the number of needed base classifiers may become very large.

In this paper, we propose a principled probabilistic framework for stream classification that is impervious to the aforementioned issues and is able to adapt the complexity of the model to the data over time. The idea is to model the data stream using a non-parametric generative model in which each concept is modeled by an incremental probabilistic classifier and concepts can emerge or die over time. Moreover, instead of the restrictive i.i.d. assumption among data of a batch, we assume that the data of a batch are exchangeable which is a much weaker assumption (refer to Section 3.1 for a detailed definition). This is realized by modeling each concept with an incremental probabilistic classifier and using the temporal Dirichlet process mixture model (TDPM) [3]. For inference, we propose a variation of forward Gibbs sampling.

To summarize, we make the following main contributions: (i) We propose a coherent generative model for stream classification. (ii) The model manages its complexity by adapting the size of the latent space and the number of classifiers over time. (iii) The proposed model handles concept drift by adapting data-concept association without unnecessary i.i.d. assumption among data of a batch. (iv) An online algorithm is proposed for inference on the non-conjugate non-parametric time-dependent model.

The remainder of this paper is organized as follows: Section 2 briefly discusses the prior art on this subject. The details of the proposed generative model are discussed in Section 3. To demonstrate the effectiveness of the proposed model, extensive experimental results on several stream datasets are reported and analyzed in Section 4. Finally, Section 5 concludes this paper and discusses paths for future research.

2   Review on Prior Art

Stream classification methods can be categorized based on different criteria. As is mentioned in [19], based on how concept drift is handled the different strategies can be categorized into informed adaptation and blind adaptation. In informed adaptation-based models, there is a separate building block that detects the drift allowing the system to act according to these triggers [8, 22]. However, blind adaptation models adapt the model without any explicit detection of concept drift [19]. In this paper, the focus is on blind adaptation.

Chu et al. proposed a probabilistic uni-model method for stream classification in [13] that uses sample weighting to handle concept drift. This method, uses a probit regression model as a classifier and adaptive density filtering (ADF) [32] to make inference on the model and update it. Probit regression, is a linear classifier with parameter w and prior distribution $N(w; \mu_0, \Sigma_0)$. After observing each new data, the posterior of w is updated and approximated by a Gaussian distribution, that is:

$$w_t \sim N(w_t; \mu_t, \Sigma_t) \tag{2}$$
$$p(y_t|x_t, w_t) = \Phi(y_t w_t^T x_t) \tag{3}$$
$$p(w_{t+1}|x_t, y_t) \propto \Phi(y_t w_t^T x_t) N(w_t; \mu_t, \Sigma_t) \tag{4}$$

In order to decrease the effect of old data, this method introduces a memory loss factor and incorporates the prior of w_t with this factor in computing the posterior of w, that is:

$$p(w_{t+1}|x_t, y_t) \propto \Phi(y_t w_t^T x_t) N(w_t; \mu_t, \Sigma_t)^\gamma \quad 0 \ll \gamma < 1 \tag{5}$$

Using this method, the effect of out-of-date data is reduced as new data arrives into the system. As it is evident in (5), this method forgets old data gradually and hence can not handle abrupt changes in the distribution of data. On the other hand, since sample selection methods only consider the selected data, they easily can handle abrupt drift but they miss the information in the old data that are relevant to the current concept.

As mentioned in Section 1, ensemble methods can be categorized into model selection and model combination methods. Model combination methods assume that each data item is generated by a linear combination of base classifiers and thereby enrich the hypothesis space [15]. There have been different methods for stream classification based on model combination [16, 34]. These methods maintain a pool of classifiers and estimate the label of each datum of batch t by combining base classifiers using:

$$\hat{y}_i^t = \arg\max_c \sum_k W_k^t I_{[h_k(x_i^t)=c]} \tag{6}$$

where W_k^t is the weight of base classifier k for batch t, which is an estimate of its accuracy relative to other classifiers. After observing the true labels of a batch, these methods update the model by adding new classifiers or removing inefficient classifiers, or changing the weights of classifiers.

The main idea of model selection methods, is to find the concept of each data item, hence reducing the problem to an online classification task. The challenge is that finding the concepts of the data is an unsupervised task. There have been different methods to tackle this issue. The simplifying assumption that is common among almost all of these methods is that all of the data of a batch are i.i.d. and hence generated from one concept. For example, [29] uses this assumption and extracts some feature from each batch and finds their concept by clustering the extracted feature vectors. This method assumes that all of the batches that lie in the same cluster, can be classified using a single classifier. This assumption may not be true in many applications, which may decrease the efficiency. Since this method finds the concepts of data by clustering the features that are extracted from the whole batch and the diversity of batches may be very high, the number of clusters may become very large and hence the model can become very complex. Hosseini et al. proposed an improved version of this method in [27]. This method introduces a new distance metric in the feature space together with pool management operations such as splitting or merging of concepts.

There is some prior work on classification using Dirichlet process mixture models [14,24,36]. All of these methods have been designed for classifying batch data and can't be applied to stream classification due to two main reasons. First, these methods does not model the temporal dependency among data and second, the inference algorithm in these methods is offline which is not suitable for stream classification. In order to solve these two issues, we proposed a time-dependent non-parametric generative model and an online inference algorithm based on forward collapsed Gibbs sampling which is an online version of Gibbs sampling [1].

3   Proposed Method

In this section, we introduce our proposed method for stream classification. In order to model concept drift, we propose a non-parametric mixture model with potentially infinite number of mixtures, in which each mixture represents a concept. This model uses a Bayesian model selection approach [31] and assumes that each data item is generated by one concept. Each concept is modeled with a generative classifier. In order to model the change in popularity of concepts over time and their emergence and death, we use TDPM. After observing the true labels of a batch, this model allows the number of concepts and the data-concept associations to be determined by inference, for which we propose an online inference algorithm based on Gibbs sampling.

For clarity, we define the problem setting and notations in Section 3.1. To make the presentation self-sufficient, TDPM is reviewed in Section 3.2. The proposed generative model is described in Section 3.3, followed by the details of the inference algorithm in Section 3.4.

3.1   Problem Setting and Notation

Consider a stream classification problem in which each data item consists of an $l-tuple$ feature vector and a label that associates it to one of C predefined classes. In this setting, data arrive in consecutive batches. The system is required to predict the labels for each batch, after which the true labels are revealed. Moreover due to limited memory, the system only has access to one previous batch of data. In summery, our goal is to classify a stream of data $(D_1, \ldots, D_T)$, where T denotes the number of batches and D_t is the batch of data that arrives at time t. Also, $D_t = (d_{ti})_{i=1}^{n_t}$ where d_{ti} is the ith data item in batch t and n_t is the number of data in that batch. Furthermore, each data item is denoted by (x_{ti}, y_{ti}), where x_{ti} is an $l-tuple$ vector with l_1 discrete features $x_{ti}^{1,\ldots,l_1}$ and l_2 continuous features $x_{ti}^{l_1+1,\ldots,l}$ and $y \in \{1, \ldots, C\}$.

For a general variable z, z_t denotes the set of all z values of batch t and $z_{t,i:j}$ denotes the corresponding z values of data i to j in batch t. Moreover, by z_t^{-i}, we mean all z values of batch t except the i'th one.

3.2   Temporal Dirichlet Process Mixture Model

Suppose that we assume that the data $(x_1, \ldots, x_N)$ are infinitely exchangeable, that is, the joint probability distribution underlying the data is invariant to permutation, then according to De Finetti theorem [26], the joint probability $p(x_1, \ldots, x_N)$ has a representation as a mixture:

$$p(x_1, \ldots, x_N) = \int \left(\prod_{i=1}^{N} p(x_i|G) \right) dP(G) \tag{7}$$

for some random variable G. The theorem needs G to range over measures, in which case $P(G)$ is a measure on measures. The Dirichlet process denoted by $DP(G_0, \alpha)$ is a measure on measures with base measure G_0 and concentration parameter α [17], and hence can be used to model exchangeable data . We write $G \sim DP(G_0, \alpha)$ if G is drawn from a Dirichlet process in which case G itself is a measure on the given parameter space θ. Integrating out G, the parameters θ follow the Chinese Restaurant Process (CRP) [9], in which the probability of redrawing a previously drawn value of θ is strictly positive which makes G a discrete probability measure with probability one; that is:

$$p(\theta_i|\theta_{1:i-1}) = \sum_k \frac{m_k}{i-1+\alpha}\delta(\phi_k) + \frac{\alpha}{i-1+\alpha}G_0 \tag{8}$$

where ϕ_ks are unique θ values and m_k is the number of θ_is having value ϕ_k. The CRP metaphor explains (8) clearly. In this metaphor, there is a Chinese restaurant with infinite number of tables. When a new customer x_i comes into the restaurant, she either sits on one of the previously occupied tables ϕ_k with probability $\frac{m_k}{i-1+\alpha}$ or sits on a new table with probability $\frac{\alpha}{i-1+\alpha}$. Using the Dirichlet process as the prior distribution of a hierarchical model, one obtains the Dirichlet process mixture model (DPM) [6]. As is evident from (8), DPM assumes the data are exchangeable. Since there are temporal dependencies among data in a stream, DPM is not appropriate for modeling.

There are several methods to incorporate temporal dependency in DPM [1,3, 11,37]. In this paper, we focus on TDPM introduced in [3], and use a variation of that in our proposed model for stream classification. TDPM assumes that the stream of data arrives in consecutive batches. Moreover, this model assumes that data are partially exchangeable, i.e., the data that belong to one batch are exchangeable but exchangeability does not hold among batches. A sample G_t drawn from $TDP(G_0, \alpha, \lambda, \Delta)$ is a time dependent probability measure over the parameter space θ:

$$G_t|\phi_{1:k}, G_0, \alpha \sim DP \left(\sum_k \frac{m'_{kt}}{\sum_l m'_{lt} + \alpha}\delta(\phi_k) + \frac{\alpha}{\sum_l m'_{lt} + \alpha}G_0, \alpha + \sum_k m'_{kt} \right) \tag{9}$$

where $\phi_{1:k}$ are the set of unique θ_i values used in recent Δ batches and m'_{kt} is the weighted number of θ_is having value ϕ_k. More formally, if m_{kt} denotes the number of θs in batch t with value ϕ_k, then:

$$m'_{kt} = \sum_{\tau=1}^{\Delta} e^{-\frac{\tau}{\lambda}} m_{k,t-\tau} \qquad (10)$$

As it is evident from Eq. (9), the data in each batch are modeled by a DP and hence it is assumed that they are exchangeable. However, the parameters of these processes evolve over time and are dependent. By marginalizing over G_t, the parameters θ follow the Recurrent Chinese Restaurant Process (RCRP) introduced in [3]:

$$\theta_{t,i}|\theta_{t-\Delta}, \theta_t^{-i}, \alpha, G_0 \propto \sum_k (m'_{kt} + m_{kt})\delta(\phi_k) + \alpha G_0 \qquad (11)$$

According to (11), when customer x_i comes to the restaurant, the probability of choosing table ϕ_k is proportional to the weighted number of customers in previous Δ batches that chose that table and the number of customers in the current batch that chose the same table. In fact, RCRP in (10), uses sample selection and sample weighting to model the evolution of the probability distribution over parameters. Moreover, this process assumes that the data in a batch are only exchangeable and doesn't force them to select the same mixture. Therefore, we use this process as the prior over the parameters of a classifier which uses model selection.

3.3 Infinite Concept Stream Classifier

In this section, we introduce our generative model for stream classification. In order to model concept drift, we propose a Bayesian model selection method [31]. Figure 1 depicts the graphical representation of the proposed generative model. In this graph, observed variables are depicted using shaded nodes and blank nodes represent latent variables. Moreover, arrows are used to represent the dependency among random variables. The plate structure is used to act as a for loop to represent repetition. As it can be seen, there are in total T batches of data, in which batch t, contains n_t data. Each data item consists of a feature vector x which is an observed variable and a latent variable label y which will be revealed to the system after it is estimated. In this model, there are potentially infinite number of concepts. Each concept is in fact a classifier with parameter set ϕ. Moreover, since it is based on model selection, it assumes that each (x_{ti}, y_{ti}) is generated by a single concept, where z_{ti} is the concept indicator. More formally, if data (x_{ti}, y_{ti}) is generated by a classifier with parameters θ_{ti}, then $\theta_{ti} = \phi_{z_{ti}}$. Since the size of data is very large in data streams, it is not possible to keep them all. Therefore, we need an incremental model for the base classifiers. The classifier model that we used in our model is a naive Bayes classifier. In this model, each continuous feature in each class is modeled by a Gaussian distribution and each discrete feature is modeled by a categorical distribution. In order to model the

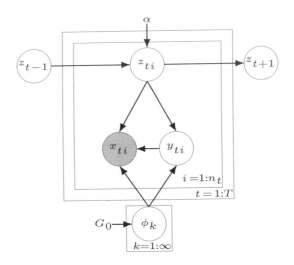

Fig. 1. The Graphical Model of the Proposed Method

temporal dependency among the data of the stream, we used $RCRP(G_0, \alpha, \lambda, \Delta)$ as the prior over concept indicators. The generative process of this model is described in Algorithm 1. According to this process, in order to generate the ith element of batch t, one may either choose one of the existing classifiers that have generated at least one data item in last Δ batches, or a new classifier. The probability of choosing each classifier is determined by RCRP. Furthermore, if a new classifier is needed to generate d_{ti}, then the parameters of the classifiers are obtained by drawing a sample from G_0. In order to make inference easier, we selected G_0 conjugate to the classifier's likelihoods. That is, Dirichlet distribution is used for the prior over class prior probabilities as well as the distribution of discrete features in each class, and Gaussian-Gamma distribution is used for the prior over continuous features in each class.

Indeed, this model assumes that the amount of activity of classifier k at batch t is proportional to the weighted number of data that this model has classified in the last Δ batches and hence uses sample selection and sample weighting concurrently to handle concept drift. Moreover, this model allows the data of a batch to select different concepts. This assumption increases the efficiency of the model in applications where the data of a batch are not necessarily i.i.d. Furthermore, unlike most existent ensemble methods that set the number and the weights of classifiers using heuristic rules, the number of classifiers and their corresponding weights are determined consistently in this method through Bayesian inference on the proposed model. The details of the inference algorithm are discussed next.

3.4 Inference

When a new batch of data arrives at the system, we need to find their labels by estimating the posterior probability of labels given all previously observed data, that is:

Algorithm 1. The proposed Generative Model

---

for all batch $t \in \{1, 2, \ldots, T\}$ **do**
    for all data $i \in \{1, \ldots, n_t\}$ **do**
        Draw $z_{t,i}|z_{1:t-1}, z_t^{-i} \sim RCRP(\alpha, \lambda, \Delta)$
        if $z_{t,i}$ is a new concept **then**
            Draw $\beta_{znew}|G_0 \sim Dir(\pi)$
            for all $c \in \{1, \ldots, C\}$ **do**
                for all $j \in \{1, \ldots, m_1\}$ **do**
                    Draw $\rho_{znew,j,c}|G_0 \sim Dir(\gamma_{j,c})$
                for all $j \in \{1, \ldots, m_2\}$ **do**
                    Draw $\lambda_{znew,j,c}|G_0 \sim Gam(a_{j,c}, b_{j,c})$
                    Draw $\mu_{znew,j,c}|G_0 \sim N(\eta_{j,c}, (\nu_{j,c}\lambda_{znew,j,c})^{-1})$
        Draw $y_{t,i} \sim Cat(y_{t,i}; \beta_{z_{t,i}})$
        Draw $x_{t,i}^{1,\ldots,l_1}|y_{t,i} \sim \prod_{j=1}^{l_1} Cat(x_{t,i}^j; \rho_{z_{t,i},j,y_{t,i}})$
        Draw $x_{t,i}^{l_1+1,\ldots,l}|y_{t,i} \sim \prod_{j=1}^{l_2} N(x_{t,i}^{j+l_1}; \mu_{z_{t,i},j,y_{t,i}}, \lambda_{z_{t,i},j,y_{t,i}}^{-1})$

$$p(y_t|x_t, x_{1:t-1}, y_{1:t-1}, G_0, \alpha, \lambda, \Delta) \qquad (12)$$

This can be done by marginalizing over concept indicators $z_{1:t}$ and concepts' parameters ϕ_ks. However, since the posterior of TDPM is not available in closed form, we need an algorithm to approximate it. Moreover, in stream classification, the algorithm only has access to one previous batch of data and hence, the inference algorithm must be online. Therefore, we approximate the posterior (12) in two phases. First, after observing the true labels of batch $t-1$, we update the model accordingly and then, after batch t is available, we approximate (12) using the updated model.

Several approximate algorithms have been introduced for inference on DPM models [21]. These methods either use Markov Chain Monte Carlo (MCMC) sampling methods [3,33] or Variational methods [10,28] to estimate the posterior distribution of desired latent variables. Moreover, online inference algorithms have been proposed for making inference on TDPM which are based on sequential Monte Carlo estimation [2] or Gibbs sampling [1]. The proposed algorithm for making inference on the proposed model is a variation of forward Gibbs sampling [1] which we explain next.

Generally, the main idea of MCMC estimation is to design a Markov Chain over the desired latent variables in which the equilibrium distribution of the Markov chain is the posterior of the variables [5]. By drawing samples from this Markov chain, one can obtain samples from the posterior of the desired random variables. Gibbs sampling is a widely used variation of MCMC sampling. If $p(z_{1:m})$ is the distribution that we want to draw samples from, then Gibbs chooses an initial value for $z_{1:m}$ and in each iteration, chooses one of the random variables z_i and replaces its value by the value drawn from $p(z_i|z_{1:m}^{-i})$. This process is repeated by iterating over z_is [20]. Gibbs sampling can not be applied to online applications such as stream classification where there is temporal dependency among latent variables. The reason is that in these models, the system doesn't have access to old data and hence iterating over all latent variables is not practical.

After observing the labels of batch t, the set of latent variables in our model is $z_{1:t}$ and ϕ_ks. In order to use Gibbs sampling to draw samples from the posterior of these variables, it is necessary to access all previous batches of data. In order to solve this issue, we use forward Gibbs sampling, an online variation of Gibbs sampling [1]. In this method, at each time step t, we estimate the posterior of new random variables z_t using the estimates of concept indicators in previous batches. In order to do so, we run batch Gibbs sampling over newly added random variables given the state of the sampler in the last batch. In fact, in this method, the value of latent variables that are set in previous batches is no longer changed and the dependency of these variables on future data is not considered. Although this causes suboptimal estimates for initial batches, the estimates will improve over time.

Formally, for inference on the proposed model, we use a collapsed Gibbs sampler, the state of which at time t is $z_{1:t}$. In order to draw a sample at time t, we collapse the concepts' parameters ϕ_ks and compute the posterior of $z_{t,i}$ given the values assigned to $z_{1:t-1}$ in previous batches. To compute this conditional distribution, we use the exchangeability among data of a batch and assume that data i is the last data of the batch. Moreover, using the independency relations among random variables, which can be inferred from the graphical model in Fig. 1, we have:

$$p(z_{ti} = k | z_{1:t}^{-(ti)}, D_{1:t}, G_0, \alpha, \lambda, \Delta) \propto \qquad (13)$$
$$p(z_{ti} = k | z_{1:t}^{-(ti)}, G_0, \alpha, \lambda, \Delta)\, p(d_{ti} | z_{ti} = k, z_{1:t}^{-(ti)}, d_{1:t}^{-(ti)})$$

According to Algorithm 1, the prior over z_{ti} obeys $RCRP(G_0, \alpha, \lambda, \Delta)$, i.e.:

$$p(z_{t,i} = k | z_{1:t}^{-(t,i)}, G_0, \alpha, \lambda, \Delta) \propto \begin{cases} m'_{k,t} + m_{k,t}, & \text{if } k \in I_{t-\Delta:t} \\ \alpha, & \text{if } k \text{ is a new concept} \end{cases} \qquad (14)$$

where $I_{t-\Delta:t}$ is the set of all concept indices that generated at least one data in the last Δ batches. Since we chose G_0 conjugate to the likelihood functions of the base classifiers, the second term in (13) can be analytically computed by marginalizing over ϕ_k; that is:

$$p(d_{t,i} | z_{t,i} = k, z_{1:t}^{-(t,i)}, d_{1:t}^{-(t,i)}) = \int p(d_{t,i} | \phi_k) p(\phi_k | \{d_{\tau,j} : z_{\tau,j} = k\}) d\phi_k \qquad (15)$$

where $p(\phi_k | \{d_{\tau,j} : z_{\tau,j} = k\})$ is the posterior of the parameters of classifier k given all data that was generated by this classifier. Since the base classifier is a naive Bayes classifier with normal likelihood for continuous features and categorical likelihood for discrete features and due to the conjugacy relationship between G_0 and classifier likelihoods, the posterior of the parameters of these classifiers can be easily computed using a few sufficient statistics.

When a new batch of data arrives to be classified, we find the labels by approximating their posteriors by:

$$p\left(y_{t+1,i}|x_{t+1}, d_{1:t}, z_{1:t}\right) \simeq p(y_{t+1,i}|x_{t+1,i}, d_{1:t}, z_{1:t}) \tag{16}$$

$$= \sum_k p(y_{t+1,i}|x_{t+1,i}, z_{t+1,i} = k, z_{1:t}, d_{1:t})p(z_{t+1,i} = k|x_{t+1,i}, z_{1:t}, d_{1:t}) \tag{17}$$

In this approximation, we have discarded the information that x_{t+1}^{-i} may have about $z_{t+1,i}$. The first term of (17) can be calculated similar to (15) and the second term is calculated by:

$$p(z_{t+1,i}|x_{t+1,i}, z_{1:t}, d_{1:t}) \propto p(x_{t+1,i}|z_{t+1,i}, z_{1:t}, d_{1:t})p(z_{t+1,i}|z_{1:t}) \tag{18}$$

where the first term is calculated using (15) and marginalizing over $y_{t+1,i}$.

4   Experimental Results

In this section, we provide experimental results and analysis regarding application of the proposed non-parametric generative model on real stream classification datasets, known as spam [29], weather [16], and electricity [25][1].

The spam dataset consists of 9324 emails extracted from the Spam Assassin Collection. Each email is represented by 500 binary features which indicate the existence of words derived using feature selection. The ratio of spam messages is approximately 20%, hence the classification problem is imbalanced. The emails are sorted according to their arrival date in to batches of 50 emails.

The weather dataset consists of 18,159 daily readings including features such as temperature, pressure, and wind speed. The data were collected by The U.S. National Oceanic and Atmospheric Administration from 1949 to 1999 in the Offutt Air Force Base in Bellevue, Nebraska which has diverse weather patterns making it a suitable dataset for evaluating concept drift. We use the same eight features as [16]. The samples belong to one of two classes: "rain" with 5698 (31%) and "no rain" with 12461 (69%) samples, and are sorted into 606 30-day batches. The model must predict the weather forecast for 30 days, after which the true forecast is revealed.

The electricity dataset consists of 45,312 samples from the Australian New South Wales Electricity Market. Each sample is described by 4 attributes, namely time stamp, day of the week, and 2 electricity demand values. The data was collected from May 1996 to December 1998, during which period the prices vary due to changes in demand and supply. The samples were taken every 30 minutes and sorted into batches of 48 samples each. The target is to predict whether the prices related to a moving average of the last 24 hours, increase or decrease.

In order to compare different stream classification methods on the above datasets, we use two well known measures, namely accuracy defined as the ratio of samples classified correctly and the κ coefficient [12] which is a robust measure of agreement that corrects for random classification. Furthermore, in order

---

[1] All codes for the experiments are available at http://ml.dml.ir/research/npsc

to evaluate accuracy over time, we use prequential accuracy with fading factor $\alpha = 0.95$ defined as [18]:

$$A_\alpha(t) = \frac{\sum_{\tau=1}^{t} \alpha^{t-\tau} a(\tau)}{\sum_{\tau=1}^{t} \alpha^{t-\tau}} \tag{19}$$

where $a(\tau)$ is the ratio of correctly classified samples in batch τ. The reason for choosing this measure is two fold: First, at time t, all previous accuracies contribute to $A_\alpha(t)$, providing an overall picture for evaluation. Second, the forgetting factor α mitigates the impact of older accuracies allowing us to observe how well the algorithm responds to concept drift.

We compare the proposed Non-Parametric Stream Classifier (NPSC) with Naive Bayes (NB) and Probit [13] as uni-model methods and with Conceptual Clustering and Prediction (CCP) [29] and Pool Management base Recurring Concepts Detection (PMRCD) [27] as state-of-the-art model selection algorithms that attempt to handle concept drift. The results are depicted in Table 1 and Fig. 2.

The parameters of Probit, CCP and PMRCD are set according to their corresponding publications. The proposed method has hyper-parameters that need to be set, namely $(G_0, \alpha, \lambda, \Delta)$. The baseline distribution G_0 can be treated as the expected distribution G_t, which is the prior distribution over the parameters of the base classifiers at time t. According to [37], it is unrealistic to assume that this parameter is constant over time. Therefore, we learn this parameter by training a single naive Bayes classifier on all observed data until time t. The precision parameter α, controls how much G_t can deviate from the baseline distribution G_0. Moreover, this parameter controls how often new classifiers emerge.This parameter was set equal to the batch size of each dataset. The parameter λ is the forgetting factor which determines how fast the effect of old data is mitigated. This parameter was set to 0.4 for all datasets. The parameter Δ can be safely set to any large value for which $e^{-\frac{\Delta}{\lambda}}$ is sufficiently small [4]; to incur less computation cost, we set Δ to 30 for all datasets.

The results show that although Probit is a uni-model method, it provides better results than CCP and PMRCD on the spam dataset. The reason is that CCP and PMRCD assume that all data in a batch belong to the same concept. This assumption coupled with the fact that initial batches in the dataset consist of data from a single class, causes their classifiers to overfit to a single class. Later batches in this dataset consist of data from different classes, which the classifiers of CCP and PMRCD can not classify correctly. We have observed that CCP and PMRCD tend to classify each batch with an accuracy similar to the ratio of the majority label. The single classifier of Probit can better handle this situation because it observes all the data and forgets older data gradually. NPSC Provides the best accuracy on this dataset, because it can use multiple classifiers without the unrealistic assumption that all data in a batch belong to the same concept.

The weather dataset exhibits recurrent and gradual concept drift, for which modeling with a finite number of concepts may be sufficient, but the assumption

Table 1. Classification Accuracy (%) And κ Measure For Different Classifiers For Different Methods

		NB	Probit	CCP	PMRCD	NPSC
Spam	Accuracy	90.7	92.4	91.6	89.7	94.5
	κ	0.8	0.8	0.76	0.7	0.85
Weather	Accuracy	73.8	73	73.2	73.0	75.5
	κ	0.31	0.41	0.37	0.32	0.41
Electricity	Accuracy	62.4	62.4	66.5	69.9	69.8
	κ	0.23	0.20	0.30	0.38	0.38

that all days in a month (one batch) belong to the same concept is still unrealistic. That may be the reason for the better performance of NPSC on this dataset (Table 1). According to Fig. 2, the ensemble methods (CCP, PMRCD, and NPSC) handle concept drift better than uni-model methods (NB and Probit), but it is hard to distinguish which performs better. This was expected due to the recurrent nature of weather which can be modeled by a finite number of concepts without the need for a complex management scheme for the pool of classifiers.

Finally, the results show that uni-model methods (NB and Probit) perform poorly on the Electricity dataset. The reason is that this dataset exhibits complex concept drift, due to the complex nature of demand and supply. Ensemble methods (CCP, PMRCD, and NPSC) perform better, because they can handle multiple concepts. On the other hand, CCP lacks a management scheme for the pool of classifiers which explains its poor performance in comparison to PMRCD and NPSC. Moreover, since each batch of data corresponds to a single day, the assumption that data in a batch belong to the same concept is not unrealistic. This explains the similar performance of PMRCD and NPSC.

5 Conclusions and Future Works

In this paper, we addressed the problem of stream classification and introduced a probabilistic framework. The proposed method handles concept drift using a nonparametric temporal model that builds a model selection based classifier via a mixture model with potentially infinite mixtures. This method finds the number of concepts and the data-concept association through inference on the proposed model. In order to make inference on the proposed model, we introduced an online algorithm which is based on Gibbs sampling.

Several directions of future research are possible. The proposed method yields accurate results using simple naive Bayes classifier. As it was mentioned in Section 4, more complex classifiers such as probit may provide better results. The challenge is that there are no conjugate priors for probit's parameters and hence, it may be necessary to use some approximate inference algorithms such as Expectation Propagation (EP) in each iteration of Gibbs sampling. Another direction is to use model combination instead of model selection. The assumption that each data is generated by one classifier may be a constraining assumption and

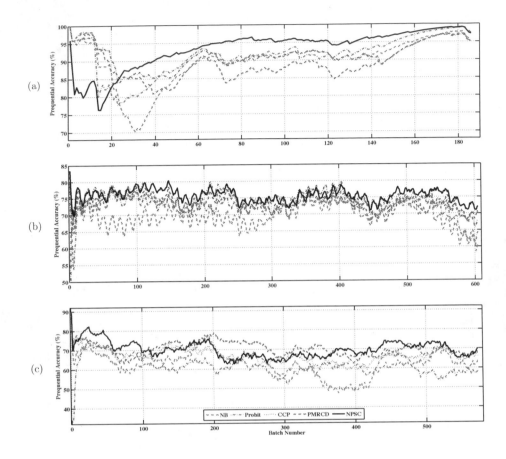

Fig. 2. Classification results of classifiers on (a) Spam, (b) Weather, and (c) Electricity

since model combination methods enrich the hypothesis space by combining different classifiers, they may increase the efficiency of the model. Furthermore, sampling based inference methods are non-deterministic and their convergence can not be verified easily. A direction we are currently pursuing is to develop an online variational inference algorithm based on the idea proposed in [28].

References

1. Ahmed, A., Low, Y., Aly, M., Josifovski, V., Smola, A.J.: Scalable distributed inference of dynamic user interests for behavioral targeting. In: Proceedings of the 17th ACM SIGKDD International Conference on Knowledge Discovery and Data Mining, pp. 114–122. ACM (2011)
2. Ahmed, A., Ho, Q., Eisenstein, J., Xing, E., Smola, A.J., Teo, C.H.: Unified analysis of streaming news. In: Proceedings of the 20th International Conference on World Wide Web, pp. 267–276. ACM (2011)

3. Ahmed, A., Xing, E.P.: Dynamic Non-Parametric Mixture Models and the Recurrent Chinese Restaurant Process: with Applications to Evolutionary Clustering. In: SDM, pp. 219–230 (2008)
4. Ahmed, A., Xing, E.P.: Timeline: A dynamic hierarchical Dirichlet process model for recovering birth/death and evolution of topics in text stream. arXiv preprint arXiv:1203.3463 (2012)
5. Andrieu, C., De Freitas, N., Doucet, A., Jordan, M.I.: An introduction to MCMC for machine learning. Machine Learning 50(1-2), 5–43 (2003)
6. Antoniak, C.E.: Mixtures of Dirichlet processes with applications to Bayesian nonparametric problems. The Annals of Statistics 2(6), 1152–1174 (1974)
7. Bifet, A., Frank, E.: Sentiment knowledge discovery in twitter streaming data. In: Pfahringer, B., Holmes, G., Hoffmann, A. (eds.) DS 2010. LNCS, vol. 6332, pp. 1–15. Springer, Heidelberg (2010)
8. Bifet, A., Pfahringer, B., Read, J., Holmes, G.: Efficient data stream classification via probabilistic adaptive windows. In: Proceedings of the 28th Annual ACM Symposium on Applied Computing, pp. 801–806. ACM (2013)
9. Blackwell, D., MacQueen, J.B.: Ferguson distributions via Plya urn schemes. The Annals of Statistics, 353–355 (1973)
10. Blei, D.M., Jordan, M.I.: Variational inference for Dirichlet process mixtures. Bayesian Analysis 1(1), 121–143 (2006)
11. Blei, D.M., Frazier, P.I.: Distance dependent Chinese restaurant processes. The Journal of Machine Learning Research 12, 2461–2488 (2011)
12. Cohen, J.: A coefficient of agreement for nominal scales. Educational and Psychological Measurement 20(1), 37–46 (1960)
13. Chu, W., Zinkevich, M., Li, L., Thomas, A., Tseng, B.: Unbiased online active learning in data streams. In: Proceedings of the 17th ACM SIGKDD International Conference on Knowledge Discovery and Data Mining, pp. 195–203. ACM (2011)
14. Davy, M., Tourneret, J.Y.: Generative supervised classification using dirichlet process priors. IEEE Transactions on Pattern Analysis and Machine Intelligence 32(10), 1781–1794 (2010)
15. Domingos, P.: Why Does Bagging Work? A Bayesian Account and its Implications. In: KDD, pp. 155–158 (1997)
16. Elwell, R., Polikar, R.: Incremental learning of concept drift in nonstationary environments. IEEE Transactions on Neural Networks 22(10), 1517–1531 (2011)
17. Ferguson, T.S.: A Bayesian analysis of some nonparametric problems. The Annals of Statistics, 209–230 (1973)
18. Gama, J., Sebastio, R., Rodrigues, P.P.: On evaluating stream learning algorithms. Machine Learning 90(3), 317–346 (2013)
19. Gama, J., Zliobaite, I., Bifet, A., Pechenizkiy, M., Bouchachia, A.: A Survey on Concept Drift Adaptation. ACM Computing Surveys 46(4) (2014)
20. Geman, S., Geman, D.: Stochastic relaxation, Gibbs distributions, and the Bayesian restoration of images. IEEE Transactions on Pattern Analysis and Machine Intelligence (6), 721–741 (1984)
21. Gershman, S.J., Blei, D.M.: A tutorial on Bayesian nonparametric models. Journal of Mathematical Psychology 56(1), 1–12 (2012)
22. Gomes, J.B., Menasalvas, E., Sousa, P.A.: Learning recurring concepts from data streams with a context-aware ensemble. In: Proceedings of the 2011 ACM Symposium on Applied Computing, pp. 994–999. ACM (2011)

23. Graepel, T., Candela, J.Q., Borchert, T., Herbrich, R.: Web-scale bayesian click-through rate prediction for sponsored search advertising in microsoft's bing search engine. In: Proceedings of the 27th International Conference on Machine Learning (ICML 2010), pp. 13–20 (2010)
24. Hannah, L.A., Blei, D.M., Powell, W.B.: Dirichlet process mixtures of generalized linear models. The Journal of Machine Learning Research 12, 1923–1953 (2011)
25. Harries, M.: Splice-2 comparative evaluation: Electricity pricing. Artificial Intelligence Group, School of Computer Science and Engineering, The University of New South Wales, Sidney, Tech.Rep. UNSW-CSE-TR-9905 (1999)
26. Heath, D., Sudderth, W.: De Finetti's theorem on exchangeable variables. The American Statistician 30(4), 188–189 (1976)
27. Hosseini, M.J., Ahmadi, Z., Beigy, H.: New management operations on classifiers pool to track recurring concepts. In: Cuzzocrea, A., Dayal, U. (eds.) DaWaK 2012. LNCS, vol. 7448, pp. 327–339. Springer, Heidelberg (2012)
28. Hoffman, M.D., Blei, D.M., Wang, C., Paisley, J.: Stochastic variational inference. The Journal of Machine Learning Research 14(1), 1303–1347 (2013)
29. Katakis, I., Tsoumakas, G., Vlahavas, I.: Tracking recurring contexts using ensemble classifiers: an application to email filtering. Knowledge and Information Systems 22(3), 371–391 (2010)
30. Klinkenberg, R.: Learning drifting concepts: Example selection vs. example weighting. Intelligent Data Analysis 8(3), 281–300 (2004)
31. Minka, T.P.: Bayesian model averaging is not model combination. Technical Report (2000)
32. Minka, T.P.: Expectation propagation for approximate Bayesian inference. In: Proceedings of the Seventeenth Conference on Uncertainty in Artificial Intelligence, pp. 362–369. Morgan Kaufmann Publishers Inc. (2001)
33. Neal, R.M.: Markov chain sampling methods for Dirichlet process mixture models. Journal of Computational and Graphical Statistics 9(2), 249–265 (2000)
34. Minku, L.L., Yao, X.: DDD: A new ensemble approach for dealing with concept drift. IEEE Transactions on Knowledge and Data Engineering 24(4), 619–633 (2012)
35. Paquet, U., Van Gael, J., Stern, D., Kasneci, G., Herbrich, R., Graepel, T.: Vuvuzelas & Active Learning for Online Classification. In: NIPS Workshop on Comp. Social Science and the Wisdom of Crowds (2010)
36. Shahbaba, B., Neal, R.: Nonlinear models using Dirichlet process mixtures. The Journal of Machine Learning Research 10, 1829–1850 (2009)
37. Zhang, J., Ghahramani, Z., Yang, Y.: A Probabilistic Model for Online Document Clustering with Application to Novelty Detection. In: NIPS, vol. 4, pp. 1617–1624 (2004)
38. Zhu, X., Zhang, P., Lin, X., Shi, Y.: Active learning from stream data using optimal weight classifier ensemble. IEEE Transactions on Systems, Man, and Cybernetics, Part B: Cybernetics 40(6), 1607–1621 (2010)

Fast Nearest Neighbor Search
on Large Time-Evolving Graphs

Leman Akoglu[1], Rohit Khandekar[2], Vibhore Kumar[3],
Srinivasan Parthasarathy[3], Deepak Rajan[4], and Kun-Lung Wu[3]

[1] Stony Brook University
leman@cs.stonybrook.edu
[2] Knight Capital Group
rkhandekar@gmail.com
[3] IBM T. J. Watson Research
{vibhorek,spartha,klwu}@us.ibm.com
[4] Lawrence Livermore National Labs
rdeepak@gmail.com

Abstract. Finding the k nearest neighbors (k-NNs) of a given vertex in
a graph has many applications such as link prediction, keyword search,
and image tagging. An established measure of vertex-proximity in graphs
is the Personalized Page Rank (PPR) score based on random walk with
restarts. Since PPR scores have long-range correlations, computing them
accurately and efficiently is challenging when the graph is too large to
fit in main memory, especially when it also changes over time. In this
work, we propose an efficient algorithm to answer PPR-based k-NN queries
in large time-evolving graphs. Our key approach is to use a divide-and-
conquer framework and efficiently compute answers in a distributed fashion. We represent a given graph as a collection of dense vertex-clusters
with their inter connections. Each vertex-cluster maintains certain information related to internal random walks and updates this information
as the graph changes. At query time, we combine this information from
a small set of relevant clusters and compute PPR scores efficiently. We
validate the effectiveness of our method on large real-world graphs from
diverse domains. To the best of our knowledge, this is one of the few
works that simultaneously addresses answering k-NN queries in possibly
disk-resident *and* time-evolving graphs.

Keywords: vertex proximity, personalized pagerank, time-evolving
graphs, disk-resident graphs, distributed pagerank, dynamic updates.

1 Introduction

Quantifying the proximity, relevance, or similarity between vertices, and more
generally finding the k nearest neighbors (k-NNs) of a given vertex in a large, time-
evolving graph is a fundamental building block for many applications. Personalized
PageRank (PPR) has proved to be a very effective proximity measure for the link
prediction and recommendation problems in such applications. Thanks to its effectiveness, there exist many algorithms in the literature that are designed to compute

T. Calders et al. (Eds.): ECML PKDD 2014, Part I, LNCS 8724, pp. 17–33, 2014.

the PPR scores of a given vertex in a graph efficiently [4, 10–13, 25]. These works, however, cannot handle graphs that are larger than a certain size, that is, they are not optimally designed for very large disk-resident graphs. Moreover, they cannot work with graphs that dynamically change over time. Other previous works deal with computing PPR queries on either disk-resident *static* graphs [21, 3] or special families of time-evolving graphs [26].

In this work we propose CLUSTERRANK, an algorithm for efficient computation of PPR queries for *both* disk-resident *and* time-evolving general graphs. Our main motivation is to build a unified framework that will enable us to tackle both of these two challenges. Our key idea is to take a divide-and-conquer approach; simply put, we split the graph into relatively small vertex-clusters and decompose the overall problem into simulating intra-cluster and inter-cluster random walks. This decomposition enables us to handle disk-resident graphs since the work is carefully split across distributed compute nodes. What is more, thanks to this modular design of our approach, our updates are local and fast when the graph changes over time. We summarize our main contributions as follows.

- *Fast query processing*: We propose a fast algorithm to answer k-NN queries on large graphs, with query response time sub-linear in input graph size.
- *Dynamic updates*: The algorithm includes fast, incremental update procedures for handling additions or deletions of edges and vertices. Thus it also works with time-evolving graphs.
- *Disk-resident graphs*: Our method can operate when the graph resides entirely on disk. Moreover, it loads only a small and relevant portion of the graph into memory to answer a query or perform dynamic updates.

We demonstrate the effectiveness and efficiency of our method, w.r.t. query accuracy and response time, on large real-world graphs from various domains.

2   Preliminaries and Overview

Vertex-Proximity. We consider finding the k-NN's of a given vertex in a graph. To calculate the k-NN's of a vertex, one needs to define a distance metric between two vertices. A widely used proximity measure that is based on random walks is Personalized Page Rank (PPR). Given a restart vertex q and a parameter $\alpha \in (0, 1)$, consider the random walk with restart starting at vertex q, such that at any step when currently present at a vertex v, it chooses any of its neighbors with equal probability α/d_v, and returns to the restart vertex q with probability $(1 - \alpha)$. The stationary probability at vertex u of the random walk with restart is defined as the PPR score of u with respect to the query vertex q. The PPR score is known to be robust under noise or small changes in the graph, in contrast to shortest paths, and favors existence of many short paths between vertices. Therefore, in this paper we consider the problem of finding the k-NN's of any given vertex in a graph, where proximity is measured by the PPR score.

Overview. The main challenge in answering a k-NN query is the computational overhead involved in simulating a random walk on a large graph that may not even fit in memory. We employ a divide-and-conquer principle to handle this

---

1. Pre-computation (Offline)

   a. Cluster the graph into low-conductance, possibly *overlapping* clusters. For each vertex v, we identify a unique cluster containing v and call it the parent of v. Store these clusters on one or more compute nodes.

   b. For each cluster, compute and store some auxiliary information relating to intra-cluster random walks, *independently* of other clusters.

2. Query processing (Online)

   a. Given a query vertex, identify the 'right' subset of clusters to consider. (If all the clusters are considered, then the final answer is exact.)

   b. Combine the auxiliary information of identified clusters to compute PPR scores.

3. Graph update processing (Online/Batch)

   a. Given an update (addition/deletion of one or more vertices/edges), identify the 'right' subset of clusters to update.

   b. Update the identified clusters and their auxiliary information.

Fig. 1. Main components of proposed framework

challenge. We cluster the graph into relatively small vertex-clusters and decompose the problem into simulating intra-cluster and inter-cluster random walks.

For a subset S of vertices, conditioned on the event that the random walk is in S, the probability that it steps out of S is proportional to its conductance—the ratio of the weight of edges crossing S to the weight of all edges incident to S. Thus a low-conductance cluster "holds" the random walk longer than a high-conductance cluster. This makes low conductance a natural choice for estimating quality of a cluster for our purposes. We allow the clusters to overlap since it is natural for a vertex to belong to multiple communities.

Consider a random walk with restart starting at q. Let the sequence of vertices the walk visits be $v_0, v_1, v_2, \ldots$. A vertex may appear several times in this sequence. The stationary distribution of this walk is the relative frequency with which different vertices appear in this walk. Now suppose $q \in S_i$. As the random walk steps through this sequence, it stays in cluster S_i for a while, then jumps to another cluster S_j. Next it stays in S_j for a while before jumping to yet another cluster, and so on. The clusters thus visited by the walk may also repeat. As a result, one can partition the walk into a sequence of contiguous blocks of vertices where each block represents a portion of the walk inside a cluster and consecutive blocks represent a jump from a cluster to another.

Now it is easy to describe how to simulate the random walk based on the clusters. For each cluster S_i and each "entry" vertex $v \in S_i$, one can compute the characteristics of the random walk inside S_i assuming it entered S_i through v. These characteristics include the probabilities with which it exits S_i to different "exit" vertices and the expected number of times it visits various vertices in S_i before exiting. Interestingly this information can be computed for each cluster S_i *independent* of other clusters. Our method pre-computes and stores this information for each cluster. At query time, it combines this information across different clusters to compute the desired PPR scores. Whenever the graph changes, due to addition/deletion of vertices/edges, it updates the relevant clusters and their information appropriately. Our overall framework is given in Figure 1.

3   Proposed Method

We propose CLUSTERRANK, a method to address the following two problems.

P1) Given a large edge-weighted graph $\mathcal{G}$, a query vertex q in $\mathcal{G}$ and an integer k, find k vertices in $\mathcal{G}$ that have highest PPR scores w.r.t. q.

P2) Given a large edge-weighted graph $\mathcal{G}(t)$ at time t, a subset $D(t)$ of existing edges in $\mathcal{G}(t)$ and a set $A(t)$ of new edges, update the graph structure and the relevant auxiliary information to delete the edges in $D(t)$ and add the edges in $A(t)$, i.e., compute $\mathcal{G}(t+1) := (\mathcal{G}(t) \setminus D(t)) \cup A(t)$.

We next describe the components in Fig. 1 in detail. To simplify the presentation, we assume the graph is unweighted. Our techniques, however, extend to graphs with non-negative edge-weights and directed graphs.[1]

3.1   Pre-computation

The pre-computation involves two steps which can be performed offline.

3.1.1. Clustering the Graph

To distribute a graph across compute nodes, one can use any top-performing known graph partitioning algorithm [1, 8, 15, 19, 23, 22]. In this work, we use [1] which finds low-conductance clusters and allows clusters to overlap. We assign each vertex v to a unique cluster S containing v that contains the maximum number of v's neighbors. We call such a cluster the *parent* cluster of v. The notion of parent clusters is used while query processing.

3.1.2. Computing Auxiliary Information for Clusters

Given the overlapping clustering computed as $\mathcal{S} = \{S_1, S_2, \ldots, S_p\}$, we next compute certain auxiliary information for each cluster $S_i \in \mathcal{S}$ independently of others clusters. Assume that the query vertex $q \notin S_i$ and assume that the random walk with restart enters S_i through a vertex $u \in S_i$. We simulate this random walk with restart till it exits cluster S_i. Suppose the random walk is at vertex $v \in S_i$. In one step, with probability $1 - \alpha$, the random walk restarts at q and hence exits S_i. With probability α, it picks a neighbor $w \in \Gamma(v)$ at random. Here $\Gamma(v)$ denotes the set of neighbors of v in $\mathcal{G}$. If $w \in S_i$, the random walk continues within S_i. If $w \notin S_i$, the random walk exits S_i to vertex w. The auxiliary information for each cluster S_i consists of two matrices, the *Count* matrix and the *Exit* matrix.

Count Matrix. The count matrix C_i is an $|S_i| \times |S_i|$ matrix defined as follows. The entry $C_i(u, v)$, for $u, v \in S_i$, equals the expected number of times a random walk with restart (restarting at $q \notin S_i$) starting at u visits v before exiting S_i. The following lemma gives a closed-form expression for C_i. Let T_i be an $|S_i| \times |S_i|$ matrix that gives transition probabilities of a random walk within S_i without restart, i.e., for $u, v \in S_i$, let $T_i(u, v) = 1/|\Gamma(u)|$ if $v \in \Gamma(u)$ and 0 otherwise.

---

[1] We modify directed graphs by adding a self-loop to each vertex, such that no vertex has out-degree zero. This ensures the random walk matrix remains a Markov chain.

Lemma 1. $C_i = (I - \alpha T_i)^{-1}$ where I is the $|S_i| \times |S_i|$ identity matrix.

Proof. It is easy to see that C_i satisfies the following relation for all $u, v \in S_i$.

$$C_i(u,v) = \begin{cases} 1 + \alpha \displaystyle\sum_{w \in \Gamma(u) \cap S} T_i(u,w)C_i(w,v), & \text{if } v = u; \\ \alpha \displaystyle\sum_{w \in \Gamma(u) \cap S} T_i(u,w)C_i(w,v), & \text{otherwise.} \end{cases} \tag{1}$$

In matrix form, the above relation can be written as $C_i = I + \alpha T_i C_i$.   □

Exit Matrix. Let $B_i = \Gamma(S_i) \setminus S_i$ denote the set of vertices not in S_i that are adjacent to vertices in S_i. The exit matrix E_i is an $|S_i| \times (|B_i| + 1)$ matrix defined as follows. The entry $E_i(u,b)$, for $u \in S_i$ and $b \in B_i$, is the probability that a random walk with restart (restarting at $q \notin S_i$) starting at u exits S_i while jumping to vertex $b \in B_i$. Since the random walk can exit S_i while jumping to a restart vertex q (assumed not to be in S_i), we have an additional column in E_i corresponding to q. Of course, we do not know the identity of the restart vertex q at the pre-computation phase. Therefore we treat q as a symbolic representative of the restart vertex. The entry $E_i(u,q)$, for $u \in S_i$, is the probability that the random walk exits S_i while jumping to the restart vertex q. The following lemma gives a closed-form expression for E_i. Let T_i^+ be an $|S_i| \times (|B_i|+1)$ matrix that gives exit probabilities of a random walk from vertices in S_i to vertices in $B_i \cup \{q\}$, i.e., for $u \in S_i$ and $b \in B_i$, let $T_i^+(u,b) = \alpha/|\Gamma(u)|$ if $b \in \Gamma(u)$ and 0 otherwise; and for $u \in S_i$, $T_i^+(u,q) = 1 - \alpha$.

Lemma 2. $E_i = (I - \alpha T_i)^{-1} T_i^+ = C_i T_i^+$ where I is the identity matrix.

Proof. It is easy to see that E_i satisfies the following relation $\forall u \in S_i$ and $\forall b \in B_i \cup \{q\}$.

$$E_i(u,b) = \begin{cases} 1 - \alpha + \alpha \displaystyle\sum_{w \in \Gamma(u) \cap S} T_i(u,w)E_i(w,v), & \text{if } b = q; \\ T_i^+(u,b) + \alpha \displaystyle\sum_{w \in \Gamma(u) \cap S} T_i(u,w)E_i(w,v), & \text{otherwise.} \end{cases} \tag{2}$$

In matrix form, the above relation can be written as $E_i = T_i^+ + \alpha T_i E_i$.   □

There are a couple of ways in which one can compute matrices C_i and E_i for a cluster. One can directly use the closed-form expressions given above. In this case, computing auxiliary information for a cluster S_i containing s vertices and containing b vertices in the neighborhood $\Gamma(S_i) \setminus S_i$ involves, computing an inverse $(I - \alpha T_i)^{-1}$ of an $s \times s$ matrix and multiplying an $s \times s$ matrix and an $s \times (b + 1)$ matrix. This takes a total of $O(s^3 + s^2 b)$ time using Gaussian elimination for inverting and textbook matrix products. One can reduce this time complexity by using Strassen's algorithm [24]. An alternative is to use the relations (1) and (2) to compute these matrices in an iterative fashion. This approach, however, is often found less effective than computing the matrix inverse.

3.2 Query Processing

The second component of CLUSTERRANK deals with query answering, and consists of two steps: (1) updating the auxiliary information for those clusters that contain the query vertex, and (2) combining such information from a subset of "relevant" clusters to compute the final PPR scores.

3.2.1. Updating the Matrices Given a Query Vertex

Given a query vertex q, we first identify its unique parent cluster S_i. We then update the count matrix C_i and the exit matrix E_i to reflect the fact that the restart vertex now lies inside the cluster S_i. We remark that the count and exit matrices corresponding to any other cluster, say S_j with $j \neq i$, containing the query vertex are not updated.

Count matrix with a query vertex inside. Given a query vertex q and its parent cluster S_i, the count matrix C_i^q is an $|S_i| \times |S_i|$ matrix defined analogously. The entry $C_i^q(u, v)$, for $u, v \in S_i$, equals the expected number of times a random walk with restart (restarting at $q \in S_i$) starting at u visits v before exiting S_i. The following lemma gives a closed-form expression for C_i^q. Let Q_q be an $|S_i| \times |S_i|$ matrix with all entries in the column q equal to 1 and all other entries zero.

Lemma 3. $C_i^q = (I - \alpha T_i - (1 - \alpha)Q_q)^{-1}$ *where I is the $|S_i| \times |S_i|$ identity.*

Proof. Recall that the random walk restarts at $q \in S_i$ at every step with probability $1 - \alpha$. Therefore C_i^q satisfies the following relation for all $u, v \in S_i$.

$$C_i^q(u, v) = \begin{cases} 1 + (1 - \alpha)C_i^q(q, v) + \alpha \sum_{w \in \Gamma(u) \cap S} T_i(u, w)C_i^q(w, v), & \text{if } v = u; \\ (1 - \alpha)C_i^q(q, v) + \alpha \sum_{w \in \Gamma(u) \cap S} T_i(u, w)C_i^q(w, v), & \text{otherwise.} \end{cases} \quad (3)$$

In matrix form, the above relation becomes $C_i^q = I + (1 - \alpha)Q_q C_i^q + \alpha T_i C_i^q$. □

Exit matrix given a query vertex inside. Given a query vertex $q \in S_i$, the exit matrix E_i^q is an $|S_i| \times |B_i|$ matrix defined analogously. The entry $E_i^q(u, b)$, for $u \in S_i$ and $b \in B_i$, is the probability that a random walk with restart (restarting at $q \in S_i$) starting at u exits S_i while jumping to vertex $b \in B_i$. Note that since the random walk does not exit S_i due to a restart, E_i^q has only $|B_i|$ columns.

The following lemma gives a closed-form expression for E_i^q. Let T_i^{+q} be an $|S_i| \times |B_i|$ matrix that gives exit probabilities of a random walk from vertices in S_i to vertices in B_i, i.e., for $u \in S_i$ and $b \in B_i$, let $T_i^+(u, b) = \alpha/|\Gamma(u)|$ if $b \in \Gamma(u)$ and 0 otherwise. This matrix is T_i^+ with the last column dropped.

Lemma 4. $E_i^q = (I - \alpha T_i - (1 - \alpha)Q_q)^{-1}T_i^{+q} = C_i^q T_i^{+q}$.

Proof. Recall that the random walk restarts at $q \in S_i$ at every step with probability $1 - \alpha$. Therefore E_i^q satisfies the following relation $\forall u \in S_i$ and $\forall b \in B_i$.

$$E_i^q(u, b) = T_i^{+q}(u, b) + (1 - \alpha)E_i^q(q, b) + \alpha \sum_{w \in \Gamma(u) \cap S} T_i(u, w)E_i^q(w, v).$$

In matrix form, we can write the above as $E_i^q = T_i^{+q} + (1 - \alpha)Q_q E_i^q + \alpha T_i E_i^q$. □

Updating the matrices using Sherman-Morrison formula. Observe, from Lemmas 1 and 3, that the expressions for C_i and C_i^q are quite similar—C_i is the inverse of a matrix and C_i^q is the inverse of the same matrix with $(1-\alpha)Q_q$ subtracted. Note also that $(1-\alpha)Q_q$ is a rank-1 matrix. We can update the inverse of a matrix efficiently when the matrix undergoes such a low-rank update. To this end, we first quote a well-known lemma.

Lemma 5 (Sherman-Morrison-Woodbury [27]). *Let n and k be any positive integers, $A \in \Re^{n \times n}, U \in \Re^{n \times k}, \Sigma \in \Re^{k \times k}, V \in \Re^{k \times n}$ be any matrices:*

$$(A + U\Sigma V)^{-1} = A^{-1} - A^{-1}U(\Sigma^{-1} + VA^{-1}U)^{-1}VA^{-1}.$$

Note that $U\Sigma V$ is a rank-k matrix. Thus after updating A with a rank-k matrix, its inverse can be computed from A^{-1} by doing 4 multiplications of $n \times n$ and $n \times k$ matrices, 2 multiplications of $n \times k$ and $k \times k$ matrices and 1 inverse of a $k \times k$ matrix. Thus overall time is $O(n^2 k)$ since $k \leq n$. This can be much more efficient than computing the inverse of an $n \times n$ matrix from scratch, especially if k is much smaller than n. If $k = 1$, the above formula reduces to what is commonly known as Sherman-Morrison formula. We refer to the formula in the above lemma as the SMW formula in the remainder of the text.

To use this approach, we have to express the rank-1 matrix $(1-\alpha)Q_q$ as $U\Sigma V$ where Σ is a 1×1 matrix, i.e., a scalar. This can be done simply by setting $\Sigma = (1-\alpha)\sqrt{|S_i|}$, U to be an $|S_i|$-size column vector with all entries $1/\sqrt{|S_i|}$ and V to be an $|S_i|$-size row vector with all entries 0 except the entry corresponding to q equal to 1. Thus C_i^q can be computed from C_i in $O(|S_i|^2)$ time. Similarly, E_i^q can be computed from C_i and E_i in $O(|S_i|^2 + |S_i||B_i|)$ time.

To simplify the notation in the following discussion, we let $\hat{C}_i$ (resp. $\hat{E}_i$) denote C_i^q (resp. E_i^q) if S_i is the parent cluster of q, and C_i (resp. E_i) otherwise.

3.2.2. Computing the PPR Scores
Recall that to compute the PPR scores, our method decomposes the random walk with restart starting from the query vertex q into intra-cluster and inter-cluster random walks. Since the information about intra-cluster random walks is already pre-computed (or appropriately updated for the parent cluster of the query vertex), we next compute the necessary information about the inter-cluster random walk. As a first step, we identify the clusters "relevant" for answering the k-NN query for q. If we want to compute PPR scores exactly, we label all the clusters as relevant. Working with all the clusters to answer a query, however, leads to excessive query response time. It turns out that one can reduce the query response time significantly by limiting the number of relevant clusters. We employ two heuristics called *1-hop* and *2-hop* to limit the relevant clusters. In the former, we label a cluster S as relevant if and only if $q \in S$. In the latter, we label a cluster S as relevant if and only if either $q \in S$ or S is the parent cluster of some vertex $b \in B_i = \Gamma(S_i) \setminus S_i$ for some cluster S_i such that $q \in S_i$. Intuitively, these heuristics quickly identify the vertices that are expected to have high PPR scores w.r.t. q.

Suppose $\mathcal{S}_q$ is the set of relevant clusters. Let $\cup \mathcal{S}_q$ denote the union of these clusters. Recall that $B_i = \Gamma(S_i) \setminus S_i$ denotes the set of vertices which the random walk inside S_i may jump to while exiting S_i. Now let $B_q = ((\cup S_q) \cap (\cup_{S_i \in S_q} B_i)) \cup \{q\}$. As the number of vertices in B_q relates to the efficiency of our method, we explicitly limit $|B_q|$; we fix a parameter β, and while using either 1-hop or 2-hop heuristic, we continue labeling the clusters relevant as long as $|B_q|$ does not exceed β or all the clusters according to the heuristic are labeled relevant.

Computing the inter-cluster random walk matrix. After identifying the relevant clusters S_q, we gather their auxiliary information to compute the inter-cluster random walk matrix. Recall that when the random walk with restart enters $u \in S_i \in S_q$, it exits S_i while jumping to some vertex $b \in B_i \cup \{q\}$ (or $b \in B_i$ if S_i is the parent cluster of q). The probability of this event is exactly given by $\hat{E}_i(u, b)$. Clearly, this vertex b can belong to multiple clusters. When the random walk jumps to b, we assume that it enters the parent cluster of b.

Thus we can think of inter-cluster jumps as a random walk on the vertices in B_q. Whenever the random walk jumps to a vertex $b \in B_i \setminus \cup S_q$ that is not in the relevant clusters, we assume that the random walk jumps back to q. The transition matrix (of dimensions $|B_q| \times |B_q|$) of this walk is as follows. For any $b_1, b_2 \in B_q$, the probability that this random walk jumps from b_1 to b_2 is

$$M_q(b_1, b_2) = \begin{cases} \hat{E}_i(b_1, b_2), & \text{if } b_2 \neq q, S_i \text{ is theparent cluster of } b_1; \\ 1 - \sum_{b \in B_q \setminus \{q\}} M_q(b_1, b), & \text{if } b_2 = q. \end{cases} \quad (4)$$

We compute M_q from the auxiliary information stored (or appropriately updated) for the relevant clusters. Recall that the random walk (or the corresponding Markov chain) is called *ergodic* if it is possible to go from every state to every other state (not necessarily in one move), and if the walk is aperiodic. Now we can assume that the given graph $\mathcal{G}$ is connected without loss of generality.[2] Thus, if we label all clusters as relevant, the resulting Markov chain M_q is ergodic (under very mild assumptions satisfied by large real-world graph topologies). Also, from the definition of 1-hop or 2-hop heuristics, the resulting Markov chain M_q is still ergodic even if we use these heuristics.

From the standard theorem of ergodic chains [9], we conclude that there is a *unique* probability row-vector $\mu \in \Re^{|B_q|}$ such that $\mu M_q = \mu$. This vector gives the expected fraction of steps the random walk spends at any vertex $b \in B_q$. This vector can be computed either by doing repeated multiplications of M_q with the starting probability distribution (which is 1 at the coordinate q and 0 elsewhere); or by computing the top eigenvector of $I - M_q^\top$ corresponding to eigenvalue 1. The eigenvector computation can be done in time $O(|B_q|^3)$.

We now "lift" this random walk back to the random walk with restart on the union of the relevant clusters $\cup S_q$. Since a cluster $S_i \in S_q$, the value $\hat{C}_i(u, v)$ gives the expected number of times the random walk with restart (starting at q) visits $v \in S_i$ before exiting S_i. Therefore, for a vertex $v \in \cup S_q$, the quantity

$$\pi_v = \sum_{S_i \in S_q : v \in S_i} \sum_{b \in B_q : S_i \text{ parent of } v} \mu_b \hat{C}_i(b, v) \quad (5)$$

---

[2] If $\mathcal{G}$ is not connected, we focus on the connected component containing q.

gives the expected number of times the random walk with restart visits $v \in \cup \mathcal{S}_q$ between consecutive inter-cluster jumps. Scaling these values so that they sum up to 1, gives the fraction of steps the random walk visits $v \in \cup \mathcal{S}_q$, i.e., $\hat{\pi}_v = \frac{\pi_v}{\sum_{u \in \cup \mathcal{S}_q} \pi_u}$. The k-NN query can then be answered by identifying k vertices with the highest values of $\hat{\pi}_v$. The following theorem is now evident.

Theorem 1. *If we label all the clusters as relevant, the computed values $\{\hat{\pi}_v \mid v \in \mathcal{G}\}$ equal the* exact PPR *values w.r.t. the query vertex q.*[3]

The k-NN query is then answered by top k vertices with the highest $\hat{\pi}_v$.

3.3 Dynamic Updates

To simplify the presentation, we describe how to handle addition of a single edge $e = \{u, v\}$. When an edge is added, the transition probability matrices T_i and T_i^+ for some clusters S_i are changed, resulting in the change of C_i and E_i according to Lemmas 1 and 2. The key observation here is that the changes in T_i and T_i^+ are low-rank. Therefore, new C_i and E_i can be computed from their old versions using the SMW formula. Let $\mathcal{S}(u)$ be the set of clusters containing u and $\mathcal{S}(v)$ containing v. We consider several cases.

Case 1: $\mathcal{S}(\mathbf{u}) = \mathcal{S}(\mathbf{v}) = \emptyset$. This case arises when both vertices u and v are new vertices. In this case, we add both these vertices to a cluster S_i with the smallest size. We also designate S_i as the parent cluster of both u and v. Note that the edge e forms a disconnected component in S_i. Therefore, the matrices C_i and E_i can be computed directly without resorting to the SMW formula.

Consider a random walk with restart (restarting at q) starting at u. Since the random walk restarts with probability $1 - \alpha$, it is easy to see that the expected number of times the random walk visits u before exiting $\{u, v\}$ is $\frac{1}{1-\alpha^2}$ and the expected number of times the random walk visits v before exiting $\{u, v\}$ is $\frac{\alpha}{1-\alpha^2}$. It is now easy to observe that if the edge $e = \{u, v\}$ is added to cluster S_i, it's count matrix can be computed from the original count matrix C_i as on the left.

$$\left[\begin{array}{c|cc} & 0 & 0 \\ C_i & \vdots & \vdots \\ & 0 & 0 \\ \hline 0 \ldots 0 & \frac{1}{1-\alpha^2} & \frac{\alpha}{1-\alpha^2} \\ 0 \ldots 0 & \frac{\alpha}{1-\alpha^2} & \frac{1}{1-\alpha^2} \end{array} \right]$$

We then use Lemma 2 to compute the new E_i as $C_i T_i^+$ where the new T_i^+ is computed as $\left[\begin{array}{c} T_i^+ \\ \hline 0 \ldots 0 \; \alpha \\ 0 \ldots 0 \; \alpha \end{array} \right]$.

Case 2: $\mathcal{S}(\mathbf{v}) = \emptyset$. In this case, vertex v is a new vertex. We add vertex v to each cluster $S_i \in \mathcal{S}(u)$ and designate some cluster picked arbitrarily among these as the parent cluster of v. We now use the SMW formula along with Lemma 1 to compute the new count matrix C_i. The probability transition matrix T_i is updated as follows. Let d_u be the degree of u before adding v. Add a new row and a new column both corresponding to vertex v to T_i and add matrix A_i of the same dimensions where A_i has all entries zero except $A_i(u, w) = -1/d_u(d_u + 1)$ for all $w \in S_i \cap \Gamma(u)$, $A_i(u, v) = 1/(d_u + 1)$, $A_i(v, u) = 1$. Since A_i has only two non-zero rows that

---

[3] For directed graphs, only vertices reachable from q by a directed path get non-zero PPR values, i.e. $\{\hat{\pi}_v > 0 \mid q \rightsquigarrow v \in \mathcal{G}\}$.

are linearly-independent, it has rank 2. Thus we compute an SVD decomposition $A_i = U \Sigma V$ where Σ is a 2×2 matrix and update C_i using Lemma 5. We again use Lemma 2 to compute the new matrix E_i as $C_i T_i^+$ where the matrix T_i^+ is updated by adding a new row corresponding to v with all entries zero and adding a matrix A_i^+ of the same dimensions. Here A_i^+ is a matrix with all zero entries except $A_i^+(u,b) = -\alpha/d_u(d_u + 1)$ for all $b \in B_i \cap \Gamma(u)$ and $A_i(v, q) = 1 - \alpha$.

Case 3: $u \in S_i$ and $v \notin S_i$. We again use the SMW formula along with Lemma 1 to compute the new count matrix C_i. The probability transition matrix T_i is now updated as follows. Let d_u be the degree of u before adding edge $e = \{u, v\}$. The matrix T_i is updated by adding a matrix A_i of the same dimensions where A_i has all entries zero except $A_i(u, w) = -1/d_u(d_u + 1)$ for all $w \in S_i \cap \Gamma(u)$. Since A_i has only one non-zero row, it has rank 1. Thus again we compute an SVD decomposition $A_i = U \Sigma V$ where Σ is a scalar and update C_i using Lemma 5.

We use Lemma 2 to compute the new E_i as $C_i T_i^+$ where T_i^+ is updated as follows. If $v \notin B_i$ currently, we add a column corresponding to v with all entries zero. Next we update T_i^+ as $T_i^+ + A_i^+$ where A_i^+ is a matrix with all zero entries except $A_i^+(u,b) = -\alpha/d_u(d_u+1)$ for all $b \in B_i \cap \Gamma(u)$ and $A_i^+(u,v) = \alpha/(d_v+1)$. The case where $u \in S_i$ and $v \notin S_i$ is analogous and is omitted.

Case 4: $u, v \in S_i$. The probability transition matrix T_i is now updated as follows. Let d_u be the degree of u and d_v be the degree of v before adding edge $e = \{u, v\}$. The matrix T_i is updated by adding a matrix A_i of the same dimensions where A_i has all entries zero except $A_i(u, w) = -1/d_u(d_u + 1)$ for all $w \in S_i \cap \Gamma(u)$, $A_i(u, v) = 1/(d_u + 1)$, $A_i(v, w) = -1/d_v(d_v + 1)$ for all $w \in S_i \cap \Gamma(v)$, $A_i(v, u) = 1/(d_v + 1)$. Since A_i has only two non-zero rows that are linearly-independent, it has rank 2. Thus again we compute an SVD decomposition $A_i = U \Sigma V$ where Σ is a 2×2 matrix and update C_i using Lemma 5.

We use Lemma 2 to compute the new E_i as $C_i T_i^+$ where T_i^+ is updated by adding a matrix A_i^+ of the same dimensions. Here A_i^+ is a matrix with all zero entries except $A_i^+(u,b) = -\alpha/d_u(d_u + 1)$ for all $b \in B_i \cap \Gamma(u)$ and $A_i^+(v,b) = -\alpha/d_v(d_v + 1)$ for all $b \in B_i \cap \Gamma(v)$.

4   Empirical Study

We evaluate our method, with respect to accuracy and efficiency, on both synthetic and real-world graphs. We first give dataset description including synthetic data generation and follow with experiment results.[4]

Synthetic Data Generation. Our graph generation algorithm is based on the planted partitions model [6]. Simply put, given the desired number of vertices in each partition we split the adjacency matrix into blocks defined by the partitioning. For each block B_{ij}, the user provides a probability p_{ij}. Using a random process based on this probability we assign a 1, i.e. an edge, to each entry in each block, and 0 otherwise. In other words, p_{ij} specifies the density of each block.

---

[4] All experiments are performed on a 3-CPU 2.8 GHz AMD Opteron 854 server with 32GB RAM. The fly-back probability α for random walks is set to 0.15.

Table 1. Graph datasets (E: #edges, N: #vertices, real graph data source: http://snap.stanford.edu/data) — C: #clusters and median conductance ϕ and size

Dataset	E	N	Description	C	med. ϕ	med. size
Synthetic	300K	909K	Planted partitions [6]	100	0.0210	3050
Web	1100K	325K	http://nd.edu links	2793	0.0625	31
Amazon	900K	262K	Product co-purchases	3739	0.1385	17
DBLP	1100K	329K	Co-authorships	4670	0.2117	27
Live Journal	21500K	2700K	Friendships	15252	0.5500	43

Using the above planted partitions model, we simulated a graph of 300K vertices, with 100 partitions of equal size. We set $p_{ii} = 10^{-3}$ and $p_i = \sum_{j, j \neq i} p_{ij} = 10^{-5}$, which yielded $909, 333$ edges in the graph.

Real Datasets. Our real graph datasets come from diverse domains such as social, Web, and co-authorship networks, and vary in size from 1 million edges to more than 20 million edges. We give a summary of our datasets in Table 1.

4.1 Pre-computation

The first phase involves clustering a given graph and then computing auxiliary information, i.e., the count and exit matrices C and E for each cluster.

There are several top-performing algorithms known for graph clustering such as [1, 8, 15, 19]. We performed experiments using both METIS [15] and the Andersen et. al. algorithm [1]. For moderate sized graphs (e.g., with 1.1 million edges), the qualities of query results obtained with either of the clustering algorithms were comparable, both in running time and accuracy. However, for large graphs (e.g., Live Journal with 21.5 million edges), METIS could not complete the clustering computation while Andersen et. al. algorithm was able to compute a clustering—it took about 35 seconds to compute each of the 15,000 clusters.

Table 1 shows the median conductance and size of the clusters found in each dataset. As shown in [17], we observe that good conductance clusters are often of small size. Moreover, graphs from different domains cluster differently, where lower conductance implies higher quality clusters.

In Figure 2, we show the precomputation time for our graphs which consists of two parts; the first (blue) bars show the average time to extract a single cluster, and the second (red) bars give the average time

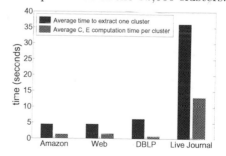

Fig. 2. Average time per cluster for two phases of pre-computation: (1) clustering and (2) C, E computation, for real graphs

to compute its corresponding C and E matrices. Note that as each graph clustered into different number of clusters, and hence we show the average time per cluster.

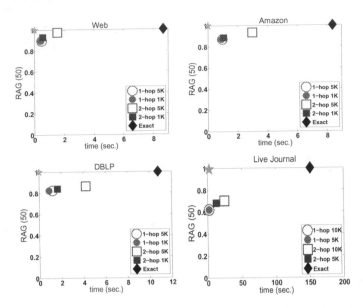

Fig. 3. Average accuracy (RAG(50)) vs. response time (sec.) for (from left to right) Web, Amazon, DBLP and Live Journal of CLUSTERRANK *1-hop* and *2-hop* heuristics compared to EXPPR. The optimal point is depicted with a star.

4.2   Query Processing

After pre-computation, our method is ready to process queries. In order to measure performance, we conducted experiments with 100 randomly chosen query vertices from each graph. We report average running time and average accuracy on all graphs. To compute accuracy, we need the "true" PPR scores. Thus, we also compute the exact PPR (EXPPR) scores using power-iterations [10].

One of the measures for accuracy is "precision at k" which can be defined as $|T_k \cap \hat{T}_k|/k \in [0, 1]$, where T_k and $\hat{T}_k$ denote the sets of top-k vertices using the exact and the test algorithm, respectively. However, precision can be excessively severe. In many real graphs, ties and near-ties in PPR scores are very common. In such a case, we would like to say that the test algorithm works well if the "true" scores of $\hat{T}_k$ are large. Therefore we use the Relative Average Goodness (RAG) at k which is defined as $\text{RAG}(k) = \frac{\sum_{v \in \hat{T}_k} p(v)}{\sum_{v \in T_k} p(v)} \in [0, 1]$ where $p(v)$ denotes the "true" PPR score of vertex v w.r.t. the query vertex.

Figure 3 shows the average RAG accuracy versus response time achieved by CLUSTERRANK for all four real-world graphs. EXPPR response time is also shown with an RAG score of 1. For the Web graph, both *1-hop* and *2-hop* average accuracy is quite close to that of the exact algorithm (0.90 and 0.97, respectively). As one might expect, the best accuracy is achieved using *2-hop* with β=5K (i.e., max. number of boundary vertices) since in that case more clusters are considered as relevant. Notice the results are similar for other graphs.

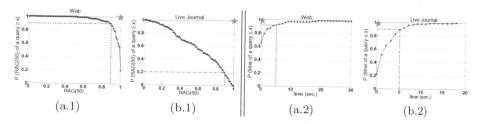

(a.1)   (b.1)   (a.2)   (b.2)

Fig. 4. NCDF distribution of RAG(50) scores for (a.1) Web, (b.1) Live Journal; and CDF distribution of query response times for (a.2) Web, (b.2) Live Journal. (using the *1-hop* heuristic and β=5K). Our method performs better on Web graph than on Live Journal, possibly due to higher quality clusters. Optimal point is depicted with a star.

We show the accuracy and response times on our synthetic graph on the right, which suggests that high accuracy is achieved for graphs with well-pronounced clusters.

	2-hop	*1-hop*
β =5K	0.9986 / 5.12	0.9865 / 2.18
β =1K	0.9892 / 2.86	0.9865 / 2.12

In Figure 4, we show the distribution of (a) accuracy scores and (b) running times of all the 100 queries in Web and Live Journal. The ideal point is also marked with a star on each figure. We observe that around 80% of the queries in Web (and around 20% in Live Journal) have an accuracy more than 0.9. Also, 90% of the queries take less than 5 seconds in both graphs.

We further study the performance of CLUSTERRANK on increasing graph sizes. We first merge clusters from our largest graph Live Journal to build a connected graph $\mathcal{G}_{1/2M}$ with half a million edges, and keep growing the number of clusters to obtain a set of increasingly larger graphs, $\{\mathcal{G}_{1M}, \mathcal{G}_{2M}, \ldots, \mathcal{G}_{21M}\}$. As before, we conduct experiments on 100 randomly chosen query vertices from $\mathcal{G}_{1/2M}$ and keep the same query set for the larger graphs to ensure that the query vertices exist in all graphs.

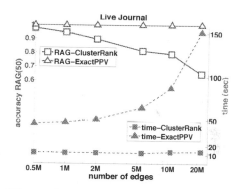

Fig. 5. Accuracy and response time for CLUSTERRANK (squares) and EXPPR (triangles) with increasing graph size

Figure 5 shows both RAG accuracy and response time versus graph size for CLUSTERRANK and EXPPR. CLUSTERRANK's accuracy remains ≈ 0.70 when the graph becomes more than 40× larger. Moreover, its response time stays almost constant at around 13 seconds across graphs with increasing size, while EXPPR's response time grows up to 150 seconds following a quadratic trend.

Next, we analyze our results qualitatively. We build the DBLP graph for years 2000-2007, and run our method on 100 randomly chosen authors. We list the top-proximity authors we found to two example authors.[5] Bold-faced authors are found to be past or future collaborators, while others are highly related with overlapping research interests (respectively, parallel computing and biomedical imaging).

'Christoph_Zenger'	'Shigehiko_Katsuragawa'
'Hans-Joachim_Bungartz'	'Joo_Kooi_Tan'
'Ralf-Peter_Mundani'	'Yoshinori_Otsuka'
'Ralf_Ebner'	'Feng_Li'
'Tobias_Weinzierl'	'Masahito_Aoyama'
'Anton_Frank'	'Shusuke_Sone'
'Ioan_Lucian_Muntean'	'Takashi_Shinomiya'
'Thomas_Gerstner'	**'Heber_MacMahon'**
'Clemens_Simmer'	'Junji_Shiraishi'
'Dirk_Meetschen'	**'Roger_Engelmann'**
'Susanne_Crewell'	'Kenya_Murase'

4.3 Dynamic Updates

To study the performance of CLUSTERRANK on dynamic updates, we use DBLP which is a time-varying graph by years. First, we build a DBLP co-authorship graph of 500K edges which spans from 1959 to 2001. Then, we perform updates to our clusters by introducing the next 1K edges in time. Note that some new edges also introduce new vertices to the graph.

Figure 6 (left) shows the distribution of the number of clusters affected per edge for the 1K new edges added, and (right) the distribution of update times. We note that more than 90% of the new edges cause fewer than 5 clusters to be updated. Moreover, 90% of the updates take less than 100 seconds, including reading/writing of the C and E matrices of the affected clusters from/to the disk.

Finally Table 2 shows CLUSTERRANK's accuracy on the time-evolving DBLP graph; i.e., after the addition of (a) 1K and (b) 500K new edges to it, which initially had 500K edges. We notice that the accuracy remains quite stable over the course of the changes to the graph, even after when the graph size doubles.

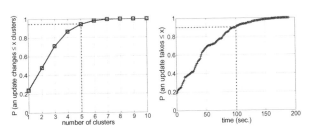

Fig. 6. Distributions of (left) number of clusters affected and (right) affected cluster update times, per new edge

---

[5] Note that direct neighbors, i.e. co-authors, are omitted from the top list as they constitute trivial nearest neighbors.

Table 2. Accuracy on DBLP with 500K, and after it grows to 501K, and 1M edges

	DBLP(500K)		DBLP(501K)		DBLP(1M)	
	2-hop	1-hop	2-hop	1-hop	2-hop	1-hop
$\beta =5K$	0.8887	0.8241	0.8717	0.8124	0.8840	0.8375
$\beta =1K$	0.8565	0.8210	0.8446	0.8095	0.8583	0.8327

5   Related Work

Scalable and efficient algorithms for exact as well as approximate computation of Page Rank (PR) scores in graphs has been studied widely [5, 7, 14, 16, 18, 20]. On the other hand, computing the *Personalized* Page Rank (PPR) scores for a given vertex, which is the central topic of our paper, is a considerably more general and harder problem than computing PR scores. The reason is that PR scores of vertices in a graph are stationary probabilities over a network-wide random walk; hence, there is a *single* (global) PR score of each vertex. On the contrary, PPR scores change as a function of the start vertex, and thus are a significant generalization of PR scores.

Designing efficient algorithms to compute PPR scores has been an active topic of relatively recent research due to its many applications including link-prediction, proximity tracking in social networks and personalized web-search. Tong et al. [25] develop fast methods for computing PPR scores. They exploit community structure by graph partitioning and correlations among partitions by low-rank approximation. However, their method is tuned for static graphs and does not address dynamic updates. The later work by Tong et al. [26] is the one which is closely related to our work; here, the authors consider proximity tracking in time-evolving *bipartite* graphs and develop matrix algebraic algorithms for efficiently answering proximity queries. The assumption that the graph is bipartite, and further, the assumption that one of its partitions is of a small size is critical to their design. For instance, these assumptions play a key role in constructing a low-rank approximation of the graph adjacency matrix, which is then perturbed to account for edge additions over time.

Sarkar and Moore [21] develop fast external memory algorithms for computing PPR scores on disk-resident graphs. They focus on suitable cluster representations in order to optimize disk accesses for minimizing query-time latency for *static* graphs. MapReduce based methods [3] optimized for disk-resident graphs also cannot deal with dynamic graphs.

The development of efficient algorithms based on linear algebraic techniques and segmented random-walks for fast computation of PPR scores on *static* graphs has also been the subject of several works. Haveliwala [12] pre-compute multiple importance scores w.r.t. various topics for each Web page towards personalization; at query time, these scores are combined to form a context-specific, composite PPR scores. Jeh and Widom [13] propose efficient algorithms to compute "partial vectors" encoding personalized views, which are incrementally used to construct the fully personalized view at query time. Fogaras et al. [10] use simulated random walk segments to approximate PPR scores by stitching the walk segments to form longer walks. Chakrabarti *et al.* [4, 11] pre-compute random walk "fingerprints" for a small fraction of the so-called hub vertices. At query time, an "active" subgraph bounded by hubs is identified where PPR scores are estimated by iterative PPV decompositions.

Most related to ours is the work by Bahmani *et al.* [2], which also uses a divide-and-conquer approach: pre-computation (random walk segments) and query-time combination of intermediate results (random walk using segments), with fast query and update times. The main difference is their underlying infrastructure; [2] needs distributed shared memory for its employed random-access Monte Carlo method, while we can work with fully distributed commodity systems—once the graph is partitioned, the compute nodes operate independently, and a dedicated node combines results only from relevant nodes.

6 Conclusion

We propose CLUSTERRANK, an efficient method for answering PPR-based k-NN queries in large time-evolving graphs. Our method addresses three major challenges associated with this problem: *(1) fast k-NN queries*; at query time, we operate on a small subset of clusters and their pre-computed information, and achieve a response time *sub-linear* in the size of the graph, *(2) efficient incremental dynamic updates*; thanks to our divide-and-conquer approach, addition or deletion of an edge/vertex triggers the update of only a small subset of clusters, which involves at most rank-2 updates to their pre-computed information, and *(3) spilling to disk*; as both query processing and dynamic updates operate on subset of clusters, only a small fraction of the graph is loaded into memory at all times while the rest sits on disk. As such, the modular design of our approach is a natural way to handle both large and time-evolving graphs simultaneously.

Acknowledgments. The authors would like to thank Christos Faloutsos for valuable discussions and the anonymous reviewers for their useful feedback. This material is based upon work supported by an ARO Young Investigator Program grant under Contract No. W911NF-14-1-0029, an ONR SBIR grant under Contract No. N00014-14-P-1155, an R&D gift from Northrop Grumman, and Stony Brook University Office of Vice President for Research. Any findings and conclusions expressed in this material are those of the authors and do not necessarily reflect the official policies or views of the funding parties.

References

1. Andersen, R., Chung, F.R.K., Lang, K.J.: Local graph partitioning using pagerank vectors. In: FOCS (2006)
2. Bahmani, B., Chowdhury, A., Goel, A.: Fast Incremental and Personalized PageRank. PVLDB 4(3), 173–184 (2010)
3. Bahmani, B., Chakrabarti, K., Xin, D.: Fast personalized PageRank on MapReduce. In: SIGMOD (2011)

4. Chakrabarti, S.: Dynamic personalized pagerank in entity-relation graphs. In: WWW 2007 (2007)
5. Chen, Y.-Y., Gan, Q., Suel, T.: Local methods for estimating pagerank values. In: CIKM 2004 (2004)
6. Condon, A., Karp, R.M.: Algorithms for graph partitioning on the planted partition model. Random Struct. Algorithms 18(2) (2001)
7. Das Sarma, A., Gollapudi, S., Panigrahy, R.: Estimating pagerank on graph streams. In: PODS (2008)
8. Ding, C.H.Q., He, X., Zha, H., Gu, M., Simon, H.D.: A min-max cut algorithm for graph partitioning and data clustering. In: ICDM 2001 (2001)
9. Feller, W.: An Introduction to Probability Theory and Its Applications. Wiley (1971)
10. Fogaras, D., Rácz, B.: Towards scaling fully personalized pagerank. In: Leonardi, S. (ed.) WAW 2004. LNCS, vol. 3243, pp. 105–117. Springer, Heidelberg (2004)
11. Gupta, M., Pathak, A., Chakrabarti, S.: Fast algorithms for top-k personalized pagerank queries. In: WWW 2008 (2008)
12. Haveliwala, T.H.: Topic-sensitive pagerank: A context-sensitive ranking algorithm for web search. In: WWW 2002 (2002)
13. Jeh, G., Widom, J.: Scaling personalized web search. In: WWW 2003 (2003)
14. Kamvar, S., Haveliwala, T., Golub, G.: Adaptive methods for the computation of pagerank. Linear Algebra Appl. 386, 51–65 (2004)
15. Karypis, G., Kumar, V.: Multilevel algorithms for multi-constraint graph partitioning. Supercomputing, 1–13 (1998)
16. Langville, A.N., Meyer, C.D.: Updating pagerank with iterative aggregation. In: WWW Alt (2004)
17. Leskovec, J., Lang, K.J., Dasgupta, A., Mahoney, M.W.: Community structure in large networks: Natural cluster sizes and the absence of large well-defined clusters. Internet Mathematics 6(1) (2009)
18. McSherry, F.: A uniform approach to accelerated pagerank computation. In: WWW 2005 (2005)
19. Ng, A.Y., Jordan, M.I., Weiss, Y.: On spectral clustering: Analysis and an algorithm. In: NIPS 2001 (2001)
20. Page, L., Brin, S., Motwani, R., Winograd, T.: The pagerank citation ranking: Bringing order to the web. Technical Report 1999-66, Stanford InfoLab (November 1999)
21. Sarkar, P., Moore, A.W.: Fast nearest-neighbor search in disk-resident graphs. In: KDD (2010)
22. Schaeffer, S.E.: Graph clustering. Computer Science Review I, 27–64 (2007)
23. Stanton, I., Kliot, G.: Streaming graph partitioning for large distributed graphs. In: KDD 2012, pp. 1222–1230 (2012)
24. Strassen, V.: Gaussian elimination is not optimal. Numer. Math. 13, 354–356 (1969)
25. Tong, H., Faloutsos, C., Pan, J.-Y.: Fast random walk with restart and its applications. In: ICDM (2006)
26. Tong, H., Papadimitriou, S., Yu, P.S., Faloutsos, C.: Proximity Tracking on Time-Evolving Bipartite Graphs. In: SDM (2008)
27. Woodbury, M.A.: Inverting modified matrices. Memorandum, Statistical Research Group

Deconstructing Kernel Machines

Mohsen Ali[1], Muhammad Rushdi[2], and Jeffrey Ho[1]

[1] Department of Computer and Information Science and Engineering, University of Florida
[2] Department of Biomedical and Systems Engineering, Cairo University, Giza, Egypt
{moali,jho}@cise.ufl.edu, mrushdi@eng.cu.edu.eg

Abstract. This paper studies the following problem: Given an SVM (kernel)-based binary classifier $\mathbf{C}$ as a black-box oracle, how much can we learn of its internal working by querying it? Specifically, we assume the feature space $\mathbb{R}^d$ is known and the kernel machine has m support vectors such that $d > m$ (or $d >> m$), and in addition, the classifier $\mathbf{C}$ is laconic in the sense that for a feature vector, it only provides a predicted label (± 1) without divulging other information such as margin or confidence level. We formulate the problem of understanding the inner working of $\mathbf{C}$ as characterizing the decision boundary of the classifier, and we introduce the simple notion of bracketing to sample points on the decision boundary within a prescribed accuracy. For the five most common types of kernel function, linear, quadratic and cubic polynomial kernels, hyperbolic tangent kernel and Gaussian kernel, we show that with $\mathbf{O}(dm)$ number of queries, the type of kernel function and the (kernel) subspace spanned by the support vectors can be determined. In particular, for polynomial kernels, additional $\mathbf{O}(m^3)$ queries are sufficient to reconstruct the entire decision boundary, providing a set of quasi-support vectors that can be used to efficiently evaluate the deconstructed classifier. We speculate briefly on the future application potential of deconstructing kernel machines and we present experimental results validating the proposed method.

Keywords: deconstruction, support vector machines, RBF.

1 Introduction

This paper proposes to investigate a new type of learning problems we called deconstructive learning. While the ultimate objective of most learning problems is the determination of classifiers from labeled training data, for deconstructive learning, the objects of study are the classifiers themselves. As its name suggests, the goal of deconstructive learning is to deconstruct a given classifier $\mathbf{C}$ by determining and characterizing (as much as possible) the full extent of its capability, revealing all of its powers, subtleties and limitations. Since classifiers in machine learning come in a variety of forms, deconstructive learning correspondingly can be formulated and posed in many different ways. This paper focuses on a family of binary classifiers based on support vector machines [1], and deconstructive learning will be formulated and studied using geometric and algebraic methods without recourse to probability and statistics. Specifically, the (continuous) feature space in which the classifier $\mathbf{C}$ is defined is assumed to be a d-dimensional vector space $\mathbb{R}^d$, and the classifier $\mathbf{C}$ is given as a binary-valued function $\mathbf{C} : \mathbb{R}^d \to \{-1, +1\}$, indicating the class assignment of each feature $\mathbf{x} \in \mathbb{R}^d$.

T. Calders et al. (Eds.): ECML PKDD 2014, Part I, LNCS 8724, pp. 34–49, 2014.

As a kernel machine, $\mathbf{C}$ is specified by a set of m support vectors $\mathbf{y}_1, \cdots, \mathbf{y}_m \in \mathbb{R}^d$ and a kernel function $\mathbf{K}(\mathbf{x}, \mathbf{y})$ such that the decision function $\mathbf{\Psi}(\mathbf{x})$ is given as the sum

$$\mathbf{\Psi}(\mathbf{x}) = \omega_1 \mathbf{K}(\mathbf{x}, \mathbf{y}_1) + \cdots \omega_m \mathbf{K}(\mathbf{x}, \mathbf{y}_m), \tag{1}$$

where $\omega_1, \cdots, \omega_m$ are the weights. With the bias $\mathbf{b}$,

$$\mathbf{C}(\mathbf{x}) = \begin{cases} +1 & \text{if } \mathbf{\Psi}(\mathbf{x}) \leq \mathbf{b}, \\ -1 & \text{if } \mathbf{\Psi}(\mathbf{x}) > \mathbf{b}. \end{cases} \tag{2}$$

The classifier $\mathbf{C}$ is also assumed to be laconic in the sense that except for the binary label, it does not divulge any other potentially useful information such as margin or confidence level. With these assumptions, we formulate the problem of deconstructing $\mathbf{C}$ through the following list of four questions (ordered in increasing difficulty)

- Can the kernel function $\mathbf{K}(\mathbf{x}, \mathbf{y})$ be determined?
- Can the subspace $\mathbf{S_Y}$ spanned by the support vectors be determined?
- Can the number m of support vectors be determined?
- Can the support vectors themselves be determined?

Without loss of generality, we will henceforth assume $\mathbf{b} = 1$. Therefore, if the support vectors and the kernel function are known, the weights ω_i can be determined completely given enough points $\mathbf{x}$ on the decision boundary

$$\mathbf{\Sigma} = \{ \mathbf{x} \mid \mathbf{x} \in \mathbb{R}^d, \ \mathbf{\Psi}(\mathbf{x}) = \mathbf{b} \ \}. \tag{3}$$

That is, a kernel machine $\mathbf{C}$ can be completely deconstructed if its support vectors and kernel function are known.

The four questions above are impossible to answer without further quantification on the type of kernel function and the number of support vectors. In this paper, we assume 1) the unknown kernel function belongs to one of the following five types: polynomial kernels of degree one, two and three (linear, quadratic and cubic kernels), hyperbolic tangent kernel and RBF kernel, and 2) the number of support vectors is less than the feature dimension, $d > m$ (or $d \gg m$) and they are linearly independent. For most applications of kernel machines, these two assumptions are not particularly restrictive since the five types of kernel are arguably among the most popular ones. Furthermore, as the feature dimensions are often very high and the support vectors are often thought to be a small number of the original training features that are critical for the given classification problem, it is generally observed that $d > m$. With these two assumptions, the method proposed in this paper shows that the first three questions can be answered affirmatively. While the last question cannot be answered for transcendental kernels, we show that using recent results on tensor decomposition (e.g., [2]), a set of quasi-support vectors can be computed for a polynomial kernel that recover the decision boundary exactly.

Given the laconic nature of $\mathbf{C}$, it seems that the only effective approach is to probe the feature space by locating points on the decision boundary $\mathbf{\Sigma}$ and to answer the above questions using local geometric features computed from these sampled points. More precisely, the proposed algorithm takes the classifier $\mathbf{C}$ and a small number of positive features in $\mathbb{R}^d$ as the only inputs. Starting with these small number of positive features, the algorithm proceeds to explore the feature space by generating new features and utilizing these new features and their class labels provided by $\mathbf{C}$ to produce points on the decision boundary. The challenge is therefore to use only a comparably small number of sampled features (i.e., queries to $\mathbf{C}$) to learn enough about Σ in order to answer the questions, and our main contribution is an algorithm that has complexity (to be defined later) linear in the dimension d of the ambient space.

Sampling points on Σ can be accomplished easily using bracketing, the same idea used in finding the root of a function (e.g., [3]). Given a pair of positive and negative features (PN-pair), the intersection of Σ and the line segment joining the two features cannot be empty, and bracketing allows at least one such point on Σ to be determined up to any prescribed precision. Using bracketing as the main tool, the first two questions can be answered by exploring the geometry of Σ in two different ways. First, the decision boundary Σ is given as the implicit surface of the multi-variate function, $\Psi(\mathbf{x}) = b$. With high-dimensional features, it is difficult to work directly with Σ or $\Psi(\mathbf{x})$; instead, the idea is to examine the intersection of Σ with a two-dimension subspace formed by a PN-pair. The locus of such intersection is in fact determined by the kernel function, and by computing such intersection, we can ascertain the kernel function on this two-dimensional subspace. For the second question, the answer is to be found in the normal vectors of the hypersurface Σ. Using bracketing, the normal vector at a given point on Σ can be determined, again in principle, up to prescribed precision. From the parametric forms of the kernel functions, it readily follows that the normal vectors of Σ are generally quite well-behaved in the sense that they either belong to the kernel subspace $\mathbf{S_Y}$ spanned by the support vectors or they are affine-translations of the kernel subspace $\mathbf{S_Y}$. For the former, a quick application of singular value decomposition immediately yields the kernel subspace $\mathbf{S_Y}$, and for the latter, the kernel subspace $\mathbf{S_Y}$ can be computed via a rank-minimization problem that can be solved (in many cases) as a convex optimization problem with the nuclear norm. If we define the complexity of the algorithm as the required number of sampled points in the feature space, it will be shown that the complexity of the proposed method is essentially $\mathbf{O}(dm)$ as it requires $\mathbf{O}(m)$ normal vectors to determine the m-dimensional kernel subspace and $\mathbf{O}(d)$ points to determine the normal vector at a point in $\mathbb{R}^d$. The constant depends on the number of steps used for bracketing, and if the features are assumed to be drawn from a bounded subset in $\mathbb{R}^d$, this constant is then independent of the dimension d.

We note that for a polynomial kernel of degree D, its decision function $\Psi(\mathbf{x})$ is a degree-D polynomial with d variables. Therefore, in principle, $\mathbf{C}$ can be deconstructed by fitting a polynomial of degree D in $\mathbb{R}^d$ given enough sampled points on Σ. However, this solution is in general not useful because it does not extend readily to transcendental kernels. Furthermore, the number of required points is in the order of d^D, and correspondingly, a direct polynomial fitting would require the inversion of a large dense (Vandermonde) matrix that is in the order of $d^D \times d^D$. With a moderate dimension of $d = 100$ and $D = 3$, this would require 10^6 points and the inversion of a $10^6 \times 10^6$ dense matrix. Our method, on the other hand, encompasses both the transcendental and polynomial kernels and at the same time, it avoids the direct polynomial fitting in $\mathbb{R}^d$ and has the overall complexity that is linear in d, making it a truly practical algorithm.

We conclude the introduction with a brief discussion on the potential usefulness of deconstructive learning, providing several examples that illustrate its significance in terms of its future prospects for theoretical development as well as practical applications. The geometric approach taken in this paper shares some visible similarities with low-dimensional reconstruction problems studied in computational geometry [4], and in fact, it is partially inspired by various 3D surface reconstruction algorithms studied in computational geometry (and computer vision) [5] [6]. However, due to the high dimensionality of the feature space, deconstructive learning offers a brand new setting that is qualitatively different from those low-dimensional spaces studied in computational geometry and various branches of geometry in mathematics. High dimensionality of the feature space has been a hallmark of machine learning, a realm that has not be actively explored by geometers, mainly for the lack of interesting examples and motivation. Perhaps deconstructive learning's emphasis on the geometry of the decision boundary in high dimensional space and its connection with machine learning could provide stimulating examples or even counterexamples unbeknown to the geometers, and therefore, provide the needed motivation for the development of new type of high-dimensional geometry [7].

On the practical side, we believe that deconstructive learning can provide a greater flexibility to the users of AI/machine learning products because it allows the users to determine the full extent of an AI/ML program/system, and therefore, create his/her own adaptation or modification of the given system for specific and specialized tasks. For example, once a kernel machine has been deconstructed, it can be subject to various kinds of improvement and upgrade in terms of its application scope, runtime efficiency and others. Imagine a kernel machine that was originally trained to recognize humans in images. By deconstructing the kernel machine and knowing its kernel type and possibly its support vectors, we can improve and upgrade it to a kernel machine that also recognizes other objects such as vehicles, scenes and other animals. The actual process of upgrading the kernel machine can be managed using existing methods such as incremental SMV or online SVM [8] [9], and at the same time, its efficiency can be improved using, for example, suitable parallelization. This provides the users with the unprecedent capability of modifying a kernel machine without access to its source code, something that to the best of our knowledge has not been studied or reported in the machine learning literature. As the kernel machines are often the main workhorse of many existing machine learning programs/systems, the ability to deconstruct a given kernel machine should have other surprising and interesting consequences and applications unforeseen at this point. Furthermore, in the context of adversarial learning [10] [11], deconstructive learning allows a kernel machine to be defeated and its deficiencies revealed. For example, how would an UAI reviewer know that a submitted binary code of a paper really does implement the algorithm proposed in the paper, not some clever implementation of a kernel machine? Deconstructive learning proposed in this paper offers a possible solution without the need to ask for the source code[1]. For more interesting examples in this direction, we leave it to the reader's imagination. Finally, perhaps the most compelling reason (to the authors) for studying deconstructive hboxlearning is

---

[1] Asking for source code is certainly not an affordable panacea for all tech problems.

inscribed by the famous motto uttered by David Hilbert more than eighty years ago: we must know and we will know! Indeed, when presented with a black-box classifier (especially the one with great repute), we have found the problem of determining the secret of its inner working by simply querying it both fascinating and challenging, a problem with its peculiar elegance and charm.

Related Work To the best of our knowledge, there is no previous work on deconstructing general kernel machines as described above. However, [10] studied the problem of deconstructing linear classifiers in a context that is slightly different from ours. This corresponds to linear kernel machines and consequently, their scope is considerably narrower than ours as (single) linear classifiers are relatively straightforward to deconstruct. Active learning (e.g., [12] [13] [14]) shares certain similarities with deconstructive learning (**DL**) in that it also has a notion of querying a source. However, the main distinction is their specificities and outlooks: for active learning, it is general and relative while for **DL**, it is specific and absolute. More precisely, for active learning, the goal is to determine a classifier from a concept class with some prescribed (PAC-like) learning error bound using samples generated from the underlying joint distribution of feature and label. In this model, the learning target is the joint distribution and the optimal learned classifier is relative to the given concept class. On the other hand, in **DL**, the learning target is a given classifier and the classifier defines an absolute partition of the feature space into two disjoint regions of positive and negative features. Furthermore, the classifier is assumed to belong to a specific concept class (e.g., kernel machines with known types of kernel function) such that the goal of **DL** is to identify the classifier within the concept class using the geometric features of the decision boundary. In this absolute setting, geometry replaces probability as the joint feature-label distribution gives way to the geometric notion of decision boundary as the main target of learning. In particular, bracketing is a fundamentally geometric notion that is generally incompatible with a probabilistic approach, and with it, **DL** possesses a much more efficient and precise tool for exploring the spatial partition of the feature space, and consequently, it allows for a direct and geometric approach without requiring much probability.

2   Preliminaries

Let $\mathbb{R}^d$ denote the feature space equipped with its standard Euclidean inner product, and for $\mathbf{x}, \mathbf{y} \in \mathbb{R}^d$, $\|\mathbf{x} - \mathbf{y}\|^2 = (\mathbf{x} - \mathbf{y})^\top (\mathbf{x} - \mathbf{y})$. For the kernel machines studied in this paper, we assume their kernel functions are of the following five types:

Linear Kernel	$\mathbf{K}(\mathbf{x}, \mathbf{y}) = \mathbf{x}^\top \mathbf{y},$
Quadratic Kernel	$\mathbf{K}(\mathbf{x}, \mathbf{y}) = (\mathbf{x}^\top \mathbf{y} + 1)^2,$
Cubic Kernel	$\mathbf{K}(\mathbf{x}, \mathbf{y}) = (\mathbf{x}^\top \mathbf{y} + 1)^3,$
Hyperbolic Tangent Kernel	$\mathbf{K}(\mathbf{x}, \mathbf{y}) = \tanh(\alpha \mathbf{x}^\top \mathbf{y} + \beta),$
Gaussian Kernel	$\mathbf{K}(\mathbf{x}, \mathbf{y}) = \exp(-\dfrac{\|\mathbf{x} - \mathbf{y}\|^2}{2\sigma^2}),$

for some constants α, β, σ. We will further refer to the three polynomial kernels and the hyperbolic tangent kernel as the Type-A kernels and the Gaussian kernel as the Type-B kernel. This particular taxonomy is based on their forms that can be generically written as

$$\mathbf{K}(\mathbf{x}, \mathbf{y}) = f(\mathbf{x}^\top \mathbf{y}), \qquad \mathbf{K}(\mathbf{x}, \mathbf{y}) = g(\|\mathbf{x} - \mathbf{y}\|^2),$$

for some smooth univariate function $f, g : \mathbb{R} \to \mathbb{R}$.

Given the forms of the kernel function, an important consequence is that the decision boundary $\boldsymbol{\Sigma}$ is determined in large part by its intersection with the kernel subspace $\mathbf{S_Y}$ spanned by the support vectors. More precisely, for $\mathbf{x} \in \mathbb{R}^d$, let $\bar{\mathbf{x}}$ denote the projection of $\mathbf{x}$ on $\mathbf{S_Y}$:

$$\bar{\mathbf{x}} = \arg \min_{y \in \mathbf{S_Y}} \|\mathbf{x} - y\|^2.$$

For Type-A kernel $\mathbf{K}(\mathbf{x}, \mathbf{y}) = f(\mathbf{x}^\top \mathbf{y})$, we have $\mathbf{K}(\mathbf{x}, \mathbf{y}_i) = \mathbf{K}(\bar{\mathbf{x}}, \mathbf{y}_i)$ for every support vector $\mathbf{y}_i$. In particular, $\bar{\mathbf{x}}$ is on the decision boundary if and only if $\mathbf{x}$ is. For Type-B kernels, we have (using Pythagorean theorem with $q^2 = \|\mathbf{x}\|^2 - \|\bar{\mathbf{x}}\|^2$)

$$\mathbf{K}(\mathbf{x}, \mathbf{y}_i) = g(\|\mathbf{x} - \mathbf{y}_i\|^2) = g(\|\bar{\mathbf{x}} - \mathbf{y}_i\|^2 + q^2),$$

and with the Gaussian kernel g, we have $g(\|\bar{\mathbf{x}} - \mathbf{y}_i\|^2 + q^2) = e^{-\frac{q^2}{2\sigma^2}} g(\|\bar{\mathbf{x}} - \mathbf{y}_i\|^2)$. It then follows that for any $\mathbf{x} \in \boldsymbol{\Sigma}$, its projection $\bar{\mathbf{x}}$ on $\mathbf{S_Y}$ must satisfy

$$\boldsymbol{\Psi}(\bar{\mathbf{x}}) = e^{\frac{q^2}{2\sigma^2}} \boldsymbol{\Psi}(\mathbf{x}) = e^{\frac{q^2}{2\sigma^2}} \mathbf{b}.$$

In other words, the decision boundary $\boldsymbol{\Sigma}$ is essentially determined by the level-sets of $\boldsymbol{\Psi}(\mathbf{x})$ on the kernel subspace $\mathbf{S_Y}$.

Since the decision boundary $\boldsymbol{\Sigma}$ is given as the implicit surface $\boldsymbol{\Psi}(\mathbf{x}) = \mathbf{b}$, a normal vector $\mathbf{n}(\mathbf{x})$ at a point $\mathbf{x} \in \mathbf{S}$ can be given as the gradient of $\boldsymbol{\Psi}(\mathbf{x})$:

$$\mathbf{n}(\mathbf{x}) = \nabla \boldsymbol{\Psi}(\mathbf{x}) = \sum_{i-1}^{m} \omega_i \nabla_{\mathbf{x}} \mathbf{K}(\mathbf{x}, \mathbf{y}_i). \tag{4}$$

For the two types of kernels we are interested in, their gradient vectors assume the following forms:

$$\nabla_{\mathbf{x}} \mathbf{K}(\mathbf{x}, \mathbf{y}) = f'(\mathbf{x}^\top \mathbf{y})\mathbf{y}, \tag{5}$$

$$\nabla_{\mathbf{x}} \mathbf{K}(\mathbf{x}, \mathbf{y}) = 2g'(\|\mathbf{x} - \mathbf{y}\|^2) (\mathbf{x} - \mathbf{y}). \tag{6}$$

Using the above formulas, it is clear that for Type-A kernels, the normal vector $\mathbf{n}(\mathbf{x})$ depends on $\mathbf{x}$ only through the coefficients in the linear combination of the support vectors, while for Type-B kernels, $\mathbf{x}$ actually contributes to the vectorial component of $\mathbf{n}(\mathbf{x})$. It will follow that an important element in the deconstruction method introduced below is to exploit this difference in how the normal vectors are computed for the two types of kernels. For example, for a polynomial kernel of degree D, a normal vector at a point $\mathbf{x} \in \boldsymbol{\Sigma}$ is

$$\mathbf{n}(\mathbf{x}) = \sum_{i=1}^{m} D \, \omega_i \, (\mathbf{x}^\top \mathbf{y}_i + 1)^{D-1} \, \mathbf{y}_i. \tag{7}$$

As a special case, for linear kernel $D = 1$, we have

$$\mathbf{n}(\mathbf{x}) = \sum_{i=1}^{m} \omega_i \mathbf{y}_i,$$

that is independent of $\mathbf{x}$. For the Gaussian kernel, we have

$$\mathbf{n}(\mathbf{x}) = \sum_{i=1}^{m} -\frac{\omega_i}{\sigma^2} \exp(-\frac{\|\mathbf{x} - \mathbf{y}_i\|^2}{2\sigma^2}) (\mathbf{x} - \mathbf{y}_i). \tag{8}$$

3 Deconstruction Method

The deconstruction algorithm requires two inputs: 1) an SVM-based binary classifier $\mathbf{\Psi}(\mathbf{x})$ that uses one of the five kernel types indicated above, and 2) a small number of positive and negative features. The algorithm uses the small number of input features to generate other pairs of positive and negative features. For a pair $\mathbf{p}, \mathbf{n}$ of positive and negative features (a PN-pair), we can be certain that the line segment joining $\mathbf{p}, \mathbf{n}$ must intersect the decision boundary in at least one point. Using bracketing, we can locate one such point $\mathbf{x}$ on the decision boundary within any given accuracy, i.e., we can use bracketing to obtain a PN-pair $\mathbf{p}, \mathbf{n}$ such that $\|\mathbf{p} - \mathbf{n}\| < \epsilon$ for some prescribed $\epsilon > 0$. With a small enough ϵ, the midpoint between $\mathbf{p}, \mathbf{n}$ can be considered approximately as a sampled point $\mathbf{x}$ on $\mathbf{\Sigma}$ and its normal vector can then be estimated. The algorithm proceeds to sample a collection of points and their normals on the decision boundary $\mathbf{\Sigma}$, and using this information, the algorithm first computes the kernel subspace $\mathbf{S_Y}$ and this step separates the Type-A kernels from the Type-B kernels (Gaussian kernel). The four Type-A kernels can further be identified by computing the intersection of $\mathbf{\Sigma}$ with a few randomly chosen two-dimensional subspaces. These two steps provide the affirmative answers to the first three questions in the introduction. For polynomial kernels, we can determine a set of quasi-support vectors that provide the exact recovery of the decision boundary $\mathbf{\Sigma}$. However, no such results for the two transcendental kernels are known at present and we leave its resolution to future research.

3.1 Bracketing

Given a PN-pair, $\mathbf{p}, \mathbf{n}$, the decision boundary must intersect the line segment joining the two features. Therefore, we can use bracketing, the well-known root-finding method (e.g., [3]), to locate the point on $\mathbf{\Sigma}$. Note that bracketing does not require the function value, only its sign. This is compatible with our classifier $\mathbf{C}$ that only gives binary values ± 1. In particular, if we bisect the interval in each step of bracketing, the length of the interval is halved at each iteration, and for a given precision requirement $\epsilon > 0$, the number of steps required to reach it is in the order of $|\log \epsilon|$. If we further assume that the features are generated from a bounded subset of $\mathbb{R}^d$ (which is often the case) with diameter less than K, then for any PN-pair $\mathbf{p}, \mathbf{n}$, bracketing terminates after at most

$$\log_2 K - \log_2 \epsilon + 1 \tag{9}$$

steps, a number that is independent of the ambient dimension d.

3.2   Estimating Normal Vectors

Given the pair $\mathbf{p}, \mathbf{n}$, let $\bar{\mathbf{p}}, \bar{\mathbf{n}}$ denote the two points near Σ after the bracketing step and $\mathbf{x}$ denote their midpoint. To estimate the normal vector at $\mathbf{x}$, we use the fact that the (unknown) kernel function is assumed to be smooth and Σ is a level-surface of the decision function $\Psi(\mathbf{x})$ that is a linear combination of smooth functions. Consequently, a randomly chosen point on Σ is almost surely non-singular [15] in that it has a small neighborhood in Σ that can be well-approximated using a linear hyperplane (its tangent space) in $\mathbb{R}^d$. Accordingly, we will estimate the normal vector at $\mathbf{x}$ by linearly fitting a set of points on Σ that belong to a small neighborhood of $\mathbf{x}$. More specifically, we chose a small $\delta > \epsilon > 0$ and generate PN-pairs on the sphere centered at $\mathbf{x}$ with radius δ. Using bracketing and the convexity of the ball enclosed by the sphere, we obtain PN-pairs that are near Σ and no more than δ away from $\mathbf{x}$. Taking the midpoint of these PN-pairs, we obtain a set of randomly generated $\mathbf{O}(d)$ points on Σ. We linearly fit a $(d-1)$-dimensional hyperplane to these points and the normal vector is then computed as the eigenvector associated to the smallest eigenvalues of the normalized covariance matrix. The result can be further sharpened by repeating the step over multiple δ and taking the (spherical) average of the unit normal vectors. However, in practice, we have observed that good normal estimates can be consistently obtained using one small $\delta \approx 10^{-3}$ (with $\epsilon = 10^{-6}$) and $2d$ sampled points[2].

3.3   Determining Kernel Subspace S_Y

To determine the kernel subspace S_Y, we will use the formulas for the normal vectors given in Equations 5 and 6. Assume that we have sampled $s > m$ points on Σ and their corresponding normal vectors. Let $\mathbb{N}, \mathbf{X}$ denote the following two matrices

$$\mathbf{X} = [\mathbf{x}_1 \, \mathbf{x}_2 \dots \mathbf{x}_s], \qquad \mathbb{N} = [\mathbf{n}_1 \, \mathbf{n}_2 \dots \mathbf{n}_s] \tag{10}$$

that horizontally stack together the points $\mathbf{x}_i$ and their normal vectors $\mathbf{n}_i$, respectively. If all $\mathbf{n}_i$ are correctly recovered (without noise), we have the following:

- For Type-A kernels, $\mathbf{n}_i \in S_Y$, i.e., $\mathbf{n}_i$ is a linear combination of the support vectors.
- For Type-B kernels, $\mathbf{n}_i \in \gamma_i \mathbf{x}_i + S_Y$, for some $\gamma_i \in \mathbb{R}$, i.e., $\mathbf{n}_i - \gamma_i \mathbf{x}_i \in S_Y$.

Note that γ_i depends on $\mathbf{x}_i$ and the two statements can be readily checked using Equations 4 - 6. Therefore, the kernel subspace S_Y can be recovered, for Type-A kernels, using Singular Value Decomposition (SVD). Specifically, let $\mathbb{N} = \mathbf{UDV}^\top$ denote the singular value decomposition of $\mathbb{N}$. There are precisely m nonzero singular values and S_Y is spanned by the first m columns of $\mathbf{U}$. For Type-B, a slight complication arises because we must determine s constants $\gamma_1, \cdots, \gamma_s$ such that the span of the following matrix is S_Y:

$$\mathbb{N} - \mathbf{X}\Gamma \equiv [\mathbf{n}_1 \, \mathbf{n}_2 \dots \mathbf{n}_s] - [\gamma_1 \mathbf{x}_1 \, \gamma_2 \mathbf{x}_2 \dots \gamma_s \mathbf{x}_s], \tag{11}$$

where Γ is a diagonal matrix with γ_i as its entries. Note that in general, $\mathbb{N}, \mathbf{X}$ are of full-rank $\min(d, s)$, and we are trying to find a set of γ_i such that the above matrix has

---

[2] We note that for sufficiently small δ, the angular error of the estimated normal is approximately in the order of $\tan^{-1}(\frac{\epsilon}{2\delta})$.

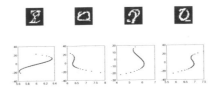

Fig. 1. Intersections of Σ and two-dimensional affine subspaces An SVM using the cubic kernel is trained on MINST dataset. **Top Row:** Midpoints of PN-pairs near the decision boundary Σ after bracketing. **Bottom Row:** Sampled polynomial curves given the intersections of the decision boundary with two-dimensional affine subspaces containing the images above.

rank $m < s$. However, for a generically chosen set of $x_1, \cdots, x_s$, the rank of $\mathbb{N} - X\Gamma$ is at least m because the support vectors are linearly independent. Therefore, γ_i can be determined via the following rank-minimization problem

$$\arg\min_{\gamma_i} \mathbf{Rank}([n_1\ n_2 \ldots n_s] - [\gamma_1 x_1\ \gamma_2 x_2 \ldots \gamma_s x_s]). \tag{12}$$

As is well-known, a convex relaxation of the above problem uses the nuclear norm $\|\cdot\|_*$ (sum of singular values) as the surrogate

$$\arg\min_{\text{diagonal } \Gamma} \|\mathbb{N} - X\Gamma\|_*, \tag{13}$$

and there are efficient algorithms for solving this type of convex optimization problem [16]. We note that for Type-A kernels, the rank is minimized at $\gamma_1 = \cdots = \gamma_s = 0$. In both cases, the span of $\mathbb{N} - X\Gamma$ gives the kernel subspace $\mathbf{S_Y}$. As the support vectors are assumed to be linearly independent, the dimension of $\mathbf{S_Y}$ then gives the number of support vectors. For noisy recovery, the above method requires the standard modification that uses the significant gap between singular values as the indicator. For Type-A kernels, this is applied to the SVD decomposition of $\mathbb{N}$ directly, and for Type-B kernels, this is applied to the SVD decomposition of $\mathbb{N} - X\Gamma$ with Γ determined by the nuclear norm minimization.

3.4 Determining Kernel Type

For determining the four Type-A kernels, we will examine the locus of the intersection of the decision boundary with a two-dimensional affine subspace containing a point close to the decision boundary. More specifically, let x_+, x_- denote a PN-pair that is sufficiently close to the decision boundary Σ. We can randomly generate a two-dimensional subspace containing x_+, x_- by, for example, taking the subspace $\mathbf{A}$ formed by x_+, x_- and the origin. For a generic two-dimensional subspace $\mathbf{A}$, its intersection with Σ is a one-dimensional curve, and the parametric form of this curve is determined by the (yet unknown) kernel function. See Figure 1. Take a polynomial kernel of degree D as an example. By its construction, the intersection of the decision boundary and the affine subspace $\mathbf{A}$ is nonempty, and the locus of the intersection formed a curve in $\mathbf{A}$ that satisfies a polynomial equation of degree D. This can be easily seen as follows: take $\mathbf{x}_+$ as the origin on $\mathbf{A}$ and choose (arbitrary) orthonormal vectors $\mathbf{U}_1, \mathbf{U}_2 \in \mathbb{R}^d$ such that the triplet $\mathbf{x}_+, \mathbf{U}_1, \mathbf{U}_2$ identifies $\mathbf{A}$ with $\mathbb{R}^2$. Therefore, any point $p \in \mathbf{A}$ can be uniquely identified with a two-dimensional vector $\mathbf{p} = [\mathbf{p}_1, \mathbf{p}_2] \in \mathbb{R}^2$ as

$$p = \mathbf{x}_+ + \mathbf{p}_1 \mathbf{U}_1 + \mathbf{p}_2 \mathbf{U}_2.$$

If $\mathbf{p} \in \mathbf{A}$ is a point in the intersection of $\mathbf{A}$ with the decision boundary $\mathbf{\Psi}(\mathbf{p}) = \mathbf{b}$, we have

$$\sum_{i=1}^{m} w_i((\mathbf{x}_+^\top \mathbf{Y}_i + \mathbf{p}_1 \mathbf{U}_1^\top \mathbf{Y}_i + \mathbf{p}_2 \mathbf{U}_2^\top \mathbf{Y}_i)1)^D = \mathbf{b}, \tag{14}$$

which is a polynomial of degree D in the two variables $\mathbf{p}_1, \mathbf{p}_2$. Therefore, to ascertain the degree of the polynomial kernel, we can (assuming $D < 4$)

- Sample at least nine points on the intersection of the decision boundary and $\mathbf{A}$.
- Fit a bivariate polynomial of degree D to the points. If the fitting error is sufficiently small, this gives an indication that the polynomial kernel is indeed of degree D.

We note that up to a multiplicative constant, a bivariate cubic polynomial in $\mathbb{R}^2$ has nine coefficients and this gives the minimum number of points required to fit a cubic polynomial. In addition, since the degree of the polynomial is invariant under any linear transform, this shows that the choice of the two basis vectors is immaterial. The advantage of the reduction from $\mathbb{R}^d$ to $\mathbb{R}^2$ is considerable as it implies that the complexity of this step is essentially independent of the ambient dimension d. For a transcendental kernel such as the hyperbolic tangent kernel, the locus of the intersection is generally not a polynomial curve and this can be detected by the curve-fitting error. Although, in principle, one affine subspace $\mathbf{A}$ is sufficient to distinguish between four Type-A kernels (as shown by the above equation), in practice, due to various issues such as possible degeneracy of the polynomial curve and the curve fitting error, we randomly sample several affine subspaces and use a majority voting scheme to determine the kernel type.

3.5 Complexity Analysis and Exact Recovery of Σ

The steps outlined above essentially aim to ascertain the parametric form of the decision boundary Σ using a (relatively) small number of sampled points on Σ. We note that the bracketing error in general can be explicitly controlled, and there are only two steps above that incur uncertainty: the normal estimate and the nuclear norm relaxation of the rank minimization problem. Our approach of using the local linear approximation to estimate the normal vector at a point is the standard one common in computational geometry and machine learning (e.g., [17,18] [19]), and the nuclear norm relaxation is the standard convex relaxation for the original NP-hard rank minimization problem [20]. A complete complexity analysis of the proposed algorithm would require detailed probabilistic estimates pertaining to these two steps, and although there are partial and related results scattered in the literature (e.g.,[20] [21]), we are unable to provide a definitive result at this point. Instead, we present a simple complexity analysis below under the assumption that these two steps can be determined exactly, i.e., the convex relaxation using the nuclear norm gives the same result as the original rank minimization problem.

The computational complexity can be defined as the number of features (not necessarily only on the decision boundary) in $\mathbb{R}^d$ sampled during the process and this number is the same as the number of queries to the classifier $\mathbf{C}$. From the above, it is clear that to determine the m-dimensional kernel subspace, at least $\mathbf{O}(m)$ sampled normals are required, i.e., $\mathbb{N}$ has at least m columns. Furthermore, to determine each normal vector at a given point $\mathbf{x}$ $\mathbf{O}(d)$ number of points are required, as the ambient dimension is d. Therefore, the total complexity is $\mathbf{O}(dm)$. The multiplicative constant here, as can be readily seen, is bounded by the maximum number of steps required for the bracketing, and this number is independent of the dimensions d, m, provided the features are drawn from a bounded subset of $\mathbb{R}^d$ (Equation 9).

Once the kernel subspace $\mathbf{S_Y}$ and the kernel type are determined, this allows us to focus on the intersection $\Sigma \cap \mathbf{S_Y}$. In the case $m << d$, this reduction from $\Sigma \subset \mathbb{R}^d$ to $\Sigma \cap \mathbf{S_Y} \subset \mathbf{S_Y}$ is computationally significant. In particular, for polynomial kernels, we can sample $\mathbf{O}(m^D)$ points on $\Sigma \cap \mathbf{S_Y}$ to reconstruct the polynomial $\mathbf{\Psi}(\mathbf{x})$ on $\mathbf{S_Y}$. At this point, $\mathbf{\Psi}(\mathbf{x})$ is a degree-D polynomial in m variables, and using recent results on tensor decomposition (e.g., [2] [22])[3], we can decompose $\mathbf{\Psi}(\mathbf{x})$ (more precisely, its homogenized version)

$$\mathbf{\Psi}(\mathbf{x}) = \sum_{i=1}^{r} \ell_i(\mathbf{x})^D, \tag{15}$$

where $\ell_1, \cdots, \ell_r$ are linear (homogeneous) polynomials. The smallest integer r for such decomposition gives the rank of the (homogeneous) polynomial (as a symmetric tensor) and in general, such decomposition is also possible for r greater than the rank. If we write the linear polynomials (after de-homogenization) as $\ell_i(\mathbf{x}) = \mathbf{z}_i^\top \mathbf{x} + 1$ for some vector $\mathbf{z}_i$, it is tempting to infer $\mathbf{z}_i$ as the support vector $\mathbf{y}_i$ from the above equation. However, because the non-uniqueness of the decomposition, $\mathbf{z}_i \neq \mathbf{y}_i$ in general. Nevertheless, $\mathbf{z}_i$ do act as if they are support vectors in the sense that the evaluation of the polynomial $\mathbf{\Psi}(\mathbf{x})$ becomes computationally trivial using the above decomposition. For polynomial kernels, the recovery of these quasi-support vectors $\mathbf{z}_i$ then determines the decision boundary Σ exactly, essentially completing the deconstruction process. Although the general algorithms for tensor decomposition [2] [22] require some mathematical machinery, the special case of quadratic kernels (degree-two polynomials) can be readily solved using eigen-decomposition of a symmetric matrix (the details are provided in the supplemental material). For transcendental kernels, no similar results are known at present. Although the reduction from $\Sigma \subset \mathbb{R}^d$ to $\Sigma \cap \mathbf{S_Y} \subset \mathbf{S_Y}$ offers the possibility of reconstructing the decision boundary in $\mathbf{S_Y}$, due to the nature of the transcendental functions, the details are considerably more difficult than the polynomial case, and we leave its resolution to future research.

4   Experiments

We present two sets of experiments in this section. The first set of experiments evaluates various components of the proposed method and the second set of experiments applies the proposed method to explicitly deconstruct a kernel machine and subsequently improve it using incremental SVM [9].

---

[3] Algorithm 5.1 in http://arxiv.org/pdf/0901.3706v2.pdf, the archived version of [2].

4.1   Evaluation of the Deconstruction Algorithm

We present two experiments using kernel machines whose support vectors are randomly generated (first experiment) and support vectors trained using real image data (second experiment). We remark that there is no qualitative differences between deconstructing kernel machines with randomly-generated support vectors and deconstructing kernel machines trained with real data since, in both cases, the kernel function and decision function (Eq 1) are the same. Using randomly-generated kernel machines allow us to study the behavior of the deconstruction algorithm over a much wider range of support vector configuration, demonstrating its accuracy and robustness. In the first set of experiments, we set the feature dimension $d = 30$, and we randomly generate 12 support vectors. For determining the kernel type, we sample 25 points close to the decision boundary Σ and at each point, we compute the intersection of Σ and a two-dimensional subspace. We fit a quadratic and then a cubic polynomial to these points, and the smallest degree giving an error below some threshold value is declared as the degree of the kernel. However, if in both cases the fitting errors are greater than the threshold value, the kernel is declared to be a Gaussian kernel at this location. This is repeated at 25 sampled locations and a majority vote is used to determine the kernel type. Once the kernel type is determined, we use SVD to determine the dimension of the kernel subspace $\mathbf{S_Y}$ and the subspace itself. For the Gaussian kernel, the nuclear-norm minimization is performed before using SVD to locate the subspace $\mathbf{S_Y}$. In this experiment, we sample $s = 100$ points on the decision boundary in order to form the matrices $\mathbb{N}$, $\mathbf{X}$ and the tolerance in the bracketing step is set at 10^{-6}. Let $\overline{\mathbf{S_Y}}$ denote the kernel subspace computed by our method. We use the principal angles [23] between the two subspaces $\mathbf{S_Y}, \overline{\mathbf{S_Y}}$ as the metric for quantifying the error.

Summary. The gap between the singular values of $\mathbb{N}$ is an important indicator of the dimension of the kernel subspace, and it is affected by the accuracy of the normals. Figure 2 shows the effect in terms of the radius δ used in computing the normals, showing the expected result that the ratio of δ/ϵ is directly related to the accuracy of the recovered normals (larger ratios provide more accuracy). For determining the kernel type, the specificity for the polynomial kernels is close to 100% with the specificity of approximately 80% for the Gaussian kernel (and hyperbolic tangent kernel). This can be attributed to the majority voting scheme used in assigning the kernel type, and we leave it as important future work for designing more robust criteria. The accuracy of the recovered kernel subspaces is shown in Figure 3 and 4a. The first figure shows the means and variances of the (cosine of) twelve principal angles, taken over one hundred randomly generated kernel machines using polynomial kernels. Note that $\cos^{-1}(0.99)$ is approximately $8°$ and this gives a good indication of the accuracy. In the second figure, the twelve principal angles computed before and after the rank-minimization are shown, indicating the correctness and necessity of performing rank-minimization. Finally, each deconstruction makes between $60,000$ and $70,000$ queries to the classifier, and on a typical 3Ghz machine, it takes no more than a few minutes to complete the deconstruction process. Since the algorithm is readily parallelizable (which would be important for deconstruction in high-dimensional feature spaces), a full parallelized and optimized implementation can be expected to shorten the running time considerably, perhaps in the range of only a few seconds.

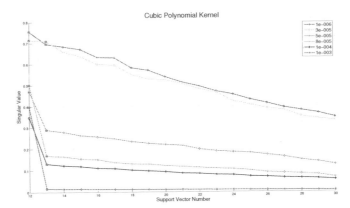

Fig. 2. Singular Values of the normal matrix $\mathbb{N}$ for different choices of δ. The kernel machine uses a polynomial cubic kernel with 12 support vectors. The expected gaps between the 12th and 13th singular values are indicated by the green markers. Note that for a fixed tolerance $\epsilon = 10^{-6}$, the optimal $\delta = 10^{-3}$. For smaller δ without changing ϵ accordingly, the estimated normals become less accurate. (**Image best viewed in color**).

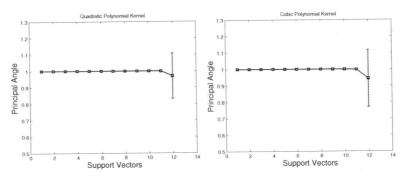

Fig. 3. Means and variances of the cosines of the twelve principal angles between $\overline{\mathbf{S}_Y}$ and $\mathbf{S}_Y$. Means and variances are taken over one hundred independent deconstruction results for kernel machines with twelve support vectors using a polynomial kernel (Quadratic kernel on the left and cubic kernel on the right). (**best viewed in color**).

In the second experiment, we train a kernel machine with cubic polynomial kernel using 1000 images from MNIST dataset [24]. The positive class consists of images of the digit 2 and the negative class consists of $0, 5, 7, 8$. The trained kernel machine has 275 support vectors. Figure 1 displays the intersections of the decision boundary with several two-dimensional affine subspaces, noticing the superpositions of the images of 2 with images of other digits. In this experiment, we randomly generate 200 two-dimensional affine subspaces and for each subspace, its vote on the type of kernel is determined as above. Figure 4b shows the distribution of votes, clearly indicating the correct result. For this experiment, the gap in the singular values of $\mathbb{N}$ indicates the correct dimension of the kernel subspace (275) and the kernel subspace is also successfully recovered.

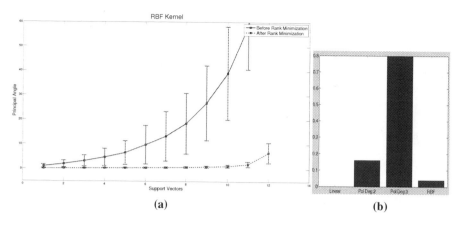

Fig. 4. Left: Means and variances of the the twelve principal angles between $\overline{S_Y}$ and S_Y. Means and variances are taken over one hundred independent deconstruction results for kernel machines with twelve support vectors using a Gaussian kernel. The principal angles before and after rank minimization are shown. (**best viewed in color**). **Right:Distribution of Votes on Kernel Type** For a cubic kernel machine trained on 1000 MNIST images, the distribution of votes on kernel type for 200 randomly sampled two-dimensional affine subspaces. The correct result is clearly indicated.

4.2   Kernel Machine Upgrade without Source Code

In the second experiment, we demonstrate the possibility of upgrading a kernel machine without access to the kernel machine's source code. As outlined in the introduction, we apply the deconstruction algorithm to deconstruct the kernel machine. This step provides us with the kernel type and quasi-support vectors (for a polynomial kernel machine). For the subsequent upgrade (or update), we use the incremental SVM algorithm [9] to retrain the kernel machine given the new training data. Specifically, we first train a kernel machine using MNIST dataset: images of digit 1 as positive samples and the negative training samples comprise the remaining digits except 8. Dimensionality reduction is applied to the images using PCA to a feature space of dimension 60. An SVM with quadratic kernel is trained on these training samples, resulting in 97.30% true positive detection rate and 99.17% true negative detection rate on the test dataset. The initial kernel machine has 48 support vectors. During deconstruction, the kernel subspace is recovered using 800 sampled normal vectors. Let $\mathbb{N}$ denote the matrix obtained by horizontally stacking together the normal vectors and $\mathbb{N} = USD$, its SVD decomposition. The plot of the singular values is shown in Figure 5b and the significant gap between the 48th and 49th singular values indicate the correct dimension (and the number of support vectors). The principle angles between the kernel subspace estimated by the first 48 columns of U and the ground-truth is shown in Figure 5a. Once the kernel subspace is recovered, we proceed to recover the quasi-support vectors. The kernel machine defined by the quasi-support vectors should be a good approximation of the original kernel machine and this is shown in Table 1a, where we compare the classification results using the recovered kernel machine and the original one. In this example, the results as expected are quite similar, with the recovered kernel machine

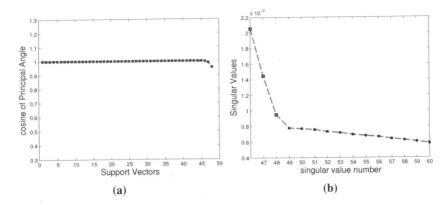

(a)   (b)

Fig. 5. Left: Cosines of the principal angles between the recovered kernel subspace and the ground-truth kernel subspace. **Right:**Singular values of the matrix $\mathbb{N}$. The gap between 48th and 49th singular values is significant as the gaps among the remaining singular values are substantially smaller. The correct dimension of the kernel subspace (and the number of support vectors) is 48.

actually performing slightly better. Once we have recovered the quasi-support vectors, we next proceed to upgrade the kernel machine. The task is to upgrade a kernel machine that recognizes only digit 1 to a kernel machine that recognizes digits 1 and 8. The classification results for the initial and upgraded kernel machines are tabulated in Table 1b. As shown in the table, before the upgrade, the original kernel machine performs poorly on the images of digit 8 and for the upgraded machine, both digits can now be successfully classified.

Table 1. Left: Confusion matrices for the original kernel machine and the kernel machine defined by the recovered quasi-support vectors. Both machines are tested on the same test dataset. **Right:**Comparisons of classification results for the original kernel machine and the upgraded kernel machine.

(a)   (b)

	Quasi-SV Machine outcome		Original Machine outcome	
	+ve	-ve	+ve	-ve
Positive	100.00%	00.00%	97.30%	2.70%
Negative	3.73%	96.27%	0.83%	99.17%

	Classification Rate	
	Original Machine	Upgraded Machine
Digit 1	97.30%	100.00%
Digit 8	00.00%	92.31%
Negative	99.17%	97.93%

5   Conclusion

We have introduced the novel notion of deconstructive learning and proposed an algorithm for deconstructing kernel machines. Preliminary experimental results have demonstrated both the viability and effectiveness of the proposed method. Although much work remains for the future, the results presented in this paper serve as a small first step in understanding the full implication and potential of deconstructive learning.

References

1. Vapnik, V.: Statistical Learning Theory. Wiley-Interscience (1998)
2. Brachat, J., Comon, P., Mourrain, B., Tsigaridas, E.: Symmetric tensor decomposition. Linear Algebra Appl. 433(11-12), 1851–1872 (2010)
3. Heath, M.: Scientific Computing. The McGraw-Hill Companies, Inc. (2002)
4. de Berg, M., Cheong, O., van Kreveld, M., Overmars, M.: Computational Geometry: Algorithms and Applications. Springer (2010)
5. Dey, T.: Curve and Surface Reconstruction: Algorithms with Mathematical Analysis. Cambridge University Press (2006)
6. Horn, B.: Robot Vision. MIT Press (2010)
7. Carlsson, G.: Topology and data. Bulletin of the American Mathematical Society 46(2), 255–308 (2009)
8. Cauwenberghs, G., Poggio, T.: Incremental and decremental support vector machine learning. In: Advances in Neural Information Processing Systems, pp. 409–415 (2001)
9. Diehl, C.P., Cauwenberghs, G.: SVM incremental learning, adaptation and optimization. In: Proceedings of the International Joint Conference on Neural Networks 2003., pp. 2685–2690. IEEE (2003)
10. Lowd, D., Meek, C.: Adversarial learning. In: Proceedings of the eleventh ACM SIGKDD International Conference on Knowledge Discovery in Data Mining, pp. 641–647. ACM (2005)
11. Torkamani, M., Lowd, D.: Convex adversarial collective classification. In: Proc. Int. Conf. Machine Learning (ICML) (2013)
12. Dasgupta, S.: Analysis of a greedy active learning strategy. In: Advances in Neural Information Processing Systems (2004)
13. Balcan, M., Beygelzimer, A., Langford, J.: Agnostic active learning. In: Proc. Int. Conf. Machine Learning (ICML) (2006)
14. Balcan, M.-F., Broder, A., Zhang, T.: Margin based active learning. In: Bshouty, N.H., Gentile, C. (eds.) COLT. LNCS (LNAI), vol. 4539, pp. 35–50. Springer, Heidelberg (2007)
15. Hirsch, M.: Differential Topology. Springer (1997)
16. Nesterov, Y., Nemirovski, A.: Some first-order algorithm for l1/nuclear norm minimization. Acta Numerica (2014)
17. Roweis, S.T., Saul, L.K.: Nonlinear dimensionality reduction by locally linear embedding. Science 290(5500), 2323–2326 (2000)
18. Belkin, M., Niyogi, P.: Semi-supervised learning on riemannian manifolds. Machine Learning 56(1-3), 209–239 (2004)
19. Rifai, S., Dauphin, Y., Vincent, P., Bengio, Y., Muller, X.: The manifold tangent classifier. In: NIPS, pp. 2294–2302 (2011)
20. Candes, E., Recht, B.: Exact matrix completion via convex optimization. Commun. ACM 55(6), 111–119 (2012)
21. Belkin, M., Niyogi, P., Sindhwani, V.: Manifold regularization: A geometric framework for learning from labeled and unlabeled examples. The Journal of Machine Learning Research 7, 2399–2434 (2006)
22. Ballico, E., Bernardi, A.: Decompoision of homogeneous polynomials with low rank. Mathematische Zeitschrift 271(3-4), 1141–1149 (2012)
23. Golub, G., Loan, C.V.: Matrix Computation. John Hopkins University Press (1996)
24. LeCun, Y., et al.: Gradient-based learning applied to document recognition. Proceedings of the IEEE 86(11), 2278–2324 (1998)

Beyond Blocks:
Hyperbolic Community Detection

Miguel Araujo[1,2], Stephan Günnemann[1], Gonzalo Mateos[1], and
Christos Faloutsos[1]

[1] Carnegie Mellon University,
5000 Forbes Avenue, Pittsburgh, PA, USA
{maraujo,sguennem,mateosg,christos}@cs.cmu.edu
[2] CRACS/INESC-TEC & Universidade do Porto,
Rua do Campo Alegre, 1021/1055, 4169-007 Porto, Portugal

Abstract. What do real communities in social networks look like? Community detection plays a key role in understanding the structure of real-life graphs with impact on recommendation systems, load balancing and routing. Previous community detection methods look for uniform blocks in adjacency matrices. However, after studying four real networks with ground-truth communities, we provide empirical evidence that communities are best represented as having an *hyperbolic* structure. We detail HYCoM - the Hyperbolic Community Model - as a better representation of communities and the relationships between their members, and show improvements in compression compared to standard methods.

We also introduce HYCoM-FIT, a fast, parameter free algorithm to detect communities with *hyperbolic* structure. We show that our method is effective in finding communities with a similar structure to self-declared ones. We report findings in real social networks, including a community in a blogging platform with over 34 million edges in which more than 1000 users established over 300 000 relations.

1 Introduction

Given a large social network, what do real communities look like? How does their size affect their structure, shape, and density[1] of connections? Are the communities' degree distributions uniform as implied by traditional community detection algorithms that look for quasi-cliques (i.e., dense rectangles or blocks of uniform density in the adjacency matrix)? One would intuitively expect that larger communities exhibit similar relational patterns to the whole graph. Accordingly, do the communities' degree distributions obey power laws?

The present paper deals with the following problems: what is the structure of communities in large, real social networks and what are suitable models to describe them? Moreover, how can one find these communities in an effective and scalable way by leveraging this particular structure and without any user-defined parameters? We analyze four real-world social networks with ground-truth communities and provide empirical evidence that communities exhibit power law

---

[1] Density equals the number of edges divided by the number of nodes squared.

T. Calders et al. (Eds.): ECML PKDD 2014, Part I, LNCS 8724, pp. 50–65, 2014.

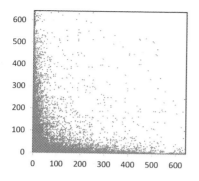 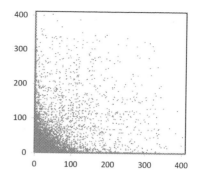

Fig. 1. Motivation for our work: Real ground-truth community

Fig. 2. Result of our work: Community found by HYCOM-FIT

degree distributions. As such, they are typically best represented as having an hyperbolic structure in the adjacency matrix, rather than rectangular (uniform) structure. We detail HYCOM - the Hyperbolic Community Model - as a better representation of communities and the relationships between their members, and introduce HYCOM-FIT as a scalable algorithm to detect communities with hyperbolic structure. To illustrate our model and algorithm, Figure 1 represents the adjacency matrix of a real (ground-truth) community externally provided when nodes are ordered by degree, and Figure 2 shows the adjacency matrix of an exemplary community found by our algorithm. Clearly, both communities do not show uniform density. In a nutshell, the main contributions of our work are:

- Introduction of the **Hyperbolic Community Model:** We provide empirical evidence that communities in large, real social graphs are better modeled using an hyperbolic model. We also show that this model is better from a compression perspective than previous models.
- **Scalability:** We develop HYCOM-FIT, an algorithm for the detection of hyperbolic communities that scales linearly with the number of edges.
- **No user-defined parameters:** HYCOM-FIT detects communities in a parameter-free fashion, transparent to the end-user.
- **Effectiveness:** We applied HYCOM-FIT on real data where we discovered communities that agree with intuition.
- **Generality:** HYCOM includes uniform block communities used by other algorithms as a special case.

2 Background and Related Work

Nodes in real-world networks organize into communities or clusters, which tend to exhibit a higher degree of 'cohesiveness' with respect to the underlying relational patterns. Group formation is natural in social networks as people organize in families, clubs and political organizations; see e.g., [19]. Communities also emerge in protein-protein interaction or gene-regulatory networks whereby genes associated to a common metabolic function tend to be more densely connected [16], or in the World Wide Web where hyperlinks between theme-related websites are more prevalent [6]. In this context, an important problem is to identify these groups of nodes from given (unlabeled) graph data.

Formally, unveiling communities in networks can be cast as a graph partitioning or clustering problem, e.g., [13]. While a fairly large number of standard methods have been proposed to this end [7], network community detection nevertheless remains a very active area of research – arguably an indicator of the problem's inherent difficulty. As discussed in [21], the threefold challenge faced is due to (c1) a lack of consensus on the structural definition of network community; (c2) the fact that node subset selection overlaid to the combinatorial structure of graphs typically leads to intractable formulations; and (c3) the lack of ground-truth to carry out an objective validation on real data.

The widespread notion of cohesiveness used to group nodes has typically reflected that community members are (i) well connected among themselves, while they are (ii) relatively well separated from the remaining nodes. Building on this intuition, methods based on adaptations of hierarchical and spectral clustering have been proposed [9,11], in addition to those relying on block-modeling [19], co-clustering or cross-associations [3]. Generative model-based approaches have been also proposed [20], while traditional methods rely on optimization of judicious criteria such as conductance and normalized cut [17], as well as modularity [14], to name a few. Similar to the proposed method, model selection approaches based on Minimum Description Length (MDL) were put forth in [2,8,18]. MDL-based algorithms are attractive since they are devoid of user-defined parameters. For a comprehensive tutorial on community detection methods and their multiple variants, the reader is referred to [7].

All previous community detection methods have been either explicitly or implicitly aimed at extracting areas of high and/or uniform density in the adjacency matrix (e.g., near cliques in the corresponding graphs). In this paper, we argue that communities in real networks do not show such a density profile but are better represented by using a hyperbolic model.

3   Empirical Observations

The goal of this section is to provide empirical evidence that real communities are not blocks of uniform density and are best represented as hyperbolic structures. We examined a collection of four real networks (Table 1) previously used in the literature [20,21] with significantly different ground-truth definitions, available in the Stanford Network Analysis Project (SNAP) collection. These datasets have *externally provided* community labels for a number of communities and, in the following, we analyze the meaning of these different community definitions and explore their underlying structure.

The YOUTUBE and LIVEJOURNAL datasets are standard friendship networks. Each node represents a user of the website and friendship relations establish links

Table 1. Summary of **real-world networks** used

Dataset	Networks with ground-truth communities				Network with node labels
	AMAZON	DBLP	YOUTUBE	LIVEJOURNAL	WIKIPEDIA
Nodes	334 863	317 081	1 134 890	3 997 962	143 508
Edges	1 851 744	2 099 732	5 975 248	34 681 889	3 753 156

between them. In these websites, users are also able to form groups that others can join. We consider each of these groups as a ground-truth community.

The DBLP dataset is a computer science co-authorship network: two authors (nodes) are connected if they published at least one paper together. Publication venues (i.e. specific journal or conference series like ECML/PKDD) define ground-truth communities. In this case, ground-truth communities roughly correspond to scientific fields.

The AMAZON dataset was collected by crawling the Amazon website and is based on the "customers who bought this item also bought" feature. Each individual node corresponds to a product and an edge exists if products i and j are frequently co-purchased. Products are organized hierarchically in categories and we view products in the same category as forming a ground-truth community. In this scenario, communities represent product similarity.

Observations. Exploring the communities in these networks allows for a better understanding of common community structures.

Density. Firstly, in Figure 3 we see that community size impacts edge density (here plotted for the DBLP data). While small communities might have any density, big communities are consistently less dense. These simple observations already indicate that blocks of uniform density are not the appropriate representation for a wide range of communities: small communities might go from small stars to full-cliques and big communities are usually not dense enough for a uniform block representation to be the most suitable. We hypothesize that nodes in big communities might play different roles and have different characteristics, in a process analogous to the differences between nodes in the global graph.

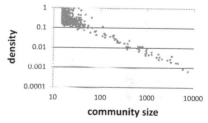

Fig. 3. Big communities are sparse: Community size vs density

Power-law degrees. One well documented relationship in real networks is the power-law between the degree of a node and its rank (i.e. position in decreasing order of degree) [5], which means the degree of a node i can be approximated as $d_i = K \cdot p_i^\alpha$, where α is the power-law exponent, K is the scaling factor correlated with number of edges and p_i the rank of node i.

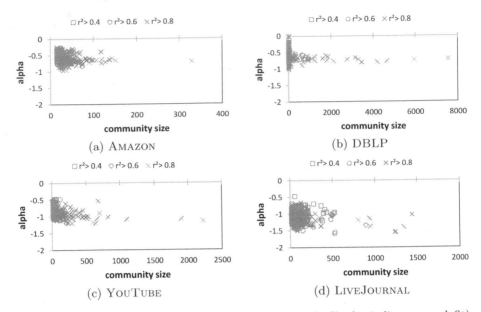

Fig. 4. **Big ground-truth communities are hyperbolic** (× indicates good fit). Community size vs α.

Our first hypothesis is that big communities follow a similar degree distribution. Figure 4 shows the calculated α values for different ground-truth communities in the 4 datasets. Communities have been marked according to their coefficient of determination (r^2) when we approximate the degree distribution within each community with a power-law. The power-law was approximated using a linear-regression in the log-log data and the coefficient was calculated using the same transformation (more details can be found in Section 5.1). It can be seen, agreeing with intuition, that power-law degree distributions represent big communities fairly well. In fact, most of the ground-truth communities do not show uniform degree distribution (which would be α = 0) but strongly skewed ones. Interestingly, α appears to decrease with community size (note the differences in the x-axis) and to be between -0.6 and -1.5 for communities with thousands of elements. Furthermore, as the frequently used uniform block model for communities indirectly assumes a uniform degree distribution, the power-law model necessarily achieves a better fit – the uniform model is a special case of the power-law model where α = 0.

Some variations between the datasets are yet to be explained but can most likely be attributed to the different community definitions. For example, some communities with uniform degree distribution in the DBLP dataset are due to anomalies such as venues with a single paper creating artificial cliques (e.g. recording errors, conference proceedings with a single entry, workshops, etc.).

Again, we want to highlight that the observations made above are based on the communities which were externally provided for these datasets ("ground-truth communities") – not based on the results of a specific algorithm.

4   Hyperbolic Community Model

The previous analysis shows that, in order to detect big communities with realistic properties, models must be able to represent non-uniform degree distributions. In this section, we first propose HyCoM, a community model that assumes communities to have a power-law degree distribution. We then detail the MDL-based formalization that will guide the community discovery process and that is used as a metric for community quality.

4.1   Community Definition

We are given an undirected network consisting of nodes $\mathcal{N}$ and edges $\mathcal{E}$. We represent this network as an adjacency matrix $\mathbf{M} \in \{0,1\}^{|\mathcal{N}| \times |\mathcal{N}|}$. As an abbreviation, we use $N = |\mathcal{N}|$. The goal is to detect Hyperbolic Communities:

Definition 1. *Hyperbolic Community*
A hyperbolic community is a triplet $C = (S, \alpha, \tau)$ with $S = [S_1, .., S_{|S|}], S_i \in \mathcal{N}$ and $S_i \neq S_j$ if $i \neq j$, representing an ordered list of nodes, $\alpha \leq 0$ being the exponent when the degree distribution of the nodes is approximated by a power-law, and $0 \leq \tau \leq 1$ a threshold that determines the number of edges represented by the community.

Given the above triplet, and knowing that the nodes in S are sorted by degree in the community, the degree of a node is $d_i \propto i^\alpha$. If we assume conditional independence given the community (i.e. we assume edge independence when we know both nodes belong to the current community), then the probability $p_{i,j}$ that the edge between nodes i and j is part of the community is also proportional to $i^\alpha \cdot j^\alpha$. Therefore, we can define the edges of an hyperbolic community to be the most probable edges given exponent α and threshold τ:

$$E(C) = \{(S_i, S_j) \in S \times S \; : \; i^\alpha \cdot j^\alpha > \tau\}.$$

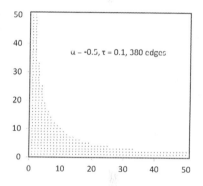

Fig. 5. Adjacency Matrix of a synthetic Hyperbolic Community

Figure 5 illustrates the adjacency matrix induced by the set $E(C)$ given a certain degree distribution and value of τ. Its characteristic shape, an hyperbola, gave name to this model.

We propose to measure the importance of a community via the principle of compression, i.e. by its ability to compress the matrix $\mathbf{M}$: if most edges of $E(C)$ are in fact part of $\mathbf{M}$, then we can compress this community easily. Finding the most important communities will lead to the best compression of $\mathbf{M}$.

More specifically, we use the MDL principle [10]. We aim to minimize the number of bits required to simultaneously encode the communities (i.e. the model) and the data (effects not captured by the model, e.g. missing edges), in a trade off between model complexity and goodness of fit. In the following, we provide details on how to compute the description cost in this setting.

4.2   MDL Description Cost

The first part of the description cost accounts for encoding the detected communities $\mathcal{C} = \{C_1, \ldots, C_n\}$ (where n is part of the optimization and not a priori given). Each community $C_i = (S_i, \alpha_i, \tau_i)$ can be described by the list S_i, the number of bits used for α_i, denoted as k_{α_i} [2], and by the number of edges $|E(C)|$ in the community. Please note that we actually do not need to encode the real-valued variable τ, but it is sufficient to encode the natural number $|E(C)|$. The coding cost for a pattern C_i is

$$L_1(C_i) = \log N + |S_i| \cdot \log N + k_{\alpha_i} + \log(|S_i|^2).$$

The first two terms encode the list of nodes, there are up to N elements in the community and we can encode each element using $\log N$ bits. The second term encodes α_i and the last term encodes the number of edges in the community. Since the number of edges is bounded by $|S_i|^2$, we can encode it with $log(|S_i|^2)$ bits. Similarly, the set of patterns $\mathcal{C} = \{C_1, \ldots, C_l\}$ can be encoded by the following number of bits:

$$L_2(\mathcal{C}) = \log^* |\mathcal{C}| + \sum_{C \in \mathcal{C}} L_1(C).$$

Since the cardinality of $\mathcal{C}$ is not known a priori, we encode it via the function $\log^*$ using the universal code length for integers [15].

The second part of the description cost accounts for encoding the actual data given the detected communities. Since one might expect to find overlapping communities, we refer to the principle of Boolean Algebra and patterns are combined by a logical disjunction: if an edge occurs in at least one of the patterns, it is also present in the reconstructed data. More formally, we reconstruct the given matrix by:

Definition 2. *Matrix reconstruction*
Given a community C, we define an indicator matrix $\mathbf{I}^C \in \{0,1\}^{N \times N}$ (using the same ordering of nodes as imposed by $\mathbf{M}$) that represents the edges of the graph encoded by community C, i.e. $\mathbf{I}^C_{x,y} = 1 \Leftrightarrow (x,y) \in E(C)$.
  Given a set of communities $\mathcal{C}$, the reconstructed network $\mathbf{M}^{\mathcal{C}}_r$ is defined as $\mathbf{M}^{\mathcal{C}}_r = \bigvee_{C \in \mathcal{C}} \mathbf{I}^C$ where $\vee$ denotes the element-wise disjunction.

Since MDL requires a lossless reconstruction of the network, the matrix $\mathbf{M}^{\mathcal{C}}_r$, however, likely does not perfectly reconstruct the data, the second part of the description cost encodes the data given the model. Here, an 'error' might be either an edge appearing in $\mathbf{M}^{\mathcal{C}}_r$ but not in $\mathbf{M}$ or vice versa. As we are considering binary matrices, the number of errors can be computed based on the squared Frobenius norm of the residual matrix, i.e. $\left\| \mathbf{M} - \mathbf{M}^{\mathcal{C}}_r \right\|^2_F$.

---

[2] The number of bits does not affect the results as the previous term is significantly bigger. We use 32 bits in our experiments.

Finally, as 'errors' correspond to edges in the graph, the description cost of the data can be computed as

$$L_3(\mathbf{M}|\mathcal{C}) = \log^* \left\|\mathbf{M} - \mathbf{M}_r^{\mathcal{C}}\right\|_F^2 + 2 \cdot \left\|\mathbf{M} - \mathbf{M}_r^{\mathcal{C}}\right\|_F^2 \cdot \log N.$$

Overall model. Given the functions L_2 and L_3, we are now able to define the communities that minimize the overall number of bits required to describe the model and the data:

Definition 3. *Finding hyperbolic communities*
Given a matrix $\mathbf{M} \in \{0,1\}^{N \times N}$, the problem of finding hyperbolic communities is defined as finding a set of patterns $\mathcal{C}^* \subseteq (\mathcal{P}(\mathcal{N}) \times \mathbb{R} \times \mathbb{R})$ such that

$$\mathcal{C}^* = \arg\min_{\mathcal{C}}[L_2(\mathcal{C}) + L_3(\mathbf{M}|\mathcal{C})].$$

Computing the optimal solution to this problem is NP-hard, given that the column reordering problem in two dimensions is NP-hard as well [12]. In the next section we present an approximate but scalable solution based on an iterative processing scheme.

5   HyCoM-FIT: Fitting Hyperbolic Communities

In this section, we introduce HyCoM-FIT, a scalable and efficient algorithm that approximates the optimal solution via an iterative method of sequentially detecting important communities. The general idea is to find in each step a single community C_i that contributes the most to the MDL-compression based on local evaluation. That is, given the already detected communities $\mathcal{C}_{i-1} = \{C_1, \ldots, C_{i-1}\}$, we are interested in finding a novel community C_i which minimizes $L_2(\{C_i\} \cup \mathcal{C}_{i-1}) + L_3(\mathbf{M}|\{C_i\} \sqcup \mathcal{C}_{i-1})$. Since $\mathcal{C}_{i-1}$ is given, this is equivalent to minimizing

$$L_1(C_i) + L_3(\mathbf{M}|\{C_i\} \cup \mathcal{C}_{i-1}). \tag{1}$$

Obviously, enumerating all possible communities is infeasible. Therefore, to detect a single community C_i, the following steps are performed:

- **Step 1: Community candidates**: We spot candidate nodes by performing a rank-1 approximation of the matrix $\mathbf{M}$. This step provides a normalized vector with the score of each node.
- **Step 2: Community construction**: The scores from the previous step are used in a hill climbing search as a bias for connectivity, while minimizing the MDL costs is used as the objective function for determining the correct community size.
- **Step 3: Matrix deflation**: Based on the current community detected, we deflate the matrix so that the rank-1 approximation is steered to find novel communities in later iterations.

In the following, we will discuss each step of the iterative procedure.

Community Candidates. As mentioned, exhaustively enumerating all possible communities is infeasible. Therefore we propose to iteratively let the communities grow. The challenge, however, is how to spot nodes which should be added to a community. For this purpose, we refer to the idea of matrix decomposition. Given the matrix $\mathbf{M}$ (or as we will explain in step 3, the deflated matrix $\mathbf{M}^{(i)}$), we compute a vector $\mathbf{a}$ such that $\mathbf{a} \cdot \mathbf{a}^T \approx \mathbf{M}$. The vector $\mathbf{a}$ reflects the community structure in the data and we treat the elements a_i as an indication of the importance of node i to this community.

Community Construction. Given the vector $\mathbf{a}$, we construct a new community. Algorithm 1 shows an overview of this step. We start by selecting an initial seed $S = \{v_1, v_2\}$ of two connected nodes with high score in $\mathbf{a}$.[3] We then let the community grow incrementally: We randomly select a neighbor v_i that is not currently part of the community, where the score vector $\mathbf{a}$ is used as the sampling bias. That is, given the current nodes S, we sample according to

$$v_i \propto \begin{cases} a_i & v_i \notin S \wedge \exists v' \in S : (v_i, v') \in \mathcal{E} \\ 0 & else \end{cases}.$$

If the MDL score (cf. Equation 1) of the new community, i.e. using the vertices $S \cup \{v_i\}$, is smaller than the MDL score using the previous community, the vertex v_i is accepted. Otherwise, a new sample is generated. This process is repeated until Δ consecutive rejections have been observed. Since the probability that an element that should have been included in the community but which was not sampled, i.e. $P(\text{"i not selected"} | \text{"i should have been selected"})$, decreases exponentially as a function of Δ and of its initial score, i.e. it can be bounded by a_i^Δ, a small value of Δ is sufficient. In our experimental analysis, a value of $\Delta = 50$ has proven to be sufficient; we consider this parameter to be general and it does not need to be defined by the user of the algorithm.

After growing the community, we then try to remove elements from the community, once again checking the change in the description cost. This alternating process is repeated until the community stabilizes. This process is guaranteed to converge as the description cost of matrix $\mathbf{M}$ is strictly decreasing.

Matrix Deflation. While the first two steps build a single community C_i, the objective of this step is to transform the matrix so that the process can be iterated in such a way that we don't get the same community repeatedly. In particular, we aim at steering the rank-1 decomposition to novel solutions.

To solve this problem we propose the principle of matrix deflation. Starting with the original matrix $\mathbf{M} =: \mathbf{M}^{(1)}$, we remove after each iteration those edges which are already described by the detected community. That is, we obtain the recursion

$$\mathbf{M}^{(i+1)} := \mathbf{M}^{(i)} - \mathbf{I}^{C_i} \circ \mathbf{M}^{(i)} \qquad [\, = \mathbf{M} - \mathbf{M}_r^{C_i} \circ \mathbf{M} \,]$$

---

[3] We tested different methods with no significant differences found in the results. Selecting the edge (i, j) with highest $min(a_i, a_j)$ provides a good initial seed.

Algorithm 1. HYCoM-FIT- Community Construction

---

function COMMUNITYCONSTRUCTION(ScoreVector a)
   S ← $initialSeed(a)$
   repeat
      $t \leftarrow 0$
      while $t < \Delta$ **do**
         $v_i \leftarrow newBiasedNode(S, a)$
         if $\mathrm{MDL}(S \cup \{v_i\}) < \mathrm{MDL}(S)$ **then** $S \leftarrow S \cup \{v_i\}$, $t \leftarrow 0$
         else $t \leftarrow t + 1$
      end while
      for all nodes n in S **do**
         if $\mathrm{MDL}(S \backslash \{n\}) < \mathrm{MDL}(S)$ **then** $S \leftarrow S \backslash \{n\}$
      end for
   until S has converged
   return S

---

where ∘ denotes the Hadamard product. As seen, the matrix $\mathbf{M}^{(i+1)}$ incorporates all communities detected so far. Using the deflated matrix, our objective in Equation 1 is replaced by

$$L_1(C_i) + L_3(\mathbf{M}^{(i)}|\{C_i\}). \tag{2}$$

Overall, the algorithm might either terminate when the matrix is fully deflated, or when a pre-defined number of communities has been found, or when some other measure of community quality (i.e. size) has not been achieved in the most recent communities.

5.1   Fast MDL Calculation

The key task of Algorithm 1 is to compute the MDL score (Equation 2) based on the current set of nodes S. Besides the set S, estimating the number of bits requires to determine the value of α, to specify a value for τ (or $|E(C)|$), and to count the number of errors made by the model. Since the MDL score is computed several times, we propose an efficient approximation for these tasks:

Approximating the Exponent of the Degree Distribution. Exhaustive test of different approximation methods is beyond the scope of this paper; for an in-depth analysis on power-law exponent estimation from empirical data we refer the reader to the review by Aaron Clauset et al. [4]. The method chosen has to be robust in degenerate situations (e.g. uniform distributions) and efficient. We opted for a linear regression of the log-log data, as it not only respects both requirements, but also because it is known to over fit to the tail of the distribution and edges between high degree nodes are already expected under the independence assumption.

Number of Edges and Value of τ. The value of $|E(C)|$ is selected as the number of edges between the nodes in S, i.e. $|(S \times S) \cap \mathcal{E}|$, since this value can efficiently be obtained by an incremental computation each time a node is added/removed from the current community. Efficiency is ensured by indexing the edges in $\mathbf{M}$ by node.

Fixing the value of $|E(C)|$, we need to derive the value of τ leading to the desired cardinality. For efficiency, we exploit the following approximation:

Lemma 1. *The value of* $|E(C)|$ *can be approximated by*

$$|E(C)| \approx (i_{start} - 1) \cdot |S| + \tau^{\frac{1}{\alpha}} \cdot (\log(i_{end}) - \log(i_{start})),$$

where $i_{start} := \max\{\lceil \tau^{\frac{1}{\alpha}} \cdot |S|^{-1} \rceil, 1\}$ *and* $i_{end} := \min\{\lfloor \tau^{\frac{1}{\alpha}} \rfloor, |S| + 1\}$.

Proof. Instead of exactly counting the number of elements $i^\alpha \cdot j^\alpha > \tau$ (cf. Figure 5), we do a continuous approximation by computing the area under the τ-isoline (intuitively: the area shaded in Figure 5). More precisely, given a specific τ (and assuming $\alpha \neq 0$), we use the isoline derived by

$$i^\alpha \cdot j^\alpha = \tau \Leftrightarrow j = \tau^{\frac{1}{\alpha}} \cdot i^{-1} =: f(i).$$

Considering the integral $\int_1^{|S|+1} f(i) \, di$ leads to an approximation of $|E(C)|$. To achieve a more accurate approximation, we consider two further improvements: (a) For each i with $f(i) < 1$, no edges are generated. Thus, we also don't need to consider the area under this part of the curve. It holds

$$f(i) \geq 1 \Rightarrow i \leq \tau^{\frac{1}{\alpha}} \quad \Rightarrow i_{end} := \min\{\lfloor \tau^{\frac{1}{\alpha}} \rfloor, |S| + 1\}.$$

The integration interval can end at i_{end}.
(b) The number of edges for each node is bounded by $|S|$. Thus, for each i with $f(i) > S$, we can restrict the function value to $|S|$. It holds

$$f(i) \leq |S| \Rightarrow i \geq \tau^{\frac{1}{\alpha}} \cdot |S|^{-1} \quad \Rightarrow i_{start} := \max\{\lceil \tau^{\frac{1}{\alpha}} \cdot |S|^{-1} \rceil, 1\}.$$

Thus, overall, given a specific τ, the value of $|E(C)|$ can be approximated by

$$\int_1^{i_{start}} |S| \, di + \int_{i_{start}}^{i_{end}} f(i) \, di = (i_{start} - 1) \cdot |S| + \tau^{\frac{1}{\alpha}} \cdot (\log(i_{end}) - \log(i_{start})).$$

$\square$

Based on Lemma 1, we find the appropriate τ by performing a binary search on the value of $\log \tau$ until the given value of $|E(C)|$ is (approximately) obtained. This step can be done in time $O(\log |S|^2)$.

Calculating the Number of Errors. Determining the number of errors can be reduced to the problem of counting the number of existing edges in $\mathbf{I}^C$. In other words, the goal is to determine how many edges $(S_i, S_j) \in (S \times S) \cap \mathcal{E}$ fulfill $i^\alpha \cdot j^\alpha > \tau$. Knowing this number, e.g. denoted as x, the number of errors is given by

$$(\mathbf{M}^{(i)} - x) + (|E(C)| - x).$$

We have to encode all edges of $\mathbf{M}^{(i)}$ as errors which are not covered by C (i.e. $\mathbf{M}^{(i)} - x$ many) and we additionally have to encode all non-existing edges which are unnecessarily included in C (i.e. $|E(C)| - x$ many).

Obviously, the value of x can be determined by simply iterating over all edges $(S \times S) \cap \mathcal{E}$ of the community, i.e. linear in the number of edges.

5.2 Complexity Analysis

Lemma 2. HyCoM-FIT *has a runtime complexity of* $O(K \cdot (|\mathcal{E}| + |S| \cdot (\log |S|^2 + E)))$, *where* K *is the number of communities we obtain,* $|\mathcal{E}|$ *is the number of edges in the network,* $|S|$ *is the average size of a community and* E *is the number of edges between the elements of* S.

Proof. Steps 1 to 3 are repeated K times, the number of communities to be obtained. Step 1, the rank-1 approximation, requires $O(|\mathcal{E}|)$ time. Step 2, the core of the algorithm, can be executed using $O(|S|)$ additions and removals to the community, each with complexity $O(\log |S|^2 + E)$ as detailed in the previous sub-section. Finally, step 3, the matrix deflation, can be done in $O(E)$ with a single pass over the edges of the community.   □

6 Experiments on Real Data

In this section, we start by evaluating the quality of the Hyperbolic Community Model using the datasets of Table 1. We subsequently evaluate HyCoM-FIT by studying its scalability and its ability to obtain empirically correct communities through the use of the node-labeled dataset.

We focus on three quality metrics: $Q1$) Model quality, $Q2$) HyCoM-FIT scalability and $Q3$) Effectiveness.

Q1) Model Quality

While Section 4 describes how to encode hyperbolic communities, it does not show whether this model is preferable over simpler models such as edge lists when encoding real communities. This aspect is not immediately clear because, even though block communities of uniform density are a special case ($\alpha = 0$) of hyperbolic, HyCoM explicitly encodes *missing* edges (i.e. errors made by the model). This observation implies that HyCoM must create dense hyperbolas to ensure that the overall cost of encoding the errors and the model is not higher than to the cost of simply encoding all edges in the graph. Since big communities are usually very sparse, it is not obvious whether better compression can be achieved by our model.

Figure 6 shows the number of bits required to encode the ground-truth communities using the hyperbolic model and the edge-list format. In this scenario, the cost of each community using HyCoM can be obtained using Definition 3 when setting $|\mathcal{C}| = 1$. As seen, the hyperbolic model consistently requires less bits to represent the ground-truth communities. While for the datasets shown in (a)-(c), the savings are substantial, the savings on the LiveJournal are less strong. In any case, though, compression based on hyperbolic structure is preferable.

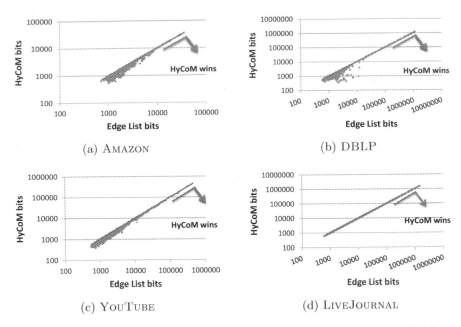

Fig. 6. Number of bits required to encode ground truth communities: **HyCoM consistently requires less bits**

Q2) HyCoM-FIT Scalability

We compared HYCOM-FIT to several popular community detection methods found in the literature: the community affiliation graph model [20], clique percolation [1] and cross-associations [3]. We obtained realistic graphs of different sizes by doing a weighted-snowball sampling[4] in the LIVEJOURNAL dataset.

Figure 7 shows the run-time of the different algorithms using their default parameters. [1] ran out of memory on a graph with 100 000 edges. HYCOM-FIT was run without any special stopping criteria (i.e. until the deflation was complete); as a consequence, bigger graphs required more communities to be fully deflated. HYCOM-FIT shows a fully linear run-time when the required number of communities is constant.

Q3) Effectiveness

In addition to the datasets with ground-truth communities previously used, we also applied HYCOM-FIT to a copy of the simple-english Wikipedia pages from March 8, 2014. In this dataset, nodes represent articles and edges represent hyperlinks between them. Unlike previous datasets, we don't consider any

---

[4] In this weighted-snowball sampling, weights correspond to the number of connections from a node to the current sample. This was done in an effort to preserve community structure.

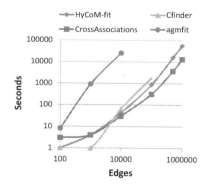

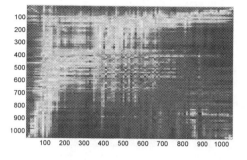

Fig. 7. HyCoM-FIT scales linearly with the number of edges

Fig. 8. Anomalous community found by HyCoM-FIT in the LiveJournal data: **HyCoM-FIT can also be used to detect anomalies**

ground-truth communities in the Wikipedia data; however, as nodes are labeled, this dataset allows us to assert the effectiveness of HyCoM-FIT.

Detecting Hyperbolic Communities. Figures 1 and 2 presented in the introduction illustrate both a ground-truth community and a community found by HyCoM-FIT in the YouTube dataset. They show not only the existence of hyperbolic communities in real data, but also the ability of our method to successfully find them. Note the similarity in the shape of both communities. Existing methods trying to find communities of uniform density would fail to detect such communities.

Anomaly Detection. HyCoM-FIT is also able to detect anomalous structures in data. Figure 8 shows a detected community from the LiveJournal dataset. We can see the adjacency matrix (here represented as an heatmap) of suspiciously highly connected accounts. Approximately 1 000 users established over 300 000 friendship relations forming a very dense community (compared to a common distribution as shown in Figure 3). Clearly, also this anomalous community shows the characteristic shape of an hyperbola.

Communities in Wikipedia. Figures 9 and 10 show two communities detected in the Wikipedia dataset. Figure 9 illustrates an hyperbolic community mostly consisting of *temporal* articles. The first 6 articles correspond to countries heavily mentioned in events (e.g. United States, France, Germany, etc.) then we have articles corresponding to months (e.g. April, July), then articles representing individual years (e.g. 2002, 1973) and finally articles corresponding to particular dates (e.g. November 25, May 13).

Figure 10 shows HyCoM-FIT's generality and its ability of detecting bipartite cores given their close resemblance to hyperbolas. In this community, the approximately 20 articles of highest degree represent articles with lists (e.g. "Country", "List of countries by area", "Members of the United Nations") while the remaining 140 articles are all individual countries.

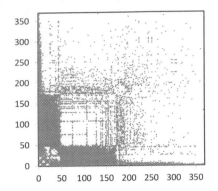

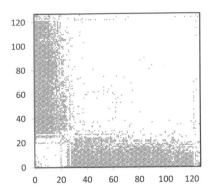

Fig. 9. Community of dates in WIKIPEDIA: **HyCoM-FIT** finds meaningful hyperbolic structures

Fig. 10. Community of countries in WIKIPEDIA: **HyCoM-FIT** also finds bipartite cores and cliques

7 Conclusions

We focused on the problem of representing communities in real graph data, and specifically on the resemblance between structure of the full graph and the structure of big communities. The main contributions are the following:

- **Hyperbolic Community Model:** We provide empirical evidence that communities in real data are better modeled using an hyperbolic model, termed HYCOM. Our model includes communities of uniform density as used by other approaches as a special case. We also show that this model is better from a compression perspective than previous models.
- **Scalability:** HYCOM-FIT is a scalable algorithm for the detection of communities fitting the HYCOM model. We leverage rank-1 decompositions and the MDL principle to guide the search process.
- **No user-defined parameters**: HYCOM-FIT detects communities in a parameter-free fashion, transparent to the end-user.
- **Effectiveness:** We applied HYCOM-FIT on various real datasets, where we discovered communities that agree with intuition.

HYCOM-FIT is available at http://cs.cmu.edu/~maraujo/hycom/.

Acknowledgments. This work is partially funded by the Fundação para a Ciência e a Tecnologia (Portuguese Foundation for Science and Technology) through the Carnegie Mellon Portugal Program under Grant SFRH/BD/52362/2013, by ERDF and FCT through the COMPETE Programme within project FCOMP-01-0124-FEDER-037281, and by a fellowship within the postdoc-program of the German Academic Exchange Service (DAAD). Research was also sponsored by the Army Research Laboratory and was accomplished under Co-operative Agreement Number W911NF-09-2-0053. Any opinions, findings, and conclusions or recommendations expressed in this material are those of the author(s) and do not necessarily reflect the views of the funding parties. The U.S. Government is authorized to reproduce and distribute reprints for Government purposes notwithstanding any copyright notation here on.

References

1. Adamcsek, B., Palla, G., Farkas, I.J., Dernyi, I., Vicsek, T.: Cfinder: locating cliques and overlapping modules in biological networks. Bioinformatics 22(8), 1021–1023 (2006)
2. Araujo, M., Papadimitriou, S., Günnemann, S., Faloutsos, C., Basu, P., Swami, A., Papalexakis, E.E., Koutra, D.: Com2: Fast automatic discovery of temporal ('comet') communities. In: Tseng, V.S., Ho, T.B., Zhou, Z.-H., Chen, A.L.P., Kao, H.-Y. (eds.) PAKDD 2014, Part II. LNCS (LNAI), vol. 8444, pp. 271–283. Springer, Heidelberg (2014)
3. Chakrabarti, D., Papadimitriou, S., Modha, D.S., Faloutsos, C.: Fully automatic cross-associations. In: KDD, pp. 79–88 (2004)
4. Clauset, A., Shalizi, C., Newman, M.: Power-law distributions in empirical data. SIAM Review 51(4), 661–703 (2009)
5. Faloutsos, M., Faloutsos, P., Faloutsos, C.: On power-law relationships of the internet topology. In: SIGCOMM, pp. 251–262 (1999)
6. Flake, G.W., Lawrence, S., Giles, C.L.: Efficient identification of web communities. In: KDD, pp. 150–160 (2000)
7. Fortunato, S.: Community detection in graphs. Physics Reports 486(35), 75–174 (2010)
8. Gionis, A., Mannila, H., Seppänen, J.K.: Geometric and combinatorial tiles in 0-1 data. In: Boulicaut, J.-F., Esposito, F., Giannotti, F., Pedreschi, D. (eds.) PKDD 2004. LNCS (LNAI), vol. 3202, pp. 173–184. Springer, Heidelberg (2004)
9. Gkantsidis, C., Mihail, M., Zegura, E.W.: Spectral analysis of internet topologies. In: INFOCOM, pp. 364–374 (2003)
10. Grünwald, P.D.: The minimum description length principle. The MIT Press (2007)
11. Günnemann, S., Färber, I., Raubach, S., Seidl, T.: Spectral subspace clustering for graphs with feature vectors. In: ICDM, pp. 231–240 (2013)
12. Johnson, D.S., Krishnan, S., Chhugani, J., Kumar, S., Venkatasubramanian, S.: Compressing large boolean matrices using reordering techniques. In: VLDB, pp. 13–23 (2004)
13. Karypis, G., Kumar, V.: Metis - unstructured graph partitioning and sparse matrix ordering system, version 2.0. Technical report (1995)
14. Newman, M.: Modularity and community structure in networks. PNAS 103(23), 8577–8582 (2006)
15. Rissanen, J.: A universal prior for integers and estimation by minimum description length. The Annals of Statistics, 416–431 (1983)
16. Sen, T., Kloczkowski, A., Jernigan, R.: Functional clustering of yeast proteins from the protein-protein interaction network. BMC Bioinformatics 7, 355–367 (2006)
17. Shi, J., Malik, J.: Normalized cuts and image segmentation. IEEE PAMI 22(8), 888–905 (2000)
18. Sun, J., Faloutsos, C., Papadimitriou, S., Yu, P.S.: Graphscope: parameter-free mining of large time-evolving graphs. In: KDD, pp. 687–696 (2007)
19. Wasserman, S., Faust, K.: Social Network Analysis: Methods and Applications. Cambridge University Press (1994)
20. Yang, J., Leskovec, J.: Community-affiliation graph model for overlapping network community detection. In: ICDM, pp. 1170–1175 (2012)
21. Yang, J., Leskovec, J.: Defining and evaluating network communities based on ground-truth. In: ICDM, pp. 745–754 (2012)

Concurrent Bandits and
Cognitive Radio Networks

Orly Avner and Shie Mannor*

Department of Electrical Engineering,
Technion - Israel Institute of Technology, Haifa, Israel

Abstract. We consider the problem of multiple users targeting the arms of a single multi-armed stochastic bandit. The motivation for this problem comes from cognitive radio networks, where selfish users need to coexist without any side communication between them, implicit cooperation or common control. Even the number of users may be unknown and can vary as users join or leave the network. We propose an algorithm that combines an ϵ-greedy learning rule with a collision avoidance mechanism. We analyze its regret with respect to the system-wide optimum and show that sub-linear regret can be obtained in this setting. Experiments show dramatic improvement compared to other algorithms for this setting.

Keywords: Bandits, Multi-user, Epsilon-greedy.

1 Introduction

In this paper we address a fundamental challenge arising in dynamic multi-user communication networks, inspired by the field of Cognitive Radio Networks (CRNs). We model a network of independent users competing over communication channels, represented by the arms of a stochastic multi-armed bandit. We begin by explaining the background, describing the general model, reviewing previous work and introducing our contribution.

1.1 Cognitive Radio Networks

Cognitive radio networks, introduced in [19], refer to an emerging field in multi-user multi-media communication networks. They encompass a wide range of challenges stemming from the dynamic and stochastic nature of these networks. Users in such networks are often divided into primary and secondary users. The primary users are licensed users who enjoy precedence over secondary users in terms of access to network resources. The secondary users face the challenge of identifying and exploiting available resources. Typically, the characteristics of the primary users vary slowly, while the characteristics of secondary users tend

---

* This research was partially supported by the CORNET consortium (www.cornet.org.il) and by an MOE Magneton grant.

to be dynamic. In most realistic scenarios, secondary users are unaware of each other. Thus, there is no reason to assume the existence of any cooperation or communication between them. Furthermore, they are unlikely to know even the *number* of secondary users in the system. Another dominant feature of CRNs is their distributed nature, in the sense that a central control does not exist.

The resulting problem is quite challenging: multiple users, coexisting in an environment whose characteristics are initially unknown, acting selfishly in order to achieve an individual performance criterion. We approach this problem from the point of view of a single secondary user, and introduce an algorithm which, when applied by all secondary users in the network, enjoys promising performance guarantees.

1.2   Multi-armed Bandits

Multi-Armed Bandits (MABs) are a well-known framework in machine learning [6]. They succeed in capturing the trade-off between exploration and exploitation in sequential decision problems, and have been used in the context of learning in CRNs over the last few years [4,5], [11]. Classical bandit problems comprise an agent (user) repeatedly choosing a single option (arm) from a set of options whose characteristics are initially unknown, receiving a certain reward based on each choice. The agent wishes to maximize the acquired reward, and in order to do so she must balance exploration of unknown arms and exploitation of seemingly attractive ones. Different algorithms have been proposed and proved optimal for the stochastic setting of this problem [2],[10], as well as for the adversarial setting [3].

We adopt the MAB framework in order to capture the challenge presented to a secondary user choosing between several unknown communication channels. The characteristics of the channels are assumed to be fixed, corresponding to a relatively slow evolution of primary user characteristics. The challenge we address in this paper arises from the fact that there are *multiple* secondary users in the network.

1.3   Multiple Users Playing a MAB

A natural extension of the CRN-MAB framework described above considers multiple users attempting to exploit resources represented by *the same* bandit. The multi-user setting leads to collisions between users, due to both exploration and exploitation; an "attractive" arm in terms of reward will be targeted by all users, once it has been identified as such. In real-life communication systems, collisions result in impaired performance. In our model, reward loss is the natural manifestation of collisions.

As one might expect, straightforward applications of classical bandit algorithms designed for the single-user case, e.g., KL-UCB [10], are hardly beneficial. The reason is that in the absence of some form of a collision avoidance mechanism, all users attempt to sample the same arm after some time. We illustrate this in Section 5.

We therefore face the problem of sharing a resource and learning its characteristics when users cannot communicate and are oblivious to each other's existence.

1.4   Related Work

Recently, considerable effort has been put into finding a solution for the multi-user CRN-MAB problem. One approach, considered in [16], is based on a Time-Division Fair Sharing (TDFS) of the best arms between all users. This policy enjoys good performance guarantees but has two significant drawbacks. First, the number of users is assumed to be fixed and known to all users, and second, the implementation of a TDFS mechanism requires pre-agreement among users to coordinate a time division schedule. Another work that deals with multi-user access to resources, but does not incorporate the MAB setting, is presented in [14]. The users reach an orthogonal configuration without pre-agreement or communication, using multiplicative updates of channel sampling probabilities based on collision information. However, this approach does not handle the learning aspect of the problem and disregards differences in the performance of different channels. Thus, it cannot be applied to our problem. The authors in [12] consider a form of the CRN-MAB problem in which channels appear different to different users, and propose an algorithm which enjoys good performance guarantees. However, their algorithm includes a negotiation phase, based on the Bertsekas auction algorithm, during which the users communicate in order to reach an orthogonal configuration. Another work that considers this form of problem and uses calibrated forecasting in order to reach an orthogonal configuration is described in [17]. The analysis is based on a game theoretic approach, and shows asymptotic convergence to an optimal strategy. Additional papers such as [9] and [13] also consider the CRN-MAB setting, offering only asymptotic performance guarantees.

The work closest in spirit to ours is [1]. The authors propose different algorithms for solving the CRN-MAB problem, attempting to lift assumptions of cooperation and communication as they go along. Their main contribution is expressed in an algorithm which is coordination and communication free, but relies on exact knowledge of the number of users in the network. In order to resolve this issue, an algorithm which is based on *estimating* the number of users is proposed. Performance guarantees for this algorithm are asymptotic, and it does not address the scenario of a time-varying number of users.

A different approach to resource allocation with multiple noncooperative users involves game theoretic concepts [20,21]. In our work we focus on cognitive, rather than strategic, users. Yet another perspective includes work on CRNs with multiple secondary users, where the emphasis is placed on collision avoidance and sensing. References such as [7] and [15] propose ALOHA based algorithms, achieving favorable results. However, these works do not consider the learning problem we are facing, and assume all channels to be known and identical.

1.5   Contribution

The main contribution of our paper is suggesting an algorithm for the multi-user CRN-MAB problem, which guarantees convergence to an optimal configuration when employed by all users. Our algorithm adheres to the strict demands imposed by the CRN environment: no communication, cooperation or coordination (control) between users, and strictly local knowledge - even the number of users is unknown to the algorithm.

Also, to the best of our knowledge, ours is the only algorithm that handles a dynamic number of users in the network successfully.

The remainder of this paper is structured as follows. Section 2 includes a detailed description of the framework and problem formulation. Section 3 presents our algorithm along with its theoretical analysis, while Section 4 discusses the setup of a dynamic number of users. Section 5 displays experimental results and Section 6 concludes our work. The proofs of our results are provided in the supplementary material.

2   Framework

Our framework consists of two components: the environment and the users. The environment is a communication system that consists of K channels with different, initially unknown, reward characteristics. We model these channels as the arms of a stochastic Multi-Armed Bandit (MAB). We denote the expected values of the reward distributions by $\boldsymbol{\mu} = (\mu_1, \mu_2, \ldots, \mu_K)$, and assume that channel characteristics are fixed. Rewards are assumed to be bounded in the interval $[0, 1]$.

The users are a group of non-cooperative, selfish agents. They have no means of communicating with each other and they are not subject to any form of central control. Unlike some of the previous work on this problem, we assume users have no knowledge of the total number of users. In Section 3 we assume the number of users is fixed and equal to N, and in Section 4 we relax this assumption; in both cases we assume $K \geq N$. Scenarios in which $K < N$ correspond to overcrowded networks and should be dealt with separately. The fact that the users share the communication network is modeled by their playing *the same* MAB. Two users or more attempting to sample the same arm at the same time will encounter a collision, resulting in a zero reward for all of them in that round. A user sampling an arm k alone at a certain time t receives a reward $r(t)$, drawn i.i.d from the distribution of arm k.

We would like to devise a policy that, when applied by all users, results in convergence to the system-optimal solution. A common performance measure in bandit problems is the expected regret, whose definition for the case of a single user is

$$\mathbb{E}\left[R\left(t\right)\right] \triangleq \mu_{k^*} t - \sum_{\tau=1}^{t} \mathbb{E}\left[r\left(\tau\right)\right],$$

where $\mu_{k^*} = \max_{k \in \{1, \ldots, K\}} \mu_k$ is the expected reward of the optimal arm.

Naturally, in the multi-user scenario not all users can be allowed to select the optimal arm. Therefore, the number of users defines a *set* of optimal arms, namely the N best arms, which we denote by K^*. Thus, the appropriate expected regret definition is

$$\mathbb{E}\left[R\left(t\right)\right] \triangleq t \sum_{k \in K^*} \mu_k - \sum_{n=1}^{N} \sum_{\tau=1}^{t} \mathbb{E}\left[r_n\left(\tau\right)\right],$$

where $r_n\left(\tau\right)$ is the reward user n acquired at time τ. We note that this definition corresponds to the expected loss due to a suboptimal sampling policy.

The socially optimal solution, which minimizes the expected regret for all users as a group, is for each to sample a different arm in K^*. Adopting such a system-wide approach makes the most sense from an engineering point of view, since it maximizes network utilization without discriminating between users.

3   Fixed Number of Users

In this section we introduce the policy applied by each of the users, described in Algorithm 1. Our policy is based on several principles:

1. Assuming an arm that experiences a collision is an "attractive" arm in terms of expected reward, we would like one of the colliding users to continue sampling it.
2. Since all users need to learn the characteristics of all arms, we would like to ensure that an arm is not sampled by a single user exclusively.
3. To avoid frequent collisions on optimal arms, we need users to back off of arms on which they have experienced collisions.
4. To avoid interfering with on-going transmissions in the steady state, we would like to prevent exploring users from "throwing off" exploiting users.

3.1   The MEGA Algorithm

The Multi-user ϵ-Greedy collision Avoiding (MEGA) algorithm is based on the ϵ-greedy algorithm introduced in [2], augmented by a collision avoidance mechanism that is inspired by the classical ALOHA protocol.

Learning is achieved by balancing exploration and exploitation through a time-dependant exploration probability. The collision avoidance mechanism is implemented using a persistence probability, p, that controls users' "determination" once a collision occurs. Its initial value is p_0, and it is incremented with each successful sample. Once a collision event begins, the persistence probability remains fixed until it ends.

A collision event ends when all users but one have "given up" and stopped sampling the arm under dispute. Upon giving up, each user resets her persistence and draws a random interval of time during which she refrains from sampling the arm under dispute. The length of these intervals increases over time in order to ensure sub-linear regret.

Algorithm 1. Multi-user ϵ-Greedy collision Avoiding (MEGA) algorithm

---

input Parameters c, d, p_0, α and β
1: **init** $p \leftarrow p_0$, $t \leftarrow 1$, $\eta(0) \leftarrow 0$, $a(0) \sim U(\{1, \ldots, K\})$, $t_{\text{next},k} \leftarrow 1 \ \forall k$
2: **note:** $\eta(t)$ is a collision indicator, $\hat{\mu}_k$ is the empirical mean of arm k's reward
3: **loop**
4:     Sample arm $a(t)$ and observe $r(t), \eta(t)$
5:     **if** $\eta(t) == 1$ **then**
6:         With probability p persist:
7:             $a(t+1) \leftarrow a(t)$
8:             **continue loop**
9:         With probability $1 - p$ give up:
10:            Mark arm as taken until time $t_{\text{next},k}$, where $t_{\text{next},k} \sim U\left([t, t + t^\beta]\right)$
11:    **else**
12:        $p \leftarrow p \cdot \alpha + (1 - \alpha)$
13:        Update $\hat{\mu}_{a(t)}$ with $r(t)$
14:    **end if**
15:    Identify available arms: $A = \{k : t_{\text{next},k} \leq t\}$
16:    **if** $A = \emptyset$ **then**
17:        $a(t+1) \leftarrow \emptyset$, i.e., refrain from transmitting in next round
18:        **continue loop**
19:    **end if**
20:    With probability $\epsilon_t = \min\left\{1, \frac{cK^2}{d^2(K-1)t}\right\}$ explore: $a(t+1) \sim U(A)$
21:    With probability $1 - \epsilon_t$ exploit: $a(t+1) \leftarrow \arg\max_{k \in A} \hat{\mu}_k$
22:    **if** $a(t+1) \neq a(t)$ **then**
23:        $p \leftarrow p_0$
24:    **end if**
25: **end loop**

Each agent executes Algorithm 1 on every round. First, she samples an arm based on her last decision and observes the reward and collision indicator. Her next decision is based on the collision indicator. If a collision occurred, she sticks to her previous decision w.p. p (line 6) or steps down and marks the arm as taken w.p. $1 - p$ (line 9). If a collision did not occur, p is incremented and the empirical mean of the sampled arm is updated (lines 12, 13).

The next arm to be sampled is chosen from the set of available (i.e., not "taken") arms. If this set is empty, the user refrains from transmitting in the following round (line 17). Otherwise, an arm is chosen according to the ϵ-greedy algorithm (lines 20, 21). The value of the persistence probability, p, is reset every time the choice of arms changes between rounds (line 23).

The parameters in Algorithm 1 are chosen so that p_0, α and β lie in the interval $(0, 1)$. In the original ϵ-greedy algorithm, the parameter d is set to be $\mu_{k^*} - \mu_{k_2}$, where μ_{k_2} is the expected reward of the second-best arm. In our case, learning the N best arms requires that d be modified and set to $\mu_{k_{N-1}} - \mu_{k_N}$. However, since the expected rewards of the arms are unknown in practice and we assume the number of users to be unknown, we use a fixed value for d. For further details see Section 5. The exploration probability, ϵ_t, is modified compared to the original ϵ-greedy algorithm [2], in order to account for the decreased efficiency of samples, caused by collisions. For our algorithm we use $\epsilon_t = \min\left\{1, \frac{cK^2}{d^2(K-1)t}\right\}$. Also, the empirical mean which determines the ranking of the arms is calculated based on the number of *successful* samples of each arm.

3.2   Analysis of the MEGA Algorithm

We now turn to a theoretical analysis of the MEGA algorithm. Our analysis shows that when all users apply MEGA, the expected regret grows at a sublinear rate, i.e., MEGA is a no-regret algorithm.

The regret obtained by users employing the MEGA algorithm consists of three components. The first component is the loss of reward due to collisions: in a certain round t, all colliding users receive zero reward. We denote the expected reward loss due to collisions by $\mathbb{E}\left[R^C(t)\right]$. The second and third components reflect the loss of reward due to sampling of suboptimal arms, i.e., arms $k \notin K^*$. Once the users have learned the ranking of the different arms, suboptimal sampling is caused either by random exploration, dictated by the ϵ-greedy algorithm, or due to the fact that all arms in K^* are marked unavailable by a user at a certain time. We denote the expected reward loss due to these issues by $\mathbb{E}\left[R^E(t)\right]$ and $\mathbb{E}\left[R^A(t)\right]$, respectively.

We begin by showing that all users succeed in learning the correct ranking of the N-best arms in finite time in Lemma 1. This result will serve as a base for the bounds of the different regret components.

Definition 1 *An ϵ-correct ranking of M arms is a sorted M-vector of empirical mean rewards of arms (i.e., $i < j \iff \hat{\mu}_i \leq \hat{\mu}_j$), such that*

$$\hat{\mu}_i \leq \hat{\mu}_j \iff \mu_i + \epsilon \leq \mu_j \;\; \forall i,j \in \{1,\ldots,M\}, i \neq j.$$

Lemma 1 *For a system of K arms and N users, $N \leq K$, in which all users employ MEGA, there exists a finite time $T = 2\frac{4K^N N}{\epsilon^2 \prod_{i=1}^{N-1}(K-i)} \log\left(\frac{2K}{\delta}\right)$ such that $\forall t > T$, all users have learned an ϵ-correct ranking of the N-best arms with a probability of at least $1 - \delta$.*
Note: we assume that $\mu_i - \mu_j \geq \epsilon$ for all $i,j \in \{1,\ldots,K\}, i \neq j$.

Proof. We prove the existence of a finite T by combining the sample complexity of stochastic MABs with the characteristics of MEGA.

First, we note that as long as $\epsilon_t = 1$, if the availability mechanism is disabled, each of the users performs uniform sampling on average. We therefore examine a slightly modified version of MEGA for the sake of this theoretical analysis.

Based on [8], a naïve algorithm that samples each arm $\ell_K = \frac{4}{\epsilon^2} \log\left(\frac{2K}{\delta}\right)$ times, identifies an ϵ-best arm with probability of at least $1 - \delta$. A loose bound on the number of samples needed in order to produce a correct *ranking* of the N-best arms of a K-armed bandit is obtained by applying an iterative procedure: sample each of the K arms ℓ_K times and select the best arm; then sample each of the remaining $K-1$ arms ℓ_{K-1} times and select the best arm; repeat the procedure

N times. Such an approach requires no more than $S = \frac{4N}{\epsilon^2} \log\left(\frac{2K}{\delta}\right)$ samples of each arm, for each user.

The collision probability of N users uniformly sampling K channels (in the absence of an availability mechanism) is given by the solution of the well-known "birthday problem" [18]:

$$\mathbb{P}\left[C\right] = 1 - \prod_{d=1}^{N-1}\left(1 - \frac{d}{K}\right).$$

As a result of the collisions, the number of samples which are "effective" in terms of learning arm statistics is reduced. For a certain arm k, sampled by a user n, the expected number of successful samples up till time t is given by

$$\mathbb{E}\left[s_{k,n}\left(t\right)\right] = \left(1 - \mathbb{P}\left[C\right]\right)\frac{t}{K} = \frac{t}{K}\prod_{i=1}^{N-1}\left(1 - \frac{i}{K}\right).$$

In order to ensure an adequate number of samples we need to choose a certain T' for which $\mathbb{E}\left[s_{k,n}\left(T'\right)\right] = S$:

$$\frac{T'}{K}\prod_{i=1}^{N-1}\left(1 - \frac{i}{K}\right) = S,$$

meaning that

$$T' = \frac{4K^N N}{\epsilon^2 \prod_{i=1}^{N-1}\left(K - i\right)} \log\left(\frac{2K}{\delta}\right).$$

Since the users' sampling is random, it is only uniform on average. By choosing $T = 2T'$, we ensure that the number of samples is sufficient with high probability:

$$\mathbb{P}\left[\mathbb{E}\left[s_{k,n}\left(2T'\right)\right] - s_{k,n}\left(2T'\right) > S\right] \leq e^{-S^2/T'} \leq \left(\frac{2K}{\delta}\right)^{-\frac{1}{2}\frac{4N}{K}\left(\frac{K-N}{K}\right)^{N-1}},$$

which is due to Hoeffding's inequality.

We note that Lemma 1 holds for a choice of the parameter c which ensures that $\epsilon_t = 1 \quad \forall t < T$:

$$c = \frac{d^2\left(K - 1\right)T}{K^2}.$$

$\square$

Based on Lemma 1 we proceed with the analysis of MEGA, incorporating the fact that for all $t > T$, all users know the correct ranking of the N best arms.

Back to our regret analysis - since the reward is bounded in $[0, 1]$, the expected regret is also bounded:

$$\mathbb{E}\left[R\left(t\right)\right] \leq \mathbb{E}\left[R^C\left(t\right)\right] + \mathbb{E}\left[R^E\left(t\right)\right] + \mathbb{E}\left[R^A\left(t\right)\right].$$

We begin by addressing the expected regret due to collisions, denoted by $\mathbb{E}\left[R^C(t)\right]$. The bound on collision regret is derived from a bound on the total number of collisions between two users on a single channel up till t, $C_p(t)$, whose expected value is bounded in Lemma 2.

Lemma 2 *The expected number of collisions between two users on a single channel up till time t is bounded:*

$$\mathbb{E}\left[C_p(t)\right] \leq 2\sqrt{\frac{1+p_0}{1-p_0}}\, t^{1-\beta/2}, \tag{1}$$

where β and p_0 are parameters of MEGA.

Once we have a bound for the pairwise, per-arm, number of collisions, we can bound the mean number of collisions for all users.

Corollary 1 *The expected number of collisions between all users over all channels up till time t is bounded:*

$$\mathbb{E}\left[C(t)\right] \leq \frac{1}{2}N(N-1)K\mathbb{E}\left[C_p(t)\right] \leq N^2 K\sqrt{\frac{1+p_0}{1-p_0}}\, t^{1-\beta/2}.$$

Since the reward is bounded in $[0,1]$, the expected regret acquired as a result of collisions up till time t is bounded by the same value:

$$\mathbb{E}\left[R^C(t)\right] \leq \frac{1}{2}N(N-1)K\mathbb{E}\left[C_p(t)\right] \leq C_1 N^2 K t^{1-\beta/2},$$

where $C_1 = \sqrt{\frac{1+p_0}{1-p_0}}$.

Corollary 1 follows from Lemma 2, since each pair of users can collide on each arm, before the dictated "quiet" period, and so we obtain a bound on the expected regret accumulated due to collisions.

Next, we examine the expected regret caused by the unavailability of arms in K^*, denoted by $\mathbb{E}\left[R^A(t)\right]$. The availability mechanism contributes to a user's regret if it marks all arms in K^* as taken, causing the user to choose an arm $k \notin K^*$ until one of the arms in K^* becomes available once again.

We compute an upper bound on the regret by analyzing the regret due to unavailability when the number of users is $N = 2$. When there are more users, the regret is bounded by the worst case, in which all of them declare an optimal arm unavailable at the same time:

$$\mathbb{E}\left[R^A(t)\right] \leq N\mathbb{E}\left[R_2^A(t)\right],$$

where $\mathbb{E}\left[R_2^A(t)\right]$ is the availability regret accumulated up till time t in the two user scenario for a single channel.

Lemma 3 *The expected regret accumulated due to unavailability of optimal arms in the interval $[T,t]$ is bounded:*

$$\mathbb{E}\left[R_2^A(t)\right] \leq C_3 t^\beta,$$

where $C_3 = 1 + 8\left(\frac{2cK^2}{d(K-1)} + \frac{7-3\alpha}{1-\alpha}\right)$ and β is a parameter of MEGA.

Corollary 2 *The expected regret contributed by the availability-detection mechanism up till time t is bounded:*

$$\mathbb{E}\left[R^A\left(t\right)\right] \leq NKT + NKC_3t^\beta.$$

Corollary 2 follows from Lemma 3, where the first term represents the assumption that the collision mechanism is disabled up until $t = T$.

Our next goal is to bound the regret due to exploration, which is dictated by the ϵ-greedy approach adopted in MEGA.

Lemma 4 *The expected regret accumulated by all users employing the MEGA algorithm due to random exploration up till time t is bounded $\forall t > m$:*

$$\mathbb{E}\left[R^E\left(t\right)\right] \leq Nm + \frac{cK^2N}{d^2K - 1}\log t,$$

where c, d are parameters of the MEGA algorithm and $m = \frac{cK^2}{d^2(K-1)}$.

Based on the lemmas and corollaries above, we have the following regret bound for the MEGA algorithm:

Theorem 1 *Assume a network consisting of N users playing a single K-armed stochastic bandit, $N \leq K$. If all users employ the policy of MEGA, the system-wide regret is bounded for all $t > \max(m, T)$ as follows:*

$$\mathbb{E}\left[R\left(t\right)\right] \leq C_1N^2Kt^{1-\beta/2} + NKT + NKC_3t^\beta + Nm + \frac{cK^2N}{d^2K - 1}\log t$$
$$= O\left(t^{1-\beta/2} + \log t + t^\beta\right).$$

The dominant term in the regret bound above depends on the value of β. For $\beta > 2/3$, the term t^β dominates the bound, while for smaller values the dominant term is $t^{1-\beta/2}$. This tradeoff is intuitive - large values of β correspond to longer "quiet" intervals, reducing the regret contributed by collisions. However, such long intervals also result in longer unavailability periods, increasing the regret contributed by the availability mechanism. Optimizing over β yields $\beta = 2/3$, and so the corresponding regret bound is

$$R\left(t\right) \leq O\left(t^{\frac{2}{3}}\right).$$

The regret bounds for the algorithms proposed in [1] and [12] are $O\left(\log t\right)$ and $O\left(\log^2 t\right)$, respectively. It is worth noting that the constants in the bound provided in [1] are very large, as they involve a binomial coefficient which depends on the numbers of users and channels. Also, the assumptions our algorithm makes are much more strict. Reference [1] requires knowing the number of users, and [12] requires ongoing communication between users, through the Bertsekas auction algorithm. Reference [1] does propose an algorithm which estimates the number of users, but its regret bound is asymptotic. In addition, the empirical results our algorithm provides are considerably better (see Section 5).

4   Dynamic Number of Users

So far, we have focused on several traits of the MEGA algorithm: it does not require communication, cooperation or coordination among users, and it does not assume prior knowledge of the number of users. Simply put, the user operates as though she were the only user in the system, and the algorithm ensures this approach does not result in the users' interfering with each other once the system reaches a "steady state".

However, communication networks like the ones we wish to model often evolve over time - users come and go, affecting the performance of other users by their mere presence or absence. As mentioned in Section 1, the algorithms proposed in [1] attempt to address scenarios similar to ours. However, they rely on either *knowing* or *estimating* the number of users. Thus, a varying number of users is beyond their scope.

It is evident from the experiments in Section 5 that the MEGA algorithm is applicable not only to a fixed number of users, but also in the case that the number of active users in the network varies over time. To the best of our knowledge, this is the *only* algorithm that is able to handle such a setup.

We defer a thorough analysis of the dynamic scenario to our future work. However, a simple performance guarantee can be obtained for the event in which a user leaves the network. Let us begin by defining the regret. Let $N(t)$ denote the number of users in the network at time t. Accordingly, $K^*(t)$ is the set of $N(t)$-best arms. The series $t_1, t_2, t_3 \ldots$ denotes change events in the number of users - arrival or departure. Time intervals during which the number of users is fixed are denoted by $T_i \triangleq [t_{i-1}, t_i - 1]$, with $t_0 \triangleq 0$. Following the definition of the regret introduced in Section 2, the regret for the dynamic scenario is

$$R(t) \triangleq \sum_{T_i} \sum_{k \in K^*(t_{i-1})} \mu_k - \sum_{\tau \in T_i} \sum_{n \in N(t_{i-1})} \mathbb{E}[r_n(\tau)],$$

where we allow a slight abuse of notation for the sake of readability.

Let us assume that a user n leaves the network at some time t, and that the number of users without him is $N(t)$. We also assume that the users had reached a steady state before this departure, i.e., the optimal configuration was being sampled with high probability. Unless user n was sampling the $N(t) + 1$-best arm, regret will start building up at this point. Based on Proposition 1, which follows directly from the definition of the MEGA algorithm, we bound the regret accumulated until the system "settles down" in the new optimal configuration, in Proposition 2.

Proposition 1 *Let K_n^* denote the set of n-optimal arms. For an arm such that $k \in K_n^*$ and also $k \in K_{n-1}^*$, if a user occupying k becomes inactive at time t, k will return to the set of regularly sampled arms within a period of no more than t^β, with a probability greater than $1 - \epsilon_t$.*

Proposition 2 *The regret accumulated in the period between user n's departure at time t and the new optimal configuration's being reached is bounded by*

$$R\left(t\right) \leq \frac{2^{(\beta+1)(N(t)-1)}-1}{2^{\beta+1}-1}t^{\beta} = O\left(t^{\beta}\right).$$

Proposition 2 follows from Proposition 1 and from the worst case analysis described in Figure 1: if the freed arm was the best one, and the user sampling the second-best arm re-occupied it, then the second-best arm would be left to be occupied, etc.. In the worst case, the time intervals before users "upgrade" their arms are back-to-back, creating a series of the form $t^{\beta}, \left(t + t^{\beta}\right)^{\beta}, \ldots$ For detailed proofs of these propositions see the supplementary material.

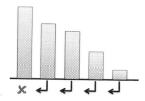

Fig. 1. Worst case occupation of new optimal configuration after a user has left the network

Proposition 2 shows that for a sufficiently low departure rate, regret remains sub-linear even in the scenario of a dynamic number of users. Clearly, frequent changes in the number of users will result in linear regret for practically any distributed algorithm, including ours.

5   Experiments

Our experiments simulate a cognitive radio network with K channels and N users. The existence of primary users is manifested in the differences in expected reward yielded by the channels (i.e., a channel taken by a primary user will yield a low reward). Over the course of the experiment, the secondary users learn the characteristics of the communication channels and settle into an orthogonal, reward-optimal transmission configuration.

The first experiments concern a fixed number of users, N, and channels, K. We assume channel rewards to be Bernoulli random variables with expected values $\boldsymbol{\mu} = (\mu_1, \mu_2, \ldots, \mu_K)$ drawn uniformly from the interval $[0, 1]$. The initial knowledge users have is only of the number of channels.

Once again, we stress that our users *do not* communicate among themselves, nor do they receive external control signals. Their only feedback is the instantaneous reward and an instantaneous collision indicator.

We begin by showing that straightforward application of classic bandit algorithms does not suffice in this case. Figure 2a and Figure 2b present simulation results for a basic scenario in which $N = K = 2$. Even in this rather simple case, the KL-UCB and ϵ-greedy algorithms fail to converge to an orthogonal configuration, and the number of collisions between users grows linearly with time. The experiment was repeated 50 times.

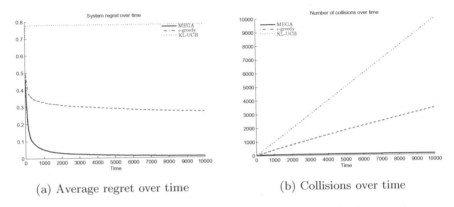

(a) Average regret over time   (b) Collisions over time

Fig. 2. KL-UCB, ϵ-greedy and MEGA performance in basic scenario

Having demonstrated the need for an algorithm tailored to our problem, we compare the performance of our algorithm, MEGA, with the ρ^{RAND} algorithm, proposed in [1]. Figure 3a displays the average regret over time and Figure 3b displays the cumulative number of collisions over time, averaged over all users. An important note is that in our experiments we provide ρ^{RAND} with the exact number of users, as it requires. The MEGA algorithm does not require this input. We did not implement the algorithm ρ^{EST} [1], as its pseudo-code was rather complicated.

The set of parameters used for MEGA was determined by cross validation: $c = 0.1, p_0 = 0.6, \alpha = 0.5, \beta = 0.8$. The value of d as dictated by the ϵ-greedy algorithm should be $d \leq \Delta = \mu_{k_{N-1}} - \mu_{k_N}$. Calculating this value requires prior knowledge of both the number of users and the channels' expected rewards. In order to avoid this issue, we set $d = 0.05$ and avoided distributions for which this condition does not hold in our experiments. The algorithm ρ^{RAND} is parameter-free, as it is a modification of the UCB1 algorithm [2].

The results in Figure 3a and Figure 3b present a scenario in which $N < K$. In the more challenging scenario of $N = K$ our algorithm's advantage is even more pronounced, as is evident from Figure 4a and Figure 4b. Here, ρ^{RAND} actually fails to converge to the optimal configuration, yielding constant average regret.

Next, we display the results of experiments in which the number of users changes over time. Initially, the number of users is 1, gradually increasing until it is equal to 4, decreasing back to 1 again. Since ρ^{RAND} needs a fixed value for

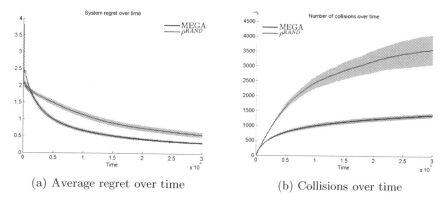

(a) Average regret over time   (b) Collisions over time

Fig. 3. Performance of MEGA compared to ρ^{RAND}. The shaded area around the line plots represents result variance over 50 repetitions. The experiment was run with $N = 6$ users and $K = 9$ channels, and the number of collisions was averaged over all users.

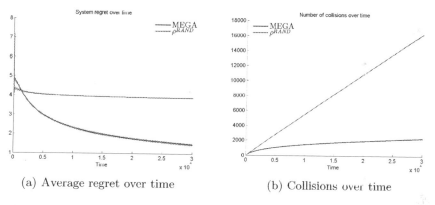

(a) Average regret over time   (b) Collisions over time

Fig. 4. Performance of MEGA compared to ρ^{RAND}. The shaded area around the line plots represents result variance over 50 repetitions. It is barely visible due to the small variance. The experiment was run with $N = 12$ users and $K = 12$ channels, and the number of collisions was averaged over all users.

the number of users, we gave it the value $N_0 = 2$, which is the average number of users in the system over time. For different values of N_0 the performance of ρ^{RAND} was rather similar; we present a single value for the sake of clarity.

As before, Figure 5a displays the average regret over time and Figure 5b displays the cumulative number of collisions over time, averaged over all users.

Clearly, MEGA exhibits better performance in terms of regret and collision rate for both scenarios. The significant improvement in the variance (represented by the shaded area around the line plots) of MEGA compared to ρ^{RAND} is also noteworthy.

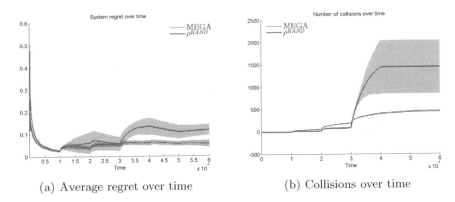

(a) Average regret over time   (b) Collisions over time

Fig. 5. Performance of MEGA compared to ρ^{RAND} in the dynamic scenario. The shaded area represents result variance over 20 repetitions. The experiment was run with $K = 12$ channels, and the number of collisions was averaged over all users.

6   Conclusion

We formulate the problem of multiple selfish users learning to split the resources of a multi-channel communication system modeled by a stochastic MAB. Our proposed algorithm, a combination of an ϵ-greedy policy with an availability detection mechanism, exhibits good experimental results for both fixed and dynamic numbers of users in the network. We augment these results with a theoretical analysis guaranteeing sub-linear regret. It is worth noting that this algorithm is subject to a very strict set of demands, as mentioned in Sections 1 and 2.

We plan to look into additional scenarios of this problem. For example, an explicit collision indication isn't always available in practice. Also, collisions may result in partial, instead of zero, reward. Another challenge is presented when different users have different views of the arms' characteristics (i.e., receive different rewards). We believe that since our algorithm does not involve communication between users, the different views might actually result in fewer collisions. We would also like to expand our theoretical analysis of the scenario in which the number of users is dynamic, deriving concrete regret bounds for it.

References

1. Anandkumar, A., Michael, N., Tang, A.K., Swami, A.: Distributed algorithms for learning and cognitive medium access with logarithmic regret. IEEE Journal on Selected Areas in Communications 29(4), 731–745 (2011)
2. Auer, P., Cesa-Bianchi, N., Fischer, P.: Finite-time analysis of the multiarmed bandit problem. Machine Learning 47(2), 235–256 (2002)
3. Auer, P., Cesa-Bianchi, N., Freund, Y., Schapire, R.E.: The nonstochastic multiarmed bandit problem. SIAM Journal on Computing 32(1), 48–77 (2002)
4. Avner, O., Mannor, S.: Stochastic bandits with pathwise constraints. In: 50th IEEE Conference on Decision and Control (December 2011)

5. Avner, O., Mannor, S., Shamir, O.: Decoupling exploration and exploitation in multi-armed bandits. In: International Conference on Machine Learning (2012)
6. Berry, D.A., Fristedt, B.: Bandit problems: sequential allocation of experiments. Chapman and Hall London (1985)
7. Choe, S.: Performance analysis of slotted aloha based multi-channel cognitive packet radio network. In: IEEE CCNC (2009)
8. Even-Dar, E., Mannor, S., Mansour, Y.: PAC bounds for multi-armed bandit and markov decision processes. In: Kivinen, J., Sloan, R.H. (eds.) COLT 2002. LNCS (LNAI), vol. 2375, p. 255. Springer, Heidelberg (2002)
9. Fang, X., Yang, D., Xue, G.: Taming wheel of fortune in the air: An algorithmic framework for channel selection strategy in cognitive radio networks. IEEE Transactions on Vehicular Technology 62(2), 783–796 (2013)
10. Garivier, A., Cappé, O.: The KL-UCB algorithm for bounded stochastic bandits and beyond. In: Conference on Learning Theory, pp. 359–376 (July 2011)
11. Jouini, W., Ernst, D., Moy, C., Palicot, J.: Multi-armed bandit based policies for cognitive radio's decision making issues. In: 2009 3rd International Conference on Signals, Circuits and Systems (SCS), pp. 1–6. IEEE (2010)
12. Kalathil, D., Nayyar, N., Jain, R.: Decentralized learning for multi-player multi-armed bandits. In: 51st IEEE Conference on Decision and Control (2012)
13. Lai, L., El Gamal, H., Jiang, H., Poor, V.H.: Cognitive medium access: Exploration, exploitation, and competition. IEEE Transactions on Mobile Computing 10(2), 239–253 (2011)
14. Leith, D.J., Clifford, P., Badarla, V., Malone, D.: WLAN channel selection without communication. Computer Networks (2012)
15. Li, X., Liu, H., Roy, S., Zhang, J., Zhang, P., Ghosh, C.: Throughput analysis for a multi-user, multi-channel ALOHA cognitive radio system. IEEE Transactions on Wireless Communications 11(11), 3900–3909 (2012)
16. Liu, K., Zhao, Q.: Distributed learning in multi-armed bandit with multiple players. IEEE Transactions on Signal Processing 58(11), 5667–5681 (2010)
17. Maghsudi, S., Stanczak, S.: Channel selection for network-assisted D2D communication via no-regret bandit learning with calibrated forecasting. CoRR, abs/1404.7061 (2014)
18. McKinney, E.H.: Generalized birthday problem. American Mathematical Monthly, 385–387 (1966)
19. Mitola, J., Maguire, G.Q.: Cognitive radio: making software radios more personal. IEEE Personal Communications 6(4), 13–18 (1999)
20. Nie, N., Comaniciu, C.: Adaptive channel allocation spectrum etiquette for cognitive radio networks. Mobile Networks and Applications 11(6), 779–797 (2006)
21. Niyato, D., Hossain, E.: Competitive spectrum sharing in cognitive radio networks: a dynamic game approach. IEEE Trans. on Wireless Communications 7, 2651–2660 (2008)

Kernel Principal Geodesic Analysis

Suyash P. Awate[1,2,*], Yen-Yun Yu[1], and Ross T. Whitaker[1]

[1] Scientific Computing and Imaging (SCI) Institute, School of Computing, University of Utah
[2] Computer Science and Engineering Department, Indian Institute of Technology (IIT) Bombay

Abstract. Kernel principal component analysis (kPCA) has been proposed as a dimensionality-reduction technique that achieves nonlinear, low-dimensional representations of data via the mapping to kernel feature space. Conventionally, kPCA relies on Euclidean statistics in kernel feature space. However, Euclidean analysis can make kPCA inefficient or incorrect for many popular kernels that map input points to a *hypersphere* in kernel feature space. To address this problem, this paper proposes a novel adaptation of kPCA, namely *kernel principal geodesic analysis* (kPGA), for hyperspherical statistical analysis in kernel feature space. This paper proposes tools for statistical analyses on the Riemannian manifold of the Hilbert sphere in the reproducing kernel Hilbert space, including algorithms for computing the sample weighted Karcher mean and eigen analysis of the sample weighted Karcher covariance. It then applies these tools to propose novel methods for (i) dimensionality reduction and (ii) clustering using mixture-model fitting. The results, on simulated and real-world data, show that kPGA-based methods perform favorably relative to their kPCA-based analogs.

1 Introduction

Kernel principal component analysis (kPCA) [47] maps points in *input space* to a (high-dimensional) *kernel feature space* where it estimates a best-fitting linear subspace via PCA. This mapping to the kernel feature space is typically denoted by $\Phi(\cdot)$. For many of the most useful and widely used kernels (e.g., Gaussian, exponential, Matern, spherical, circular, wave, power, log, rational quadratic), the input data x gets mapped to a *hypersphere*, or a *Hilbert sphere*, in the kernel feature space. Such a mapping also occurs when using (i) kernel normalization, which is common, e.g., in pyramid match kernel [28], and (ii) polynomial and sigmoid kernels when the input points have constant l^2 norm, which is common in digit image analysis [46]. This special structure arises because for these kernels $k(\cdot, \cdot)$, the self similarity of any data point x equals unity (or some constant), i.e., $k(x, x) = 1$. The kernel defines the inner product in the kernel feature space $\mathcal{F}$, and thus, $\langle \Phi(x), \Phi(x) \rangle_{\mathcal{F}} = 1$, which, in turn, equals the distance of the mapped point $\Phi(x)$ from the origin in $\mathcal{F}$. Thus, all of the mapped points $\Phi(x)$ lie on a Hilbert sphere in kernel feature space. Figure 1(a) illustrates this behavior.

The literature shows that for many high-dimensional real-world datasets, where the data representation uses a large number of dimensions, the intrinsic dimension is often quite small, e.g., between 5–20 in [18,43,24,29,42]. The utility of kPCA lies in capturing the intrinsic dimension of the data through the few principal (linear) modes of variation in kernel feature space. This paper proposes a novel extension of kPCA to model

* We thank NIH support via NCRR CIBC P41-RR12553 and NCBC NAMIC U54-EB005149.

T. Calders et al. (Eds.): ECML PKDD 2014, Part I, LNCS 8724, pp. 82–98, 2014.

distributions on the Hilbert sphere manifold in kernel feature space. Manifold-based statistical analysis explicitly models data to reside in a lower dimensional subspace of the ambient space, representing variability in the data more efficiently (fewer degrees of freedom). In this way, the proposed method extends kPCA to (i) define more meaningful modes of variation in kernel feature space by explicitly modeling the data on the Hilbert sphere in kernel feature space, (ii) represent variability using fewer modes, and (iii) reduce curvature of distributions by modeling them explicitly on the Hilbert sphere, instead of modeling them in the ambient space, to avoid artificially large measurements of variability observed in the ambient space. Figure 1(b) illustrates this idea.

Typically, Euclidean PCA of spherical data introduces one additional (unnecessary) component, aligned orthogonally to the sphere and proportional to the sectional curvature. In practice, however, PCA in high-dimensional spaces (e.g., kernel feature space) is known to be unstable and prone to error [4], which interacts with the curvature of the Hilbert sphere on which the data resides. Thus, our empirical results demonstrate that the actual gains in our hyperspherical analysis in kernel feature space surpass what we would expect for the low-dimensional case.

While several works in the literature [3,21,23,27,46] address the properties and uses of kernel feature spaces, these works do *not* systematically explore this special structure of kernel feature space and its implications for PCA in kernel feature space; that is the focus of this paper. Recently, [21] have, in an independent development, examined the use of the Karcher mean in kernel feature spaces, but they propose a different estimation strategy and they do *not* formulate, estimate, or demonstrate the use of principle components on the sphere, which is the main purpose of this work.

This paper makes several contributions. It proposes new formulations and algorithms for computing the sample Karcher mean on a Hilbert sphere in reproducing kernel Hilbert space (RKHS). To analyze sample Karcher covariance, this paper proposes a kernel-based PCA on the Hilbert sphere in RKHS, namely, *kernel principal geodesic analysis* (kPGA). It shows that just as kPCA leads to a standard eigen-analysis problem, kPGA leads to a generalized eigen-analysis problem. This paper evaluates the utility of kPGA for (i) nonlinear dimensionality reduction and (ii) clustering with a Gaussian mixture model (GMM) and an associated expectation maximization (EM) algorithm on the Hilbert sphere in RKHS. Results on simulated and real-world data show that kPGA-based methods perform favorably with their kPCA-based analogs.

2   Related Work

There are several areas of related work that inform the results in this paper. The Karcher mean and associated covariance have recently become important tools for statistical analysis [39]. The algorithm for the Karcher mean proposed in [17] is restricted to analyzing the intrinsic mean and does *not* address how to capture covariance for data lying on spheres, even in finite-dimensional spaces. Other algorithms for the Karcher mean exist and may be more efficient numerically [34]. To capture covariance structure on Riemannian manifolds, Fletcher et al. [25] propose PGA and an associated set of algorithms. Likewise, a small body of work relies on the local geometric structure of Riemannian spaces of covariance matrices for subsequent statistical analysis [7,20,50].

Because many RKHSs are infinite dimensional, we must acknowledge the problem of modeling distributions in such spaces [30] and the corresponding theoretical problems [16]. Of course, these same theoretical concerns arise in kPCA, and other well-known kernel methods, and thus the justification for this work is similar. First, we may assume or assert that the covariance operator of the mapped data is of trace class or, even more strongly, restricted to a finite-dimensional manifold defined by the cardinality of the input data. Second, the proposed methods are intended primarily for data analysis rather than statistical estimation, and, thus, we intentionally work in the subspace defined by the data (which is limited by the data sample size).

In addition to the dimensionality structure, the Hilbert sphere imposes its own structure and has an associated geometry with underlying theoretical implications. The proposed approach in this paper extends PGA [25] to the Hilbert sphere in RKHS. The important geometrical properties of the sphere for the proposed extension concern (i) the geodesic distance between two points, which depends on the arc cosine of their dot product, and (ii) the existence and formulation of tangent spaces [11,15,31].

The work in [21] is more directly related to the proposed method, because it uses logarithmic and exponential maps on the Hilbert sphere in RKHS for data analysis. However, [21] does *not* define a mean or a covariance on the Hilbert sphere in RKHS; it also requires the solution of the ill-posed preimage problem. Unlike [21], we define covariance and its low-dimensional approximations on the Hilbert sphere, represented in terms of the Gram matrix of the data, and incorporate this formulation directly into novel algorithms for dimensionality reduction and clustering via EM [22], including geodesic Mahalanobis distance on the Hilbert sphere in RKHS.

We apply the proposed method for (i) dimensionality reduction for machine-learning applications and (ii) mixture modeling. This builds on the work in kPCA [47], and therefore represents an alternative to other nonlinear mapping methods, such as Sammon's nonlinear mapping [45], Isomap [51] and other kernel-based methods [35,52]. For applications to clustering, the proposed approach generalizes kernel k-means [47] and kernel GMMs [53], where we use formulations of means and/or covariances that respect the hyperspherical geometry of the mapped points in RKHS.

3   Geometry of the Hilbert Sphere in Kernel Feature Space

Many popular kernels are associated with a RKHS that is infinite dimensional. Thus, the analysis in this paper focuses on such spaces. Nevertheless, analogous theory holds for other important kernels (e.g., normalized polynomial) where the RKHS is finite dimensional.

Let X be a random variable taking values x in *input space* $\mathcal{X}$. Let $\{x_n\}_{n=1}^N$ be a set of observations in input space. Let $k(\cdot, \cdot)$ be a real-valued Mercer kernel with an associated map $\Phi(\cdot)$ that maps x to $\Phi(x) := k(\cdot, x)$ in a RKHS $\mathcal{F}$ [6,46]. Consider two points in RKHS: $f := \sum_{i=1}^I \alpha_i \Phi(x_i)$ and $f' := \sum_{j=1}^J \beta_j \Phi(x_j)$. The inner product $\langle f, f' \rangle_{\mathcal{F}} := \sum_{i=1}^I \sum_{j=1}^J \alpha_i \beta_j k(x_i, x_j)$. The norm $\|f\|_{\mathcal{F}} := \sqrt{\langle f, f \rangle_{\mathcal{F}}}$. When $f, f' \in \mathcal{F} \setminus \{0\}$, let $f \otimes f'$ be the rank-one operator defined as $f \otimes f'(h) := \langle f', h \rangle_{\mathcal{F}} f$. Let $Y := \Phi(X)$ be the random variable taking values y in RKHS.

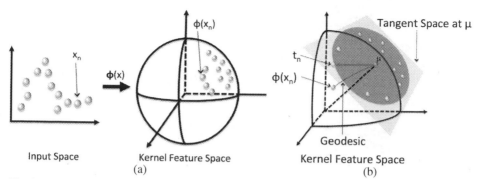

Fig. 1. Kernel Principal Geodesic Analysis (kPGA). **(a)** Points in input space get mapped, via several popular Mercer kernels, to a hypersphere or a Hilbert sphere in kernel feature space. **(b)** Principal geodesic analysis on the Hilbert sphere in kernel feature space.

Assuming Y is bounded and assuming the expectation and covariance operators of Y exist and are well defined, kPCA uses observations $\{y_n := \Phi(x_n)\}_{n=1}^N$ to estimate the eigenvalues, and associated eigenfunctions, of the covariance operator of Y [14,47]. The analysis in this paper applies to kernels that map points in input space to a Hilbert sphere in RKHS, i.e., $\forall x : k(x,x) = \kappa$, a constant (without loss of generality, we assume $\kappa = 1$). For such kernels, the proposed kPGA modifies kPCA using statistical modeling on the Riemannian manifold of the unit Hilbert sphere [5,10] in RKHS.

Consider a and b on the unit Hilbert sphere in RKHS represented, in general, as $a := \sum_n \gamma_n \Phi(x_n)$ and $b := \sum_n \delta_n \Phi(x_n)$. The *logarithmic map*, or Log map, of a with respect to b is the vector

$$\mathrm{Log}_b(a) = \frac{a - \langle a, b \rangle_{\mathcal{F}} b}{\|a - \langle a, b \rangle_{\mathcal{F}} b\|_{\mathcal{F}}} \arccos(\langle a, b \rangle_{\mathcal{F}}) = \sum_n \zeta_n \Phi(x_n), \text{ where } \forall n : \zeta_n \in \mathbb{R}. \tag{1}$$

Clearly, $\mathrm{Log}_b(a)$ can always be written as a weighted sum of the vectors $\{\Phi(x_n)\}_{n=1}^N$. The *tangent vector* $\mathrm{Log}_b(a)$ lies in the *tangent space*, at b, of the unit Hilbert sphere. The tangent space to the Hilbert sphere in RKHS inherits the same structure (inner product) as the ambient space and, thus, is also a RKHS. The geodesic distance between a and b is $d_g(a,b) = \|\mathrm{Log}_b(a)\|_{\mathcal{F}} = \|\mathrm{Log}_a(b)\|_{\mathcal{F}}$.

Now, consider a tangent vector $t := \sum_n \beta_n \Phi(x_n)$ lying in the tangent space at b. The *exponential map*, or Exp map, of t with respect to b is

$$\mathrm{Exp}_b(t) = \cos(\|t\|_{\mathcal{F}})b + \sin(\|t\|_{\mathcal{F}}) \frac{t}{\|t\|_{\mathcal{F}}} = \sum_n \omega_n \Phi(x_n), \text{ where } \forall n : \omega_n \in \mathbb{R}. \tag{2}$$

Clearly, $\mathrm{Exp}_b(t)$ can always be written as a weighted sum of the vectors $\{\Phi(x_n)\}_{n=1}^N$. $\mathrm{Exp}_b(t)$ maps a tangent vector t to the unit Hilbert sphere, i.e., $\|\mathrm{Exp}_b(t)\|_{\mathcal{F}} = 1$.

4   PCA on the Hilbert Sphere in Kernel Feature Space

This section proposes the kPGA algorithm for PCA on the unit Hilbert sphere in RKHS.

4.1 Sample Karcher Mean

The sample Karcher mean on Riemannian manifolds is a consistent estimator of the theoretical Karcher mean of the underlying random variable [12,13]. The sample weighted Karcher mean of set of observations $\{y_m\}_{m=1}^M$, on the unit Hilbert sphere in RKHS, with associated weights $\{p_m \in \mathbb{R}^+\}_{m=1}^M$ is defined as

$$\mu := \arg\min_{\nu} \sum_m p_m d_g^2(\nu, y_m). \qquad (3)$$

The existence and uniqueness properties of the Karcher mean on the Riemannian manifold of the unit Hilbert sphere are well studied [1,32,33]; a study on finite-dimensional Hilbert spheres appears in [17]. The sample Karcher mean on a Hilbert sphere exists and is unique if the pointset is contained within (i) an open convex Riemannian ball of radius $\pi/2$ [33], i.e., an open hemisphere, or (ii) a similar closed ball if one of the points lies in its interior [17]. Thus, the sample Karcher mean exists and is unique for all kernels that map points within a single orthant of the Hilbert sphere in RKHS; this is true for all positive-valued kernels, e.g., the Gaussian kernel.

Clearly, a Karcher mean μ must lie within the space spanned by $\{y_m\}_{m=1}^M$; if not, we could project the assumed "mean" ν' onto the span of $\{y_m\}_{m=1}^M$ and reduce all distances $d_g(y_m, \nu')$ on the Hilbert sphere because of the spherical Pythagoras theorem, thereby resulting in a more-optimal mean ν'' with $d_g(y_m, \nu'') < d_g(y_m, \nu'), \forall m$ and a contradiction to the initial assumption. Therefore, if the points y_m are represented using another set of points $\{\Phi(x_n)\}_{n=1}^N$, i.e., $\forall m, y_m := \sum_n w_{mn}\Phi(x_n)$, then the mean μ can be represented as $\mu = \sum_n \xi_n \Phi(x_n)$, where $\forall n : \xi_n \in \mathbb{R}$.

We propose the following gradient-descent algorithm to compute the mean μ.

1. **Input:** A set of points $\{y_m\}_{m=1}^M$ on the unit Hilbert sphere in RKHS. Weights $\{p_m\}_{m=1}^M$. As described previously, we assume that, in general, each y_m is represented using another set of points $\{\Phi(x_n)\}_{n=1}^N$ and weights w_{mn} on the unit Hilbert sphere in RKHS, i.e., $y_m := \sum_n w_{mn}\Phi(x_n)$.
2. Initialize iteration count: $i = 0$. Initialize the mean estimate to

$$\mu^0 = \frac{\sum_m p_m y_m}{\|\sum_m p_m y_m\|_{\mathcal{F}}} = \sum_n \xi_n \Phi(x_n), \text{ where } \xi_n = \frac{\sum_m p_m w_{mn}}{\|\sum_m p_m y_m\|_{\mathcal{F}}}. \qquad (4)$$

3. Iteratively update the mean estimate, until convergence, by (i) taking the Log maps of all points with respect to the current mean estimate, (ii) performing a weighted average of the resulting tangent vectors, and (iii) taking the Exp map of the weighted average scaled by a step size τ^i, i.e.,

$$\mu^{i+1} = \text{Exp}_{\mu^i}\left(\frac{\tau^i}{M}\sum_m p_m \text{Log}_{\mu^i}(y_m)\right), \text{ where } \tau^i \in (0,1). \qquad (5)$$

4. **Output:** Mean μ lying on the unit Hilbert sphere in RKHS.

In practice, we use a gradient-descent algorithm with an adaptive step size τ^i such that the algorithm (i) guarantees that the objective-function value is non increasing every iteration and (ii) increases/decreases the step size each iteration to aid faster convergence.

We detect convergence as the point when the objective function cannot be reduced using any non-zero step size. Typically, in practice, a few iterations suffice for convergence.

The convergence of gradient descent for finding Karcher means has been studied [2,17]. In certain conditions, such as those described earlier when the sample Karcher mean on a Hilbert sphere is unique, the objective function becomes convex [19], which leads the gradient descent to the global minimum.

4.2 Sample Karcher Covariance and Eigen Analysis

Given the sample weighted Karcher mean μ, consider a random variable $Z := \text{Log}_\mu(Y)$ taking values in the tangent space at μ. Assuming that both the expectation and covariance operators of Z exist and are well defined (this follows from the similar assumption on Y), the sample weighted Karcher covariance operator, in the tangent space at μ, is

$$C := (1/M) \sum_m p_m z_m \otimes z_m, \text{ where } z_m := \text{Log}_\mu(y_m). \tag{6}$$

Because the tangent space is a RKHS, the theoretical analysis of covariance in RKHS in standard kPCA [14,48] applies to C as well (note that the set $\{z_m\}_{m=1}^M$ is empirically centered by construction; i.e., $\sum_m z_m = 0$). Thus, as the sample size $M \to \infty$, the partial sums of the empirically-computed eigenvalues converge to the partial sums of the eigenvalues of the theoretical covariance operator of Z.

Using the Log map representation in Section 3, $z_m = \sum_{n'} \beta_{n'm} \Phi(x_{n'})$ leading to

$$C = \sum_{n'} \sum_{n''} E_{n'n''} \Phi(x_{n'}) \otimes \Phi(x_{n''}), \text{ where } E_{n'n''} = \frac{1}{M} \sum_m p_m \beta_{n'm} \beta_{n''m}. \tag{7}$$

If λ is a positive eigenvalue of C and v is the corresponding eigenfunction, then

$$v = \frac{Cv}{\lambda} = \frac{1}{\lambda} \sum_{n'} \sum_{n''} E_{n'n''} \Phi(x_{n'}) \otimes \Phi(x_{n''}) v = \sum_{n'} \alpha_{n'} \Phi(x_{n'}),$$

$$\text{where } \alpha_{n'} = \sum_{n''} \frac{E_{n'n''}}{\lambda} \langle \Phi(x_{n''}), v \rangle_{\mathcal{F}}. \tag{8}$$

Thus, any eigenfunction v of C lies within the span of the set of points $\{\Phi(x_n)\}_{n=1}^N$ used to represent $\{y_m\}_{m=1}^M$. For any $\Phi(x_\eta) \in \{\Phi(x_n)\}_{n=1}^N$ and the eigenfunction v,

$$\langle \Phi(x_\eta), Cv \rangle_{\mathcal{F}} = \lambda \langle \Phi(x_\eta), v \rangle_{\mathcal{F}}. \tag{9}$$

Thus, $\langle \Phi(x_\eta), \sum_{n'} \sum_{n''} E_{n'n''} \Phi(x_{n'}) \otimes \Phi(x_{n''}) \sum_{n'''} \alpha_{n'''} \Phi(x_{n'''}) \rangle_{\mathcal{F}} =$

$$\lambda \langle \Phi(x_\eta), \sum_{n'''} \alpha_{n'''} \Phi(x_{n'''}) \rangle_{\mathcal{F}}. \tag{10}$$

Thus, $\sum_{n'''} \left(\sum_{n'} K_{\eta n'} \sum_{n''} E_{n'n''} K_{n''n'''} \right) \alpha_{n'''} = \lambda \sum_{n'''} K_{\eta n'''} \alpha_{n'''}, \tag{11}$

$$\text{where } K_{ij} := \langle \Phi(x_i), \Phi(x_j) \rangle_{\mathcal{F}}$$

is the element in row i and column j of the Gram matrix K. Considering E and K as $N \times N$ real matrices and defining $F := EK$ and $G := KF$ leads to

$$\sum_{n''} E_{n'n''} K_{n''n'''} = F_{n'n'''} \text{ and } \sum_{n'} K_{\eta n'} \sum_{n''} E_{n'n''} K_{n''n'''} = G_{\eta n'''}. \qquad (12)$$

Therefore, the left hand side of Equation 9 equals $G_{\eta \bullet} \alpha$, where (i) $G_{\eta \bullet}$ is the ηth row of the $N \times N$ matrix G and (ii) α is the $N \times 1$ column vector with the nth component as α_n. Similarly, the right hand side of Equation 9 equals $K_{\eta \bullet} \alpha$, where $K_{\eta \bullet}$ is the ηth row of the $N \times N$ matrix K. Using Equation 9 to form one equation for all $\eta = 1, \cdots, N$, gives the following generalized eigen-analysis problem

$$G\alpha = \lambda K \alpha. \qquad (13)$$

If $k(\cdot, \cdot)$ is a symmetric positive-*definite* (SPD) Mercer kernel and the points $\{\Phi(x_n)\}_{n=1}^N$ are *distinct*, then K is SPD (hence, invertible) and the generalized eigen-analysis problem reduces to the standard eigen-analysis problem

$$EK\alpha = \lambda \alpha. \qquad (14)$$

Thus, (i) the eigenvalues $\{\lambda_n\}_{n=1}^N$ are same as the eigenvalues of the sample covariance operator C and (ii) each eigenvector α gives one eigenfunction of C through Equation 8. Note that standard kPCA requires eigen decomposition of the (centralized) matrix K.

The definition of the sample covariance operator C implies that the rank of C is upper bounded by the sample size M. Because the eigenvalues of C are the same as those for EK or for the pair (G, K), if $M < N$, then the rank of the $N \times N$ matrices EK and G are also upper bounded by M. While K is an $N \times N$ symmetric positive (semi) definite matrix of rank at-most N, E is an $N \times N$ symmetric positive (semi) definite matrix of rank at-most M because $E = BPB^T$ where (i) B is a $N \times M$ matrix where $B_{nm} = \beta_{nm}$ and (ii) P is an $M \times M$ diagonal matrix where $P_{mm} = p_m/M$.

4.3   Kernel Principal Geodesic Analysis (kPGA) Algorithm

We summarize the proposed **kPGA** algorithm below.

1. **Input:** (i) A set of points $\{y_m\}_{m=1}^M$ on the unit Hilbert sphere in RKHS. (ii) Weights $\{p_m\}_{m=1}^M$. As described previously, we assume that, in general, each y_m is represented using another set of points $\{\Phi(x_n)\}_{n=1}^N$ and weights w_{mn} on the unit Hilbert sphere in RKHS, i.e., $y_m := \sum_n w_{mn} \Phi(x_n)$.
2. Compute the Gram matrix K.
3. Compute the Karcher mean μ using the algorithm in Section 4.1.
4. Compute the matrix E or $G = KEK$ as described in Section 4.2.
5. To analyze the Karcher covariance, perform eigen analysis for the linear system $G\alpha = \lambda K\alpha$ or $EK\alpha = \lambda \alpha$ to give eigenvalues $\{\lambda_n\}_{\eta=1}^N$ (sorted in non-increasing order) and eigenvectors $\{\alpha_\eta\}_{\eta=1}^N$.
6. **Output:** (i) Mean μ lying on the unit Hilbert sphere in RKHS. (ii) Principal components or eigenfunctions $\{v_n = \sum_{n'} \alpha_{\eta n'} \Phi(x_{n'})\}_{n=1}^N$ in the tangent space at μ. (iii) Eigenvalues $\{\lambda_n = \lambda_\eta\}_{n=1}^N$ capturing variance along principal components.

5   Applications

This section proposes kPGA-based algorithms for (i) nonlinear dimensionality reduction and (ii) clustering using a mixture model fitted using EM.

5.1   Nonlinear Dimensionality Reduction

First, we propose the following algorithm for dimensionality reduction using kPGA.

1. **Input:** A set of points $\{x_n\}_{n=1}^N$ along with their maps $\{\Phi(x_n)\}_{n=1}^N$ on the unit Hilbert sphere in RKHS. Weights $\{p_n = 1\}_{n=1}^N$.
2. Apply the kPGA algorithm in Section 4.2 to the observed sample $\{\Phi(x_n)\}_{n=1}^N$ to compute mean μ, eigenvalues $\{\lambda_n\}_{n=1}^N$ (sorted in non-increasing order), and corresponding eigenfunctions $\{v_n\}_{n=1}^N$.
3. Select the largest $Q < N$ eigenvalues $\{\lambda_q\}_{q=1}^Q$ that capture a certain fraction of energy in the eigenspectrum. Select the corresponding subspace $\mathbb{G}_Q = <v_1, \cdots, v_Q>$.
4. Project the Log map of each point $\Phi(x_n)$ on the subspace $\mathbb{G}_Q$ to give the embedding coordinates $e_{nq} := \langle \mathrm{Log}_\mu \Phi(x_n), v_q \rangle_{\mathcal{F}}$ and projected tangent vectors $t_n = \sum_q e_{nq} v_q$ in the tangent space at the mean μ.
5. Take the Exp map of projections $\{t_n\}_{n=1}^N$ to produce $\{y_n = \mathrm{Exp}_\mu(t_n)\}_{n=1}^N$ lying within a Q-dimensional subsphere on the unit Hilbert sphere in RKHS.
6. **Output**: Embedding subspace (lower dimensional) $\mathbb{G}_Q$, embedding coordinates $\{(e_{n1}, \cdots, e_{nQ})\}_{n=1}^N$, and (re)mapped points on the Hilbert subsphere $\{y_n\}_{n=1}^N$.

5.2   Clustering Using Mixture Modeling and Expectation Maximization

We now propose an algorithm for clustering a set of points $\{x_n\}_{n=1}^N$, into a fixed number of clusters, by fitting a mixture model on the unit Hilbert sphere in RKHS.

The proposed approach entails mixture modeling in a finite-dimensional subsphere of the unit Hilbert sphere in RKHS, after the dimensionality reduction of the points $\{\Phi(x_n)\}$ to a new set of points $\{y_n\}$ (as in Section 5.1). Modeling PDFs on Hilbert spheres entails fundamental trade-offs between model generality and the viability of the underlying parameter estimation. For instance, although Fisher-Bingham PDFs on $\mathbb{S}^d$ are able to model generic anisotropic distributions (anisotropy around the mean) using $O(d^2)$ parameters, their parameter estimation may be intractable [9,37,40]. On the other hand, parameter estimation for the $O(d)$-parameter von Mises-Fisher PDF is tractable [9], but that PDF can only model isotropic distributions. We take another approach that uses a tractable approximation of a normal law on a Riemannian manifold [41], allowing modeling of anisotropic distributions through its covariance parameter in the tangent space at the mean. Thus, the proposed PDF evaluated at $\Phi(x)$ is $P(\Phi(x)|\mu, C) \doteq \exp\left(-0.5 d_g^2(\mu, \Phi(x); C)\right) / ((2\pi)^{Q/2}|C|^{1/2})$, where $|C| = \Pi_{q=1}^Q \lambda_q$ and $d_g(\mu, \nu; C)$ is the *geodesic Mahalanobis distance* between the point $\Phi(x)$ and mean μ, given covariance C.

The geodesic Mahalanobis distance relies on a regularized sample inverse-covariance operator [38] $C^{-1} := \sum_{q=1}^Q (1/\lambda_q) v_q \otimes v_q$, where λ_q is the qth sorted eigenvalue of

C, v_q is the corresponding eigenfunction, and $Q \leq \min(M, N)$ is a regularization parameter. Then, the corresponding square-root inverse-covariance operator is $C^{-1/2} := \sum_q (1/\sqrt{\lambda_q})v_q \otimes v_q$ and the geodesic Mahalanobis distance of the point ν from mean μ is $d_g(\nu, \mu; C) := (\langle C^{-1/2}t, C^{-1/2}t \rangle_{\mathcal{F}})^{0.5}$ where $t := \mathrm{Log}_\mu(\nu)$.

Let Y be a random variable that generates the N independent and identically-distributed data points $\{y_n\}_{n=1}^N$ as follows. For each n, we first draw a cluster number $l \in \{1, 2, \cdots, L\}$ with probability w_l (where $\forall l : w_l > 0$ and $\sum_l w_l = 1$) and then draw y_n from $P(Y|\mu_l, C_l)$. Thus, the probability of observing y_n is $P(y_n) = \sum_l w_l P(y_n|\mu_l, C_l)$.

The *parameters* for $P(Y)$ are $\theta = \{w_l, \mu_l, C_l\}_{l=1}^L$. We solve for the maximum-likelihood estimate of θ via EM. Let $\{S_n\}_{n=1}^N$ be *hidden* random variables that give, for each n, the cluster number $s_n \in \{1, \cdots, L\}$ that generated data point y_n.

EM performs iterative optimization. Each EM iteration involves an E step and an M step. At iteration i, given parameter estimates θ^i, the E step defines a function $\mathcal{Q}(\theta|\theta^i) := E_{P(\{S_n\}_{n=1}^N|\{y_n\}_{n=1}^N, \theta^i)}[\log P(\{S_n, y_n\}_{n=1}^N|\theta)]$. For our mixture model,

$$\mathcal{Q}(\theta|\theta^i) = \sum_n \sum_l P(s_n = l|y_n, \theta^i) \left(\log w_l - 0.5 \log |C_l| - 0.5 d_g^2(\mu_l, y_n; C_l)\right)$$

$$+ \text{ constant, where} \tag{15}$$

$$P(s_n = l|y_n, \theta^i) = \frac{P(s_n = l|\theta^i)P(y_n|s_n = l, \theta^i)}{P(y_n|\theta^i)} = \frac{w_l^i P(y_n|\mu_l^i, C_l^i)}{\sum_l w_l^i P(y_n|\mu_l^i, C_l^i)}. \tag{16}$$

We denote $P(s_n = l|y_n, \theta^i)$ in shorthand by the class membership P_{nl}^i. We denote $\sum_n P_{nl}^i$ in shorthand by P_l^i. Simplifying gives

$$\mathcal{Q}(\theta|\theta^i) = \sum_l P_l^i \left(\log w_l - 0.5 \log |C_l|\right) - 0.5 \sum_n \sum_l P_{nl}^i d_g^2(\mu_l, y_n; C_l) + \text{ constant.} \tag{17}$$

The M step maximizes $\mathcal{Q}(\theta)$, under the constraints on w_l, using the method of Lagrange multipliers, to give the optimal values and, hence, the updates, for parameters θ.

Thus, the proposed clustering algorithm is as follows.

1. **Input:** A set of points $\{\Phi(x_n)\}_{n=1}^N$ on the unit Hilbert sphere in RKHS with all associated weights p_n set to unity.
2. Reduce the dimensionality of the input using the algorithm in Section 5.1 to give points $\{y_n\}_{n=1}^N$ on a lower-dimensional subsphere of the Hilbert sphere in RKHS.
3. Initialize iteration count $i := 0$. Initialize parameters $\theta^0 = \{w_l^0, \mu_l^0, C_l^0\}_{l=1}^L$ as follows: run farthest-point clustering [26] (with kernel-based distances; with randomly-selected first point) to initialize kernel k means [47] that, in turn, initializes μ_l^0 and C_l^0 to be the mean and covariances of cluster l, respectively, and w_l^0 equal to the number of points in cluster l divided by N.
4. Iteratively update the parameter estimates, until convergence, as follows.
5. Evaluate probabilities $\{P_{nl}^i\}$ using current parameter estimates θ^i.

6. Update means $\mu_l^{i+1} = \arg\min_\mu \sum_n P_{nl}^i d_g^2(\mu_l, y_n; C_l)$ using a gradient-descent algorithm similar to that used in Section 4.1 for the sample weighted Karcher mean.
7. Update covariances $C_l^{i+1} = \sum_n (P_{nl}^i / P_l^i) \text{Log}_{\mu_l^{i+1}}(y_n) \otimes \text{Log}_{\mu_l^{i+1}}(y_n)$.
8. Update probabilities $w_l^{i+1} = P_l^i / (\sum_l P_l^i)$.
9. **Output:** Parameters: $\theta = \{w_l, \mu_l, C_l\}_{l=1}^L$. Labeling: Assign $\Phi(x_n)$ to the cluster l that maximizes $P(y_n | \mu_l, C_l)$.

6   Results and Discussion

This section shows results on simulated data, real-world face images from the Olivetti Research Laboratory (ORL) [44], and real-world data from the University of California Irvine (UCI) machine learning repository [8].

6.1   Nonlinear Dimensionality Reduction

We employ kPCA and the proposed kPGA for nonlinear dimensionality reduction on simulated and real-world databases. To evaluate the quality of dimensionality reduction, we use the co-ranking matrix [36] to compare rankings of pairwise distances between (i) data points in the original high-dimensional space (i.e., without any dimensionality reduction) and (ii) the projected data points in the lower-dimensional embedding found by the algorithm. Based on this motivation, a standard measure to evaluate the quality of dimensionality-reduction algorithms is to average, over all data points, the fraction of other data points that remain inside a κ neighborhood defined based on the original distances [36]. For a fixed number of reduced dimensions, an ideal dimensionality-reduction algorithm would lead to this quality measure being 1 for every value of $\kappa \in \{1, 2, \cdots, N - 1\}$, where N is the total number of points in the dataset.

Simulated Data – Points on a High-Dimensional Unit Hilbert Sphere. We generate $N = 200$ data points lying on the unit Hilbert sphere in $\mathbb{R}^{100}$. We ensure the intrinsic

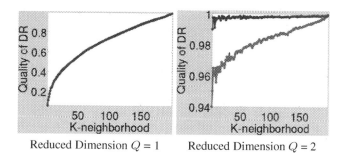

Reduced Dimension $Q = 1$ Reduced Dimension $Q = 2$

Fig. 2. Nonlinear Dimensionality Reduction on Simulated Data. The performance for the proposed kPGA is in blue and that for the standard kPCA is in red. The horizontal axis shows values of κ in the κ neighborhood [36]. The quality measure on the vertical axis indicates the preservation of κ-sized neighborhoods based on distances in the original space (see text). For a fixed number of reduced dimensions Q, the ideal performance is a quality measure of 1 for all κ.

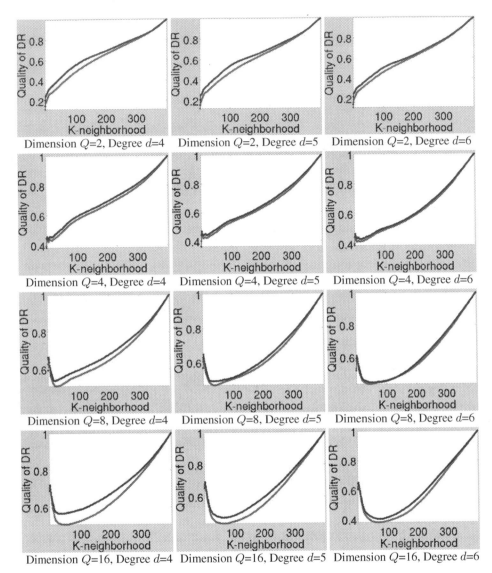

Fig. 3. Nonlinear Dimensionality Reduction on ORL Face Images. The **blue** curves represent the **proposed kPGA** and the **red** curves represent standard kPCA. Each subfigure plots quality measures (on vertical axis) for reduced-dimension values $Q = 2, 4, 8, 16$ and polynomial-kernel-parameter values $d = 4, 5, 6$. Within each subfigure (on horizontal axis), $\kappa = 1, \cdots, 399$. See Figure 4 for additional results with reduced-dimension values $Q = 32, 64, 128, 256$.

dimensionality of the dataset to be 2 by considering a subsphere $\mathbb{S}^2$ of dimension 2 and sampling points from a von Mises-Fisher distribution on $\mathbb{S}^2$ [37]. We set the kernel as $k(x, y) := \langle x, y \rangle$ that reduces the map $\Phi(\cdot)$ to identity (i.e., $\Phi(x) := x$) and, thereby, performs the analysis on the original data that lies on a Hilbert sphere in input space. Figure 2 shows the results of the dimensionality reduction using kPCA and kPGA.

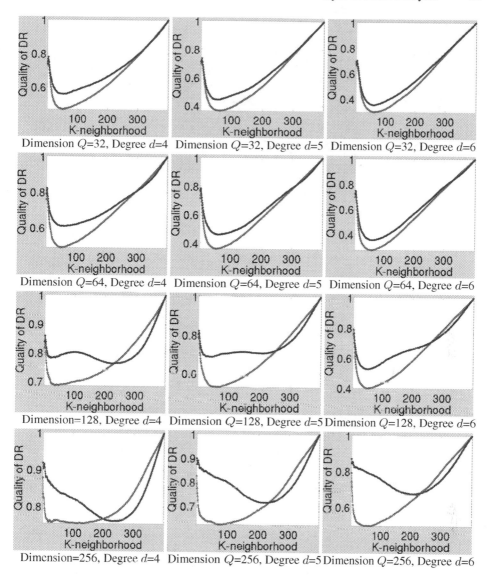

Fig. 4. Nonlinear Dimensionality Reduction on ORL Face Images. Continued from Figure 3.

When the reduced dimensionality is forced to be 1, which we know is suboptimal, both kPCA and kPGA perform comparably. However, when the reduced dimensionality is forced to 2 (which equals the intrinsic dimension of the data), then kPGA clearly outperforms kPCA; kPGA preserves the distance-based κ neighborhoods for almost every value of $\kappa \in \{1, \cdots, 199\}$. The result in Figure 2 is also consistent with the covariance eigenspectra produced by kPCA and kPGA. Standard kPCA, undesirably, gives 3 non-zero eigenvalues $(0.106, 0.0961, 0.0113)$ that reflect the dimensionality of the data representation for points on $\mathbb{S}^2$. On the other hand, the proposed kPGA gives

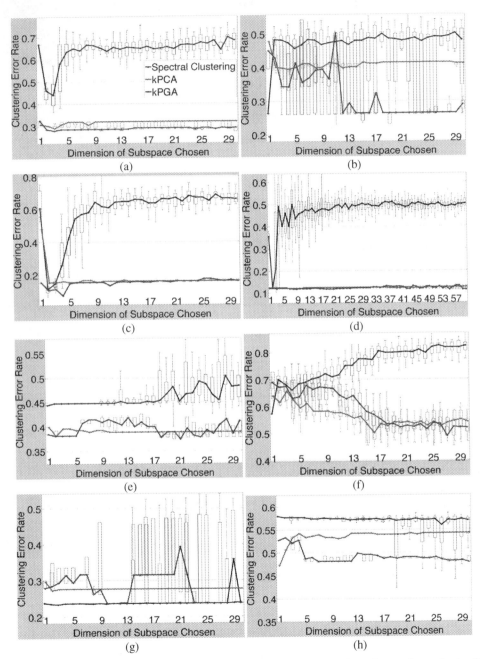

Fig. 5. Clustering on UCI Datasets. Box plots of error rates from clustering random subsets of the dataset. We use a Gaussian kernel. **(a)–(h)** show results on Wine, Haberman, Iris, Vote, Heart, Ecoli, Blood, and Liver datasets, respectively.

only 2 non-zero eigenvalues $(0.1246, 0.1211)$ that reflect the intrinsic dimension of the data. Thus, kPGA needs fewer components/dimensions to represent the data.

Real-World Data – ORL Face Image Database. The ORL database [44] comprises $N = 400$ face images of size 112×92 pixels. To measure image similarity, a justifiable kernel is the polynomial kernel $k(x, y) := (\langle x, y \rangle)^d$ after normalizing the intensities in each image x (i.e., subtract mean and divide by standard deviation) so that $\langle x, x \rangle = 1 = k(x, x)$ [46]. Figure 3 and Figure 4 show the results of nonlinear dimensionality reduction using standard kPCA and the proposed kPGA. For a range of values of the reduced dimension (i.e., $2, 4, 8, 16, 32, 64, 128, 256$) and a range of values of the polynomial kernel degree d (i.e., $d = 4, 5, 6$), the proposed kPGA outperforms standard kPCA with respect to the κ-neighborhood based quality measure.

6.2 Clustering Using Mixture Modeling and Expectation Maximization

We use the UCI repository to evaluate clustering in RKHS. Interestingly, for all but 2 of the UCI datasets used in this paper, the number of modes in kPCA (using the Gaussian kernel) capturing 90% of the spectrum energy ranges from 3–15 (mean 8.5, standard deviation 4.5). For only 2 datasets is the corresponding number of modes more than 20. This number is usually close to the intrinsic dimension of the data.

Real-World Data – UCI Machine Learning Repository. We evaluate clustering algorithms by measuring the error rate in the assignments of data points to clusters; we define error rate as the fraction of the total number of points in the dataset assigned to the incorrect cluster. We evaluate clustering error rates on a wide range of subspace dimensions $Q \in \{1, \cdots, 30\}$. For each Q, we repeat the following process 50 times: we randomly select 70% points from each cluster, run the clustering algorithm, and compute the error rate. We use the Gaussian kernel $k(x_i, x_j) = \exp(-0.5\|x_i - x_j\|_2^2/\sigma^2)$ and set σ^2, as per convention, to the average squared distance between all pairs (x_i, x_j).

Figures 5 compares the performance of spectral clustering [49], standard kPCA, and the proposed kPGA. In Figures 5(a)–(f), kPGA gives the lowest error rates (over all Q) and outperforms spectral clustering. In Figures 5(a)–(d), kPGA performs better or as well for almost all choices of Q. In Figure 5(g), kPGA performs as well as spectral clustering (over all Q). In Figure 5(h), kPGA performs slightly worse than kPCA (over all Q), but kPGA performs the best whenever $Q > 2$.

7 Conclusion

This paper addresses the hyperspherical geometry of points in kernel feature space, which naturally arises from many popular kernels and kernel normalization. This paper proposes kPGA to perform PGA on the Hilbert sphere manifold in RKHS, through algorithms for computing the sample weighted Karcher mean and the eigenvalues and eigenfunctions of the sample weighted Karcher covariance. It leverages kPGA to propose methods for (i) nonlinear dimensionality reduction and (ii) clustering using mixture-model fitting on the Hilbert sphere in RKHS. The results, on simulated and real-world data, show that kPGA-based methods perform favorably with their kPCA-based analogs.

References

1. Afsari, B.: Riemannian L^p center of mass: Existence, uniqueness, and convexity. Proc. Am. Math. Soc. 139(2), 655–673 (2011)
2. Afsari, B., Tron, R., Vidal, R.: On the convergence of gradient descent for finding the Riemannian center of mass. SIAM J. Control and Optimization 51(3), 2230–2260 (2013)
3. Ah-Pine, J.: Normalized kernels as similarity indices. In: Proc. Pacific-Asia Conf. Advances in Knowledge Discovery and Data Mining, vol. 2, pp. 362–373 (2010)
4. Ahn, J., Marron, J.S., Muller, K., Chi, Y.Y.: The high-dimension, low-sample-size geometric representation holds under mild conditions. Biometrika 94(3), 760–766 (2007)
5. Amari, S., Nagaoka, H.: Methods of Information Geometry. Oxford Univ. Press (2000)
6. Aronszajn, N.: Theory of reproducing kernels. Trans. Amer. Math. Soc. 68(3), 337–404 (1950)
7. Arsigny, V., Fillard, P., Pennec, X., Ayache, N.: Log-Euclidean metrics for fast and simple calculus on diffusion tensors. Mgn. Reson. Med. 56(2), 411–421 (2006)
8. Bache, K., Lichman, M.: UCI machine learning repository (2013), http://archive.ics.uci.edu/ml
9. Banerjee, A., Dhillon, I., Ghosh, J., Sra, S.: Clustering on the unit hypersphere using von Mises-Fisher distributions. J. Mach. Learn. Res. 6, 1345–1382 (2005)
10. Berger, M.: A Panoramic View of Riemannian Geometry. Springer (2007)
11. Berman, S.: Isotropic Gaussian processes on the Hilbert sphere. Annals of Probability 8(6), 1093–1106 (1980)
12. Bhattacharya, R., Patrangenaru, V.: Large sample theory of intrinsic and extrinsic sample means on manifolds. I. Annals Stats. 31(1), 1–29 (2005)
13. Bhattacharya, R., Patrangenaru, V.: Large sample theory of intrinsic and extrinsic sample means on manifolds. II. Annals Stats. 33(3), 1225–1259 (2005)
14. Blanchard, G., Bousquet, O., Zwald, L.: Statistical properties of kernel principal component analysis. Machine Learning 66(3), 259–294 (2007)
15. Boothby, W.M.: An introduction to differentiable manifolds and Riemannian geometry, vol. 120. Academic Press (1986)
16. Bühlmann, P., Van De Geer, S.: Statistics for high-dimensional data: methods, theory and applications. Springer (2011)
17. Buss, S., Fillmore, J.: Spherical averages and applications to spherical splines and interpolation. ACM Trans. Graph. 20(2), 95–126 (2001)
18. Carter, K., Raich, R., Hero, A.: On local intrinsic dimension estimation and its applications. IEEE Trans. Signal Proc. 58(2), 650–663 (2010)
19. Charlier, B.: Necessary and sufficient condition for the existence of a Frechet mean on the circle. ESAIM: Probability and Statistics 17, 635–649 (2013)
20. Cherian, A., Sra, S., Banerjee, A., Papanikolopoulos, N.: Jensen-Bregman logDet divergence with application to efficient similarity search for covariance matrices. IEEE Trans. Pattern Anal. Mach. Intell. 35(9), 2161–2174 (2012)
21. Courty, N., Burger, T., Marteau, P.-F.: Geodesic analysis on the Gaussian RKHS hypersphere. In: Flach, P.A., De Bie, T., Cristianini, N. (eds.) ECML PKDD 2012, Part I. LNCS, vol. 7523, pp. 299–313. Springer, Heidelberg (2012)
22. Dempster, A., Laird, N., Rubin, D.: Maximum likelihood from incomplete data via the EM algorithm. J. Royal Statistical Society B(39), 1–38 (1977)
23. Eigensatz, M.: Insights into the geometry of the Gaussian kernel and an application in geometric modeling. Master thesis. Swiss Federal Institute of Technology (2006)
24. Felsberg, M., Kalkan, S., Krueger, N.: Continuous dimensionality characterization of image structures. Image and Vision Computing 27(6), 628–636 (2009)

25. Fletcher, T., Lu, C., Pizer, S., Joshi, S.: Principal geodesic analysis for the study of nonlinear statistics of shape. IEEE Trans. Med. Imag. 23(8), 995–1005 (2004)

26. Gonzalez, T.: Clustering to minimize the maximum intercluster distance. Theor. Comp. Sci. 38, 293–306 (1985)

27. Graf, A., Smola, A., Borer, S.: Classification in a normalized feature space using support vector machines. IEEE Trans. Neural Networks 14(3), 597–605 (2003)

28. Grauman, K., Darrell, T.: The pyramid match kernel: Efficient learning with sets of features. Journal of Machine Learning Research 8, 725–760 (2007)

29. Hein, M., Audibert, J.Y.: Intrinsic dimensionality estimation of submanifolds in R^d. In: Int. Conf. Mach. Learn., pp. 289–296 (2005)

30. Hoyle, D.C., Rattray, M.: Limiting form of the sample covariance eigenspectrum in PCA and kernel PCA. In: Int. Conf. Neural Info. Proc. Sys. (2003)

31. Kakutani, S., et al.: Topological properties of the unit sphere of a Hilbert space. Proceedings of the Imperial Academy 19(6), 269–271 (1943)

32. Karcher, H.: Riemannian center of mass and mollifier smoothing. Comn. Pure Appl. Math. 30(5), 509–541 (1977)

33. Kendall, W.S.: Probability, convexity and harmonic maps with small image I: uniqueness and fine existence. Proc. Lond. Math. Soc. 61, 371–406 (1990)

34. Krakowski, K., Huper, K., Manton, J.: On the computation of the Karcher mean on spheres and special orthogonal groups. In: Proc. Workshop Robotics Mathematics, pp. 1–6 (2007)

35. Lawrence, N.: Probabilistic non-linear principal component analysis with Gaussian process latent variable models. J. Mach. Learn. Res. 6, 1783–1816 (2005)

36. Lee, J.A., Verleysen, M.: Quality assessment of dimensionality reduction: Rank-based criteria. Neurocomputing 72, 1432–1433 (2009)

37. Mardia, K., Jupp, P.: Directional Statistics. Wiley (2000)

38. Mas, A.: Weak convergence in the function autoregressive model. J. Multiv. Anal. 98, 1231–1261 (2007)

39. Nielsen, F., Bhatia, R.: Matrix Information Geometry. Springer (2013)

40. Peel, D., Whiten, W., McLachlan, G.: Fitting mixtures of Kent distributions to aid in joint set identification. J. Amer. Stat. Assoc. 96, 56–63 (2001)

41. Pennec, X.: Intrinsic statistics on Riemannian manifolds: Basic tools for geometric measurements. J. Mathematical Imaging and Vision 25(1), 127–154 (2006)

42. Raginsky, M., Lazebnik, S.: Estimation of intrinsic dimensionality using high rate vector quantization. In: Proc. Adv. Neural Information Processing Systems, pp. 1–8 (2005)

43. de Ridder, D., Kouropteva, O., Okun, O., Pietikainen, M., Duin, R.: Supervised locally linear embedding. In: Kaynak, O., Alpaydın, E., Oja, E., Xu, L. (eds.) ICANN/ICONIP 2003. LNCS, vol. 2714, pp. 333–341. Springer, Heidelberg (2003)

44. Samaria, F., Harter, A.: Parameterisation of a stochastic model for human face identification. In: Proc. IEEE Workshop on Applications of Computer Vision, pp. 138–142 (1994)

45. Sammon, J.W.: A nonlinear mapping for data structure analysis. IEEE Trans. Computers 18(5), 401–409 (1969)

46. Scholkopf, B., Smola, A.: Learning with Kernels. MIT Press (2002)

47. Scholkopf, B., Smola, A., Muller, K.R.: Nonlinear component analysis as a kernel eigenvalue problem. Neural Computation 10, 1299–1319 (1998)

48. Shawe-Taylor, J., Williams, C., Cristianini, N., Kandola, J.: On the eigenspectrum of the Gram matrix and the generalisation error of kernel PCA. IEEE Trans. Info. Th. 51(7), 2510–2522 (2005)

49. Shi, J., Malik, J.: Normalized cuts and image segmentation. IEEE Trans. Pattern Anal. Mach. Intell. 22(8), 888–905 (2000)
50. Sommer, S., Lauze, F., Hauberg, S., Nielsen, M.: Manifold valued statistics, exact principal geodesic analysis and the effect of linear approximations. In: Daniilidis, K., Maragos, P., Paragios, N. (eds.) ECCV 2010, Part VI. LNCS, vol. 6316, pp. 43–56. Springer, Heidelberg (2010)
51. Tenenbaum, J.B., De Silva, V., Langford, J.C.: A global geometric framework for nonlinear dimensionality reduction. Science 290(5500), 2319–2323 (2000)
52. Walder, C., Schölkopf, B.: Diffeomorphic dimensionality reduction. In: Int. Conf. Neural Info. Prof. Sys., pp. 1713–1720 (2008)
53. Wang, J., Lee, J., Zhang, C.: Kernel trick embedded Gaussian mixture model. In: Gavaldá, R., Jantke, K.P., Takimoto, E. (eds.) ALT 2003. LNCS (LNAI), vol. 2842, pp. 159–174. Springer, Heidelberg (2003)

Attributed Graph Kernels Using the Jensen-Tsallis q-Differences

Lu Bai[1], Luca Rossi[2], Horst Bunke[3], and Edwin R. Hancock[1,*]

[1] Department of Computer Science, University of York
Deramore Lane, Heslington, York, YO10 5GH, UK
[2] School of Computer Science, University of Birmingham
Edgbaston, Birmingham, B15 2TT, UK
[3] Institute of Computer Science and Applied Mathematics
University of Bern, Switzerland

Abstract. We propose a family of attributed graph kernels based on mutual information measures, i.e., the Jensen-Tsallis (JT) q-differences (for $q \in [1,2]$) between probability distributions over the graphs. To this end, we first assign a probability to each vertex of the graph through a continuous time quantum walk (CTQW). We then adopt the tree-index approach [1] to strengthen the original vertex labels, and we show how the CTQW can induce a probability distribution over these strengthened labels. We show that our JT kernel (for $q - 1$) overcomes the shortcoming of discarding non-isomorphic substructures arising in the R-convolution kernels. Moreover, we prove that the proposed JT kernels generalize the Jensen-Shannon graph kernel [2] (for $q = 1$) and the classical subtree kernel [3] (for $q = 2$), respectively. Experimental evaluations demonstrate the effectiveness and efficiency of the JT kernels.

Keywords: Graph kernels, tree-index method, continuous-time quantum walk, Jensen-Tsallis q-differences.

1 Introduction

There has recently been an increasing interest in evolving graph kernels into kernel machines (e.g., a Support Vector Machine) for graph classification [4, 5]. A graph kernel is usually defined in terms of a similarity measure between graphs. Most of the recently introduced graph kernels are in fact instances of the generic R-convolution kernel proposed by Haussler [6]. For a pair of graphs $G_1(V_1, E_1)$ and $G_2(V_2, E_2)$, suppose $\{\mathcal{S}_{1;1}, \ldots, \mathcal{S}_{1;n_1}, \ldots, \mathcal{S}_{1;N_1}\}$ and $\{\mathcal{S}_{2;1}, \ldots, \mathcal{S}_{2;n_2}, \ldots, \mathcal{S}_{q;N_2}\}$ are the sets of the substructures of G_1 and G_2 respectively. An R-convolution kernel k_R between G_1 and G_2 can be defined as

$$k_R(G_1, G_2) = \sum_{n_1=1}^{N_1} \sum_{y=1}^{N_2} \delta(\mathcal{S}_{1;n_1}, \mathcal{S}_{2;n_2}), \tag{1}$$

* Edwin R. Hancock is supported by a Royal Society Wolfson Research Merit Award.

T. Calders et al. (Eds.): ECML PKDD 2014, Part I, LNCS 8724, pp. 99–114, 2014.
© Springer-Verlag Berlin Heidelberg 2014

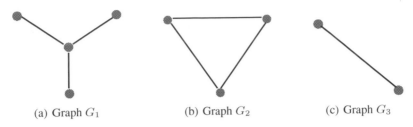

(a) Graph G_1 (b) Graph G_2 (c) Graph G_3

Fig. 1. Example Graphs

where we have that

$$\delta(\mathcal{S}_{1;n_1}, \mathcal{S}_{2;n_2}) = \begin{cases} 1 \text{ if } \mathcal{S}_{1;n_1} \simeq \mathcal{S}_{2;n_2}, \\ 0 \text{ otherwise,} \end{cases} \tag{2}$$

δ is the Dirac kernel, i.e., it is 1 if the arguments are equal and 0 otherwise, and $\mathcal{S}_{1;n_1} \simeq \mathcal{S}_{2;n_2}$ indicates that $\mathcal{S}_{1;n_1}$ is isomorphic to $\mathcal{S}_{2;n_2}$. k_R is a positive definite kernel.

The existing R-convolution graph kernels can be categorized into three classes, namely the graph kernels based on comparing all pairs of a) walks (e.g., the random walk kernel [7]), b) paths (e.g., the shortest path kernel [8]), and c) restricted subgraph or subtree structures (e.g., the subtree or subgraph kernel [9–11]). Unfortunately, there are two main problems arising in the R-convolution kernels. First, Eq.(1) indicates that the R-convolution kernels only enumerate the pairs of isomorphic substructures. As a result, the substructures which are not isomorphic are discarded. For instance, for the three graphs shown in Fig.1 (a), (b) and (c), the pair of graphs G_1 and G_3 and the pair of graphs G_2 and G_3 both share three same pairs of isomorphic substructures (i.e., pairwise vertices connected by an edge). The R-convolution kernels only count the number of pairs of isomorphic substructures. As a result, the kernel value for G_1 and G_3 is the same as that for G_2 and G_3, though the graphs G_1 and G_2 are structurally different. Second, Eq.(2) indicates that the R-convolution kernels simply record whether two substructures are isomorphic. As a result, the kernels do not reflect any other potential information between these substructures. These drawbacks clearly limit the accuracy of the similarity measure (i.e., the kernel value) between a pair of graphs.

Recently, Bai and Hancock [2] have developed an alternative kernel, namely the Jensen-Shannon (JS) graph kernel, by measuring the Jensen-Shannon divergence (JSD) for a pair of graphs. The JSD is a dissimilarity measure between probability distributions in terms of the nonextensive entropy difference associated with them. The JSD between a pair of graphs is defined in terms of the difference between the entropy of a composite graph and the entropies of the individual graphs. Unlike the R-convolution kernels, the entropy associated with a probability distribution of an individual graph can be computed without decomposing the graph. As a result, the computation of the JS graph kernel avoids the burdensome computation of comparing all the substructure pairs. Unfortunately, the JS graph kernel only captures the global similarity between a pair of graphs, and thus lacks information on the interior topology of the graphs. Moreover, the required composite entropy is computed from a composite structure which does not reflect the correspondence information between the original graphs. Finally, this kernel is restricted to non-attributed graphs. As a summary of the existing kernels (i.e., the R-convolution and JS graph kernels), it is fair to say that developing efficient and effective graph kernels still remains an open challenge.

To overcome the shortcomings of existing graph kernels, we aim to propose novel kernels for attributed graphs where we make use of the JT q-differences. In information theory, the JT q-difference [12] is a nonextensive measure between probability distributions over structure data. Moreover, the JT q-differences generalize the Hadamard-Schur (element wise) product (for $q = 0$), the classical Jensen-Shannon divergence (JSD) (for $q = 1$), and the inner product (for $q = 2$). To compute the JT q-differences between attributed graphs, we commence by performing a CTQW on each graph to assign each vertex a time-averaged probability. The reasons for using the CTQW are twofold (see details in Sec. 2.2). First, the CTQW reduces the tottering effect arising in classical random walks [13]. Second, the CTQW offers us a richer structure than the classical random walk. We then apply the tree-index (TI) method [1] on each graph to strengthen the vertex labels. At each iteration h, we compute the probability of a strengthened label by summing the probabilities of the vertices having the same label, and then obtain a probability distribution for each graph. With the probability distributions for a pair of attributed graphs to hand, the JT kernels between the graphs are computed in terms of the JT q-differences. As a result, our kernels reflect the similarity between the probability distributions over the global graphs, while also capturing the local correspondence information between the substructures.

Note that, in this paper, we only consider $q = 1$ or 2 for the JT q-differences. The reasons for this are twofold. First, we show that our JT kernel (for $q = 1$) not only generalizes the JS graph kernel [2], but it also overcomes the shortcomings of the JS graph kernel. We also show that the JT kernel overcomes the shortcoming of discarding non-isomorphic substructures that arises in R-convolution kernels. Finally, we show that our JT kernel (for $q - 2$) generalizes the R-convolution kernels based on subtrees. The remainder of this paper is organized as follows: Section 2 reviews the TI method and the CTQW, Section 3 gives the definition of our new kernels, Section 4 provides experimental evaluation and Section 5 concludes the work.

2   Preliminary Concepts

In this section, we review some preliminary concepts which will be used in this work. We commence by introducing a TI method for strengthening the vertex label. Finally, we show how to assign a probability to a vertex of a graph by performing the CTQW.

2.1   A TI Method for Strengthening Vertex Labels

In this subsection, we introduce the TI method of Dahm et al. [1] for strengthening the vertex label of a graph. Given an attributed graph $G(V, E)$, let the label of a vertex $v \in V$ be denoted as $f(v)$. Using the TI method, the new strengthened label for v at the iteration h is defined as

$$TI_h(v) = \begin{cases} f(v) & \text{if } h = 0, \\ \cup_u \{TI_{h-1}(u)\} & \text{otherwise,} \end{cases} \quad (3)$$

where $u \in V$ is adjacent to v. At each iteration h, the TI method takes the union of neighbouring vertex label lists from the last iteration as a new label list for v (the initial step is identical to listing). This creates an iteratively deeper list corresponding to a subtree rooted at v of height h. An example of how the TI method defined in Eq.(3) strengthens the vertex label is shown in Fig.2. In this example, the initialized vertex labels for vertices A to E are their corresponding vertex degrees, i.e., 1, 2, 3, 2 and 2 respectively. Using the TI method, the second iteration indicates the strengthened labels for vertices A to E as $\{\{1,3\}\}$, $\{\{2\},\{2,2,2\}\}$, $\{\{1,3\},\{2,3\},\{2,3\}\}$,$\{\{2,2,2\},\{2,3\}\}$, and $\{\{2,2,2\},\{2,3\}\}$ respectively.

Unfortunately, Fig.2 indicates that the above procedure clearly leads to a rapid explosion of the labels length. Moreover, strengthening a vertex label by only taking the union of the neighbouring label lists also ignores the original label information of the vertex. To overcome these problems, at each iteration h we propose to strengthen the label of a vertex as a new label list by taking the union of both the original vertex label and its neighbouring vertex labels. We use a Hash function to compress the strengthened label list into a new short label. The pseudocode of the re-defined TI algorithm is shown in Algorithm 1, where the neighbourhood of a vertex $v \in V$ is denoted as $\mathcal{N}(v) = \{u|(v,u) \in E\}$.

Algorithm 1. Vertex labels strengthening procedure

1: Initialization.

- Input an attributed graph $G(V, E)$.
- Set $\boldsymbol{h=0}$. For a vertex $v \in V$, assign the original label $f(v)$ as the initial label $\mathcal{L}_h(v)$.

2: Update the label for each vertex.

- Set $\boldsymbol{h=h+1}$. For each vertex $v \in G$, assign it a new strengthened label list as

$$\mathcal{L}_h(v) = \cup_{u\in\mathcal{N}(v)}\{\mathcal{L}_{h-1}(u), \mathcal{L}_{h-1}(v)\}. \tag{4}$$

Note that, $\mathcal{L}_{h-1}(v)$ is at the end of the label list $\mathcal{L}_h(v)$, and $\mathcal{L}_{h-1}(u)$ is arranged as ascending order.

3: For each vertex, compress its strengthened label list into a new short label.

- Using the Hash function $\mathbf{H} : \mathcal{L} \to \Sigma$, compress the label list $\mathcal{L}_h(v)$ into a new short label for each vertex v as

$$\mathcal{L}_h(v) = \mathbf{H}(\mathcal{L}_h(v)). \tag{5}$$

4: Check h.

- Check h. Repeat steps 2, 3 and 4 until the iteration h achieves an expected value.

Note that, in step 4 we use the same function $\mathbf{H}$ for any graph. This guarantees that all the identical labels of different graphs are mapped into the same number.

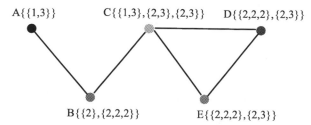

Fig. 2. Example of strengthened labels

2.2 Vertex Probabilities from The CTQW

In this subsection, we show how to assign a probability to a vertex of a graph by performing a CTQW. The CTQW is the quantum analogue of the classical continuous-time random walk (CTRW) [14]. Similarly to the CTRW, the state space of the CTQW is the vertex set V of a graph $G(V, E)$. However, unlike the CTRW, where the state vector is real-valued and the evolution is governed by a doubly stochastic matrix, the state vector of the CTQW is complex-valued and its evolution is governed by a time-varying unitary matrix. Hence the evolution of the CTQW is reversible, which implies that the CTQW is non-ergodic and does not possess a limiting distribution. As a result, the CTQW possesses a number of interesting properties not exhibited in the CTRW. One notable consequence of this is that the problem of tottering of the CTRW is naturally reduced [13]. Furthermore, the quantum walk has been shown to successfully capture the topological information in a graph structure [15–18]. Note also that, contrary to the CTRW and its non-backtracking counterparts [19–21], the CTQW is not dominated by the low frequency of the Laplacian spectrum, and thus is potentially able to discriminate better among different graph structures.

Using the Dirac notation, we define the basis state corresponding to the CTQW being at a vertex $u \in V$ as $|u\rangle$, where $|.\rangle$ denotes an orthonormal vector in a n-dimensional complex-valued Hilbert space $\mathcal{H}$. A general state of the CTQW is a complex linear combination of the basis states $|u\rangle$, i.e., an amplitude vector, such that the state of the CTQW at time t is

$$|\psi_t\rangle = \sum_{u \in V} \alpha_u(t)\, |u\rangle, \tag{6}$$

where the amplitudes $\alpha_u(t) \in \mathbb{C}$. The probability of the CTQW visiting a vertex $u \subset V$ at time t is

$$\Pr(X^t = u) = \alpha_u(t)\alpha_u^*(t), \tag{7}$$

where $\alpha_u^*(t)$ is the complex conjugate of $\alpha_u(t)$. For all $u \in V$, $t \in \mathbb{R}^+$, we have $\sum_{u \in V} \alpha_u(t)\alpha_u^*(t) = 1$ and $\alpha_u(t)\alpha_u^*(t) \in [0, 1]$. Note that there is no restriction on the sign or phase of the amplitudes and this allows destructive and constructive interference to take place during the evolution of the walk. These interference patterns are responsible for the faster hitting times [13] and the ability of the CTQW to capture the presence of particular structural patterns in the graph [16]. Note also that when the quantum walk backtracks on an edge it does so with opposite phase, thus creating destructive inter fence which reduces the problem of tottering of classical random walks [13].

Let A be the adjacency matrix of G, then the degree matrix D is a diagonal matrix whose elements are given by $D(u, u) = d_u = \sum_{v \in V} A(u, v)$, where d_u is the degree of u. We compute the Laplacian matrix as $L = D - A$. The spectral decomposition $L = \Phi^\top \Lambda \Phi$ of the Laplacian matrix L is given by the diagonal matrix $\Lambda = diag(\lambda_1, \lambda_2, ..., \lambda_{|V|})$ with the ordered eigenvalues as elements ($\lambda_1 < \lambda_2 < ... < \lambda_{|V|}$) and the matrix $\Phi = (\phi_1 | \phi_2 | ... | \phi_{|V|})$ with the corresponding ordered orthonormal eigenvectors as columns.

Given an initial state $|\psi_0\rangle$, the Schrödinger equation gives us the state of the walk at time t, i.e.,

$$|\psi_t\rangle = \Phi^\top e^{-i\Lambda t} \Phi |\psi_0\rangle . \tag{8}$$

In this work we propose to let the initial amplitude be proportional to $D(u, u)$, i.e.,

$$\alpha_u(0) = \sqrt{D(u, u) / \sum D(u, u)}. \tag{9}$$

Note that, on the other hand, choosing a uniform distribution over the vertices of G would result in the system remaining stationary, as the initial state vector would be an eigenvector of the Laplacian and thus a stationary state of the walk.

In quantum mechanics, a state $|\psi_t\rangle$ is called a pure state. In general, however, we deal with mixed states, i.e., a statistical ensemble of pure states $|\psi_t\rangle$, each with probability p_t. The density matrix associated with such a system is defined as $\rho = \sum_t p_t |\psi_t\rangle \langle \psi_t|$, where $|\psi_t\rangle \langle \psi_t|$ denotes the outer product between $|\psi_t\rangle$ and its conjugate transpose. Let $|\psi_t\rangle$ be the state corresponding to the CTQW that has evolved from $|\psi_0\rangle$ defined as in Eq.(9) until a time t. Using Eq.(8), we can define the time-averaged density matrix (i.e., mixed density matrix) ρ_G^T for G as

$$\rho_G^T = \frac{1}{T} \int_0^T \Phi^\top e^{-i\Lambda t} \Phi |\psi_0\rangle \langle \psi_0| \Phi^\top e^{i\Lambda t} \Phi \, \mathrm{d}t. \tag{10}$$

Let ϕ_{ra} and ϕ_{cb} denote the (ra)-th and (cb)-th elements of the matrix of eigenvectors Φ of the Laplacian matrix L. When we let $T \to \infty$, Rossi et al. [16] have shown that the (r, c)-th element of ρ_G^∞ can be computed as

$$\rho_G^\infty(r, c) = \sum_{\lambda \in \tilde{\Lambda}} \sum_{a \in B_\lambda} \sum_{b \in B_\lambda} \phi_{ra} \phi_{cb} \bar{\psi}_a \bar{\psi}_b, \tag{11}$$

where $\tilde{\Lambda}$ is the set of distinct eigenvalues of the Laplacian matrix L, and B_λ is a basis of the eigenspace associated with λ. Eq.(11) indicates that the mixed density matrix ρ_G^T relies on computing the eigendecomposition of G, and thus has time complexity $O(|V|^3)$.

The mixed density matrix ρ_G^∞ is a $|V| \times |V|$ matrix. Note that, for a graph G, the time-averaged probability of the CTQW to visit a vertex $v \in V$ at time $T \to \infty$ is

$$p_Q(v) = \rho_G^\infty(v, v). \tag{12}$$

Interestingly, despite the non-stationary behaviour of the CTQW, we have that as $T \to \infty$ the time-averaged probability converges. Also, Eq.(8) indicates that the evolution of the CTQW relies on the spectral decomposition of the graph Laplacian, which in turn encapsulates rich interior graph information. Thus, the probability $p_Q(v)$ reflects the interior topology information of the graph.

3 Graph Kernels from JT q-differences

In this section, we define a family of novel graph kernels from JT q-differences (for $q = 1$ and 2). The JT q-difference is a dissimilarity measure between probability distributions [12]. Consider the graphs $G_y(V_y, E_y)$ where $y \in Y = \{1, 2\}$ and $x \in X = \{l_1, \ldots, l_{|X|}\}$ is the label of a vertex $v_y \in V_y$. X is a label set which contains any possible vertex label. Let $\mathbf{P}_y = \{p_{y1}, \ldots, p_{y|X|}\}$, where $p_{yx} = P(X = x|Y = y)$ is the probability distribution over vertex labels of G_y. $\pi_y \in \pi = \{\pi_1, \pi_2\}$ is the weight associated with $\mathbf{P}_y$, such that $\pi_y \geq 0$ and $\sum_{y \in Y} \pi_y = 1$. $P(X|Y)$ is the joint probability distribution for the two variables X and Y. P_X and P_Y are two probability distributions over $x \in X$ and $y \in Y$ respectively. We define the JT q-differences (for q=$\{1,2\}$) between the probability distributions $\mathbf{P}_1$ and $\mathbf{P}_2$ (i.e., $\mathbf{P}_{y=1}$ and $\mathbf{P}_{y=2}$) as

$$JT_q^\pi(\mathbf{P}_1, \mathbf{P}_2) = S_q(X) - \int_{y \in Y} \pi_y^q S_q(X|Y) = S_q(X) - S_q(X|Y), \quad (13)$$

where $S_q(.)$ is the Tsallis entropy [22] and

$$S_q(X) = \frac{k}{q-1}\left(1 - \sum_{x \in X} 2P_X(x)^q\right), \quad (14)$$

and

$$S_q(X|Y) = S_q(X) + S_q(Y) - (q-1)S_q(X)S_q(Y). \quad (15)$$

By letting $k = 1$ [12], Eq.(14) can be simplified as

$$S_q(X) = -\sum_{x \in X} P_X(x)^q \ln_q P_X(x), \quad (16)$$

where $\ln_q(P_X(x)) = (P_X(x)^{(1-q)} - 1)/(1 - q)$ is the q-logarithm function introduced by Tsallis [22]. Let $\pi_1 = \pi_2 = 1/2$. We define the JT q-difference for the pair of graphs G_1 and G_2 by re-writing Eq.(13) as

$$JT_q(G_1, G_2) = JT_q^{1/2,1/2}(\mathbf{P}_1, \mathbf{P}_2) = S_q\left(\frac{\mathbf{P}_1 + \mathbf{P}_2}{2}\right) - \frac{S_q(\mathbf{P}_1) + S_q(\mathbf{P}_2)}{2^q}. \quad (17)$$

Note that Furuichi [23] has defined the Tsallis MI as $I_q(X;Y) = S_q(X) - S_q(X|Y)$ $= I_q(Y;X)$, which is a nonextensive measure. Thus, for the graphs G_1 and G_2, the JT q-difference $JT_q(G_1, G_2)$ not only reflects the dissimilarity between their probability distributions $\mathbf{P}_1$ and $\mathbf{P}_2$, but also measures the MI (i.e., the mutual dependence) between the graphs and their probability distributions. In other words, $JT_q(G_1, G_2)$ reflects the dissimilarity over the global graphs. By contrast, the R-convolution kernels measure the (dis)similarity in terms of the substructures.

3.1 The JT Graph Kernel

In this subsection, we define a family of JT graph kernels using the JT q-difference (for $q = 1$ and 2). For a graph $G(V, E)$, we commence by performing a CTQW (for $T = \infty$)

on G, and we associate with each vertex $v \in V$ the time-averaged probability $p_Q(v)$. At each iteration h, we strengthen the vertex labels using Algorithm 1. The probability of the label $x \in X$ for G at iteration h is $p_x^h = \sum_{v \in V} p_Q(v)$, where each vertex v satisfies $\mathcal{L}_h(v) = x$. We thus obtain a probability distribution over the vertex labels of G as $\mathbf{P^h} = \{p_1^h, \ldots, p_{|X|}^h\}$. For a pair of graphs $G_1(V_1, E_1)$ and $G_2(V_2, E_2)$, we compute the probability distributions $\mathbf{P_1^h} = \{p_{11}^h, \ldots, p_{1|X|}^h\}$ and $\mathbf{P_2^h} = \{p_{21}^h, \ldots, p_{2|X|}^h\}$ over the vertex labels through the CTQW. Based on Eq.(17), the JT q-difference for G_1 and G_2 at iteration h is defined as

$$JT_q^h(G_1, G_2) = JT_q^{1/2, 1/2}(\mathbf{P_1^h}, \mathbf{P_2^h}). \tag{18}$$

Definition (Jensen-Tsallis Graph Kernel). The kernel $k_{JT}^{(q,H)} : G_1 \times G_2 \longrightarrow R^+$ for the graphs G_1 and G_2 is

$$k_{JT}^{(q,H)}(G_1, G_2) = \sum_{h=0}^{H} \exp\{-\lambda JT_q^h(G_1, G_2)\} = \sum_{h=0}^{H} \exp\{-\lambda JT_q^{1/2, 1/2}(\mathbf{P_1^h}, \mathbf{P_2^h})\}, \tag{19}$$

where H is the largest number of TI iterations, $q = 1$ or 2, and $0 < \lambda \le 1$ is a decay factor. Here, λ is used to ensure that the large value dose not tend to dominate the kernel value. Based on our experimental evaluation, the different values of λ do not influence the performance of our JT kernels. Thus, in this work we decide to set λ to 1. $\square$

Lemma. *The JT graph kernel is positive definite (**pd**).*

Proof. This follows the definition in [24]. In fact, if a similarity or dissimilarity measure $s_G(G_1, G_2)$ between a pair of graphs G_1 and G_2 is symmetrical, then a diffusion kernel $k_s = \exp(\lambda s_G(G_1, G_2))$ or $k_s = \exp(-\lambda s_G(G_1, G_2))$ associated with the similarity or dissimilarity measure $s_G(G_1, G_2)$ is **pd**. As a result, the JT kernel $k_{JT}^{(q,H)}$ is the sum of several **pd** kernels in terms of the exponentiated JT q-difference, and is also **pd**. $\square$

Reisen and Bunke [24] observed that in a diffusion kernel the exponentiation enhances the (dis)similarity between the graphs. Thus, the kernel $k_{JT}^{(q,H)}$ enhances the similarity measure between the graphs.

3.2   Relation to State of the Art Graph Kernels

Proposition 1. When $q = 2$, the JT graph kernel generalizes the R-convolution kernels based on subtrees. $\square$

Proof. We verify this proposition by revealing the relationship between the JT kernel and the classical subtree kernel [3]. For a graph $G(V, E)$, the strengthened label $\mathcal{L}_h(v)$, which is defined in Eq.(5), corresponds to a subtree of height h rooted at a vertex $v \in V$. For a pair of graphs $G_1(V_1, E_1)$ and $G_2(V_2, E_2)$, let $v_1 \in V_1$ and $v_2 \in V_2$ denote a pair of vertices. If $\mathcal{L}_h(v_1) = \mathcal{L}_h(v_2)$, the subtrees of height h rooted at v_1 and v_2 are isomorphic. Thus, a subtree kernel k_{st}^H for G_1 and G_2 can be defined as

$$k_{st}^H(G_1, G_2) = \sum_{h=0}^{H} k_{st}^h(G_1, G_2) = \sum_{h=0}^{H} \sum_{v_1 \in V_1} \sum_{v_2 \in V_2} \delta\{\mathcal{L}_h(v_1), \mathcal{L}_h(v_2)\}, \tag{20}$$

where H is the largest iteration h for Algorithm 1. δ is a Dirac kernel, that is, it is 1 when its arguments are equal and 0 otherwise. Clearly, the subtree kernel k_{st}^H counts the number of all the pairwise isomorphic subtrees identified by the TI method. k_{st}^h is a base subtree kernel counting the number of pairwise isomorphic subtrees of height h.

As a result, the base subtree kernel k_{st}^h can be defined by an inner product $\langle \cdot, \cdot \rangle$, i.e., a linear kernel. For the graph G, let $n^h(x)$ denote the number of vertices having the same label $x \in X$ at iteration h. Thus G can be represented by the feature vector $FV^h = \{n^h(l_1), \ldots, n^h(l_{|X|})\}^\top$. Given $FV_1^h = \{n_1^h(l_1), \ldots, n_1^h(l_{|X|})\}^\top$ and $FV_2^h = \{n_2^h(l_1), \ldots, n_2^h(l_{|X|})\}^\top$, the kernel k_{st}^h between G_1 and G_2 at iteration h can be defined as

$$k_{st}^h(G_1, G_2) = \langle FV_1^h, FV_2^h \rangle. \tag{21}$$

Martin et al. [12] have observed that the JT q-difference is related to the inner product when $q = 2$, i.e., $JT_2^{1/2,1/2}(\mathbf{P_1^h}, \mathbf{P_2^h}) = 1/2 - 1/2\langle \mathbf{P_1^h}, \mathbf{P_2^h}\rangle$. Simply, we have $JT_2^{1/2,1/2}(\mathbf{P_1^h}, \mathbf{P_2^h}) = -\langle \mathbf{P_1^h}, \mathbf{P_2^h}\rangle$. As a result, the JT kernel $k_{JT}^{(2,H)}$ ($\lambda = 1$) can be re-written as

$$k_{JT}^{(2,H)}(G_1, G_2) = \sum_{h=0}^{H} \exp\{\langle \mathbf{P_1^h}, \mathbf{P_2^h}\rangle\}, \tag{22}$$

For a graph G, the probability of a label $x \in X$ at the iteration h is $p_x^h = \sum_{v \in V} p_Q(v)$, where each vertex v satisfies $\mathcal{L}_h(v) = x$ and $p_Q(v)$ is the time averaged probability of a CTQW visiting v. Given G and its associated feature vector FV^h with elements $n^h(x)$, if we compute the probability of a label x instead of computing the frequency $n^h(x)$ of x (i.e., we compute the probability for a class of isomorphic subtrees which correspond to the label x, instead of computing the number of these subtrees), we re-write FV^h as $FV_p^h = \mathbf{P}^h$. For the graphs G_1 and G_2, we thus have

$$k_{JT}^{(2,H)}(G_1, G_2) = \sum_{h=0}^{H} \exp\{\langle FV_{p1}^h, FV_{p2}^h \rangle\} = \sum_{h=0}^{H} \exp\{k_{st}^h(G_1, G_2)\}. \tag{23}$$

As a result, the kernel $k_{JT}^{(2;H)}$ can be seen as a generalization of the subtree kernel k_{st}^H where we assign a probability to each class of isomorphic subtrees and then exponentiate the base subtree kernel k_{st}^h at each iteration h. $\square$

Discussions. Prop. 1 and its proof make two interesting observations on the JT kernel $k_{JT}^{(2,H)}$. First, for each pair of graphs the kernel $k_{JT}^{(2,H)}$ computes the probability of a vertex label by summing the probabilities of the vertices having the same label. The probabilities of these vertices are computed by means of a CTQW, whose propagation depends on the interior connections of the graph and thus reflects its interior topology information. For a pair of graphs, the kernel measures the similarity between their pairwise probabilities in terms of matching labels, where the labels correspond to classes of isomorphic subtrees in the graphs. The kernel thus reflects more information among these subtrees rather than only the isomorphism. Second, the $k_{JT}^{(2,H)}$ exponentiates the base subtree kernel k_{st}^h (i.e., the similarity measure between pairwise subtrees of height

h). Thus, the kernel $k_{JT}^{(2,H)}$ enhances the similarity measure between the graphs by exponentiating the kernel k_{st}^h.

Unfortunately, the JT kernel $k_{JT}^{(2,H)}$ also suffers from the drawback of discarding non-isomorphic subtrees arising in the subtree kernel. This can be observed from Eq.(23). For a pair of graphs, if a class of isomorphic subtrees are only contained in one graph, the probability for a label corresponding to these subtrees in the other graph is 0. As a result, Eq.(23) cannot reflect the similarity between pairwise probabilities of the labels for the graphs. Below, we will show how the kernel $k_{JT}^{(1,H)}$ solves this problem.

Proposition 2. When $q = 1$, the JT graph kernel generalizes the JS graph kernel.   □

Proof. We verify the proposition by revealing the relationship between the JT kernel and the JS graph kernel [2]. For a graph $G(V, E)$ and its degree matrix D, the probability of a classical steady state random walk (CSSRW) visiting a vertex $v \in V$ is

$$p_v = \frac{D(v,v)}{\sum_v D(v,v)} = \frac{\sum_{u \in V} A(u,v)}{\sum_{u \in V} \sum_{v \in V} A(u,v)}. \tag{24}$$

Thus, we can associate to the vertices of G the distribution $\mathbf{P^c} = \{p_1^c, \ldots, p_{|V|}^c\}$. For a pair of graphs $G_1(V_1, E_1)$ and $G_2(V_2, E_2)$, we compute the probability distributions $\mathbf{P_1^c} = \{p_{11}^c, \ldots, p_{1|V|}^c\}$ and $\mathbf{P^c} = \{p_{21}^c, \ldots, p_{2|V|}^c\}$ associated with the CSSRW. The JS graph kernel for G_1 and G_2 is defined as

$$k_{JS}(G_1, G_2) = \log 2 - D_{JS}(\mathbf{P_1^c}, \mathbf{P_2^c}), \tag{25}$$

where $D_{JS}(G_1, G_2)$ is the JSD and is defined as

$$D_{JS}(\mathbf{P_1^c}, \mathbf{P_2^c}) = H_S(\mathbf{P_U^c}) - \frac{H_S(\mathbf{P_1^c}) + H_S(\mathbf{P_2^c})}{2}. \tag{26}$$

Here, $H_S(\mathbf{P_1^c})$ is a Shannon entropy associated with the probability distribution $\mathbf{P_1^c}$ and is defined as

$$H_S(\mathbf{P_1^c}) = -\sum_{v \in V} p_{1v}^c \log p_{1v}^c. \tag{27}$$

Moreover, $H_S(\mathbf{P_U^c})$ is the Shannon entropy of the probability distribution from an union graph of G_1 and G_2 (i.e., the composite graph of G_1 and G_2). The union graph can be either a product graph or a disjoint union graph of G_1 and G_2 [2, 25].

For the JT kernel $k_{JT}^{(1,H)}$, Martin et al. [12] have observed that the JT q-difference is related to the JSD when $q = 1$, i.e., $JT_1^{1/2,1/2}(\mathbf{P_1^h}, \mathbf{P_2^h}) = D_{JS}(\mathbf{P_1^h}, \mathbf{P_2^h})$, where

$$D_{JS} = H_S(\frac{\mathbf{P_1^h} + \mathbf{P_2^h}}{2}) - \frac{H_S(\mathbf{P_1^h}) + H_S(\mathbf{P_2^h})}{2}. \tag{28}$$

Thus, the JT kernel $k_{JT}^{(1,H)}$ ($\lambda = 1$) can be re-written as

$$k_{JT}^{(1,H)}(G_1, G_2) = \sum_{h=0}^{H} \exp\{-D_{JS}(\mathbf{P_1^h}, \mathbf{P_2^h})\}$$

$$= \sum_{h=0}^{H} \exp\{H_S(\mathbf{P_1^h}) + H_S(\mathbf{P_2^h}) - H_S(\frac{\mathbf{P_1^h} + \mathbf{P_2^h}}{2})\}. \tag{29}$$

Here, $H_S(\frac{\mathbf{P_1^h}+\mathbf{P_2^h}}{2})$ is the Shannon entropy of the composite probability distribution $\frac{\mathbf{P_1^h}+\mathbf{P_2^h}}{2}$, and is defined as

$$H_S\Big(\frac{\mathbf{P_1^h} + \mathbf{P_2^h}}{2}\Big) = -\sum_{x=1}^{|X|} \frac{p_{1x}^h + p_{2x}^h}{2} \log \frac{p_{1x}^h + p_{2x}^h}{2}. \tag{30}$$

As a result, the JT kernel $k_{JT}^{(1,H)}$ and the JS graph kernel k_{JS} are both related to the JSD. Through Eq.(25), Eq.(26) and Eq.(29), we observe that the kernel $k_{JT}^{(1,H)}$ generalizes the kernel k_{JS} by three computational steps, i.e., a) assigning each vertex a probability using the CTQW instead of using the CSSRW, b) computing the probability distribution over the vertex labels at each iteration h instead of computing the probability distribution associated with the CSSRW, and c) computing the composite Shannon entropy using the composite probability distribution $\frac{\mathbf{P_1^h}+\mathbf{P_2^h}}{2}$ instead of using the probability distribution $\mathbf{P_U^c}$ from the union graph. Moreover, the JT kernel $k_{JT}^{(1,H)}$ also exponentiates the JSD measure for graphs at each iteration h. $\square$

Discussions. Compared to the JS graph kernel k_{JS}, Prop. 2 and its proof reveal four advantages for the JT kernel $k_{JT}^{(1,H)}$. First, from Eq.(30) we observe that there is correspondence between pairwise discrete probabilities p_{1x}^h and p_{2x}^h through the same label x. As a result, the kernel $k_{JT}^{(1,H)}$ overcomes the shortcoming of lacking correspondence information between pairwise discrete probabilities that arises in the JS graph kernel k_{JS}. Second, since a pair of identical strengthened labels for a pair of vertices correspond to a pair of isomorphic subtrees rooted at the vertices. The kernel $k_{JT}^{(1,H)}$ encapsulates correspondence information between pairwise isomorphic substructures. By contrast, the JS graph kernel k_{JS} does not reflect substructure correspondence. Third, the kernel $k_{JT}^{(1,H)}$ overcomes the restriction on non-attributed graphs for the JS graph kernel k_{JS}. Fourth, similar to the kernel $k_{JT}^{(2,H)}$, the kernel $k_{JT}^{(1,H)}$ also enhances the similarity measure for graphs by exponentiating the JSD (i.e.,the dissimilarity measure between graphs). Furthermore, the evolution of the CTQW relies on the topology information of graphs. Thus, the kernel $k_{JT}^{(1,H)}$ also reflects rich interior topology information of graphs, relying on the CTQW. By contrast, the probability distribution associated with the CSSRW (i.e., the vertex degree distribution) only reflects limited topology information, because the vertex degree of a graph is structurally simple and reflects limited topology information.

Note finally that, through Eq.(29) and Eq.(30) we also observe that all the probabilities of the different labels are used to compute the entropies. We have known that each label corresponds to a subtree. All the strengthened labels of a graph will be used to compute the Shannon entropy. As a result, unlike existing R-convolution kernels which count the number of pairwise isomorphic substructures, the kernel $k_{JT}^{(1,H)}$ incorporates all the identified subtrees into the computation. The kernel $k_{JT}^{(1,H)}$ thus overcomes the shortcoming of discarding non-isomorphic substructures. In other words, the JT kernel $k_{JT}^{(1,H)}$ may distinguish different classes of graphs better than the R-convolution kernels.

3.3 Computational Analysis

For N graphs (each graph has n vertices) and their label set X, computing the $N \times N$ kernel matrix using the JT kernels $k_{JT}^{(1,H)}$ and $k_{JT}^{(2,H)}$ requires time complexity $O(HN^2 n^2 + HN^3 n + Nn^3)$ and $O(HN^2 n^2 + Nn^3)$, respectively. The reasons are explained as follows. a) For both of the kernels $k_{JT}^{(1,H)}$ and $k_{JT}^{(2,H)}$, computing the compressed strengthened labels for a graph at each iteration h ($0 \le h \le H$) needs to visit all the n^2 entries of the adjacency matrix, and thus requires time complexity $O(Hn^2)$ for all the H iterations. b) For both of the kernels $k_{JT}^{(1,H)}$ and $k_{JT}^{(2,H)}$, computing the probabilities for all the vertices of a graph using the CTQW requires time complexity $O(n^3)$, because the CTQW relies on the eigen-decomposition of the graph Laplacian. Computing the probability distribution for a graph requires time complexity $O(HNn^2)$ (for the worst case, i.e., each vertex label for the N graphs at all the H iterations are all different and there thus are NHn different labels in X), because it needs to visit all the HNn entries in X for the n vertices. c) For the kernel $k_{JT}^{(1,H)}$, computing the $N \times N$ kernel matrix requires time complexity $O(HN^3 n)$, because the Tsallis entropy $S_q(.)$ for each pair of graphs requires time complexity $O(HNn)$. On the other hand, for the kernel $k_{JT}^{(2,H)}$, computing the $N \times N$ kernel matrix only requires time complexity $O(HN^2)$, because Eqs.(22) and (23) indicate that $k_{JT}^{(2,H)}$ can directly compute the kernel value for a pair of graphs by computing the inner product of their probability distributions. In other words, $k_{JT}^{(2,H)}$ does not need to compute the Tsallis entropy for each pair of graphs. As a result, the complete time complexities for the JT kernels $k_{JT}^{(1,H)}$ and $k_{JT}^{(2,H)}$ are $O(HN^2 n^2 + HN^3 n + Nn^3)$ and $O(HN^2 n^2 + Nn^3)$, respectively.

4 Experimental Results

In this section, we empirically evaluate the performance of our JT kernels on standard attributed graphs from bioinformatics. Furthermore, we also compare our new kernels with several state of the art graph kernels.

Table 1. Information of the Graph-based Datasets

Datasets	MUTAG	NCI1	NCI109	ENZYMES	PPIs	PTC(MR)
Max # vertices	28	111	111	126	238	109
Min # vertices	10	3	4	2	3	2
Mean # vertices	17.93	29.87	29.68	32.63	109.63	25.60
# graphs	188	4110	4127	600	219	344
# classes	2	2	2	6	5	2

4.1 Graph Datasets

We evaluate our kernels on standard graph datasets. These datasets include: MUTAG, NCI1, NCI109, ENZYMES, PPIs and PTC(MR). More details are shown in Table.1.
MUTAG: The MUTAG dataset consists of graphs representing 188 chemical compounds, and aims to predict whether each compound possesses mutagenicity.

NCI1 and NCI109: The NCI1 and NCI109 datasets consist of graphs representing two balanced subsets of datasets of chemical compounds screened for activity against non-small cell lung cancer and ovarian cancer cell lines respectively. There are 4110 and 4127 graphs in NCI1 and NCI109 respectively.

ENZYMES: The ENZYMES dataset consists of graphs representing protein tertiary structures consisting of 600 enzymes from the BRENDA enzyme. The task is to correctly assign each enzyme to one of the 6 EC top-level.

PPIs: The PPIs dataset consists of protein-protein interaction networks (PPIs). The graphs describe the interaction relationships between histidine kinase in different species of bacteria. Histidine kinase is a key protein in the development of signal transduction. If two proteins have direct (physical) or indirect (functional) association, they are connected by an edge. There are 219 PPIs in this dataset and they are collected from 5 different kinds of bacteria (i.e., a) *Aquifex*4 and *thermotoga*4 PPIs from *Aquifex aelicus* and *Thermotoga maritima*, b) *Gram-Positive*52 PPIs from *Staphylococcus aureus*, c) *Cyanobacteria*73 PPIs from *Anabaena variabilis*, d) *Proteobacteria*40 PPIs from *Acidovorax avenae*, and e) *Acidobacteria*46 PPIs). Note that, unlike the experiment in [26] that only uses the *Proteobacteria*40 and the *Acidobacteria*46 PPIs as the testing graphs, we use all the PPIs as the testing graphs in this paper. As a result, the experimental results for some kernels are different on the PPIs dataset.

PTC: The PTC (The Predictive Toxicology Challenge) dataset records the carcinogenicity of several hundred chemical compounds for Male Rats (MR), Female Rats (FR), Male Mice (MM) and Female Mice (FM). These graphs are very small (i.e.,20 − 30 vertices, and 25 − 40 edges) and sparse. We select the graphs of MR for evaluation.

4.2   Experiments on Graph Classification

We evaluate the performance of our JT kernels $k_{JT}^{(q,H)}$ for $q = 1$ (JT1) and $q = 2$ (JT2). Moreover, we also compare our kernels with several alternative state of the art graph kernels. These graph kernels for comparison include: 1) the Jensen-Shannon graph kernel (JSGK) associated with the CSSRW [2], 2) the unaligned and aligned quantum Jensen-Shannon kernels (QJSK and QJSKA) [26], 3) the Weisfeiler-Lehman subtree kernel (WLSK) [9], 4) the shortest path graph kernel (SPGK) [8], 5) the graphlet count graph kernel with graphlet of size 3 (GCGK3) [11], 6) the backtrackeless random walk kernel using the Ihara zeta function based cycles (BRWK) [19], 7) the random walk graph kernel (RWGK) [3].

For each kernel, we compute the kernel matrix on each graph dataset. We perform 10-fold cross-validation using the C-Support Vector Machine (C-SVM) Classification to compute the classification accuracies, using LIBSVM. We use nine samples for training and one for testing. All the C-SVMs were performed along with their parameters optimized on each dataset. We report the average classification accuracies (± standard error) and the runtime for each kernel in Table 2. The runtime is measured under Matlab R2011a running on a 2.5GHz Intel 2-Core processor (i.e. i5-3210m). Note that, both our JT kernel and the WLSK kernel are related to a TI method. In this work, we set the parameter H (i.e., the maximum number of TI iteration) to 10, i.e., we vary h from 1 to 10 for both our kernel and the WLSK kernel. The reasons of setting $H = 10$ are twofold. First, for most of the datasets, the strengthened vertex labels of the graphs tend to be all different after $h = 10$. In other words, after $h = 10$, there are nearly no isomorphic subtrees, i.e., we achieve maximum discrimination. Second, in our experiments we observe that the classification performance tends to be more stable after $h = 10$. As a result, for each dataset we compute 10 different kernel matrices for both our kernel and the WLSK kernel. The classification accuracy is then the average accuracy over the 10 kernel matrices. Note that, the experimental results (for the WLSK kernel) on some datasets are different from those in [27], since the authors of [27] set $H = 3$. Finally, recall that our JT kernel, and the WLSK and SPGK kernels are all able to handle attributed graphs. However, the graphs in the PPIs dataset are unattributed graphs, thus we decided to use the vertex degree as a vertex label.

Table 2. Classification Accuracy (In % ± Standard Error) and Runtime for Various Kernels

Datasets	MUTAG	NCI1	NCI109	ENZYMES	PPIs	PTC(MR)
JT1	85.10 ± 0.64	$\mathbf{86.35 \pm 0.12}$	$\mathbf{87.00 \pm 0.15}$	$\mathbf{57.41 \pm 0.53}$	87.28 ± 0.61	$\mathbf{60.16 \pm 0.50}$
JT2	$\mathbf{85.50 \pm 0.55}$	85.32 ± 0.14	85.79 ± 0.13	56.41 ± 0.42	$\mathbf{88.47 \pm 0.47}$	58.50 ± 0.39
JSGK	83.11 ± 0.80	62.50 ± 0.33	63.00 ± 0.35	20.81 ± 0.29	34.57 ± 0.54	57.29 ± 0.41
QJSK	82.72 ± 0.44	69.09 ± 0.20	70.17 ± 0.23	36.58 ± 0.46	65.61 ± 0.77	56.70 ± 0.49
QJSKA	82.83 ± 0.50	–	–	24.31 ± 0.27	61.09 ± 0.98	57.39 ± 0.46
WLSK	82.88 ± 0.57	84.77 ± 0.13	84.49 ± 0.13	52.75 ± 0.44	88.09 ± 0.41	58.26 ± 0.47
SPGK	83.38 ± 0.81	74.21 ± 0.30	73.89 ± 0.28	41.30 ± 0.68	59.04 ± 0.44	55.52 ± 0.46
GCGK3	82.04 ± 0.39	63.72 ± 0.12	62.33 ± 0.13	24.87 ± 0.22	46.61 ± 0.47	55.41 ± 0.59
BRWK	77.50 ± 0.75	60.34 ± 0.17	59.89 ± 0.15	20.56 ± 0.35	–	53.97 ± 0.31
RWGK	80.77 ± 0.72	–	–	22.37 ± 0.35	41.29 ± 0.89	55.91 ± 0.37

Datasets	MUTAG	NCI1	NCI109	ENZYMES	PPIs	PTC(MR)
JT1	14"	$7h21'$	$7h24'$	$11'30"$	$3'20"$	$1'10"$
JT2	3"	$10'50"$	$10'55"$	30"	$1'43"$	8"
JSGK	1"	1"	1"	1"	1"	1"
QJSK	20"	$2h55'$	$2h55'$	$4'23"$	$3'24"$	$1'46"$
QJSKA	$1'30"$	> 1 day	> 1 day	$1h10'$	$1h54'$	$16'40"$
WLSK	3"	$2'31"$	$2'37"$	20"	20"	9"
SPGK	1"	16"	16"	4"	22"	1"
GCGK3	1"	5"	5"	2"	4"	1"
BRWK	11"	$6'49"$	$6'49"$	$3'5"$	> 1 day	29"
RWGK	14"	> 1 day	> 1 day	$9'52"$	$4'26"$	$2'35"$

Results and Discussions. In terms of classification accuracy, our JT kernels overcome the alternative kernels on most of the datasets. Only the accuracy of the WLSK kernel on the PPIs dataset is a little higher than our kernel $k_{JT}^{(2,H)}$. The reasons of the effectiveness of our kernels are explained as follows. a) Compared to the JSGK kernel, our JT kernels overcome the restriction on non-attributed graphs. Moreover, our JT kernels also overcome the shortcoming of lacking correspondence information. Finally, since each strengthened label of the JT kernels correspondences to a subtree, the JT kernels reflect richer interior topology information than the JSGK kernel. By contrast, the JSGK kernel only reflects limited information in terms of the vertex degree distribution. b) Compared to the QJSK and QJSKA kernels, our JT kernels also overcome the restriction on non-attributed graphs arising in the two quantum kernels. Moreover, Bai et al. [26] show that the QJSK kernel requires the computation of an additional mixed state where the system has equal probability of being in each of the two original quantum states. Unless this quantum kernel takes into account the correspondences between the vertices of the two graphs, it can be shown that this kernel is not permutation invariant. While our JT kernels are not only permutation invariant but also reflect the correspondence information between pairwise probabilities computed from the CTQW. c) Compared to the BRWK, RWGK and GCGK3 kernels, our JT kernels reflect richer topology information in terms of the subtrees identified by the TI method (i.e.,the strengthened labels). The reason for this is that the subtree based strategy can overcomes the shortcoming of structurally simple problem arising in the path and walk based strategies. Moreover, the CTQW required for our kernels also possess more interesting properties than the BRWK and RWGK based on the classical random walk. d) Compared to the WLSK kernel, our JT kernel $k_{JT}^{(2,H)}$ generalizes the subtree based kernel and reflects richer information. Moreover, our JT kernel $k_{JT}^{(1,H)}$ overcomes the shortcoming of discarding non-isomorphic subtrees arising in the WLSK kernel. Finally, the performance of the kernel $k_{JT}^{(1,H)}$ is a little better than that of the kernel $k_{JT}^{(2,H)}$. The reason for this is that the kernel $k_{JT}^{(2,H)}$ also suffers from the problem of discarding non-isomorphic substructures, while the kernel $k_{JT}^{(1,H)}$ can overcome this shortcoming. e) In terms of the runtime, our JT kernel $k_{JT}^{(2,H)}$ is fast and is competitive to the fast subtree kernel WLSK. Furthermore, the computational efficiency of our JT kernel $k_{JT}^{(1,H)}$ is obviously slower, but it can still finish the computation in a polynomial time on any dataset. By contrast, some kernels cannot finish the computation on some datasets in one day.

5   Conclusions

In this paper, we develop a family of JT kernels for attributed graphs. For a graph, we use a TI method to strengthen the vertex labels and then compute the probability distribution over the vertex labels through the CTQW. For a pair of graphs, the JT kernels are computed by measuring the JT q-difference between their probability distributions. We show that the JT kernels not only generalize some state of the art kernels but also overcome the shortcomings arising in these kernels. The experiments demonstrate the effectiveness and efficiency of the JT kernels. Our future work is to investigate the relationship between state of the art kernels and our JT kernel with other q values (e.g., $q = 0$). Furthermore, we are also interested in extending this work on financial analysis.

Acknowledgments. We thank Dr. Chaoyan Wang for the insights of future extension in financial analysis.

References

1. Dahm, N., Bunke, H., Caelli, T., Gao, Y.: A Unified Framework for Strengthening Topological Node Features and Its Application to Subgraph Isomorphism Detection. In: Kropatsch, W.G., Artner, N.M., Haxhimusa, Y., Jiang, X. (eds.) GbRPR 2013. LNCS, vol. 7877, pp. 11–20. Springer, Heidelberg (2013)
2. Bai, L., Hancock, E.R.: Graph Kernels from the Jensen-Shannon Divergence. Journal of Mathematical Imaging and Vision 47(1-2), 60–69 (2013)
3. Gärtner, T., Flach, P.A., Wrobel, S.: On Graph Kernels: Hardness Results and Efficient Alternatives. In: Schölkopf, B., Warmuth, M.K. (eds.) COLT/Kernel 2003. LNCS (LNAI), vol. 2777, pp. 129–143. Springer, Heidelberg (2003)

4. Bunke, H., Riesen, K.: A Family of Novel Graph Kernels for Structural Pattern Recognition. In: Rueda, L., Mery, D., Kittler, J. (eds.) CIARP 2007. LNCS, vol. 4756, pp. 20–31. Springer, Heidelberg (2007)
5. Jebara, T., Kondor, R.I., Howard, A.: Probability Product Kernels. Journal of Machine Learning Research 5, 819–844 (2004)
6. Haussler, D.: Convolution Kernels on Discrete Structures. Technical Report UCS-CRL-99-10 (1999)
7. Kashima, H., Tsuda, K., Inokuchi, A.: Marginalized Kernels Between Labeled Graphs. In: Proc. ICML, pp. 321–328 (2003)
8. Borgwardt, K.M., Kriegel, H.P.: Shortest-path Kernels on Graphs. In: Proc. ICDM, pp. 74–81 (2005)
9. Shervashidze, N., Borgwardt, K.M.: Fast Subtree Kernels on Graphs. In: Proc. NIPS, pp. 1660–1668 (2009)
10. Costa, F., Grave, K.D.: Fast Neighborhood Subgraph Pairwise Distance Kernel. In: Proc. ICML, pp. 255–262 (2010)
11. Shervashidze, N., Vishwanathan, S.V.N., Petri, T., Mehlhorn, K., Borgwardt, K.M.: Efficient Graphlet Kernels for Large Graph Comparison. Journal of Machine Learning Research 5, 488–495 (2009)
12. Martins, A.F.T., Smith, N.A., Xing, E.P., Aguiar, P.M.Q., Figueiredo, M.A.T.: Nonextensive Information Theoretic Kernels on Measures. Journal of Machine Learning Research 10, 935–975 (2009)
13. Julia, K.: Quantum Random Walks: An Introductory Overview. Contemporary Physics 44(4), 307–327 (2003)
14. Farhi, E., Gutmann, S.: Quantum Computation and Decision Trees. Physical Review A 58, 915 (1998)
15. Aubry, M., Schlickewei, U., Cremers, D.: The Wave Kernel Signature: A Quantum Mechanical Approach to Shape Analysis. In: Proc. ICCV Workshops, pp. 1626–1633 (2011)
16. Rossi, L., Tosello, A., Hancock, E.R., Wilson, R.C.: Characterizing Graph Symmetries through Quantum Jensen-Shannon Divergence. Physical Review E 88(3-1), 032806 (2013)
17. Suau, P., Hancock, E.R., Escolano, F.: Graph Characteristics from the Schrödinger Operator. In: Kropatsch, W.G., Artner, N.M., Haxhimusa, Y., Jiang, X. (eds.) GbRPR 2013. LNCS, vol. 7877, pp. 172–181. Springer, Heidelberg (2013)
18. Cottrell, S., Hillery, M.: Finding Structural Anomalies in Star Graphs Using Quantum Walks. Physical Review Letters 112(3), 030501 (2014)
19. Aziz, F., Wilson, R.C., Hancock, E.R.: Backtrackless Walks on A Graph. IEEE Transactions on Neural Networks and Learning System 24(6), 977–989 (2013)
20. Mahé, P., Ueda, N., Akutsu, T., Perret, J., Vert, J.: Extensions of marginalized graph kernels. In: Proc. ICML (2004)
21. Alon, N., Benjamini, I., Lubetzky, E., Sodin, S.: Non-backtracking Random Walks Mix Faster. Communications in Contemporary Mathematics 9(4), 585–603 (2007)
22. Tsallis, C.: Possible Generalization of Boltzman-Gibbs Statistacs. J. Stats. Physics 52, 479–487 (1988)
23. Furuichi, S.: Information Theoretical Properities of Tsallis Entropies. Journal of Math. Physics 47, 2 (2006)
24. Riesen, K., Bunke, H.: Graph Classification and Clustering based on Vector Space Embedding. World Scientific Publishing (2010)
25. Bai, L., Hancock, E.R., Ren, P.: Jensen-Shannon Graph Kernel using Information Functionals. In: Proc. ICPR, pp. 2877–2880 (2012)
26. Bai, L., Rossi, L., Torsello, A., Hancock, E.R.: A Quantum Jensen-Shannon Kernel for Unattributed Graphs. To appear in Pattern Recognition (2014)
27. Kriege, N., Mutzel, P.: Subgraph Matching Kernels for Attributed Graphs. In: Proc. ICML (2012)

Sub-sampling for Multi-armed Bandits

Akram Baransi[1], Odalric-Ambrym Maillard[1], and Shie Mannor[1]

Department of Electrical Engineering,
Technion - Israel Institute of Technology, Haifa, Israel
abaransi@tx.technion.ac.il,
odalric-ambrym.maillard@ens-cachan.org, shie@ee.technion.ac.il

Abstract. The stochastic multi-armed bandit problem is a popular model of the exploration/exploitation trade-off in sequential decision problems. We introduce a novel algorithm that is based on sub-sampling. Despite its simplicity, we show that the algorithm demonstrates excellent empirical performances against state-of-the-art algorithms, including Thompson sampling and KL-UCB. The algorithm is very flexible, it does need to know a set of reward distributions in advance nor the range of the rewards. It is not restricted to Bernoulli distributions and is also invariant under rescaling of the rewards. We provide a detailed experimental study comparing the algorithm to the state of the art, the main intuition that explains the striking results, and conclude with a finite-time regret analysis for this algorithm in the simplified two-arm bandit setting.

Keywords: Multi-armed Bandits, Sub-sampling, Reinforcement Learning.

1 Introduction

In sequential decision making under uncertainty, the main dilemma that a decision maker faces is to explore, or not to explore. One of these problems is the popular stochastic multi-armed bandit problem, termed in reference to the 19th century gambling game and introduced by [23, 20]. It illustrates the fundamental trade-off between *exploration*, that is, making decisions that improve the knowledge of the environment, and *exploitation*, that is, choosing the decision that has maximized the previous payoff. Classically, each decision is referred to as an "arm". There is a finite set of arms and each arm, when pulled, returns a real value, called the *reward*, which is independently and identically drawn from an unknown distribution. At each time step the decision maker chooses an arm based on the sequence of rewards that has been observed so far, pulls this arm and observes a new sample from the corresponding unknown underlying distribution. The objective is to find a policy for choosing the next arm to be pulled, that maximizes the sum of the expected rewards, or equivalently minimize the expected *regret*, that is the loss caused by not pulling the best arm at each time step. If $\mathcal{A}$ denotes the set of arms and $\{\mu_a\}_{a \in \mathcal{A}}$ the mean reward of the distribution of each arm, we denote $\star \in \mathrm{argmax}_{a \in \mathcal{A}} \mu_a$ an optimal arm and the (expected) regret of an algorithm that pulls arms $a_1, \ldots, a_T$ up to time T is classically defined as

$$\mathfrak{R}_t = \mathbb{E} \left[\sum_{t=1}^{T} (\mu_\star - \mu_{a_t}) \right]. \tag{1}$$

T. Calders et al. (Eds.): ECML PKDD 2014, Part I, LNCS 8724, pp. 115–131, 2014.
© Springer-Verlag Berlin Heidelberg 2014

Previous Work. Since the formulation of the problem by Robbins (1952), the regret, that measures the cumulative loss resulting from pulling sub-optimal arms, has been a popular criterion for assessing the quality of a strategy. Gittins index based policies ([13, 14, 12]), which were initially introduced by Gittins in 1979, is a family of Bayesian-optimal policies that based on indices that fully characterize each arm given the current history of the game, and at each time step the arm with the highest index will be pulled. However, the high computational cost of Gittins indices and the fact that they are practically limited to a specific set of distributions, arose the need of modifying the policies and make them more efficient. In [8], extending the seminal work of [18], the authors characterized the achievable performance. They showed that under suitable conditions on the possible distributions associated to each arm, any policy that is "admissible" (that is, not grossly under-performing, see [18] for details) must satisfy the following asymptotic lower-performance bound

$$\liminf_{T\to\infty} \frac{\mathfrak{R}_T}{\log(T)} \geqslant \sum_{a:\mu_a<\mu_\star} \frac{\mu_\star - \mu_a}{\mathcal{K}_{\inf}\left(\nu_a; \mu_\star\right)} \, , \qquad (2)$$

where $\mathcal{K}_{\inf}\left(\nu_a; \mu_\star\right)$ is an information-theoretic quantity which measures the minimal Kullback-Leibler divergence between ν_a and distributions in the model that have expectations larger than $\mu_\star$. In the same papers, [18], [10], [8] suggested that Gittins indices can be approximated by quantities that can be interpreted as upper bounds of confidence intervals.

In [1], the generic class of index policies termed **UCB** (Upper Confidence Bounds) was introduced, together with an asymptotic analysis of their performance. [5] provided the first finite time analysis for a particular variant of **UCB** based on Hoeffding's inequality, showing that the regret grows logarithmically with the time horizon T. A few algorithms from the **UCB** family have been recently introduced such as **UCB**-V ([4]), **MOSS** ([3]), Improved-**UCB** ([6]), as well as the recent Kullback-Leibler-based algorithms **DMED** ([15]), **Kinf** ([19]), **kl-UCB** ([11]) and **KL-UCB** ([9]), that were shown to be first-order optimal.

Besides Gittins index and the **UCB**-type algorithms, another important class of algorithms is that introduced by Thompson ([23, 24]), and called **Thompson sampling**. The algorithm assumes that the arms' distributions belong to a parametric family of distributions $\mathcal{P} = \{p(.|\theta), \theta \in \Theta\}$ where $\Theta \subseteq R$, it starts by putting a prior distributions on each one of the arms' parameters, and at each time step a posterior distribution is maintained according to the rewards observed so far. In practice each different $\mathcal{P}$ leads to a different implementation of the algorithm. At each time step, this Bayesian algorithm draws one sample for each arm from its posterior, then pulls the arm that maximizes the expected reward given that parameter. Recently, the analysis developed in [9] enabled to tackle the first frequentist optimal bound for the Thompson-sampling algorithm ([16]) in case of a family of Bernoulli distributions, thus proving that this algorithm also achieves optimality with a the regret that grows logarithmically with T. See also [2], as well as the recent extension to another class of distributions in [17].

Contribution. In this paper we introduce a novel algorithm called **BESA** (**B**est **E**mpirical **S**ampled **A**verage) for the stochastic multi-armed bandit problem (see Section 2). The

---

Algorithm 1. BESA(a,b) for a two-arm bandit

Require: Two arms a, b, current time t.

1. Sample $I_t^a \sim \mathrm{Wr}(N_t(a); N_t(b))$ and $I_t^b \sim \mathrm{Wr}(N_t(b); N_t(a))$.
2. Define $\tilde{\mu}_{t,a} = \hat{\mu}(X_{1:N_t(a)}^a(I_t^a))$ and $\tilde{\mu}_{t,b} = \hat{\mu}(X_{1:N_t(b)}^b(I_t^b))$.
3. Choose (break ties by choosing the arm with the smaller N_t)

$$a_t = \underset{a' \in \{a,b\}}{\mathrm{argmax}} \, \tilde{\mu}_{t,a'} \,.$$

---

algorithm has a different flavor than previously introduced algorithms. It is *not* based on the computation of an empirical confidence bounds but rather on the sampling of some specific quantity. It is for that reason related in spirit to the Thompson sampling algorithm. However, unlike Thompson sampling, **BESA** does not rely on a parametric set of distribution (or a prior) and is instead fully *non-parametric*. In Section 3, we compare the performance of the algorithm against state-of-the-art algorithms, including Thompson sampling and KL-UCB, in several scenarios with different types of reward distributions and show that the algorithm demonstrates excellent empirical performances against them. In Section 4, we provide a possible explanation for the strong performance of **BESA**, and then discuss its properties; Perhaps the most important property of **BESA** is its flexibility, since the same implementation can be used for any type of reward distributions, contrary to Thompson sampling or KL-UCB whose implementations differ according to the considered set of distribution. Finally in Section 5, we provide a finite-time regret bound for this algorithm in the two-arm bandit problem. We show with a rough analysis that the expected regret of the algorithm in this case is $O(\log(T))$ where T is the time horizon. The focus of the paper is to introduce and report the striking empirical performance of this *simple* and *flexible* algorithm.

Setup and Notations. We consider a multi-armed bandit setting with finitely many arms $\mathcal{A}$ and respective reward distributions $\{\nu_a\}_{a \in \mathcal{A}}$, where $\nu_a \in \mathcal{P}([0,1])$ and $\mathcal{P}([0,1])$ denotes the set of probability measures with support in $[0,1]$. We denote $\mu_a \in [0,1]$ the mean of the distribution ν_a, and $X_{1:n}^a = (X_1^a, \ldots, X_n^a)$ a sample of size n, i.i.d. from ν_a. In the sequel, we use the short-hand notation $[n]$ for the set of integer $\{1, \ldots, n\}$. For a set of indices $I \subset [n]$ of size m, say $I = (i_1, \ldots, i_m)$, we write $X_{1:n}^a(I) = (X_{i_1}^a, \ldots, X_{i_m}^a)$ for the corresponding sub-sampled set. A sample of size m drawn *without replacement* from the set $[n]$ is written $I(n; m) \sim \mathrm{Wr}(n; m)$. Here $\mathrm{Wr}(n; m)$ denotes a distribution over sets of integers (with the convention that $\mathrm{Wr}(n; m) = \delta_{[n]}$ if $m > n$, where δ refers to a Dirac distribution). Finally, for a sample S of real values, we denote $\hat{\mu}(S)$ the average of the sample components. For instance we have $\hat{\mu}(X_{1:n}^a) = \frac{1}{n} \sum_{i=1}^{n} X_i^a$. Let $\star \in \mathcal{A}$ denote an arm with maximal mean $\mu_\star$. The regret of an algorithm that pulls arms $\{a_t\}_{t \in [T]}$ up to time T is defined by (1), where the expectation is taken with respect to all sources of randomness. We also denote the number of pulls of an arm $a \in \mathcal{A}$ up to time t by $N_t(a) = \sum_{t'=1}^{t-1} \mathbb{I}\{a_{t'} = a\}$.

2   The BESA Algorithm

The main contribution of this paper is to introduce a novel algorithm, called **BESA** (**B**est **E**mpirical **S**ampled **A**verage) that uses a sub-sampling procedure in order to compare

---

Algorithm 2. BESA($\mathcal{A}$) for a multi-armed bandit

Require: Set of arms $\mathcal{A}$ of size A, current time t.

1. **if** $\mathcal{A} = \{a\}$ **then**
2.   Choose $a_t = a$.
3. **else**
4.   Choose $a_t = \mathbf{BESA}_t(\mathbf{BESA}_t(\{a_i\}_{1 \leqslant i < \lceil A/2 \rceil}), \mathbf{BESA}_t(\{a_i\}_{\lfloor A/2 \rfloor < i \leqslant A}))$
5. **end if**

---

between the empirical values of two arms. The pseudo-code of the algorithm, for two arms is provided in Algorithm 1. The version for the more general multi-armed bandit uses a tournament strategy described in Algorithm 2.

The main idea of the algorithm is to make a fair comparison between the arms: Given two arms a and b that has been pulled $n_a = N_t(a)$ and $n_b = N_t(b) > n_a$ times respectively at time t, comparing the empirical averages of the arms is not a fair comparison since a has not gotten the same number of opportunities as b to show its abilities. **BESA** compensates for this situation by sub-sampling uniformly n_a rewards out of the n_b rewards of arm b. **BESA** then compares the empirical average of the rewards from a, to the empirical average of the rewards sub-sampled from b. It finally chooses b if its computed value is larger than the one of a. We provide in Algorithm 1 a more formal and unified presentation of this strategy. If $n_b > n_a$, the sampled set (line 2) is $I_t^b \sim \mathtt{Wr}(N_t(b); N_t(a))$ (indeed $I_t^a \sim \mathtt{Wr}(N_t(a); N_t(b))$ is the full set $[N_t(a)]$ in this case). Then (line 3,4) the compared values become $\widehat{\mu}(X_{1:N_t(a)}^a)$ and $\widehat{\mu}(X_{1:N_t(b)}^b(I_t^b))$.

BESA, FTL and Thompson Sampling. At first sight, **BESA** seems close to a version of the standard **F**ollow **T**he **L**eader (FTL) algorithm. This algorithm selects $\operatorname{argmax}_{a \in \mathcal{A}} \widehat{\mu}(X_{1:N_t(a)}^a)$, that is the best empirical arm (with no sub-sampling). **FTL** is known to be a *bad* strategy in the bandit setting as it can lead to a linear regret in a number of situations. It is thus *a priori* striking that **BESA** can be any reasonable. On the other hand, **BESA** uses a sampling strategy in order to select the subset used to compute the sub-sampled mean. This is in this respect related in spirit to the **Thompson sampling** strategy, that is known to be both Bayesian optimal, and frequentist optimal, achieving the state-of-the-art for the bandit setting with Bernoulli distribution of rewards ([16]), or more recently distributions in the one-dimension exponential family ([17]). Note that **Thompson sampling** actually refers to a collection of algorithms whose implementation depend on the prior we have on reward distributions. Thus **Thompson sampling** for Bernoulli distributions is for instance different than **Thompson sampling** for exponential distributions. In contrast, **BESA** keeps the same form regardless of the distribution on rewards.

A Tournament for Many Arms. We extend Algorithm 1 written for two arms to the more general case of a finite set $\mathcal{A}$ of arms by using a divide-and-conquer style algorithm (see Algorithm 2). This intuitively corresponds to organizing a tournament between arms. To avoid relying too much on a specific ordering of the arms that may bias the final result (and look arbitrary), we randomly shuffle the set of arms before each decision. That is, at time t, we create a copy $\tilde{\mathcal{A}}_t$ of $\mathcal{A}$ that is obtained by shuffling $\mathcal{A}$ uniformly at random, and then output the arm $\mathbf{BESA}(\tilde{\mathcal{A}}_t)$.

3   Numerical Experiments

Our findings show that, surprisingly, **BESA** is a strong competitor of the state-of-the-art strategies from the bandit literature. Before providing one possible explanation for this striking performance, we now report these intriguing results more precisely.

In this section, the **BESA** algorithm is compared against well-known optimal algorithms such as **KL-UCB**, **KL-UCB+** ([11, 19, 9]), **Thompson sampling** ([23, 24]) with prior Beta(1,1), **KL-UCB**-exp ([9]) and **UCB**-tuned ([5]), on different scenarios with different set of arms. To avoid implementation bias, we use the open-source code available on-line at http://mloss.org/software/view/415 for the implementation of these algorithms. Note that these algorithms do not need parameter tuning. In each one of the scenarios detailed below, the time horizon is set to $T = 20,000$, and the scenario is run on 50,000 independent experiments. In each run, so as to have a fair comparison, the rewards of the arms are all drawn in advance, and all the algorithms are run on the same set of drawn rewards. In other words, on each of the 50,000 runs, $\forall n \in \{1, \ldots, 20000\}, a \in \mathcal{A}$ all the algorithms will observe the same reward on the nth pull of arm a. This enables us to measure the percentage of runs on which one algorithm is better than a reference one, thus providing another measure of performance, besides the empirical average cumulative. We systematically report below in Table 1, ..., 7 this percentage using **BESA** as a reference. In Figures 1, ..., 7, the dark gray represents the plot quartiles, while the light gray represents the upper 5 percents quantile. Finally, in section 5, we introduce the so-called balance function $\alpha_{1/2}$ (see definition 1) that acts as a complexity parameter. For clarity, we report the scaling of this function in most of the following scenarios as well.

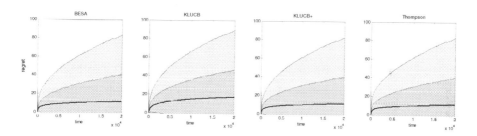

Fig. 1. Regret against time for the two-arm scenario, with $\mu_\star = 0.9$ and $\mu_a = 0.8$

Table 1. Performance measures for $T = 20,000$ in the two-arm scenario, with $\mu_\star = 0.9$ and $\mu_a = 0.8$. Complexity $\alpha_{1/2}(M, 1) = O(0.9^M)$.

	BESA	KL-UCB	KL-UCB+	Thompson sampling
Average regret at T	11.83	17.48	11.54	11.3
Beat **BESA**	--	1.82%	41.6%	58.28%
Average running time	2.86X	2.7X	3.12X	X

3.1   Scenario 1: Two Bernoulli Arms

In this scenario we consider the case of two Bernoulli arms $\mathcal{A} = \{\star, a\}$, with expectations $\mu_\star = 0.9$ and $\mu_a = 0.8$, respectively. The empirical average cumulative regret of each algorithm is shown in the first raw of Table 1, while the second raw shows the percentage of the runs on which the algorithm gave a lower regret than **BESA**, and the third shows the average run time where X denotes the average run time of the fastest algorithm. In Figure 1 the average regret is shown as a function of time. The same scenario has been considered in [11]. On one hand, from figure 1 in [11] one can conclude that the average cumulative regret of **UCB**-V is larger than 50, while all the other algorithms but **KL-UCB** have average cumulative regret between 21 and 36. On the other hand, as reported in Table 1, the average cumulative regret of **BESA** is 11.38. Thus **BESA** outperforms all the algorithms considered in [11] such as e.g. **UCB**-tuned, **DMED**, MOSS on this scenario, including **KL-UCB**. Note that **KL-UCB**+ does get a slightly lower expected regret, but does not beat **BESA** more than 50 per cent of the time. **Thompson sampling** here slightly outperforms **BESA**.

3.2   Scenario 2: Bernoulli with a Small Δ

This scenario is similar to scenario 1 but with a smaller gap Δ: We consider the case of two Bernoulli arms, with expectations $\mu_1 = 0.81$ and $\mu_2 = 0.8$ respectively. Similarly to scenario 1 the average regret, the percentage of experiments on which **BESA** was beaten and the average run time are shown in Table 2, and the cumulative regret as a function of time is shown in Figure 2. Note that the average regret of **BESA** is close to that of **KL-UCB**+ and smaller than that of **KL-UCB** and **Thompson sampling**. Interestingly enough, in addition the percentage of runs on which **BESA** is beaten by any of the state-of-the-art algorithms is smaller than 37%.

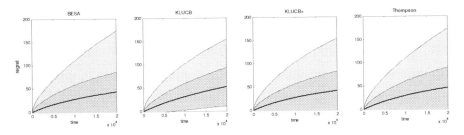

Fig. 2. Regret against time for the two-arm scenario, with $\mu_\star = 0.81$ and $\mu_a = 0.8$.

Table 2. Performance measures for $T = 20,000$ for the two-arm scenario, with $\mu_\star = 0.81$ and $\mu_a = 0.8$. Complexity $\alpha_{1/2}(M, 1) = O(0.9^M)$.

	BESA	KL-UCB	KL-UCB+	Thompson sampling
Average regret at T	42.6	52.34	41.71	46.14
Beat **BESA**	--	25.61%	36.86%	35.2%
Average running time	4.56X	2.78X	3.47X	X

3.3   Scenario 3: Bernoulli with Low Means

In this scenario we consider the scenario used in [11], and inspired by a situation, frequent in applications like marketing or Internet advertising, where the mean reward of each arm is very low. More precisely we consider a harder case which has ten Bernoulli arms, the best arm has expectation 0.1, three arms have expectation 0.05, three arms expectation 0.02, and the rest with expectation 0.01. Table 3 summarizes the results of this experiment, and the regret as a function of time is shown in Figure 3. As can be seen from Table 3 the average regret of **BESA** is much smaller than **KL-UCB** and **Thompson sampling** regrets, and it is beaten by **KL-UCB** only in 1.57% of the runs and by **Thompson sampling** only in 3.09% of the runs. It is also beaten by **KL-UCB**+ less than 36% of the time. As can be seen in figure 2 of [11] the regrets of all the algorithms but **DMED**+ and **KL-UCB**+, which include e.g. CP-**UCB**, **DMED**, **UCB**-Tuned, are between 100 and 400. Thus we can conclude that **BESA**'s average is smaller than the average of those algorithms.

3.4   Scenario 4: All Half but One

In this scenario we consider a case with ten Bernoulli arms, considered as being hard: The optimal arm has expectation 0.51 while all the others have expectation 0.5. The results of this experiment are shown in Table 4 and Figure 4. We note that **BESA** gets a smaller average regret than its competitors, and is not beaten by them more than 42% of the time. Thus **BESA** performs best in this hard setting.

3.5   Scenario 5: Truncated Exponential

In order to further demonstrate the flexibility of the **BESA** algorithm, we consider in this scenario the case of rewards coming from an exponential distribution. Five arms

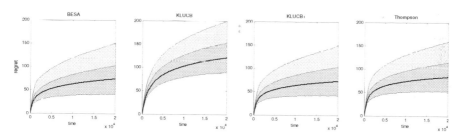

Fig. 3. Regret against time for scenario 3

Table 3. Performance measures for $T = 20,000$ for scenario 3. $\alpha_{1/2}(M, 1) = O(0.5025^M)$.

	BESA	KL-UCB	KL-UCB+	Thompson sampling
Average regret at T	74.41	121.21	72.84	83.36
Beat **BESA**	−−	1.57%	35.41%	3.09%
Average running time	13.85X	2.83X	3.08X	X

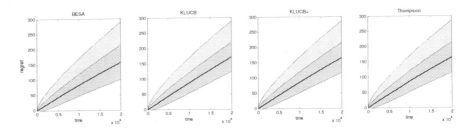

Fig. 4. Regret against time for scenario 4

Table 4. Performance measures for $T = 20,000$ for scenario 4. $\alpha_{1/2}(M, 1) = O(0.75^M)$.

	BESA	KL-UCB	KL-UCB+	Thompson sampling
Average regret at T	156.7	170.82	165.28	165.08
Beat **BESA**	--	41.36%	41.57%	40.78%
Average running time	19.64X	2.78X	2.96X	X

were considered with parameters $\left\{\frac{1}{5}, \frac{1}{4}, \frac{1}{3}, \frac{1}{2}, 1\right\}$, truncated at 10 then divided by 10 (thus they are bounded in $[0, 1]$). The results of this experiment are shown in Table 5 and Figure 5. Note that the regret of **KL-UCB**-exp, which is the version of **KL-UCB** specifically tuned for exponential families and achieving the state-of-the-art for this case, is lower than that of **BESA** only on 5.72% of the runs. Note that **BESA** need not know that the distributions are exponential, that is, we use exactly the same algorithm. Now, as can be seen in the figure of **BESA** the graph is not smooth: the reason is that **BESA** misses the optimal arm if the first reward that it gives is too low. In order to get a smoother behavior, we ran a slightly modified version of **BESA** to skip these corner cases: The modified algorithm is called **BESAT**, and simply pulls each arm ten times before starting running the regular **BESA**. As can be seen in the results this improved the regret dramatically. Now **KL-UCB**-exp beats **BESAT** only on 1.38% of the runs, and similar numbers is achieved for **UCB**-tuned. In [11] a similar scenario is considered, with the difference that they didn't divided the reward by 10. It is actually easy to prove that both **BESA** and **BESAT** are actually invariant by rescaling, that is they pull the same arms in the same order wither we divide the reward by 10 or not. Thus, running the algorithms with the same runs without dividing the rewards by 10, the regret of **BESA** is 532.6 and the one of **BESAT** is 314.1, which is still better than the regrets reported in [11] at Figure 3 (they are above 600).

3.6   Scenario 6: Truncated Poisson

In this scenario we consider the case of Poisson rewards, six Poisson arms are considered with parameters $0.5 + \frac{i}{3}$, where $1 \leqslant i \leqslant 6$, truncated at 10 then divided by 10. A similar scenario was considered in [9] where **KL-UCB**-Poisson is the leading algorithm. From Table 6, **BESA** and **BESAT** outperform **KL-UCB** and **KL-UCB**-Poisson on 95.95% of the runs for **BESA** and 97.99% for **BESAT**, with a much smaller average regret.

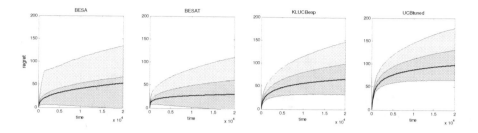

Fig. 5. Regret against time for scenario 5

Table 5. Performance measures for $T = 20,000$ for scenario 5

	BESA	**BESAT**	**KL-UCB**-exp	**UCB**-tuned
Average regret at T	53.26	31.41	65.67	97.6
Beat **BESA**	$--$	40.59%	5.72%	4.33%
Beat **BESAT**	59.41%	$--$	1.38%	0.85%
Average running time	6.01X	7.09X	2.76X	X

3.7 Scenario 7: Uniform Distributions

In this experiment, we consider a challenging setting where arms are uniformly distributed with $X_a \sim \mathcal{U}([0.2, 0.4])$ and $X_\star \sim \mathcal{U}([0, 1])$. Note that there is no natural **KL-UCB** nor **Thompson sampling** algorithm to deal with such family of distribution. In such a scenario, we note that $\alpha_{1/2}(M, 1)$ does not decay exponentially to 0 with M, indicating that this is a difficult scenario for **BESA**. However, it holds that

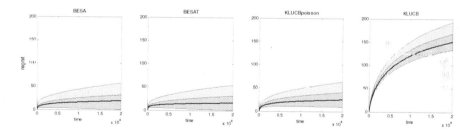

Fig. 6. Regret against time for scenario 6

Table 6. Performance measures for $T = 20,000$ for scenario 6

	BESA	**BESAT**	**KL-UCB**-Poisson	**KL-UCB**
Average regret at T	19.37	16.72	25.05	150.56
Beat **BESA**	$--$	39.92%	4.05%	0.72%
Beat **BESAT**	59.51%	$--$	2.01%	0.17%
Average running time	3.53X	3.49X	1.15X	X

Table 7. Performance measures for $T = 20,000$ for scenario 7

Alg.	UCB	KL-UCB	Thompson sampling
Average regret	21.23	20.72	13.18
Beat **BESA** $n_0 = 0$	24.26%	24.28%	24.7%
Beat **BESA** $n_0 = 3$	7.27%	7.3%	7.83%
Beat **BESA** $n_0 = 7$	1.56%	1.58%	1.76%
Beat **BESA** $n_0 = 10$	0.62%	0.63%	0.74%

BESA	$n_0 = 0$	$n_0 = 3$	$n_0 = 6$	$n_0 = 7$	$n_0 = 8$	$n_0 = 9$	$n_0 = 10$
Average regret	920.12	213.44	50.6	35.38	25.88	17.85	15.42

$\alpha_{1/2}(1, n) = O(\beta^n)$ with $\beta = 0.2$. According to Theorem 1, we should initialize **BESA** by pulling n_0 times each arm before applying **BESA** (that is, run **BESAT**), where for $T = 20,000$, $n_0 \simeq 6.15$. We ran **BESAT** with different number of initialization pulls $n_0 \in \{0, 3, 6, 7, 8, 9, 10\}$, and we also ran **UCB KL-UCB** and **Thompson sampling** (with Beta(1,1) prior) on the same set of arms. The average regrets of each of the algorithms in addition to the percentage on which each non-**BESA** algorithm beats **BESA** with different n_0 are provided in Table 7. The average regret of **BESA** improves with increasing n_0 as expected, and as can be seen from the table, **BESA** with $n_0 = 10$ gave a lower average regret than **UCB** and **KL-UCB** and a bit higher than **Thompson sampling** and it was beaten by the other algorithms on less than 0.8% of the runs.

3.8   Summary of the Experimental Results

From the first four numerical experiments, we deduce that **BESA** is able to compete with the state-of-the-art bandit algorithms in the Bernoulli case. It becomes especially good when the gaps are *small*, or when the Bernoulli parameters are small, which are two main cases of practical interest (especially in web-advertising). Scenario 5, 6 and 7 highlight the flexibility of the algorithm: the same algorithm competes favorably against one of the best algorithm for exponential distributions as well as for Poisson distributions. Using a slight modification, we can beat them by an even larger margin.

4   Intuition and Properties

In this section, we provide an explanation for the striking performance of the **BESA** algorithm, and discuss further its advantages and drawbacks. In the next section, we use this intuition to derive a regret bound for the **BESA** algorithm.

4.1   Why Does It Work?

To give intuition why **BESA** works, let us focus on the two-arm bandit problem. The heuristic idea behind the algorithm is that a comparison between two empirical mean estimates built on a very different number of samples n_a and n_b is not really "fair", and that it seems more natural to compare empirical means based on the same number of samples. Thus the algorithm we introduce is based on sub-sampling. In the rich sub-sampling literature, the works of [7] and [21], show that using sub-sampling without replacement ensures convergence guarantees in a strictly broader setting than sub-sampling with replacement (a.k.a bootstrap). This may provide some informal support for the soundness of the method. However we now provide a more direct justification for the striking performance of **BESA**.

On the theoretical side, one can justify the intuition by looking at the probability of repeatedly choosing a wrong action. If $\mu_b > \mu_a$ and the number of plays of each arm satisfies $n_a > n_b$, the probability that **BESA** chooses a wrong action is approximatively

$$\mathbb{P}\left[\widehat{\mu}\left(X^a_{1:n_a}\left(I(n_a;n_b)\right)\right) > \widehat{\mu}\left(X^b_{1:n_b}\right)\right], \qquad (3)$$

where $I(n_a;n_b) \sim \mathtt{Wr}(n_a;n_b)$, and the probability of making M consecutive mistakes is essentially

$$\mathbb{P}\left[\forall m \in [M] \; \widehat{\mu}\left(X^a_{1:n_a^{(m)}}\left(I_m(n_a^{(m)};n_b)\right)\right) > \widehat{\mu}\left(X^b_{1:n_b}\right)\right], \qquad (4)$$

where for all $m \leqslant M$, $I_m(n_a;n_b) \sim \mathtt{Wr}(n_a;n_b)$, and where we introduced for convenience the short-hand notation $n_a^{(m)} = n_a + m - 1$.

Now, for deterministic[1] n_a, n_b, (3) typically scales with $\exp\left(-2n_b(\mu_b - \mu_a)^2\right)$, by a standard Hoeffding inequality since n_b samples are involved. On the other hand, (4) can decrease at a much faster rate, intuitively of order $\exp\left(-2n_b\tilde{M}(\mu_b - \mu_a)^2\right)$ where $\tilde{M}$ is the number of non-overlapping sub-samples of size n_b. Indeed if $I_{m'}(n_a^{(m)};n_b) \cap I_m(n_a^{(m)};n_b) = \emptyset$ for all $m \neq m' \in \mathcal{M} \subset [M]$ where $|\mathcal{M}| = \tilde{M}$, then the corresponding empirical means are independent from each other, which leads to the intuitive improvement. Using sub-sampling, the later event is of high probability for a reasonable $\tilde{M}$ provided that n_a/n_b is large enough. Note that in the case when we do not resort to sub-sampling, such a phenomenon will not happen, due to the strong dependency between the samples at two consecutive time steps. Thus $\tilde{M}$ is essentially 1 which means that the probability of committing M successive mistakes, will stay big, of the order of $\exp\left(-2n_b(\mu_b - \mu_a)^2\right)$. Now in an ideal case, with only $M = n_a/n_b$ subsets, we might get a mistake error scaling with $\exp\left(-2n_a(\mu_b - \mu_a)^2\right)$, even though b has only been pulled n_b times. As long as $n_a/n_b \ll n_a - n_b$, then we need less trials than the procedure based only on confidence interval estimates before accurately discarding the wrong arm with the same probability. This intuitive idea is formalized and captured by Lemma 1, which we consider to be the key for the current analysis.

4.2   Properties of the Algorithm

We now highlight some of the main properties of **BESA**.

---

[1] The exact argument needs to deal with the fact that $n_a = N_t(a)$ and $n_b = N_t(b)$ are both random stopping times.

Simplicity. We first note the simplicity of the **BESA** compared to previous methods, such as **KL-UCB** for instance that requires some fancy linear program in order to compute the upper confidence bound, or **Thompson sampling** that requires to be able to find an appropriate conjugate prior and implement the update of its parameters, or even the **UCB**-type strategies that generally require some free parameter to be adjusted. Here, **BESA** requires no parameter tuning, and no complicated prior/posterior relation is needed either. Thus the algorithm is directly applicable in a broad range of situation.

Flexibility. A striking property of the algorithm is its flexibility to adapt to various situations. For instance, note that **BESA** does not need to know the support of the distributions, and is also invariant under rescaling. This is not the case of most bandit algorithms that explicit use the knowledge of the support $[a, b]$ of distributions. We believe this can be a serious advantage in some situation. Moreover, both **Thompson sampling** and **KL-UCB** are dependent on a considered parametric set of distributions: in practice a different set of distribution leads to a different implementation. **BESA** does not need such parameters, keeps the same form in all situations, and more importantly still achieves excellent performance in a number of situations, as detailed in Section 3.

Efficiency. Finally, one might wonder about the computationally efficiency of **BESA** due to the use of a sub-sampling method, that is generally not memory less. We were a bit worried about this fact, and thus we implemented the algorithm in a naive way and reported the computational cost of the algorithm in each table for completeness. We conclude from these results that the computational cost of the algorithm is essentially not a problem. Moreover, note that, due to the i.i.d. nature of the data, one may use fancier but more efficient sub-sampler techniques. A naive implementation needs to save all the received rewards. In case the rewards take only finitely many values one can use instead a counter for each possible value. To avoid distracting the reader from the main message, we do not discuss possible tricks that could be used to improve further the numerical efficiency of the method.

5   Regret Upper-Bound

In this section, we provide a simple regret analysis of the **BESA** algorithm[2]. Formalizing further the heuristic intuition of Section 4, it is actually possible to derive a non-trivial regret bound for the **BESA** algorithm, given in Theorem 1. In order to characterize the difficulty of a bandit problem, we now define the following problem-dependent quantity

Definition 1. *For integers M, n and $\lambda \in [0, 1]$, we define the balance function of the distributions $(\nu_a, \nu_\star)$ as*

$$\alpha_\lambda(M, n) = \mathbb{E}_{Z \sim \nu_{\star,n}}\left[\left(1 - F_{\nu_{a,n}}(Z) + \lambda \nu_{a,n}(Z)\right)^M\right],$$

where $\nu_{a,n}$ is the distribution of $\sum_{i=1}^{n} X_i^a$ with $X_i^a \overset{i.i.d}{\sim} \nu_a$ and F_ν is the cdf of ν (that is, $F_\nu(x) = \mathbb{P}_{X \sim \nu}(X \leqslant x)$).

---

[2] We study a slightly modified version, that break ties uniformly at random.

Let us provide some intuition on two examples. Note that α_λ is not increasing both in M and n. First, let us consider two Bernoulli arms with $X^a \sim \mathcal{B}(\mu_a)$ and $X^\star \sim \mathcal{B}(\mu_\star)$. We can compute easily that $\alpha_\lambda(M,1) = (\lambda \mu_a)^M \mu_\star + \big(\lambda + \mu_a(1-\lambda)\big)^M (1 - \mu_\star)$. Now, since $\mu_a < \mu_\star \leqslant 1$, we deduce that $\alpha_{1/2}(M,1) \overset{M \to \infty}{\to} 0$, and more precisely that

$$\alpha_{1/2}(M,1) = O\left(\left(\frac{\mu_a \vee (1 - \mu_a)}{2} \right)^M \right).$$

Thus it converges exponentially fast to 0. Note, however that $\alpha_1(M,1) \overset{M \to \infty}{\to} 1 - \mu_\star$, which is non zero unless $\mu_\star = 1$. Second, let us consider the case of two Uniform arms $X^a \sim \mathcal{U}([0.2, 0.4])$ and $X^\star \sim \mathcal{U}([0, 1.])$. For all λ it holds that $\alpha_\lambda(M,n) \overset{M \to \infty}{\to} 0.2^n$. Thus there is no exponential decay with M. However, it holds that $\alpha_\lambda(1,n) = O(0.2^n)$.

We now prove the following

Theorem 1. *Let $\mathcal{A} = \{\star, a\}$ be a two-armed bandit with bounded rewards in $[0,1]$, and $\Delta = \mu^\star - \mu_a$ be the mean gap. Let us moreover assume that there exists $\alpha \in (0,1)$ and $c > 0$ such that $\alpha_{1/2}(M,1) \leqslant c\alpha^M$. Then the regret of **BESA** at time T is controlled by*

$$\mathfrak{R}_T \leqslant \frac{11 \log(T)}{\Delta} + C'_{\nu_a, \nu_\star} + O(1),$$

where $C'_{\nu_a, \nu_\star}$ depends on the parameters of the problem α, c and Δ, but not on T. Moreover if there exists some $\beta \subset (0,1)$ and $c > 0$ such that $\alpha_{1/2}(1,n) \leqslant c\beta^n$. Let us define

$$n_{0,T} = \left\lceil \frac{\ln(T) - \ln\big((1-\beta)C\big)}{\ln(1/\beta)} \right\rceil.$$

*Then if **BESA** is initialized with $n_{0,T}$ pulls of each arm, then its regret at time T is controlled by*

$$\mathfrak{R}_T \leqslant \frac{11 \log(T)}{\Delta} + n_{0,T} + \tilde{C}_{\nu_a, \nu_\star} + O(1).$$

where $\tilde{C}_{\nu_a, \nu_\star}$ depends on C and on the parameters β, c and Δ, but not on T.

Remark 1. Up to Lemma 1 (see below) which is a purely probabilistic result, and is independent on the bandit setting, the proof is arguably simpler than the typical ones used for Thompson sampling. In particular, we do not need to resort to a fancy "Bernoulli-Beta" trick that is used in classical proofs of Thompson sampling and does not extend easily to general distributions (see for instance [17] that is entirely devoted to the extension to exponential families of dimension one).

Remark 2. Since one needs not use empirical confidence intervals in the analysis, but simply confidence intervals, one can hope to derive much tighter results in the future, using Kullback-Leibler-based Chernoff bounds, or event the sharpest Sanov bounds.

We provide the full proof of this result in the appendix. It mainly follows the proof of [16] that provides a sharp analysis of the **Thompson sampling** algorithm, with some simplifications: first, we consider only two arms, which enables us to skip a recurrence argument (but the same technique could be used to extend our analysis to the case of K-arms); then, we only use mean-based arguments essentially for clarity of exposure. We believe it is more important at this point to provide a clear intuition about why the algorithm works than to provide a tight analysis based on Kullback-Leibler concentration results that are trickier to catch. Now for clarity, we summarize in the next Lemma what we believe is the key result for the regret analysis of **BESA**. This purely probabilistic result is specific to the properties of the sub-sampling procedure.

The sketch of proof is as follows: As usual, we express the regret in terms of the expected number of pulls of sub-optimal arms, that are further decomposed according to the event that the optimal arm has been pulled enough or not. Under the event that the optimal arm is pulled enough, we control easily the probability of mistake resorting to standard proof techniques based on concentration bounds. One difference with respect to standard bounds is that we use here a Serfling-Hoeffding ([22]) concentration inequality. This gives the first term of the regret. The next and difficult step is to show, as usual, that the optimal arm is indeed pulled enough with high probability. To that end, we borrow a proof technique considered in [16]: we introduce the random times τ_j between the jth and $(j+1)^{th}$ pull of the optimal arm, and show that they cannot be too large for too many j, that is to show that the number of *consecutive* mistakes made by the algorithm must be small with high probability. This is one key of the regret analysis of **Thompson sampling** that enables to derive an optimal performance bound. The novelty that we introduce for the analysis of **BESA** is to relate this number of consecutive mistakes to the probability that many sub-samples of small size do not overlap, as explained in Section 4. The precise lemma that covers this part, and that eventually leads to our regret bound is the following:

Lemma 1 (Maximal non-overlapping sub-samples). *Let* $\mathcal{M} = \{p, \ldots, q\} \subset \mathbb{N}$ *be some interval. Let* $j, M \in \mathbb{N}$ *be such that* $p \geqslant 2j$, *and* $M \leqslant |\mathcal{M}|$. *For all* $s \in \mathcal{M}$, *we introduce the random variable* $I_s(s-j;j) \sim \mathtt{WR}(s-j;j)$. *Then the function defined by*

$$f_{\mathcal{M}}(M, j) = 1 - \mathbb{P}\Big[\exists s_1 < \cdots < s_M \in \mathcal{M},$$

$$\forall m \neq m' \in [M] : I_{s_m}(s_m - j; j) \cap I_{s_{m'}}(s_{m'} - j; j) = \emptyset\Big]$$

is decreasing with p. *Moreover, for a sequence of intervals* $\mathcal{M}_t = \{p_t, \ldots, q_t\}$, *such that* $\lim_{t \to \infty} \frac{q_t - p_t}{t} = C > 0$ *and integers* $M_{t,j}$ *such that* $M_{t,j} j = O(\ln(q_t))$, *we have*

$$f_{\mathcal{M}_t}(M_{t,j}, j) = o(t^{-1}). \tag{5}$$

One way to show this lemma is by studying $f_{\mathcal{M}}(M, j)$ formally, and trying to see how it behaves with the different parameters. This however turns out to be tedious and

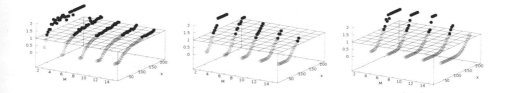

Fig. 7. α as a function of x and M for $j = 3$ (left), $j = 8$ (middle) and $j = 15$ (right). (Note that for $\Delta = 0.3$ and $T = 20,000$, then $\frac{1}{2\Delta} \log(T) \simeq 16.5$.) The black circles indicate when $\alpha_x > 1$ and the white circles when $\alpha_x < 1$.

very technical. On the other hand, we can take advantage of the fact that this function is problem-independent and thus can be computed off-line. It can actually be simulated, and since it is decreasing with $|\mathcal{M}|$, it is enough to study its behavior for small $|\mathcal{M}|$. For our purpose in the analysis, we only need to look at small values of $M = O(\log(T))$ (note that for a time horizon $T = 20,000$, then $\log(T) < 10$). Similarly we use small value for $n_0 \leqslant j \leqslant u_T = O(\log(T))$ as well. It is not difficult to simulate $f_{\mathcal{M}}(M, j)$ for $\mathcal{M}_x = [2j + 1, x]$ with various values of M and x. We are interested in the ratio $\alpha = -\log(f_{\mathcal{M}_x}(M, j))/\log(x)$, and observe numerically that this ratio is increasing with x and quickly becomes larger than 1, that is $f_{\mathcal{M}_x}(M, j) < x^{-1}$ for large enough $x \geqslant C_M = o(M)$ that is slowly increasing with M. In the analysis of Theorem 1, we use $x = O(t/\log(t))$, and $M = O(\log(T))$, and thus as soon as $t \geqslant C' \log(t)$, which happens for $t \geqslant C$ for some numerical constant C, then $f_{\mathcal{M}_t}(M, j)$ starts decaying faster than t^{-1}, that is $f_{\mathcal{M}_t}(\log(t)/j, j) = o(t^{-1})$. In figure 7, we plot the function α in terms of x and M, and for different values of j. Each point is the result obtained via 5,000 replications. We report especially in blue the regions when α becomes larger than 1, which is the critical value to ensure that the lemma holds.

6 Discussion and Conclusion

In this paper, we introduced a novel algorithm for the stochastic multi-armed bandit that is based on *sub-sampling*. We provided a careful experimental analysis of the **BESA** algorithm, by comparing it with the optimized versions of the state-of-the-art algorithms known in each situation. We demonstrated the advantage of **BESA** specifically in the case of Bernoulli distributions, including the case of small parameters and small gaps, as well as exponential and Poisson distributions. For completeness, we reported three measures of performance of the algorithm: plots of the cumulative reward, included quantiles, the percentage it is beaten by other standard algorithms and the numerical complexity with respect to the fastest method.

The algorithm has several striking properties: it is simple to implement, and is very flexible. It does not need to know a set of distributions in advance, unlike **Thompson sampling** or **KL-UCB** and does not even need to know the support, unlike **UCB** or **kl-UCB**. It is also invariant under rescaling of the rewards. This is thus a fully non-parametric algorithm, that competes favorably against standard algorithms.

We finally provide a regret analysis for **BESA**, which shows that the regret of the algorithm is logarithmic (we believe that the constants are not tight). More importantly, we provided a novel proof technique that we believe conveys the core intuition why the algorithm is working, and can lead to much tighter bounds in the future.

Now that we have introduced this algorithm and shown its flexibility, it seems natural to try to extend the **BESA** to other settings. One first direction of research is to consider the *contextual*-bandit problem, another one is to consider the *adversarial* multi-armed bandit setting.

Acknowledgements. This work was supported by the European Community's Seventh Framework Programme (FP7/2007-2013) under grant agreement 306638 (SUPREL) and the Technion.

References

[1] Agrawal, R.: Sample mean based index policies with O(log n) regret for the multi-armed bandit problem. Advances in Applied Probability 27(4), 1054–1078 (1995)

[2] Agrawal, S., Goyal, N.: Further optimal regret bounds for thompson sampling. In: International Conference on Artificial Intelligence and Statistics, Scottsdale, AZ, US. JMLR W&CP, vol. 31 (2013)

[3] Audibert, J.-Y., Bubeck, S.: Minimax policies for adversarial and stochastic bandits

[4] Audibert, J.-Y., Munos, R., Szepesvári, C.: Exploration-exploitation trade-off using variance estimates in multi-armed bandits. Theoretical Computer Science 410, 1876–1902 (2009)

[5] Auer, P.: Using confidence bounds for exploitation-exploration trade-offs. Journal of Machine Learning Research 3, 397–422 (2003)

[6] Auer, P., Ortner, R.: UCB revisited: Improved regret bounds for the stochastic multi-armed bandit problem. Periodica Mathematica Hungarica 61(1-2), 55–65 (2010)

[7] Bickel, P.J., Sakov, A.: On the choice of m in the m out of n bootstrap and confidence bounds for extrema. Statistica Sinica 18, 967–985 (2008)

[8] Burnetas, A.N., Katehakis, M.N.: Optimal adaptive policies for sequential allocation problems. Adv. Appl. Math. 17(2), 122–142 (1996)

[9] Cappé, O., Garivier, A., Maillard, O.-A., Munos, R., Stoltz, G.: Kullback–leibler upper confidence bounds for optimal sequential allocation. Ann. Statist. 41(3), 1516–1541 (2013)

[10] Chang, F., Lai, T.L.: Optimal stopping and dynamic allocation. Advances in Applied Probability 19(4), 829–853 (1987)

[11] Garivier, A., Cappé, O.: The KL-UCB algorithm for bounded stochastic bandits and beyond. In: Proceedings of the 24th annual Conference on Learning Theory, COLT 2011 (2011)

[12] Gittins, J.C.: Bandit processes and dynamic allocation indices. Journal of the Royal Statistical Society. Series B (Methodological) 41(2), 148–177 (1979)

[13] Gittins, J.C., Jones, D.M.: A dynamic allocation index for the discounted multiarmed bandit problem. Biometrika 66(3), 561–565 (1979)

[14] Gittins, J.C., Weber, R., Glazebrook, K.: Multi-armed Bandit Allocation Indices. Wiley (1989)

[15] Honda, J., Takemura, A.: An asymptotically optimal bandit algorithm for bounded support models, pp. 67–79

[16] Kaufmann, E., Korda, N., Munos, R.: Thompson sampling: an asymptotically optimal finite-time analysis. In: Bshouty, N.H., Stoltz, G., Vayatis, N., Zeugmann, T. (eds.) ALT 2012. LNCS (LNAI), vol. 7568, pp. 199–213. Springer, Heidelberg (2012)

[17] Korda, N., Kaufmann, E., Munos, R.: Thompson sampling for 1-dimensional exponential family bandits. In: Burges, C.J.C., Bottou, L., Ghahramani, Z., Weinberger, K.Q. (eds.) NIPS, Lake Tahoe, Nevada, United States, vol. 26, pp. 1448–1456 (2013)

[18] Lai, T.L., Robbins, H.: Asymptotically efficient adaptive allocation rules. Advances in Applied Mathematics 6(1), 4–22 (1985)

[19] Maillard, O.-A., Munos, R., Stoltz, G.: Finite-time analysis of multi-armed bandits problems with kullback-leibler divergences. In: Proceedings of the 24th Annual Conference on Learning Theory, COLT 2011 (2011)

[20] Robbins, H.: Some aspects of the sequential design of experiments. Bulletin of the American Mathematics Society 58, 527–535 (1952)

[21] Romano, J.P., Shaikh, A.M.: On the uniform asymptotic validity of subsampling and the bootstrap. The Annals of Statistics 40(6), 2798–2822 (2012)

[22] Serfling, R.J.: Probability inequalities for the sum in sampling without replacement. The Annals of Statistics 2(1), 39–48 (1974)

[23] Thompson, W.R.: On the likelihood that one unknown probability exceeds another in view of the evidence of two samples. Biometrika 25, 285–294 (1933)

[24] Thompson, W.R.: On the theory of apportionment. American Journal of Mathematics 57, 450–456 (1935)

Knowledge-Powered Deep Learning for Word Embedding

Jiang Bian, Bin Gao, and Tie-Yan Liu

Microsoft Research
{jibian,bingao,tyliu}@microsoft.com

Abstract. The basis of applying deep learning to solve natural language processing tasks is to obtain high-quality distributed representations of words, i.e., word embeddings, from large amounts of text data. However, text itself usually contains incomplete and ambiguous information, which makes necessity to leverage extra knowledge to understand it. Fortunately, text itself already contains well-defined morphological and syntactic knowledge; moreover, the large amount of texts on the Web enable the extraction of plenty of semantic knowledge. Therefore, it makes sense to design novel deep learning algorithms and systems in order to leverage the above knowledge to compute more effective word embeddings. In this paper, we conduct an empirical study on the capacity of leveraging morphological, syntactic, and semantic knowledge to achieve high-quality word embeddings. Our study explores these types of knowledge to define new basis for word representation, provide additional input information, and serve as auxiliary supervision in deep learning, respectively. Experiments on an analogical reasoning task, a word similarity task, and a word completion task have all demonstrated that knowledge-powered deep learning can enhance the effectiveness of word embedding.

1 Introduction

With rapid development of deep learning techniques in recent years, it has drawn increasing attention to train complex and deep models on large amounts of data, in order to solve a wide range of text mining and natural language processing (NLP) tasks [4, 1, 8, 13, 19, 20]. The fundamental concept of such deep learning techniques is to compute distributed representations of words, also known as word embeddings, in the form of continuous vectors. While traditional NLP techniques usually represent words as indices in a vocabulary causing no notion of relationship between words, word embeddings learned by deep learning approaches aim at explicitly encoding many semantic relationships as well as linguistic regularities and patterns into the new embedding space.

Most of existing works employ generic deep learning algorithms, which have been proven to be successful in the speech and image domains, to learn the word embeddings for text related tasks. For example, a previous study [1] proposed a widely used model architecture for estimating neural network language model; later some studies [5, 21] employed the similar neural network architecture to learn word embeddings in order to improve and simplify NLP applications. Most recently, two models [14, 15] were

T. Calders et al. (Eds.): ECML PKDD 2014, Part I, LNCS 8724, pp. 132–148, 2014.

proposed to learn word embeddings in a similar but more efficient manner so as to capture syntactic and semantic word similarities. All these attempts fall into a common framework to leverage the power of deep learning; however, one may want to ask the following questions: *Are these works the right approaches for text-related tasks? And, what are the principles of using deep learning for text-related tasks?*

To answer these questions, it is necessary to note that text yields some unique properties compared with other domains like speech and image. Specifically, while the success of deep learning on the speech and image domains lies in its capability of discovering important signals from noisy input, the major challenge for text understanding is instead the missing information and semantic ambiguity. In other words, image understanding relies more on the information contained in the image itself than the background knowledge, while text understanding often needs to seek help from various external knowledge since text itself only reflects limited information and is sometimes ambiguous. Nevertheless, most of existing works have not sufficiently considered the above uniqueness of text. Therefore it is worthy to investigate how to incorporate more knowledge into the deep learning process.

Fortunately, this requirement is fulfillable due to the availability of various text-related knowledge. First, since text is constructed by human based on morphological and grammatical rules, it already contains well defined morphological and syntactic knowledge. Morphological knowledge implies how a word is constructed, where morphological elements could be syllables, roots, or affix (prefix and suffix). Syntactic knowledge may consist of part-of-speech (POS) tagging as well as the rules of word transformation in different context, such as the comparative and superlative of an adjective, the past and participle of a verb, and the plural form of a noun. Second, there has been a rich line of research works on mining semantic knowledge from large amounts of text data on the Web, such as WordNet [25], Freebase [2], and Probase [26]. Such semantic knowledge can indicate entity category of the word, and the relationship between words/entities, such as synonyms, antonyms, *belonging-to* and *is-a*. For example, Portland *belonging-to* Oregon; Portland *is-a* city. Given the availability of the morphological, syntactic, and semantic knowledge, the critical challenge remains as how to design new deep learning algorithms and systems to leverage it to generate high-quality word embeddings.

In this paper, we take an empirical study on the capacity of leveraging morphological, syntactic, and semantic knowledge into deep learning models. In particular, we investigate the effects of leveraging morphological knowledge to define new basis for word representation and as well as the effects of taking advantage of syntactic and semantic knowledge to provide additional input information and serve as auxiliary supervision in deep learning. In our study, we employ an emerging popular continuous bag-of-words model (CBOW) proposed in [14] as the base model. The evaluation results demonstrate that, knowledge-powered deep learning framework, by adding appropriate knowledge in a proper way, can greatly enhance the quality of word embedding in terms of serving syntactic and semantic tasks.

The rest of the paper is organized as follows. We describe the proposed methods to leverage knowledge in word embedding using neural networks in Section 2. The experimental results are reported in Section 3. In Section 4, we briefly review the related work on word embedding using deep neural networks. The paper is concluded in Section 5.

2   Incorporating Knowledge into Deep Learning

In this paper, we propose to leverage morphological knowledge to define new basis for word representation, and we explore syntactic and semantic knowledge to provide additional input information and serve as auxiliary supervision in the deep learning framework. Note that, our proposed methods may not be the optimal way to use those types of knowledge, but our goal is to reveal the power of knowledge for computing high-quality word embeddings through deep learning techniques.

2.1   Define New Basis for Word Representation

Currently, two major kinds of basis for word representations have been widely used in the deep learning techniques for NLP applications. One of them is the 1-of-v word vector, which follows the conventional bag-of-word models. While this kind of representation preserves the original form of the word, it fails to effectively capture the similarity between words (i.e., every two word vectors are orthogonal), suffers from too expensive computation cost when the vocabulary size is large, and cannot generalize to unseen words when it is computationally constrained.

Another kind of basis is the letter n-gram [11]. For example, in letter tri-gram (or tri-letter), a vocabulary is built according to every combination of three letters, and a word is projected to this vocabulary based on the tri-letters it contains. In contrast to the first type of basis, this method can significantly reduce the training complexity and address the problem of word orthogonality and unseen words. Nevertheless, letters do not carry on semantics by themselves; thus, two words with similar set of letter n-grams may yield quite different semantic meanings, and two semantically similar words might share very few letter n-grams. Figure 1 illustrates one example for each of these two word representation methods.

Representation	Example
1-of-v word vector	Crocodile: $\{w_1, w_2, \ldots, crocodile, \ldots, w_{N-1}, w_N\}$ $\langle 0, 0, \ldots, 1, \ldots, 0, 0 \rangle$
Letter n-gram vector	Crocodile={#cr,cro,roc,oco,cod,odi,dil,ile,le#}: $\{abc, \ldots, \#cr, \ldots, def, \ldots, cro, \ldots, roc, \ldots, oco, \ldots, cod, \ldots, xyz\}$ $\langle 0, \ldots, 1, \ldots, 0, \ldots, 1, \ldots, 1 \ldots, 1, \ldots, 1, \ldots, 0 \rangle$

Fig. 1. An example of how to use 1-of-v word vector and letter n-gram vector as basis to represent a word

To address the limitations of the above word representation methods, we propose to leverage the morphological knowledge to define new forms of basis for word representation, in order to reduce training complexity, enhance capability to generalize to new emerging words, as well as preserve semantics of the word itself. In the following, we will introduce two types of widely-used morphological knowledge and discuss how to use them to define new basis for word representation.

Root/Affix. As an important type of morphological knowledge, root and affix (prefix and suffix) can be used to define a new space where each word is represented as a vector of root/affix. Since most English words are composed by roots and affixes and both roots and affixes yield semantic meaning, it is quite beneficial to represent words using the vocabulary of roots and affixes, which may not only reduce the vocabulary size, but also reflect the semantics of words. Figure 2 shows an example of using root/affix to represent a word.

Knowledge	Examples	
Root/Affix	Crocodile={croc; ile}:	{an,...,croc,...,dis,...,ile,...,in,...,pre,...,zoo} ⟨0,...,1,...,0,...,1,...,0 ...,0,...,0 ⟩
Syllable	Crocodile={croc; o; dile}:	{aba,...,croc,...,dile,...,epi,...,ink,...,o,...,zip} ⟨ 0,...,1,...,1,...,0,...,0 ...,1,...,0 ⟩

Fig. 2. An example of how to use root/affix and syllable to represent a word

Syllable. Syllable is another important type of morphological knowledge that can be used to define the word representation. Similar to root/affix, using syllable can significantly reduce the dimension of the vocabulary. Furthermore, since syllables effectively encodes the pronunciation signals, they can also reflect the semantics of words to some extent (considering that human beings can understand English words and sentences based on their pronunciations). Meanwhile, we are able to cover any unseen words by using syllables as vocabulary. Figure 2 presents an example of using syllables to represent a word.

2.2 Provide Additional Input Information

Existing works on deep learning for word embeddings employ different types of data for different NLP tasks. For example, Mikolov *et al* [14] used text documents collected from Wikipedia to obtain word embeddings; Collobert and Weston [4] leveraged text documents to learn word embeddings for various NLP applications such as language model and chunking; and, Huang *et al* [11] applied deep learning approaches on queries and documents from click-through logs in search engine to generate word representations targeting the relevance tasks. However, those various types of text data, without extra information, can merely reflect partial information and usually cause semantic ambiguity. Therefore, to learn more effective word embeddings, it is necessary to leverage additional knowledge to address the challenges.

In particular, both syntactic and semantic knowledge can serve as additional inputs. An example is shown in Figure 3. Suppose the 1-of-v word vector is used as basis for word representations. To introduce extra knowledge beyond a word itself, we can use entity categories or POS tags as the extension to the original 1-of-v word vector. For example, given an entity knowledge graph, we can define an entity space. Then, a word will be projected into this space such that some certain elements yield non-zero values if the word belongs to the corresponding entity categories. In addition, relationship between words/entities can serve as another type of input information. Particularly, given

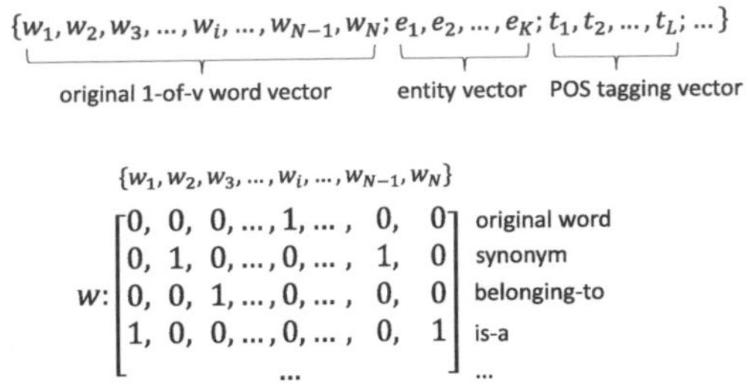

{w1, w2, w3, …, wi, …, wN−1, wN; e1, e2, …, eK; t1, t2, …, tL; …}
original 1-of-v word vector
entity vector
POS tagging vector
{w1, w2, w3, …, wi, …, wN−1, wN}
w:
0, 0, 0, …, 1, …, 0, 0
0, 1, 0, …, 0, …, 1, 0
0, 0, 1, …, 0, …, 0, 0
1, 0, 0, …, 0, …, 0, 1
…
original word
synonym
belonging-to
is-a
…

Fig. 3. An example of using syntactic or semantic knowledge, such as entity category, POS tags, and relationship, as additional input information

various kinds of syntactic and semantic relations, such as *synonym*, *antonym*, *belonging-to*, *is-a*, etc., we can construct a relation matrix $\mathcal{R}_w$ for one word w (as shown in Figure 3), where each column corresponds to a word in the vocabulary, each row encodes one type of relationship, and one element $\mathcal{R}_w(i,j)$ has non-zero value if w yield the i-th relation with the j-th word.

2.3 Serve as Auxiliary Supervision

According to previous studies on deep learning for NLP tasks, different training samples and objective functions are suitable for different NLP applications. For example, some works [4, 14] define likelihood based loss functions, while some other work [11] leverages cosine similarity between queries and documents to compute objectives. However, all these loss functions are commonly used in the machine learning literature without considering the uniqueness of text.

Main Objective
Auxiliary Tasks
Syntactic Knowledge
Semantic Knowledge
Word Embedding

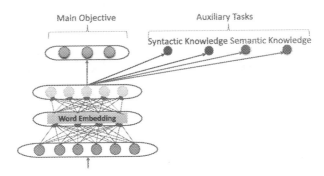

Fig. 4. Using syntactic and semantic knowledge as auxiliary objectives

Text related knowledge can provide valuable complement to the objective of the deep learning framework. Particularly, we can create auxiliary tasks based on the knowledge to assist the learning of the main objective, which can effectively regularize the learning of the hidden layers and improve the generalization ability of deep neural networks so as to achieve high-quality word embedding. Both semantic and syntactic knowledge can serve as auxiliary objectives, as shown in Figure 4.

Note that this multi-task framework can be applied to any text related deep learning technique. In this work, we take the continuous bag-of-words model (CBOW) [14] as a specific example. The main objective of this model is to predict the center word given the surrounding context. More formally, given a sequence of training words $w_1, w_2, \cdots,$ w_X, the main objective of the CBOW model is to maximize the average log probability:

$$\mathcal{L}_M = \frac{1}{X} \sum_{x=1}^{X} \log p(w_x | \mathcal{W}_x^d) \tag{1}$$

where $\mathcal{W}_x^d = \{w_{x-d}, \cdots, w_{x-1}, w_{x+1}, \cdots, w_{x+d}\}$ denotes a $2d$-sized training context of word w_x.

To use semantic and syntactic knowledge to define auxiliary tasks to the CBOW model, we can leverage the entity vector, POS tag vector, and relation matrix (as shown in Figure 3) of the center word as the additional objectives. Below, we take entity and relationship as two examples for illustration. Specifically, we define the objective for entity knowledge as

$$\mathcal{L}_E = \frac{1}{X} \sum_{x=1}^{X} \sum_{k=1}^{K} \mathbf{1}(w_x \in e_k) \log p(e_k | \mathcal{W}_x^d) \tag{2}$$

where K is the size of entity vector; and $\mathbf{1}(\cdot)$ is an indicator function, $\mathbf{1}(w_x \in e_k)$ equals 1 if w_x belongs to entity e_k, otherwise 0; note that the entity e_k could be denoted by either a single word or a phrase. Moreover, assuming there are totally R relations, i.e., there are R rows in the relation matrix, we define the objective for relation as:

$$\mathcal{L}_R = \frac{1}{X} \sum_{x=1}^{X} \sum_{r=1}^{R} \lambda_r \sum_{n=1}^{N} r(w_x, w_n) \log p(w_n | \mathcal{W}_x^d) \tag{3}$$

where N is vocabulary size; $r(w_x, w_n)$ equals 1 if w_x and w_n have relation r, otherwise 0; and λ_r is an empirical weight of relation r.

3   Experiments

To evaluate the effectiveness of the knowledge-powered deep learning for word embedding, we compare the quality of word embeddings learned with incorporated knowledge to those without knowledge. In this section, we first introduce the experimental settings, and then we conduct empirical comparisons on three specific tasks: a public analogical reasoning task, a word similarity task, and a word completion task.

3.1   Experimental Setup

Baseline Model. In our empirical study, we use the continuous bag-of-words model (CBOW) [14] as the baseline method. The code of this model has been made publicly available[1]. We use this model to learn the word embeddings on the above dataset. In

---

[1] http://code.google.com/p/word2vec/

the following, we will study the effects of different methods for adding various types of knowledge into the CBOW model. To ensure the consistency among our empirical studies, we set the same number of embedding size, i.e. 600, for both the baseline model and those with incorporated knowledge.

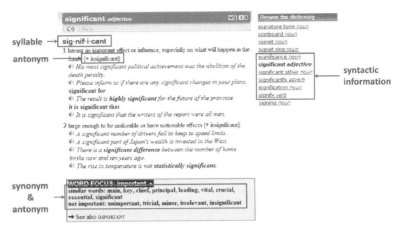

Fig. 5. Longman Dictionaries provide several types of morphological, syntactic, and semantic knowledge

Table 1. Knowledge corpus used for our experiments (Type: MOR-morphological; SYN-syntactic; SEM-semantic)

Corpus	Type	Specific knowledge	Size
Morfessor	MOR	root, affix	200K
Longman	MOR/SYN /SEM	syllable, POS tagging, synonym, antonym	30K
WordNet	SYN/SEM	POS tagging, synonym, antonym	20K
Freebase	SEM	entity, relation	1M

Applied Knowledge. For each word in the Wikipedia dataset as described above, we collect corresponding morphological, syntactic, and semantic knowledge from four data sources: Morfessor [23], Longman Dictionaries[2], WordNet [25], and Freebase[3].

Morfessor provides a tool that can automatically split a word into roots, prefixes, and suffixes. Therefore, this source allows us to collect morphological knowledge for each word existed in our training data.

Longman Dictionaries is a large corpus of words, phrases, and meaning, consisting of rich morphological, syntactic, and semantic knowledge. As shown in Figure 5, Longman Dictionaries provide word's syllables as morphological knowledge, word's syntactic transformations as syntactic knowledge, and word's synonym and antonym as semantic knowledge. We collect totally 30K words and their corresponding knowledge from Longman Dictionaries.

---

[2] http://www.longmandictionariesonline.com/
[3] http://www.freebase.com/

WordNet is a large lexical database of English. Nouns, verbs, adjectives, and adverbs are grouped into sets of cognitive synonyms (synsets), each expressing a distinct concept. Synsets are interlinked by means of conceptual-semantic and lexical relations. Note that WordNet interlinks not just word forms (syntactic information) but also specific senses of words (semantic information). WordNet also labels the semantic relations among words. Therefore, WordNet provides us with another corpus of rich semantic and syntactic knowledge. In our experiments, we sample 15K words with 12K synsets, and there are totally 20K word-senses pairs.

Freebase is an online collection of structured data harvested from many online sources. It is comprised of important semantic knowledge, especially the entity and relation information (e.g., categories, *belonging-to*, *is-a*). We crawled 1M top frequent words and corresponding information from Freebase as another semantic knowledge base.

We summarize these four sources in Table 1[4].

3.2   Evaluation Tasks

We evaluate the quality of word embeddings on three tasks.

1. Analogical Reasoning Task:

The analogical reasoning task was introduced by Mikolov *et al* [16, 14], which defines a comprehensive test set that contains five types of semantic analogies and nine types of syntactic analogies[5]. For example, to solve semantic analogies such as *Germany : Berlin = France : ?*, we need to find a vector x such that the embedding of x, denoted as vec(x) is closest to vec(*Berlin*) - vec(*Germany*) + vec(*France*) according to the cosine distance. This specific example is considered to have been answered correctly if x is *Paris*. Another example of syntactic analogies is *quick : quickly = slow : ?*, the correct answer of which should be *slowly*. Overall, there are 8,869 semantic analogies and 10,675 syntactic analogies.

In our experiments, we trained word embeddings on a publicly available text corpus[6], a dataset about the first billion characters from Wikipedia. This text corpus contains totally 123.4 million words, where the number of unique words, i.e., the vocabulary size, is about 220 thousand. We then evaluated the overall accuracy for all analogy types, and for each analogy type separately (i.e., semantic and syntactic).

2. Word Similarity Task:

A standard dataset for evaluating vector-space models is the WordSim-353 dataset [7], which consists of 353 pairs of nouns. Each pair is presented without context and associated with 13 to 16 human judgments on similarity and relatedness on a scale from 0 to 10. For example, (*cup*, *drink*) received an average score of 7.25, while (*cup*, *substance*) received an average score of 1.92. Overall speaking, these 353 word pairs reflect more semantic word relationship than syntactic relationship.

In our experiments, similar to the Analogical Reasoning Task, we also learned the word embeddings on the same Wikipedia dataset. To evaluate the quality of learned

---

[4] We plan to release all the knowledge corpora we used in this study after the paper is published.

[5] http://code.google.com/p/word2vec/source/browse/trunk/
questions-words.txt

[6] http://mattmahoney.net/dc/enwik9.zip

word embedding, we compute Spearman's ρ correlation between the similarity scores computed based on learned word embeddings and human judgments.

3. Sentence Completion Task:

Another advanced language modeling task is Microsoft Sentence Completion Challenge [27]. This task consists of 1040 sentences, where one word is missing in each sentence and the goal is to select word that is the most coherent with the rest of the sentence, given a list of five reasonable choices. In general, accurate sentence completion requires better understanding on both the syntactic and semantics of the context.

In our experiments, we learn the 600-dimensional embeddings on the 50M training data provided by [27], with and without applied knowledge, respectively. Then, we compute score of each sentence in the test set by using each of the sliding windows (window size is consistent with the training process) including the unknown word at the input, and predict the corresponding central word in a sentence. The final sentence score is then the sum of these individual predictions. Using the sentence scores, we choose the most likely sentence to answer the question.

3.3   Experimental Results

Effects of Defining Knowledge-Powered Basis for Word Representation. As introduced in Section 2.1, we can leverage morphological knowledge to design new basis for word representation, including root/affix-based and syllable-based bases. In this experiment, we separately leverage these two types of morphological basis, instead of the conventional 1-of-v word vector and letter n-gram vector, in the CBOW framework (as shown in Figure 6). Then, we compare the quality of the newly obtained word embeddings with those computed by the baseline models. Note that, after using root/affix, syllable, or letter n-gram as input basis, the deep learning framework will directly generate the embedding for each root/affix, syllable, or letter n-gram; the new embedding of a word can be obtained by aggregating the embeddings of this word's morphological elements.

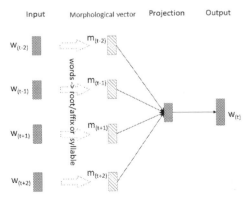

Fig. 6. Define morphological elements (root, affix, syllable) as new bases in CBOW

Table 2 shows the accuracy of analogical questions by using baseline word embeddings and by using those learned from morphological knowledge-powered bases, respectively. As shown in Table 2, different bases yield various dimensionalities; and,

Table 2. The accuracy of analogical questions by using word embeddings learned with different bases for word representation

Representation	Dimensionality	Semantic Accuracy	Syntactic Accuracy	Overall Accuracy	Overall Relative Gain
Original words	220K	16.62%	34.98%	26.65%	-
Root/affix	24K	14.27%	44.15%	30.59%	14.78%
Syllable	10K	2.67%	18.72%	11.44%	-57.07%
Letter 3-gram	13K	0.18%	9.12%	5.07%	-80.98%
Letter 4-gram	97K	17.29%	32.99%	26.89%	0.90%
Letter 5-gram	289K	16.03%	34.27%	26.00%	-2.44%

using root/affix to represent words can significantly improve the accuracy with about 14% relative gain, even with a much lower input dimensionality than the original 1-of-v representation.

However, syllable and letter 3-gram lead to drastically decreasing accuracy, probably due to their low dimensionalities and high noise levels. In addition, as the average word length of the training data is 4.8, using letter 4-gram and 5-gram is very close to using 1-of-V as basis. Therefore, as shown in Table 2, letter 4-gram and 5-gram can perform as good as baseline.

Table 3 illustrate the performance for the word similarity task by using word embeddings trained from different bases. From the table, we can find that, letter 4-gram and 5-gram yields similar performances to the baseline; however, none of root/affix, syllable, and letter tri-gram can benefit word similarity task.

Table 3. Spearman's ρ correlation on WordSim-353 by using word embeddings learned with different bases

Model	$\rho \times 100$	Relative Gain
Original words	60.1	-
Root/affix	60.6	0.83%
Syllable	17.9	-70%
3-gram	14.2	-76%
4-gram	60.3	0.33%
5-gram	60.0	-0.17%

For the sentence completion task, Table 4 compares the accuracy by using word embeddings trained with different bases. Similar to the trend of the first task, except Root/affix that can raise the accuracy by 3-4%, other bases for word representation have little or negative influence on the performance.

Table 4. Accuracy of different models on the Microsoft Sentence Completion Challenge

Model	Accuracy	Relative gain
Original words	41.2%	-
Root/affix	42.7%	3.64%
Syllable	40.0%	-2.91%
3-gram	41.3%	0.24%
4-gram	40.8%	-0.97%
5-gram	41.0%	-0.49%

Effects of Providing Additional Knowledge-Augmented Input Information. In this experiment, by using the method described in Section 2.2, we add syntactic and semantic knowledge of each input word as additional inputs into the CBOW model (as shown in Figure 7). Then, we compare the quality of the newly obtained word embeddings with the baseline.

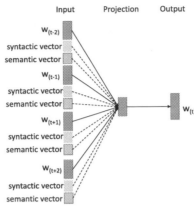

Fig. 7. Add syntactic and semantic knowledge of input word as additional inputs in CBOW

For the analogical reasoning task, Table 5 reports the accuracy by using wording embeddings learned from the baseline model and that with knowledge-augmented inputs, respectively. From the table, we can find that using syntactic knowledge as additional input can benefit syntactic analogies significantly but drastically hurt the semantic accuracy, while semantic knowledge gives rise to an opposite result. This table also illustrates that using both semantic and syntactic knowledge as additional inputs can lead to about 24% performance gain.

Table 5. The accuracy of analogical questions by using word embeddings learned with different additional inputs

Raw Data	Semantic Accuracy	Relative Gain	Syntactic Accuracy	Relative Gain	Total Accuracy	Relative Gain
Original words	16.62%		34.98%		26.65%	
+ Syntactic knowledge	6.12%	−63.18%	46.84%	33.90%	28.67%	7.58%
+ Semantic knowledge	49.16%	195.78%	17.96%	−48.66%	31.38%	17.74%
+ both knowledge	27.37%	64.68%	36.33%	3.86%	33.22%	24.65%

Furthermore, Table 6 illustrates the performance of the word similarity task on different models. From the table, it is clear to see that using semantic knowledge as additional inputs can cause a more than 4% relative gain while syntactic knowledge brings little influence on this task.

Table 6. Spearman's ρ correlation on WordSim-353 by using word embeddings learned with different additional input

Model	$\rho \times 100$	Relative Gain
Original words	60.1	-
+ Syntactic knowledge	60.6	0.83%
+ Semantic knowledge	62.6	4.16%
+ both knowledge	60.9	1.33%

Table 7. Accuracy of different models on the Microsoft Sentence Completion Challenge

Model	Accuracy	Relative Gain
Original words	41.2%	-
+ Syntactic knowledge	43.7%	6.07%
+ Semantic knowledge	44.1%	7.04%
+ Both knowledge	43.8%	6.31%

Moreover, Table 7 shows the accuracy of the sentence completion task by using models with different knowledge-augmented inputs. From the table, we can find that using either semantic or syntactic knowledge as additional inputs can benefit the performance, with more than 6% and 7% relative gains, respectively.

Effects of Serving Knowledge as Auxiliary Supervision. As introduced in Section 2.3, in this experiment, we use either separate or combined syntactic and semantic knowledge as auxiliary tasks to regularize the training of the CBOW framework (as shown in Figure 8). Then, we compare the quality of the newly obtained word embeddings with those computed by the baseline model.

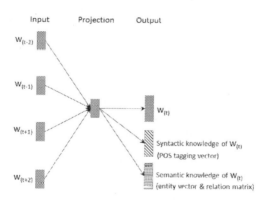

Fig. 8. Use syntactic and semantic knowledge as auxiliary objectives in CBOW

Table 8 illustrates the accuracy of analogical questions by using word embeddings learned from the baseline model and from those with knowledge-regularized objectives, respectively. From the table, we can find that leveraging either semantic or syntactic knowledge as auxiliary objectives results in quite little changes to the accuracy, and using both of them simultaneously can yield 1.39% relative improvement.

Furthermore, Table 9 compares different models' performance on the word similarity task. From the table, we can find that using semantic knowledge as auxiliary objective can result in a significant improvement, with about 5.7% relative gain, while using syntactic knowledge as auxiliary objective cannot benefit this task. And, using both knowledge can cause more than 3% improvement.

Moreover, for the sentence completion task, Table 10 shows the accuracy of using different knowledge-regularized models. From the table, we can find that, while syntac-

Table 8. The accuracy of analogical questions by using word embeddings learned from baseline model and those with knowledge-regularized objectives

Objective	Semantic Accuracy	Relative Gain	Syntactic Accuracy	Relative Gain	Total Accuracy	Relative Gain
Original words	16.62%		34.98%		26.65%	
+ Syntactic knowledge	17.09%	2.83%	34.74%	−0.69%	26.73%	0.30%
+ Semantic knowledge	16.43%	−1.14%	35.33%	1.00%	26.75%	0.38%
+ both knowledge	17.59%	5.84%	34.86%	−0.34%	27.02%	1.39%

Table 9. Spearman's ρ correlation on WordSim-353 by using baseline model and the model trained by knowledge-regularized objectives

Model	$\rho \times 100$	Relative Gain
Original words	60.1	-
+ Syntactic knowledge	59.8	-0.50%
+ Semantic knowledge	63.5	5.66%
+ both knowledge	62.1	3.33%

tic knowledge does not cause much accuracy improvement, using semantic knowledge as auxiliary objectives can significantly increase the performance, with more than 9% relative gain. And, using both knowledge as auxiliary objectives can lead to more than 7% improvement.

Table 10. Accuracy of different models on the Microsoft Sentence Completion Challenge

Model	Accuracy	Relative Gain
Original words	41.2%	-
+ Syntactic knowledge	41.9%	1.70%
+ Semantic knowledge	45.2%	9.71%
+ both knowledge	44.2%	7.28%

3.4 Discussions

In a summary, our empirical studies investigate three ways (i.e., new basis, additional inputs, and auxiliary supervision) of incorporating knowledge into three different text related tasks (i.e., analogical reasoning, word similarity, and sentence completion), and we explore three specific types of knowledge (i.e., morphological, syntactic, and semantic). Figure 9 summarizes whether and using which method each certain type of knowledge can benefit different tasks, in which a tick indicates a relative gain of larger than 3% and a cross indicates the remaining cases. In the following of this section, we will take further discussions to generalize some guidelines for incorporating knowledge into deep learning.

Different Tasks Seek Different Knowledge. According to the task descriptions in Section 3.2, it is clear to see that the three text related tasks applied in our empirical studies are inherently different to each other, and such differences further decide each task's sensitivity to different knowledge.

Specifically, the analogical reasoning task consists of both semantic questions and syntactic questions. As shown in Figure 9, it is beneficial to applying both syntactic and semantic knowledge as additional input into the learning process. Morphological knowledge, especially root/affix, can also improve the accuracy of this task, because root/affix plays a key role in addressing some of the syntactic questions, such as *adj : adv*, *comparative : superlative*, the evidence of which can be found in Table 2 that illustrates using root/affix as basis can improve syntactic accuracy more than semantic accuracy.

Task \ Knowledge type	Morphological		Syntactic	Semantic
Analogical reasoning	✓ Root/affix	✓	✓ Additional input	✓ Additional input
	✗ Syllable			
	✗ Letter n-gram		✗ Auxiliary objective	✗ Auxiliary objective
Word similarity	✗ Root/affix	✗	✗ Additional input	✓ Additional input
	✗ Syllable			
	✗ Letter n-gram		✗ Auxiliary objective	✓ Auxiliary objective
Sentence completion	✓ Root/affix	✓	✓ Additional input	✓ Additional input
	✗ Syllable			
	✗ Letter n-gram		✗ Auxiliary objective	✓ Auxiliary objective

Fig. 9. A summary of whether and using which method each certain type of knowledge can benefit different tasks

As aforementioned, the goal of the word similarity task is to predict the semantic similarity between two words without any context. Therefore only semantic knowledge can enhance the learned word embeddings for this task. As shown in Table 6 and 9, it is clear to see that using semantic knowledge as either additional input or auxiliary supervision can improve the word similarity task.

As a sentence is built to represent certain semantics under human defined morphological and syntactic rules, sentence completion task requires accurate understanding on the semantics of the context, the syntactic structure of the sentence, and the morphological rules for key words in it. Thus, as shown in Figure 9, all three types of knowledge can improve the accuracy of this task if used appropriately.

Effects of How to Incorporate Different Knowledge. According to our empirical studies, syntactic knowledge is effective to improve analogical reasoning and sentence completion only when it is employed as additional input into the deep learning framework, which implies that syntactic knowledge can provide valuable input information but may not be suitable to serve as regularized objectives. Our empirical studies also demonstrate that, using semantic knowledge as either additional input or regularized objectives can improve the performance of the word similarity task and sentence completion tasks. Furthermore, comparing Table 9 and 10 with Table 6 and 7, we can find that applying semantic knowledge as auxiliary objectives can achieve slightly better performance than using it as additional input. However, for the analogical reasoning task, semantic knowledge is effective only when it is applied as additional input.

4 Related Work

Obtaining continious word embedding has been studied for a long time [9]. With the progress of deep learning, deep neural network models have been applied to obtain word embeddings. One of popular model architectures for estimating neural network language model (NNLM) was proposed in [1], where a feed-forward neural network with a linear projection layer and a non-linear hidden layer was used to learn jointly the word embedding and a statistical language model. Many studies follow this approach to improve and simplify text mining and NLP tasks [4–6, 8, 11, 19, 22, 20, 17, 10]. In these studies, estimation of the word embeddings was performed using different model architectures and trained on various text corpora.

For example, Collobert et al [5] proposed a unified neural network architecture to learn adequate internal representations on the basis of vast amounts of mostly unlabeled training data, to deal with various natural language processing tasks. In order to adapt the sequential property of language modeling, a recurrent architecture of NNLM was present in [13], referred as RNNLM, where the hidden layer at current time will be recurrently used as input to the hidden layer at the next time. Huang et al [11] developed a deep structure that project queries and documents into a common word embedding space where the query-document similarity is computed as the cosine similarity. The word embedding model is trained by maximizing the conditional likelihood of the clicked documents for a given query using the click-through data. Mikolov et al [14, 15] proposed the continuous bag-of-words model (CBOW) and the continuous skip-gram model (Skip-gram) for learning distributed representations of words from large amount of unlabeled text data. Both models can map the semantically or syntactically similar words to close positions in the learned embedding space, based on the principal that the context of the similar words are similar.

Recent studies have explored knowledge related word embedding, the purpose of of which is though quite different. For example, [3] focused on learning structured embeddings of knowledge bases; [18] paid attention to knowledge base completion; and [24] investigated relation extraction from free text. They did not explicitly study how to use knowledge to enhance word embedding. Besides, Luong et al [12] proposed to apply morphological information to learn better word embedding. But, it did not explore other ways to leverage various types of knowledge.

5 Conclusions and Future Work

In this paper, we take an empirical study on using morphological, syntactic, and semantic knowledge to achieve high-quality word embeddings. Our study explores these types of knowledge to define new basis for word representation, provide additional input information, and serve as auxiliary supervision in deep learning framework. Evaluations on three text related tasks demonstrated the effectiveness of knowledge-powered deep learning to produce high-quality word embeddings in general, and also reveal the best way of using each type of knowledge for a given task.

For the future work, we plan to explore more types of knowledge and apply them into the deep learning process. We also plan to study the co-learning of high-quality word embeddings and large-scale reliable knowledge.

References

1. Bengio, Y., Ducharme, R., Vincent, P., Janvin, C.: A neural probabilistic language model. The Journal of Machine Learning Research 3, 1137–1155 (2003)
2. Bollacker, K., Evans, C., Paritosh, P., Sturge, T., Taylor, J.: Freebase: a collaboratively created graph database for structuring human knowledge. In: Proceedings of the 2008 ACM SIGMOD International Conference on Management of Data, pp. 1247–1250. ACM (2008)
3. Bordes, A., Weston, J., Collobert, R., Bengio, Y., et al.: Learning structured embeddings of knowledge bases. In: AAAI (2011)
4. Collobert, R., Weston, J.: A unified architecture for natural language processing: Deep neural networks with multitask learning. In: Proceedings of the 25th International Conference on Machine Learning, ICML 2008, pp. 160–167. ACM, New York (2008)
5. Collobert, R., Weston, J., Bottou, L., Karlen, M., Kavukcuoglu, K., Kuksa, P.: Natural language processing (almost) from scratch. Journal of Machine Learning Research 12, 2493–2537 (2011)
6. Deng, L., He, X., Gao, J.: Deep stacking networks for information retrieval. In: ICASSP, pp. 3153–3157 (2013)
7. Finkelstein, L., Gabrilovich, E., Matias, Y., Rivlin, E., Solan, Z., Wolfman, G., Ruppin, E.: Placing search in context: The concept revisited. ACM Transactions on Information Systems (2002)
8. Glorot, X., Bordes, A., Bengio, Y.: Domain adaptation for large-scale sentiment classification: A deep learning approach. In: Proceedings of the Twenty-eight International Conference on Machine Learning, ICML (2011)
9. Hinton, G.E., McClelland, J.L., Rumelhart, D.E.: Distributed representations. In: Parallel Distributed Processing: Explorations in the Microstructure of Cognition, vol. 3, pp. 1137–1155. MIT Press (1986)
10. Huang, E., Socher, R., Manning, C., Ng, A.: Improving word representations via global context and multiple word prototypes. In: Proc. of ACL (2012)
11. Huang, P.-S., He, X., Gao, J., Deng, L., Acero, A., Heck, L.: Learning deep structured semantic models for web search using clickthrough data. In: Proceedings of the 22Nd ACM International Conference on Conference on Information & Knowledge Management, CIKM 2013, pp. 2333–2338. ACM, New York (2013)
12. Luong, M.-T., Socher, R., Manning, C.D.: Better word representations with recursive neural networks for morphology. CoNLL-2013, 104 (2013)
13. Mikolov, T.: Statistical Language Models Based on Neural Networks. PhD thesis, Brno University of Technology (2012)
14. Mikolov, T., Chen, K., Corrado, G., Dean, J.: Efficient estimation of word representations in vector space. CoRR, abs/1301.3781 (2013)
15. Mikolov, T., Sutskever, I., Chen, K., Corrado, G.S., Dean, J.: Distributed representations of words and phrases and their compositionality. In: Burges, C.J.C., Bottou, L., Ghahramani, Z., Weinberger, K.Q. (eds.) NIPS, pp. 3111–3119 (2013)
16. Mikolov, T., Yih, W.-T., Zweig, G.: Linguistic regularities in continuous space word representations. In: Proceedings of NAACL-HLT, pp. 746–751 (2013)
17. Mnih, A., Hinton, G.E.: A scalable hierarchical distributed language model. In: NIPS, pp. 1081–1088 (2008)

18. Socher, R., Chen, D., Manning, C.D., Ng, A.: Reasoning with neural tensor networks for knowledge base completion. In: Advances in Neural Information Processing Systems, pp. 926–934 (2013)
19. Socher, R., Lin, C.C., Ng, A.Y., Manning, C.D.: Parsing natural scenes and natural language with recursive neural networks. In: Proceedings of the 26th International Conference on Machine Learning, ICML (2011)
20. Tur, G., Deng, L., Hakkani-Tur, D., He, X.: Towards deeper understanding: Deep convex networks for semantic utterance classification. In: ICASSP, pp. 5045–5048 (2012)
21. Turian, J.P., Ratinov, L.-A., Bengio, Y.: Word representations: A simple and general method for semi-supervised learning. In: ACL, pp. 384–394 (2010)
22. Turney, P.D.: Distributional semantics beyond words: Supervised learning of analogy and paraphrase. In: Transactions of the Association for Computational Linguistics (TACL), pp. 353–366 (2013)
23. Virpioja, S., Smit, P., Grnroos, S., Kurimo, M.: Morfessor 2.0: Python implementation and extensions for morfessor baseline. In: Aalto University Publication Series SCIENCE + TECHNOLOGY (2013)
24. Weston, J., Bordes, A., Yakhnenko, O., Usunier, N.: Connecting language and knowledge bases with embedding models for relation extraction. arXiv preprint arXiv:1307.7973 (2013)
25. WordNet. "about wordnet". Princeton university (2010)
26. Wu, W., Li, H., Wang, H., Zhu, K.: Probase: A probabilistic taxonomy for text understanding. In: Proc. of SIGMOD (2012)
27. Zweig, G., Burges, C.: The microsoft research sentence completion challenge. Microsoft Research Technical Report MSR-TR-2011-129 (2011)

Density-Based Subspace Clustering in Heterogeneous Networks

Brigitte Boden[1], Martin Ester[2], and Thomas Seidl[1]

[1] RWTH Aachen University, Aachen, Germany
{boden,seidl}@cs.rwth-aachen.de
[2] Simon Fraser University, Burnaby, BC, Canada
ester@cs.sfu.ca

Abstract. Many real-world data sets, like data from social media or bibliographic data, can be represented as heterogeneous networks with several vertex types. Often additional attributes are available for the vertices, such as keywords for a paper. Clustering vertices in such networks, and analyzing the complex interactions between clusters of different types, can provide useful insights into the structure of the data. To exploit the full information content of the data, clustering approaches should consider the connections in the network as well as the vertex attributes. We propose the density-based clustering model TCSC for the detection of clusters in heterogeneous networks that are densely connected in the network as well as in the attribute space. Unlike previous approaches for clustering heterogeneous networks, TCSC enables the detection of clusters that show similarity only in a subset of the attributes, which is more effective in the presence of a large number of attributes.

1 Introduction

In many applications, data of various kinds are available, and there is a need for analyzing such data. Clustering, the task of grouping objects based on their similarity, is one of the most important data mining tasks, and clustering algorithms for different kinds of data exist. Graph clustering aims at grouping the vertices of a network into clusters such that many edges between vertices of the same cluster exist, i.e. the vertices are densely connected. While most graph clustering methods are constrained to homogeneous networks (networks with a single vertex type), real-world data can often better be represented by *heterogeneous* networks with several vertex types. For example, bibliographic data can be represented as a network with the vertex types "paper" and "author". When we consider heterogeneous networks, a novel challenge for clustering arises: Besides detecting clusters, clustering approaches should also analyze the interactions between clusters of different types, e.g. "which groups of authors are interested in which groups of papers?" Furthermore, real-world data often contains additional information ("attributes") about the vertices of a graph. E.g., a "paper" vertex can be further described by a vector of keywords. To exploit the full information content of the data, the similarity of the vertex attributes should be considered

T. Calders et al. (Eds.): ECML PKDD 2014, Part I, LNCS 8724, pp. 149–164, 2014.

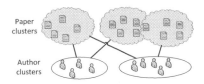

Fig. 1. Example author-paper clustering

for the clustering, as well as the connections in the network. An important aspect is that not all of the attributes may be relevant for a cluster. E.g., for a cluster of papers on similar topics, we would expect the papers to have *some*, but not *all* keywords in common. Thus, we aim at detecting clusters of vertices in heterogeneous networks that are densely connected and also show similarity in a subset of the attributes (called *subspace*), similar to the principle of *subspace clustering* for vector data [8]. To the best of our knowledge, there exists no previous approach for subspace clustering in heterogeneous networks.

In principle, it would be possible to project the network to a homogeneous network just containing vertices of one of the types (e.g. build a co-author network by connecting authors with common papers). However, by doing this information about the other types (e.g. the topics of the papers) is lost. Furthermore, an important aspect in our work is analyzing the connections between clusters of different types, which would not be possible at all in such a setting. In our experimental section, we show the superiority of our approach over a baseline using such a projection.

In our work, we consider heterogeneous networks that contain edges between vertices of different types (e.g. a paper is connected to its authors), but can also contain edges between vertices of the same type (e.g. citations between papers). Furthermore, for each of the vertex types there can be additional attributes. In such networks, we want to cluster the vertices of each type such that the clusters of different types interact with each other. An important challenge is how to model the interactions between the clusters. Intuitively, two clusters (of different types) are connected if the vertices of each cluster are densely connected *via the vertices of the other cluster*. Consider the example in Fig. 1. Here we observe two author clusters and three paper clusters (two of which are overlapping). The connections between the clusters indicate that the vertices of those clusters are connected by many edges. Naturally, a group of authors can publish papers about different topics, and also different groups of authors publish papers on the same topic. Thus it makes sense that an author cluster can be connected to several paper clusters and vice versa. Each cluster can interact with a different number of clusters. Therefore, just connecting each cluster to a specified number of other clusters would be problematic. Thus, in our approach the number of connections of a cluster is not restricted.

We observe that there are different ways to represent a data set as a heterogeneous network. Information about entities (e.g. words contained in papers) can be modeled in different ways: Either as an additional vertex type or as an attribute of another vertex type. In our work, we model only those information types as vertex types that we want to cluster. Other types of information are modeled as attributes. Furthermore, we want to highlight that there is no unique definition of the clustering problem in heterogeneous networks. The existing approaches vary greatly in their clustering objectives. Some approaches (e.g. [15]) cluster only the vertices of one type, while the vertices of other types are considered as "attribute types" of the clustered type. Other approaches (e.g. [13]) aim at clustering the vertices of all types such that each cluster contains vertices of different types. In Fig. 1, those approaches would either partition the authors into three clusters or merge two of the paper clusters in order to find a clustering of both types. Other approaches ([16], [17]) partition the vertices of each type separately, with the aim that the group membership of two vertices determines the probability of an edge between them. In this paper[1], we present the cluster model TCSC (Typed Combined Subspace Cluster), which belongs to the last, most general, category. In contrast to the previous approaches, TCSC additionally considers the similarity of vertices in subspaces of their attributes and allows the clusters to overlap, which makes sense in many applications. However, redundancy in the clustering result due to too much overlap is avoided by using a redundancy model. We introduce the algorithm HSC (Heterogeneous Subspace Clustering) for detecting TCSC clusters and evaluate it in experiments on real-world data sets.

2   Related Work

Combined Clustering of Graph and Attribute Data. Recently, several clustering approaches have been proposed that consider (homogeneous) graphs with vertex attributes. These approaches can be seen as a combination of graph clustering and vector clustering approaches. However, they mostly rely on fullspace-clustering on the vertex attributes (e.g. [12,20,19]) or only consider binary attributes [1]. The approaches [10,6,7] propose the combination of subspace clustering and graph clustering, aiming at finding clusters of vertices that are densely connected and as well show similar attribute values in a subset of their attributes. However, none of them considers heterogeneous networks.

Clustering in Heterogeneous Networks. The existing approaches for clustering in heterogeneous networks vary greatly in the types of networks that they consider, as well as in their clustering objectives. Several approaches [4,18] consider graphs with a single vertex type and multiple edge types. In some cases, such networks are called multi-dimensional [18] or multislice [11] networks. [4] considers graphs with multiple edge types and edge attributes. In such graphs, densely connected clusters are detected that also have similar attribute values. Other approaches [2,16] consider bipartite graphs: [2] defines a null model for modularity which considers bipartite networks and detects communities based on this measure. [16] proposes a new modularity measure for bipartite graphs, resulting in one partition of the vertices for each vertex type.

---

[1] The contents of this paper are also included in the first author's PhD thesis [3].

There also exist approaches that can handle graphs with an arbitrary number of vertex types: [5,15] cluster star-structured heterogeneous networks with one central type, where only the clustering of the vertices of the central type is optimized. In [14], networks with several vertex types are considered, which are not restricted to star-structured networks. The user has to specify a target type, i.e. a vertex type that should be clustered. The other types are called "feature types" and are used like attributes of the target type vertices. [13] considers a heterogeneous network with incomplete vertex attributes. The authors mention that only a subset of the attributes may be relevant for the clustering (similar to the idea of subspace clustering). However, in this approach the user has to specify the relevant set of attributes. In the resulting clustering, each cluster can contain vertices of every type. The clustering is mostly based on the attributes, while the links are only used to ensure a "structural consistency" (i.e. connected vertices are clustered together with higher probability). However, the resulting clusters do not have to be dense or even connected in the network. [9] propose a random-walk based approach for community detection in heterogeneous networks, which aims at finding a single community based on a set of seed items.

The most similar approach to our work is [17]. This approach considers evolving multi-mode networks, i.e. networks with different types of vertices that evolve over time. The vertices of each type are clustered simultaneously, with the aim that the group membership of two vertices determines their interaction. However, the approach does not provide information about connections between groups. Furthermore, clusters should evolve smoothly over the time steps. The proposed method only considers multi-partite networks, but extensions for considering edges between vertices of the same type and vertex attributes are mentioned. In our experimental section, we compare this approach (with the extension for using attributes) with our approach.

3   The TCSC Clustering Model

In this section, we introduce our TCSC model for clustering in heterogeneous networks. Basically, a cluster consists of a set of vertices of the same type that are densely connected via the vertices of the other types and show similar attribute values in a subset of their dimensions (this subset is called the *subspace* of the cluster). The model is partly based on the DB-CSC model [7] for homogeneous networks with vertex attributes. A new important challenge for heterogeneous networks is the detection of interactions between the clusters of different types.

For defining the clusters, we adopt the principle of density-based clustering, which allows the detection of dense clusters without restricting them to certain shapes or sizes. Basically, in density-based clustering clusters are defined as dense regions in the data space that are separated by sparse regions. In our work, we aim at detecting clusters that are not only dense considering their attribute values, but also are densely connected in the network via the vertices of their interacting clusters of other types. Thus, the clusters in our model correspond to dense regions in the graph as well as in a subspace of the attribute space. Therefore, we define the local neighborhood of a vertex such that it represents the graph neighborhood as well as the neighborhood in the attribute space.

Formally, the input for our model is a vertex-labeled graph with T different vertex types. Formally, $G = (V, E, t, l)$ with vertices V, edges E, a type indicator function $t : V \rightarrow \{1, \ldots, T\}$ and a vertex labeling function l. Let V_i denote the set of vertices of type i: $V_i = \{v \in V : t(v) = i\}$ and Dim_i the set of dimensions for type i, then $l : V_i \rightarrow \mathbb{R}^{|Dim_i|}$.

Neighborhood Definitions. For the clustering, we do not only consider vertices as neighbors that are directly connected by an edge, but also vertices that are connected via other vertices. For example, in a paper-author network we would consider two authors as neighbors if they are co-authors of a common paper, i.e. they are connected via two hops in the network. Therefore, we use the k-neighborhood of a vertex to define its local neighborhood in the graph. Formally, the graph k-neighborhood is defined as follows:

Definition 1 (Graph k-neighborhood). *A vertex u is k-reachable from a vertex v (over a set of vertices V) if $\exists v_1, \ldots, v_k \in V : v_1 = v \wedge v_k = u \wedge \forall j \in \{1, \ldots, k-1\} : (v_j, v_{j+1}) \in E$. The graph k-neighborhood of vertex $v \in V$ is given by: $N_k^V(v) = \{u \in V \mid u$ is j-reachable from v (over V) $\wedge j \leq k\} \cup \{v\}$.*

Furthermore, we define the ϵ-neighborhood of a vertex in the attribute space. Naturally, this neighborhood can only contain vertices of the same type:

Definition 2 (ϵ-neighborhood). *The distance between two vertices x and y of type i w.r.t. a subspace $S \subseteq Dim_i$ is defined as the maximum norm[2] $dist^S(x, y) = \max_{d \in S} |x[d] - y[d]|$ with the special case $dist^\emptyset(x, y) = 0$. The ϵ-neighborhood of $v \subset V$ for a subspace $S \subseteq Dim_{t(v)}$ is defined as: $N_{\epsilon,S}^V(v) = \{u \in V_{t(v)} \mid dist^S(l(u), l(v)) \leq \epsilon\}$*

As we want to consider the connections in the graph and the similarity in the attribute space simultaneously, we define a combined local neighborhood:[3]

Definition 3 (Combined local neighborhood). *The combined neighborhood of $v \in V$ w.r.t. a subspace $S \subseteq Dim_{t(v)}$ is defined as: $N_S^V(v) = N_k^V(v) \cap N_{\epsilon,S}^V(v)$*

This neighborhood contains only vertices of the same type. This makes sense for the clustering as each cluster should only contain vertices of a single type. In Fig. 2, the combined neighborhood for $k = 2, \epsilon = 1$ of vertex A considering only dimension 1 would be $N_{\{1\}}^V(A) = \{A, B, C, D, E\}$. For dimension 2, $N_{\{2\}}^V(A) = \{A, C, D, E, F\}$.

Modeling Clusters and their Interactions. In our cluster definition, we have to ensure that all objects in a cluster are dense w.r.t. the combined neighborhood

---

[2] We choose the maximum norm because we want to consider two vertices similar only if they are similar in *all* of the dimensions in S.

[3] Another idea would be to combine the graph and attribute distance into a unified distance function. However, in that case a very small distance in the graph could even out a larger distance in the attribute space and vice versa. Instead, we want the vertices in the combined neighborhood to be similar in both regards.

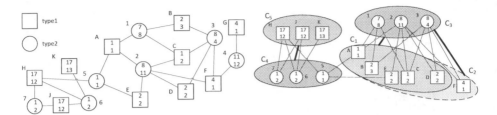

Fig. 2. Example with two vertex types   **Fig. 3.** Clustering for the example network

(ensured by property (1) in Def. 4) and the cluster is density-connected via the neighborhoods (property (2)). Furthermore, we want to detect the interactions between clusters of different types. Intuitively, two clusters of different types should be "connected" to each other if many edges exist between the vertices of these clusters. In other words, the connections to the vertices of the other cluster should induce density in a cluster. Therefore, we define a cluster C of vertices of type i *w.r.t. a set of clusters of the other types.* We call this set of clusters the *adjacent clusters* of C, denoted by $A(C)$. C has to be dense w.r.t. the union of the adjacent clusters. Therefore, the combined neighborhood is computed using only the vertices of the cluster itself and the union of the adjacent clusters:

Definition 4 (Typed Combined Subspace Cluster). *A typed combined subspace cluster $C = (O, S)$ of type i in a graph $G = (V, E, t, l)$ w.r.t. the parameters k_i, ϵ_i and $minPts_i$ and a set of adjacent clusters $A(C) = \{(O_j, S_j) \mid O_j \subseteq (V \setminus V_i), j = 1, \ldots, |A(C)|\}$ consists of a set of vertices $O \subseteq V_i$ and a set of relevant dimensions $S \subseteq Dim_i$*[4] *that fulfill the following properties:*

(1) *density:* $\forall v \in O : |N_S^{O \cup (\cup_{1 \leq j \leq a_C} O_j)}(v)| \geq minPts_i$

(2) *connectivity:* $\forall u, v \in O : \exists w_1, \ldots, w_l \in O : w_1 = u \wedge w_l = v \wedge \forall i \in \{1, \ldots, l-1\} : w_i \in N_S^{O \cup (\cup_{1 \leq j \leq a_C} O_j)}(w_{i+1})$

(3) *density w.r.t. all adjacent clusters:* $\forall (O_j, S_j) \in A : \exists W \subseteq O : (W, S)$ *forms a cluster w.r.t. the set $A(W) = \{(O_j, S_j)\}$ (for non-bipartite graphs: ignoring the edges $(u, v) \in E : u, v \in O$)*

(4) *reciprocity:* $\forall C_j \in A(C) : C \in A(C_j)$

(5) *maximality:* $\neg \exists O' \supset O : O'$ *fulfills (1) and (2)*

To avoid adding unrelated clusters, we require that each adjacent cluster has to induce density in at least a subset of C (property (3)). If there exist edges between vertices of the same type, we ignore them for testing this density (else, the cluster could be dense considering those edges alone, and thus the cluster definition would be fulfilled w.r.t. arbitrary other clusters). To avoid adding

---

[4] Generally, we require the subspace S to be non-empty, such that the vertices of the cluster show similarity in at least one dimension. However, in a heterogeneous network it is possible that not all of the vertex types have attributes. In this case, it makes sense to allow the detection of clusters with empty subspace. Thus, the user can choose if clusters with empty subspace should be included in the result.

clusters that just incidentally induce density in a small subset of C, we require a reciprocity of the adjacency between clusters (property (4)). Please note that we do *not* require a maximality property on $A(C)$ and redundant clusters can be removed from $A(C)$ later. Thus, a cluster C can fulfill the cluster definition for different sets $A(C)$. How to finally select $A(C)$ for the clustering result is discussed below.

The example network in Fig. 2 contains the following clusters (shown in Fig. 3) for the parameters $k_1 = k_2 = 2, \epsilon_1 = \epsilon_2 = 1, minPts_1 = minPts_2 = 3$:

- $C_1 = (\{A, B, C, D, E\}, \{1, 2\})$, connected to C_3
- $C_2 = (\{A, B, C, D, E, F\}, \{2\})$, connected to C_3
- $C_3 = (\{1, 2, 3\}, \{1\})$, connected to C_1, C_2
- $C_4 = (\{H, J, K\}, \{1, 2\})$, connected to C_5
- $C_5 = (\{5, 6, 7\}, \{1, 2\})$, connected to C_4

Interestingness of a TCSC Cluster. As we detect clusters in different subspaces, we can possibly find quite similar clusters in similar subspaces, like C_1 and C_2. To avoid redundancy in the result, we have to be able to decide which of the clusters is more interesting for our clustering result. Generally, we consider clusters with many vertices as interesting. However, a higher dimensionality also makes a cluster more interesting. Therefore, we introduce an interestingness function for clusters that considers both criteria. The interestingness function for a typed cluster is normalized by the overall number of vertices and the dimensionality of the corresponding type:

Definition 5 (Interestingness Measure). *The interestingness of a TCSC cluster $C = (O, S)$ of type i is defined as $Q(C) = \frac{|O| \cdot |S|}{|V_i| \cdot |Dim_i|}$ if $|Dim_i| > 0$, and $Q(C) = \frac{|O|}{|V_i|}$ else.*

In our example, $Q(C_1) = 0.35$ and $Q(C_2) = 0.3$. Thus, the two-dim. cluster C_1 is preferred as it is only slightly smaller than the similar one-dim. cluster C_2.

Parameters. Our model requires several parameters: ϵ, $minPts$ and k have to be set for each type. Setting these parameters for each type separately leads to a greater flexibility of the model: Especially if the number of vertices for the different types strongly differs, we should not expect the clusters of each type to fulfill e.g. the same $minPts$ value. In Section 5, we present a method for finding good parameter settings for ϵ and $minPts$. For k, a suitable setting can be directly obtained from the structure of the graph. E.g. for a bipartite paper-author network, $k_{author} = 2$ is a good choice as we need two hops to reach one author from another author. If there exist intra-type edges, we can consider them additionally by setting $k_{author} = 3$. For networks with more types, the distances between vertices of the same type can be larger, if vertices of different types have to be traversed. In this case, higher k values are required. Unlike other approaches, our model does not require the number of clusters as a parameter.

Selecting the Final Clustering Result. In order to avoid redundant clusters in the result, we first define a binary redundancy relation between two clusters of the same type, adopting the definition from the DB-CSC model [7]. A cluster is considered redundant w.r.t. another cluster if its quality is lower and the clusters show a high overlap in their vertex sets as well as their relevant subspaces:

Definition 6 (Redundancy between clusters). *Given the redundancy parameters* $r_{obj}, r_{dim} \in [0,1]$, *the binary redundancy relation* $\prec_{red}$ *is defined by:*
For all clusters $C = (O, S), C' = (O', S')$: $C \prec_{red} C' \Leftrightarrow Q(C) < Q(C') \wedge$
$\frac{|O \cap O'|}{|O|} \geq r_{obj} \wedge \frac{|S \cap S'|}{|S|} \geq r_{dim}$

In Fig. 3, $C_2 \prec_{red} C_1$ (e.g. for $r_{obj} = r_{dim} = 0.5$).

Now we can define the final result set, which should not contain clusters that are redundant w.r.t. each other and has to be maximal w.r.t. this property. Furthermore, we ensure a maximality property for the set of adjacent clusters for a cluster C: If $C, A(C)$ together form a cluster according to Def. 4, then in the TCSC clustering the set of adjacent clusters of C must contain all the clusters in $A(C)$ except those that are redundant w.r.t. another cluster in $A(C)$.

Definition 7 (TCSC clustering). *Given the set Clusters of all TCSC clusters, the resulting TCSC clustering Result* $\subseteq$ *Clusters fulfills*

- *redundancy-freeness:* $\neg \exists C_i, C_j \in Result : C_i \prec_{red} C_j$
- *maximality:* $\forall C_i \in Clusters \setminus Result : \exists C_j \in Result : C_i \prec_{red} C_j$
- *maximality for adjacent clusters:* $\forall C : \forall C_i \in \{C_x \mid C \text{ fulfills Def. 4 w.r.t.}$ $A(C) \cup \{C_x\}\} \setminus A(C) : \exists C_j \in A(C) : C_i \prec_{red} C_j$

Furthermore, we have to consider the connections between clusters of different types. A cluster $C \in Result$ may be adjacent to a cluster of another type that is excluded from the result due to redundancy. In this case, by just deleting this cluster we would lose the information about this connection. For solving this problem, we use the following theorem:

Theorem 1. *For the clustering defined in Def. 7 it holds:* $\forall C \in Result : \forall C_A \in A(C) : (C_A \in Result \vee \exists C_{A'} \in Result : C_A \prec_{red} C_{A'})$.

Proof Assume $\exists C \in Result, C_A \in A(C), C_A \in Clusters \setminus Result$. Following the maximality property in Def. 7 it holds $\exists C_{A'} \in Result : C_A \prec_{red} C_{A'}$.

I.e., if an adjacent cluster C_A is excluded from the result, there exists a similar cluster $C_{A'}$ that is contained in the result. In our implementation, we "reconnect" the cluster C to $C_{A'}$, if this connection does not yet exist. In Fig. 3, the result is $\{C_1, C_3, C_4, C_5\}$. $\square$

4   The HSC Algorithm

In this section we give a short overview of the HSC algorithm. While HSC is partly based on DB-CSC [7], for heterogeneous networks novel challenges arise, which we discuss in this section. First, we explain the overall processing of the algorithm, followed by a detailed description of the refinement of a cluster.

```
method: main()
 1  Result = ∅ // current result set
 2  queue = ∅ // priority queue, sorted by quality
 3  for i = 1, · · · , T do A(Vᵢ) = {Vⱼ | 1 ≤ j ≤ T, j ≠ i}
 4  Detect set netclus of network-only clusters
 5  if network-only clusters are allowed then
 6  ⌊  add all network clusters to queue

 7  for C ∈ netclus, d ∈ Dim_{t(O)} do  DFS_trav(O, {d})
 8  repeat
 9  │   Sort queue ascendingly by dimensionality
10  │   for C = (O, S) ∈ queue : A(C) has changed do
11  │   ⌊  refine_cluster(C)

12  until adjacency between clusters converges
13  while queue ≠ ∅ do
14  │   remove first cluster Clus from queue
15  │   if ∃C = (O, S) ∈ Result : Clus ≺_red C  then
16  │   │   "reconnect" Clus's connections to C
17  │   ⌊   goto line 13 // discard redundant cluster

18  ⌊  Result = Result ∪ {Clus}

19  return Result
method: DFS_trav(vertex set O of type t, subspace S)
20  foundClusters = refine_cluster(C = (O, S))
21  add foundClusters to queue
22  for Cᵢ = (Oᵢ, S) ∈ foundClusters do
23  │   for d ∈ {max{S} + 1, . . . , Dim_t} do
24  │   ⌊   DFS_trav(Oᵢ, S ∪ {d}) // check subsets of Oᵢ
```

Algorithm 1. Pseudo-Code for the HSC algorithm

The pseudo-code for HSC is given in Algorithm 1. The final result set *Result* is initialized as an empty set (line 1), which is then filled iteratively by the algorithm until it contains the final, non-redundant clustering result defined in Def. 7. However, when a TCSC cluster C is detected during the processing, it can not directly be decided if C should be added to the clustering result, as a higher-quality cluster C' could be detected later such that $C \prec_{red} C'$ holds. Therefore, all detected clusters are temporarily stored in a priority queue *queue* (initialized in line 2) that is sorted according to the interestingness of the clusters.

From the definition of the combined neighborhood and the subspace distance, it follows that our TCSC clusters fulfill an anti-monotonicity property w.r.t. the subspaces, i.e. if there exists a cluster $C = (O, S)$ in subspace S, then for each subspace $S' \subset S$ there exists a vertex set $O' \supset O$ such that (O', S') also forms a TCSC cluster. This property can be exploited algorithmically: In order to detect clusters in higher-dimensional subspaces, the algorithm only has to check subsets of the vertex sets of the detected lower-dimensional clusters. Therefore, HSC starts by detecting clusters based only on the network information (i.e. clusters with the empty subspace). However, for heterogeneous networks, already the detection of these "network-only" clusters is a challenging problem. Because a cluster is defined based on its adjacent clusters of the other types, we can not simply determine the clusters for each type separately. Thus, the algorithm has to iteratively update the clusters and connections. A small example for the iterative processing on a paper-author network is illustrated in Fig. 4. If the

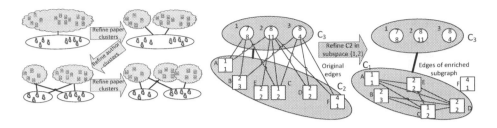

Fig. 4. Example for the detection of network clusters   **Fig. 5.** Example for typed enriched subgraph

user allows clusters with empty subspace in the result, the network clusters are added to the *queue* (line 6). Based on the detected "network-only" clusters, HSC can now detect clusters in higher-dimensional subspaces (line 7). Based on these clusters, a depth-first search is performed in the subspaces[5] (line 20 – 24). Here, $max\{S\}$ denotes the dimension with maximal ID in subspace S.

Also for the higher-dimensional clusters, we encounter the challenge that the clusters of the different types depend on each other. During the DFS traversal, each cluster was refined based on the adjacent clusters that were currently known at the time of refinement. For many clusters in the queue, the set of adjacent clusters may have changed. Therefore, these clusters are refined based on the updated set of adjacent clusters (line 10 – 11). This process is repeated until the sets of adjacent clusters do not change anymore (line 12). For this step, the clusters are sorted ascendingly by dimensionality, as clusters of lower dimensionality tend to be connected to more clusters of the other types. After the set of clusters and adjacencies has converged, HSC detects the final clustering by processing the priority queue (sorted by quality), discarding clusters that are redundant w.r.t. a cluster already in the result set and adding non-redundant clusters to the result (line 18).

Refinement Based on Adjacent Clusters. Given a cluster candidate (i.e. a set of vertices O, a set of adjacent clusters and a subspace), the refinement method for the detection of TCSC clusters returns the set of all valid TCSC clusters with vertex sets $O' \subseteq O$ based on this subspace and adjacent clusters. The refinement method is based on a structure named *typed enriched subgraph*, which represents the similarity of the attributes in the considered subspace as well as the connectedness of the vertices via the adjacent clusters. In the typed enriched subgraph for a vertex set O, each vertex is connected to all vertices from its combined neighborhood (Def. 3), which is determined based on a given subspace S and using the connections via the vertices of O itself and of a vertex set O_A, which represents the vertices of all adjacent clusters:

---

[5] In practice, we can save computations and avoid the detection of some redundant clusters by using a technique from [7] that avoids the traversal of subspaces where probably only redundant clusters will be detected. Its details can be found in [7].

```
method: refine_cluster(cand. C = (O, S) of type t)
  1   foundClusters = ∅, prelimClusters = {C}
  2   for C_A ∈ A(C) do remove C from A(C_A)
  3   while prelimClusters ≠ ∅ do
  4        remove first C_x = (O_x, S) from prelimClusters
  5        compute adj. vertex set O_A = ∪_(O_i,S_i)∈A(C_x) O_i
  6        generate typed enriched subgraph G^(O_x)_(S,O_A)
  7        find (minPts_t − 1)-cores Cores = {O'_1, . . . , O'_m}
  8        for each core O'_i determine adjacent clusters A(O'_i)
  9        if |Cores| = 1 ∧ O'_1 = O_x then
 10        └─  foundClusters = foundClusters ∪ {(O'_1, S)}
 11        else prelimClusters = prelimClusters ∪ Cores
 12   return foundClusters
```

Algorithm 2. Method for refining a single cluster

Definition 8 (Typed Enriched Subgraph). [6] *Given a set of vertices $O \subseteq V_i$, a subspace S, the original graph $G = (V, E, l)$ and a vertex set $O_A \subseteq V \setminus V_i$, the enriched subgraph $G^O_{S,O_A} = (V', E')$ w.r.t. O_A is defined by $V' = O$ and $E' = \{(u, v) \mid v \in N^{O \cup O_A}_S(u) \wedge v \neq u\}$ using the distance function $dist^S$.*

To fulfill the density property, each vertex in a TCSC cluster of type t has to have at least $minPts_t$ vertices in its combined neighborhood (which also contains the vertex itself). In the enriched subgraph, a TCSC cluster thus corresponds to a $(minPts_t − 1)$-core. In [7], it has been shown that the combined clusters can be detected by iteratively detecting $(minPts − 1)$-cores. Our method for finding TCSC clusters works in a similar fashion. The pseudo-code for the refinement method is given in Algorithm 2. If a candidate $C = (O, S)$ is refined, first the connection to C is removed from its adjacent clusters, which will later be connected to the new clusters detected in subsets of O (line 2). Then, HSC iteratively detects clusters in subsets of O (line 5 − 7). If O was not changed by the core-detection (line 9), the refinement has converged and the found vertex set is a cluster. However, if one or several smaller cores were detected, they cannot directly be output as clusters, because their adjacent clusters may have changed. Thus, the procedure is repeated (line 11) until convergence.

Fig. 5 shows an example for the graph from Fig. 2. Assume C_2 (in subspace $\{2\}$) and C_3 and their connection have already been detected. Now, we want to refine C_2 for the subspace $\{1, 2\}$. Thus, we construct the typed enriched subgraph for the vertices of C_2 based on this subspace and the vertices of C_3, and obtain the new cluster $\{A, B, C, D, E\}$ for the subspace $\{1, 2\}$.

5   Experiments

In this section we evaluate the performance of HSC (implemented in Java). We compare HSC to the algorithm of Tang et al. [17], the only existing algorithm for

---

[6] Please note that the typed enriched subgraph of a vertex set contains only vertices of the same type. However, it is determined using the combined neighborhood, which takes the connections to the adjacent clusters into account.

clustering heterogeneous networks that also separates vertices of different types into different clusters and (in its extension) considers vertex attributes.[7]

Data Sets. For our experiments, we use two heterogeneous real-world data sets that each contain several vertex attributes. Yelp is a website where users can rate and review businesses. Our yelp network was extracted from the yelp academic data set (http://www.yelp.com) and has three vertex types: "User", "Business" and "Review". The network contains all the businesses from the academic data set that belong to the categories "Restaurant" and "Pizza", all the users who rated at least one of these businesses and all of the corresponding reviews. Overall, there are 6931 user vertices, 283 business vertices and 8584 review vertices. A review vertex is connected to the user who submitted this review as well as to the business it rates. Furthermore, the network also contains intra-type edges and thus is not tripartite: Two businesses are connected by an edge if they are located close to each other (up to 300m apart). For all vertex types, the data set provides additional attribute information: For the businesses, we have the attributes "review count" (number of ratings received) and "average rating" (from 1 to 5 stars). For the users, we also have "review count" (number of ratings submitted) and "average rating" (of the ratings by this user). Furthermore, user vertices have the attributes "funny", "useful" and "cool", which correspond to attributes given to the reviews of a user by other users. Review vertices have the single attribute "stars". All values were normalized to $[0, 1]$.

Our second network was extracted from the DBLP (http://dblp.uni-trier.de) database and has the vertex types "Author" and "Paper". It contains all papers of selected conferences from the database and data mining area from the years 2000 to 2004. Each paper is connected to the vertices representing its authors. Authors vertices do not contain attributes. The papers have binary attributes indicating the occurrence of certain keywords in the title. To avoid irrelevant keywords, only words that occurred in at least 100 papers are represented. Overall, we have 5497 author vertices and 3354 paper vertices with 208 attributes.

Experiments on Yelp Data. As no ground truth for the clustering is available, we can not evaluate the clustering quality directly. Therefore, we divide the edges of the Yelp data set into a training set (95% of the edges) and a test set (5% of the edges). Using these data sets, we obtain an accuracy measure that measures how well the test edges are predicted by the result of HSC, i.e. which percentage of the test edges connect vertices from adjacent clusters in our result. We also create a test set of edges between vertices that are not connected in the network and use this set to obtain a "false positive" rate. Please note that we can not expect to reach perfect or nearly perfect accuracy values using this measure, as this would only be possible for graphs where the clusters correspond to fully connected cliques and no edges between clusters exist, which does not hold for real graphs. However, we can use this measure to compare how well different clustering results capture the structure of the graph. Using this method, we analyze the influence

---

[7] We extended the implementation from the authors homepage for attribute values as described in [17] and treat our networks as networks with a single time stamp.

of parameters on the result. In each experiment, we measure the percentage of correctly predicted edges and of correctly predicted non-existing edges. This method can be used for finding good parameter setting. For each parameter, we choose the value maximizing the minimum of both accuracy values. For HSC, we vary the values for ϵ and $minPts$ for each type. The parameter k is discrete and is set as discussed in Section 3. For the method from [17], the number of clusters for each type has to be given as a parameter, thus we vary these values. However, for this method computing the accuracy is problematic, as it does not produce binary connections between clusters. Instead, we consider the group interaction matrix A that is used by [17] and interpret positive entries as "connection" and negative entries as "no connection".

To analyze the advantage of our heterogeneous clustering method over clustering methods for homogeneous graphs, we test a baseline that projects the heterogeneous network to one homogeneous network for each vertex type (e.g. connecting two users if they are connected to the same business in the heterogeneous network) and then uses DB-CSC [7] on each network separately. To enable a comparison with the results of HSC, this baseline detects connections between the clusters from the different networks in a post-processing step. To analyze the influence of subspace clustering on our results, we also test a fullspace version of HSC that detects only clusters that show similarity in all dimensions.

The results for the "user" vertex type are shown in Fig. 6. The experiments for the other vertex types show similar results (not printed here due to space limitations). For the parameter $minPts_{user}$, increasing values lead to a lower accuracy for the test edges and a higher accuracy for the non-existing test edges (Fig. 6(a)). This is due to the fact that for higher $minPts$ values, less and smaller TCSC clusters are detected due to the stricter density criterion. Therefore, less correct edges, but also less non-existing edges are predicted by the clustering result. According to this results, we set $minPts_{user} = 50$. The baseline clustering the homogeneous projections of the network separately reaches very similar values in the accuracy for test edges. However, the accuracy for non-existing test edges is considerably worse than for HSC. This is due to the fact that this method cannot use the information about the clustering structure of the other vertex type, and thus also clusters vertices together that are connected via noise vertices of the other type. Therefore, the resulting clusters are supersets of the TCSC clusters and the predicted connections show worse accuracy for non-existing test edges. We also evaluate the fullspace version of HSC in this experiment. We observe that only few clusters can be found in the fullspace and thus the accuracy value for existing edges is very low. For values greater than 25, the fullspace version detects no clusters at all. Therefore, the fullspace version is not used in the following experiments. The runtimes for all versions (Fig. 6(b)) decrease for increasing $minPts$ values, as less clusters are detected. The runtimes of HSC are slightly higher than those of the fullspace version of HSC, as the fullspace version does not have to look for clusters in different subspaces. The method clustering homogeneous networks shows far higher runtimes, as it

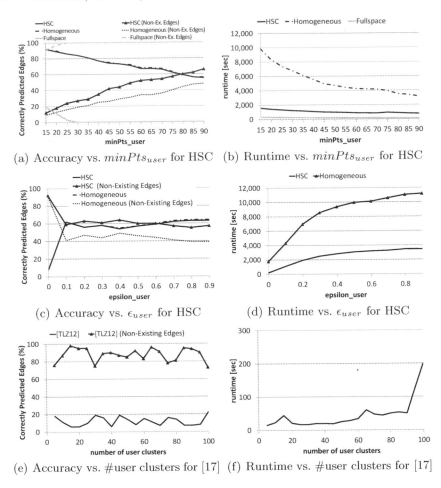

(a) Accuracy vs. $minPts_{user}$ for HSC   (b) Runtime vs. $minPts_{user}$ for HSC

(c) Accuracy vs. ϵ_{user} for HSC   (d) Runtime vs. ϵ_{user} for HSC

(e) Accuracy vs. #user clusters for [17]   (f) Runtime vs. #user clusters for [17]

Fig. 6. Experimental results on the Yelp data set with 3 vertex types

has to cluster all networks separately and cannot use the information about the clustering structure of the respective other vertex types for pruning.

For an increasing ϵ_{user}, both accuracy values for HSC remain relatively stable (Fig. 6(c)). Like in the previous experiment, the homogeneous clustering variant shows similar behavior in the accuracy for test edges and considerably worse values for the accuracy for non-existing test edges. The runtime (Fig. 6(d)) increases quickly for increasing ϵ-values until it reaches a plateau, as for higher ϵ-values larger vertex neighborhoods have to be considered. Again, the runtime of the homogeneous variant is far higher than that of HSC.

For the method from [17], we do not observe a trend in the accuracy for an increasing number of user clusters (Fig. 6(e)). In contrast to HSC, in this partitioning method each vertex is grouped in exactly one cluster, thus the trend described above does not occur here. The accuracy for test edges is considerably lower than that of HSC, while a high accuracy for non-existing edges is obtained.

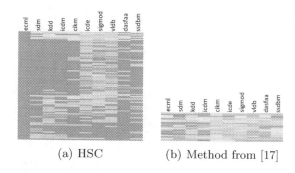

(a) HSC   (b) Method from [17]

Fig. 7. Distribution of paper clusters over conferences for the DBLP data set (heat map color gradient from Green= "0%" to Red= "100% of the papers in the cluster belong to this conference")

This shows that this approach predicts less connections between clusters for the 3-type network. Overall, the method shows far lower runtimes than HSC (Fig. 6(f)), as it does not consider subspaces of the attribute space and does not exclude outliers. However, HSC reaches better accuracy values: We can find parameter settings such that both accuracy values for HSC are about 60%.

Experiments on DBLP Data. On DBLP, HSC detects 14 author clusters with an average size of 59 and 98 paper clusters with an average size of 10 and average dimensionality of 1.05, i.e. most of the paper clusters have one or two keywords in common. The method from [17] detects 20 paper clusters with an average size of 168 and 20 author clusters with an average size of 275. As this method does not consider subspace clusters, there is no information about the relevant keywords for the clusters; the papers in a cluster do not necessarily have common keywords at all. However, considering subspaces and connections between clusters also causes higher runtimes: The runtime was 5 sec. for the method from [17] and 278 sec. for HSC. To provide an impression of the detected clusters, we depict the distribution of the detected paper clusters over the conferences in Fig. 7. Each row in the diagrams corresponds to a paper cluster. For most of the clusters detected by HSC, the papers in the cluster belong to a small set of conferences. Particularly, most clusters show a clear tendency either to the database area or the data mining area. For the method from [17], the tendency is less clear.

6 Conclusion

We propose the clustering model TCSC for the clustering of vertices in heterogeneous networks, which takes into account the connections in the network as well as the vertex attributes. Furthermore, TCSC detects interactions between clusters of different types. TCSC is the first clustering model which considers subspace clustering in heterogeneous networks. We introduce the algorithm HSC for computing the TCSC clustering.

Acknowledgements. This work has been supported by the B-IT Research School of the Bonn-Aachen International Center for Information Technology.

References

1. Akoglu, L., Tong, H., Meeder, B., Faloutsos, C.: Pics: Parameter-free identification of cohesive subgroups in large attributed graphs. In: Proceedings of the Twelfth SIAM International Conference on Data Mining, pp. 439–450 (2012)
2. Barber, M.: Modularity and community detection in bipartite networks. Phys. Rev. E 76(6), 066102 (2007)
3. Boden, B.: Combined Clustering of Graph and Attribute Data. Ph.D. thesis, RWTH Aachen University, Aachen (2014)
4. Boden, B., Günnemann, S., Hoffmann, H., Seidl, T.: Mining coherent subgraphs in multi-layer graphs with edge labels. In: SIGKDD, pp. 1258–1266 (2012)
5. Gao, B., Liu, T., Zheng, X., Cheng, Q., Ma, W.: Consistent bipartite graph co-partitioning for star-structured high-order heterogeneous data co-clustering. In: SIGKDD, pp. 41–50 (2005)
6. Günnemann, S., Färber, I., Boden, B., Seidl, T.: Subspace clustering meets dense subgraph mining: A synthesis of two paradigms. In: ICDM (2010)
7. Günnemann, S., Boden, B., Seidl, T.: Finding density-based subspace clusters in graphs with feature vectors. DMKD 25(2), 243–269 (2012)
8. Kriegel, H.P., Kröger, P., Zimek, A.: Clustering high-dimensional data: A survey on subspace clustering, pattern-based clustering, and correlation clustering. TKDD 3(1), 1–58 (2009)
9. Li, X., Ng, M.K., Ye, Y.: Multicomm: Finding community structure in multi-dimensional networks. TKDE 99(PrePrints), 1 (2013)
10. Moser, F., Colak, R., Rafiey, A., Ester, M.: Mining cohesive patterns from graphs with feature vectors. In: SDM, pp. 593–604 (2009)
11. Mucha, P.J., Richardson, T., Macon, K., Porter, M.A., Onnela, J.P.: Community structure in time-dependent, multiscale, and multiplex networks. Science 328(5980), 876–878 (2010)
12. Shiga, M., Takigawa, I., Mamitsuka, H.: A spectral clustering approach to optimally combining numerical vectors with a modular network. In: SIGKDD, pp. 647–656 (2007)
13. Sun, Y., Aggarwal, C., Han, J.: Relation strength-aware clustering of heterogeneous information networks with incomplete attributes. VLDB 5(5), 394–405 (2012)
14. Sun, Y., Norick, B., Han, J., Yan, X., Yu, P., Yu, X.: Integrating meta-path selection with user-guided object clustering in heterogeneous information networks. In: SIGKDD, pp. 1348–1356 (2012)
15. Sun, Y., Yu, Y., Han, J.: Ranking-based clustering of heterogeneous information networks with star network schema. In: SIGKDD, pp. 797–806 (2009)
16. Suzuki, K., Wakita, K.: Extracting multi-facet community structure from bipartite networks. In: CSE, vol. 4, pp. 312–319 (2009)
17. Tang, L., Liu, H., Zhang, J.: Identifying evolving groups in dynamic multimode networks. TKDE 24(1), 72–85 (2012)
18. Tang, L., Wang, X., Liu, H.: Community detection via heterogeneous interaction analysis. DMKD 25(1), 1–33 (2012)
19. Yang, J., McAuley, J.J., Leskovec, J.: Community detection in networks with node attributes. In: ICDM, pp. 1151–1156 (2013)
20. Zhou, Y., Cheng, H., Yu, J.X.: Graph clustering based on structural/attribute similarities. PVLDB 2(1), 718–729 (2009)

Open Question Answering with Weakly Supervised Embedding Models

Antoine Bordes[1], Jason Weston[2], and Nicolas Usunier[1]

[1] Université de Technologie de Compiègne – CNRS,
Heudiasyc UMR 7253, Compiègne, France
[2] Google, 111 8th avenue, New York, NY, USA
{bordesan,nusunier}@utc.fr, jaseweston@gmail.com

Abstract. Building computers able to answer questions on any subject is a long standing goal of artificial intelligence. Promising progress has recently been achieved by methods that learn to map questions to logical forms or database queries. Such approaches can be effective but at the cost of either large amounts of human-labeled data or by defining lexicons and grammars tailored by practitioners. In this paper, we instead take the radical approach of learning to map questions to vectorial feature representations. By mapping answers into the same space one can query any knowledge base independent of its schema, without requiring any grammar or lexicon. Our method is trained with a new optimization procedure combining stochastic gradient descent followed by a fine-tuning step using the weak supervision provided by blending automatically and collaboratively generated resources. We empirically demonstrate that our model can capture meaningful signals from its noisy supervision leading to major improvements over PARALEX, the only existing method able to be trained on similar weakly labeled data.

Keywords: natural language processing, question answering, weak supervision, embedding models.

1 Introduction

This paper addresses the challenging problem of open-domain question answering, which consists of building systems able to answer questions from any domain. Any advance on this difficult topic would bring a huge leap forward in building new ways of accessing knowledge. An important development in this area has been the creation of large-scale Knowledge Bases (KBs), such as FREEBASE [4] and DBPEDIA [16] which store huge amounts of general-purpose information. They are organized as databases of triples connecting pairs of entities by various relationships and of the form (left entity, relationship, right entity). Question answering is then defined as the task of retrieving the correct entity or set of entities from a KB given a query expressed as a question in natural language.

The use of KBs simplifies the problem by separating the issue of collecting and organizing information (i.e. *information extraction*) from the one of searching

T. Calders et al. (Eds.): ECML PKDD 2014, Part I, LNCS 8724, pp. 165–180, 2014.

through it (i.e. *question answering* or *natural language interfacing*). However, open question answering remains challenging because of the scale of these KBs (billions of triples, millions of entities and relationships) and of the difficulty for machines to interpret natural language. Recent progress [6,3,13,10] has been made by tackling this problem with semantic parsers. These methods convert questions into logical forms or database queries (e.g. in SPARQL) which are then subsequently used to query KBs for answers. Even if such systems have shown the ability to handle large-scale KBs, they require practitioners to hand-craft lexicons, grammars, and KB schema for the parsing to be effective. This non-negligible human intervention might not be generic enough to conveniently scale up to new databases with other schema, broader vocabularies or other languages than English.

In this paper, we instead take the approach of converting questions to (un-interpretable) vectorial representations which require no pre-defined grammars or lexicons and can query any KB independent of its schema. Following [10], we focus on answering simple factual questions on a broad range of topics, more specifically, those for which single KB triples stand for both the question and an answer (of which there may be many). For example, (`parrotfish.e`, `live-in.r`, `southern-water.e`) stands for *What is parrotfish's habitat?* and `southern-water.e` and (`cantonese.e`, `be-major-language-in.r`, `hong-kong.e`) for *What is the main language of Hong-Kong?* and `cantonese.e`. In this task, the main difficulties come from i) lexical variability (rather than from complex syntax), ii) having multiple answers per question, and iii) the absence of a supervised training signal.

Our approach is based on learning low-dimensional vector embeddings of words and of KB triples so that representations of questions and correspond-ing answers end up being similar in the embedding space. Unfortunately, we do not have access to any human labeled (query, answer) supervision for this task. In order to avoid transferring the cost of manual intervention to the one of labeling large amounts of data, we make use of weak supervision. We show empirically that our model is able to take advantage of noisy and indirect su-pervision by (i) automatically generating questions from KB triples and treating this as training data; and (ii) supplementing this with a data set of questions collaboratively marked as paraphrases but with no associated answers. We end up learning meaningful vectorial representations for questions involving up to 800k words and for triples of an mostly automatically created KB with 2.4M entities and 600k relationships. Our method strongly outperforms previous re-sults on the WikiAnswers+ReVerb evaluation data set introduced by [10]. Even if the embeddings obtained after training are of good quality, the scale of the optimization problem makes it hard to control and to lead to convergence. Thus, we propose a method to fine-tune embedding-based models by carefully optimizing a matrix parameterizing the similarity used in the embedding space, leading to a consistent improvement in performance.

The rest of the paper is organized as follows. Section 2 discusses some previous work and Section 3 introduces the problem of open question answering. Then, Section 4 presents our model and Section 5 our experimental results.

2   Related Work

Large-scale question answering has a long history, mostly initiated via the TREC tracks [23]. The first successful systems transformed the questions into queries which were fed to web search engines, the answer being subsequently extracted from top returned pages or snippets [14,1]. Such approaches require significant engineering to hand-craft queries and then parse and search over results.

The emergence of large-scale KBs, such as FREEBASE [4] or DBPEDIA [16], changed the setting by transforming open question answering into a problem of querying a KB using natural language. This is a challenging problem, which would require huge amount of labeled data to be tackled properly by purely supervised machine learning methods because of the great variability of language and of the large scale of KBs. The earliest methods for open question-answering with KBs, based on hand-written templates [26,22], were not robust enough to such variability over possibly evolving KBs (addition/deletion of triples and entities). The solution to gain more expressiveness via machine learning comes from distant or indirect supervision to circumvent the issue of labeled data. Initial works attempting to learn to connect KBs and natural language with less supervision have actually been tackling the information extraction problem [17,12,15,20].

Recently, new systems for learning question answering systems with few labeled data have been introduced based on semantic parsers [6,3,13,11]. Such works tend to require realistic amounts of manual intervention via labeled examples, but still need vast efforts to carefully design lexicons, grammars and the KB. In contrast, [10] proposed a framework for open question answering requiring little human annotation. Their system, PARALEX, answers questions with more limited semantics than those introduced in [3,13], but does so at a very large scale in an open-domain manner. It is trained using automatically and collaboratively generated data and using the KB REVERB [9]. In this work, we follow this trend by proposing an embedding-based model for question answering that is also trained under weak and cheap supervision.

Embedding-based models are getting more and more popular in natural language processing. Starting from the neural network language model of [2], these methods have now reached near state-of-the-art performance on many standard tasks while usually requiring less hand-crafted features [7,21]. Recently, some embedding models have been proposed to perform a connection between natural language and KBs for word-sense disambiguation [5] and for information extraction [25]. Our work builds on these approaches to instead learn to perform open question answering under weak supervision, which to our knowledge has not been attempted before.

3   Open-Domain Question Answering

In this paper, we follow the question answering framework of [10] and use the same data. Hence, relatively little labeling or feature engineering has been used.

3.1   Task Definition

Our work considers the task of question answering as in [10]: given a question q, the corresponding answer is given by a triple t from a KB. This means that we consider questions for which a set of triples t provides an interpretation of the question and its answer, such as:

- q: *What environment does a dodo live in ?*
  t: (dodo.e, live-in.r, makassar.e)

- q: *What are the symbols for Hannukah ?*
  t: (menorah.e, be-for.r, hannukah.e)

- q: *What is a laser used for?*
  t: (hologram.e,be-produce-with.r,laser.e)

Here, we only give a single t per question, but many can exist. In the remainder, the KB is denoted $\mathcal{K}$ and its set of entities and relationships is $\mathcal{E}$. The word vocabulary for questions is termed $\mathcal{V}$. n_v and n_e are the sizes of $\mathcal{V}$ and $\mathcal{E}$ respectively.

Our model consists in learning a function $S(\cdot)$, which can score question-answer triple pairs (q, t). Hence, finding the top-ranked answer $\hat{t}(q)$ to a question q is directly carried out by:

$$\hat{t}(q) = \arg\max_{t' \in \mathcal{K}} S(q, t') \ .$$

To handle multiple answer, we instead present the results as a ranked list, rather than taking the top prediction, and evaluate that instead.

Using the scoring function $S(\cdot)$ allows the model to directly query the KB without needing to define an intermediate structured logical representation for questions as in semantic parsing systems. We aim at learning $S(\cdot)$, with no human-labeled supervised data in the form (question, answer) pairs, but only by indirect supervision, generated either automatically or collaboratively. We detail in the rest of this section our process for creating training data.

3.2   Training Data

Our training data consists of two sources: an automatically created KB, RE-VERB, from which we generate questions and a set of pairs of questions collaboratively labeled as paraphrases from the website WIKIANSWERS.

Knowledge Base. The set of potential answers $\mathcal{K}$ is given by the KB REVERB [9]. REVERB is an open-source database composed of more than 14M triples, made of more than 2M entities and 600k relationships, which have been automatically extracted from the ClueWeb09 corpus [18]. In the following, entities are denoted with a .e suffix and relationships with a .r suffix.

Table 1. Examples of triples from the KB REVERB

```
left entity, relationship, right entity
churchill.e, be-man-of.r, great-accomplishment.e
churchill-and-roosevelt.e, meet-in.r, cairo.e
churchill.e, reply-on.r, may-19.e
crick.e, protest-to.r, churchill.e
churchill.e, leave-room-for.r, moment.e
winston-churchill.e, suffer-from.r, depression.e
churchill.e, be-prime-minister-of.r, great-britain.e
churchill.e, die-in.r, winter-park.e
winston-churchill.e, quote-on.r, mug.e
churchill.e, have-only.r, compliment.e
```

REVERB contains broad and general knowledge harvested with very little human intervention, which suits the realistically supervised setting. But, as a result, REVERB is ambiguous and noisy with many useless triples and entities as well as numerous duplicates. For instance, `winston-churchill.e`, `churchill.e` and even `roosevelt-and-churchill.e` are all distinct entities. Table 3.2 presents some examples of triples: some make sense, some others are completely unclear or useless.

In contrast to highly curated databases such FREEBASE, REVERB has more noise but also many more relation types (FREEBASE has around 20k). So for some types of triple it has much better coverage, despite the larger size of FREEBASE; for example FREEBASE does not cover verbs like afraid-of or suffer-from.

Questions Generation. We have no available data of questions q labeled with their answers, i.e. with the corresponding triples $t \in \mathcal{K}$. Following [10], we hence decided to create such question-triple pairs automatically. These pairs are generated using the 16 seed questions displayed in Table 2. At each round, we pick a triple at random and then generate randomly one of the seed questions. Note only triples with a `*-in.r` relation (denoted `r-in` in Table 2) can generate from the pattern *where did e r ?*, for example, and similar for few other constraints. Otherwise, the pattern is chosen randomly. Except for these exceptions, we used all 16 seed questions for all triples hence generating approximately $16 \times 14M$ questions stored in a training set we denote $\mathcal{D}$.

The generated questions are imperfect and noisy and create a weak training signal. Firstly, their syntactic structure is rather simplistic, and real questions as posed by humans (such as in our actual test) can look quite different to them. Secondly, many generated questions do not correspond to semantically valid English sentences. For instance, since the type of entities in REVERB is unknown, a pattern like *who does e r ?* can be chosen for a triple where the type of ? in (?, r, e) is not a person, and similar for other types (e.g. *when*). Besides, for the strings representing entities and relationships in the questions, we simply used their names in REVERB, replacing - by spaces and stripping off their suffixes, i.e. the string representing `winston-churchill.e` is simply *winston*

Table 2. Patterns for generating questions from REVERB triples following [10]

KB Triple	Question Pattern	KB Triple	Question Pattern
(?, r, e)	*who r e ?*	(?, r, e)	*what is e's r ?*
(?, r, e)	*what r e ?*	(e, r, ?)	*who is r by e ?*
(e, r, ?)	*who does e r ?*	(e, r-in, ?)	*when did e r ?*
(e, r, ?)	*what does e r ?*	(e, r-on, ?)	*when did e r ?*
(?, r, e)	*what is the r of e ?*	(e, r-in, ?)	*when was e r ?*
(?, r, e)	*who is the r of e ?*	(e, r-on, ?)	*when was e r ?*
(e, r, ?)	*what is r by e ?*	(e, r-in, ?)	*where was e r ?*
(?, r, e)	*who is e's r ?*	(e, r-in, ?)	*where did e r ?*

churchill. While this is often fine, this is also very limited and caused many incoherences in the data. Generating questions with a richer KB than REVERB, such as FREEBASE or DBPEDIA, would lead to better quality because typing and better lexicons could be used. However, this would contradict one of our motivations which is to train a system with as little human intervention as possible (and hence choosing REVERB over hand-curated KBs).

Paraphrases. The automatically generated examples are useful to connect KB triples and natural language. However, they do not allow for a satisfactory modeling of English language because of their poor wording. To overcome this issue, we again follow [10] and supplement our training data with an indirect supervision signal made of pairs of question paraphrases collected from the WIKIANSWERS website.

On WIKIANSWERS, users can tag pairs of questions as rephrasing of each other. [10] harvested a set of 18M of these question-paraphrase pairs, with 2.4M distinct questions in the corpus. These pairs have been labeled collaboratively. This is cheap but also causes the data to be noisy. Hence, [10] estimated that only 55% of the pairs were actual paraphrases. The set of paraphrases is denoted $\mathcal{P}$ in the following. By considering all words and tokens appearing in $\mathcal{P}$ and $\mathcal{D}$, we end up with a size for the vocabulary $\mathcal{V}$ of more than 800k.

4   Embedding-Based Model

Our model ends up learning vector embeddings of symbols, either for entities or relationships from REVERB, or for each word of the vocabulary.

4.1   Question-KB Triple Scoring

Architecture. Our framework concerns the learning of a function $S(q, t)$, based on embeddings, that is designed to score the similarity of a question q and a triple t from $\mathcal{K}$.

Our scoring approach is inspired by previous work for labeling images with words [24], which we adapted, replacing images and labels by questions and triples. Intuitively, it consists of projecting questions, treated as a bag of words (and possibly n-grams as well), on the one hand, and triples on the other hand, into a shared embedding space and then computing a similarity measure (the dot product in this paper) between both projections. The scoring function is then:

$$S(q,t) = \boldsymbol{f}(q)^\top \boldsymbol{g}(t)$$

with $\boldsymbol{f}(\cdot)$ a function mapping words from questions into $\mathbb{R}^k$, $\boldsymbol{f}(q) = \boldsymbol{V}^\top \Phi(q)$. $\boldsymbol{V}$ is the matrix of $\mathbb{R}^{n_v \times k}$ containing all word embeddings $\boldsymbol{v}$ that will be learned, where k is a hyperparameter specifying the embedding dimension. $\Phi(q)$ is the (sparse) binary representation of q ($\in \{0,1\}^{n_v}$) indicating absence or presence of words. Similarly, $\boldsymbol{g}(\cdot)$ is a function mapping entities and relationships from KB triples into $\mathbb{R}^k$, $\boldsymbol{g}(t) = \boldsymbol{W}^\top \Psi(t)$. $\boldsymbol{W}$ is the matrix of $\mathbb{R}^{n_e \times k}$ containing all entity and relationship embeddings $\boldsymbol{w}$, that will also be learned. $\Psi(t)$ is the (sparse) binary representation of t ($\in \{0,1\}^{n_e}$) indicating absence or presence of entities and relationships.

Representing questions as a bag of words might seem too limited, however, in our particular setup, syntax is generally simple, and hence quite uninformative. A question is typically formed by an interrogative pronoun, a reference to a relationship and another one to an entity. Besides, since lexicons of relationships and entities are rather disjoint, even a bag of words representation should lead to decent performance, up to lexical variability. There are counter-examples such as *What are cats afraid of ?* vs. *What are afraid of cats ?* which require different answers, but such cases are rather rare. Future work could consider adding parse tree features or semantic role labels as input to the embedding model.

Contrary to previous work modeling KBs with embeddings (e.g. [25]), in our model, an entity does not have the same embedding when appearing in the left-hand or in the right-hand side of a triple. Since, $\boldsymbol{g}(\cdot)$ sums embeddings of all constituents of a triple, we need to use 2 embeddings per entity to encode for the fact that relationships in the KB are not symmetric and so that appearing as a left-hand or right-hand entity is different.

This approach can be easily applied at test time to score any (question, triple) pairs. Given a question q, one can predict the corresponding answer (a triple) $\hat{t}(q)$ with:

$$\hat{t}(q) = \arg\max_{t' \in \mathcal{K}} S(q,t') = \arg\max_{t' \in \mathcal{K}} \left(\boldsymbol{f}(q)^\top \boldsymbol{g}(t') \right).$$

Training by Ranking. Previous work [24,25] has shown that this kind of model can be conveniently trained using a ranking loss. Hence, given our data set $\mathcal{D} = \{(q_i, t_i), i = 1, \ldots, |\mathcal{D}|\}$ consisting of (question, answer triple) training pairs, one could learn the embeddings using constraints of the form:

$$\forall i, \ \forall t' \neq t_i, \ \ \boldsymbol{f}(q_i)^\top \boldsymbol{g}(t_i) > 0.1 + \boldsymbol{f}(q_i)^\top \boldsymbol{g}(t') \ ,$$

where 0.1 is the margin. That is, we want the triple that labels a given question to be scored higher than other triples in $\mathcal{K}$ by a margin of 0.1. We also enforce a constraint on the norms of the columns of $\boldsymbol{V}$ and $\boldsymbol{W}$, i.e. $\forall i, ||\boldsymbol{v}_i||_2 \leq 1$ and $\forall j, ||\boldsymbol{w}_j||_2 \leq 1$.

To train our model, we need positive and negative examples of (q, t) pairs. However, $\mathcal{D}$ only contains positive samples, for which the triple actually corresponds to the question. Hence, during training, we use a procedure to corrupt triples. Given $(q, t) \in \mathcal{D}$, we create a negative triple t' with the following method: pick another random triple t_{tmp} from $\mathcal{K}$, and then, replace with 66% chance each member of t (left entity, relationship and right entity) by the corresponding element in t_{tmp}. This heuristic creates negative triples t' somewhat similar to their positive counterpart t, and is similar to schemes of previous work (e.g. in [7,5]).

Training the embedding model is carried out by stochastic gradient descent (SGD), updating W and V at each step, including projection to the norm constraints. At the start of training, the parameters of $f(\cdot)$ and $g(\cdot)$ (the $n_v \times k$ word embeddings in V and the $n_e \times k$ entities and rel. embeddings in W) are initialized to random weights (mean 0, standard deviation $\frac{1}{k}$). Then, we iterate the following steps to train them:

1. Sample a positive training pair (q_i, t_i) from $\mathcal{D}$.
2. Create a negative triple t'_i ensuring that $t'_i \neq t_i$.
3. Make a stochastic gradient step to minimize $\left[0.1 - f(q_i)^\top g(t_i) + f(q_i)^\top g(t'_i)\right]_+$.
4. Enforce the constraint that each embedding vector is normalized.

The learning rate of SGD is updated during the course of learning using ADA-GRAD [8]. $\left[x\right]_+$ is the positive part of x.

Multitask Training with Paraphrases Pairs. We multitask the training of our model by training on pairs of paraphrases of questions (q_1, q_2) from $\mathcal{P}$ as well as training on the pseudolabeled data constructed in $\mathcal{D}$. We use the same architecture simply replacing $g(\cdot)$ by a copy of $f(\cdot)$. This leads to the following function that scores the similarity between two questions:

$$S_{\mathrm{prp}}(q_1, q_2) = f(q_1)^\top f(q_2) \ .$$

The matrix W containing embeddings of words is shared between S and S_{prp}, allowing it to encode information from examples from both $\mathcal{D}$ and $\mathcal{P}$. Training of S_{prp} is also conducted with SGD (and ADAGRAD) as for S, but, in this case, negative examples are created by replacing one of the questions from the pair by another question chosen at random in $\mathcal{P}$.

During our experiments, W and V were learned by alternating training steps using S and S_{prp}, switching from one to another at each step. The initial learning rate was set to 0.1 and the dimension k of the embedding space to 64. Training ran for 1 day on a 16 core machine using HOGWILD [19].

4.2   Fine-Tuning the Similarity between Embeddings

The scale of the problem forced us to keep our architecture simple: with $n_e \approx$ 3.5M (with 2 embeddings for each entity) and $n_v \approx$ 800k, we have to learn around 4.3M embeddings. With an embedding space of dimension $k = 64$, this leads to around 275M parameters to learn. The training algorithm must also

Table 3. Performance of variants of our embedding models and Paralex [10] for reranking question-answer pairs from the WIKIANSWERS+REVERB test set

Method	F1	Prec	Recall	MAP
PARALEX *(No. 2-arg)*	0.40	**0.86**	0.26	0.12
PARALEX	0.54	0. 77	0.42	0.22
Embeddings	0.68	0.68	0.68	0.37
Embeddings *(no paraphrase)*	0.60	0.60	0.60	0.34
Embeddings *(incl. n-grams)*	0.68	0.68	0.68	0.39
Embeddings+fine-tuning	**0.73**	0.73	**0.73**	**0.42**

stay simple to scale on a training set of around 250M of examples ($\mathcal{D}$ and $\mathcal{P}$ combined); SGD appears as the only viable option.

SGD, combined with ADAGRAD for adapting the learning rate on the course of training, is a powerful algorithm. However, the scale of the optimization problem makes it very hard to control and conduct properly until convergence. When SGD stops after a pre-defined number of epochs, we are almost certain that the problem is not fully solved and that some room for improvement remains: we observed that embeddings were able to often rank correct answers near the top of the candidates list, but not always in the first place.

In this paper, we introduce a way to fine-tune our embedding-based model so that correct answers might end up more often at the top of the list. Updating the embeddings involves working on too many parameters, but ultimately, these embeddings are meant to be used in a dot-product that computes the similarity between q and t. We propose to learn a matrix $M \in \mathbb{R}^{k \times k}$ parameterizing the similarity between words and triples embeddings. The scoring function becomes:

$$S_{\text{ft}}(q, t) = f(q)^{\top} M g(t) .$$

M has only k^2 parameters and can be efficiently determined by solving the following convex problem (fixing the embedding matrices W and V):

$$\min_M \frac{\lambda}{2} \parallel M \parallel_F^2 + \frac{1}{m} \sum_{i=1}^{m} \left[1 - S_{\text{ft}}(q_i, t_i) + S_{\text{ft}}(q_i, t_i') \right]_+^2 ,$$

where $\parallel X \parallel_F$ is the Frobenius norm of X. We solve this problem in a few minutes using L-BFGS on a subset of $m = 10M$ examples from $\mathcal{D}$. We first use 4M examples to train and 6M as validation set to determine the value of the regularization parameter λ. We then retrain the model on the whole 10M examples using the selected value, which happened to be $\lambda = 1.7 \times 10^{-5}$.

This fine-tuning is related to learning a new metric in the embedding space, but since the resulting M is not symmetric, it does not define a dot-product. Still, M is close to a constant factor times identity (as in the original score $S(\cdot)$). The fine-tuning does not deeply alter the ranking, but, as expected, allows for a slight change in the triples ranking, which ends in consistent improvement in performance, as we show in the experiments.

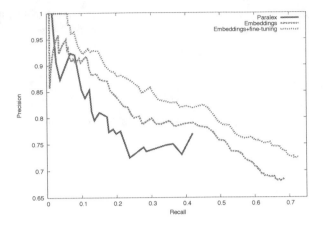

Fig. 1. Precision-recall curves of our embedding model and PARALEX [10] for reranking question-answer pairs from the WIKIANSWERS+REVERB test set

5   Experiments

5.1   Evaluation Protocols

We first detail the data and metrics which were chosen to assess the quality of our embedding model.

Test Set. The data set WIKIANSWERS+REVERB contains no labeled examples but some are needed for evaluating models. We used the test set which has been created by [10] in the following way: (1) they identified 37 questions from a held-out portion of WIKIANSWERS which were likely to have at least one answer in REVERB, (2) they added all valid paraphrases of these questions to obtain a set of 691 questions, (3) they ran various versions of their PARALEX system on them to gather candidate triples (for a total of 48k), which they finally hand-labeled.

Reranking. We first evaluated different versions of our model against the PAR-ALEX system in a reranking setting. For each question q from the WIKIAN-SWERS+REVERB test set, we take the provided candidate triples t and rerank them by sorting by the score $S(q,t)$ or $S_{ft}(q,t)$ of our model, depending whether we use fine-tuning or not. As in [10], we then compute the precision, recall and F1-score of the highest ranked answer as well as the mean average precision (MAP) of the whole output, which measures the average precision over all levels of recall.

Full Ranking. The reranking setting might be detrimental for PARALEX because our system simply never has to perform a full search for the good answer among the whole REVERB KB. Hence, we also conducted an experiment where, for each of the 691 questions of the WIKIANSWERS+REVERB test set, we ranked all 14M triples from REVERB. We labeled the top-ranked answers ourselves and computed precision, recall and F1-score.

Table 4. Performance of our embedding model for retrieving answers for questions from the WIKIANSWERS+REVERB test set, among the whole REVERB KB (14M candidates)

Method	F1
Embeddings	0.16
Embeddings+fine-tuning	0.22
Embeddings +string-matching	0.48
Embeddings+fine-tuning+string-matching	**0.57**

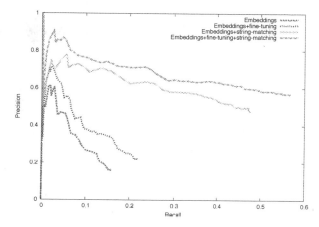

Fig. 2. Precision-recall curves for retrieving answers for questions from the WIKIAN-SWERS+REVERB test set, among the whole REVERB KB (14M candidates)

5.2 Results

This section now discusses our empirical performance.

Reranking. Table 3 and Figure 1 present the results of the reranking experiments. We compare various versions of our model against two versions of PARALEX, whose results were given in [10].

First, we can see that multitasking with paraphrase data is essential since it improves F1 from 0.60 to 0.68. Paraphrases allow for the embeddings to encode a richer connection between KB constituents and words, as well as between words themselves. Note that the WIKIANSWERS data provides word alignment between paraphrases, which we did not use, unlike PARALEX. We also tried to use n-grams (2.5M most frequent) as well as the words to represent the question, but this did not bring any improvement, which might at first seem counter-intuitive. We believe this is due to two factors: (1) it is hard to learn good embeddings for n-grams since their frequency is usually very low and (2) our automatically generated questions have a poor syntax and hence, many n-grams in this data set do not make sense. We actually conducted experiments with several variants of our model, which tried to take the word ordering into account (e.g. with convolutions), and they all failed to outperform our best performance without

Table 5. Examples of nearest neighboring entities and relationships from REVERB for some words from our vocabulary. The prefix L:, resp. R:, indicates the embedding of an entity when appearing in **left-hand** side, resp. **right-hand** side, of triples.

Word	Closest entities or relationships from REVERB in the embedding space
get rid of	get-rid-of.r be-get-rid-of.r rid-of.r can-get-rid-of.r will-get-rid-of.r should-get-rid-of.r have-to-get-rid-of.r want-to-get-rid-of.r will-not-get-rid-of.r help-get-rid-of.r
useful	be-useful-for.r be-useful-in.r R:wide-range-of-application.e can-be-useful-for.r be-use-extensively-for.r be-not-very-useful-for.r R:plex-or-technical-algorithm.e R:internal-and-external-use.e R:authoring.e R:good-or-bad-purpose.e
radiation	R:radiation.e L:radiation.e R:gamma-radiation.e L:gamma-radiation.e L:x-ray.e L:gamma-ray.e L:cesium-137.e R:electromagnetic-radiation.e L:external-beam-radiation.e L:visible-light.e
barack-obama	L:president-elect-barack-obama.e L:barack-obama.e R:barack-obama.e L:president-barack-obama.e L:obama-family.e L:sen.-barack-obama.eL:president-elect-obama.e R:president-barack-obama.e L:democratic-presidential-candidate-barack-obama.e L:today-barack-obama.e
iphone	R:iphone.e L:iphone.e R:t-mobile.e R:apple-iphone.e L:lot-of-software.e L:hotmail.e R:windows-mobile-phone.e L:skype.e R:smartphone.e R:hd-dvd-player.e

word order, once again perhaps because the supervision is not clean enough to allow for such elaborated language modeling. Fine-tuning the embedding model is very beneficial to optimize the top of the list and grants a bump of 5 points of F1: carefully tuning the similarity makes a clear difference.

All versions of our system greatly outperform PARALEX: the fine-tuned model improves the F1-score by almost 20 points and, according to Figure 1, is better in precision for all levels of recall. PARALEX works by starting with an initial lexicon mapping from the KB to language and then gradually increasing its coverage by iterating on the WIKIANSWERS+REVERB data. Most of its predictions come from automatically acquired templates and rules: this allows for a good precision but it is not flexible enough across language variations to grant a satisfying recall. Most of our improvement comes from a much better recall.

However, as we said earlier, this reranking setting is detrimental for PARALEX because PARALEX was evaluated on the task of reranking some of its own predictions. The results provided for PARALEX, while not corresponding to those of a full ranking among all triples from REVERB (it is still reranking among a subset of candidates), concerns an evaluation setting more complicated than for our model. Hence, we also display the results of a full ranking by our system in the following.

Full Ranking. Table 4 and Figure 2 display the results of our model to rank all 14M triples from REVERB. The performance of the plain models is not good (F1 = 0.22 only for S_{ft}) because the ranking is degraded by too many candidates. But most of these can be discarded beforehand.

We hence decided to filter out some candidates before ranking by using a simple string matching strategy: after pos-tagging the question, we construct a set of candidate strings containing (i) all noun phrases in the question that appear less than 1,000 times in REVERB, (ii) all proper nouns in the question, if any, otherwise the least frequent noun phrase in REVERB. This set of strings is then augmented with the singular form of plural nouns, removing the final "s", if any. Then, only the triples containing at least one of the candidate strings are

Table 6. Performance of our embedding model for retrieving answers for questions from the WEBQUESTIONS test set, among the whole REVERB KB (14M candidates)

Method	Top-1	Top-10	F1
Emb.	0.025	0.094	0.025
Emb.+fine-tuning	0.032	0.106	0.032
Emb. +string-match.	0.085	0.246	0.068
Emb.+fine-tuning+string-match.	**0.094**	**0.270**	**0.076**

scored by the model. On average, about 10k triples (instead of 14M) are finally ranked for each question, making our approach much more tractable.

As expected, string matching greatly improves results, both in precision and recall, and also significantly reduces evaluation time. The final F1 score obtained by our fine-tuned model is even better than the result of PARALEX in reranking, which is pretty remarkable, because this time, this setting advantages PARALEX quite a lot.

Embeddings. Table 5 displays some examples of nearest neighboring entities from REVERB for some words from our vocabulary. As expected, we can see that verbs or adverbs tend to correspond to relationships while nouns refer to entities. Interestingly, the model learns some synonymy and hyper/hyponymy. For instance, *radiation* is close to x-ray.e and *iphone* to smartphone.e. This happens thanks to the multitasking with paraphrase data, since in our automatically generated (q, t) pairs, the words *radiation* and *iphone* are only used for entities with the strings radiation and iphone respectively in their names.

5.3   Evaluation on WebQuestions

Our initial objective was to be able to perform open-domain question answering. In this last experimental section, we tend to evaluate how generic our learned system is. To this end, we propose to ask our model to answer questions coming from another dataset from the literature, but without retraining it with labeled data, just by directly using the parameters learned on WIKIANSWERS+REVERB.

We chose the data set WEBQUESTIONS [3], which consists of natural language questions matched with answers corresponding to entities of FREEBASE: in this case, no triple has to be returned, only a single entity. We used exact string matching to find the REVERB entities corresponding to the FREEBASE answers from the test set of WEBQUESTIONS and obtained 1,538 questions labeled with REVERB out of the original 2,034.

Results of different versions of our model are displayed in Table 6. For each test question, we record the rank of the first REVERB triple containing the answer entity. Top-1 and Top-10 are computed on questions for which the system returned at least one answer (around 1,000 questions using string matching), while F1 is computed for all questions. Of course, performance is not great and can not be directly compared with that of the best system reported in

[3] (more than 0.30 of F1). One of the main reasons is that most questions of WEBQUESTIONS, such as *Who was vice-president after Kennedy died?*, should be represented by multiple triples, a setting for which our system has not been designed. Still, for a system trained with almost no manual annotation nor prior information on another dataset, with an other –very noisy– KB, the results can be seen as particularly promising. Besides, evaluation is broad since, in REVERB, most entities actually appear many times under different names as explained in Section 3. Hence, there might be higher ranked answers but they are missed by our evaluation script.

6   Conclusion

This paper introduces a new framework for learning to perform open question answering with very little supervision. Using embeddings as its core, our approach can be successfully trained on imperfect labeled data and indirect supervision and significantly outperforms previous work for answering simple factual questions. Besides, we introduce a new way to fine-tune embedding models for cases where their optimization problem can not be completely solved.

In spite of these promising results, some exciting challenges remain, especially in order to scale up this model to questions with more complex semantics. Due to the very low supervision signal, our work can only answer satisfactorily simple factual questions, and does not even take into account the word ordering when modeling them. Further, much more work has to be carried out to encode the semantics of more complex questions into the embedding space.

Acknowledgments. Part of this work was carried out in the framework of the Labex MS2T (ANR-11-IDEX-0004-02), and was funded by the French National Agency for Research (EVEREST-12-JS02-005-01).

References

1. Banko, M., Brill, E., Dumais, S., Lin, J.: Askmsr: Question answering using the worldwide web. In: Proceedings of 2002 AAAI Spring Symposium on Mining Answers from Texts and Knowledge Bases (2002)
2. Bengio, Y., Ducharme, R., Vincent, P., Jauvin, C.: A neural probabilistic language model. Journal of Machine Learning Research 3, 1137–1155 (2003)
3. Berant, J., Chou, A., Frostig, R., Liang, P.: Semantic parsing on Freebase from question-answer pairs. In: Proceedings of the 2013 Conference on Empirical Methods in Natural Language Processing (October 2013)
4. Bollacker, K., Evans, C., Paritosh, P., Sturge, T., Taylor, J.: Freebase: a collaboratively created graph database for structuring human knowledge. In: Proceedings of the 2008 ACM SIGMOD International Conference on Management of Data. ACM (2008)
5. Bordes, A., Glorot, X., Weston, J., Bengio, Y.: Joint learning of words and meaning representations for open-text semantic parsing. In: Proc. of the 15th Intern. Conf. on Artif. Intel. and Stat., vol. 22. JMLR W&CP (2012)

6. Cai, Q., Yates, A.: Large-scale semantic parsing via schema matching and lexicon extension. In: Proceedings of the 51st Annual Meeting of the Association for Computational Linguistics. Long Papers, vol. 1 (August 2013)

7. Collobert, R., Weston, J., Bottou, L., Karlen, M., Kavukcuoglu, K., Kuksa, P.: Natural language processing (almost) from scratch. Journal of Machine Learning Research 12, 2493–2537 (2011)

8. Duchi, J., Hazan, E., Singer, Y.: Adaptive subgradient methods for online learning and stochastic optimization. The Journal of Machine Learning Research 12 (2011)

9. Fader, A., Soderland, S., Etzioni, O.: Identifying relations for open information extraction. In: Proceedings of the Conference of Empirical Methods in Natural Language Processing (EMNLP 2011), Edinburgh, Scotland, UK, July 27-31 (2011)

10. Fader, A., Zettlemoyer, L., Etzioni, O.: Paraphrase-driven learning for open question answering. In: Proceedings of the 51st Annual Meeting of the Association for Computational Linguistics, Sofia, Bulgaria, pp. 1608–1618 (2013)

11. Fader, A., Zettlemoyer, L., Etzioni, O.: Open question answering over curated and extracted knowledge bases. In: KDD (2014)

12. Hoffmann, R., Zhang, C., Ling, X., Zettlemoyer, L., Weld, D.S.: Knowledge-based weak supervision for information extraction of overlapping relations. In: Proceedings of the 49th Annual Meeting of the Association for Computational Linguistics: Human Language Technologies, vol. 1 (2011)

13. Kwiatkowski, T., Choi, E., Artzi, Y., Zettlemoyer, L.: Scaling semantic parsers with on-the-fly ontology matching. In: Proceedings of the 2013 Conference on Empirical Methods in Natural Language Processing (October 2013)

14. Kwok, C., Etzioni, O., Weld, D.S.: Scaling question answering to the web. ACM Transactions on Information Systems (TOIS) 19(3) (2001)

15. Lao, N., Subramanya, A., Pereira, F., Cohen, W.W.: Reading the web with learned syntactic-semantic inference rules. In: Proceedings of the 2012 Joint Conference on Empirical Methods in Natural Language Processing and Computational Natural Language Learning (2012)

16. Lehmann, J., Isele, R., Jakob, M., Jentzsch, A., Kontokostas, D., Mendes, P.N., Hellmann, S., Morsey, M., van Kleef, P., Auer, S., Bizer, C.: DBpedia - a large-scale, multilingual knowledge base extracted from wikipedia. Semantic Web Journal (2014)

17. Mintz, M., Bills, S., Snow, R., Jurafsky, D.: Distant supervision for relation extraction without labeled data. In: Proc. of the Conference of the 47th Annual Meeting of ACL (2009)

18. Pomikálek, J., Jakubícek, M., Rychlý, P.: Building a 70 billion word corpus of english from clueweb. In: LREC, pp. 502–506 (2012)

19. Recht, B., Ré, C., Wright, S.J., Niu, F.: Hogwild!: A lock-free approach to parallelizing stochastic gradient descent. In: Advances in Neural Information Processing Systems (NIPS 24) (2011)

20. Riedel, S., Yao, L., McCallum, A., Marlin, B.M.: Relation extraction with matrix factorization and universal schemas. In: Proceedings of NAACL-HLT (2013)

21. Socher, R., Perelygin, A., Wu, J.Y., Chuang, J., Manning, C.D., Ng, A.Y., Potts, C.: Recursive deep models for semantic compositionality over a sentiment treebank. In: Proceedings of the Conference on Empirical Methods in Natural Language Processing (EMNLP) (2013)

22. Unger, C., Bühmann, L., Lehmann, J., Ngonga Ngomo, A.C., Gerber, D., Cimiano, P.: Template-based question answering over rdf data. In: Proceedings of the 21st International Consference on World Wide Web (2012)

23. Voorhees, E.M., Tice, D.M.: Building a question answering test collection. In: Proceedings of the 23rd Annual International ACM SIGIR Conference on Research and Development in Information Retrieval. ACM (2000)
24. Weston, J., Bengio, S., Usunier, N.: Large scale image annotation: learning to rank with joint word-image embeddings. Machine Learning 81(1) (2010)
25. Weston, J., Bordes, A., Yakhnenko, O., Usunier, N.: Connecting language and knowledge bases with embedding models for relation extraction. In: Proceedings of the 2013 Conference on Empirical Methods in Natural Language Processing, pp. 1366–1371 (2013)
26. Yahya, M., Berberich, K., Elbassuoni, S., Ramanath, M., Tresp, V., Weikum, G.: Natural language questions for the web of data. In: Proceedings of the Conference on Empirical Methods in Natural Language Processing (2012)

Towards Automatic Feature Construction for Supervised Classification

Marc Boullé

Orange Labs,
2 avenue Pierre Marzin, 22300 Lannion, France
marc.boulle@orange.com
http://perso.rd.francetelecom.fr/boulle

Abstract. We suggest an approach to automate variable construction for supervised learning, especially in the multi-relational setting. Domain knowledge is specified by describing the structure of data by the means of variables, tables and links across tables, and choosing construction rules. The space of variables that can be constructed is virtually infinite, which raises both combinatorial and over-fitting problems. We introduce a prior distribution over all the constructed variables, as well as an effective algorithm to draw samples of constructed variables from this distribution. Experiments show that the approach is robust and efficient.

Keywords: supervised learning, relational learning, feature construction, feature selection, regularization.

1 Introduction

In a data mining project, the data preparation phase aims at constructing a data table for the modeling phase [19,6]. The data preparation is both time consuming and critical for the quality of the mining results. It mainly consists in a search of an effective data representation, based on variable construction and selection. Variable selection has been extensively studied in the literature [12]. Two main approaches, filter and wrapper, have been proposed. Filter methods consider the correlation between the input variables and the output variable as a pre-processing step, independently of the chosen classifier. Wrapper methods search the best subset of variables for a given classification technique, used as a black box. Wrapper methods, which are time consuming, are restricted to the modeling phase of data mining, as a post-optimization of a classifier. Filter methods are better suited to the data preparation phase, since they can be combined with any data modeling approach and can deal with large numbers of input variables. In this paper, we focus on the filter approach, in the context of supervised classification.

Variable construction [18] has been less studied than variable selection in the literature. It implies a large amount of work for the data analyst and heavily relies on domain knowledge to construct new potentially informative variables. In practice, the initial raw data usually originate from relational databases. As most

T. Calders et al. (Eds.): ECML PKDD 2014, Part I, LNCS 8724, pp. 181–196, 2014.
© Springer-Verlag Berlin Heidelberg 2014

classification techniques need a flat input data table with instances × variables tabular format, such relational data cannot be directly analyzed.

Learning from relational data has recently received an increasing attention in the literature. The term Multi-Relational Data Mining (MRDM) was initially introduced in [13] to address novel knowledge discovery techniques from multiple relational tables. The common point between these techniques is that they need to transform the relational representation. In Inductive Logic Programming (ILP) [9], data is recoded as logic formulas. In 1BC method [16] and its successor 1BC2 [17], simple predicates are used together with a naive Bayes classifier. More expressive approaches cause scalability problems especially with large-scale data. Other methods named by propositionalisation [14] try to flatten the relational data by constructing new variables. These variables aggregate the information contained in non target tables in order to obtain a classical tabular format. For example, the RELAGGS method [15] uses functions such as mean, median, min, max to summarize numerical variables from secondary tables in zero to many relationship, or counts per value for the categorical variables. The TILDE method [2,24] aims at constructing complex variables based on conjunctions of selection conditions of records in secondary tables. However, the expressiveness of such methods faces the following problems: complex parameter setting, combinatorial explosion of the number of potentially constructed variables and risk of over-fitting that increases with the number of constructed variables.

In this paper, we suggest an approach aiming at the automation of variable construction, with the three-fold following objective: simplicity of parameters, efficient control of the combinatorial search in the space of variable construction and robustness w.r.t. over-fitting. Section 2 presents a formal description of a variable construction domain. Section 3 introduces an evaluation criterion of the constructed variables exploiting a Bayesian approach, by suggesting a prior distribution over the space of variables that can be constructed. Section 4 studies the problem of drawing a sample from this space and describes an efficient and computable algorithm for drawing samples of constructed variables of given size. Section 5 evaluate the approach on several datasets. Finally, Section 6 gives a summary and discusses future work.

2   Specification of Variable Construction Domain

We suggest a formal specification of a variable construction domain in order to efficiently drive the construction of variables for supervised classification. The objective is not to propose a new expressive and general formalism for describing domain knowledge, but simply to clarify the framework exploited by the variable construction algorithms presented in Section 4. This framework consists in two parts: description of the data structure and choice of the construction rules.

2.1   Data Structure

The simplest data structure is the tabular one. Data instances are represented by a list of variables, each defined by its name and type. The standard types, numerical or categorical, can be extended to other specialized types, such as date, time or text. As real data usually comes from relational databases, extending tabular format to multi-table looks natural. We suggest to describe these structures similarly to structured or object-oriented programming languages. The statistical unit (root instance) belongs to a *root* table. A root instance is then defined by a list of variables, whose type can be simple (numerical, categorical...) as in the tabular case, or structured: one record of a secondary table in zero to one relationship or several records of a secondary table in zero to many relationship. In the case of supervised classification, the output variable is a categorical variable in the root table. Figure 1 presents an example of the use of this formalism. The root instance is a Customer, with secondary records Usages in zero to many relationship. The variables are either of simple type (Cat, Num or Date) or structured type (Table(Usage)). The identifier variables (prefixed by #) are mainly used for practical purposes, in order to establish a matching with a relational database; they are not considered as input variables.

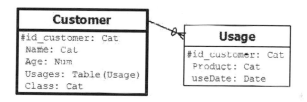

Fig. 1. Data structure for a problem of customer relationship management

2.2   Variable Construction Rules

A variable construction rule is similar to a function (or method) in a programming language. It is defined by its name, the list of its operands and its return value. The operands and the return value are typed, with the types defined in Section 2.1. For example, the *YearDay(Date)→Num* rule builds a numerical variable from a date variable. The operands can originate from an *original* variable (in the initial data representation), from the return value of another rule, or from values coming from a train dataset. In this paper, the construction rules used in the experiments of Section 5 are the following ones:

- *Selection(Table, Num)→Table*: selection of records from the table according to a conjunction of selection terms (membership in a numerical interval or in group of categorical values),
- *Count(Table)→Num*: count of records in a table,
- *Mode(Table, Cat)→Cat*: most frequent value of a variable in a table,
- *CountDistinct(Table, Cat)→Num*: number of distinct values,
- *Mean(Table, Num)→Num*: mean value,
- *Median(Table, Num)→Num*: median value,
- *Min(Table, Num)→Num*: min value,
- *Max(Table, Num)→Num*: max value,

- *StdDev(Table, Num)→Num*: standard deviation,
- *Sum(Table, Num)→Num*: sum of values.

Using the data structure presented in Figure 1 and the previous construction rules (plus the *YearDay* rule for date variables), one can construct the following variables to enrich the description of a customer:

- *MainProduct = Mode(Usages, Product)*,
- *LastUsageYearDay = Max(Usages, YearDay(useDate))*,
- *NbUsageProd1FirstQuarter = Count(Selection(Usages, YearDay(useDate) ∈ [1;90] and Product = "Prod1"))*.

3   Evaluation of Constructed Variables

The issue is to exploit the variable construction domain in order to efficiently drive the construction of variables which are potentially informative for the prediction of the output variable. In the framework introduced in Section 2, the data structure can have several level of depth or even have a graph structure. For example, a molecule is a graph where the vertices are the atoms and the edges are the bounds between atoms. The constructed rules can be used as operands of other rules, leading to computation formulas of any length. The space of constructed variables is then of potentially infinite size. This raises the two major following problems:

1. combinatorial explosion for the exploration of this construction space,
2. risk of over-fitting.

We suggest to solve these problems by introducing an evaluation criterion of the constructed variables according to a Bayesian approach in order to penalize complex variables. For this purpose, we propose a prior distribution on the space of all variables and an efficient sampling algorithm of the space of variables according to their prior distribution.

3.1   Evaluation of a Variable

Variable construction aims to enrich the root table with new variables that will be taken as input of a classifier. As usual classifiers take as input only numerical or categorical variables, only these variables need to be evaluated.

Supervised Preprocessing. The MODL supervised preprocessing methods [3,4] consist in partitioning a numerical variable into intervals or a categorical variable into groups of values, with a piecewise constant class conditional density estimation. The parameters of a specific preprocessing model are the number of parts, the partition and the multinomial distribution of the classes within each part. In the MODL approach, supervised preprocessing is turned into a model selection problem and solved in a Bayesian way. A prior distribution is proposed on this model space. This prior exploits the hierarchy of the parameters, with a uniform distribution at each stage of the hierarchy. The methods exploit a maximum a posteriori (MAP) technique to select the most probable preprocessing model given the input data. Taking the negative log of probabilities that are no other than coding lengths [22] in the minimum description length (MDL) approach [20], this amounts to the description length of a preprocessing model $M_P(X)$ (using a supervised partition) of a variable X plus the description length of the output data D_Y given the model and the input data D_X.

$$cost_P(X) = L(M_P(X)) + L(D_Y|M_P(X), D_X). \tag{1}$$

We asymptotically have $cost_P(X) \approx Nent(Y|X)$ where N is the number of train instances and $ent(Y|X)$ the conditional entropy [7] of the output given the input variable. Formula (1) and the related optimisation algorithms are fully detailed in [4] for supervised discretization and [3] for supervised value grouping.

Null Model and Variable Filtering. The null model $M_P(\emptyset)$ corresponds to the case of a preprocessing model with one single part (interval or group of values) and thus to the direct modeling of the output values using a multinomial distribution, without using the input variable. The value of criterion $cost_P(\emptyset)$ amounts to a direct coding of the output values: the null cost is $cost_P(\emptyset) \approx Nent(Y)$, where $ent(Y)$ is the entropy of Y. The evaluation criterion of a variable is then exploited according to a filter approach [12]: only variables whose evaluation is better than the null cost are considered informative and retained at the end of the data preparation phase.

Accounting for the Variable Construction Process. When the number of original or constructed variables increases, the chance for a variable to be wrongly considered as informative becomes critical. In order to prevent this risk of over-fitting, we suggest in this paper to exploit the space of constructed variables described in Section 2 by proposing a prior distribution over the set of all variable construction models $M_C(X)$. We then get a Bayesian regularization of the constructed variables, which allows to penalize the most "complex" variables. This translates into an additional construction cost $L(M_C(X))$ in the evaluation criterion of the variables, which becomes that of Formula (2).

$$cost_{CP}(X) = L(M_C(X)) + L(M_P(X)) + L(D_Y|M_P(X), D_X). \tag{2}$$

$L(M_C(X))$ is the negative log of the prior probability (coding length) of an original or constructed variable X, defined below.

3.2   Prior Distribution of the Original and Constructed Variables

A variable to evaluate is a numerical or categorical variable in the root table, either original or built using construction rules recursively. The space of such variables being of virtually infinite size, defining a prior probability on this space raises many problems and involves many choices. To guide these choices, we will stick to the following general principles:

1. taking into account the constructed variables has a minimal impact on the original variables,
2. in order to have a minimum bias, the prior is as uniform as possible,
3. the prior exploits at best the variable construction domain.

Case of Original Variables. In the case where no variable can be constructed, the problem reduces to the choice of an original variable to evaluate among the K numerical or categorical variables of the root table. Using a uniform prior for this choice, we obtain $p(M_C(X)) = 1/K$, thus $L(M_C(X)) = \log K$.

Case of Constructed Variables. In the case where variables can be constructed, one must first choose whether to use an original variable or to construct a new variable. Using a uniform prior $(p = 1/2)$ for this choice implies an additional cost of $\log 2$, which violates the principle of minimal impact on the original variables. We then suggest to consider the choice of constructing a new variable as an additional possibility beyond the K original variables. The cost of an original variable becomes $L(M_C(X)) = \log(K + 1)$, with an additional cost of $\log(1 + 1/K) \approx 1/K$ w.r.t. the case of original variables only.

Constructing a new variable then relies on the following hierarchy of choices:

- choice of constructing a new variable,
- choice of the construction rule among the R applicable rules (with the required return value type and available operands of the required types),
- for each operand of the rule, choice of using an original variable or to construct a new variable with a rule whose return value is compatible with the expected operand type.

Using a hierarchical prior, uniform at each level of the hierarchy, the cost of a constructed variable is decomposed on the operands of the used construction rule according to the recursive Formula (3), where the variables X_{op} are the original or constructed variables used as operands op of the rule $\mathcal{R}$.

$$L(M_C(X)) = \log(K + 1) + \log R + \sum_{op \in \mathcal{R}} L(M_C(X_{op})). \tag{3}$$

Case of the *Selection* Rule. The case of the *Selection* rule that extracts records from a secondary table according to a conjunction of selection terms is treated similarly. The hierarchy of choices is extended in the following way: number of selection operands, list of selection variables (original or constructed) and for each selection variable, choice of the selection part (numerical interval or group of categorical values). The selection part is itself chosen hierarchically with first a choice of granularity of the partitioned variable into a set of quantiles and second the index of the quantile in the partition. In definitions 1 and 2, we precisely define quantile partitions both for numerical and categorical variables.

Definition 1 (Numerical quantile partition). *Let D be a dataset of N instances and X a numerical variable. Let $x_1, x_2, \ldots, x_N$ be the N sorted values of X in dataset D. For a given number of parts P, the dataset is divided into P equal frequency intervals* $]-\infty, x_{\lfloor 1+\frac{N}{P} \rfloor}[, \; [x_{\lfloor 1+\frac{N}{P} \rfloor}, x_{\lfloor 1+2\frac{N}{P} \rfloor}[, \; \ldots,$ $[x_{\lfloor 1+i\frac{N}{P} \rfloor}, x_{\lfloor 1+(i+1)\frac{N}{P} \rfloor}[, \; \ldots, [x_{\lfloor 1+(P-1)\frac{N}{P} \rfloor}, +\infty[.$

Definition 2 (Categorical quantile partition). *Let D be a dataset of N instances and X a categorical variable with V values. For a given number of parts P, let $N_P = \lceil \frac{N}{P} \rceil$ be the expected minimum frequency per part. The categorical quantile partition into (at most) P parts is defined by singleton parts for each value of X with frequency above the threshold frequency N_P and a "garbage" part consisting of all values of X below the threshold frequency.*

The number of selection terms is chosen according to the universal prior for integer numbers of Rissanen [21]. This prior distribution is as flat as possible, with larger probabilities for small integer numbers. Each selection variable (original or constructed) is distributed using the prior defined previously in this section. As for the granularities, we consider only powers of two $2^0, 2^1, 2^2, \ldots 2^p, \ldots$ for the sizes of the partitions, with the exponent p distributed according to the universal prior for integer numbers. Finally, the index of each quantile is distributed uniformly among the 2^p parts.

Whereas all the other rules exploit only the data structure and the set of construction rules, the *Selection* rule exploits the values of the train dataset to build the actual definition of the selection parts. This requires one reading step of each secondary table to instantiate the formal definition of each part (granularity and part index) into an actual definition, with numerical boundaries for intervals and categorical values for groups of values.

Synthesis. Figure 2 presents an example of such a prior distribution over the set of variables that can be built using the construction rules *Mode, Min, Max* and *YearDay*, in the case of the customer relationship management dataset of Figure 1. For example, the cost of selecting the original variable *Age* is

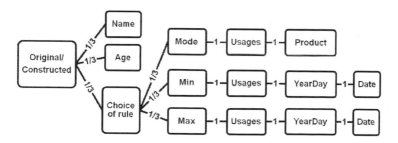

Fig. 2. Prior distribution of variable construction in the case of the customer relationship management dataset

$L(M_C(Age)) = \log 3$. That of constructing the variable with formula $Min($ $Usages, YearDay(Date))$ exploits of a hierarchy of choices leading to

$$L(M_C(Min(Usages, YearDay(Date)))) = \log 3 + \log 3 + \log 1 + \log 1 + \log 1.$$

This prior distribution on the space of variable construction corresponds to a Hierarchy of Multinomial Distributions with potentially Infinite Depth (HM-DID). The original variables are obtained from the first level of hierarchy of the prior, whereas the constructed variables get all the more lower prior probabilities as they exploit deeper parts of the HMDID prior with complex formulas.

4   Building a Random Sample of Variables

The objective is to build a given number of variables, potentially informative for supervised classification, in order to create an input tabular representation,. We suggest to build a sample of variables by drawing them according to their prior distribution. We present a first "natural" algorithm for building samples of variables, and demonstrate that it is neither efficient nor even computable. We then propose a second algorithm that solves the problem.

4.1   Successive Random Draws

Algorithm 1. Successive random draws

Require: K {Number of draws}
Ensure: $\mathcal{V} = \{V\}, |\mathcal{V}| \leq K$ {Sample of constructed variables}
 1: $\mathcal{V} = \emptyset$
 2: **for** $k = 1$ to K **do**
 3:     Draw V according to HMDID prior
 4:     Add V in $\mathcal{V}$
 5: **end for**

Algorithm 1 consists in successively drawing K according to the HMDID prior. Each draw starts from the root of the prior and goes down in the hierarchy until obtaining an original or constructed variable, which corresponds to a leaf in the prior hierarchy. This natural algorithm cannot be used in the general case, because it is neither efficient nor computable, as we demonstrate below.

Algorithm 1 Is not Efficient. Let us consider a construction domain with V original variables that can be evaluated in the root table and no construction rule. The HMDID prior reduces to a standard multinomial distribution with V equidistributed values. If K draws are performed according to this multinomial distribution, the expectation of the number of distinct obtained variables is $V(1-e^{-K/V})$ [11]. In the case where $K = V$, this corresponds to the size of a bootstrap sample, that is $(1 - 1/e) \approx 63\%$ variables obtained using V draws. To obtain 99% of the original variables, one needs $K \approx 5V$ draws, which is not efficient. Furthermore, in the case with construction rules, the multinomial at the root of the HMDID consists now into $K + 1$ equidistributed values. The draws result in the construction of a new variable in only $\frac{1}{K+1}$ of the cases. It is noteworthy that this problem of inefficiency occurs at all levels of depth of the HMDID prior for the draw of the operands of the rules under construction.

Algorithm 1 Is not Computable. Let us consider a construction domain with one single numerical variable x and one single construction rule $f(Num, Num) \rightarrow Num$. The variables that can be constructed are x, $f(x,x)$, $f(x, f(x,x))$, $f(f(x,x), f(x,x))$, $f(f(x,x), f(f(x,x),x))$... In combinatorial mathematics, the Catalan number $C_n = \frac{(2n)!}{(n+1)!n!} \approx \frac{4^n}{\pi n^{3/2}}$ counts the number of such expressions. C_n is the number of different ways $n+1$ factors can be completely parenthesized or the number of full binary trees with $n + 1$ leaves. Each variable represented by a binary construction tree with n leaves (repetitions of x in the formula) comes into C_{n-1} formally distinct copies, each with a prior probability of $2^{-(2n-1)}$ according to the HMDID prior. The expectation of the size $s(V)$ of a constructed variable V (size defined by the number of leaves in the binary construction tree) can then be computed. Using the above approximation of the Catalan number, Formula (4) states that the expectation of the size of the variable is infinite.

$$E(s(V)) = \sum_{n=1}^{\infty} n2^{-(2n-1)}C_{n-1} = \infty. \tag{4}$$

This means that if one draws a random variable according to the HMDID prior among all expressions involving f and x, the drawing process will never stop on average. Algorithm 1 is thus not computable in the general case.

4.2   Simultaneous Random Draws

As variables cannot be drawn individually as in Algorithm 1, we suggest to draw a complete sample using several draws simultaneously. For a multinomial distribution $(n; p_1, p_2, \ldots, p_K)$ with n draws and K values, the probability that a sample results in counts $n_1, n_2, \ldots, n_K$ per value is:

$$\frac{n!}{n_1!n_2!\ldots n_K!}p_1^{n_1}p_2^{n_2}\ldots p_K^{n_K}. \tag{5}$$

The most probable sample is obtained by maximizing Formula (5), which results into counts $n_k = p_k n$ per value according to maximum likelihood. For example, in the case of an equidistributed multionomial distribution with $p_k = 1/K$ and $n = K$ draws, Formula (5) is maximized for $n_k = 1$. As a consequence, all the values are drawn, which solves the inefficiency problem described in Section 4.1. Algorithm 2 exploits this drawing process using maximum likelihood recursively.

Algorithm 2. Simultaneous random draws

Require: K {Number of draws}
Ensure: $\mathcal{V} = \{V\}, |\mathcal{V}| \leq K$ {Sample of constructed variables}
 1: $\mathcal{V} = \emptyset$
 2: Start from the root of the hierarchy of the HMDID prior
 3: Compute the number of draws K_i per branch of the prior tree (original variable, rule, operand...)
 4: **for all** branch of the prior tree **do**
 5:     **if** terminal leaf of the prior tree (original variable or variable constructed with a complete formula) **then**
 6:         Add V in $\mathcal{V}$
 7:     **else**
 8:         Propagate the construction process recursively by assigning the K_i draws of each branch on the multinomial distribution at the sub-level of the prior tree
 9:     **end if**
10: **end for**

The draws are assigned on the original or constructed variables at each level of depth of the HMDID prior, which results in a number of draws that decreases with the depth of the prior hierarchy. In case of even choices (for example, one single draw among K variables), the draw is chosen randomly uniformly, with a priority for original variables when both original and constructed variables are possible. By assigning recursively the draws according to multinomial distributions at each branch of the hierarchy of the HMDID prior, with numbers of draws that decrease with the depth of the hierarchy, Algorithm 2 is both efficient and computable.

In Algorithm 2, the number of draws may be greater than 1 in some leaves of the prior hierarchy. This implies that the number of obtained variables can be inferior to the number of initial draws. To reach a given number of variables K, Algorithm 2 is first called with K draws, then called again successively with twice the number of draws at each call, until the number of required variables is reached or until no additional variable is built in the last call.

5   Experiments

The proposed method is evaluated by focusing on the following points: ability to construct large numbers of variables without problem of combinatorial explosion, robustness w.r.t. over-fitting and contribution to the classification performance.

5.1   Experimental Setup

The experiment performs comparisons with alternative relational data mining methods based on propositionalisation and with inductive logic programming Bayesian classifiers. The compared methods are:

- MODL is the method described in this paper. It exploits the following construction rules (cf. Section 2.2): *Selection, Count, Mode, CountDistinct, Mean, Median, Min, Max, StdDev, Sum*. The only parameter (see Section 4.2) is the number of variables to construct: 1, 3, 10, 30, 100, 300, 1000, 3000 and 10000 in the experiments. The variables are constructed using Algorithm 2 then filtered using Formula (2), which accounts both for construction and preprocessing cost. The filtered variables are used as input of a selective naive Bayes classifier with variable selection and model averaging (SNB) [1] [5], which is both robust and accurate in the case of very large numbers of variables.
- RELAGGS is a method similar to the Relaggs propositionalisation method [15]. It exploits the same construction rules as MODL and exhaustively constructs all the possible variables, except for the *Selection* rule that raises combinatorial problems. Instead, RELAGGS constructs all the rules based on counts per categorical value in the secondary tables. The data preprocessing and the SNB classifier are the same as for MODL, without accounting the construction of the variables (Formula (1) is used for filtering).
- 1BC is the first-order Bayesian classifier described in [16]. It can be considered as a propositionalisation method, with one variable per value in a secondary table. To preprocess the numerical values of each table, all numerical variables are discretized into equal frequency intervals. In the experiments, we use discretisation into 1, 2, 5, 10, 20, 50, 100 and 200 bins.
- 1BC2 is the successor of 1BC described in [17]. While 1BC applies propositionalisation, 1BC2 is a true first-order classifier. [2]

Fourteen relational datasets are considered in the experiments. The Auslan, CharacterTrajectories, JapaneseVowels, OptDigit and SpliceJunction datasets come from the UCI repository [1] and are related to the recognition of Australian sign language, characters from pen tip trajectories, Japanese speakers from cepstrum coefficients of two uttered vowels, handwritten digits from a matrix of 32*32 black and white pixels, and boundaries between intron and exon in gene sequences (DNA). These sequential or time series datasets are represented with one root table and a secondary table in zero to many relationship. The Diterpenes [10], Musk1, Musk2 [8] and Mutagenesis [23] datasets are related to molecular chemistry. The Mutagenesis dataset is a graph with molecules (*lumo, logp* plus the class variable) in a root table, atoms (*element, type, charge*) as vertices and bonds (*bondtype*) as edges. The Miml dataset [25] is related to image recognition, with five different target variables. Table 1 gives a summary of these datasets.

All the experiments are performed using a stratified 10-fold cross validation. In each train fold, the variables are constructed and selected and the classifier is trained, while the test accuracy is evaluated in the test fold.

---

[1] The SNB classsifier is available as a shareware at http://www.khiops.com.

[2] 1BC et 1BC2 are available at
http://www.cs.bris.ac.uk/Research/MachineLearning/1BC/. I am grateful to Nicolas Lachiche for providing support and advices regarding their use.

Table 1. Relational datasets: number of instances, records in the secondary tables, categorical and numerical variables, classes, and accuracy of the majority class

Dataset	Instances	Records	Cat.vars	Num.vars	Classes	Maj.
Auslan	2,565	146,949	1	23	96	0.011
CharacterTrajectories	2,858	487,277	1	4	20	0.065
Diterpenes	1,503	30,060	2	1	23	0.298
JapaneseVowels	640	9961	1	13	9	0.184
MimlDesert	2,000	18,000	1	15	2	0.796
MimlMountains	2,000	18,000	1	15	2	0.771
MimlSea	2,000	18,000	1	15	2	0.710
MimlSunset	2,000	18,000	1	15	2	0.768
MimlTrees	2,000	18,000	1	15	2	0.720
Musk1	92	476	1	166	2	0.511
Musk2	102	6,598	1	166	2	0.618
Mutagenesis	188	10,136	3	4	2	0.665
OptDigits	5,620	5,754,880	1	3	10	0.102
SpliceJunction	3,178	191,400	2	1	3	0.521

As for computational efficiency, the overhead of the construction algorithm is negligible w.r.t. the overall training time. Actually, Algorithm 2 consists in drawing a sample of constructed variables with their construction formulas. This algorithm mainly relies on the exploration of the construction domain (data structure and set of construction rules). The *Selection* rule requires one reading step of each secondary table to build the actual selection operands. This reading step dominates the time of the variable construction process, and is itself dominated by the data preparation and modeling steps of the classifier.

5.2   Results

The mean test accuracy versus the number of constructed variables per dataset is reported in Figure 3, with the standard deviation represented by error bars. The baseline (horizontal gray dashed line) is the accuracy of the majority classifier. The MODL performance is reported for each number of actually constructed variables. The RELAGGS performance is reported only once, with the number of constructed variables resulting from an exhaustive application of the construction rules. The 1BC and 1BC2 performance are reported for each bin number of the discretization preprocessing, with a number of constructed variables based on the total number of actually different values per variable.

Control of Variable Construction. The RELAGGS, 1BC and 1BC2 methods have little control on the size of the constructed representation space, which strongly varies with the complexity of the data structure and the number of values in the dataset. For example, the size of the representation space goes from a

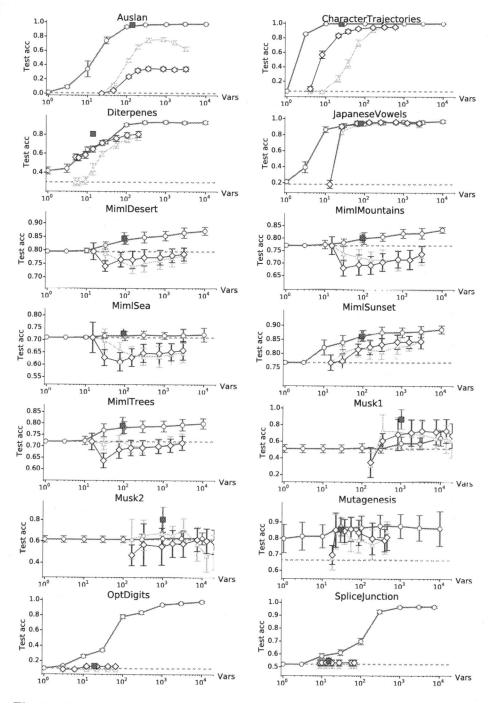

Fig. 3. Test accuracy versus number of constructed variables per dataset.
MODL: ○ red   RELAGGS: ■ green   1BC: △ cyan   1BC2: ◇ blue.

few tens of variables (SpliceJunction dataset) to around 20,000 variables (Musk2 dataset) for the 1BC method. The MODL method is much more expressive than the alternative methods, with the *Selection* rule which can build conjunctions of selection terms at any granularity. Still, Algorithm 2 allows to control the combinatorial exploration of this huge space and to construct the requested number of variables, as shown in Figure 3.

Examples of Constructed Variables. For the Mutagenesis dataset, in one train fold, the null cost ($\approx NEnt(Y)$) is 115.08 for the encoding of the classes of the 177 train instances. Among the 10,000 generated variables, only 618 are identified as informative. The original variable *lumo* in the root table has a low construction cost of 2.08. The simplest informative constructed variable involves one construction rule having one operand: *Count(Atoms)* with a construction cost of 5.08. The most complex informative constructed variable involves three construction rules, including the *Selection* rule with two selection terms. This rule *Sum(Selection(Atoms, (type ≤ 23.5 and charge ≤ -0.0685)), Count-Distinct(AdjacentBonds, bondtype))* has a construction cost of 17.66, which is not negligible compared to the null cost for this small dataset. This illustrates the ability of Algorithm 2 to build rather complex variables and of Formula (2) to filter the constructed variables.

Test Accuracy. The 1BC and 1BC2 often obtain similar performance, except for Auslan where 1BC is better and CharacterTrajectories and Diterpenes where 1BC2 dominates 1BC. Both methods get generally better performance as the number of bins increases, but they suffer from over-fitting, especially in the Miml image datasets where their performance is under the baseline. The RELAGGS propositionalisation method is used together with the SNB classifier (same as for MODL) and inherits from its accuracy and robustness. It always dominates the 1BC and 1BC2 methods. In the Musk1 and Musk2 datasets, the MODL method is not better than the baseline. Actually, these datasets are very small (around 100 instances), have a complex data structure (166 variables in the secondary table) and the classes are not well separable. Altogether, the construction penalization in Formula (2) results in rejecting almost all constructed variables. More instances would be necessary to learn the concept. In all other datasets, the MODL method benefits from its expressive construction language and obtains better test accuracy than the alternative methods. Remarkably, it often achieves good test accuracy with fewer variables than the alternative methods, and its performance never decreases with the number of constructed variables, which is the only user parameter in the approach.

Robustness. In order to evaluate the robustness of our approach, the classes have been randomly shuffled in each dataset before performing the experiment again. Two experiments are performed, one using criterion $cost_{CP}$ of Formula (2) (accounting for the construction cost of the variables), the other using $cost_P$ of Formula (1) (not accounting for the construction cost). The number of selected variables is collected in both cases. The used preprocessing methods [3,4] are very robust. However, when 10,000 variables are constructed, on average 5 variables per dataset are wrongly selected, with more than 20 variables for the JapaneseVowels dataset. When the construction regularization is used (criterion $cost_{CP}$), the method is extremely robust: the overall 1.4 millions of constructed variables over all the datasets and folds of the cross-validation are all identified as information-less, without any exception.

6 Conclusion

In this paper, we have suggested a framework aiming at automating variable construction for supervised classification. On the basis of a description of a multi-table schema and a set of construction rules, we have suggested a prior distribution over the space of all variables that can be constructed. We have demonstrated that drawing variables according to this prior distribution raises critical problems of inefficiency and non-computability, then proposed an efficient algorithm that is able to perform many simultaneous draws in order to build a tabular representation with the required number of variables. The experiments indicate that the proposed method solves the problem of combinatorial explosion that usually limits the approaches which construct variables by a systematic application of construction rules, and the problem of over-fitting that occurs in case of representations with very large numbers of variables. The obtained classification performance are promising. In future work, we plan to extend the description of the variable construction domain by providing additional construction rules with potential specialization per application domain. Another research direction consists in drawing constructed variables according to their posterior distribution rather than their prior distribution. Finally, accounting for correlations between the constructed variables so as to avoid the risk of constructing many variants of the same variables raises another challenge.

References

1. Bache, K., Lichman, M.: UCI machine learning repository (2013), http://archive.ics.uci.edu/ml
2. Blockeel, H., De Raedt, L., Ramon, J.: Top-Down Induction of Clustering Trees. In: Proceedings of the Fifteenth International Conference on Machine Learning, pp. 55–63 (1998)
3. Boullé, M.: A Bayes optimal approach for partitioning the values of categorical attributes. Journal of Machine Learning Research 6, 1431–1452 (2005)
4. Boullé, M.: MODL: a Bayes optimal discretization method for continuous attributes. Machine Learning 65(1), 131–165 (2006)
5. Boullé, M.: Compression-based averaging of selective naive Bayes classifiers. Journal of Machine Learning Research 8, 1659–1685 (2007)
6. Chapman, P., Clinton, J., Kerber, R., Khabaza, T., Reinartz, T., Shearer, C., Wirth, R.: CRISP-DM 1.0: step-by-step data mining guide. Tech. rep., The CRISP-DM consortium (2000)

7. Cover, T., Thomas, J.: Elements of information theory. Wiley-Interscience, New York (1991)
8. De Raedt, L.: Attribute-Value Learning Versus Inductive Logic Programming: The Missing Links (Extended Abstract). In: Page, D.L. (ed.) ILP 1998. LNCS, vol. 1446, pp. 1–8. Springer, Heidelberg (1998)
9. Džeroski, S., Lavrač, N.: Relational Data Mining. Springer-Verlag New York, Inc. (2001)
10. Džeroski, S., Schulze-Kremer, S., Heidtke, K.R., Siems, K., Wettschereck, D., Blockeel, H.: Diterpene Structure Elucidation From 13C NMR Spectra With Inductive Logic Programming. Applied Artificial Intelligence, Special Issue on First-Order Knowledge Discovery in Databases 12(5), 363–383 (1998)
11. Efron, B., Tibshirani, R.: An introduction to the bootstrap. Monographs on Statistics and Applied Probability, vol. 57. Chapman & Hall/CRC, New York (1993)
12. Guyon, I., Gunn, S., Nikravesh, M., Zadeh, L. (eds.): Feature Extraction: Foundations And Applications. Springer (2006)
13. Knobbe, A.J., Blockeel, H., Siebes, A., Van Der Wallen, D.: Multi-Relational Data Mining. In: Proceedings of Benelearn 1999 (1999)
14. Kramer, S., Flach, P.A., Lavrač, N.: Propositionalization approaches to relational data mining. In: Džeroski, S., Lavrač, N. (eds.) Relational Data Mining, ch. 11, pp. 262–286. Springer (2001)
15. Krogel, M.-A., Wrobel, S.: Transformation-based learning using multirelational aggregation. In: Rouveirol, C., Sebag, M. (eds.) ILP 2001. LNCS (LNAI), vol. 2157, pp. 142–155. Springer, Heidelberg (2001)
16. Lachiche, N., Flach, P.: Ibc: A first-order bayesian classifier. In: Proceedings of the 9th International Workshop on Inductive Logic Programming, pp. 92–103. Springer (1999)
17. Lachiche, N., Flach, P.A.: 1bc2: A true first-order bayesian classifier. In: Matwin, S., Sammut, C. (eds.) ILP 2002. LNCS (LNAI), vol. 2583, pp. 133–148. Springer, Heidelberg (2003)
18. Liu, H., Motoda, H.: Feature Extraction, Construction and Selection: A Data Mining Perspective. Kluwer Academic Publishers (1998)
19. Pyle, D.: Data preparation for data mining. Morgan Kaufmann Publishers, Inc., San Francisco (1999)
20. Rissanen, J.: Modeling by shortest data description. Automatica 14, 465–471 (1978)
21. Rissanen, J.: A universal prior for integers and estimation by minimum description length. Annals of Statistics 11(2), 416–431 (1983)
22. Shannon, C.: A mathematical theory of communication. Tech. Rep. 27, Bell Systems Technical Journal (1948)
23. Srinivasan, A., Muggleton, S., King, R., Sternberg, M.: Mutagenesis: ILP experiments in a non-determinate biological domain. In: Wrobel, S. (ed.) Proceedings of the 4th International Workshop on Inductive Logic Programmin (ILP 1994). GMD-Studien, vol. 237, pp. 217–232 (1994)
24. Vens, C., Ramon, J., Blockeel, H.: Refining aggregate conditions in relational learning. In: Fürnkranz, J., Scheffer, T., Spiliopoulou, M. (eds.) PKDD 2006. LNCS (LNAI), vol. 4213, pp. 383–394. Springer, Heidelberg (2006)
25. Zhou, Z.H., Zhang, M.L.: Multi-instance multi-label learning with application to scene classification. In: Schölkopf, B., Platt, J., Hofmann, T. (eds.) Advances in Neural Information Processing Systems (NIPS 2006), vol. i, pp. 1609–1616. MIT Press, Cambridge (2007)

Consistency of Losses for Learning from Weak Labels

Jesús Cid-Sueiro[1,*], Darío García-García[2], and Raúl Santos-Rodríguez[3]

[1] Universidad Carlos III de Madrid, Spain
jcid@tsc.uc3m.es
[2] OMNIA Team, Commonwealth Bank of Australia
dario.garcia@cba.com.au
[3] Image Processing Lab., Univ. de Valencia, Spain
raulsantosrodriguez@gmail.com

Abstract. In this paper we analyze the consistency of loss functions for learning from weakly labelled data, and its relation to properness. We show that the consistency of a given loss depends on the mixing matrix, which is the transition matrix relating the weak labels and the true class. A linear transformation can be used to convert a conventional classification-calibrated (CC) loss into a *weak* CC loss. By comparing the maximal dimension of the set of mixing matrices that are admissible for a given CC loss with that for proper losses, we show that classification calibration is a much less restrictive condition than properness. Moreover, we show that while the transformation of conventional proper losses into a weak proper losses does not preserve convexity in general, conventional convex CC losses can be easily transformed into weak and convex CC losses. Our analysis provides a general procedure to construct convex CC losses, and to identify the set of mixing matrices admissible for a given transformation. Several examples are provided to illustrate our approach.

1 Introduction

The analysis of the conditions required to any loss function in order to estimate posterior class probabilities, and those required to optimize classification performance, has been an important area of research for many years, both in binary and multiclass settings [7, 19, 21], providing insights on the influence of loss functions for learning. Most of this work has focused on the standard fully or semi-supervised cases. However, many interesting real-world problems do not fit into those categories. This has spanned a broad literature on non-standard learning paradigms, being weak supervision one of the most widespread of them. In a weakly supervised learning problem, each instance is assumed to be labelled as belonging to one of several candidate categories, at most one of them being true. This paper studies the asymptotic properties of loss functions in this more general learning setting.

Weakly supervised learning has attracted recent interest due to its suitability to model several scenarios in bioinformatics or computer vision. We trace this formulation back

---

* This research has been partly supported by the spanish Ministry of Economy and Competitiveness (MINECO) under projects TEC2011-22480, PRI-PIBIN-2011- 1266 and LIFE-VISION TIN2012-38102-C03-01.

T. Calders et al. (Eds.): ECML PKDD 2014, Part I, LNCS 8724, pp. 197–210, 2014.

to [3, 12, 13] for the noisy label case, and also [8] and [11] for more general scenarios were samples may have multiple labels (including the true class). In the last decade several authors addressed this and related problems under different names, including partial labels [6, 8, 17], ambiguous labels [10], multiple labels [11], or crowd learning [18]. It is also a particular case of the more general problems of learning from soft labels [5], learning from measurements [14] or learning from candidate labelling sets [16].

There are fundamental limitations to learning from weak labels. If the statistical mixing model relating the true class to the weak labels is unknown, there is no way to infer it from a sample set with weak labels only, at least without making additional assumptions. This is a well-known problem in semi-supervised learning. However, it can be shown that the full knowledge of the mixing process is actually not necessary for learning purposes. In [4] a general procedure was proposed to transform a standard (i.e. fully-supervised) proper loss into a *weak loss* that is also proper, in the sense that posterior class probabilities can be estimated provided that the mixing processes is restricted to lie in certain linear subspace. Unfortunately, the proposed method scales exponentially with the number of classes. Moreover, in general, the resulting weak losses are non convex.

Since posterior probability estimation in weakly labelled scenarios requires very strong conditions on losses and mixing matrices, in this paper we focus on the analysis of losses for classification. Up to our knowledge, there is no general approach to the consistency problem for learning from weak labels in the literature.

Specifically our contributions in this paper are the following:

1. We provide a general theoretical analysis of consistency. We present if-and-only-if conditions for classification calibration, and we show that, in general, the underlying assumptions about the mixing process are relaxed (with respect to those of probability estimation) if only classification consistency is required. We show that consistent losses can be obtained by a linear transformation of any conventional consistent loss. This means that a machine learning practitioner can keep working with the same type of losses she/he is familiar with (Section 3.1).

2. Properness is significantly more restrictive than consistency. If one is worried about classification errors, and not about posterior probability estimation, the dimension of the maximal set of mixing matrices that is covered by a consistent loss is higher than that covered by a proper loss (Section 3.2).

3. If one has no information at all about the mixing matrix, one can assume independent labels. For that case, a straightforward approach is provided (Section 4).

4. Making a difference with the case of weak proper losses, convexity can still be preserved when transforming a conventional loss into a weak consistent loss (Section 5).

2   Formulation

2.1   Notation

Vectors are written in boldface, matrices in boldface capital and sets in calligraphic letters. For any integer n, $\mathbf{e}_i^n$ is a n-dimensional unit vector with all zero components apart from the i-th component which is equal to one, and $\mathbb{1}_n$ is a n-dimensional all-ones vector. Superindex $\intercal$ denotes transposition. We will use $\Psi()$ to denote a loss based on weak labels (for brevity, "weak loss"), and $\tilde{\Psi}$ to losses based on the true class. The number of classes is c, and the number of possible weak label vectors is $d \leq 2^c$. $|\mathbf{z}|$ is the number of nonzero elements in $\mathbf{z}$. The set of all $d \times c$ matrices with stochastic columns is $\mathcal{M} = \{\mathbf{M} \in [0,1]^{d \times c} : \mathbf{M}^\intercal \mathbb{1}_d = \mathbb{1}_c\}$.

2.2   Learning from Weak Labels

Let $\mathcal{X}$ be a sample space, $\mathcal{Y} = \{\mathbf{e}_j^c, j = 0, 1, \ldots, c-1\}$ a set of labels, and $\mathcal{Z} = \{\mathbf{b}_1, \ldots, \mathbf{b}_d\} \subset \{0,1\}^c$ a set of weak or partial label vectors. Sample $(\mathbf{x}, \mathbf{z}) \in \mathcal{X} \times \mathcal{Z}$ is drawn from an unknown distribution P.

Weak label vector $\mathbf{z} \in \mathcal{Z}$ is a noisy version of the true class $\mathbf{y} \in \mathcal{Y}$. A common assumption [1, 6, 9, 11] is that the true class is always present in $\mathbf{z}$, i.e., $z_j = 1$ when $y_j = 1$, but this assumption is not required in our setting, which admits noisy label scenarios (as, for instance, in [18]). Without loss of generality, we assume that $\mathcal{Z}$ contains only weak labels with nonzero probability (i.e. $P\{\mathbf{z} = \mathbf{b}\} > 0$ for any $\mathbf{b} \in \mathcal{Z}$).

The dependency between $\mathbf{z}$ and $\mathbf{y}$ is modelled through an arbitrary $d \times c$ conditional mixing probability matrix $\mathbf{M}(\mathbf{x}) \in \mathcal{M}$ with components

$$m_{ij}(\mathbf{x}) = P\{\mathbf{z} = \mathbf{b}_i | y_j = 1, \mathbf{x}\} \tag{1}$$

where $\mathbf{b}_i \in \mathcal{Z}$ is the i-th element of $\mathcal{Z}$. Defining posterior probability vectors $\mathbf{p}(\mathbf{x})$ and $\boldsymbol{\eta}(\mathbf{x})$ with components $p_i = P\{\mathbf{z} = \mathbf{b}_i | \mathbf{x}\}$ and $\eta_j = P\{\mathbf{y} = \mathbf{e}_j^c | \mathbf{x}\}$, we can write $\mathbf{p}(\mathbf{x}) = \mathbf{M}(\mathbf{x})\boldsymbol{\eta}(\mathbf{x})$. In general, the dependency with $\mathbf{x}$ will be omitted and we will write, for instance,

$$\mathbf{p} = \mathbf{M}\boldsymbol{\eta}. \tag{2}$$

In general, the mixing matrix could depend on $\mathbf{x}$, though a constant mixing matrix is a common assumption [1, 9, 11, 18], as well as the statistical independence of the incorrect labels [1, 9, 11]. Assuming a constant matrix is not required in our analysis. Any property derived for $\mathbf{M}$ can be by extended to a property that must be satisfied by $\mathbf{M}(\mathbf{x})$ for all $\mathbf{x}$.

The goal is to infer $\mathbf{y}$ given $\mathbf{x}$ without knowing model P. To do so, a set of i.i.d. weakly labelled samples, $\mathcal{S} = \{(\mathbf{x}_k, \mathbf{z}_k), k = 1, \ldots, K\} \sim P$ is available. True classes $\mathbf{y}_k$ are not observed.

2.3   Classification Calibration, Ranking Calibration and Properness

The goal of the learning algorithm is to find an accurate class predictor using a weakly labelled training set. Our focus is to determine consistent predictors that would eventually find the Bayesian predictor if the size of the training set becomes infinity.

Let $\mathcal{F}$ be a function class and $\Psi(\mathbf{z}, \mathbf{f})$ with $\mathbf{f} \in \mathcal{F}$ be a loss function based on weak label $\mathbf{z}$, $\boldsymbol{\Psi} = (\Psi(\mathbf{b}_1, \mathbf{f}), \ldots, \Psi(\mathbf{b}_d, \mathbf{f}))^\intercal$, where $\mathbf{b}_i$ is the i-th element in $\mathcal{Z}$ (according to some arbitrary ordering). The $\boldsymbol{\Psi}$-risk of a function $\mathbf{f}$ from the function class $\mathcal{F}$ is

$$R_{\boldsymbol{\Psi}}(\mathbf{f}) = \mathbb{E}_{\mathcal{X}\mathcal{Z}}\{\Psi(\mathbf{z}, \mathbf{f}(\mathbf{x}))\} \tag{3}$$

and the minimum $\boldsymbol{\Psi}$-risk is

$$R_{\boldsymbol{\Psi}}^* = \inf_{\mathbf{f} \in \mathcal{F}} R_{\boldsymbol{\Psi}}(\mathbf{f}) \tag{4}$$

Note that our definition of the risk functions differs from the conventional setting in that it is a function of the weak labels. We are interested in the risk of $\mathbf{f}$ as a predictor of the true class through some function $\text{pred}(\mathbf{x}) \in \text{argmax}_j \{f_j(\mathbf{x})\}$. This risk is defined as

$$R(\mathbf{f}) = \mathbb{E}_{\mathcal{X}\mathcal{Y}}\{\mathbf{1}[\text{pred}(\mathbf{f}(\mathbf{x})) \neq \mathbf{y}]\} \tag{5}$$

where $\mathbf{1}[.]$ is the indicator function. It is well known that the minimum possible risk is $R^* = \mathbb{E}\{1 - \max_{\mathbf{y}}\{P(\mathbf{y}|\mathbf{x})\}\}$, and it is achieved by the Bayesian predictor $\text{argmax}_{\mathbf{y}} P(\mathbf{y}|\mathbf{x})$. If $\hat{\mathbf{f}}_K$ minimizes the empirical $\boldsymbol{\Psi}$-risk for a training set of size K, one would expect $R_{\boldsymbol{\Psi}}(\hat{\mathbf{f}}_K)$ to converge to $R_{\boldsymbol{\Psi}}^*$ in probability as K goes to infinity. When that makes $R(\hat{\mathbf{f}}_K)$ converge to R^* in probability, we say that $\boldsymbol{\Psi}$ is consistent. It turns out that consistency is strongly related to the notion of classification calibration [2] [19] (also related to Fisher consistency [15] or infinite-sample consistency [21]). To simplify the discussion, our definitions assume that $\inf_{\mathbf{f}} \mathbb{E}_{\mathbf{z}}\{\boldsymbol{\Psi}(\mathbf{z}, \mathbf{f})\}$ is reachable by some minimizer $\mathbf{f}^*$.

Definition 1 (Classification calibration). *Weak loss* $\boldsymbol{\Psi}(\mathbf{z}, \mathbf{f})$ *is classification calibrated (CC) to predict* $\mathbf{y}$ *from* $\mathbf{f}$ *if* $\mathbf{f}^* \in \text{arg min}_{\mathbf{f}} \mathbb{E}_{\mathbf{z}}\{\boldsymbol{\Psi}(\mathbf{z}, \mathbf{f})\}$ *satisfies* $(\eta_i > \max_{j \neq i} \eta_j \Rightarrow f_i^* > \max_{j \neq i} f_j^*)$.

Zhang [21] and Tewari [19] have shown that classification calibration is essentially equivalent to consistency in the multiclass fully supervised setting (i.e., when $\mathcal{Z} = \mathcal{Y}$ and $\mathbf{z} = \mathbf{y}$). For our analysis, two stronger conditions will be helpful. The first one imposes that the predictor preserves the order of all classes, and not only the predicted one.

Definition 2 (Ranking calibration). *Weak loss* $\boldsymbol{\Psi}(\mathbf{z}, \mathbf{f})$ *is ranking calibrated (RC) to predict* $\mathbf{y}$ *from* $\mathbf{f}$ *if* $\mathbf{f}^* \in \text{arg min}_{\mathbf{f}} \mathbb{E}_{\mathbf{z}}\{\boldsymbol{\Psi}(\mathbf{z}, \mathbf{f})\}$ *satisfies* $(\eta_i > \eta_j \Rightarrow f_i^* > f_j^*)$.

The second one imposes that the predictor is a good estimator of the posterior class probabilities.

Definition 3 (Properness). *Weak loss* $\boldsymbol{\Psi}(\mathbf{z}, \mathbf{f})$ *is proper to predict* $\mathbf{y}$ *from* $\mathbf{f}$ *if* $\eta \in \text{arg min}_{\mathbf{f}} \mathbb{E}_{\mathbf{z}}\{\boldsymbol{\Psi}(\mathbf{z}, \mathbf{f})\}$. *The loss is strictly proper if* η *is the unique minimizer.*

Proper losses in the weakly labelled scenario were examined in [4]. Our main focus in the following sections is to study classification consistency and comparatively discuss the potential benefits of avoiding the estimation of posterior class probabilities.

3   Weak Loss Characterization

3.1   Virtual Label Representation

Classification calibration, ranking calibration or properness of a given loss depend on the mixing matrix $\mathbf{M}$. To make this dependency explicit, we will say that $\boldsymbol{\Psi}$ is M-CC,

M-RC or M-proper if it is CC, RC or proper for a mixing matrix $\mathbf{M}$, respectively. Additionally, given a set $\mathcal{Q} \subset \mathcal{M}$ of mixing matrices (remember that $\mathcal{M}$ is the set of all left-stochastic matrices), we will say that $\boldsymbol{\Psi}$ is $\mathcal{Q}$-CC, $\mathcal{Q}$-RC or $\mathcal{Q}$-proper if it has the corresponding property for any $\mathbf{M} \in \mathcal{Q}$.

According to definitions 1, 2 and 3, a loss $\boldsymbol{\Psi}$ will be CC, RC or proper depending solely on the characteristics of the expected value, $\mathbb{E}_{\mathbf{z}}\{\Psi(\mathbf{z}, \mathbf{f})\}$. Therefore, the properties of any loss $\boldsymbol{\Psi}$ will be the same than that of any other loss with the same expected value. In particular, let us define the *equivalent* loss

$$\tilde{\Psi}(\mathbf{y}, \mathbf{f}) = \mathbf{y}^{\mathsf{T}} \mathbf{M}^{\mathsf{T}} \boldsymbol{\Psi} \tag{6}$$

or, equivalently, in vector form,

$$\tilde{\boldsymbol{\Psi}}(\mathbf{f}) = \mathbf{M}^{\mathsf{T}} \boldsymbol{\Psi}(\mathbf{f}) \tag{7}$$

Note that $\tilde{\Psi}$ depends on the true classes, and is only useful for fully supervised scenarios. Using (2) and (7) it is straightforward to show that $\mathbb{E}_{\mathbf{z}}\{\Psi(\mathbf{z}, \mathbf{f})\} = \mathbb{E}_{\mathbf{z}}\{\mathbf{z}^{\mathsf{T}} \boldsymbol{\Psi}\} = \mathbb{E}_{\mathbf{z}}\{\mathbf{y}^{\mathsf{T}} \tilde{\boldsymbol{\Psi}}\}$. Thus, $\boldsymbol{\Psi}$ and $\tilde{\boldsymbol{\Psi}}$ have the same expected value. This connection between a loss and its equivalent loss is used in [4] to derive the properness of losses in a weakly labelled scenario, and can be used in a straightforward way to prove the following extension:

Theorem 1. *Consider a weak loss* $\Psi : \mathcal{Z} \times \mathbb{R}^c \to \mathbb{R}^+$ *and a mixing matrix* $\mathbf{M} \in \mathcal{M}$, *and let the* equivalent *loss* $\tilde{\Psi} : \mathcal{Y} \times \mathbb{R}^c \to \mathbb{R}^+$ *be given by (6).*

- Ψ *is (strictly)* $\mathbf{M}$-*proper iff* $\tilde{\Psi}$ *is (strictly) proper.*
- Ψ *is* $\mathbf{M}$-*RC iff* $\tilde{\Psi}$ *is RC.*
- Ψ *is* $\mathbf{M}$-*CC iff* $\tilde{\Psi}$ *is CC.*

Theorem 1 states an iff connection between the CC of a weak loss and the CC of its equivalent loss. Additionally, since the consistency is also a function of the conditional risk, the consistency of a weak loss is also equivalent to the consistency of the equivalent loss. Thus, we can conclude that the equivalence between classification calibration and consistency in supervised learning can be extended to weak losses.

Eq. (7) suggests a way to generate a suitable partial loss from a conventional loss: let $\tilde{\mathbf{Y}}$ be a left-inverse of $\mathbf{M}$ (i.e. $\tilde{\mathbf{Y}}\mathbf{M} = \mathbf{I}$). If $\tilde{\boldsymbol{\Psi}}$ is a CC, RC or strictly proper loss, then the partial loss

$$\boldsymbol{\Psi}(\mathbf{f}) = \tilde{\mathbf{Y}}^{\mathsf{T}} \tilde{\boldsymbol{\Psi}}(\mathbf{f}) \tag{8}$$

is CC, RC or strictly proper, respectively (because $\mathbf{M}^{\mathsf{T}} \boldsymbol{\Psi}(\mathbf{f}) = \mathbf{M}^{\mathsf{T}} \tilde{\mathbf{Y}}^{\mathsf{T}} \tilde{\boldsymbol{\Psi}}(\mathbf{f}) = \tilde{\boldsymbol{\Psi}}(\mathbf{f})$, and the conditions of Th. 1 are satisfied). The analysis of losses based on (8) is the main issue of the following sections.

Note that, if $\mathbf{z} = \mathbf{b}_i$, $\Psi(\mathbf{z}, \mathbf{f}) = (\mathbf{e}_i^d)^{\mathsf{T}} \boldsymbol{\Psi}(\mathbf{f})$ and, using (8), we get $\Psi(\mathbf{z}, \mathbf{f}) = (\mathbf{e}_i^d)^{\mathsf{T}} \tilde{\mathbf{Y}}^{\mathsf{T}} \tilde{\boldsymbol{\Psi}}(\mathbf{f}) = (\tilde{\mathbf{Y}} \mathbf{e}_i^d)^{\mathsf{T}} \mathbf{M}^{\mathsf{T}} \boldsymbol{\Psi}(\mathbf{f})$. Comparing this expression with (6), we can interpret the columns of $\tilde{\mathbf{Y}}$ in (8) as *virtual* labels, to be used instead of $\mathbf{y}$ when the true class is unknown. Thus, $\tilde{\mathbf{Y}}$ is a matrix of virtual labels.

3.2 Maximal Sets of Mixing Matrices

Eqs. (6) and (8) show that a weak loss may be CC, RC or proper for a wide set of mixing matrices. In particular, if $\mathcal{W}$ is the subset of all right-inverse matrices of $\tilde{\mathbf{Y}}$ that are in $\mathcal{M}$, $\boldsymbol{\Psi}(\mathbf{f})$ is $\mathcal{W}$-CC, $\mathcal{W}$-RC or $\mathcal{W}$-proper provided that $\tilde{\boldsymbol{\Psi}}$ is CC, RC or proper, respectively. This has a practical relevance because the mixing matrix may be partially unknown, or may depend on $\mathbf{x}$, and we could be interested in finding weak losses with the desired property for a large set of mixing matrices.

Our main interest in this section is to find the largest set of mixing matrices having the desired property. For a given matrix of virtual labels, $\tilde{\mathbf{Y}}$, we will define the maximal sets $\mathcal{Q}_{cc}(\tilde{\mathbf{Y}})$, $\mathcal{Q}_{rc}(\tilde{\mathbf{Y}})$ and $\mathcal{Q}_p(\tilde{\mathbf{Y}})$ as the largest set of mixing matrices for which a weak loss given by (8) is CC, RC or strictly proper, respectively, for any CC, RC or strictly proper loss $\tilde{\boldsymbol{\Psi}}$, respectively. In this section we show that, in general, $\mathcal{Q}_{cc}(\tilde{\mathbf{Y}}) = \mathcal{Q}_{rc}(\tilde{\mathbf{Y}}) \supset \mathcal{Q}_p(\tilde{\mathbf{Y}}) = \mathcal{W}$ and, thus, there is a penalty in the size of the maximal set if accurate posterior class probability estimates are required.

We start showing that $\mathcal{Q}_p(\tilde{\mathbf{Y}})$ is equal to the set of all admissible right-inverses of the virtual label matrix.

Theorem 2. *Given a strictly proper loss $\tilde{\boldsymbol{\Psi}}(\mathbf{f}, \mathbf{y})$ and a virtual (full rank) label matrix $\tilde{\mathbf{Y}}$, the weak loss $\boldsymbol{\Psi}(\mathbf{f}) = \tilde{\mathbf{Y}}^{\mathsf{T}} \tilde{\boldsymbol{\Psi}}(\mathbf{f})$ is strictly $\mathcal{W}$-proper, for $\mathcal{W} = \{\mathbf{M} | \tilde{\mathbf{Y}}\mathbf{M} = \alpha\mathbf{I}, \alpha > 0\} \cap \mathcal{M}$, and $\mathcal{W}$ is maximal.*

Moreover, the dimension of the maximal set is $D_p \leq dc - c^2 - c + 1$.

Proof. See the Appendix.

Now, we characterize $\mathcal{Q}_{rc}(\tilde{\mathbf{Y}})$.

Theorem 3. *Given a RC loss $\tilde{\boldsymbol{\Psi}}(\mathbf{f}, \mathbf{y})$ and a full rank matrix $\tilde{\mathbf{Y}}$, the weak loss $\boldsymbol{\Psi}(\mathbf{f}) = \tilde{\mathbf{Y}}^{\mathsf{T}} \tilde{\boldsymbol{\Psi}}(\mathbf{f})$ is $\mathcal{W}$-RC, for $\mathcal{W} = \{\mathbf{M} | \tilde{\mathbf{Y}}\mathbf{M} \in \mathcal{V}\} \cap \mathcal{M}$, where $\mathcal{V}$ is the set of matrices in the form $\lambda\mathbf{I} + \mathbb{1}_c\mathbf{v}^{\mathsf{T}}$ (for arbitrary $\lambda \in \mathbb{R}^+$ and $\mathbf{v} \in \mathbb{R}^c$).*

Moreover, $\mathcal{W}$ is maximal, and its dimension is $D_{rc} \leq dc - c^2 + 1$.

Proof. See the Appendix.

The above theorem shows that the matrices of the maximal set can be written in the form

$$\mathbf{M} = \mathbf{M}_0(\lambda\mathbf{I} + \mathbb{1}_c\mathbf{v}^{\mathsf{T}}) \tag{9}$$

where $\mathbf{M}_0$ is an arbitrary right inverse of the virtual label matrix and $\lambda > 0$ and $\mathbf{v}$ are arbitrary parameters. As we will see later, these extra parameters may have a significant value: for some virtual label matrices, the resulting loss is not proper for any mixing matrix, while classification calibration is preserved for a large set of mixing matrices.

The proof of Theorem 3 can be used step by step to prove an identical result for classification calibrated losses.

Theorem 4. *Given a CC loss $\tilde{\boldsymbol{\Psi}}(\mathbf{f}, \mathbf{y})$ and a full rank matrix $\tilde{\mathbf{Y}}$, the weak loss $\boldsymbol{\Psi}(\mathbf{f}) = \tilde{\mathbf{Y}}^{\mathsf{T}} \tilde{\boldsymbol{\Psi}}(\mathbf{f})$ is $\mathcal{W}$-CC, for $\mathcal{W} = \{\mathbf{M} | \tilde{\mathbf{Y}}\mathbf{M} \in \mathcal{V}\} \cap \mathcal{M}$, where $\mathcal{V}$ is the set of matrices in the form $\lambda\mathbf{I} + \mathbb{1}_c\mathbf{v}^{\mathsf{T}}$ (for arbitrary $\lambda \in \mathbb{R}^+$ and $\mathbf{v} \in \mathbb{R}^c$).*

Moreover, $\mathcal{W}$ is maximal, and its dimension is $D_{cc} \leq dc - c^2 + 1$.

Proof. See Appendix.

Note that, although classification calibration is a less restrictive condition than ranking calibration, $\mathcal{Q}_{cc}(\tilde{\mathbf{Y}}) = \mathcal{Q}_{rc}(\tilde{\mathbf{Y}})$. This, however, does not imply that any CC loss is also RC: if $\tilde{\boldsymbol{\Psi}}$ is CC but not RC, weak loss $\tilde{\mathbf{Y}}\tilde{\boldsymbol{\Psi}}$ will be CC but not RC. But Theorem 4 shows that by relaxing the ranking calibration to classification calibration, the set of admissible mixing matrices does not change.

4 Losses for Independent Labels

The main drawback for the application of conventional CC, RC or proper losses to construct weak losses is that, in general, the mixing matrix is unknown, and we may have no a priori information about a (small enough) set $\mathcal{Q}$ containing the true mixing matrix.

As an alternative, we can construct consistent labels starting from some simplifying assumptions on the mixing matrix. An appealing choice consists in assuming that the noisy labels are statistically independent, that is,

$$P(\mathbf{z}|\mathbf{y} = \mathbf{e}_m^c) = \alpha_m^{z_m}(1 - \alpha_m)^{1-z_m}\beta_m^{|\mathbf{z}|-1}(1 - \beta_m)^{c-|\mathbf{z}|} \tag{10}$$

for some parameters $\alpha_m \subset [0, 1]$ and $\beta_m \subset [0, 1]$. Recall that $|\mathbf{z}|$ is the number of nonzero elements in $\mathbf{z}$. Unfortunately, as noted in [4] there is no proper loss for general independent labels: If $\mathbf{M}_{\alpha\beta}$ is the mixing matrix resulting from (10), and $\mathcal{Q} = \{\mathbf{M}_{\alpha\beta}, \boldsymbol{\alpha} \in [0, 1]^c, \boldsymbol{\beta} \in [0, 1]^c\}$, no $\mathcal{Q}$-proper loss exists.

However, there does exist a proper weak loss for *quasi-independent* label models. In particular consider the conditional probability model given by

$$P(\mathbf{z}|\mathbf{y} = \mathbf{e}_m^c) = \begin{bmatrix} z_m\beta_{m,|\mathbf{z}|} & |\mathbf{z}| < c \\ 0 & |\mathbf{z}| = c \text{ or } |\mathbf{z}| = 0 \end{bmatrix} \tag{11}$$

where coefficients $\beta_{m,n}$ satisfy the linear constraint

$$\sum_{n=1}^{c} \binom{c-1}{n-1} \beta_{m,n} = 1 \tag{12}$$

for any m (so that probabilities sum up to one). Let $\mathbf{M}_\beta$ be the corresponding mixing matrix, and $\mathcal{Q}' = \{\mathbf{M}_\beta\}$. Taking $\beta_{m,|\mathbf{z}|} \propto \beta_m^{|\mathbf{z}|-1}(1 - \beta_m)^{c-|\mathbf{z}|}$ (which satisfies the given constraints), the model (11) is equivalent to (10) with $\alpha_m = 1$, unless for the fact that a zero probability is given to $\mathbf{z} = \mathbb{1}_c$. The set $\mathcal{Q}' = \{\mathbf{M}_\beta\}$ additionally includes many other non-independent label models. The probability of a given partial label vector depends on the true class, m, and the number of noisy labels ($|\mathbf{z}| - 1$) but it is independent of the specific choice of the noisy labels.

The advantage of (11) is that there do exist consistent losses for this model, that do not depend on the particular value of coefficients $\beta_{m,|\mathbf{z}|}$.

Theorem 5. *Consider the virtual labels given by*

$$\tilde{y}_j = \begin{bmatrix} 1 & z_j = 1 \\ -\frac{|\mathbf{z}|-1}{c-|\mathbf{z}|} & z_j = 0 \end{bmatrix} \tag{13}$$

(the case $|\mathbf{z}| = c$ is ignored). If $\tilde{\boldsymbol{\Psi}}$ is a CC, RC or (strictly) proper loss, then $\Psi(\mathbf{z}, \mathbf{f}) = \tilde{\mathbf{y}}^{\mathsf{T}}\tilde{\boldsymbol{\Psi}}(\mathbf{f})$ is Q'-CC, RC or (strictly) proper, respectively.

Proof. Let $\tilde{Y}$ be the virtual label matrix such that its i-th column contains the virtual label corresponding to weak label vector $\mathbf{b}_i$. Then, using (11) and (13),

$$(\tilde{Y}\mathbf{M}_\beta)_{ij} = \sum_{k=1}^{d} \tilde{y}_{ik} P\{\mathbf{z} = \mathbf{b}_k | \mathbf{y} = \mathbf{e}_j^c\}$$

$$= \sum_{k=1}^{d} \tilde{y}_{ik} b_{kj} \beta_{j,|\mathbf{b}_k|}$$

$$= \sum_{k=1}^{d} \left(b_{ki} - (1 - b_{ki}) \frac{|\mathbf{b}_k| - 1}{c - |\mathbf{b}_k|} \right) b_{kj}$$

$$= \sum_{k=1}^{d} \left(b_{ki} \left(1 + \frac{|\mathbf{b}_k| - 1}{c - |\mathbf{b}_k|} \right) - \frac{|\mathbf{b}_k| - 1}{c - |\mathbf{b}_k|} \right) b_{kj} \beta_{j,|\mathbf{b}_k|}$$

$$= \sum_{k=1}^{d} b_{ki} b_{kj} \beta_{j,|\mathbf{b}_k|} \left(1 + \frac{|\mathbf{b}_k| - 1}{c - |\mathbf{b}_k|} \right) - \sum_{k=1}^{d} b_{kj} \beta_{j,|\mathbf{b}_k|} \frac{|\mathbf{b}_k| - 1}{c - |\mathbf{b}_k|} \tag{14}$$

Consider first the case $i \neq j$. Noting that there are $\binom{c-2}{n-2}$ weak label vectors $\mathbf{b}_k$ with n nonzero elements and $b_{ki} = b_{kj} = 1$, and there are $\binom{c-1}{n-1}$ weak label vectors $\mathbf{b}_k$ with n nonzero elements and $b_{ki} = 1$ we can write

$$(\tilde{Y}\mathbf{M}_\beta)_{ij} = \sum_{n=2}^{c} \binom{c-2}{n-2} \beta_{j,n} \left(1 + \frac{n-1}{c-n} \right) - \sum_{n=1}^{c} \binom{c-1}{n-1} \beta_{j,n} \frac{n-1}{c-n}$$

$$= \sum_{n=2}^{c} \left(\binom{c-2}{n-2} \left(1 + \frac{n-1}{c-n} \right) - \binom{c-1}{n-1} \frac{n-1}{c-n} \right) \beta_{j,n} = 0 \tag{15}$$

Finally, for $i = j$,

$$(\tilde{Y}\mathbf{M}_\beta)_{jj} = \sum_{k=1}^{c} b_{kj} \beta_{j,|\mathbf{b}_k|} \left(1 + \frac{|\mathbf{b}_k| - 1}{c - |\mathbf{b}_k|} \right) - \sum_{k=1}^{c} b_{kj} \beta_{j,|\mathbf{b}_k|} \frac{|\mathbf{b}_k| - 1}{c - |\mathbf{b}_k|}$$

$$= \sum_{n=1}^{c} \binom{c-1}{n-1} \beta_{j,n} \left(1 + \frac{n-1}{c-n} \right) - \sum_{k=1}^{c} \binom{c-1}{n-1} \beta_{j,n} \frac{n-1}{c-n}$$

$$= \sum_{n=1}^{c} \binom{c-1}{n-1} \beta_{j,n} = 1 \tag{16}$$

where we have used (12) in the last step. Therefore $\tilde{Y}\mathbf{M}_\beta = \mathbf{I}$, which completes the proof.

The following are some possible choices for a loss based on different choices of $\tilde{\Psi}$.

Example 1 (Pairwise comparison).

Let

$$\tilde{\Psi}(\mathbf{y}, \mathbf{f}) = \sum_{j=0}^{c-1} \phi(\mathbf{y}^\mathsf{T}\mathbf{f} - f_j) \tag{17}$$

This is a general form of the multiclass support vector machine proposed by Weston and Watkins [20], whose consistency was proven by Zhang in [21] for a differentiable non-negative non-increasing $\phi()$. The weak loss based on (13) and (17) is

$$\Psi(\mathbf{z}, \mathbf{f}) = \sum_{k:z_k=1} \sum_{j=0}^{c-1} \phi(f_k - f_j)$$
$$- \frac{|\mathbf{z}| - 1}{c - |\mathbf{z}|} \sum_{k:z_k=0} \sum_{j=0}^{c-1} \phi(f_k - f_j) \tag{18}$$

Example 2 (One versus all).

Let

$$\tilde{\Psi}(\mathbf{y}, \mathbf{f}) = \phi(\mathbf{y}^\mathsf{T}\mathbf{f}) + \sum_{j=0}^{c-1}(1 - y_j)\phi(-f_j) \tag{19}$$

where ϕ is convex, bounded below and differentiable with $\phi(f) < \phi(-f)$ when $f > 0$. The partial loss based on (13) and (19) is

$$\Psi(\mathbf{z}, \mathbf{f}) = \sum_{k:z_k=1} (\phi(f_k) - \phi(-f_k))$$
$$+ \frac{|\mathbf{z}| - 1}{c - |\mathbf{z}|} \sum_{k:z_k=0} (\phi(f_k) - \phi(-f_k))$$
$$+ (2|\mathbf{z}| - 1) \sum_{j=0}^{c-1} \phi(-f_j) \tag{20}$$

Example 3 (Unconstrained Background Discriminative Method).

Let

$$\tilde{\Psi}(\mathbf{y}, \mathbf{f}) = \phi(\mathbf{y}^\mathsf{T}\mathbf{f}) + s\left(\sum_{j=0}^{c-1} t(f_j)\right) \tag{21}$$

where ϕ, s and t convex differentiable functions [21]. The weak loss based on (13) and (21) is

$$\Psi(\mathbf{z}, \mathbf{f}) = \sum_{k:z_k=1} \phi(f_k) + \frac{|\mathbf{z}| - 1}{c - |\mathbf{z}|} \sum_{k:z_k=0} \phi(f_k)$$
$$+ (2|\mathbf{z}| - 1)s\left(\sum_{j=0}^{c-1} t(f_j)\right) \tag{22}$$

4.1 Classification Calibrated Losses for Independent Labels

Despite there is no weak proper loss for independent labels, there may exist CC losses. Consider, for example, the virtual label vector given by $\tilde{y}_i = z_i$ (i.e., the columns of the virtual label matrix $\tilde{\mathbf{Y}} = \mathbf{Z}$ are all possible label vectors (i.e. the i-th column of $\mathbf{Z}$ is $\mathbf{b}_i$).

Note that, for $d > c$, since the virtual label matrix is non-negative, its right-inverse has negative components and, therefore, there is no stochastic matrix satisfying $\mathbf{Z}\mathbf{M} = \mathbf{I}$. Therefore, unless for trivial cases with $d = c$, the loss $\boldsymbol{\Psi}(\mathbf{f}) = \mathbf{Z}\tilde{\boldsymbol{\Psi}}$ is not $\mathbf{M}$-proper for any mixing matrix.

However, there exist CC losses for this case.

Theorem 6. *Given a CC loss* $\tilde{\boldsymbol{\Psi}}$, *weak loss* $\boldsymbol{\Psi}(\mathbf{f}) = \mathbf{Z}\tilde{\boldsymbol{\Psi}}(\mathbf{f})$ *is CC for any mixing matrix satisfying the general model*

$$P(\mathbf{z}|\mathbf{y} = \mathbf{e}_m^c) = \alpha^{z_m}(1 - \alpha)^{1 - z_m}\beta^{|\mathbf{z}| - 1}(1 - \beta)^{c - |\mathbf{z}|} \tag{23}$$

for any $\alpha > \beta$.

Proof. If $\mathbf{V} = \mathbf{Z}\mathbf{M}$, we have

$$v_{ij} = \sum_{k=1}^{d} b_{ki} m_{kj}$$

$$= \sum_{k=1}^{d} b_{ki}\alpha^{b_{kj}}(1 - \alpha)^{1 - b_{kj}}\beta^{|\mathbf{b}_k| - 1}(1 - \beta)^{c - |\mathbf{b}_k|} \tag{24}$$

For $i = j$, we get

$$v_{ij} = \alpha \sum_{k=1}^{d} b_{ki}\beta^{|\mathbf{b}_k| - 1}(1 - \beta)^{c - |\mathbf{b}_k|}$$

$$= \alpha \sum_{n=1}^{c} \binom{c - 1}{n - 1}\beta^{n-1}(1 - \beta)^{c - n} = \alpha \tag{25}$$

and, for $i \neq j$,

$$v_{ij} = \sum_{k=1}^{d} b_{ki}\alpha^{b_{kj}}(1 - \alpha)^{1 - b_{kj}}\beta^{|\mathbf{b}_k| - 1}(1 - \beta)^{c - |\mathbf{b}_k|}$$

$$= \alpha \sum_{n=2}^{c} \binom{c - 2}{n - 2}\beta^{n-1}(1 - \beta)^{c - n}$$

$$+ (1 - \alpha) \sum_{n=1}^{c-1} \binom{c - 2}{n - 1}\beta^{n}(1 - \beta)^{c - n - 1} = \beta \tag{26}$$

Therefore

$$\mathbf{V} = (\alpha - \beta)\mathbf{I} + \beta \mathbb{1}_c \mathbb{1}_c^{\mathsf{T}} \tag{27}$$

which has the form (41) with $\lambda = \alpha - \beta > 0$ and $\mathbf{v} = \beta \mathbb{1}_c$.

5 Convexity

For optimization purposes, we may be interested in using virtual label matrices in such a way that the transformation (8) preserves convexity. This way, convexity of the weak losses can be guaranteed provided that the conventional loss $\tilde{\boldsymbol{\Psi}}$ has convex components.

It is not difficult to show that a necessary and sufficient condition for the components of $\boldsymbol{\Psi}(\mathbf{f}) = \tilde{\mathbf{Y}}\tilde{\boldsymbol{\Psi}}(\mathbf{f})$ to be convex for any convex $\tilde{\Psi}(\mathbf{z}, \mathbf{f})$ is that $[\tilde{\mathbf{Y}}]_{ji} \geq 0$ for any $1 \leq j \leq c, 1 \leq i \leq d$.

Unfortunately, restricting the virtual matrix to have non-negative components constitutes a strong limitation for the design of weak convex proper losses: for instance, if all components of virtual matrix $\tilde{\mathbf{Y}}$ are strictly positive, its right inverse has at least one negative component and, thus, it is not stochastic. Therefore, $\mathcal{Q}_p(\tilde{\mathbf{Y}}) = \emptyset$.

Moreover, though the right inverse of a matrix with all non-negative elements may be non-negative, each zero element in $\tilde{\mathbf{Y}}$ imposes strong constraints on the number of non-zero elements in its right inverse. In particular, it is easy to see that, if $\tilde{\mathbf{Y}}$ does not have negative components and $\boldsymbol{\Psi}(\mathbf{f}) = \tilde{\mathbf{Y}}\tilde{\boldsymbol{\Psi}}(\mathbf{f})$ is M-proper, then $[\tilde{\mathbf{Y}}]_{ji} > 0$ implies $[\mathbf{M}]_{ik} = 0$ for any $k \neq j$. In summary, properness is a limiting factor to preserve convexity in a practical design of weak losses: even for a convex $\tilde{\Psi}$, Ψ is in general non-convex.

However, as we have seen in the previous example, non-negative virtual label matrices can be used to construct CC or RC losses. The following are examples of convex classification calibrated weak loss functions constructed from conventional losses used for multiclass classification.

Example 4 (Pairwise comparison.).
    If

$$\tilde{\Psi}(\mathbf{y}, \mathbf{f}) = \sum_{j=0}^{c-1} \phi(\mathbf{y}^\mathsf{T}\mathbf{f} - f_j) \tag{28}$$

The weak loss $\boldsymbol{\Psi}(\mathbf{f}) = \mathbf{B}\tilde{\boldsymbol{\Psi}}(\mathbf{f})$ is

$$\Psi(\mathbf{z}, \mathbf{f}) = \tilde{\mathbf{y}}^\mathsf{T}\tilde{\boldsymbol{\Psi}}(\mathbf{f}) = \sum_{k=0}^{c-1} z_k \tilde{\psi}_k(\mathbf{f})$$

$$= \sum_{k=0}^{c-1} z_k \sum_{j=0}^{c-1} \phi(f_k - f_j) \tag{29}$$

Example 5 (One versus all.).
    Let

$$\tilde{\Psi}(\mathbf{y}, \mathbf{f}) = \phi(\mathbf{y}^\mathsf{T}\mathbf{f}) + \sum_{j \neq \text{ind}(\mathbf{y})} \phi(-f_j) \tag{30}$$

The weak loss $\boldsymbol{\Psi}(\mathbf{f}) = \mathbf{B}\tilde{\boldsymbol{\Psi}}(\mathbf{f})$ is

$$\Psi(\mathbf{z}, \mathbf{f}) = \sum_{k=0}^{c-1} z_k \left(\phi(f_k) + \sum_{j \neq k} \phi(-f_j) \right) \tag{31}$$

6   Conclusions

In this paper we have analyzed conditions on the conditional probability model relating weak labels and true classes to guarantee that a loss is classification-calibrated. As expected, we have found that classification calibration imposes less constraints on the mixing matrix. Moreover, we show a straighforward way to construct a weak loss from a conventional loss that preserves convexity and also classification calibration for a wide set of mixing matrices including independent label models.

Appendix

Proof of Theorem 2

Assume that $\Psi(\mathbf{z}, \mathbf{f})$ is strictly $\mathbf{M}$-proper, but $\mathbf{M} \notin \mathcal{W}$. Then we have

$$\mathbb{E}_{\mathbf{z}}\{\Psi(\mathbf{z}, \mathbf{f})\} = \mathbf{p}^{\mathsf{T}}\Psi(\mathbf{f}) = \boldsymbol{\eta}^{\mathsf{T}}\mathbf{M}^{\mathsf{T}}\tilde{\mathbf{Y}}^{\mathsf{T}}\tilde{\boldsymbol{\Psi}}(\mathbf{f}) = \boldsymbol{\eta}^{\mathsf{T}}\mathbf{V}^{\mathsf{T}}\tilde{\boldsymbol{\Psi}}(\mathbf{f}) \tag{32}$$

where $\mathbf{V} = \tilde{\mathbf{Y}}\mathbf{M}$ is a $c \times c$ matrix. We consider two cases: (i) $\mathbf{V}\boldsymbol{\eta}$ has non-negative components, an (ii) at least one component of $\mathbf{V}\boldsymbol{\eta}$ is negative.

In case (i), since $\tilde{\boldsymbol{\Psi}}$ is strictly proper, (32) is minimum at

$$\mathbf{f}^* = \frac{\mathbf{V}\boldsymbol{\eta}}{\mathbb{1}_c^{\mathsf{T}}\mathbf{V}\boldsymbol{\eta}} \tag{33}$$

Since $\mathbf{M} \notin \mathcal{W}$, we have $\mathbf{V} \neq \alpha\mathbf{I}$ for any α and, thus, $\mathbf{f}^* \neq \boldsymbol{\eta}$ (at least for some $\boldsymbol{\eta}$), which is in contradiction with the fact that $\boldsymbol{\Psi}$ is strictly proper.

The basic idea of the proof of case (ii) is to show that, if $\mathbf{V}\boldsymbol{\eta}$ has some negative components, then the unique minimizer of $\boldsymbol{\Psi}(\mathbf{f})$ (the weak loss) for $\mathbf{f}$ in the probability simplex $\mathcal{P}$, must lie in the boundary of $\mathcal{P}$. In such case, for any $\boldsymbol{\eta}$ in the interior of $\mathcal{P}$, $\boldsymbol{\eta} \notin \arg\min_{\mathbf{f}} \boldsymbol{\eta}^{\mathsf{T}}\boldsymbol{\Psi}(\mathbf{f})$ and, thus, $\boldsymbol{\Psi}$ is not proper. A proof that the minimizer of $\boldsymbol{\Psi}(\mathbf{f})$ must be in the boundary of $\mathcal{P}$ follows.

First, note that, if $\tilde{\boldsymbol{\Psi}}$ is strictly proper, it is invertible. Otherwise, if $\tilde{\boldsymbol{\Psi}}(\boldsymbol{\eta}_1) = \tilde{\boldsymbol{\Psi}}(\boldsymbol{\eta}_2)$, $\boldsymbol{\eta}_1^{\mathsf{T}}\tilde{\boldsymbol{\Psi}}(\boldsymbol{\eta}_1) = \boldsymbol{\eta}_1^{\mathsf{T}}\tilde{\boldsymbol{\Psi}}(\boldsymbol{\eta}_2)$, which is in contradiction with $\boldsymbol{\eta}_1$ being the unique minimizer of $\boldsymbol{\eta}_1^{\mathsf{T}}\tilde{\boldsymbol{\Psi}}(\mathbf{f})$. Also, the image set $\mathcal{S} = \{\tilde{\boldsymbol{\Psi}}(\mathbf{f}), \mathbf{f} \in \mathcal{P}\}$ is a convex manifold in $\mathbb{R}^c$ (i.e., it is in the boundary of its convex hull). To see this, since $\tilde{\boldsymbol{\Psi}}$ is strictly proper, $\boldsymbol{\eta}\tilde{\boldsymbol{\Psi}}(\boldsymbol{\eta}) \leq \boldsymbol{\eta}\tilde{\boldsymbol{\Psi}}(\mathbf{f})$, thus, $\boldsymbol{\eta}$ is the normal vector of a supporting hyperplane of $\mathcal{S}$ in $\tilde{\boldsymbol{\Psi}}(\boldsymbol{\eta})$ and, thus, $\tilde{\boldsymbol{\Psi}}(\boldsymbol{\eta})$ is a boundary point in the convex hull of $\mathcal{S}$.

Additionally, for $\mathbf{a} = \mathbf{V}\boldsymbol{\eta}$, if $\boldsymbol{\eta} = \arg\min_{\mathbf{f}} \mathbf{a}\tilde{\boldsymbol{\Psi}}(\mathbf{f})$, then $\mathbf{a}$ is also a normal vector of a supporting hyperplane of $\mathcal{S}$ in $\tilde{\boldsymbol{\Psi}}(\boldsymbol{\eta})$. But, if $\mathcal{S}$ is strictly convex, it has a single supporting hyperplane at almost every point and, thus, for almost every $\boldsymbol{\eta} \in \mathcal{P}$, $\mathbf{V}\boldsymbol{\eta} = \alpha\boldsymbol{\eta}$, for some $\alpha \neq 0$. Since $\mathbf{V}\boldsymbol{\eta}$ has some negative components, we must have $\alpha < 0$ and, thus $\boldsymbol{\eta} = \arg\min_{\mathbf{f}} \alpha\boldsymbol{\eta}^{\mathsf{T}}\tilde{\boldsymbol{\Psi}}(\mathbf{f}) = \arg\max_{\mathbf{f}} \boldsymbol{\eta}^{\mathsf{T}}\tilde{\boldsymbol{\Psi}}(\mathbf{f})$, which is in contradiction with $\boldsymbol{\eta}$ being a minimizer of $\boldsymbol{\eta}^{\mathsf{T}}\tilde{\boldsymbol{\Psi}}(\mathbf{f})$

Note that equation $\tilde{\mathbf{Y}}\mathbf{M} = \alpha\mathbf{I}$ states a set of $c^2 - 1$ linear constraints. This, along with $\mathbf{M}\mathbb{1}_d = \mathbb{1}_c$ shows that the dimension of the maximal set is at most $dc - c^2 - c + 1$.

Proof of Theorem 3

Let $\mathbf{f}^*$ be the minimum of $\mathbb{E}_{\mathbf{z}}\{\Psi(\mathbf{z},\mathbf{f})\} = \boldsymbol{\eta}^{\mathsf{T}}\mathbf{V}^{\mathsf{T}}\tilde{\boldsymbol{\Psi}}(\mathbf{f})$ with $\mathbf{V} = \tilde{\mathbf{Y}}\mathbf{M}$. Also, let $\boldsymbol{\eta}$ such that $\eta_i = \mu + \epsilon$ and $\eta_j = \mu - \epsilon$, for some small $\epsilon > 0$. Since $\Psi(\mathbf{z},\mathbf{f})$ is RC and $\eta_i > \eta_j$, we have $f_i^* > f_j^*$. Also, since $\tilde{\boldsymbol{\Psi}}$ is ranking-consistent, we have

$$\mathbf{e}_i^c\mathbf{V}\boldsymbol{\eta} \leq \mathbf{e}_j^c\mathbf{V}\boldsymbol{\eta} \tag{34}$$

therefore

$$v_{ii}(\mu + \epsilon) + v_{ij}(\mu - \epsilon) + \sum_{k \notin \{i,j\}} v_{ik}\eta_k$$
$$\leq v_{ji}(\mu + \epsilon) + v_{jj}(\mu - \epsilon) + \sum_{k \notin \{i,j\}} v_{jk}\eta_k \tag{35}$$

Since this must be true for arbitrary small ϵ, we have

$$(v_{ii} + v_{ij})\mu + \sum_{k \not\subset \{i,j\}} v_{ik}\eta_k \leq (v_{ji} + v_{jj})\mu + \sum_{k \notin \{i,j\}} v_{jk}\eta_k \tag{36}$$

Alternatively, taking $\eta_i = \mu - \epsilon$ and $\eta_j = \mu + \epsilon$, we can conclude that the opposite inequality is also true. Therefore, the above inequality can be replaced by an equality. Since this must be true for any μ and $\{\eta_k\}$, we get

$$v_{ii} + v_{ij} = v_{jj} + v_{ji} \tag{37}$$
$$v_{ik} = v_{jk}, \qquad , k \neq i, k \neq j \tag{38}$$

Since this must be true for any pair i, j of classes, Eq. (38) implies that the non-diagonal values of each column must be equal, while (37) imposes a restriction on the diagonal element. Both conditions and the inequality relations are equivalent to claiming that matrix $\mathbf{V}$ must have the general form $\mathbf{V} = \lambda\mathbf{I} + \mathbb{1}_c\mathbf{v}^{\mathsf{T}}$ for some constant $\lambda > 0$ and vector $\mathbf{v}$. Also, we can write the above constraints as

$$(\mathbf{e}_i^c - \mathbf{e}_j^c)^{\mathsf{T}}\tilde{\mathbf{Y}}\mathbf{M}(\mathbf{e}_i^c + \mathbf{e}_j^c) = 0 \tag{39}$$
$$(\mathbf{e}_i^c - \mathbf{e}_j^c)^{\mathsf{T}}\tilde{\mathbf{Y}}\mathbf{M}\mathbf{e}_k^c = 0, \qquad , k \neq i, k \neq j \tag{40}$$

This states a set of at most $c^2 - c - 1$ linear constraints over $\mathbf{M}$. This along with $\mathbf{M}\mathbb{1}_d = \mathbb{1}_c$ shows that the dimension of the maximal set is at most $dc - c^2 + 1$. To see that $\mathcal{Q}$ is maximal, note that, for any $\mathbf{M} \in \mathcal{Q}$, we have that $\mathbb{E}_{\mathbf{z}}\{\Psi(\mathbf{z},\mathbf{f})\} = \boldsymbol{\eta}^{\mathsf{T}}\mathbf{V}^{\mathsf{T}}\tilde{\boldsymbol{\Psi}}(\mathbf{f})$. Noting that

$$\mathbf{V}\boldsymbol{\eta} = \lambda\boldsymbol{\eta} + (\mathbf{v}^{\mathsf{T}}\boldsymbol{\eta})\mathbb{1}_c \tag{41}$$

we can see that $\boldsymbol{\eta}^* = \mathbf{V}\boldsymbol{\eta}$ is an order-preserving transformation. Thus $\Psi(\mathbf{z},\mathbf{f})$ is RC.

Proof. of Theorem 4

The proof follows the same steps than Th. 3. The only difference is that we must take μ and η_k such that $\mu > \eta_k$, for all $k \neq i, k \neq j$.

References

[1] Ambroise, C., Denoeux, T., Govaert, G., Smets, P.: Learning from an imprecise teacher: probabilistic and evidential approaches. In: Applied Stochastic Models and Data Analysis, vol. 1, pp. 100–105 (2001)

[2] Bartlett, P.L., Jordan, M.I., McAuliffe, J.D.: Convexity, classification, and risk bounds. Journal of the American Statistical Association 101(473), 138–156 (2006)

[3] Chittineni, C.: Learning with imperfectly labeled patterns. Pattern Recognition 12(5), 281–291 (1980)

[4] Cid-Sueiro, J.: Proper losses for learning from partial labels. In: Advances in Neural Information Processing Systems 25, pp. 1574–1582 (2012)

[5] Côme, E., Oukhellou, L., Denux, T., Aknin, P.: Mixture model estimation with soft labels. In: Dubois, D., Lubiano, M., Prade, H., Gil, M., Grzegorzewski, P., Hryniewicz, O. (eds.) Soft Methods for Handling Variability and Imprecision. AISC, vol. 48, pp. 165–174. Springer, Heidelberg (2008)

[6] Cour, T., Sapp, B., Taskar, B.: Learning from partial labels. Journal of Machine Learning Research 12, 1225–1261 (2011)

[7] Devroye, L., Györfi, L., Lugosi, G.: A Probabilistic Theory of Pattern Recognition, Applications of Mathematics, vol. 31. Springer (1997)

[8] Grandvalet, Y.: Logistic regression for partial labels. In: 9th Information Processing and Management of Uncertainty in Knowledge-based System, pp. 1935–1941 (2002)

[9] Grandvalet, Y., Bengio, Y.: Learning from partial labels with minimum entropy (2004)

[10] Hüllermeier, E., Beringer, J.: Learning from ambiguously labeled examples. Intell. Data Anal. 10(5), 419–439 (2006)

[11] Jin, R., Ghahramani, Z.: Learning with multiple labels. In: Advances in Neural Information Processing Systems 15, pp. 897–904 (2002)

[12] Krishnan, T.: Efficiency of learning with imperfect supervision. Pattern Recognition 21(2), 183–188 (1988)

[13] Krishnan, T., Nandy, S.C.: Discriminant analysis with a stochastic supervisor. Pattern Recognition 20(4), 379–384 (1987)

[14] Liang, P., Jordan, M., Klein, D.: Learning from measurements in exponential families. In: Proceedings of the 26th Annual International Conference on Machine Learning, pp. 641–648. ACM (2009)

[15] Lin, Y.: A note on margin-based loss functions in classification. Statistics & Probability Letters 68(1), 73–82 (2004)

[16] Luo, J., Orabona, F.: Learning from candidate labeling sets. In: Advances in Neural Information Processing Systems 23, pp. 1504–1512 (2010)

[17] Nguyen, N., Caruana, R.: Classification with partial labels. In: Proceedings of the 14th ACM SIGKDD International Conference on Knowledge Discovery and Data Mining, pp. 551–559. ACM, New York (2008)

[18] Raykar, V.C., Yu, S., Zhao, L.H., Valadez, G.H., Florin, C., Bogoni, L., Moy, L.: Learning from crowds. Journal of Machine Learning Research 99, 1297–1322 (2010)

[19] Tewari, A., Bartlett, P.L.: On the consistency of multiclass classification methods. Journal of Machine Learning Research 8, 1007–1025 (2007)

[20] Weston, J., Watkins, C.: Support vector machines for multi-class pattern recognition. In: Proceedings of the Seventh European Symposium on Artificial Neural Networks, vol. 4, pp. 219–224 (1999)

[21] Zhang, T.: Statistical analysis of some multi-category large margin classification methods. Journal of Machine Learning Research 5, 1225–1251 (2004)

Active Learning for Support Vector Machines with Maximum Model Change

Wenbin Cai[1], Ya Zhang[1], Siyuan Zhou[1], Wenquan Wang[1],
Chris Ding[2], and Xiao Gu[1]

[1] Shanghai Key Laboratory of Multimedia Processing & Transmissions
Shanghai Jiao Tong University, Shanghai, China
{cai-wenbin,ya_zhang,zhousiyuan,wangwenquan,gugu97}@sjtu.edu.cn
[2] University of Texas at Arlington, Arlington, Texas, USA
chqding@uta.edu

Abstract. Margin-based strategies and model change based strategies represent two important types of strategies for active learning. While margin-based strategies have been dominant for Support Vector Machines (SVMs), most methods are based on heuristics and lack a solid theoretical support. In this paper, we propose an active learning strategy for SVMs based on Maximum Model Change (MMC). The model change is defined as the difference between the current model parameters and the updated parameters obtained with the enlarged training set. Inspired by Stochastic Gradient Descent (SGD) update rule, we measure the change as the gradient of the loss at a candidate point. We analyze the convergence property of the proposed method, and show that the upper bound of label requests made by MMC is smaller than passive learning. Moreover, we connect the proposed MMC algorithm with the widely used *simple margin* method in order to provide a theoretical justification for margin-based strategies. Extensive experimental results on various benchmark data sets from UCI machine learning repository have demonstrated the effectiveness and efficiency of the proposed method.

Keywords: Active Learning, Maximum Model Change, SVMs.

1 Introduction

In supervised learning, a large amount of labeled data is usually required to obtain a high quality model. A widely used method for data collection is passive learning, where the training examples are randomly selected according to a certain underlying distribution and annotated by human editors. However, in many practical applications, there might not be sufficient labeled data examples due to the high cost associated with data annotation. To solve this problem, active learning aims at selectively labeling the most informative instances with the goal of maximizing the accuracy of the model trained. In a typical active learning framework, the learner iteratively chooses informative data examples from a large unlabeled set (denoted as pool $\mathcal{U}$) with a predefined sampling function,

T. Calders et al. (Eds.): ECML PKDD 2014, Part I, LNCS 8724, pp. 211–226, 2014.

and then labels them. This data sampling process is repeated until a certain performance expectation is achieved or a certain labeling budget is used up. Active learning is well-motivated in many supervised learning tasks where unlabeled data may be abundant but labeled data examples are expensive to obtain [8,9].

Support vector machines (SVMs), which have arisen from statistical learning theory, play a significant role in the machine learning community with solid mathematical and statistical foundation [11,12]. Due to many desired properties including excellent generalization performance, robustness to the noise, and capability to deal with high dimensional data, SVMs have been successfully applied to many learning applications. As a result, active learning for SVMs has recently drawn a great deal of attention. In previous studies, several active learning algorithms have been specifically proposed for SVMs [5,10,13,14,17,18]. Most of them are derived with the notion of margin, i.e. preferring the points located in the margin. For example, *simple margin* [18], the most widely adopted strategy for SVMs, selects the examples that are closest to the decision boundary as the most informative ones. Although the margin-based active learning heuristics are fairly straightforward and natural for SVMs, these popular approaches lack a solid theoretical justification, i.e. how can we guarantee that margin-based active sampling performs better than passive learning.

In this paper, we introduce a new interpretation for the margin-based active learning by bridging it with the idea of model change. In particular, we attempt to provide theoretical justifications for the margin-based methods. We consider the capability of examples to change the model, and accordingly propose a novel margin-based active learning strategy for SVMs called Maximum Model Change (MMC), which is to choose the examples leading to the maximal change to the current model. The change is quantified as the difference between the current model parameters and the new parameters obtained with the expanded training set. Inspired by the well-studied work on the Stochastic Gradient Descent (SGD) update rule [15,16,19], where the parameters are updated repeatedly according to the negative gradient of the objective function at each single training example, we use the gradient of the loss at a candidate example to approximate the model change. Under the model change principle, the instances lying in the margin are proven to be the ones having the capability to change the model. We further analyze the convergence property of the proposed MMC method, and show that 1) MMC is guaranteed to converge, and 2) the upper bound of label requests made by MMC is smaller than that of passive learning. We further connect MMC with simple margin to provide a uniform view to these two methods. The property holds for other well-known SVMs active learning methods as well. We validate our algorithm with various benchmark data sets from UCI machine learning repository. Extensive experimental results have demonstrated the effectiveness and efficiency of the proposed active learning approach.

The main contributions of this paper are summarized as follows.

- Focusing on SVMs as the base learner, we introduce a novel interpretation for margin-based active learning with model change, and propose a new sampling algorithm called Maximum Model Change (MMC).

- We theoretically analyze the convergence property of the proposed approach, and compare the sampling bound against passive learning.
- We connect MMC with the widely adopted simple margin heuristic in order to provide a uniform view to these two active learning methods.

The rest of this paper is structured as follows: Section 2 briefly reviews the related work. Our active learning approach for SVMs, Maximum Model Change (MMC), is presented in Section 3. Section 4 provides the theoretical justification of the convergence property for the proposed approach, and compare the sampling bound with that of passive learning. Section 5 explores the relationship between MMC and simple margin. Section 6 presents the experimental results. Finally, we conclude the paper in Section 7.

2   Related Work

The goal of active learning is to train a high quality model using as few labeled training set as possible, therefore minimizing the labeling cost. In this section, we first briefly review several general active learning strategies, and then summarize existing margin-based active learning methods for SVMs.

2.1   Active Learning

Various active learning strategies have been proposed in the literature. Here we briefly review the typical active learning strategies:

1. **Uncertainty Sampling (US)**: The US approach aims to choose the examples whose labels the current classifier is most uncertain about. This strategy is usually straightforward to implement for probabilistic models. Take binary classification as an instance, US aims to query the data point whose posterior probability is most close to 0.5 [22]. For multi-class classification problems, examples with the smallest margin between the first and second most probable class labels are selected [1].
2. **Query By Committee (QBC)**: The QBC strategy generates a committee of model members and select unlabeled instances about which the models disagree the most [4]. A popular function to quantify the disagreement is vote entropy. To efficiently generate the committee, popular ensemble learning methods, such as Bagging and Boosting, have been employed [2].
3. **Expected Error Reduction (EER)**: The EER strategy aims to minimize the generalization error of the model. Roy et al. [20] proposed an optimal active sampling method to choose the example that leads to the lowest generalization error on the future test set once labeled and added to the training set. The weakness is that the computational cost of this method is extremely high. Instead of choosing the example yielding the smallest generalization error, Nguyen et al. [7] suggested to query the instance that has the largest contribution to the current error. Cohn et al. [21] proposed a statistically optimal active learning approach, which aims to choose the examples minimizing the output variance to reduce the generalization error.

4. **Expected Model Change (EMC)**: This strategy is to select data points that are expected to incur a large model change once added to the training set. Settles et al. [23] proposed an algorithm for logistic regression, and the change is quantified as the gradient length of the objective function obtained by the enlarged training set. Donmez et al. [3] presented a sampling approach for ranking tasks, which measures the change as the difference between the current model and the additional model trained with the selected examples. Recently, Cai et al. [24] applied this strategy to regression tasks.

There are several other active learning strategies proposed. A comprehensive active learning survey can be found in [6].

2.2   Active Learning for SVMs

Support vector machines (SVMs), built on solid mathematical and statistical foundation, play an important role in supervised learning. Many active learning algorithms, especially margin-based algorithms, have been specifically proposed for SVMs. We summarize existing margin-based active learning for SVMs as follows:

1. **Simple Margin** [18]: The simple margin algorithm is one of the most widely adopted active learning strategy when employing SVMs as the base learner, which chooses the examples that are closest to the separating hyperplane.
2. **MaxMin Margin** [18]: This active learning method aims to select the data instances that equally split the version space once labeled and added to the training set.
3. **Ratio Margin** [18]: This sampling approach is an extension of MaxMin Margin by taking particular consideration of the shape of version space.
4. **Representative Sampling** [5]: This sampling algorithm selects the most representative points within the margin using the clustering techniques.
5. **Multi-criteria-based Sampling** [14]: This approach simultaneously considers multiple criteria for sampling, and queries the data examples that are both informative and representative.
6. **Diversity-based Sampling** [13]: This strategy extends the simple margin to batch mode active learning by incorporating a diversity measure, which is calculated by their angles, to enforce the selected points to be diverse.
7. **Confidence-based Sampling** [10]: This active sampling algorithm can be regarded as a variant of the simple margin, which measures the uncertainty value of each sample as its conditional error.

As listed above, a common feature among the margin-based active learning methods is that they tend to pick the data examples located in the margin. Although existing margin-based active learning strategies are quite straightforward for SVMs, one limitation is that they lack solid theoretical support. In the next sections, we propose a new active learning algorithm for SVMs, together with theoretical justification.

3   Maximum Model Change for SVMs

In this section, we first provide a brief introduction to SVMs, focusing on the model fitting with the Stochastic Gradient Descent (SGD) rule. Then, the details of the proposed active learning algorithm, Maximum Model Change (MMC), for SVMs are provided. Finally, we analyze the computational complexity of the proposed algorithm.

3.1   Training SVMs with Stochastic Gradient Descent

For simplicity, we concentrate on the binary classification problem in this paper. It is straightforward to generalize the proposed method to multi-class problems. Given a training set $\mathcal{L} = \{x_i, y_i\}_{i=1}^n$, where $x_i \in \mathbb{R}^d$ is a d-dimensional feature vector and $y_i \in \{1, -1\}$ is a class label, the separation hyperplane of linear SVM model is represented as:

$$f(x) = w^\mathrm{T} x + b = 0, \tag{1}$$

where w denotes the weight vector parameterizing the model. For simplicity, we omit the bias term b throughout this study, which is commonly used in practice. In fact, it is easy to employ the bias by padding extra dimension of all 1's.

Building a SVM classifier is to solve the following Quadratic Programming (QP) problem:

$$\begin{aligned} \min_{w} \quad & \frac{1}{2}||w||^2 + C\sum_{i=1}^{n}\xi_i \\ \text{s.t.} \quad & y_i w^\mathrm{T} x_i \geq 1 - \xi_i, \\ & \xi_i \geq 0, \qquad i = 1, ..., n, \end{aligned} \tag{2}$$

where ξ_i is a slack variable. The above QP problem can be equivalently rewritten as an unconstrained problem by re-arranging the constraints and substituting the parameter C with $C = \frac{1}{\lambda}$ as follows:

$$\min_{w} \frac{\lambda}{2}||w||^2 + \sum_{i=1}^{n}[1 - y_i w^\mathrm{T} x_i]_+, \tag{3}$$

where the subscript indicates the positive part. The first term is the regularizer, and the second term represents the standard Hinge loss. More generally, the soft margin loss is adopted with a margin parameter $\gamma \geq 0$, which treats the margin as a variable [19]. Thus, the SVM optimization problem can be reformulated as:

$$\min_{w} \frac{\lambda}{2}||w||^2 + \sum_{i=1}^{n}[\gamma - y_i w^\mathrm{T} x_i]_+, \tag{4}$$

where the second term $[...]+$ is the so-called soft margin loss. When $\gamma = 1$, the second term is the Hinge loss.

To find w minimizing the objective function, a widely used search approach is Stochastic Gradient Descent (SGD), which updates the parameter w repeatedly according to the negative gradient of the objective function with respect to each training example:

$$w \leftarrow w - \alpha \frac{\partial \mathcal{O}_w(x_i)}{\partial w}, \qquad i = 1, 2, ..., n, \tag{5}$$

where $\mathcal{O}_w(x_i)$ and α are the objective function and the learning rate, respectively. With the particular of objective function (4), the update rule can be written as:

$$w \leftarrow \begin{cases} (1 - \alpha\lambda)w + \alpha y_i x_i, & \text{if } y_i w^{\mathrm{T}} x_i < \gamma, \\ (1 - \alpha\lambda)w, & \text{otherwise.} \end{cases} \tag{6}$$

In the literatures, several SGD-based learning algorithms have been well studied for solving the SVMs optimization problems [15,19]. They share the same update rule (6) with different scheduling of the learning rate.

3.2   Model Change Computation

Here, we consider the SGD rule in the active learning cases. Suppose a candidate example x^+ is added to the training set with a given class label y^+, the objective function on the expanded training set $\mathcal{L}^+ = \mathcal{L} \cup (x^+, y^+)$ then becomes:

$$\min_w \ \frac{\lambda}{2}\|w\|^2 + \sum_{i=1}^{n}[\gamma - y_i w^{\mathrm{T}} x_i]_+ + \underbrace{[\gamma - y^+ w^{\mathrm{T}} x^+]_+}_{:=\ell_w(x^+)}. \tag{7}$$

As a result, the parameter w is changed due to the inclusion of the new example (x^+, y^+). We estimate the effect of adding the new point on the training loss to approximate the change, and hence the model change can be approximated with the gradient of the loss function at the example (x^+, y^+):

$$\mathcal{C}_w(x^+) = \triangle w \approx \alpha \frac{\partial \ell_w(x^+)}{\partial w}. \tag{8}$$

The derivative of the loss at the candidate point (x^+, y^+) is calculated as:

$$\frac{\partial \ell_w(x^+)}{\partial w} = \begin{cases} -y^+ x^+, & \text{if } y^+ w^{\mathrm{T}} x^+ < \gamma, \\ 0, & \text{otherwise.} \end{cases} \tag{9}$$

Clearly, the model updates its weight based on solely those examples that satisfy the inequality $y^+ w^{\mathrm{T}} x^+ < \gamma$, which is straightforward for SVMs.

The goal of MMC is to query the example that maximally changes the current model. According to (8) and (9), only the set $\Psi = \{x : y^+ w^{\mathrm{T}} x^+ < \gamma\} \subseteq \mathcal{U}$ has the ability to change the model, and hence only this set needs to be considered in active learning. The sampling criteria can be expressed as:

$$x^*_{\mathrm{MMC}} = \arg \max_{x^+ \in \Psi} \|\mathcal{C}_w(x^+)\|. \tag{10}$$

Algorithm 1. MMC active learning for SVMs

Input: The labeled data set $\mathcal{L}=\{(x_i,y_i)\}_{i=1}^{n}$, the unlabeled pool set $\mathcal{U}$, the parameter γ, the SVM classifier $f(x)$ trained with $\mathcal{L}$.
1: **for** each x^+ in $\mathcal{U}$ **do**
2:    **if** $|w^{\mathrm{T}}x^+| < \gamma$ **then**
3:        $\mathbb{E}_{y^+}\{||\mathcal{C}_w(x^+)||\} \leftarrow ||x^+||$.
4:    **end if**
5: **end for**
Output: $x^* \leftarrow \arg\max_{x^+} \mathbb{E}_{y^+}\{||\mathcal{C}_w(x^+)||\}$

In practice, the true label y^+ of the example x^+ is unknown in advance. With $y \in \{1,-1\}$, we have

$$\Omega = \{x : |w^{\mathrm{T}}x^+| < \gamma\} \subseteq \{x : y^+w^{\mathrm{T}}x^+ < \gamma, \ y \in \{1,-1\}\}.$$

We hence rewrite the inequality constraint $y^+w^{\mathrm{T}}x^+ < \gamma$ as $|w^{\mathrm{T}}x^+| < \gamma$. Meanwhile, we take the expected model change over each possible class labels $y^+ \in Y = \{1,-1\}$ to approximate the true change. Suppose that the learning rate α for each candidate point is identical, the final sampling criteria can be reformulated as:

$$\begin{aligned}
x_{\mathrm{MMC}}^* &= \arg\max_{x^+ \in \Omega} \mathbb{E}_{y^+}\{||\mathcal{C}_w(x^+)||\} \\
&= \arg\max_{x^+ \subset \Omega} \sum_{y^+ \in Y} \hat{P}(y^+|x^+)|| - y^+x^+|| \\
&= \arg\max_{x^+ \in \Omega} \sum_{y^+ \in Y} \hat{P}(y^+|x^+)||x^+|| \\
&= \arg\max_{x^+ \in \Omega} ||x^+||, \quad\quad\quad\quad\quad\quad (11)
\end{aligned}$$

where $\hat{P}(y^+|x^+)$ represents the conditional probability of label y^+ given example x^+ estimated by the current classifier. The last step above follows from the fact that $\hat{P}(y^+ = +1|x^+) + \hat{P}(y^+ = -1|x^+) = 1$. An intuitive explanation for MMC is that the data examples maximally changing the current classifier are expected to result in faster convergence to the optimal model. The corresponding pseudo-code is given in Algorithm 1.. Based on the above derivation, MMC can be deemed as a margin-based active learning strategy as well because it shares the common feature of preferring examples within the margin, i.e. $\{x : |w^{\mathrm{T}}x^+| < \gamma\}$.

3.3   Computational Complexity

Assume that there are n labeled examples in the training set, and m unlabeled instances in the pool set. There are three main operations in the MMC method: SVM training, sample filtering, and sample selection.

SVM training typically needs $O(n^2)$ calculation. For the sample filtering, the main operation is to calculate the inner product, which has a time complexity of

$O(d)$. Therefore, the total time complexity is $O(md)$ at this step. For the sample selection, most time is spent on computing the norm with a complexity of $O(d)$, and hence the total time complexity is $O(kd))$ if there are k eligible examples. Summing up, the total time complexity for MMC is $O(n^2 + (m + k)d)$, which is promising for real-world tasks.

4   Theoretical Analysis

The goal of a learning model is to minimize the generalization error on future data. Clearly, the generalization error is changed if and only if the model is changed. Thus, active learning only needs to select the samples that change the current model, which is support vectors for SVMs. A nice feature of SVMs is that support vectors usually represent a tiny portion of the training data.

We have shown that points within the margin are the ones having the ability to change the current model. In this section, we attempt to provide a theoretical backup behind our strategy by analyzing the convergence property. Assume that $\{\exists \epsilon, x : y^+ w^T x^+ \leq \gamma - \epsilon\} = \{x : y^+ w^T x^+ < \gamma\}$. Since the scaling factors is to scale the derived bound by some fixed constant, which does not affect the convergence property, for clarity, we drop the scaling factors in the update rule:

$$w \leftarrow \begin{cases} w + y_i x_i, & \text{if } y_i w^T x_i < \gamma, \\ w, & \text{otherwise.} \end{cases} \tag{12}$$

Theorem. (Convergence property) *Suppose that $||x_j|| \leq R$ for all $x_j \in \mathcal{L} \cup \mathcal{U}$. Let the current solution be w^c, and further suppose that there exists an optimal solution w^* such that $y_j(w^*)^T x_j \geq \gamma$ for all examples x_j. Let $||w^c|| = M$ and $||w^*|| = N$. Then, the total number of label requests $\mathcal{A}$ made by MMC is at most*

$$O\left(\frac{N}{\gamma} \left(M + N + \frac{N(R^2 - \epsilon)}{\gamma} \right) \right).$$

Proof. The proposed MMC algorithm chooses the data points only that change the current model, which implies that

$$y^+ w^T x^+ < \gamma \Leftrightarrow y^+ w^T x^+ \leq \gamma - \epsilon. \tag{13}$$

According to the SGD update rule in Eq. (12), we have

$$w^{(t+1)} \leftarrow w^{(t)} + y^+ x^+, \qquad t = 1, 2, ..., \mathcal{A}. \tag{14}$$

where $w^{(t=1)}$ stands for the current solution, i.e. $w^{(t=1)} = w^c$. According to the above update rule in Eq. (14), we get:

$$\begin{aligned} ||w^{(t+1)}||^2 &= ||w^{(t)} + y^+ x^+||^2 \\ &= ||w^{(t)}||^2 + ||x^+||^2 + 2y^+ (w^{(t)})^T x^+ \\ &\leq ||w^{(t)}||^2 + ||x^+||^2 + 2(\gamma - \epsilon) \\ &\leq ||w^{(t)}||^2 + R^2 + 2(\gamma - \epsilon). \end{aligned} \tag{15}$$

and

$$(w^{(t+1)})^{\mathrm{T}}w^* = (w^t)^{\mathrm{T}}w^* + y^+(x^+)^{\mathrm{T}}w^*$$
$$\geq (w^t)^{\mathrm{T}}w^* + \gamma. \tag{16}$$

Through iterative deduction of the above two equations, we have

$$||w^{(\mathcal{A}+1)}||^2 \leq ||w^c||^2 + \mathcal{A}R^2 + 2\mathcal{A}(\gamma - \epsilon)$$
$$= M^2 + \mathcal{A}R^2 + 2\mathcal{A}(\gamma - \epsilon). \tag{17}$$

and

$$(w^{(\mathcal{A}+1)})^{\mathrm{T}}w^* \geq (w^c)^{\mathrm{T}}w^* + \mathcal{A}\gamma. \tag{18}$$

Because $(w^c)^{\mathrm{T}}w^* = ||w^c|| \cdot ||w^*|| \cos\phi$, where ϕ is the angle between w^c and w^*, we have:

$$(w^{(\mathcal{A}+1)})^{\mathrm{T}}w^* \geq \mathcal{A}\gamma - ||w^c|| \cdot ||w^*||$$
$$= \mathcal{A}\gamma - MN. \tag{19}$$

According to the Cauchy-Schwartz inequality, we see that

$$(w^{(\mathcal{A}+1)})^{\mathrm{T}}w^* \leq ||(w^{(\mathcal{A}+1)})|| \cdot ||w^*||. \tag{20}$$

Putting together Eq. (17) and Eq. (19) we get

$$\mathcal{A}\gamma - MN \leq \sqrt{M^2 + \mathcal{A}R^2 + 2\mathcal{A}(\gamma - \epsilon)}N. \tag{21}$$

Hence, we get

$$\mathcal{A} \leq \frac{N}{\gamma}\left(2(M+N) + \frac{N(R^2 - 2\epsilon)}{\gamma}\right)$$
$$= O\left(\frac{N}{\gamma}\left(M + N + \frac{N(R^2 - \epsilon)}{\gamma}\right)\right). \tag{22}$$

$\square$

Corollary. *Suppose that $||x_j|| \leq R$ for all $x_j \in \mathcal{L} \cup \mathcal{U}$. Let the current solution be w^c, and further suppose that there exists an optimal solution w^* such that $y_j(w^*)^{\mathrm{T}}x_j \geq \gamma$ for all examples x_j. Let $||w^c|| = M$ and $||w^*|| = N$. Suppose the probability of selecting the points satisfying the inequality $y^+w^{\mathrm{T}}x^+ < \gamma$ is P_a. Then the total number of label requests made by passive learning is at most*

$$O\left(\frac{N}{\gamma P_a}\left(M + N + \frac{N(R^2 - \epsilon)}{\gamma}\right)\right).$$

Proof. This corollary can be directly derived from the above theorem, and hence we skip the proof and only present the result. $\square$

According to the theoretical justifications provided by the above convergence theorem and corollary, we get the following conclusions: (1) because $0 < P_a < 1$, the upper bound of label requests made by MMC is proven to be smaller than random selection, demonstrating that the margin-based strategy is expected to outperform passive learning, and (2) the convergence property guarantees that MMC converges with the maximal label requests derived above.

5   Linkage between MMC and Simple Margin

As discussed before, one of the most widely used SVM active learning solution is simple margin, which chooses the points that are closet to the decision boundary. The distance between a point x and the boundary $w^{\mathrm{T}}x = 0$ is computed as:

$$\mathrm{Dist}(w, x) = \frac{|w^{\mathrm{T}}x|}{||w||}, \tag{23}$$

and the sampling function can be written as:

$$x_{\mathrm{SM}}^* = \arg\min_{x^+ \in \mathcal{U}} \mathrm{Dist}(w, x^+) = \arg\min_{x^+ \in \mathcal{U}} |w^{\mathrm{T}}x^+|. \tag{24}$$

Although it achieves good practical performance, it still lacks of reasonable theoretical justifications.

Here, we attempt to explore the connection between MMC and simple margin to provide a potentially theoretical justification. Let $x_{(j)}^+$ be the j-th close-to-boundary example in the pool, e.g. $x_{(1)}^+ = x_{\mathrm{SM}}^*$. Assume there are m unlabeled examples in the pool. According to Eq. (24), we have

$$|w^{\mathrm{T}}x_{\mathrm{SM}}^*| = |w^{\mathrm{T}}x_{(1)}^+| < |w^{\mathrm{T}}x_{(2)}^+| < \cdots < |w^{\mathrm{T}}x_{(m)}^+|. \tag{25}$$

Now, let us consider the inequality $|w^{\mathrm{T}}x^+| < \gamma$ used for sample filtering. If we restrict the margin parameter γ as:

$$|w^{\mathrm{T}}x_{\mathrm{SM}}^*| < \gamma \leq |w^{\mathrm{T}}x_{(2)}^+|, \tag{26}$$

it is clear to see that there will be only one point, i.e. the one most close to boundary, satisfying this inequality. Hence we have

$$x_{\mathrm{SM}}^* = \Omega = \{x : |w^{\mathrm{T}}x^+| < \gamma\} \Rightarrow x_{\mathrm{SM}}^* = x_{\mathrm{MMC}}^*. \tag{27}$$

Thus, simple margin can be viewed as a special case of MMC, and the theoretical results derived above is applicable to this popular method as well.

6   Experiments

6.1   Data Sets and Experimental Settings

To validate the performance of the proposed algorithm, we use eight benchmark data sets of various sizes from the UCI machine learning repository[1]: **Biodeg**,

---

[1] http://archive.ics.uci.edu/ml/

Table 1. The information of the eight binary-class data sets from UCI repository

Data set		# Examples	# Features	Class distribution
Biodeg		1055	41	356/699
Ionosphere		351	34	225/126
Parkinsons		195	22	147/48
WDBC		569	30	357/212
Letter	**D-vs-P**	1608	16	805/803
	E-vs-F	1543	16	768/775
	M-vs-N	1575	16	792/783
	U-vs-V	1577	16	813/764

Ionosphere, Parkinsons, WDBC, Letter. For **Letter**, a multi-class data set, we select four pairs of letters (i.e. **D-vs-P, E-vs-F, M-vs-N, U-vs-V**) that are relatively difficult to distinguish, and construct a binary-class data set for each pair. Table 1 shows the information of the eight binary-class data sets.

Each data set is randomly divided into three disjoint subsets: the base labeled training set (denoted as $\mathcal{L}$), the unlabeled pool set (denoted as $\mathcal{U}$), and the test set (denoted as $\mathcal{T}$). We use the base labeled set $\mathcal{L}$ as the small labeled data set to train the initial SVM models. The pool set $\mathcal{U}$ is used as a large size unlabeled data set to select the most informative examples, and the separate test set $\mathcal{T}$ is used to evaluate different active learning algorithms. More specifically, the active learning scenario for each data set is constructed as: $\mathcal{L}(5\%)+\mathcal{U}(75\%)+\mathcal{T}(20\%)$. We normalize the features with the function below:

$$f_{(i,j)}^N = \frac{f_{(i,j)} - \min_{i \in n}\{f_{(i,j)}\}}{\max_{i \in n}\{f_{(i,j)}\} - \min_{i \in n}\{f_{(i,j)}\}}, \tag{28}$$

where n denotes the number of examples in each of data set, and $f_{(i,j)}$ represents the j-th feature from the i-th example.

The optimal margin parameter γ is determined by the standard 5-fold cross validation. In this study, the active learning process iterates 10 rounds. In each round of data selection, 3% of the whole examples are selected from $\mathcal{U}$. These examples are then added to the training set, and SVM classifiers are re-trained and tested on the separate test set $\mathcal{T}$.

6.2   Comparison Methods and Evaluation Metric

To test the effectiveness of the proposed active learning algorithm, we compare it against the following four competitors including three state-of-art active learning for SVMs methods, and one baseline random selection: (1) S-MARGIN [18]: the simple margin algorithm, (2) CLUSTER [5]: the clustering-based representative sampling approach, (3) QUIRE [14]: the multi-criteria-based sampling, and (4) RAND: the random sampling. A detailed description of each of these algorithms

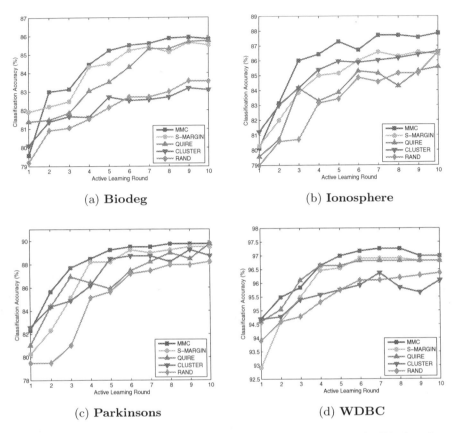

Fig. 1. Comparison results of different active learning algorithms on the **Biodeg, Iono-sphere, Parkinsons**, and **WDBC** data sets

is provided in Section 2. For evaluation, the classification accuracy is adopted to measure the performance on the test set:

$$\text{Accuracy} = \frac{1}{|\text{T}|} \sum_{i=1}^{|\text{T}|} \mathbf{1}\{f(x_i) = y_i\}, \tag{29}$$

where $|\text{T}|$ stands for the size of the test set, and y_i and $f(x_i)$ are the ground truth and prediction of x_i, respectively. $\mathbf{1}\{.\}$ is the indicator function. To avoid random fluctuation, each experiment is repeated 10 times by varying the base-pool-test sets, and the averaged classification accuracy is reported.

6.3   Comparison Results and Discussions

The comparison results of the five data selection algorithms on these eight UCI benchmark data sets are presented in Figure 1 and Figure 2. The X-axis denotes

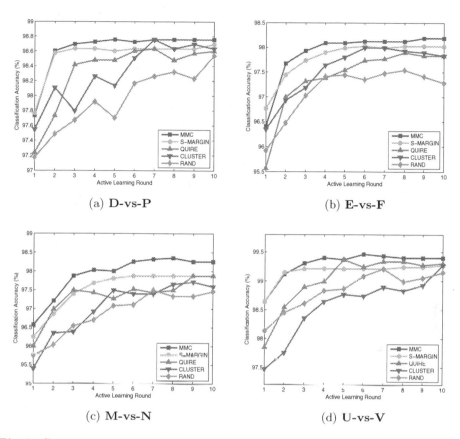

Fig. 2. Comparison results of different active learning algorithms on the **D-vs-P**, **E-vs-F**, **M-vs-N**, and **U-vs-V** data sets

the number of iterations for the active learning process, and the Y-axis represents the classification accuracy. Several general observations as shown in these figures are explained as follows.

(1) For all five algorithms, the classification accuracy generally increases with the iterations of active learning, which matches the intuition that model's performance is positively correlated with the amount of training set available.

(2) The proposed MMC algorithm is observed to perform the best among the five approaches in most cases during the entire data selection process, demonstrating that the proposed active learning method is more effective in choosing the most informative examples to improve the model quality. This is likely due to the reason that MMC quantifies the model change as the gradient of the loss, which is highly correlated with the objective function used to evaluate the SVM models. Therefore, the examples selected by MMC are more likely to contribute positively to improve the model. In addition, we observe that MMC converges much faster than the competitors on several data sets (e.g. **D-vs-P**, **E-vs-F**, **M-**

Table 2. The p-value of Wilcoxon signed rank test of MMC versus S-MARGIN, QUIRE, CLUSTER and RAND on the UCI data sets

Data sets	vs. S-MARGIN	vs. QUIRE	vs. CLUSTER	vs. RAND
Biodeg	p<0.1	p<0.1	p<0.05	p<0.05
Ionosphere	p<0.05	p<0.05	p<0.05	p<0.05
Parkinsons	p<0.05	p<0.05	p<0.05	p<0.05
WDBC	p<0.05	p<0.05	p<0.05	p<0.05
D-vs-P	p<0.05	p<0.05	p<0.05	p<0.05
E-vs-F	p<0.1	p<0.05	p<0.05	p<0.05
M-vs-N	p<0.05	p<0.05	p<0.05	p<0.05
U-vs-V	p<0.05	p<0.05	p<0.05	p<0.05

vs-N, U-vs-V), i.e. the highest classification accuracy is achieved with much less examples added to the training set. This agrees with the intuitive explanation that the data examples greatly changing the current classifier are expected to produce faster convergence to the optimal model.

(3) We see that the performance of CLUSTER is inconsistent. It works well on some data sets, but performs poorly on the others. This phenomena may be explained as follows. The CLUSTER method utilizes a clustering technique to choose the representative data points, which may fail if there is no clear cluster structure in the data. On the contrary, QUIRE is observed to yield relatively good performance on most data sets. The success of QUIRE may be attributed to the principle of choosing examples that are both informative and representative.

(4) To better validate the effectiveness of the proposed approach, we conduct the significance test on the comparisons. Table 2 presents the results of Wilcoxon signed rank test of MMC versus S-MARGIN, QUIRE, CLUSTER and RAND strategies on the benchmark UCI data sets. The comparison results with $p > 0.05$ are underlined. It shows that the proposed method performs statistically better ($p < 0.05$) than S-MARGIN, QUIRE, CLUSTER and RAND on most data sets. We also perform the 2-tailed paired T-test to further examine the effectiveness of MMC. Due to the space limitation, the p-values according to the 2-tailed T-test are not reported here, and the results show that MMC significantly outperforms ($p < 0.05$) the competitors in most cases during the sample selection process.

6.4   Efficiency Comparison

In this subsection, we compare the CPU running time taken by MMC versus the competitors. All algorithms were implemented using MATLAB on a standard desktop computer with 2.53 GHz CPU and 8 GB of memory.

Table 3 shows the comparison results, together with the information of the pool set. As shown in the table, the time complexity of MMC is slightly higher than that of S-MARGIN, but much more efficient than the other two strategies, i.e. QUIRE and CLUSTER. This is due to the reason that QUIRE involves

Table 3. The CPU running time (seconds), together with the information of pool set

Data sets	# Ex. × Features ($\mathcal{U}$)	MMC	S-MARGIN	QUIRE	CLUSTER
Biodeg	791 × 41	0.04	0.01	100.85	1.12
Ionosphere	263 × 34	0.02	0.01	3.79	0.21
Parkinsons	146 × 22	0.00	0.00	0.84	0.15
WDBC	427 × 30	0.01	0.00	16.38	0.44
D-vs-P	1206 × 16	0.07	0.02	341.94	1.03
E-vs-F	1157 × 16	0.02	0.01	301.87	0.78
M-vs-N	1181 × 16	0.02	0.02	324.70	1.02
U-vs-V	1183 × 16	0.05	0.01	219.10	1.00

calculating the inverse of a large scale matrix, and CLUSTER requires considerable efforts on clustering. In summary, the proposed MMC method is quite efficient in computational complexity, and is promising for real-world applications.

7 Conclusions

In this paper, focusing on SVMs, we introduce a new interpretation for margin-based active learning with the idea of expected model change, and accordingly propose a novel margin-based active learning algorithm named Maximum Model Change (MMC), which is to choose the examples leading to the maximal change in the current classifier. The change is measured as the difference between the current model parameters and the updated parameters trained with the accumulated training set. Inspired by the SGD rule for solving the SVMs optimization problems, the change is approximated as the gradient of the loss at a candidate point. In addition, we provide a theoretical analysis of the convergence property for the proposed algorithm, and compare the derived sampling bound against passive learning. The comparison shows that the upper bound of sample requests made by MMC is smaller than passive learning. We further connect the proposed approach with the widely adopted simple margin approach to provide a theoretical justification for this popular algorithm. Substantial experimental results on various benchmark UCI data sets have demonstrated that the proposed strategy is highly effective in selecting informative examples, and efficient in computation time.

Acknowledgments. This research was supported by National Natural Science Foundation of China (No. 61003107 & No. 61221001) and the High Technology Research and Development Program of China (2012AA011702).

References

1. Settles, B., Craven, M.: An analysis of active learning strategies for sequence labeling tasks. In: Proc. of EMNLP 2008, pp. 1070–1079 (2008)

2. Abe, N., Mamitsuka, H.: Query learning strategies using boosting and bagging. In: Proc. of ICML 1998, pp. 1–10 (1998)
3. Donmez, P., Carbonell, J.G.: Optimizing estimated loss reduction for active sampling in rank learning. In: Proc. of ICML 2008, pp. 248–255 (2008)
4. Freund, Y., Seung, H.S., Shamir, E., Tishby, N.: Selective sampling using the query by committee algorithm. Machine Learning, 133–168 (1997)
5. Xu, Z., Yu, G., Tresp, V., Xu, X., Wang, J.: Representative sampling for text classification using support vector machines. In: Sebastiani, F. (ed.) ECIR 2003. LNCS, vol. 2633, pp. 393–407. Springer, Heidelberg (2003)
6. Settles, B.: Active learning. Morgan & Claypool (2012)
7. Nguyen, H.T., Smeulders, A.: Active learning using pre-clustering. In: Proc. of ICML 2004, pp. 623–630 (2004)
8. Tang, M., Luo, X., Roukos, S.: Active learning for statistical natural language parsing. In: Proc. of ACL 2002, pp. 120–127 (2002)
9. Li, L., Jin, X., Pan, S., Sun, J.: Multi-domain active learning for text classification. In: Proc. of KDD 2012, pp. 1086–1094 (2012)
10. Li, M., Sethi, I.K.: Confidence-based acitve learning. IEEE Trans. Pattern Analysis and Machine Intelligence, 1251–1261 (2006)
11. Vapnik, V.: The nature of statistical learning Theory. Springer (1999)
12. Burges, C.: A tutorial on support vector machines for pattern recognition. Data Mining and Knowledge Discovery, 121–167 (1998)
13. Brinker, K.: Incorporating diversity in active learning with support vector machines. In: Proc. of ICML 2003 (2003)
14. Huang, S., Jin, R., Zhou, Z.: Active learning by querying informative and representative examples. In: Proc. of NIPS 2010, pp. 892–900 (2010)
15. Shwartz, S., Singer, Y., Srebro, N.: Pegasos: primal estimated sub-gradient solver for SVM. In: Proc. of ICML 2007, pp. 807–814 (2007)
16. Wang, Z., Crammer, K., Vucetic, S.: Breaking the curse of kernelization: budgeted stochastic gradient descent for large-scale SVM training. Journal of Machine Learning Research, 3103–3131 (2012)
17. Shen, D., Zhang, J., Su, J., Zhou, G., Tan, C.L.: Multi-criteria-based active learning for named entity recognition. In: Proc. of ACL 2004, pp. 589–596 (2004)
18. Tong, S., Koller, D.: Support vector machine active learning with applications to text classification. Journal of Machine Learning Research, 45–66 (2001)
19. Kivinen, J., Smola, A., Williamson, R.: Online learning with kernels. IEEE Trans. Signal Processing, 2165–2176 (2004)
20. Roy, N., McCallum, A.: Toward optimal active learning through sampling estimation of error reduction. In: Proc. of ICML 2001, pp. 441–448 (2001)
21. Chon, A., Ghahramani, Z., Jordan, M.I.: Active learning with statisitical models. Journal of Machine Learning Research, 129–145 (1996)
22. Lewis, D., Gale, W.: A sequential algorithm for training text classifiers. In: Proc. of SIGIR 1994, pp. 3–12 (1994)
23. Settles, B., Craven, M., Ray, S.: Multiple-instance active learning. In: Proc. of NIPS 2008, pp. 1289–1296 (2008)
24. Cai, W., Zhang, Y., Zhou, J.: Maximizing expected model change for active learning in regression. In: Proc. of ICDM 2008, pp. 51–60 (2013)

Anomaly Detection with Score Functions Based on the Reconstruction Error of the Kernel PCA

Laetitia Chapel[1] and Chloé Friguet[2]

[1] Univ. Bretagne-Sud, UMR 6074, IRISA, F-56000 Vannes, France
`laetitia.chapel@univ-ubs.fr`
[2] Univ. Bretagne-Sud, UMR 6205, LMBA, F-56000 Vannes, France
`chloe.friguet@univ-ubs.fr`

Abstract. We propose a novel non-parametric statistical test that allows the detection of anomalies given a set of (possibly high dimensional) sample points drawn from a nominal probability distribution. Our test statistic is the distance of a query point mapped in a feature space to its projection on the eigen-structure of the kernel matrix computed on the sample points. Indeed, the eigenfunction expansion of a Gram matrix is dependent on the input data density f_0. The resulting statistical test is shown to be uniformly most powerful for a given false alarm level α when the alternative density is uniform over the support of the null distribution. The algorithm can be computed in $O(n^3 + n^2)$ and testing a query point only involves matrix vector products. Our method is tested on both artificial and benchmarked real data sets and demonstrates good performances $w.r.t.$ competing methods.

1   Introduction

Anomaly detection [1], also called novelty detection or one-class classification, aims at declaring a query point $\boldsymbol{\eta}$ as "normal" or not with respect to a nominal model. The underlying d-dimensional nominal probability distribution $f_0(\boldsymbol{x})$ over the input space $\mathcal{X} \subset \mathbb{R}^d$ is unknown but is described by a set of n independently and identically distributed ($i.i.d.$) nominal data points, gathered in the training set $\mathcal{S} = \{\boldsymbol{x}_1, \cdots, \boldsymbol{x}_n\}$ where $\boldsymbol{x}_k$ are d-dimensional vectors. The anomaly detection problem can be formulated as a statistical test of the hypothesis that the query point $\boldsymbol{\eta}$ comes from the nominal distribution f_0:

$$\begin{cases} \mathcal{H}_0: & \boldsymbol{\eta} \sim f_0 \quad \text{i.e. } \boldsymbol{\eta} \text{ is consistent with nominal data} \\ \mathcal{H}_1: & \boldsymbol{\eta} \nsim f_0 \quad \text{i.e. } \boldsymbol{\eta} \text{ deviates from nominal data} \end{cases} \tag{1}$$

while controlling the type-I error ($i.e.$ the probability to declare a nominal point as anomaly) under a fixed level α:

$$p(\boldsymbol{\eta}) = \mathbb{P}(\text{reject } \mathcal{H}_0 | \mathcal{H}_0 \text{ is true}) \leq \alpha \tag{2}$$

Conversely, the type-II error is the probability to declare a true anomaly as nominal. If the p-value $p(\boldsymbol{\eta}) \leq \alpha$, the null hypothesis is rejected, and $\boldsymbol{\eta}$ is declared anomaly $w.r.t.$ f_0.

T. Calders et al. (Eds.): ECML PKDD 2014, Part I, LNCS 8724, pp. 227–241, 2014.

The standard approach in anomaly detection consists in declaring as anomaly a point lying in a low density region, considering a threshold t for the nominal density f_0 such that $\boldsymbol{\eta}$ is declared as anomaly if $f_0(\boldsymbol{\eta}) < t$. The probabilistic interpretation can be obtained by considering the cumulative distribution function F_0 associated to the contour $f_0(\boldsymbol{\eta}) = t$ to get the p-value. Indeed, if $\boldsymbol{\eta}$ is drawn from the nominal density f_0, it is expected to lie outside the anomaly threshold with probability $p(\boldsymbol{\eta}) = 1 - F_0(t)$. Thus, the standard approach in anomaly detection is related to density level set estimation. In practice, a ranking of $\boldsymbol{\eta}$ among all nominal points in $\mathcal{S}$ can be used to estimate the p-value function $p(\boldsymbol{\eta})$:

$$\hat{p}(\boldsymbol{\eta}) = \frac{1}{n} \sum_{k=1}^{n} \left(\mathbb{1}\{F(\boldsymbol{\eta}) \leq F(\boldsymbol{x}_k)\} \right) \tag{3}$$

where $\mathbb{1}\{.\}$ denotes the indicator function. F would be chosen to be an estimate of the nominal cumulative distribution function F_0 so that $\hat{p}(\boldsymbol{\eta})$ approximates accurately $p(\boldsymbol{\eta})$.

The estimation of the nominal density f_0 is a challenging task and classical approaches are parametric in nature, *i.e.* the key assumption is that the family of the nominal distribution f_0 is known. Such methods include Hidden Markov Models or Gaussian Mixture Modeling (see [2] for a review). On the other hand, performances of non-parametric methods do not depend of an assumed distribution.

Instead of directly estimate the nominal density, we propose in this paper to consider a surrogate of f_0. Our proposal is based on the properties of the eigen-decomposition of the Gram matrix of a given kernel $\boldsymbol{K}$, such as extracted by kernel-PCA. An accurate surrogate of the density function can then be derived, even in the high-dimensional setting, which is a challenge in anomaly detection [3].

The key principle of our approach is to consider that nominal points are projected with low reconstruction errors on a KPCA space, and that anomalies lie far from the projected space. The proposed test statistic is then derived from the reconstruction error of a new point projected on a KPCA space, namely the square distance between the original point mapped in the feature space and its projection on the KPCA space, computed from a set of nominal points $\mathcal{S}$.

The score function maps the data from the feature space to the $[0, 1]$ interval, and the associated p-values have a uniform distribution when test points are drawn according to $\mathcal{H}_0$. When the score function has a value lower than a given threshold t, the point is labeled as anomaly. The algorithm has the following properties: (i) it performs well in high dimensional spaces as it is based on manifold assumptions (ii) it is non-parametric and there is no complicated parameters to tune (iii) it provides p-value estimates, allowing for type-I error control and (iv) it converges to the uniformly most powerful test when anomalies are drawn from a mixture of a nominal density f_0 and a uniform density.

The rest of this paper is organized as follows. Section 2 describes the KPCA framework for density estimation and introduces the reconstruction error. The anomaly detection test procedure, based on the reconstruction error, is presented in Sect. 3. Our proposal is related to some other works in anomaly detection, covariate shift and selective sampling which are evoked in Sect. 4. Section 5 reports experiments to illustrate our approach on both artificial and real datasets in high-dimensional spaces. Finally, we draw some conclusions and give some perspectives.

2   Kernel Eigen-Decomposition and Density Estimation

Kernel-PCA and Notations The aim of methods related to Singular Value Decomposition (SVD), such as linear Principal Component Analysis (PCA), is to identify and extract structures from the data by computing linear functions. In PCA, the subspace spanned by the first eigenvectors is used both to give a low-dimensional model with minimal residual and to provide a low-dimensional representation of the data. However, as nonlinear structures often occur, a kernelized version of the PCA (Kernel-PCA or KPCA for short) has been developed to deal with nonlinear structures in the input space by reducing them to linear ones in a feature space: data are mapped into a higher-dimensional space in which the information about their mutual positions is used for further analyzes.

We suppose an unknown distribution $f_0(\boldsymbol{x})$ on the input space $\mathcal{X}$. Kernel methods map $\boldsymbol{x}$ into a feature space $\mathcal{F}$ through the non-linear embedding map $\psi : \boldsymbol{x} \in \mathcal{X} \mapsto \psi(\boldsymbol{x}) \in \mathcal{F}$. Pairwise inner products can be computed efficiently directly from the original data points using a kernel function $\kappa(\boldsymbol{x}; \boldsymbol{x}') = \langle \psi(\boldsymbol{x}), \psi(\boldsymbol{x}') \rangle$.

The kernel or Gram matrix $\boldsymbol{K}$ is defined as the matrix whose entries are $\boldsymbol{K}_{kj} = \kappa(\boldsymbol{x}_k, \boldsymbol{x}_j), \forall \boldsymbol{x}_k, \boldsymbol{x}_j \in \mathcal{X}$.

Eigen-Decomposition of the Gram Matrix The Gram matrix $\boldsymbol{K}$ can be decomposed as

$$\boldsymbol{K} = \boldsymbol{V} \Lambda \boldsymbol{V}^T \qquad (4)$$

where Λ denotes the diagonal matrix of eigenvalues λ_i, sorted in decreasing order for convenience, and the eigen-vectors are stored column-wise in matrix $\boldsymbol{V} = \left[\phi_i\right]_{1 \le i \le \infty}$. Performing a KPCA comes down to the previous decomposition of $\boldsymbol{K}$ as the optimal space is the one spanned by $\boldsymbol{V}$.

The input data density f_0 is related to the kernel function $\kappa(\boldsymbol{x}, \boldsymbol{y})$; indeed, the eigenfunction expansion is dependent on f_0 [4]:

$$\int_{\mathcal{X}} \kappa(\boldsymbol{x}, \boldsymbol{y}) f_0(\boldsymbol{x}) \phi_i(\boldsymbol{x}) d\boldsymbol{x} = \lambda_i \phi_i(\boldsymbol{y}). \qquad (5)$$

This continuous problem can be approximated on a finite sample $\mathcal{S}$ when the $\boldsymbol{x}_k$ are sampled from f_0 by the following empirical average:

$$\int_{\mathcal{X}} \kappa(\boldsymbol{x}, \boldsymbol{y}) f_0(\boldsymbol{x}) \phi_i(\boldsymbol{x}) d\boldsymbol{x} \simeq \frac{1}{n} \sum_{k=1}^{n} \kappa(\boldsymbol{x}_k; \boldsymbol{y}) \phi_i(\boldsymbol{x}_k). \qquad (6)$$

According to [4], the continuous eigenfunctions and eigen-spectrum are therefore reliably asymptotically estimated from the eigenvalue decomposition of the kernel matrix $\boldsymbol{K}(\mathcal{S})$ computed on the random sample of data.

Going through each sample $\boldsymbol{x}_j$ in $\mathcal{S}$ gives a set of n linear equations:

$$\frac{1}{n} \sum_{k=1}^{n} \kappa(\boldsymbol{x}_k, \boldsymbol{x}_j)\phi_i(\boldsymbol{x}_k) = \lambda_i \phi_i(\boldsymbol{x}_j) \quad ; \quad j \in [1; n] \tag{7}$$

which can be rewritten as eigen-decomposition:

$$\boldsymbol{K}(\mathcal{S})\widehat{\boldsymbol{V}} = n\widehat{\boldsymbol{V}}\widehat{\boldsymbol{\Lambda}} \tag{8}$$

where $\widehat{\boldsymbol{V}}$ has columns with the empirical eigen-vectors $\widehat{\boldsymbol{\phi}}_i$ and $\widehat{\boldsymbol{\Lambda}}$ is a diagonal matrix containing the empirical eigenvalues $\widehat{\lambda}_i$, $1 \leq i \leq n$. It gives an estimate of the system of eigenfunction associated with (5) and converges towards their asymptotical counterparts with $n \to \infty$ [5]. According to (7), the Nyström approximation [6] of the ith eigenfunction can be calculated for a new point $\boldsymbol{\eta}$ by

$$\phi_i(\boldsymbol{\eta}) = \frac{1}{n\lambda_i} \sum_{k=1}^{n} \kappa(\boldsymbol{x}_k; \boldsymbol{\eta})\phi_i(\boldsymbol{x}_k). \tag{9}$$

This can then be related to KPCA. In order to be the eigen-vector of the correlation matrix

$$C(\mathcal{S}) = \frac{1}{n}\psi(\mathcal{S}) \cdot \psi(\mathcal{S})^T \tag{10}$$

where $\psi(\mathcal{S}) = \left[\psi(\boldsymbol{x}_1); \dots; \psi(\boldsymbol{x}_n)\right]^T$, the previous eigen-vectors have to be normalized

$$\phi_i^*(\boldsymbol{\eta}) = \frac{1}{n\sqrt{\lambda_i}} \sum_{k=1}^{n} \kappa(\boldsymbol{x}_k; \boldsymbol{\eta})\phi_i(\boldsymbol{x}_k). \tag{11}$$

Reconstruction Error of the KPCA. If a new sample $\boldsymbol{\eta}$ is drawn according to the same probability distribution as $\mathcal{S}$, it is reasonable to compute the KPCA projection of $\boldsymbol{\eta}$ onto the subspace spanned by $\psi(\mathcal{S})$ [5] as

$$r(\boldsymbol{\eta} \to \psi(\mathcal{S})) = \kappa(\mathcal{S}; \boldsymbol{\eta}) \cdot \boldsymbol{V} \cdot \Lambda^{-1/2} \tag{12}$$

where $\kappa(\mathcal{S}; \boldsymbol{\eta}) = \left[\kappa(\boldsymbol{x}_1; \boldsymbol{\eta}); \dots; \kappa(\boldsymbol{x}_n; \boldsymbol{\eta})\right]$. The quality of the projection is assessed by computing the square norm of the residual, *i.e.* the error resulting from using the projection $r(\boldsymbol{\eta} \to \psi(\mathcal{S}))$ rather than the actual vector $\psi(\boldsymbol{\eta})$. The reconstruction error $\tau_{\mathcal{S}}(\boldsymbol{\eta})$ of the sample in the feature space is then given by

$$\begin{aligned} \tau_{\mathcal{S}}(\boldsymbol{\eta}) &= \|\psi(\boldsymbol{\eta}) - r(\boldsymbol{\eta} \to \psi(\mathcal{S}))\|^2 \\ &= \|\psi(\boldsymbol{\eta})\|^2 - \|r(\boldsymbol{\eta} \to \psi(\mathcal{S}))\|^2 \\ &= \kappa(\boldsymbol{\eta}; \boldsymbol{\eta}) - \kappa(\mathcal{S}; \boldsymbol{\eta}) \cdot \boldsymbol{K}(\mathcal{S})^{-1} \cdot \kappa(\mathcal{S}; \boldsymbol{\eta})^T. \end{aligned} \tag{13}$$

Note that if $\boldsymbol{\eta} \subset \mathcal{S}$, nothing is lost in the projection and hence $\tau_{\mathcal{S}}(\boldsymbol{\eta}) = 0$. Besides, if a RBF Gaussian kernel is considered, note also that $\kappa(\boldsymbol{\eta}; \boldsymbol{\eta}) = 1$ and $\tau_{\mathcal{S}}(\boldsymbol{\eta}) \in [0; 1]$.

The extent to which the projection value captures new data according to the same distribution f_0 as the training data is a critical question and has been discussed in order to assess the KPCA performances [4, 7, 5]. Proposition 1 states that good capture of the new data can be expected, as the expectation of the empirical value $\tau_{\mathcal{S}}(\boldsymbol{\eta})$ computed from $\boldsymbol{K}(\mathcal{S})$ converges to the true expectation of $\tau(\boldsymbol{\eta})$ computed from $\boldsymbol{K}$. We first suppose that the support of the distribution under $\mathcal{H}_0$ is bounded in a ball of radius R in $\mathcal{F}$.

Proposition 1. *Under $\mathcal{H}_0$, $\tau_{\mathcal{S}}(\boldsymbol{\eta}) \to 0$ with high probability, provided that the empirical eigen-spectrum decays sufficiently fast.*

In other words, as long as the percentage of variance captured by low-dimensional eigen-spaces is concentrated, $\tau(\boldsymbol{\eta})$ can be reliably estimated from $\tau_{\mathcal{S}}(\boldsymbol{\eta})$ and $\tau_{\mathcal{S}}(\boldsymbol{\eta}) \to 0$ as n grows. The proof is given in [5] or [8].

Impact of the Choice of the Kernel. The decay rate of the eigenvalues depends on the connection between the kernel and the distribution. The choice of the kernel is hence crucial; discussion about the decay of the spectrum in the case of a Gaussian or translation-invariant kernels is provided in [9]. Even if expressing the connection in closed form is in general impossible, the decay rate can be checked *a posteriori* in order to assess the validity of the results.

3   Anomaly Detection Test Based on the Reconstruction Error of the KPCA

3.1   General Principle of the Proposed Testing Procedure

Anomaly detection is related to density level set estimation: once the nominal density level set is learnt from the nominal data distribution, a point that falls outside the level set, in the low density region, is declared as an anomaly. In the KPCA framework, as stated in [7], the eigen-decomposition of the Gram matrix extracted with KPCA provides an accurate estimate of the nominal density function so that we propose to use the eigen-decomposition to derive a testing strategy for anomaly detection. Hereafter, instead of directly estimating the (unknown) f_0, we propose to consider the reconstruction error defined in (13) as a surrogate of F_0 in (3).

Indeed, $\tau(\boldsymbol{\eta})$ can be viewed as a measure that gives an insight of how far the query point is from the true kernel matrix $\boldsymbol{K}$. If the query point $\boldsymbol{\eta}$ is drawn from the nominal density f_0, its empirical eigenvalues and eigenvectors should coincide with those of the underlying process. Therefore, $\psi(\boldsymbol{\eta})$ and $\psi(\mathcal{X})$ lie roughly in the same space and the reconstruction error $\tau(\boldsymbol{\eta})$ is "small", or more precisely, it tends to 0 with high probability. On the contrary, if $\boldsymbol{\eta}$ is not drawn from the nominal density f_0, the reconstruction error $\tau(\boldsymbol{\eta})$ is "high". This behaviour makes $\tau(.)$ a suitable surrogate to rank the querry point, ordering it with respect to the target nominal density contour on which it lies and hence the level set of τ is close to the target level set of f_0 everywhere.

The statistical test then comes down to thresholding the τ values. Interestingly enough, it is the Uniformly Most Powerful (UMP) test: indeed, in density level set, it is deduced from the Neymann-Pearson test theory that the optimal test with type-I error control of level α is obtained by thresholding f_0 when the density under the alternative hypothesis is the uniform density over the support of the null distribution.

3.2   Calibrating the Test

To compute the critical region for a given level α, the test procedure relies on the definition of the null cumulative distribution function (cdf) G_0 of the test statistic. Unfortunately, the limiting cdf of τ_S under the null hypothesis is unknown but can be determined in various ways (see for instance [10]). We here consider a resampling technique to calibrate the test, which usually leads to a fast and accurate estimation of the cdf.

We take advantage of the fact that the training set S is drawn according to $\mathcal{H}_0$ to construct an empirical cdf G. As $\forall \boldsymbol{x}_k \in S, \tau_S(\boldsymbol{x}_k) = 0$, an alternative strategy has to be defined. Proposition 2 describes the procedure that can be used to obtain an empirical null cdf G and states that, with high probability, it converges to the null cdf G_0.

Proposition 2. *Given a set S drawn according to the nominal distribution f_0, the estimate of the empirical null cumulative distribution function G_0 of τ_S is given by:*

$$G(t) = \frac{1}{n} \sum_{k=1}^{n} \mathbb{1}(\tau_S^{-k}(\boldsymbol{x}_k) \leq t) \tag{14}$$

where

$$\tau_S^{-k}(\boldsymbol{x}_k) = \kappa(\boldsymbol{x}_k; \boldsymbol{x}_k) - \kappa(\boldsymbol{x}_k; S) \cdot \boldsymbol{K}(S \backslash \{\boldsymbol{x}_k\})^{-1} \cdot \kappa(\boldsymbol{x}_k; S)^T \tag{15}$$

is the reconstruction error of sample $\boldsymbol{x}_k$ computed on $S \backslash \{\boldsymbol{x}_k\}$, the training set deprived of $\boldsymbol{x}_k$. By the strong law of large number, G converges almost surely to G_0 for every value of t:

$$sup_{t \in \mathbb{R}} |G(t) - G_0(t)| \mapsto 0 \tag{16}$$

We can now define our anomaly detection test in the following proposition.

Proposition 3. *The anomaly detection test is based on the computation of estimated p-values defined by:*

$$\hat{p}(\boldsymbol{\eta}) = \frac{1}{n} \sum_{\boldsymbol{x}_k \in S} \mathbb{1}\left\{\tau_S(\boldsymbol{\eta}) \leq \tau_S^{-k}(\boldsymbol{x}_k)\right\}. \tag{17}$$

Algorithm 1. Anomaly detection with score function based on the error recon-struction of the kernel-PCA

Input: $\mathcal{S} = \{x_1; \cdots ; x_n\}$, the n-set of nominal points, drawn i.i.d from the underlying density f_0
Input: η, the query point
Input: α, the significance level
  Training phase
  for x_k in $\mathcal{S}$ **do**
    compute $\tau_{\mathcal{S}}^{-k}(x_k)$ according to (15), where $K(\mathcal{S}\backslash\{x_k\})^{-1}$ is computed from $K(\mathcal{S})^{-1}$ (18)
  end for
  Test phase
  compute $\tau_{\mathcal{S}}(\eta)$ according to (13)
  if $\hat{p}(\eta)$ defined in (17) $\leq \alpha$ **then**
    declare η as anomaly
  else
    declare η as nominal
  end if

The main steps of the proposed procedure are described in Algorithm 1. It computes the estimated nominal values of $\tau_{\mathcal{S}}(x_k)$ from $\mathcal{S}$ and query point η is then detected as anomaly if its projection on the feature space is too high, that is to say $\hat{p}(\eta) \leq \alpha$. Complexity of computing all the different $\tau_{\mathcal{S}}^{-k}(x_k)$ is $O(n^3 + n^2)$. Testing a query point only involves matrix vector products.

3.3 Implementation Issues and Practical Solutions

Computing the empirical cdf under the null hypothesis involves the inversion of n Gram matrices $K(\mathcal{S}\backslash\{x_k\})^{-1}$, $x_k \in \mathcal{S}$ of size $(n-1) \times (n-1)$. This computational burden can be alleviated by computing $K(\mathcal{S})^{-1}$ once and deducing all $K(\mathcal{S}\backslash\{x_k\})^{-1}$ matrices by considering the following algebra decomposition that uses the Schur complement of matrix $K(\mathcal{S})^{-1}$:

$$K(\mathcal{S}\backslash\{x_k\})^{-1} = K(\mathcal{S})_{-k,-k}^{-1} - \frac{K(\mathcal{S})_{-k,k}^{-1} \cdot K(\mathcal{S})_{k,-k}^{-1}}{K(\mathcal{S})_{k,k}^{-1}} \tag{18}$$

where $K(\mathcal{S})_{k,k}^{-1}$ corresponds to the kth column and k^{th} row of matrix $K(\mathcal{S})^{-1}$, the minus sign meaning that the matrix is deprived from its kth column and/or its kth row.

For some numerical analysis issues, the matrix $K(\mathcal{S})$ may not be invertible (in theory, it should not be the case as it is a Gram matrix). Yet, it is always possible to compute its pseudo-inverse or to consider regularization techniques that find a close invertible matrix.

In addition, the regularization can also be used in order to avoid over-fitting and high sensitivity to features noise. We can consider the case of Tikhonov regularization where a regularized version of $K(\mathcal{S})$ is used:

$$\widetilde{K}(\mathcal{S}) = K(\mathcal{S}) + \lambda I \tag{19}$$

where the small positive constant $\lambda > 0$ is the Tikhonov factor.

An alternative strategy to reduce the impact of feature noise would be to keep only the $q < n$ first eigen-vectors. Indeed, an improvement of the nominal density estimation can be made by considering a truncation of the eigen-decomposition, considering an estimate of f_0 based on the most important components extracted with KPCA. [4] show that, in the particular case of the RBF Gaussian kernel, the dominant eigen-values are accurately estimated, the smaller ones being more poorly estimated. Considering a density estimate based on the first components of the Gram matrix decomposition do not suffer from the estimation step.

4   Related Work

Our work is related to some other works in anomaly detection, covariate shift and selective sampling.

[11] states that the reconstruction error defined in (13), but computed only on the first q dimensions, gives a good insight on its own to the belonging of a point to a probability distribution. The threshold above which a point is declared as anomaly has to be set empirically using the ROC curve. Experiments demonstrate higher performance on several datasets than one-class SVM and Parzen window density estimator. In the present work, the same statistic is considered but the threshold allows controlling the type-I error while minimizing the type-II error. Discussion in [11] about complexity of the algorithm, as well as sensibility to outliers hence holds here and are not recalled.

The $\tau_\mathcal{S}$ statistic is also equal to one minus the surrogate kernel of $K(\mathcal{S})$ on sample η [12]: it corresponds to the projection of a kernel matrix K from η (or possibly a new set of sample) to $\mathcal{S}$, while preserving the nominal eigen-structures. They use it as a bridge to compare different kernel matrices. One key assumption of this work is that there was a covariate shift, which is the main difference with our work. They then use the surrogate kernel in order to match data distributions and then compensate for the covariate shift in the Hilbert space.

[13] use the τ statistic in the computer graphics domain. Instead of viewing it as a measure for anomaly, they instead used it as a measure of importance of a sample in characterizing a distribution: points η with high $\tau_\mathcal{S}(\eta)$ values are considered important to be added to the shape as they are different from the other points. They hence provide a sampling scheme in order to use $\ell \ll n$ points to compute $K(\mathcal{S})$. In the same line, [14] state that τ provides a quality measure for Nyström method and derive a selective sampling scheme based on its value. Such schemes can be used in complement of our approach, in order to reduce the computational cost of extracting the eigen-functions of $K(\mathcal{S})$ and of testing, but it is beyond the scope of this paper.

5 Experiments

Non-parametric methods for outlier detection usually rely on the assessment of neighborhoods, where the kth-nearest neighbor distances are used to derive an outlier score (*e.g.* see [15], [16] or [17]). For a sake of simplicity, the following experiments are limited to the comparison of the proposed algorithm with the k-lpe algorithm [15], which has been shown to be asymptotically optimal and allows the control of the type-I error, and one-class SVM [18] which is probably the most classical anomaly detection algorithm, albeit not being able to control the type-I error.

Outlier detection can be evaluated using Receiver Operating Characteristics (ROC) curves, that are numerically compared with the area under these curves (AUC). For each method, we perform a grid search in order to select the best set of parameters (see Tab. 1 for a summary) and report the best averaged AUC values over 100 repetitions. We only consider the Tikhonov regularization of the Gram matrix as it leads to similar or better results than the truncated spectrum. With a Gaussian kernel, the one-class SVM is a non-parametric estimator of a level set of density governing the training set $\mathcal{S}$, with parameter ν defining the corresponding level. Varying the ν value hence allows the definition of all the density level sets, and then the construction of the ROC curve. Note that one-class SVM does not always provide an entire ROC curve: in that case, we artificially add the $(0,0)$ and $(1,1)$ points in the curve in order to compute the AUC. For k-lpe and our method, we vary the threshold α in order to obtain the empirical ROC curves. However, as the AUC does not allow the assessment of the score information [19], we also report the type-I and type-II errors associated to different thresholds. We are unaware of a method for controlling the type-I error for SVM, and we then report errors for the proposed method and k-lpe only. For both our algorithm and one-class SVM, a Gaussian kernel is considered.

Table 1. Parameters and values taken for the experiments for the three tested methods

Method	Param.	Values
oc-SVM	γ	$1/\{0.01, 0.05, 0.1, 0.2, \cdots, 0.9, 0.95, 0.99\}$-quantiles of $\|x - y\|^2$ [20]
	ν	$\{0.01, \cdots, 1\}$ by 0.01
k-lpe	k	$\{2, 4, 6, 8, 10, 12\}$
our	γ	$1/\{0.01, 0.05, 0.1, 0.2, \cdots, 0.9, 0.95, 0.99\}$-quantiles of $\|x - y\|^2$
	λ	$\{10^{-2}, \cdots, 10^{-8}\}$ by 10^{-1}, 0

The data used for the experiments are common benchmarks of various dimensions for classification, from IDA [21] or libSVM [22] datasets, picking a nominal class and considering the other ones as anomalies. Table 2 summarises the main characteristics of the datasets.

Table 2. Description of the datasets used for the evaluation of the anomaly detection methods

Dataset	Classes		# instances	# features
	Nominal (#)	Anomaly(#)		
Banana	-1 (2 924)	+1 (2 376)	5 300	2
Diabetes	-1 (500)	+1 (268)	768	8
Thyroid	-1 (150)	+1 (65)	215	5
Mushroom	-1 (3 916)	+1 (4 208)	8 124	112
Sonar	-1 (111)	+1 (97)	60	208
USPS	digit 2 (1 553)	other digits (7 745)	9 298	256

5.1   Evaluation of the Proposed Method with a Toy Example

Distribution of the p-values. We first test our algorithm on the benchmark dataset banana, randomly picking $n = 500$ training points in the -1 class as nominal data (see Fig. 1(a)). We then consider a test set of $n_N = 1000$ nominal points (*i.e.* of label -1) and $n_A = 1000$ anomalies, drawn according to a bivariate uniform density on the interval $[0; 1]^2$. We choose the median values of the indices of considered ranges of parameters γ and λ. Figure 1(b) represents the empirical density of the p-values computed as described in Algo. 1. The p-values associated to the n_N nominal query points appear to be approximately uniformly distributed while scores for the other points are highly concentrated around 0. This illustrates the fact that the type-I error is controlled while the type-II error is minimized.

Sensitivity to the Training Set Size. Figure 2 and Tab. 3 shows the variation of the type-I, type-II and AUC values in fonction of increasing training set sizes and for different thresholds α. We notice that type-I errors are close to the expected α-values and that both type-II errors and AUC values slowly decrease along with the training set size, accurate results being already obtained for small training set sizes of 50 or 100 nominal training points.

Table 3. Sensitivity to the training set size: AUC values for different input sample sizes n

n	50	100	200	300	500	700	900
AUC (%)	88.49	89.72	90.51	90.75	91.23	91.37	91.51

5.2   Simulations on Several Datasets

Supervised Setting. When some anomalies are identified as well as nominal data, a grid search over the parameter space can be performed. In this framework, AUC values, type-I and type-II errors are respectively given in Tab 4, 5 and 6. We first note that one-class SVM always has the lowest AUC values. For low dimensional

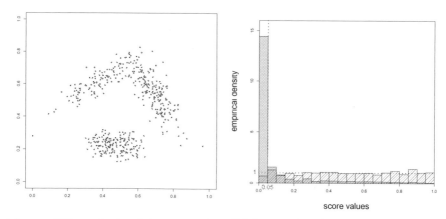

(a) $n = 500$ nominal points are used to compute the different τ_S values.

(b) Score values for the remaining nominal test points (in red) and 1000 points drawn according to a bivariate uniform distribution (in gray). Horizontal line represents an empirical density value of 1; vertical line a score value of $\alpha = 0.05$.

Fig. 1. Toy example

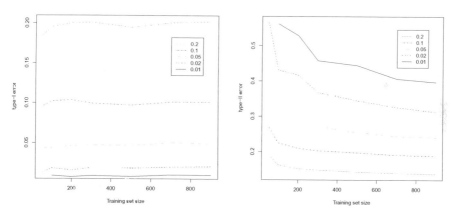

(a) Type-I error for different input sample sizes and for different thresholds (α)

(b) Type-II error for different input sample sizes and for different thresholds (α)

Fig. 2. Sensitivity to the training set size

datasets, k-lpe and the proposed procedure has similar performances in terms of the AUC values or the type-II errors. For higher dimensional dataset, we notice that k-lpe may exhibit inconsistent type-I errors (mushroom dataset), high type-II errors (mushroom and USPS datasets) or lower AUC values (sonar dataset): indeed, it is well-known that nearest-neighbors based approaches may suffer the dimensionality curse [23]. In comparison, the anomaly detector based on the reconstruction error of the KPCA exhibits the highest AUC values more often, as well as consistent type-I errors and low type-II errors.

Table 4. Averaged AUC (%) over 100 repetitions - Best results are reported boldfaced

Dataset	$n = 50$			$n = 100$			$n = 500$		
	our	k-lpe	oc-svm	our	k-lpe	oc-svm	our	k-lpe	oc-svm
Banana	**88.04**	87.90	83.50	**90.47**	89.77	86.82	**92.53**	92.48	91.54
Diabetes	73.21	73.44	67.66	74.11	**74.75**	67.36	-	-	-
Thyroid	**98.15**	97.94	96.12	**99.04**	98.52	97.10	-	-	-
Mushroom	**97.80**	97.23	75.35	**99.04**	98.68	84.48	**99.83**	99.58	94.52
Sonar	**70.34**	64.86	60.77	**73.12**	70.34	61.79	-	-	-
USPS	**97.71**	96.07	90.50	**97.95**	97.05	93.49	**98.73**	97.80	95.96

Table 5. Averaged type-I (%) over 100 repetitions - Inconsistent values are reported boldfaced

Dataset	α	$n = 50$		$n = 100$		$n = 500$	
		our	k-lpe	our	k-lpe	our	k-lpe
Banana	2%	1.91	1.35	1.75	1.86	1.91	1.87
	5%	6.09	5.17	4.72	5.11	4.77	4.91
	20%	22.02	20.77	20.30	21.01	20.01	20.10
Diabetes	2%	1.51	1.52	2.04	1.92	-	-
	5%	5.28	5.77	5.02	4.70	-	-
	20%	20.06	20.53	20.74	20.38	-	-
Thyroid	2%	2.00	2.09	1.80	2.06	-	-
	5%	5.64	6.06	4.80	5.10	-	-
	20%	21.41	21.41	20.28	21.64	-	-
Mushroom	2%	2.00	2.27	1.63	1.24	1.76	**40.49**
	5%	5.47	5.97	4.20	7.59	4.88	**57.18**
	20%	21.54	**28.22**	20.60	**50.51**	20.02	**57.18**
Sonar	2%	1.85	1.87	1.91	1.91	-	-
	5%	6.10	6.00	4.54	5.64	-	-
	20%	21.49	21.98	18.82	22.18	-	-
USPS	2%	1.86	1.66	1.90	1.66	1.87	1.95
	5%	5.24	5.30	4.75	4.71	4.90	5.10
	20%	20.20	21.06	19.95	20.33	19.88	20.11

Table 6. Averaged type-II (%) over 100 repetitions - Best results are reported bold-faced

Dataset	α	$n = 50$		$n = 100$		$n = 500$	
		our	k-lpe	our	k-lpe	our	k-lpe
Banana	2%	**74.53**	81.38	**64.51**	71.63	**45.18**	50.77
	5%	**44.21**	47.92	**38.82**	42.52	**28.49**	30.00
	20%	**17.48**	18.27	15.15	**14.57**	11.77	**11.47**
Diabetes	2%	95.18	**95.12**	**93.79**	94.49	-	-
	5%	85.52	**83.96**	**85.20**	87.49	-	-
	20%	50.76	**49.94**	**47.00**	50.26	-	-
Thyroid	2%	16.91	**14.72**	15.80	**14.32**	-	-
	5%	6.31	**5.65**	6.63	**6.07**	-	-
	20%	**0.88**	1.37	**0.40**	0.68	-	-
Mushroom	2%	53.47	**35.40**	**20.32**	49.73	0.03	**0.01**
	5%	**6.47**	12.22	**0.19**	7.86	**0.00**	0.01
	20%	**0.01**	0.29	0.01	**0.00**	**0.00**	0.01
Sonar	2%	**98.75**	99.14	**99.18**	99.23	-	-
	5%	**90.39**	93.68	**89.05**	90.80	-	-
	20%	**63.03**	69.90	61.60	**61.36**	-	-
USPS	2%	**45.43**	62.79	**34.18**	54.38	**19.00**	39.29
	5%	**13.02**	25.51	**11.94**	18.99	**3.60**	8.30
	20%	**0.32**	2.33	**0.10**	1.15	**0.02**	0.40

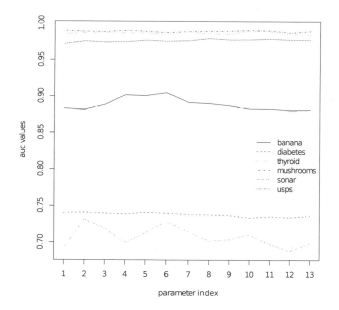

Fig. 3. AUC values along with parameter γ indices

Sensitivity to Parameter γ. In Proposition 1, there is no condition on the parameters values but on the decay speed of the empirical eigen-spectrum. Figure 3 shows the AUC values obtained for the proposed range of γ values (from the lowest to the highest) for the tested datasets, given a fixed value of $\lambda = 10^{-4}$. We note that, at least for the considered datasets, all the tested values within the proposed range give similar results for the AUC.

6   Conclusion

We have defined a new statistical test for detecting anomalies from a sample of possibly high-dimensional nominal data. The distribution of the data is described by an embedded feature space thanks to KPCA and the test statistic is the reconstruction error of a query point on this feature space. The reconstruction error cumulative distribution function is estimated from the nominal data points, which is then used for deriving a threshold that minimizes the type-I error, while minimizing the type-II error. This work can be viewed as an extension of [11], providing a scheme to compute the threshold above which points are declared anomaly. Future works aim at deriving the true asymptotic distribution of our test statistic. In addition, some computational schemes can be used in order to reduce the computational cost of calculating the test statistic and of testing. Improvement of KPCA for large-scale data is indeed an active research area and different approaches can be considered, such as selective sampling (see Section 4) or factorization of $\boldsymbol{K}(\mathcal{S})$ (for example incomplete Cholesky factorisation as in [9]).We now aim at putting in place such schemes in order to have a computational competing method.

Acknowledgements. This work has been supported in part by French Agence Nationale de la Recherche (ANR) through Asterix project (reference ANR-13-JS02-0005-01).

References

[1] Chandola, V., Banerjee, A., Kumar, V.: Anomaly detection: A survey. ACM Comput. Surv. 41(3), 15:1–15:58 (2009)

[2] Markou, M., Singh, S.: Novelty detection: a review - part 1: statistical approaches. Signal Processing 83(12), 2481–2497 (2003)

[3] Zimek, A., Schubert, E., Kriegel, H.P.: A survey on unsupervised outlier detection in high-dimensional numerical data. Statistical Analysis and Data Mining 5(5), 363–387 (2012)

[4] Williams, C., Seeger, M.: The effect of the input density distribution on kernel-based classifiers. In: Proceedings of the 17th International Conference on Machine Learning, pp. 1159–1166 (2000)

[5] Shawe-Taylor, J., Williams, C.K., Cristianini, N., Kandola, J.: On the eigenspectrum of the Gram matrix and the generalization error of kernel-PCA. IEEE Transactions on Information Theory 51(7), 2510–2522 (2005)

[6] Baker, C.T.: The numerical treatment of integral equations, vol. 13. Clarendon Press, Oxford (1977)

[7] Girolami, M.: Orthogonal series density estimation and the kernel eigenvalue problem. Neural Comput 14(3), 669–688 (2002)

[8] Blanchard, G., Bousquet, O., Zwald, L.: Statistical properties of kernel principal component analysis. Mach. Learn. 66(2-3), 259–294 (2007)

[9] Bach, F.R., Jordan, M.I.: Kernel independent component analysis. The Journal of Machine Learning Research 3, 1–48 (2003)

[10] Harchaoui, Z., Bach, F., Cappé, O., Moulines, E.: Kernel-Based Methods for Hypothesis Testing: A Unified View. IEEE Signal Processing Magazine 30(4), 87–97 (2013)

[11] Hoffmann, H.: Kernel PCA for novelty detection. Pattern Recognition 40(3), 863–874 (2007)

[12] Zhang, K., Zheng, V.W., Wang, Q., Kwok, J.T., Yang, Q., Marsic, I.: Covariate shift in hilbert space: A solution via surrogate kernels. In: Proceedings of the 30th International Conference on Machine Learning, vol. 28, pp. 388–395 (2013)

[13] Öztireli, A.C., Alexa, M., Gross, M.: Spectral sampling of manifolds. ACM Transactions on Graphics (TOG) 29(6), 168 (2010)

[14] Liu, R., Jain, V., Zhang, H.: Sub-sampling for efficient spectral mesh processing. In: Nishita, T., Peng, Q., Seidel, H.-P. (eds.) CGI 2006. LNCS, vol. 4035, pp. 172–184. Springer, Heidelberg (2006)

[15] Zhao, M., Saligrama, V.: Anomaly detection with score functions based on nearest neighbor graphs. In: Advances in Neural Information Processing Systems 22, pp. 2250–2258 (2009)

[16] Sricharan, K., Hero, A.: Efficient anomaly detection using bipartite k-nn graphs. In: Advances in Neural Information Processing Systems, pp. 478–486 (2011)

[17] Knorr, E.M., Ng, R.T., Tucakov, V.: Distance-based outliers: algorithms and applications. The VLDB Journal 8(3-4), 237–253 (2000)

[18] Schölkopf, B., Platt, J.C., Shawe-Taylor, J., Smola, A.J., Williamson, R.C.: Estimating the support of a high-dimensional distribution. Neural Computation 13(7), 1443–1471 (2001)

[19] Schubert, E., Wojdanowski, R., Zimek, A., Kriegel, H.P.: On evaluation of outlier rankings and outlier scores. In: Proceedings of the 2012 SIAM International Conference on Data Mining, pp. 1047–1058 (2012)

[20] Caputo, B., Sim, K., Furesjo, F., Smola, A.: Appearance-based object recognition using SVMs: which kernel should I use? In: NIPS Workshop on Statistical Methods for Computational Experiments in Visual Processing and Computer Vision (2002)

[21] Müller, K.R., Mika, S., Rätsch, G., Tsuda, K., Schölkopf, B.: An introduction to kernel-based learning algorithms. IEEE Transactions on Neural Networks 12, 181–201 (2001)

[22] Chang, C.C., Lin, C.J.: LIBSVM: A library for support vector machines. ACM Transactions on Intelligent Systems and Technology 2, 27:1–27:27 (2011), http://www.csie.ntu.edu.tw/~cjlin/libsvm

[23] Beyer, K., et al.: When is "nearest neighbor" meaningful? In: Beeri, C., Bruneman, P. (eds.) ICDT 1999. LNCS, vol. 1540, pp. 217–235. Springer, Heidelberg (1998)

Fast Gaussian Pairwise Constrained Spectral Clustering[*]

David Chatel[1], Pascal Denis[1], and Marc Tommasi[1,2]

[1] INRIA Lille
[2] Lille University

Abstract. We consider the problem of spectral clustering with partial supervision in the form of must-link and cannot-link constraints. Such pairwise constraints are common in problems like coreference resolution in natural language processing. The approach developed in this paper is to learn a new representation space for the data together with a distance in this new space. The representation space is obtained through a constraint-driven linear transformation of a spectral embedding of the data. Constraints are expressed with a Gaussian function that locally reweights the similarities in the projected space. A global, non-convex optimization objective is then derived and the model is learned via gradient descent techniques. Our algorithm is evaluated on standard datasets and compared with state of the art algorithms, like [14,18,31]. Results on these datasets, as well on the CoNLL-2012 coreference resolution shared task dataset, show that our algorithm significantly outperforms related approaches and is also much more scalable.

1 Introduction

Clustering is the task of mapping a set of points into groups (or "clusters") in such a way that points which are assigned to the same group are more similar to each others than they are to points assigned to other groups. Clustering algorithms have a large range of applications in data mining and related fields, from exploratory data analysis to well-known partitioning problems like noun phrase coreference resolution to more recent problems like community detection in social networks.

Over the recent years, various approaches to clustering have relied on spectral decomposition of the graph representing the data, whether the data inherently come in the form of a graph (e.g., a social network) or the graph is derived from the data (e.g., a similarity graph between data vectors). One way to understand spectral clustering is to view it as a continuous relaxation of the NP-complete normalized- or ratio-cut problems [28,22,21]. Spectral clustering has important advantages over previous approaches like k-means, one being that it does not

---

[*] This work was supported by the French National Research Agency (ANR). Project Lampada ANR-09-EMER-007.

T. Calders et al. (Eds.): ECML PKDD 2014, Part I, LNCS 8724, pp. 242–257, 2014.

make strong assumptions on the shape (e.g., convexity) of the underlying clusters. Spectral clustering first consists in computing the first k eigenvectors associated with the smallest eigenvalues of the graph Laplacian. Discrete partitions are then obtained by running k-means on the space spanned by these eigenvectors. This leads to approximations of different optimal cuts of the graphs, which are known to be potentially quite loose [10,11]. Spectral clustering can also be understood in terms of the spectral embedding of the graph, the change of representation of the data represented by nodes. Indeed, the spectral decomposition of the graph Laplacian gives a projection of the data in a new feature space in which Euclidean distance corresponds to a similarity given by the graph (e.g., the resistance distance [15,27]).

In practice, it is often the case that the space spanned by the first k eigenvectors is not rich enough to single out the correct partition. Running k-means in a transformation of this space may yield a better partition than the one found in the original space. We propose to exploit pairwise constraints to guide the process of finding such a transformation. From this perspective, our work builds upon and extends previous attempts at incorporating constraints in spectral clustering [30,16,34,14,5,19,32,18,32,26]. While clustering is often performed in a unsupervised way, there are many situations in which some form of supervision is available or can easily be acquired. For instance, part of the domain knowledge in natural language processing problems, like noun phrase coreference resolution, naturally translates into constraints. For instance, gender and number mismatches between noun phrases (e.g., *Bill Clinton* vs. *she/they*) give strong indication that these noun phrases should not appear in the same cluster.

In this paper, we consider the setting wherein supervision is only partial, which is arguably more realistic setting when annotation is costly. Partial supervision takes the form of *pairwise constraints*, whereby two points are assigned to identical (*must-link*) or different clusters (*cannot-link*), irrespective of the clusters labels. All must-link constraints can be satisfied in polynomial time using a simple transitive closure. In some problems, constraints may be inconsistent, due to noisy preprocessing of the data for instance, and satisfying all cannot-link constraints is NP-complete for $k > 2$, see [7]. These constraints can contradict the unconstrained cuts of the graph. For example, two nodes close in graph could be constrained as cannot-link and conversely two nodes far away in the graph could be constrained as must-link. One open research question is how does one best integrate this type of partial supervision into the clustering algorithm.

In this paper, we propose to learn a linear transformation $\mathbf{X}$ of the spectral embedding of the graph with the partial supervision given by the constraints. Our algorithm also learns a similarity in order to find a partition such that similar nodes are in the same cluster, dissimilar nodes are in different clusters, and the maximum number of pairwise constraints are satisfied. When two nodes must link (respectively cannot link), their similarity is constrained to be close to 1 (respectively close to 0). In the learning step, the similarity is locally distorted around constrained nodes using a Gaussian function applied on the Euclidean distance in the feature space obtained by $\mathbf{X}$. In order to increase the gap between

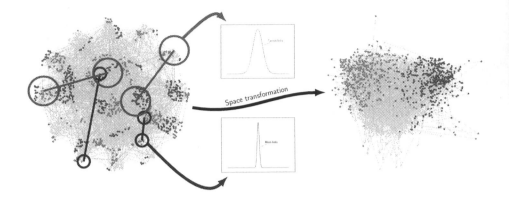

Fig. 1. This figure shows intuitively the process behind FGPWC. From a spectral embedding of a graph, Gaussian functions distort the distance between constrained pairs of nodes such that it become smaller or larger depending depending on the quality (must-link or cannot-link) attributed to the constraint. Gaussian functions act as a new similarity for the pair of nodes and it should be close to 1 if the pair must link and close to 0 if the pair cannot link.

must-link and cannot-link constraints, we use two Gaussian functions of different variances. As illustrated in Figure 1, this technique ensures that the distance in the new feature space between nodes in cannot-link constraints is significantly larger than the distance between nodes that must link. From this modeling, we derive a non-convex optimization problem to learn the transformation $\mathbf{X}$. We solve this problem using a gradient descent approach with an initialization for $\mathbf{X}$ that coincides with the unconstrained solution of the problem.

Our algorithm, FGPWC (for Fast Gaussian PairWise Clustering), is evaluated empirically on a large variety of datasets, corresponding either to genuine network data or to vectorial data converted into graphs. Two sets of experiments are conducted: the first one involves classification task, using commonly used data sets in the field. Empirical results place our algorithm above competing systems on most of the data sets. The second one involves a real task in the field of natural language processing: namely, the noun phrase coreference resolution task as described in the CoNLL-2012 shared task [25]. Our results show our algorithm compares favorably with the unconstrained spectral clustering approach of [6], outperforming it on medium-size and large clusters.

2   Background and Notation

Let $G = (\mathcal{V}, \mathcal{E}, \mathbf{W})$ be an undirected connected graph with node set $\mathcal{V} = \{v_1, \ldots, v_n\}$, edge set $\mathcal{E} \subseteq \mathcal{V} \times \mathcal{V}$ and non-negative similarity matrix $\mathbf{W}$, such that $\mathbf{W}_{ij}$ is the weight on the edge (v_i, v_j). Let $(\lambda_1, \boldsymbol{u}_1), \ldots, (\lambda_n, \boldsymbol{u}_n)$ be eigenvalue/eigenvectors pairs of the graph Laplacian $\mathbf{L_{sym}} = \mathbf{I} - \mathbf{D}^{-1/2}\mathbf{W}\mathbf{D}^{-1/2}$,

such that $\lambda_1 \leq \lambda_2 \leq \cdots \leq \lambda_n$. The matrix $\mathbf{U} = \left(\sqrt{\frac{1}{\lambda_1}} \boldsymbol{u}_1 \; \sqrt{\frac{1}{\lambda_2}} \boldsymbol{u}_2 \cdots \; \sqrt{\frac{1}{\lambda_n}} \boldsymbol{u}_n \right)$
is a spectral embedding of the graph. It can be thought as an Euclidean feature
space where each node v_i is represented by a data point whose coordinates in
this space are the components of the vector $\boldsymbol{v}_i$ equal to the ith row of the matrix
$\mathbf{U}$. The first eigenvector $\boldsymbol{u}_1$ is the constant vector $\mathbb{1}$ biased by the degrees of
the nodes, $\boldsymbol{u}_1 = \mathbf{D}^{1/2}\mathbb{1}$ and can be dropped from the feature space, as it does
not provide any information for characterizing nodes. Eigenvectors $\boldsymbol{u}_2, \ldots, \boldsymbol{u}_n$
are functions that map the manifold of the graph to real lines. If $\boldsymbol{f}$ is such a
function, then $\boldsymbol{f}^\top \mathbf{L_{sym}} \boldsymbol{f} = \frac{1}{2} \sum_{i,j=1}^n \mathbf{W}_{ij}(\boldsymbol{f}_i - \boldsymbol{f}_j)^2$ provides an estimate of how
far nearby points will be mapped by $\boldsymbol{f}$ [3]. As m increases, the space spanned by
$\boldsymbol{u}_2, \ldots, \boldsymbol{u}_m$ with $m n$ will describe smaller and smaller details in the data. In the
following, we consider a spectral embedding $\mathbf{V}_m = \left(\boldsymbol{u}_2 \cdots \boldsymbol{u}_m \right) = \left(\boldsymbol{v}_1 \ldots \boldsymbol{v}_n \right)^\top$.
To each each node of the graph v_i correspond a vector $\boldsymbol{v}_i$ that lives in this space.

Pairwise constraints are defined as follows. Let $\mathcal{M}, \mathcal{C} \subset \mathcal{V} \times \mathcal{V}$ be two sets of
pairs of nodes, describing must-link and cannot-link constraints. Let K be the
total number of constraints. If $(v_i, v_j) \in \mathcal{M}$, then v_i and v_j should be in the
same cluster, and if $(v_i, v_j) \in \mathcal{C}$ then v_i and v_j should be in different clusters.
We introduce the $K \times m$ matrices $\mathbf{A}$, $\mathbf{B}$ and the K-dimensional vector $\boldsymbol{q}$:

$$\mathbf{A} = \begin{pmatrix} \boldsymbol{v}_{i_1} \\ \vdots \\ \boldsymbol{v}_{i_K} \end{pmatrix} \qquad \mathbf{B} = \begin{pmatrix} \boldsymbol{v}_{j_1} \\ \vdots \\ \boldsymbol{v}_{j_K} \end{pmatrix} \qquad \boldsymbol{q}_k = \begin{cases} 1 & \text{if } (v_{i_k}, v_{j_k}) \in \mathcal{M} \\ 0 & \text{if } (v_{i_k}, v_{j_k}) \in \mathcal{C} \end{cases}$$

where $(\boldsymbol{v}_{i_k}, \boldsymbol{v}_{j_k})$ are vectors describing the kth pair of nodes (v_{i_k}, v_{j_k}) in $\mathcal{M} \cup \mathcal{C}$.

3   Problem Formulation

We propose to learn a linear transformation ϕ of the feature space $\mathbf{V}_m$ that
best satisfies the constraints. Let $\phi(\boldsymbol{v}_i) = \boldsymbol{v}_i \mathbf{X}$ where $\mathbf{X}$ is a $m \times m$ matrix
describing the transformation of the space. We want to find a projection of the
feature space $\phi(\boldsymbol{v}_i)$ such that the clusters are dense and far away from each
other. Ideally, if nodes $(v_i, v_j) \in \mathcal{M}$ then the distance between $\phi(\boldsymbol{v}_i)$ and $\phi(\boldsymbol{v}_j)$
should equal zero and if nodes $(v_i, v_j) \in \mathcal{C}$ then the distance between $\phi(\boldsymbol{v}_i)$
and $\phi(\boldsymbol{v}_j)$ should be very large. We introduce two Gaussian functions to locally
distort the similarities for constrained pairs. Gaussian parameters σ_m and σ_c
are chosen such that $\sigma_m \leq \sigma_c$. The similarity between two nodes v_i and v_j
is $\exp^{-\|\boldsymbol{v}_i - \boldsymbol{v}_j\|^2 / \sigma_m}$ if $(v_i, v_j) \in \mathcal{M}$ and $\exp^{-\|\boldsymbol{v}_i - \boldsymbol{v}_j\|^2 / \sigma_c}$ if $(v_i, v_j) \in \mathcal{C}$ where
$\|\cdot\|$ is the Frobenius norm. Therefore, we want to ensure that $\mathbf{X}$ is such that
$\exp^{-\|\boldsymbol{v}_i - \boldsymbol{v}_j\|^2 / \sigma_m}$ is close to 1 if $(v_i, v_j) \in \mathcal{M}$ and $\exp^{-\|\boldsymbol{v}_i - \boldsymbol{v}_j\|^2 / \sigma_c}$ is close to 0 if
$(v_i, v_j) \in \mathcal{C}$. We now encode the set of all constraints in a matrix form. Let us
first consider the K-dimensional vector $\boldsymbol{\sigma} \in \{1/\sigma_m, 1/\sigma_c\}^K$

Let $\mathbb{1}$ be the m-dimensional vector of all ones. Notice that $[(\mathbf{A} - \mathbf{B})\mathbf{X}]^2 \mathbb{1}$,
is the vector whose components are equal to the distance between pairs of con-
strained nodes in the transformed space. Let $\circ$ be the Hadamard product. Then

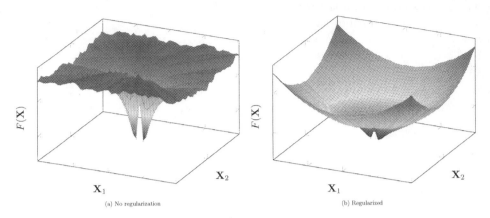

(a) No regularization   (b) Regularized

Fig. 2. Normalization effect on a simple example. 900 data points in $\mathbb{R}^2$ were drawn using a normal distribution $\mathcal{N}(0,1)$. Only 1‰ must and cannot-links have been uniformly drawn to separate data in two groups of positive and negative points. These figures plot $F(\mathbf{X})$ in the neighborhood of $\mathbf{X}^\star$. The two dimensions of $\mathbf{X}^\star$ in this example are referred by $\mathbf{X}_1$ and $\mathbf{X}_2$.

$\exp^{-[(\mathbf{A}-\mathbf{B})\mathbf{X}]^2\mathbb{1}\circ\sigma}$ is the vector whose components equal the corresponding must-link or cannot-link similarity depending on whether the associated pairs of nodes are in $\mathcal{M}$ or $\mathcal{C}$. The values in $\mathbf{X}$ are not bounded in this expression. So, we propose to add a regularization term on $\mathbf{X}$. This gives the optimization problem:

$$\min_{\mathbf{X}} F(\mathbf{X}) = \left\|\exp^{-[(\mathbf{A}-\mathbf{B})\mathbf{X}]^2\mathbb{1}\circ\sigma} - \boldsymbol{q}\right\|^2 + \gamma \left\|\mathbf{X}\right\|^2 \tag{1}$$

The effect of this regularization step is depicted in Figures 2a and 2b. In this toy example, data points where drawn using a normal distribution with mean 0. Constraints are added in order to separate positive and negative points in two clusters. Only 1‰ must and cannot-links have been uniformly drawn. We can see that in both non regularized and regularized cases, global optimums are identical. However, Figure 2a shows that far away from the global optimum, the non regularized objective function is not smooth. The regularization handles this issue, see figure 2b.

3.1   Algorithm

Our algorithm for learning the transformation $\mathbf{X}$ is presented in Algorithm 1. It takes as input a weighted adjacency matrix of a graph, and two matrices for must-link and cannot-link constraints. Parameters are the number k of clusters as usual in k-MEANS, but also the widths of the Gaussian functions σ_m and σ_c and the dimension m of $\mathbf{X}$.

The target dimension m is related to the amount of contradiction between the graph and the constraints. Remember that eigenvectors of $\mathbf{L_{sym}}$ are functions which maps nodes from the manifold of the graph to real lines and the associated eigenvalues provides us with an estimate of how far apart these functions maps nearby points [3]. When the pairwise constraints do not contradict the manifold of the graph, i.e. must-link pairs are already close on the manifold and cannot-link pairs are already far apart, m does not need to be large, because the eigenvectors associated with smallest eigenvalues will provide eigenmaps which do not contradict the constraints. Hence, a solution can be found in the very first eigenvectors. However, when the pairwise constraints contradict the manifold of the graph: must-links that are initially far apart on the manifold or cannot-links that are close, we need to consider a larger number of eigenvectors m, because the eigenvectors providing the eigenmaps that will not contradict the constraints will be later dimensions of the embedded space, describing smaller details.

Our algorithm is a typical gradient descent and its initialization can be at random. However, we propose to initialize it close to unconstrained spectral clustering $\mathbf{X}_0 = (\mathbf{V}_m^\top \mathbf{L_{sym}} \mathbf{V}_m)^{-1/2}$. We stop the descent after imax iterations or when the Frobenius norm of the partial derivative $\frac{\partial F(\mathbf{X})}{\partial \mathbf{X}}$ is less than ϵ.

Algorithm 1. FGPWC

Input: $\mathbf{W} \in \mathbb{R}^{n \times n}, \mathbf{M} \in \mathbb{R}^{n \times n}, \mathbf{C} \in \mathbb{R}^{n \times n}, m, k, \sigma_m, \sigma_c$
Output: $\mathbf{X}^\star \in \mathbb{R}^{m \times d}, \mathcal{P}$ partition of $\mathcal{V}$
1 **begin**
2      $\mathbf{L_{sym}} \leftarrow \mathbf{I} - \mathbf{D}^{-1/2}\mathbf{W}\mathbf{D}^{-1/2}$
3      $\mathbf{V}_m \leftarrow$ first m smallest eigenvectors of $\mathbf{L_{sym}}$
4      $\mathbf{X}_0 \leftarrow (\mathbf{V}_m^\top \mathbf{L_{sym}} \mathbf{V}_m)^{-1/2}$
5      $i \leftarrow 0, \alpha \leftarrow 1$
6   **repeat**
7          $i \leftarrow i + 1, \mathbf{X}_i \leftarrow \mathbf{X}_{i-1} - \alpha \partial F(\mathbf{X}_{i-1})/\partial \mathbf{X}$
8   **if** $F(\mathbf{X}_i) >= F(\mathbf{X}_{i-1})$ **then**
9             $\alpha \leftarrow \alpha/2$
10   **else**
11             $\mathbf{X}^\star \leftarrow \mathbf{X}_i$
12   **until** $\|\partial F(\mathbf{X}_i)/\partial \mathbf{X}\|^2 < \epsilon$ **or** $i > $ imax
13      $\mathcal{P} \leftarrow k\text{-MEANS}(\mathbf{V}_m \mathbf{X}, k)$
14   **return** $\mathcal{P}$

---

4   Related Work

The use of supervision in clustering tasks has been addressed in many ways. A first related approach is that of [33], which is inspired by distance learning. Constraints are given through a set of data point pairs that should be close. The authors then consider the problem of learning a weighted matrix of similarities.

They derive an optimization problem of high complexity, which they solve by doing alternate gradient ascent on two objectives, one bringing closer points that are similar and the other putting off the other points. Similarly, in [13] learning spectral clustering is the problem of finding weighted matrix or the spectrum of the Gram matrix given a known partition. A related field is supervised clustering [9], the problem of training a clustering algorithm to produce desirable clusterings: given sets of items and complete clusterings over these sets, we learn how to cluster future sets of items.

Another set of related approaches are constrained versions of the k-means clustering algorithm. In [30], it is proposed that, at each step of the algorithm, each point is assigned to the closest centroid provided that must-link and cannot-link constraints are not violated. It is not clear how the choice of the ordering on points affects the clustering. Moreover, constraints are considered as hard constraints which makes the approach prone to noise effects. Kulis et al improve on the work of [30] in [16]. Their algorithm relies on weighted kernel k-means ([8]). The authors build a kernel matrix $\mathbf{K} = \sigma\mathbf{I} + \mathbf{W} + \mathbf{S}$, where $\mathbf{W}$ is a similarity matrix, $\mathbf{S}$ is a supervision matrix such that $\mathbf{S}_{ij}$ is positive (respectively negative) when nodes i and j must link (respectively cannot link) or zero when unconstrained. The addition of $\sigma\mathbf{I}$ ensures the positive semi-definiteness of $\mathbf{K}$ (otherwise, $\mathbf{K}$ would not be a kernel, would not have any latent Euclidean space, a requirement for k-means to converge and for theoretical justification).

Introducing constraints in spectral clustering has received a lot of attention in the last decade ([34,14,5,19,32]). In many cases, the proposed approaches rely on a modification of the similarity matrix and then the resolution of the associated approximated normalized cut. For instance, in [14], weights in the similarity matrix are forced to 0 or 1 following must-link and cannot-link constraints. But this kind of weights may have a limited impact on the result of the clustering, in particular when the considered two nodes have many paths that link them together. [34] consider three kinds of constraint and cast them into an optimization problem including membership constraints in a 2-partitioning graph problem. To guarantee a smooth solution, they reformulate the optimization problem so that it involves computing the eigen decomposition of the graph Laplacian associated with the data. The approach relies on an optimization procedure that includes nullity of the flow from labeled nodes in cluster 1, to labeled nodes in cluster 2. The algorithm closely resembles the semi-supervised harmonic Laplacian approach developed for instance in [35]. But this approach is also limited to the binary case. In [19], pairwise constraints are used to propagate affinity information to the other edges in the graph. A closed form of the optimal similarity matrix can be computed but its computation requires one matrix inversion per cannot-link constraint.

In [18], constrained clustering is done by learning a transformation of the spectral embedding into another space defined by a kernel. The algorithm attempts to project data points representing nodes onto the bound of a unit-hypersphere. The inner product between vectors describing nodes that must link is close to 0, and the inner product between vectors describing nodes that cannot-link is close to 1. That way, if a node v_i belongs to the cluster j, then the vector $\boldsymbol{v}_i$ describing v_i will be projected to $\mathbb{1}_j$ where e_j is a vector of length k full of zeros except on the jth component where it is equal to 1. The number of dimensions of the hypersphere is directly related to the ability to separate clusters. One drawback is that the algorithm uses semidefinite programs whose size is quadratic in that number of dimensions.

Recently, [31,32] propose to include constraints by modifying directly the optimization problem rather than modifying the Laplacian. In their algorithm called CSP, they introduce a matrix $\mathbf{Q}$ where $\mathbf{Q}_{ij}$ is 1 if i and j must-link, -1 if i and j cannot-link and 0 otherwise. Then, a constraint $\mathbf{f}^\top \mathbf{Q}\mathbf{f} > \alpha$ is added to the normalized cut objective considered in unconstrained spectral clustering. Parameter α is considered as a way to soften constraints. Their approach outperforms previous approaches such as the one based on kernel k-means defined in [16]. An original approach based on tight relaxation of graph cut ([11]) is presented in [26]. The approach deals with must and cannot-links but in the two clusters case. It guarantees that no constraints are violated as long as they are consistent. For problems with more than two clusters, hierarchical clustering is proposed. Unfortunately in this case, the algorithm loses most of its theoretical guarantees.

5   Experiments

We conducted two sets of experiments. In the first experiments, we evaluate our algorithm on a variety of well-known clustering and classification datasets, and compare it to four related constrained clustering approaches: CCSR [18], SL [14], CSP [32] and COSC [26]. CCSR also seeks a projection of space in which constraints are satisfied. SL modifies the adjacency matrix and puts 0 for cannot-link pairs and 1 for must-link pairs. CSP modifies the minimum cut objective function introducing a term for solving a part of the constraints. COSC is based on a tight relaxation of the constrained normalized cut into a continuous optimization problem.

In a second set of experiments, we apply our algorithm to the problem of noun phrase coreference resolution, a very important problem in Natural Language Processing. The task consists in determining for a given text which noun phrases (e.g., proper names, pronouns) refer to the same real-world entity (e.g., Bill Clinton). This problem can be easily recast as a (hyper-)graph partitioning problem [24,6]. We evaluate our algorithm on the CoNLL-2012 English dataset and compare it to the unconstrained spectral clustering approach of [6], a system that ranked among the top 3 systems taking part in the CoNLL-2012 shared task.

5.1   Clustering on UCI and Network Data Sets

Dataset and Preprocessing. We first consider graphs built from UCI datasets and networks. Table 1 summarizes their properties and the characteristics of the associated clustering problem. Graph construction uses a distance that is computed based on features. First, continuous features are normalized between 0 and 1 and nominal features are converted into binary features. Second, given feature vectors x and x' associated with two datapoints, we consider two kinds of similarities: either RBF kernels of the form $\exp(-\|x - x'\|^2 / 2\sigma^2)$ or cosine similarity $x \cdot x' / (\|x\| \times \|x'\|)$. In the case of cosine similarity we also apply k-NN and weight edges with similarity. For instance, from the imdb movie dataset we extract records in which Brad Pitt, Harrison Ford, Robert De Niro and Sylvester Stallone have played. The task is to determine which of the four actors played in which movie. The movies in which more than one of these actors have played are not part of the dataset so that classes do not overlap. We have collected all the actors (except for the four actors that serve as classes) who played in 1606 movies. Each movie is described by binary features representing the presence or absence of an actor in its cast. The similarity measure between movies is the cosine similarity.

Evaluation Metric. We use Adjusted Rand Index (ARI) [12] as our main evaluation measure. The standard Rand Index compares two clusterings by counting correctly classified pairs of elements. It is defined as: $\mathcal{R}(\mathcal{C}, \mathcal{C}') = \frac{2(TP+TN)}{n(n-1)}$ where n is the number of nodes in the graph and TP, TN are true positive and true negative pairs. By contrast, the Adjusted Rand Index which is the normalized difference of the Rand Index and its expected value under the null hypothesis. This index has an expected value of zero for independant clusterings and maximum value 1 for identical clusterings. We report the mean over the 10 runs corresponding to 10 sets of constraints of the ARI computed against the ground truth. As an additional measure, we also report the number of violated constraints in the computed partition and the computation time for each algorithm.

System Settings. For each dataset, 10 different sets of constraints were selected at random. The number of constraints is chosen to avoid trivial solutions. Indeed, if the number of must-link constraints is high, a transitive closure quickly gives a perfect solution. So, the interesting cases are when only a few number of constraints is considered. Given a graph with n nodes, a set of pairs is added to the set of constraints with probability $1/n$. A pair forms a must-link constraint if the two nodes have the same class and a cannot-link constraint otherwise.

All algorithms (except COSC) rely on a k-means step which is non deterministic. So, we repeat 30 times each execution and select the partitions that violates a minimal number of constraints. The results evaluated on unconstrained pairs are averaged considering the 10 different sets of constraints.

All experiments were conducted using octave with openblas. For CCSR and COSC, we use the code provided by the authors on their webpages. We are using k-MEANS with smart initialization [1]. Finally, note that we found that initializing gradient descent so that it is close to unconstrained spectral clustering performs better than random initialization.

Results and Discussion. Results for the first set of experiments for 22 datasets are presented in Table 1. Empty cells corresponds to the case where the algorithm did not terminate after 15 minutes.

Table 1. Summary of data sets. First 5 columns show the data set properties: number of nodes in the graph, number of classes, how they have been constructed and number of dimensions in the spectral embedding used for the experiments. The following columns report performances for the various algorithms. Columns CSP and SL report poor results. This is mainly due to the fact that the supervision by must-link constraints is very weak. They do not fully exploit the cannot-link constraints. In our experiments, graphs are not sparse but constraints are sparse. COSC is expecting a sparse graph as an input and satisfy all the constraints when the number of clusters is equal to 2. When the number of clusters is greater than two, COSC looses its guarantees. Moreover, when constraints are very sparse, there is many different ways to satisfy them, and the hierarchical 2-way clustering COSC is performing for more than two clusters can achieve very poor results when the earliest cuts are wrong.

Dataset	size	k	Similarity	m	FGPWC tuning viols.		FGPWC no tuning viols.		SL viol.		CSP viol.		COSC viol.		CCSR viol.	
breasttissue	106	6	RBF	20	0.3088	3	0.327-	2	-0.0050	35	0.1339	52	0.0695	9	0.2104	5
glass	214	6	RBF	20	0.2552	16	0.1461	23	0.0115	73	0.0182	124	0.0347	20	0.1872	26
hayes-roth	132	3	Cosine	20	0.2783	3	0.1736	13	-0.0146	35	0.0170	78	0.0079	12	0.0842	21
hepatitis	80	2	RBF	10	0.1910	10	0.1220	11	0.0822	17	0.0106	42	0.0184	0	-0.0127	17
imdb	1606	4	Cosine	400	0.1385	93	0.1553	74	-0.0001	648	-	-	-0.0181	298	-	-
interlaced circles	900	3	RBF	60	0.6458	53	0.3023	131	0.1260	208	0.0002	574	0.0110	172	-	-
ionosphere	351	2	RBF	50	0.5041	37	0.4037	11	0.0045	68	0.0045	172	0.0889	19	-	-
iris	150	3	RBF	10	0.9410	1	0.8841	2	0.5657	16	0.0142	68	0.0797	0	0.8485	4
moons	900	2	RBF	10	0.9215	19	0.9045	22	0.0643	231	0.0000	468	-	-	0.6684	72
phoneme	4509	5	RBF	200	0.7073	126	0.0461	746	-0.0002	1842	-	-	-	-	-	-
promoters	106	2	Cosine	10	0.7182	3	0.43C7	3	0.0007	21	0.0043	70	0.0341	0	0.5946	8
spam	4601	2	RBF	20	0.9783	21	0.00C2	1127	0.0002	1067	-	-	-	-	0.9783	26
tic-tac-toe	958	2	RBF	200	1.0000	0	0.9541	5	0.0037	242	0.0056	404	-	-	-	-
vehicles	846	4	RBF	100	0.3175	55	0.3456	92	0.0001	316	0.0000	728	0.0038	116	-	-
wdbc	569	2	RBF	10	0.8568	14	0.8699	19	0.0024	126	0.0024	264	-	-	0.7255	35
webkb-cornell	195	5	Cosine	10	0.4868	13	0.1166	2	-0.0021	77	-0.0079	134	0.0577	13	0.3317	13
webkb-texas	187	5	Cosine	10	0.4705	11	0.2525	4	-0.0087	68	0.0045	122	0.0707	9	0.2848	25
webkb-wisconsin	265	5	Cosine	10	0.6719	21	0.10-8	10	0.0131	77	0.0072	164	0.0226	23	0.3346	32
wikipedia	835	3	Network	10	0.6298	49	0.0105	23	0.4621	111	0.0001	474	0.6960	33	0.5409	76
wine	178	3	RBF	10	0.9649	0	0.9040	1	0.0004	70	0.0031	84	0.0091	41	0.8566	10
xor	900	2	RBF	10	1.0000	0	1.0000	0	-0.0011	223	0.0000	470	-	-	1.0000	0
zoo	101	7	Cosine	10	0.9218	0	0.6536	0	0.1326	25	0.0092	50	0.1447	1	0.7025	2

The column FGPWC "no tuning" is the case where hyperparameters have been set to the following values: $\sigma_m = .15$, $\sigma_c = 1.5$ and m equals the number of eigenvalues lower than .9. The complete spectral embedding of the graph is row normalized, thus the original space is bounded by the unit-hypersphere. Consequently, in the spectral embedding before transformation, distances between data points are less than one. In the column FGPWC "tuning", we tune the σ_m and σ_c parameters using an exhaustive search in the interval $[0.01, 1]$ for σ_m and in the interval $[0.01, 2]$ for σ_c both uniformly splited in 10 equal-size parts.

Without tuning hyperparameters any further, we obtain better results than other approaches in 12 cases. We can also see that our approach is capable of returning a result within a few minutes, whereas some other methods will not within 15 minutes on large data sets. When we tune hyperparameters, we observe that FGPWC outperforms all methods on all datasets while keeping a reasonnable computational time.

We can see that COSC is able to return partitions with 0 violated constraints when the number of clusters $k = 2$, however, the partitions are not necessarily close to the ground-truth partition. An explanation of this phenomenon is that we are providing very few constraints to the different algorithms. Hence, there are many different ways to fullfill the constraints. Columns CSP and SL give poor results. This is mainly due to the fact that the supervision by must-link constraints is very weak. They do not fully exploit the cannot-link constraints. In our experiments, graphs are not sparse but constraints. COSC is expecting a sparse graph as an input and satisfy all the constraints when the number of clusters is equal to 2. When the number of clusters is greater than two, COSC looses its guarantees. Moreover, when constraints are very sparse, there are many different ways to satisfy them, and the hierarchical 2-way clustering COSC is performing for more than two clusters can achieve very poor results when the earliest cuts are wrong. It is particularly interesting to compare FGPWC to CCSR, since the the approaches developped in the two algorithms are both based on a change of representation of the spectral embedding. CCSR is competitive with FGPWC w.r.t. the ARI measure in many cases. However, we can see that CCSR becomes intractable as the size of the embedding m increases, while this is not a problem for FGPWC. This is also confirmed by the computational time.

Small graphs can be harder if constraints contradict the similarity $\mathbf{W}$, because in this case m needs to be larger, but for a large enough m, our algorithm will over-fit. It is related to the degree of freedom in solving a system of K equations, where K is the fixed number of constraints, with more and more variables (as m increases).

5.2   Noun Phrase Coreference Resolution

Dataset and Preprocessing. For the coreference resolution task, we use the English dataset used for the CoNLL-2012 shared task [25]. Recall that the task consists, for each document, in partitioning a set of noun phrases (aka *mentions*) into classes of equivalence that denote real-wold *entities*. This task is illustrated on the following small excerpt from CoNLL-2012:

*Was Sixty Minutes unfair to [Bill Clinton]$_1$ in airing Louis Freeh's charges
against [him]$_1$?*

In this case, noun phrases "Bill Clinton" and "him" both refer to the same entity
(i.e. William Jefferson Clinton), encoded here by the fact that they share the
same index[1]. The English CoNLL-2012 corpus contains over over 2K documents
(1.3M words) that fall into 7 categories, corresponding to different domains (e.g.,
newsiwre, weblogs, telephone conversation). We used the official train/dev/test
splits that come with the data. Since we were specifically interested in comparing
approaches rather than developing the best end-to-end system, we used the *gold
mentions*; that is, we clustered only the noun phrases that we know were part
of ground-truth entities.

The mention graphs are built from a model of pairwise similarity, which is
trained on the training section of CoNLL-2012. The similarity function is learned
using logistic regression, each pair of mentions being described by a set of features. We re-use features that are commonly used for mention pair classification (see e.g., [23],[4]), including grammatical type and subtypes, string and
substring matches, apposition and copula, distance (number of separating men-
tions/sentences/words), gender and number match, synonymy/hypernym and
animacy (based on WordNet), family name (based on closed lists), named entity
types, syntactic features and anaphoricity detection.

Evaluation Metrics. The systems' outputs are evaluated using the three standard coreference resolution metrics: MUC [29], B^3 [2], and Entity-based CEAF
(or CEAF$_e$) [20]. Following the convention used in CoNLL-2012, we report a
global F1-score (henceforth, CoNLL score), which corresponds to an unweighted
average of the MUC, B^3 and CEAF$_e$ F1 scores. Micro-averaging is used throughout when reporting scores for the entire CoNLL-2012 test. Additionally, we are
reporting the adjusted rand index.

In order to analyze performance for different cluster sizes, we also computed
per-cluster precision and recall scores. Precision p_i and recall r_i are computed
for each reference entity class C_i for all documents. Then, the micro-averaged
F1-score score is computed as follows:

$$\bar{p} = \sum_i \frac{|C_i|\, p_i}{\sum_j |C_j|} \qquad \bar{r} = \sum_i \frac{|C_i|\, r_i}{\sum_j |C_j|} \qquad F1 = \frac{2\bar{p}\bar{r}}{\bar{p} + \bar{r}}$$

System Settings. Following the approach in [6], we first create for each document a fully connected similarity[2] graph between mentions and then run our
clustering algorithm on this graph. Compared to the tasks on the UCI dataset,
the main difficulties are the determination of the number of clusters and the fact

---

[1] Note that noun phrases like "Sixty Minutes" and "Louis Freeh" also denote entities,
  but such singleton entities are not part of the CoNLL annotations.

[2] Pamameter estimation for this pairwise mention model was performed using Limited-memory BFGS implemented as part of the Megam package http://www.umiacs.umd.edu/~hal/megam/version0_3/. Default settings were
  used.

that we have to deal with many small graphs (documents contain between 1 and 300 mentions).

The same defaut values were used for the σ_m and σ_c parameters, as in the previous experiments (that is, 0.15 and 1.5, respectively). In our aglorithm we need to fix parameter m. We fix a value that is a tradeoff between the dimension of $\mathbf{L_{sym}}$ and the number of constraints. Indeed, we want to keep structural information comming from the graph through the eigendecomposition of $\mathbf{L_{sym}}$. Also, we reject the situations where m is much larger than the number of constraints because they can lead to solutions that are non satisfactory. In that latter case, the optimization problem can be solved without any impact on non-constrained pairs and therefore without any generalization based on the given constraints. Because the multiplicity of eigenvalue 1 is large in this dataset, m is estimated by $m = |\{\lambda_i : \lambda_i \leq 0.99\}|$ where λ_i are the eigenvalues of $\mathbf{L_{sym}}$. The number of clusters k is estimated by $k = |\{\lambda_i : \lambda_i \geq 10^{-5}\}|$ where λ_i are the eigenvalues of $\mathbf{X}^\top \mathbf{X}$.

As for the inclusion of constraints, we experimented with two distinct settings. In the first setting, we automatically extracted based on domain knowledge (setting (c) in the results below). Must-link constraints were generated for pairs of mention that have the same character string. For cannot-link constraints, we used number, gender, animacy, and named entity type dismatches (e.g., noun phrases with different values for gender cannot corefer). These constraints are similar to some of the deterministic rules used in [17] and overlap with the information already in the features. This first constraint extraction generates a lot of constraints (usually, more than 50% of all available constraints for a document), but it is also noisy. Some of the constraints extracted this way are incorrect as they are based on information that is not necessarily in the dataset (e.g., gender and number are predicted automatically). The precision of these constraints is usually higher than 95%. In a second simulate interactive setting, we extracted a smaller set of must-link and cannot-link constraints directly from the ground-truth partitions, by drawing coreferential and non-coreferential mention pairs at random according to a uniform law (setting (b) below). In turn, all of these constraints are correct. Each mention pair has a probability $1/n$ to be drawn, with n the mention count.

Results and Discussion. We want to show that FGPWC works better on large graphs and larger clusters. We perform per-cluster evaluation, this is summarized in Figure 3. All plots represent the F1-score, averaged on runs all documents per cluster size. Plot (a) reports results for the unconstrained spectral clustering approach of [6]. Their method uses a recursive 2-way spectral clustering algorithm. The parameter used to stop the recursion has been tuned on a development set. The other plots are obtained using (b) FGPWC with constraints generated uniformly at random from an oracle and (c) FGPWC with constraints derived automatically from text based on domain knowledge.

In the latter case (c) FGPWC has not been able to improve the results obtained by (a). We think that constraints extracted from text does not add new information but change the already optimized measure in the similarity graph.

However, even adding less constraints at random from an oracle using a uniform distribution is more informative. When we are using constraints that do not comes from the features used for the similarity construction step, we see that FGPWC outperform other methods for clusters larger than 5. However, we can see that FGPWC can degrade smallest clusters. There are two explanations for this: we obtain better performance on larger clusters because the way we select random constraints. Using a uniform distribution, there is more chance to add constraints for larger clusters. And moreover, clusters with few or no constraints, in our case: small clusters, are usually scattered around the space, because FG-PWC globally transforms the space to fit the constraints. We can also see that (b) outperforms (c) on small clusters. Probably because more constraints are being added for small clusters in (b). All of this supports the idea that constraints in this kind of task should be generated from another set of features applicable to all mentions, regardless of the size of the clusters they belong to.

Overall, we obtain a CoNLL score of 0.71 (0.80 MUC, 0.75 B^3, 0.57 $CEAF_e$, 0.48 ARI), for [6], 0.56 (0.76 MUC, 0.57 B^3, 0.36 $CEAF_e$, 0.31 ARI) using our method along with extracted constraints and 0.58 (0.67 MUC, 0.58 B^3, 0.49 $CEAF_e$, 0.40 ARI) with ground-truth random constraints. That is, we see a clear drop of performance when using the constraints, be they noisy or not. Closer examination reveals that this decrease stems from poor performance on small clusters, while these clusters are the most representative in this task.

The F1-score is lower than for the state of the art. But interestingly, in presence of uniformly distributed pairwise constraints, our algorithm can significantly improve clustering results on clusters larger than 5, compared to the state of the art [6]. This suggests that active methods can lead to dramatic improvements

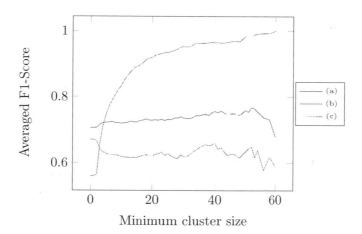

Fig. 3. Averaged F1-score vs minimum cluster size for FGPWC with CoNLL 2012 data set: (a) method in [6], (b) FGPWC uniformly distributed from reference; (c) FGPWC All extracted must/cannot-links

and our algorithm easily supports that through the introduction of pairwise constraints. Moreover, our method can be used to detect larger clusters, and leave the smaller cluster to another method.

6   Conclusion

We proposed a novel constrained spectral clustering framework to handle must-link and cannot-link constraints. This framework can handle both 2 clusters and more than 2 clusters cases using the exact same algorithm. Unlike previous methods, we can cluster data which require more eigenvectors in the analysis. We can also handle cannot-link constraints without giving up on computational complexity. We carried out experiments on UCI and network data sets. We also provide an experiment on the real task of noun-phrase coreference and discuss the results. We discuss the relationship between Laplacian eigenmaps and the constraints, that can explain why adding constraints can degrade clustering results. We empirically show that our method, that involves a simple and fast gradient descent, outperforms several state of the art algorithms on various data sets. For noun-phrase coreference, the challenge ahead will be to find rules to generate constraints from the text which are more uniformly distributed. We also want to find a way to better handle small clusters. A step in that direction is to investigate better adapted cut criteria and active learning methods.

References

1. Arthur, D., Vassilvitskii, S.: K-means++: The advantages of careful seeding. In: Proc. of SODA, pp. 1027–1035 (2007)
2. Bagga, A., Baldwin, B.: Algorithms for scoring coreference chains. In: Proc. of TREC, vol. 1, pp. 563–566 (1998)
3. Belkin, M., Niyogi, P.: Laplacian eigenmaps for dimensionality reduction and data representation. Neural Computation 15, 1373–1396 (2002)
4. Bengtson, E., Roth, D.: Understanding the value of features for coreference resolution. In: Proc. of EMNLP, pp. 294–303 (2008)
5. De Bie, T., Suykens, J.A.K., De Moor, B.: Learning from General Label Constraints. In: Fred, A., Caelli, T.M., Duin, R.P.W., Campilho, A.C., de Ridder, D. (eds.) SSPR&SPR 2004. LNCS, vol. 3138, pp. 671–679. Springer, Heidelberg (2004)
6. Cai, J., Strube, M.: End-to-end coreference resolution via hypergraph partitioning. In: Proc. of COLING, pp. 143–151 (2010)
7. Davidson, I., Ravi, S.S.: Clustering with constraints: Feasibility issues and the k-means algorithm. In: Proc. of SIAM Data Mining Conference (2005)
8. Dhillon, I.S., Guan, Y., Kulis, B.: A unified view of kernel k-means, spectral clustering and graph cuts. Technical report, UTCS (2004)
9. Finley, T., Joachims, T.: Supervised clustering with support vector machines. In: Proc. of ICML, pp. 217–224. ACM Press (2005)
10. Guattery, S., Miller, G.L.: On the quality of spectral separators. SIAM Journal on Matrix Analysis and Applications 19(3), 701–719 (1998)
11. Hein, M., Setzer, S.: Beyond spectral clustering - tight relaxations of balanced graph cuts. In: Proc. of NIPS, pp. 2366–2374 (2011)

12. Hubert, L., Arabie, P.: Comparing partitions. Journal of Classification 2, 193–218 (1985)
13. Jordan, M., Bach, F.: Learning spectral clustering. In: Proc. of NIPS (2004)
14. Kamvar, S.D., Klein, D., Manning, C.D.: Spectral learning. In: Proc. of IJCAI, pp. 561–566 (2003)
15. Klein, D.J., Randic, M.: Resistance distance. Journal of Mathematical Chemistry 12, 81–95 (1993)
16. Kulis, B., Basu, S., Dhillon, I.S., Mooney, R.J.: Semi-supervised graph clustering: a kernel approach. In: Proc. of ICML, pp. 457–464 (2005)
17. Lee, H., Peirsman, Y., Chang, A., Chambers, N., Surdeanu, M., Jurafsky, D.: Stanford's multi-pass sieve coreference resolution system at the conll-2011 shared task. In: Proc. of CoNLL: Shared Task, pp. 28–34 (2011)
18. Li, Z., Liu, J., Tang, X.: Constrained clustering via spectral regularization. In: Proc. of CVPR, pp. 421–428 (2009)
19. Lu, Z., Carreira-Perpinan, M.A.: Constrained spectral clustering through affinity propagation. In: Proc. of CVPR (2008)
20. Luo, X.: On coreference resolution performance metrics. In: Proc. of HLT-EMNLP, pp. 25–32 (2005)
21. Luxburg, U.: A tutorial on spectral clustering. Statistics and Computing 17(4), 395–416 (2007)
22. Ng, A.Y., Jordan, M.I., Weiss, Y.: On spectral clustering: Analysis and an algorithm. In: Proc. of NIPS, pp. 849–856. MIT Press (2001)
23. Ng, V., Cardie, C.: Improving machine learning approaches to coreference resolution. In: Proc. of ACL, pp. 104–111 (2002)
24. Nicolae, C., Nicolae, G.: Bestcut: A graph algorithm for coreference resolution. In: Proc. EMNLP, pp. 275–283 (2006)
25. Pradhan, S., Moschitti, A., Xue, N., Uryupina, O., Zhang, Y.: Conll-2012 shared task: Modeling multilingual unrestricted coreference in ontonotes. In: Joint Conference on EMNLP and CoNLL-Shared Task, pp. 1–40 (2012)
26. Rangapuram, S.S., Hein, M.: Constrained 1-spectral clustering. In: Proc. of AISTATS, pp. 1143–1151 (2012)
27. Saerens, M., Fouss, F., Yen, L., Dupont, P.E.: The principal components analysis of a graph, and its relationships to spectral clustering. In: Boulicaut, J.-F., Esposito, F., Giannotti, F., Pedreschi, D. (eds.) ECML 2004. LNCS (LNAI), vol. 3201, pp. 371–383. Springer, Heidelberg (2004)
28. Shi, J., Malik, J.: Normalized cuts and image segmentation. IEEE Transactions on Pattern Analysis and Machine Intelligence 22(8), 888–905 (2000)
29. Vilain, M., Burger, J., Aberdeen, J., Connolly, D., Hirschman, L.: A model-theoretic coreference scoring scheme. In: Proc. of the Conference on Message Understanding, pp. 45–52 (1995)
30. Wagstaff, K., Cardie, C., Rogers, S., Schroedl, S.: Constrained K-means Clustering with Background Knowledge. In: Proc. of ICML, pp. 577–584 (2001)
31. Wang, X., Davidson, I.: Flexible constrained spectral clustering. In: Proc. of KDD, pp. 563–572 (2010)
32. Wang, X., Qian, B., Davidson, I.: On constrained spectral clustering and its applications. In: Data Mining and Knowledge Discovery (2012)
33. Xing, E.P., Ng, A.Y., Jordan, M.I., Russell, S.: Distance Metric Learning, with Application to Clustering with Side-information. In: Proc. of NIPS (2002)
34. Yu, S.X., Shi, J.: Grouping with Bias. In: Proc. of NIPS (2001)
35. Zhu, X., Ghahramani, Z., Lafferty, J.: Semi-supervised learning using gaussian fields and harmonic functions. In: Proc. ICML, p. 912 (2003)

A Generative Bayesian Model
for Item and User Recommendation in Social
Rating Networks with Trust Relationships

Gianni Costa, Giuseppe Manco, and Riccardo Ortale

ICAR-CNR
Via Bucci 41c
87036 Rende (CS)
{costa,manco,ortale}@icar.cnr.it

Abstract. A Bayesian generative model is presented for recommending interesting items and trustworthy users to the targeted users in social rating networks with asymmetric and directed trust relationships. The proposed model is the first unified approach to the combination of the two recommendation tasks. Within the devised model, each user is associated with two latent-factor vectors, i.e., her susceptibility and expertise. Items are also associated with corresponding latent-factor vector representations. The probabilistic factorization of the rating data and trust relationships is exploited to infer user susceptibility and expertise. Statistical social-network modeling is instead used to constrain the trust relationships from a user to another to be governed by their respective susceptibility and expertise. The inherently ambiguous meaning of unobserved trust relationships between users is suitably disambiguated. An intensive comparative experimentation on real-world social rating networks with trust relationships demonstrates the superior predictive performance of the presented model in terms of RMSE and AUC.

1 Introduction

The growing popularity gained by various online services for social networking has led to the increasing availability of online social rating networks [14], i.e., environments in which users rate items and establish connections to real-world acquaintances within their social networks. In particular, the presence of explicit trust relationships between users makes such environments an appealing setting for the development of realistic recommendation processes, in which the targeted users turn to their social networks for decision making and are more strongly influenced by (directly or indirectly) trusted real-world acquaintances.

Two fundamental tasks in social rating networks with trust relationships are item recommendation and user recommendation The former consists in taking advantage of the trust relationships to suggest unrated items, that are expected to be of interest to the targeted users. The latter instead consists in taking advantage of the trust relationships to suggest users having no relationships with the targeted users and still expected to be trusted by them.

T. Calders et al. (Eds.): ECML PKDD 2014, Part I, LNCS 8724, pp. 258–273, 2014.

Each individual task has been extensively studied in the literature in isolation. Previous research on rating prediction for item recommendation in social rating networks can be divided into two major areas of focus reflecting the nature of relationships in the underlying social networks, i.e., *unilateral* relationships (such as, e.g., trust) or *cooperative and mutual* relationships (such as, e.g., in the case of friends, classmates, colleagues, relatives and so forth) [19]. Rating prediction in trust networks has been the subject of several studies such as, e.g., [14, 17, 18]. A variety of other research efforts including [10, 19, 23, 27, 31] has instead focused on social rating networks with mutual relationships. Instead, the existing approaches to link prediction can be classified into two distinct classes, i.e., *temporal* and *structural* approaches. The temporal approaches predict links between the nodes of a graph, whose evolution involves new links and new nodes with respective ties. The structural approaches assume graphs with fixed sets of nodes and, thus, they are concerned only with the prediction of new links between already observed nodes. Temporal and structural approaches can be further divided into *unsupervised* or *supervised*. Unsupervised approaches [15] do not involve a learning phase. Rather, they compute predefined scores based on graph topology alone. On the contrary, link prediction is treated as a binary classification task in supervised approaches [1, 2, 6–8, 20, 21, 30, 32], which essentially learn some suitable model with which to predict scores for pairs of nodes [20]. Certain supervised approaches also allow the optional exploitation of side information on the nodes, e.g., [21, 20]. Rating and link prediction are both instances of *dyadic prediction*, which is the more general problem of predicting a label for unobserved interactions between pairs of entities [13, 20]. Nonetheless, modeling and studying them jointly has been so far unexplored.

In this paper, to the best of our knowledge, we propose the first unified approach to trust-aware recommendation of both items and users in social rating networks. The devised approach consists in a Bayesian nonparametric hierarchical model, in which the interactions from users to users as well as between users and items are assumed to be explained by some suitable number of latent factors. More precisely, each user is associated with two real-valued latent-factor vectors, namely, her *susceptibility* and *expertise*, similarly to [3]. The entries of the susceptibility vector are the degree to which the user is sensible to the corresponding latent factors. The entries of the expertise vector are the extent to which the user can meet the susceptibility requirements of other users on the corresponding latent factors. Additionally, each item is associated with a real-valued latent-factor vector, whose entries indicate the degree to which the item is characterized by the corresponding latent factors.

The proposed model combines ideas from Bayesian probabilistic matrix factorization [25] and statistical social-network modeling to infer and exploit the foresaid latent-factor vector representations of users and items. Specifically, the seminal Bayesian approach in [25] is extended to infer the susceptibility and expertise of each user as well as the latent-factor vector representation of every item through the probabilistic factorization of the user-rating and trust-relationship matrices. Statistical social-network modeling is instead employed for a twofold purpose. On one hand, it is used to model trust relationships governed by the susceptibility and expertise of the trusting and trusted users, respectively. On the other hand, it is leveraged to properly deal with the inherent ambiguity of the unobserved trust relationships. Therein, a missing trust relationship between two users may mean either actual lack of trust or lack of awareness. Such possibilities are mixed together across the unobserved trust relationships of the social rating network at hand and, in general, cannot be distinguished beforehand. An especially interesting and novel aspect of the devised model is that each unobserved trust relationship is associated with a respective binary latent variable, whose inferred value allows to suitably account for its actual meaning.

Unlike previous approaches to item recommendation, the devised model infers the susceptibility and expertise of the individual users by accounting for both the available ratings as well as the trust relationships. Such representations are shared across rating and link prediction, which enables performing both tasks jointly. Moreover, differently from existing approaches to link prediction, the establishment of a link from a user to another is ruled only by their respective susceptibility and expertise. Yet, unobserved trust relationships are treated by drawing from research in one-class collaborative filtering (e.g., [22, 28]).

The presented model is comparatively investigated over real-world social rating networks. The empirical evidence demonstrates the superiority of its predictive performance in terms of both RMSE and AUC.

The contents of this paper are organized as follows. Section 2 introduces notation and some preliminary concepts. Section 3 covers the proposed model. Section 4 develops approximate posterior inference within the proposed model. Section 5 presents the empirical results of an intensive comparative evaluation of the proposed model against state-of-the-art competitors on real-world social rating networks. Finally, Section 6 concludes and highlights future research.

2   Preliminaries and Problem Statement

A social rating network [14] can be formalized as a tuple $\mathcal{N} = \langle N, A, \mathcal{I}, R \rangle$ where N is a set of n users and $A \subseteq N \times N \times \{0, 1\}$ is a set of directed links between users. The underlying graph $\mathcal{G} = \langle N, A \rangle$ represents trust relationships between users. In particular, a positive link $u \xrightarrow{1} v$ means that u trusts v and, dually, a negative link $u \xrightarrow{0} v$ denotes lack of u's trust in v. We will generically use matrix notation $A_{u,v}$ to succinctly denote $u \to v$. Clearly, $A_{u,v}$ is either 0 or 1, according to whether the link $u \to v$ is negative or positive. In the following, we assume to be aware only of positive links and, thus, a missing link from u to v can denote either lack of trust, or lack of awareness (i.e., u is not aware of v).

$\mathcal{G}$ can be viewed as a graph with attributes by also accounting for additional node information. We focus on the degrees of preference (or ratings) assigned by the individual users from N to the elements of a set $\mathcal{I}$ of m items. Such preference degrees are summarized into the ratings $R \subseteq N \times \mathcal{I} \times \mathcal{V}$, whose generic entry $\langle u, i, r \rangle$ denotes the rating $r \in \mathcal{V} = \{1, \ldots, V\}$ assigned by user $u \in N$ to item $i \in \mathcal{I}$. Hereafter, we denote the rating r relative to the entry $\langle u, i, r \rangle$ as $R_{u,i}$.

We assume that trust relationships between users as well as their ratings to items can be explained in terms of a number of latent (i.e., unobserved and unknown) factors, that also contribute to characterize the individual items. More precisely, each user is associated with an extent of *susceptibility* and *expertise* with respect to the individual latent factors. A rating is governed by the combination of the susceptibility and expertise of a user with the extent to which the targeted item is characterized by each latent factor. A trust relationship from a user to another is determined by their respective susceptibility and expertise. Given a generic social rating network $\mathcal{N}$, we aim to infer a probabilistic model from the trust relationships observed in $\mathcal{G}$, that allows to recommend both interesting items and further trustworthy users to the targeted users within the network. The recommendation of interesting items is essentially a rating prediction task. Given a user $u \in N$ and an item $i \in \mathcal{I}$ such that $R_{u,i}$ is unknown, the degree of u's preference for i is predicted using $\mathcal{G}$ and R. In particular, if the trusted neighbors of u in $\mathcal{G}$ enjoyed i, then $R_{u,i}$ should be predicted accordingly. Analogously, the recommendation of trustworthy users is a trust prediction task. Given a pair of users $u, v \in N$ such that $u \to v \notin A$, the trust of u in v is again predicted using $\mathcal{G}$ and R. Specifically, if the trusted neighbors of u in $\mathcal{G}$ trust v because of her ratings, then u should trust v as well and, hence, a trust relationship should be established in $\mathcal{G}$ from u to v, i.e., the positive link $u \xrightarrow{1} v$ should be added to A. Instead, if trusted neighbors of u do not trust v, or if v's ratings significantly differ from u's known ratings, then a negative link $u \xrightarrow{0} v$ should be established. Trust relationships A and ratings R are the only observed data in $\mathcal{N}$. All other aforementioned aspects of interest cannot be measured directly.

3   The Devised Bayesian Generative Model

We propose a Bayesian hierarchical model, that combines probabilistic matrix factorization and network modeling for the recommendation of items and users in a social rating network $\mathcal{N}$. Specifically, matrix factorization is exploited to explicitly capture the latent factors governing both trust relationships and item ratings. Network modeling contributes to determine user susceptibility and expertise. Probabilistic matrix factorization and statistical network modeling are seamlessly integrated for performing collaborative filtering, in order to suggest interesting items and establish missing relationships with trustable users.

In the following K is the overall number of latent factors behind the observed trust relationships in $\mathcal{G}$. Each user $u \in N$ is associated with two column vectors $\mathbf{P}_u, \mathbf{F}_u \in \mathbb{R}^K$. The generic k-th entry of $\mathbf{P}_u$ indicates the susceptibility of u to the latent topic k. Analogously, the k-th entry of $\mathbf{F}_u$ denotes the degree of expertise exhibited by u with regard to k. The susceptibility and expertise of all users in $\mathcal{G}$ are collectively denoted by means of matrices $\mathbf{P}$ and $\mathbf{F}$, respectively, where $\mathbf{P}, \mathbf{F} \in \mathbb{R}^{K \times M}$. A representation based on the latent factors is also adopted for the items in the set $\mathcal{I}$. The generic item $i \in \mathcal{I}$ is associated with one column vector $\mathbf{Q}_i \in \mathbb{R}^K$, whose k-th entry is the extent at which the latent factor k characterizes the item i. The latent factor representations of all items are collectively represented by the matrix $\mathbf{Q}$, where $\mathbf{Q} \in \mathbb{R}^{K \times N}$. Ratings $R_{u,i}$ for all $u \in N$ and $i \in \mathcal{I}$ are considered as random variables ranging in the set $\mathcal{V}$ of admissible values. Thus, in the proposed model the data likelihood, i.e., the conditional distribution over the observed data in R and A is given by

$$\Pr(R|\mathbf{P}, \mathbf{Q}, \mathbf{F}, \alpha) = \prod_{u \in N} \prod_{i \in \mathcal{I}} \mathcal{N}(R_{u,i}; \theta_{u,i}, \alpha^{-1})^{\delta_{u,i}} \tag{3.1}$$

$$\Pr(A|\mathbf{P}, \mathbf{Q}, \mathbf{F}, \mathbf{Z}, \beta) = \prod_{u \to v \in A} \mathcal{N}(A_{u,v}; \vartheta_{u,v}, \beta^{-1}) \tag{3.2}$$

where

$$\theta_{u,i} = (\mathbf{P}_u + \mathbf{F}_u)' \mathbf{Q}_j \text{ and } \vartheta_{u,v} = \mathbf{P}_u' \mathbf{F}_v$$

and $\mathcal{N}(x|\mu, \alpha^{-1})$ is the Gaussian distribution with mean μ and precision α. In particular, the observed links are centered around the dot product between the susceptibility of the start user and the expertise of the end user, which can be interpreted as the capability of the latter of satisfying the requirements of the former. Ratings involve the dot product of the sum of user susceptibility and expertise with the latent-factor representation of items, which entirely captures the interaction between users and items. Function $\delta_{u,i}$ is instead a binary indicator, which equals 1 if $R_{u,i} > 0$ (i.e., if u actually rated i) and 0 otherwise.

The representations in terms of latent-factors associated with users (i.e., their susceptibility and expertise) as well as items are drawn from prior distributions, which are assumed to be Gaussian with parameters $\Theta_\mathbf{P} = \{\mu_\mathbf{P}, \Lambda_\mathbf{P}\}$, $\Theta_\mathbf{Q} = \{\mu_\mathbf{Q}, \Lambda_\mathbf{Q}\}$ and $\Theta_\mathbf{F} = \{\mu_\mathbf{F}, \Lambda_\mathbf{F}\}$, respectively. In addition, Gaussian-Wishart prior distributions (denoted $\mathcal{NW}$ in the following) are placed on such parameters. For a generic parameter set $\Theta = \{\mu, \Lambda\}$, we have

$$\Pr(\Theta|\Theta_0) = \mathcal{N}\left(\mu; \mu_0, [\beta_0 \Lambda]^{-1}\right) \cdot \mathcal{W}(\Lambda; \nu_0, \mathbf{W}_0)$$

where $\Theta_0 = \{\mu_0, \beta_0, \nu_0, \mathbf{W}_0\}$ is the set of hyperparameters for the prior distribution placed on $\Theta = \{\mu, \Lambda\}$ and $\mathcal{W}(\Lambda; \nu_0, \mathbf{W}_0)$ is the Wishart distribution.

The overall generative process is graphically represented in Fig. 1, and can be devised as in Fig. 2. Notice that $A_{u,v}$ is a binary random variable and that its value is sampled from a continuous distribution. This is essentially accomplished by choosing the value of $A_{u,v}$ that is nearest to the mean $\mathbf{P}_u' \mathbf{F}_v$ of the distribution. More precisely, the discretization procedure looks at the densities $\Pr(A_{u,v} = 1|\mathbf{P}_u' \mathbf{F}_v, \beta^{-1})$ and $\Pr(A_{u,v} = 0|\mathbf{P}_u' \mathbf{F}_v, \beta^{-1})$ (whose sum differs from 1). Then, $A_{u,v}$ is set to 1 if $\Pr(A_{u,v} = 1|\mathbf{P}_u' \mathbf{F}_v, \beta^{-1}) > \Pr(A_{u,v} = 0|\mathbf{P}_u' \mathbf{F}_v, \beta^{-1})$ or 0 if $\Pr(A_{u,v} = 0|\mathbf{P}_u' \mathbf{F}_v, \beta^{-1}) > \Pr(A_{u,v} = 1|\mathbf{P}_u' \mathbf{F}_v, \beta^{-1})$. To elucidate, $A_{u,v} = 1$ in the case of Fig. 3(a) being closest to $\mathbf{P}_u' \mathbf{F}_v$. Instead, $A_{u,v} = 0$ in the case of Fig. 3(b), since this value is closest to $\mathbf{P}_u' \mathbf{F}_v$.

Predicting u's interest $R_{u,i}^*$ in an unrated item i or a missing trust relationship $A_{u,v}^*$ from u to v in the context of the Bayesian hierarchical model described so far requires, respectively, the predictive distributions $\Pr(R_{uj}^*|R, A, \Xi)$ and

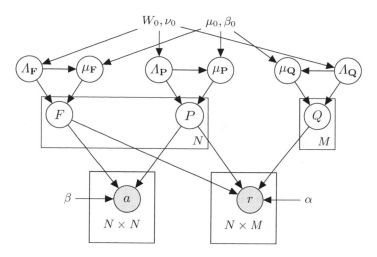

Fig. 1. Graphical representation of the proposed Bayesian hierarchical model

1. Sample

$$\Theta_{\mathbf{P}} \sim \mathcal{NW}(\Theta_0)$$
$$\Theta_{\mathbf{Q}} \sim \mathcal{NW}(\Theta_0)$$
$$\Theta_{\mathbf{F}} \sim \mathcal{NW}(\Theta_0)$$

2. For each item $i \in \mathcal{I}$ sample

$$\mathbf{Q}_i \sim \mathcal{N}(\mu_{\mathbf{Q}}, \Lambda_{\mathbf{Q}}^{-1})$$

3. For each user $u \in N$ sample

$$\mathbf{P}_u \sim \mathcal{N}(\mu_{\mathbf{P}}, \Lambda_{\mathbf{P}}^{-1})$$
$$\mathbf{F}_u \sim \mathcal{N}(\mu_{\mathbf{F}}, \Lambda_{\mathbf{F}}^{-1})$$

4. For each pair $\langle u, v \rangle \in N \times N$ sample

$$A_{u,v} \sim \mathcal{N}\left(\left(\mathbf{P}'_u \mathbf{F}_v\right), \beta^{-1}\right)$$

5. For each pair $\langle u, i \rangle \in N \times \mathcal{I}$ sample

$$R_{u,i} \sim \mathcal{N}\left(\left(\mathbf{P}_u + \mathbf{F}_u\right)\mathbf{Q}'_j, \alpha^{-1}\right)$$

Fig. 2. Generative process for the proposed Bayesian hierarchical model

$\Pr(A^*_{uv} | R, A, \boldsymbol{\Xi})$ relative to the prior $\boldsymbol{\Xi} = \{\Theta_0, \beta, \alpha\}$. Exact inference consists in computing these predictive distributions as reported at Eq. 3.3 and Eq. 3.4, where we set $\boldsymbol{\Theta} = \{\mathbf{P}, \Theta_{\mathbf{P}}, \mathbf{F}, \Theta_{\mathbf{F}}, \mathbf{Q}, \Theta_{\mathbf{Q}}\}$ for readability sake.

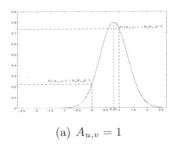 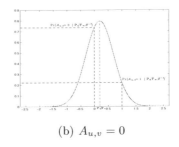

(a) $A_{u,v} = 1$ (b) $A_{u,v} = 0$

Fig. 3. The procedure to sample a binary $A_{u,v}$ value from a Gaussian with mean $\mathbf{P}'_u \mathbf{F}_v$

$$\Pr(R^*_{u,i}|R, A, \boldsymbol{\Xi}) = \int \Pr(R^*_{u,i}|\mathbf{P}_u, \mathbf{F}_u, \mathbf{Q}_i, \alpha) \Pr(\boldsymbol{\Theta}|A, R, \boldsymbol{\Xi}) \, \mathrm{d}\boldsymbol{\Theta} \qquad (3.3)$$

$$\Pr(A^*_{u,v}|A, R, \boldsymbol{\Xi}) = \int \Pr(A^*_{u,v}|\mathbf{P}_u, \mathbf{F}_v, \beta) \Pr(\boldsymbol{\Theta}|A, R, \boldsymbol{\Xi}) \, \mathrm{d}\boldsymbol{\Theta} \qquad (3.4)$$

However, the initial assumption that A only contains positive links introduces a severe bias in the model, as clearly no negative trust can be directly inferred through the posterior $\Pr(\boldsymbol{\Theta}|A, R, \boldsymbol{\Xi})$. If we consider A as an adjacency matrix, the latter generally tends to be extremely sparse. Therefore, only a very small percentage of its entries are labeled as positive, and ambiguity arises in the interpretation of all other entries, since in such cases actual lack of trust and lack of awareness cannot be distinguished. To handle this, we explicitly model awareness through a binary latent variable $Y_{u,v}$ relative to a pair (u, v) such that $u \to v \notin A$. The value $Y_{u,v} = 1$ denotes confidence in the lack of u's trust in v, whereas $Y_{u,v} = 0$ indicates confidence in the fact that u is not aware of v. The matrix of all variables is denoted by $\mathbf{Y}$ in the following.

The latent variables $Y_{u,v}$ for all the pairs $u \to v \notin A$ are drawn from a Bernoulli distribution with parameter $\epsilon_{u,v}$.

$$\Pr(Y_{u,v}) = \epsilon_{u,v}^{Y_{u,v}} \left(1 - \epsilon_{u,v}\right)^{1-Y_{u,v}} \qquad (3.5)$$

Again, we can provide a full Bayesian treatment by placing a Beta prior distribution with hyperparameter $\boldsymbol{\gamma} = \{\gamma_1, \gamma_2\}$ on the individual parameters $\epsilon_{u,v}$:

$$\Pr(\epsilon_{u,v}|\boldsymbol{\gamma}) = \frac{1}{B(\gamma_1, \gamma_2)} \epsilon_{u,v}^{\gamma_1 - 1} \left(1 - \epsilon_{u,v}\right)^{\gamma_2 - 1} \qquad (3.6)$$

The adoption of the latent variables $\mathbf{Y}$ allows us to provide an unbiased estimate of the posterior $\Pr(\boldsymbol{\Theta}|A, R, \boldsymbol{\Xi})$ as

$$\Pr(\boldsymbol{\Theta}|A, R, \boldsymbol{\Xi}) = \int \sum_{\mathbf{Y}} \Pr(\boldsymbol{\Theta}, \mathbf{Y}, \epsilon|A, R, \boldsymbol{\Xi}, \gamma) \, \mathrm{d}\epsilon, \qquad (3.7)$$

which can be plugged directly into equations 3.3 and 3.4. Also, we can further decompose the posterior as follows:

$$\Pr(\boldsymbol{\Theta}, \mathbf{Y}, \epsilon | A, R, \boldsymbol{\Xi}, \gamma) \propto \Pr(R | \boldsymbol{\Theta}, \alpha) \Pr(A | \boldsymbol{\Theta}, \mathbf{Y}, \beta)$$
$$\cdot Pr(\boldsymbol{\Theta} | \boldsymbol{\Theta}_0) \Pr(\mathbf{Y} | \epsilon) \Pr(\epsilon | \gamma)$$

where finally the term $\Pr(A | \boldsymbol{\Theta}, \mathbf{Y}, \beta)$ can be devised as

$$\Pr(A | \mathbf{P}, \mathbf{F}, \mathbf{Y}, \beta) = \prod_{u \to v \in A} \mathcal{N}(1; \vartheta_{u,v}, \beta^{-1}) \cdot \prod_{u \to v \notin A} \mathcal{N}(0; \vartheta_{u,v}, \beta^{-1})^{Y_{u,v}} \quad (3.8)$$

4 Inference

The exact computation of both Eq. 3.3 and Eq. 3.4 is analytically intractable, because of the complexity of the posterior $\Pr(\boldsymbol{\Theta}, \mathbf{Y}, \epsilon | A, R, \boldsymbol{\Xi}, \gamma)$. Therefore, we resort to *Monte-Carlo* approximation that allows to estimate the predictive distributions by averaging over samples of the model parameters:

$$\Pr(R_{u,i}^* | R, A, \boldsymbol{\Xi}, \gamma) \approx \frac{1}{H} \sum_{h=1}^{H} \Pr(R_{u,i}^* | \mathbf{P}_u^{(h)}, \mathbf{F}_u^{(h)}, \mathbf{Q}_i^{(h)}, \alpha) \quad (4.1)$$

$$\Pr(\Lambda_{u,v}^* | A, R, \boldsymbol{\Xi}, \gamma) \approx \frac{1}{H} \sum_{h=1}^{H} \Pr(A_{u,v}^* | \mathbf{P}_u^{(h)}, \mathbf{F}_v^{(h)}, \beta). \quad (4.2)$$

Here, the matrices $\mathbf{P}^{(h)}$, $\mathbf{F}^{(h)}$ and $\mathbf{Q}^{(h)}$ are sampled by running a Markov chain, whose stationary distribution approaches the posterior $\Pr(\boldsymbol{\Theta}, \mathbf{Y}, \epsilon | A, R, \boldsymbol{\Xi}, \gamma)$. In particular, we exploit the Gibbs sampling technique, that provides simple inference algorithms even when the underlying model has a very large number of hidden variables. The Markov chain is built by sequentially considering a variable $\varphi \in \{\mathbf{P}_u, \mathbf{F}_u, \mathbf{Q}_i, Y_{u,v}, \epsilon_{u,v}\}_{u,v \in N, i \in \mathcal{I}}$ and sampling according to the probability $\Pr(\varphi | Rest)$, where $Rest$ represents all remaining variables in $\{\mathbf{P}_u, \mathbf{F}_u, \mathbf{Q}_i, Y_{u,v}, \epsilon_{u,v}\}_{u,v \in N, i \in \mathcal{I}}$. Thus, inference in the context of our probabilistic model involves computing the full conditional distributions of the latent variables, which are discussed in the following.

Sampling $\mathbf{P}$, $\mathbf{F}$ *and* $\mathbf{Q}$. By exploiting conjugacy, the full conditional of each factor can be expressed as a multivariate gaussian. For example, for $\mathbf{P}$, we can observe that

$$\Pr(\mathbf{P}_u | Rest) \propto \Pr(\mathbf{P}_u | \boldsymbol{\Theta}_{\mathbf{P}}) \prod_{i \in \mathcal{I}} \Pr(\mathbf{R}_{u,i} | \mathbf{P}_u, \mathbf{F}_u, \mathbf{Q}_i, \alpha)^{\delta_{u,i}}$$
$$\cdot \prod_{v:u \to v \in A} \Pr(1 | \mathbf{P}_u, \mathbf{F}_u, \beta) \prod_{v:u \to v \notin A} \Pr(0 | \mathbf{P}_u, \mathbf{F}_u, \beta)^{Y_{u,v}},$$

which results in

$$\mathbf{P}_u \sim \mathcal{N}\left(\mu_P^{*(u)}, \left[\Lambda_P^{*(u)}\right]^{-1}\right)$$

with

$$\Lambda_P^{*(u)} = \Lambda_{\mathbf{P}} + \alpha \sum_{i \in \mathcal{I}} \delta_{u,i} \mathbf{Q}_i \mathbf{Q}_i' + \beta \sum_{v \in N} \tilde{Y}_{u,v} \mathbf{F}_v \mathbf{F}_v'$$

and

$$\mu_P^{*(u)} = \left[\Lambda_P^{*(u)}\right]^{-1} \left[\alpha \sum_{i \in \mathcal{I}} \delta_{u,i} \mathbf{Q}_j R_{u,i} - \alpha \left(\sum_{i \in \mathcal{I}} \delta_{u,i} \mathbf{Q}_i \mathbf{Q}_i' \right) \mathbf{F}_u + \beta \sum_{v:u \to v \in A} \mathbf{F}_v + \Lambda_{\mathbf{P}} \mu_{\mathbf{P}} \right]$$

Here, $\tilde{Y}_{u,v} = 1$ if either $u \to v \in A$ or $Y_{uv} = 1$ (that is to say, $\tilde{Y}_{u,v}$ models awareness of u for v). Similarly, we have

$$\mathbf{F}_u \sim \mathcal{N} \left(\mu_F^{*(u)}, \left[\Lambda_F^{*(u)}\right]^{-1} \right)$$

$$\mathbf{Q}_i \sim \mathcal{N} \left(\mu_Q^{*(i)}, \left[\Lambda_Q^{*(i)}\right]^{-1} \right)$$

where

$$\Lambda_F^{*(u)} = \Lambda_F + \alpha \sum_{i \in \mathcal{I}} \delta_{u,i} \mathbf{Q}_i \mathbf{Q}_i' + \beta \sum_{v \in N} \tilde{Y}_{u,v} \mathbf{P}_v \mathbf{P}_v'$$

$$\Lambda_Q^{*(i)} = \Lambda_{\mathbf{Q}} + \alpha \sum_{u \in N} \delta_{u,i} \left(\mathbf{P}_u + \mathbf{F}_u \right) \left(\mathbf{P}_u + \mathbf{F}_u \right)'$$

and

$$\mu_F^{*(u)} = \left[\Lambda_F^{*(u)}\right]^{-1} \left[\alpha \sum_{i \in \mathcal{I}} \delta_{u,i} \mathbf{Q}_i R_{u,i} - \alpha \left(\sum_{i \in \mathcal{I}} \delta_{u,i} \mathbf{Q}_i \mathbf{Q}_i' \right) \mathbf{P}_u + \beta \sum_{v:u \to v \in A} \mathbf{P}_v + \Lambda_{\mathbf{F}} \mu_{\mathbf{F}} \right]$$

$$\mu_Q^{*(i)} = \left[\Lambda_Q\right]^{*(i)^{-1}} \left[\alpha \sum_{u \in N} \left(\mathbf{P}_u + \mathbf{F}_u \right) \delta_{u,i} R_{u,i} + \Lambda_{\mathbf{Q}} \mu_{\mathbf{Q}} \right]$$

Sampling $\mathbf{Y}$ *and* ϵ. For each pair (u,v) such that $u \to v \notin A$, we can express the full conditional likelihood as

$$\Pr(Y_{u,v}|\epsilon_{u,v}, A, \mathbf{P}_u, \mathbf{F}_v, \beta) \propto \Pr(0|\mathbf{P}_u, \mathbf{F}_v, \beta)^{Y_{u,v}} \cdot \Pr(Y_{u,v}|\epsilon_{u,v}).$$

which yields the equation

$$\Pr(Y_{u,v}|\epsilon_{u,v}, A, \mathbf{P}_u, \mathbf{F}_v, \beta) = \frac{\exp\left\{ -\beta/2 \left(\mathbf{P}_u' \mathbf{F}_v \right)^2 + \eta_{uv} \right\}}{\exp\left\{ -\beta/2 \left(\mathbf{P}_u' \mathbf{F}_v \right)^2 + \eta_u \right\} + 1} \qquad (4.3)$$

with $\eta_{uv} = \log \epsilon_{u,v}/(1 - \epsilon_{u,v})$.

The distribution over the individual $\epsilon_{u,v}$ (for each (u,v) such that $u \to v \notin A$) can be obtained by conditioning on their respective Markov blanket. By exploiting conjugacy, we obtain

$$\Pr\left(\epsilon_{u,v}|Y_{u,v}, \gamma \right) = \frac{\gamma_1 + Y_{u,v}}{\gamma_1 + \gamma_2 + 1} \qquad (4.4)$$

Sampling $\Theta_\mathbf{P}$, $\Theta_\mathbf{Q}$ *and* $\Theta_\mathbf{F}$. Again, the conjugacy of the Gaussian-Wishart to the multivariate normal distribution provides a simplification of the full conditional into a Gaussian-Wishart [9, pp. 178]. In general, for a multivariate normal sample $\mathbf{X} \equiv \mathbf{x}_1, \ldots, \mathbf{x}_n$, the posterior $\Pr(\Theta|\mathbf{X}, \Theta_0)$ results into a $\mathcal{NW}(\Theta; \Theta_n)$ where $\Theta_n = \{\mu_n, \beta_n, \nu_n, \mathbf{W}_n\}$ and

$$\mu_n = \frac{\beta_0 \mu_0 + n}{\beta_0 + n}, \qquad \beta_n = \beta_0 + n, \qquad \nu_n = \nu_0 + n$$

$$[\mathbf{W}_n]^{-1} = \mathbf{W}_0^{-1} + \mathbf{S}_\mathbf{X} + \frac{\beta_0 n}{\beta_0 + n}(\mu_0 - \overline{\mathbf{x}})(\mu_0 - \overline{\mathbf{x}})'$$

with $\overline{\mathbf{x}} = 1/n \sum_i \mathbf{x}_i$ and $\mathbf{S}_\mathbf{X} = \sum_i (\mathbf{x}_i - \overline{\mathbf{x}})(\mathbf{x}_i - \overline{\mathbf{x}})'$.

Thus, the posteriors for $\Theta_\mathbf{P}$, $\Theta_\mathbf{F}$ and $\Theta_\mathbf{Q}$ are obtained by updating the respective statistics from which the corresponding hyperparameters depend.

The Gibbs sampling algorithm for approximate inference. Fig. 4 illustrates the Gibbs sampler used to perform approximate inference within the devised model. An execution of the sampler essentially consists in the repetition of a certain number of iterations (lines 3-20). The generic iteration h divides into two stages. The first stage is devoted to sampling hyperparameters $\Theta_\mathbf{P}^{(h)}$, $\Theta_\mathbf{F}^{(h)}$ and $\Theta_\mathbf{Q}^{(h)}$ and $\epsilon_{u,v}$ (lines 4-9). Model parameters $\mathbf{Y}$, $\mathbf{P}_u$, $\mathbf{F}_u$ and $\mathbf{Q}_j$ are then sampled at the second stage (lines 10-19).

Notice that running the Markov chain to its equilibrium through a maximum number of iteration is a widely-adopted convergence-criterion [16]. The overall number of iterations must be carefully set, so that to the probability of transitions of the sampler between latent states converges to a stationary distribution after a preliminary burn-in period. This permits to gather samples drawn after convergence for prediction (as discussed in Sec.4), while discarding burn-in samples which are sensible to the initialization of the sampler.

Also, concerning $\mathbf{Y}$, we do not sample the whole set of pairs (u, v) such that $u \to v \notin A$. This is a crucial efficiency issue. In practice, we are assuming $\mathbf{Y}$ contains several unknown values, and hence only a limited amount of unconnected pairs in a corresponding set U has to be considered. The underlying assumption is that the number $|U|$ of pairs to sample is the result of a prior Poisson process, fixed in the beginning and not reported here for lack of space.

5   Experimental Evaluation

The joint modeling of users' trust networks and ratings provides a powerful framework to detect and understand different patterns within the input social rating network. In this section we analyze the application of the proposed model to real-world social rating networks. More specifically, we are interested in evaluating the effectiveness of our approach in three respects.

- Firstly, we measure its accuracy in rating prediction.
- Secondly, we evaluate the accuracy in predicting trust between pairs of users by measuring the AUC of the proposed model.

```
GIBBS SAMPLING(N, Θ₀ = {μ₀, β₀, ν₀, W₀}, γ, α, β)
 1:  Sample a subset U ⊆ N × N such that u → v ∉ A;
 2:  Initialize P⁽⁰⁾, F⁽⁰⁾, Q⁽⁰⁾, Y⁽⁰⁾;
 3:  for h = 1 to H do
 4:     Sample Θ_P⁽ʰ⁾ ∼ NW(Θₙ) where Θₙ is computed by updating Θ₀ with P̄, S_P;
 5:     Sample Θ_F⁽ʰ⁾ ∼ NW(Θₙ) where Θₙ is computed by updating Θ₀ with F̄, S_F;
 6:     Sample Θ_F⁽ʰ⁾ ∼ NW(Θₙ) where Θₙ is computed by updating Θ₀ with Q̄, S_Q
 7:     for each (u, v) ∈ U do
 8:        Sample ε_{u,v}⁽ʰ⁾ according to Eq. 4.4;
 9:     end for
10:     for each (u, v) ∈ U do
11:        Sample Y_{uv}⁽ʰ⁾ according to Eq. 4.3;
12:     end for
13:     for each u ∈ N do
14:        Sample P_u ∼ N (μ_P^{*(u)}, [Λ_P^{*(u)}]⁻¹) ;
15:        Sample F_u ∼ N (μ_F^{*(u)}, [Λ_F^{*(u)}]⁻¹);
16:     end for
17:     for each i ∈ I do
18:        Sample Q_i ∼ N (μ_Q^{*(i)}, [Λ_Q^{*(i)}]⁻¹);
19:     end for
20:  end for
```

Fig. 4. The scheme of Gibbs sampling algorithm in pseudo code

Table 1. Summary of the chosen social rating networks

	Ciao	Epinions
Users	7,375	49,289
Trust Relationships	111,781	487,181
Items	106,797	139,738
Ratings	282,618	664,823
InDegree (Avg/Median/Min/Max)	15.16/6/1/100	9.8/2/1/2589
OutDegree (Avg/Median/Min/Max)	16.46/4/1/804	14.35/3/1/1760
Ratings on items (Avg/Median/Min/Max)	2.68/1/1/915	4.75/1/1/2026
Ratings by Users (Avg/Median/Min/Max)	38.32/18/4/1543	16.55/6/1/1023

- Thirdly, we analyze the structure of the model and investigate the properties that can be derived, such as relationships among factors and propensities of users within given factors.

Datasets. We conducted experiments on two datasets representing social rating networks from the popular product review sites *Epinions* and *Ciao*, described in [29]. Users in these sites can share their reviews about products. Also they can establish their trust networks from which they may seek advice to make decisions. Both sites employ a 5-star rating system. Some statistics of the datasets are shown in Table 1 and in Fig. 5. We can notice that both the trust relationships and the rating distributions are heavy-tailed. *Epinions* exhibits a larger number of users, as well as a larger sparsity coefficient on A.

Evaluation Setting. We chose some state-of-the-art baselines from the current literature. For rating prediction, we compared our approach against *SocialMF* [14]. The metric used here is the standard RMSE. We exploited the implementation

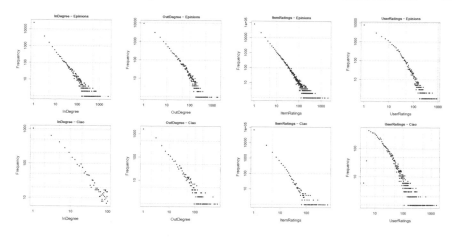

Fig. 5. Distributions of trust relationships and ratings in *Epinions* and *Ciao*

of *SocialMF* made available at `http://mymedialite.net`. For trust prediction, we adapted the framework described in [20]. For each user, we considered the ratings as user features and we trained the factorization model which minimizes the AUC loss. We exploited the implementation made available by the authors at `http://cseweb.ucsd.edu/~akmenon/code`. We refer to this method as *AUC-MF* in the following. In addition, we considered a further comparison in terms of both RMSE and AUC against a basic matrix factorization approach based on SVD named *Joint SVD (JSVD)* [11]. We computed a low-rank factorization of the joint adjacency/feature matrix $\mathbf{X} = [A\ R]$ as $\mathbf{X} \approx \mathbf{U} \cdot diag(\sigma_1, \ldots, \sigma_K) \cdot \mathbf{V}^T$, where K is the rank of the decomposition and $\sigma_1, \ldots, \sigma_K$ are the square roots of the K greatest eigenvalues of $\mathbf{X}^T\mathbf{X}$. The matrices $\mathbf{U}$ and $\mathbf{V}$ resemble the roles of $\mathbf{P}$, $\mathbf{F}$ and $\mathbf{Q}$: The term $U_{u,k}$ can be interpreted as the tendency of u to trust users, relative to factor k. Analogously, $V_{u,k}$ represents the tendency of u to be trusted, and $V_{i,k}$ represents the rating tendency of item i in k. The score can be hence computed as [26] $score(u,x) = \sum_{k=1}^{K} U_{u,k}\sigma_k V_{x,k}$, where x denotes either a user v or an item i.

In all the experiments, we performed a Monte-Carlo Cross Validation, by performing 5 training/test splits. Within the partitions, 70% of the data were retained as training, and the remaining 30% as test. The splitting was accomplished for the sole data upon which to measure the performance (i.e., ratings for the RMSE and links for the AUC).

Concerning the AUC, it is worth noticing that *Epinions* and *Ciao* only contain positive trust relationships, and the computation of the AUC relies on the presence of negative values. Negative values are indeed crucial in the approach [20], since the latter relies on a loss function which penalizes situations where the score of negative links is higher than the score of positive links. In principle, we can consider all links in the test-set as positive examples, and all non-existing links as negative example. However, the sparsity of the networks poses a major tractability issue, as it would make the computation of the AUC infeasible. A better estimation strategy in [2, 26] consists in narrowing the negative examples

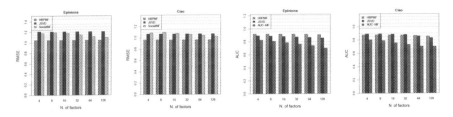

Fig. 6. Prediction results

to all the 2-hops non-existing links, i.e., all triplets (u, v, w) where both (u, v) and (v, w) exhibit a trust relationship in A, but (u, w) does not.

Results. Fig. 6 reports the averaged results of the evaluation. We ran the experiments on a variable number of latent factors, ranging from 4 to 128. We can notice that the proposed hierarchical model, denoted as *HBPMF*, achieves the minimum RMSE on both datasets. There is a tendency of the RMSE to progressively decrease. However, this tendency is more evident on *SocialMF*, while the other two methods exhibit negligible differences.

The opposite trend is observed in trust prediction. Here, all methods tend to prefer a low number of factors, as the best results are achieved with $K = 4$. The devised *HBPMF* model achieves the maximum AUC on the *Epinions* dataset, and results comparable to *JSVD* on *Ciao*. The detailed results are shown in Fig. 7, where the ROC curves are reported. In general, the predictive accuracy of the Bayesian hierarchical model is stable with regards to the number of factors. This is a direct result of the Bayesian modeling, which makes the model robust to the growth of the model complexity. Fig. 8 also shows how the accuracy varies according to the distributions which characterize the data. We can notice a correlation between accuracy and node degrees, as well as the number of ratings provided by a user or received by an item.

To evaluate the effects of the joint modeling of both the trust relationships and the ratings, we conducted some further experiments with $K = 4$. In a first experiment, we performed the sampling without considering the trust relationships. More precisely, we performed a simple *BPMF* (as described in [25]). Dually, we discarded the rating matrix and performed the sampling by only considering the trust relationships. The first graph of Fig. 9 shows the comparison between the

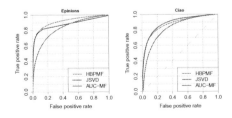

Fig. 7. ROC curves on trust prediction for $K = 4$

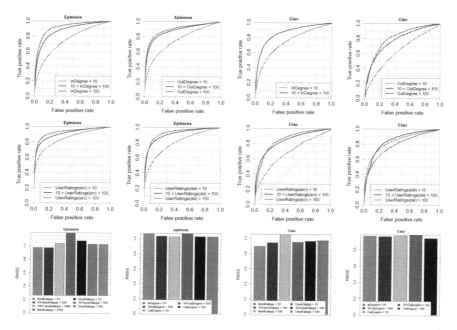

Fig. 8. Data distribution vs. AUC and rating prediction

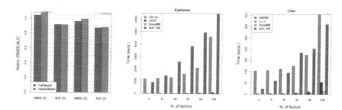

Fig. 9. (a) Effects of the joint modeling. (1 denotes *Epinions*, and 2 denotes *Ciao*). (b) Average running time for iteration (JSVD reports the total time).

results of these partial models against those achieved through the full *HBPMF* model. The effects of the joint modeling can be appreciated on the RMSE: in practice, the additional information provided by the trust relationships refines the modeling of the data, thus lowering the RMSE. By contrast, the effects of the joint modeling on the AUC do not highlight substantial improvements.

Finally, the last two graphs of Fig. 9 report the running times relative to the methods. For the *HBPMF*, we achieved stable results for the RMSE after 100 iterations, whereas the AUC result was stable after 20 iterations. Both *SocialMF* and *AUC-MF* exhibited stable results with 20 iterations. The computational overhead of the Gibbs Sampling procedure plays a crucial role here. Therein, it would be interesting to investigate alternative inference strategies based on variational approximation, which are known to guarantee fast convergence.

6   Conclusions and Future Research

We presented the first unified approach to the recommendation of interesting items and trustworthy users in social rating networks with trust relationships. The key intuition is that the interactions from users to users as well as between users and items are explained by the same latent factors, which ultimately allows to combine user and item recommendation into a simple and intuitive Bayesian generative model. A comparative experimentation over real-world social rating networks confirmed such an intuition: the devised model was shown to deliver a superior predictive performance in terms of both RMSE and AUC.

Future research will focus on two major directions. We planned to study an extension of our model in which the Indian Buffet Process [12] is exploited to automatically infer the most appropriate number of latent factors from the input social rating network. In addition, variational approximate inference and related learning algorithms will be studied to improve the computational efficiency. Finally, a further line of research is relative to how the proposed models can be adapted to support recommendation tasks behind rating prediction [24, 5, 4].

References

1. Airoldi, E.M., Blei, D.M., Fienberg, S.E., Xing, E.P.: Mixed membership stochastic blockmodels. The Journal of Machine Learning Research 9, 1981–2014 (2008)
2. Backstrom, L., Leskovec, J.: Supervised random walks: Predicting and recommending links in social networks. In: Proc. ACM WSDM Conf., pp. 635–644 (2011)
3. Barbieri, N., Bonchi, F., Manco, G.: Cascade-based community detection. In: Proc. of ACM WSDM Conf., pp. 33–42 (2013)
4. Barbieri, N., Manco, G.: An analysis of probabilistic methods for top-n recommendation in collaborative filtering. In: Gunopulos, D., Hofmann, T., Malerba, D., Vazirgiannis, M. (eds.) ECML PKDD 2011, Part I. LNCS (LNAI), vol. 6911, pp. 172–187. Springer, Heidelberg (2011)
5. Barbieri, N., Manco, G., Ortale, R., Ritacco, E.: Balancing prediction and recommendation accuracy: Hierarchical latent factors for preference data. In: Proc. of SIAM Int. Conf. on Data Mining, pp. 1035–1046 (2012)
6. Costa, G., Ortale, R.: A bayesian hierarchical approach for exploratory analysis of communities and roles in social networks. In: Proc. of the IEEE/ACM ASONAM Conf., pp. 194–201 (2012)
7. Costa, G., Ortale, R.: Probabilistic analysis of communities and inner roles in networks: Bayesian generative models and approximate inference. Social Network Analysis and Mining 3(4), 1015–1038 (2013)
8. Costa, G., Ortale, R.: A Unified Generative Bayesian Model for Community Discovery and Role Assignment based upon Latent Interaction Factors. In: Proc. of the IEEE/ACM ASONAM Conf. (2014)
9. DeGroot, M.: Optimal Statistical Decisions. McGraw-Hill (1970)
10. Delporte, J., Karatzoglou, A., Matuszczyk, T., Canu, S.: Socially enabled preference learning from implicit feedback data. In: Blockeel, H., Kersting, K., Nijssen, S., Železný, F. (eds.) ECML PKDD 2013, Part II. LNCS (LNAI), vol. 8189, pp. 145–160. Springer, Heidelberg (2013)
11. Gong, N.Z., et al.: Joint link prediction and attribute inference using a social-attribute network. ACM TIST 5(2) (2014)

12. Griffiths, T.L., Ghahramani, Z.: The indian buffet process: An introduction and review. The Journal of Machine Learning Research 12, 1185–1224 (2011)
13. Hofman, T., Puzicha, J., Jordan, M.I.: Learning from dyadic data. In: Proc. NIPS Conf., pp. 466–472 (1999)
14. Jamali, M., Ester, M.: A matrix factorization technique with trust propagation for recommendation in social networks. In: Proc. of ACM RECSYS Conf., pp. 135–142 (2010)
15. Liben-Nowell, D., Kleinberg, J.: The link-prediction problem for social networks. Journal of the American Society for Information Science and Technology 58(7), 1019–1031 (2007)
16. Liu, J.S.: Monte Carlo Strategies in Scientific Computing. Springer (2001)
17. Ma, H., King, I., Lyu, M.R.: Learning to recommend with social trust ensemble. In: Proc. of Int. ACM SIGIR Conf., pp. 203–210 (2009)
18. Ma, H., Yang, H., Lyu, M.R., King, I.: Sorec: Social recommendation using probabilistic matrix factorization. In: Proc. of ACM CIKM Conf., pp. 931–940 (2008)
19. Ma, H., Zhou, D., Liu, C., Lyu, M.R., King, I.: Recommender systems with social regularization. In: Proc. of ACM WSDM Conf., pp. 287–296 (2011)
20. Menon, A.K., Elkan, C.: Link prediction via matrix factorization. In: Gunopulos, D., Hofmann, T., Malerba, D., Vazirgiannis, M. (eds.) ECML PKDD 2011, Part II. LNCS (LNAI), vol. 6912, pp. 437–452. Springer, Heidelberg (2011)
21. Miller, K.T., Griffiths, T.L., Jordan, M.I.: Nonparametric latent feature models for link prediction. In: Proc. NIPS Conf., pp. 1276–1284 (2009)
22. Pan, R., Zhou, Y., Cao, B., Liu, N.N., Lukose, R.M., Scholz, M., Yang, Q.: One-class collaborative filtering. In: Proc. IEEE ICDM Conf., pp. 502–511 (2008)
23. Purushotham, S., Liu, Y., Kuo, C.C.J.: Collaborative topic regression with social matrix factorization for recommendation systems. In: Proc. ICML Conf., pp. 759–766 (2012)
24. Rendle, S., Christoph, F., Zeno, G., Lars, S.: Bpr: Bayesian personalized ranking from implicit feedback. In: Proc. UAI Conf. (2009)
25. Salakhutdinov, R., Mnih, A.: Bayesian probabilistic matrix factorization using markov chain monte carlo. In: Proc. ICML Conf., pp. 880–887 (2008)
26. Badrul, M.: Sarwar et al. Application of dimensionality reduction in recommender system – a case study. In: ACM WEBKDD Workshop (2000)
27. Shen, Y., Jin, R.: Learning personal+social latent factor model for social recommendation. In: Proc. of ACM SIGKDD Conf., pp. 1303–1311 (2012)
28. Sindhwani, V., Bucak, S.S., Hu, J., Mojsilovic, A.: One-class matrix completion with low-density factorizations. In: Proc. of IEEE ICDM Conf., pp. 1055–1060 (2010)
29. Tang, J., Gao, H., Liu, H.: mtrust: Discerning multi-faceted trust in a connected world. In: Proc. ACM WSDM Conf., pp. 93–102 (2012)
30. Yang, S.-H., et al.: Like like alike: Joint friendship and interest propagation in social networks. In: Proc. WWW Conf., pp. 537–546 (2011)
31. Yang, X., Steck, H., Liu, Y.: Circle-based recommendation in online social networks. In: Proc. ACM SIGKDD Conf., pp. 1267–1275 (2012)
32. Zhu, J.: Max-margin nonparametric latent feature models for link prediction. In: Proc. NIPS Conf., pp. 719–726 (2012)

Domain Adaptation
with Regularized Optimal Transport

Nicolas Courty[1], Rémi Flamary[2], and Devis Tuia[3]

[1] Université de Bretagne Sud, IRISA, Vannes, France
[2] Université de Nice, Lab. Lagrance UMR CNRS 7293, OCA, Nice, France
[3] EPFL, LASIG, Lausanne, Switzerland

Abstract. We present a new and original method to solve the domain adaptation problem using optimal transport. By searching for the best transportation plan between the probability distribution functions of a source and a target domain, a non-linear and invertible transformation of the learning samples can be estimated. Any standard machine learning method can then be applied on the transformed set, which makes our method very generic. We propose a new optimal transport algorithm that incorporates label information in the optimization: this is achieved by combining an efficient matrix scaling technique together with a majoration of a non-convex regularization term. By using the proposed optimal transport with label regularization, we obtain significant increase in performance compared to the original transport solution. The proposed algorithm is computationally efficient and effective, as illustrated by its evaluation on a toy example and a challenging real life vision dataset, against which it achieves competitive results with respect to state-of-the-art methods.

1 Introduction

While most learning methods assume that the test data $\mathbf{X}_t = (\mathbf{x}_i^t)_{i=1,\ldots,N_t}$, $\mathbf{x}_i \in \mathbb{R}^d$ and the training data $\mathbf{X}_s = (\mathbf{x}_i^s)_{i=1,\ldots,N_s}$ are generated from the same distributions $\mu_t = \mathcal{P}(\mathbf{X}_t)$ and $\mu_s = \mathcal{P}(\mathbf{X}_s)$, real life data often exhibit different behaviors. Many works study the generalization capabilities of a classifier allowing to transfer knowledge from a labeled source domain to an unlabeled target domain: this situation is referred to as transductive transfer learning [1]. In our work, we assume that the source and target domains are by nature different, which is usually referred to as domain adaptation. In the classification problem, the training data are usually associated with labels corresponding to C different classes. We consider the case where only the training data are associated with a label $\mathbf{Y}_s = (\mathbf{y}_i)_{i=1,\ldots,N_s}$, $\mathbf{y}_i \in \{1, \ldots, C\}$, yielding an **unsupervised domain adaptation** problem, since no labelled data is available in the target domain. In this acceptation, the training (resp. testing) domain is usually referred to as source (resp. target) distribution.

Domain adaptation methods seek to compensate for inter domain differences by exploiting the similarities between the two distributions. This compensation is

T. Calders et al. (Eds.): ECML PKDD 2014, Part I, LNCS 8724, pp. 274–289, 2014.

usually performed by reweighing the contribution of each samples in the learning process (*e.g.* [2]) or by means of a global data transformation that aligns the two distributions in some common feature space (*e.g.* [3]). Our work departs from these previous works by assuming that there exists a non-rigid transformation of the distribution that can account for the non-linear transformations occurring between the source and target domains. This transformation is conveniently expressed as a transportation of the underlying probability distribution functions thanks to optimal transport (OT). The OT problem has first been introduced by the French mathematician Gaspard Monge in the middle of the 19th century as the way to find a minimal effort solution to the transport of a given mass of dirt into a given hole. The problem reappeared later in the work of Kantorovitch [4], and found recently surprising new developments as a polyvalent tool for several fundamental problems [5]. In the domain of machine learning, OT has been recently used for computing distances between histograms [6] or label propagation in graphs [7].

Contributions. Our contributions are twofold: *i)* First, we show how to transpose the optimal transport problem to the domain adaptation problem, and we propose experimental validations of this idea. To the best of our knowledge, this is the first time that optimal transport is considered in the domain adaptation setting. *ii)* Second, we propose an elegant group-based regularization for integrating label information, which has the effect of regularizing the transport by adding inter-class penalties. The resulting algorithm exploits a proven efficient optimization approaches and will benefit from any advances in this domain. The proposed optimal transport with label regularization (**OT-reglab**) allows to achieve competitive state-of-the-art results on challenging datasets.

2   Related Work

Two main strategies have been considered to tackle the domain adaptation problem: on the one hand, there are approaches considering the transfer of instances, mostly via sample re-weighting schemes based on density ratios between the source and target domains [2,8]. By doing so, authors compare the data distributions in the input space and try to make them more similar by weighting the samples in the source domain.

On the other hand, many works have considered finding a common feature representation for the two (or more) domains, or a latent space, where a classifier using only the labeled samples from the source domain generalize well on the target domains [9,10]. The representation transfer can be performed by matching the means of the domains in the feature space [10], aligning the domains by their correlations [11] or by using pairwise constraints [12]. In most of these works, the common latent space is found via feature extraction, where the dimensions retained summarize the information common to the domains. In computer vision, methods exploiting a gradual alignment of sets of eigenvectors have been proposed: in [13], authors start from the hypothesis that domain adaptation can be better approached if comparing gradual distortions and therefore use intermediary projections of both domains along the Grassmannian geodesic connecting the source and target observed eigenvectors. In [14,15], authors propose to obtain all sets of transformed intermediary domains by using a geodesic-flow kernel instead of sampling a fixed number of projections along the geodesic. While these methods have the advantage of providing easily computable out-of-sample extensions (by projecting unseen samples onto the latent space eigenvectors), the transformation defined is global and applied the same way to the whole target domain.

An approach combining the two logics is found in [3], where authors extend the sample re-weighing reasoning to similarity of the distributions in the feature space by the use of surrogate kernels. By doing so, a linear transformation of the domains is found, but, as for the feature representation approaches above, it is the same for all samples transferred.

Our proposition strongly differs from those reviewed above, as it defines a local transportation plan for each sample in the source domain. In this sense, the domain adaptation problem can be seen as a graph matching problem for all samples to be transported, where their final coordinates are found by mapping the source samples to coordinates matching the marginal distribution of the target domain. In the authors knowledge, this is the first attempt to use optimal transportation theory in domain adaptation problem

3   Optimal Transportation

In this Section, we introduce the original formulation of optimal transport through the Monge-Kantorovitch problem and its discrete formulation. Then, regularized versions of the optimal transport are exposed.

3.1   The Monge-Kantorovitch Problem and Wasserstein Space

Let us first consider two domains Ω_1 and Ω_2 (in the following, we will assume without further indication that $\Omega_1 = \Omega_2 = \mathbb{R}^d$). Let $\mathcal{P}(\Omega_i)$ be the set of all the probability measures over Ω_i. Let $\mu \in \mathcal{P}(\Omega_1)$, and $\mathbf{T}$ be an application from $\Omega_1 \rightarrow \Omega_2$. The *image measure* of μ by $\mathbf{T}$, noted $\mathbf{T}\#\mu$, is a probability measure over Ω_2 which verifies:

$$\mathbf{T}\#\mu(y) = \mu(\mathbf{T}^{-1}(\mathbf{y})), \quad \forall \mathbf{y} \in \Omega_2. \tag{1}$$

Let $\mu_s = \mathcal{P}(\Omega_1)$ and $\mu_t = \mathcal{P}(\Omega_2)$ be two probability measures from the two domains. $\mathbf{T}$ is said to be a transport if $\mathbf{T}\#\mu_s = \mu_t$. The cost associated to this transport is

$$C(\mathbf{T}) = \int_{\Omega_1} c(\mathbf{x}, \mathbf{T}(\mathbf{x})) d\mu(\mathbf{x}), \tag{2}$$

where the cost function $c : \Omega_1 \times \Omega_2 \rightarrow \mathbb{R}^+$ can be understood as a regular distance function, but also as the energy required to move a mass $\mu(\mathbf{x})$ from $\mathbf{x}$

to $\mathbf{y}$. It is now possible to define the **optimal transport** $\mathbf{T}_0$ as the solution of the following minimization problem:

$$\mathbf{T}_0 = \arg\min_{\mathbf{T}} \int_{\Omega_1} c(\mathbf{x}, \mathbf{T}(\mathbf{x})) d\mu(\mathbf{x}), \quad \text{s.t. } \mathbf{T} \# \mu_s = \mu_t \tag{3}$$

which is the original Monge transportation problem. The equivalent Kantorovitch formulation of the optimal transport [4] seeks for a probabilistic coupling $\gamma \in \mathcal{P}(\Omega_1 \times \Omega_2)$ between Ω_1 and Ω_2:

$$\gamma_0 = \arg\min_{\gamma} \int_{\Omega_1 \times \Omega_2} c(\mathbf{x}, \mathbf{y}) d\gamma(\mathbf{x}, \mathbf{y}), \quad \text{s.t. } \mathrm{P}^{\Omega_1} \# \gamma = \mu_s, \mathrm{P}^{\Omega_2} \# \gamma = \mu_t, \tag{4}$$

where P^{Ω_i} is the projection over Ω_i. In this formulation, γ can be understood as a joint probability measure with marginals μ_s and μ_t. γ_0 is the unique solution to the optimal transport problem. It allows to define the **Wasserstein distance** between μ_s and μ_t as:

$$\mathbf{W}_2(\mu_s, \mu_t) = \inf_{\gamma} \int_{\Omega_1 \times \Omega_2} c(\mathbf{x}, \mathbf{y}) d\gamma(\mathbf{x}, \mathbf{y}), \quad \text{s.t. } \mathrm{P}^{\Omega_1} \# \gamma = \mu_s, \mathrm{P}^{\Omega_2} \# \gamma = \mu_t, \tag{5}$$

This distance, also known as the Earth Mover Distance in computer vision community [16], defines a metric over the space of integrable squared probability measure.

3.2   Optimal Transport of Discrete Distributions

Usually one does not have a direct access to μ_s or μ_t but rather to collections of samples from those distributions. It is then straightforward to adapt the optimal transport problem to the discrete case. The two distributions can be written as

$$\mu_s = \sum_{i-1}^{n_s} p_i^s \delta_{\mathbf{x}_i^s}, \quad \mu_t = \sum_{i=1}^{n_t} p_i^t \delta_{\mathbf{x}_i^t} \tag{6}$$

where $\delta_{\mathbf{x}_i}$ is the Dirac at location $\mathbf{x}_i \in \mathbb{R}^d$. p_i^s and p_i^t are probability masses associated to the i-th sample, and belong to the probability simplex, i.e. $\sum_{i=1}^{n_s} p_i^s = \sum_{i=1}^{n_t} p_i^t = 1$. The set of probabilistic coupling between those two distributions is then the set of doubly stochastic matrices $\mathcal{P}$ defined as

$$\mathcal{P} = \left\{ \gamma \in (\mathbb{R}^+)^{n_s \times n_t} \mid \gamma \mathbf{1}_{n_t} = \mu_s, \gamma^T \mathbf{1}_{n_s} = \mu_t \right\} \tag{7}$$

where $\mathbf{1}_d$ is a d-dimensional vector of ones. The Kantorovitch formulation of the optimal transport [4] reads:

$$\gamma_0 = \arg\min_{\gamma \in \mathcal{P}} \langle \gamma, \mathbf{C} \rangle_F \tag{8}$$

where $\langle ., . \rangle_F$ is the Frobenius dot product and $\mathbf{C} \geq 0$ is the cost function matrix of term $C(i, j)$ related to the energy needed to move a probability mass from $\mathbf{x}_i^s$ to $\mathbf{x}_j^t$. This cost can be chosen for instance as the Euclidian distance between the two locations, i.e. $C(i, j) = ||\mathbf{x}_i^s - \mathbf{x}_j^t||^2$, but other types of metric could be considered, such as Riemannian distances over a manifold [5].

Remark 1. When $n_s = n_t = n$ and when $\forall i, j \; p_i^s = p_j^t = 1/n$, the γ_0 is simply a permutation matrix

Remark 2. In the general case, it can be shown that γ_0 is a sparse matrix with at most $n_s + n_t - 1$ non zero entries (rank of constraints matrix).

This problem can be solved by linear programming, with combinatorial algorithms such as the simplex methods and its network variants (transport simplex, network simplex, etc.). Yet, the computational complexity was shown to be $\mathcal{O}(n^2)$ in practical situations [17] for the network simplex (while being $\mathcal{O}(n^3)$ in theory) which leverages the utility of the method to handle big data. However, the recent regularization of Cuturi [6] allows a very fast transport computation as discussed in the next Section.

3.3   Regularized Optimal Transport

When the target and source distributions are high-dimensional, or even in presence of numerous outliers, the optimal transportation plan may exhibit some irregularities, and lead to incorrect transport of points. While it is always possible to enforce *a posteriori* a given regularity in the transport result, a more theoretically convincing solution is to regularize the transport by relaxing some of the constraints in the problem formulation of Eq.(8). This possibility has been explored in recent papers [18,6].

In [18], Ferradans and colleagues have explored the possibility of relaxing the mass conservation constraints of the transport, *i.e.* slightly distorting the marginals of the coupling γ_0. Technically, this boils down to solving the same minimization problem but with inequality constraints on the marginals in Eq.(7). As a result, elements of the source and target distributions can remain still. Yet, one major problem of this approach is that it converts the original linear program into more computationally demanding optimizations impractical for large sets.

In a recent paper [6], Cuturi proposes to regularize the expression of the transport by the entropy of the probabilistic coupling. The regularized version of the transport γ_0^λ is then the solution of the following minimization problem:

$$\gamma_0^\lambda = \arg\min_{\gamma \in \mathcal{P}} \langle \gamma, \mathbf{C} \rangle_F - \frac{1}{\lambda} h(\gamma), \tag{9}$$

where $h(\gamma) = -\sum_{i,j} \gamma(i,j) \log \gamma(i,j)$ computes the entropy of γ. The intuition behind this form of regularization is the following: since most of the elements of γ_0 should be zero with high probability, one can look for a smoother version of the transport by relaxing this sparsity through an entropy term. As a result, and contrary to the previous approach, more couplings with non-nul weights are allowed, leading to a denser coupling between the distributions. An appealing result of this formulation is the possibility to derive a computationally very efficient algorithm, which uses the scaling matrix approach of Sinkhorn-Knopp [19]. The optimal regularized transportation plan is found by iteratively computing two scaling vectors $\mathbf{u}$ and $\mathbf{v}$ such that:

$$\gamma_0^\lambda = \mathrm{diag}(\mathbf{u}) \exp(-\lambda \mathbf{C}) \mathrm{diag}(\mathbf{v}), \tag{10}$$

where the exponential exp(.) operator should be understood element-wise.

Note that while these regularizations allow the inclusion of additional priors in the optimization problem, they do not take into account the fact that the elements of the source distribution belong to different classes. This idea is the core of our regularization strategy.

4   Domain Adaptation with Label Regularized Optimal Transport

From the definitions above, the use of optimal transport for domain adaptation is rather straightforward: by computing the optimal transport from the source distribution μ_s to the target distribution μ_t, one defines a transformation of the source domain to the target domain. This transformation can be used to adapt the training distribution by means of a simple interpolation. Once the source labeled samples have been transported, any classifier can be used to predict in the target domain. In this section, we present our optimal transport with label regularization algorithm (**OT-labreg**) and derive a new efficient algorithm to solve the problem. We finally discuss how to interpolate the training set from this regularized transport.

4.1   Regularizing the Transport with Class Labels

Optimal transport aims at minimizing a transport cost linked to a metric between distributions. It does not include any information about the particular nature of the elements of the source domain (*i.e.* the fact that those samples belong to different classes). However, this information is generally available, as labeled samples are used in the classification step following adaptation. Our proposition to take advantage of label information is to penalize couplings that match together samples with different labels. This is illustrated in Figure 1.c, where one can see that samples belonging to the same classes are only associated to points associated to the same class, contrarily to the standard and regularized versions of the transport (Figures 1.a and 1.b).

Principles of the Label Regularization. Over each column of γ, we want to concentrate the transport information on elements of the same class c. This is usually done by using $\ell_p - \ell_q$ mixed-norm regularization, among which the $\ell_1 - \ell_2$ known as as group-lasso is a favorite. The main idea is that, even if we do not know the class of the target distribution, we can promote group sparsity in the columns of γ such that a given target point will be associated with only one of the classes.

Promoting group sparsity leads to a new term in the cost function (9), which now reads:

$$\gamma_0 = \arg\min_{\gamma \in \mathcal{P}} \langle \gamma, \mathbf{C} \rangle_F - \frac{1}{\lambda} h(\gamma) + \eta \sum_j \sum_c ||\gamma(\mathcal{I}_c, j)||_q^p, \qquad (11)$$

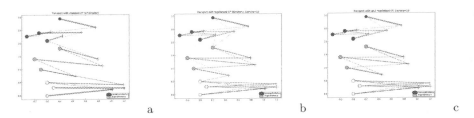

a   b   c

Fig. 1. Illustration of the transport for two simple distributions depicted in the image. The colored disks represent 3 different classes. The transport solution is depicted as blue lines whose thickness relate to the strength of the coupling. (a) Solution of the original optimal transport solution (**OT-ori**); (b) using the Sinkhorn transport (**OT-reg** [6]); (c) using our class-wise regularization term (**OT-reglab**).

where $\mathcal{I}_c$ contains the index of the lines such that the class of the element is c, $\gamma(\mathcal{I}_c, j)$ is a vector containing coefficients of the jth column of γ associated to class c and $|| \cdot ||_q^p$ denotes the ℓ_q norm to the power of p. η is a regularization parameter that weights the impact of the supervised regularization.

The choice of the p, q parameters is particularly sensitive. For $p \geq 1$ and $q \geq 1$ the regularization term is convex. The parameters $p = 1, q = 2$ lead to the classical group-lasso that is used, for instance, for joint features selection in multitask learning. The main problem of using the group-lasso in this case is that it makes the optimization problem much more difficult. Indeed, when using an ℓ_2 norm in the objective function, the efficient optimization procedure proposed in [6] cannot be used anymore. Moreover there is no particular reason to choose the ℓ_2 norm for regularizing coefficients of a transport matrix. Those coefficients being all positive and associated to probabilities, we propose to use $q = 1$ that will basically sum the probabilities in the groups. When $q = 1$, one needs to carefully chose the p coefficient in order to promote group sparsity. In this work we propose to use $p = 1/2 < 1$. This parameter is a common choice for promoting sparsity, as the square root is non-differentiable in zero and has been used recently for promoting non-grouped sparsity in compressed sensing [20]. An additional advantage of our proposal is that, despite the fact that the proposed regularization is non-convex, a simple approach known as reweighted ℓ_1 can be performed for its optimization, as detailed below.

4.2   Majoration Minimization Strategy

The optimization problem with a $\ell_p - \ell_1$ regularization boils down to optimizing

$$\gamma_0 = \arg\min_{\gamma \in \mathcal{P}} J(\gamma) + \eta \Omega(\gamma), \qquad (12)$$

with $J(\gamma) = \langle \gamma, C \rangle_F - \frac{1}{\lambda} h(\gamma)$ and $\Omega(\gamma) = \sum_j \sum_c ||\gamma(\mathcal{I}_c, j)||_1^p$. We want to be able to use the optimization in [6] to solve the left term, as it is very efficient.

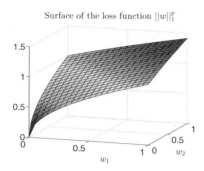

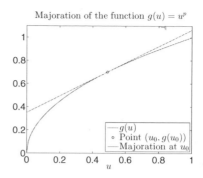

Fig. 2. Illustration of the regularization term loss for a 2D group (left). Illustration of the convexity of $g(\cdot)$ and its linear majoration (right).

First, note that the regularization term can be reformulated as

$$\Omega(\gamma) = \sum_j \sum_c g(\|\gamma(\mathcal{I}_c, j)\|_1) \tag{13}$$

where $g(\cdot) = (\cdot)^p$ is a concave function of a positive variable $(\forall \gamma \geq 0)$. A classical approach to address this problem is to perform what is called Majorization-minimization [21]. This can be done because the $\ell_p - \ell_1$ regularization term is concave in the positive orthant as illustrated in the left part of figure 2. It is clear from this Figure that the surface can be majored by an hyperplane. For a given group of variable, one can use the concavity of g to majorize it around a given vector $\hat{\mathbf{w}} > 0$

$$g(\mathbf{w}) \leq g(\|\hat{\mathbf{w}}\|_1) + \nabla g(\|\hat{\mathbf{w}}\|_1)^\top (\mathbf{w} - \hat{\mathbf{w}}) \tag{14}$$

with $\nabla g(\|\hat{\mathbf{w}}\|_1) = p(\|\hat{\mathbf{w}}\|_1)^{p-1}$ for $\hat{\mathbf{w}} > 0$. An illustration of the majoration of $g(\cdot)$ can be seen in the right part of Figure 2. For each group, the regularization term can be majorized by a linear approximation. In other words, for a fixed $\hat{\gamma}$

$$\Omega(\gamma) \leq \tilde{\Omega}(\gamma) = \langle \gamma, \mathbf{G} \rangle_F + cst \tag{15}$$

where the matrix $\mathbf{G}$ has components

$$\mathbf{G}(\mathcal{I}_c, j) = p(\|\hat{\gamma}(\mathcal{I}_c, j)\| + \epsilon)^{p-1}, \quad \forall c, j \tag{16}$$

Note that we added a small $\epsilon > 0$ that helps avoiding numerical instabilities, as discussed in [20]. Finally, solving problem (11) can be performed by iterating the two steps illustrated in Algorithm 1. This iterative algorithm is of particular interest in our case as it consists in iteratively using an efficient Sinkhorn-Knopp matrix scaling approach. Moreover this kind of MM algorithm is known to converge in a small number of iterations.

Algorithm 1. Majoration Minimization for $\ell_p - \ell_1$ regularized Optimal Transport

---

Initialize $\mathbf{G} = \mathbf{0}$
Initialize $\mathbf{C}_0$ as in Equation (8)
repeat
    $\mathbf{C} \leftarrow \mathbf{C}_0 + \mathbf{G}$
    $\gamma \leftarrow$ Solve problem (9) with $\mathbf{C}$
    $\mathbf{G} \leftarrow$ Update $\mathbf{G}$ with Equation (16)
until Convergence

4.3 Interpolation of the Source Domain

Once the transport γ_0 has been defined using either Equations (8), (9) or (11), the source samples must be transported in the target domain using their transportation plan. One can seek the interpolation of the two distributions by following the geodesics of the Wasserstein metric [5] (parameterized by t). This allows to define a new distribution μ_t such that:

$$\mu_t = \arg\min_{\mu}(1 - t)W_2(\mu_s, \mu)^2 + tW_2(\mu_t, \mu)^2. \tag{17}$$

One can show that this distribution is:

$$\mu_t = \sum_{i,j} \gamma_0(i,j)\delta_{(1-t)\mathbf{x}_i^s + t\mathbf{x}_j^t}. \tag{18}$$

In our approach, we suggest to compute directly the image of the source samples as the result of this transport, *i.e.* for $t = 1$. Those images can be expressed through γ_0 as barycenters of the target samples. Let $\mathbf{T}_{\gamma_0} : \mathbb{R}^d \to \mathbb{R}^d$ be the mapping induced by the optimal transport coupling. This map transforms the source elements $\mathbf{X}_s$ in a target domain dependent version $\hat{\mathbf{X}}_s$. The mapping $\mathbf{T}_{\gamma_0}$ can be conveniently expressed as:

$$\hat{\mathbf{X}}_s = \mathbf{T}_{\gamma_0}(\mathbf{X}_s) = \mathrm{diag}((\gamma_0 \mathbf{1}_{n_t})^{-1})\gamma_0 \mathbf{X}_s. \tag{19}$$

We note that $\mathbf{T}_{\gamma_0}$ is fully invertible and can be also used to compute an adaptation from the target domain to the source domain by observing that $\mathbf{T}_{\gamma_0}^{-1} = \mathbf{T}_{\gamma_0^T}$. Let us finally remark that similar interpolation methods were used in the domain of color transfer [18].

5 Experimental Validation

In this Section, we validate the proposed algorithm in two domain adaptation examples. On the first one, we study the behavior of our approach on a simple toy dataset. The second one considers a challenging computer vision dataset, used for a comparison with state-of-the-art methods. In every experiment, the original optimal transport (**OT-ori**) is computed with a network simplex approach [17]. The Sinkhorn transport, which corresponds to the regularized version of the optimal transport (**OT-reg**) described in Section 3.3, was implemented following the algorithm proposed in [6]. Our approach, **OT-reglab**, follows the Algorithm 1. As expected, these last two methods are generally one order of magnitude faster than the network simplex approach.

As for the choice of the weights of Eq. (6), the problem can be cast as an estimation of a probability mass function of a discrete variable on the sample space of the source and target distributions. A direct and reasonable choice is to take an uniform distribution, *i.e.* $p_i^s = \frac{1}{n_s}$ and $p_t^s = \frac{1}{n_t}$. This choice gives the same value for every samples in the two discrete distributions. Alternatively, one can seek to strengthen the weights of samples that are in a high density region, and lower weights for samples in low density regions. This way, outliers should be associated with lower masses. A possible solution relies on a discrete variant of the Nadaraya-Watson estimator [22] where one enforces the sum-to-1 property:

$$p_i^s = \frac{\sum_{j=1}^{n_s} k_\sigma(\mathbf{x}_i^s, \mathbf{x}_j^s)}{\sum_{j=1}^{n_s} \sum_{i=1}^{n_s} k_\sigma(\mathbf{x}_i^s, \mathbf{x}_j^s)} \tag{20}$$

where $k_\sigma(\cdot, \cdot)$ is a gaussian kernel of bandwidth σ. The drawback of such an estimator is that it adds an hyper parameter to the method. Yet, while standard approaches [22] can be used to estimate this parameter, we observed in our experiments, and for large number of samples, that this parameter exerts little influence over the final result (less than a standard deviation) for a large range of values.

5.1   Toy Dataset

In this first experiment, the behavior of the optimal transport is examined on a simple two-dimensional dataset. We consider a two-class distribution by sampling independently for each class $c1$ and $c2$ following the normal distributions $\mathcal{N}_1^s$ and $\mathcal{N}_2^s$. The set of all those samples constitute the source domain. The target domain samples are then obtained by sampling the mixture $\mathcal{N}_1^t + \mathcal{N}_2^t$. The target distributions $\mathcal{N}_i^t$, $(i = 1, 2)$ are deduced from $\mathcal{N}_i^s$, $(i = 1, 2)$ by changing both the scale and translating the distribution mean. The produced domain transformation is thus non-linear and cannot be expressed by a simple $2D$ transformation of the input space. This makes the problem particularly interesting with respect to our initial assumptions on the nature of the domain change. We then sample randomly from these distributions n_1^s, n_2^s, n_1^t and n_2^t samples from $\mathcal{N}_1^s, \mathcal{N}_2^s, \mathcal{N}_1^t$ and $\mathcal{N}_2^t$ to form the corresponding learning and test sets. An illustration of this toy dataset is given in Figure 3.a for $n_1^s + n_2^s = 100$ samples in the source distribution (red and white circles) and $n_1^t + n_2^t = 200$ samples in the target one (blue crosses). Note that the size of the points in the Figure is proportional to its weight p_i and reflects the density of the distribution.

Figure 3.b presents the result of the optimal transport **OT-ori** coupling as a set of non-nul connections (red and black arcs) between the source and the

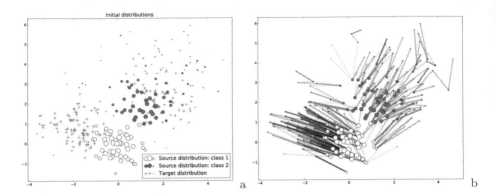

Fig. 3. Illustration of the transport **OT-ori** on a simple toy dataset. The initial distributions are depicted in the right image (a) The source distribution is depicted in white and red for respectively class 1 and 2, the target distributions are in blue. In image (b), we show the optimal transport couplings, depicted as links colored with respect to the source class label.

target distributions. The color of those connections is related to the magnitude of the coupling (up to a global scaling factor). As expected, the coupling matrix γ_0 contains less than $100 + 200 - 1 = 299$ non-nul entries. One can see that some white and red elements are clearly misled by the transport, but the overall adaptation remains coherent with the test distribution.

Figure 4 illustrates the results obtained on this dataset with the regularized versions of the transport **OT-reg** and **OT-reglab** for a regularization parameter value of $\lambda = 1$. The γ_0 matrix of **OT-reg**, on the left of the first row of Figure 4, is indeed sparse, but much less than the corresponding one in **OT-ori**. This can be assessed by comparing the denser connections issued from **OT-reg** (left panel of the second row of Figure 4) with respect to those observed for **OT-ori** (right panel of Figure 3). In the proposed **OT-reglab** (right column of Figure 4), the sparsity is clearly enforced per class (the rows of the coupling matrix are sorted by class), which yields a sparser coupling matrix with block structure. In the last row of Figure 4 we show the result of the adaptation of the source distribution following the procedure described in Section 4.3. Two additional interesting behaviors are observed in the regions highlighted by red squares, where some of the incoherencies observed in **OT-ori** and **OT-reg** of the transport are resolved by the label regularization proposed with **OT-reglab**.

Classification Measures. We now consider performances of a classifier trained on the source samples adapted to the target distribution. In those experiments, we use a SVM classifier with a Gaussian kernel. The hyperparameters of the classifier are computed for each trial by a 2-fold cross validation over a grid of potential values. For every setting considered, the data generation / adaptation / classification was conducted 20 times to leverage the importance of the sampling. When informative, we provide the standard deviation of the result.

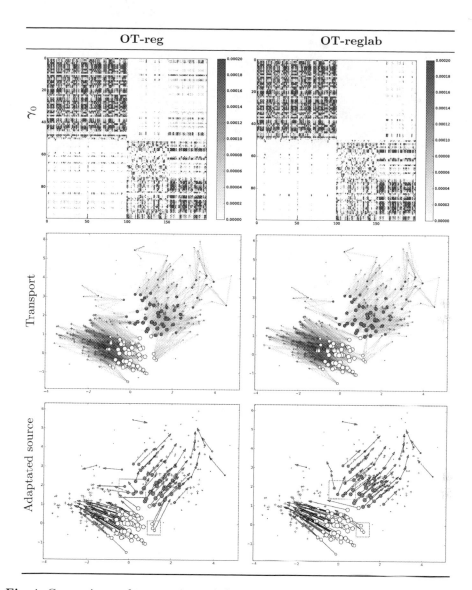

Fig. 4. Comparisons of two versions of the regularized transport: Sinkhorn transport (**OT-reg**, left column) and Sinkhorn transport with the label regularization (**OT-reglab**, right column). The first row shows the transport coupling matrices γ_0, the second row their equivalent graphical representations, with connections colored by the source node label. The third row is the adaptation of the source samples induced by γ_0 using Equation (19).

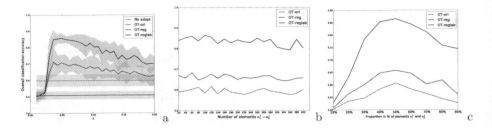

Fig. 5. Classification results for the toy dataset example: (a) influence of the regularization parameter λ; (b) influence of the proportions of samples between the source and the target distributions; (c) influence of the balance of classes on the overall performance of the adaptation.

In the first experiment, we examine the importance of the regularization parameter λ over the overall classification accuracy (Figure 5.a). In this case, we set $n_1^s = n_2^s = n_1^t = n_2^t = 100$. We confirm that the use of the transport for domain adaptation increases the performances significantly (by 8%) over a classification conducted directly with the source distribution as learning set. When varying the λ regularization parameter and using **OT-reglab**, another very significant increase is achieved (up to 25% for $\lambda = 0.04$), which demonstrates the relevance of our transport regularization. In the second experiment, we set $n_1^s = n_2^s = 100$ and we increase the number of elements in the source target n_1^t and n_2^t equivalently. For this experiment and for the next one, λ is set by a standard cross-validation method. In this case, the standard deviation is omitted as it is constant over the experiments and no informative. One can observe that the performances of the classification are *i)* consistent with the first experiment and *ii)* constant over the volume of samples in the target domain as long as the proportions are conserved. In the third experiment, we set n_1^t and n_2^t to the value of 100 samples each and we vary the proportion of the classes through a parameter $p \in [0,1]$ with $n_1^s = p * 100$ and $n_2^s = (1-p) * 100$. This parameter allows to control the proportion of elements in class 1 and in class 2 in the source distribution. As shown in Figure 5.c, the best result is achieved when the proportion of each class samples is similar in the source and target distributions (at 50%). This somehow highlights one limit of the method: the mass equivalent to each class should match in proportions for both distributions to get the best adaption result. Nevertheless, we can see from Figure 5.c that a variation of $\pm 15\%$ between the source and target distribution still leads to significant performance improvements.

5.2   Visual Adaptation Dataset

We now evaluate our method on a challenging real world dataset coming from the computer vision community. The objective is now a visual recognition task of several categories of objects, studied in the following papers [23,13,14,15]. The dataset contains images coming from four different domains: *Amazon* (online

Table 1. Overall recognition accuracies in % and standard deviation on the domain adaptation of visual features

	Methods							
	without labels						with	
	no adapt.	SuK [3]	SGF [13]	GFK [14]	OT-ori	OT-reg	GFK-lab [14]	OT-reglab
C→A	20.8 ± 0.4	32.1 ± 1.7	36.8 ± 0.5	36.9 ± 0.4	30.6 ± 1.6	41.2 ± 2.9	40.4 ± 0.7	**43.5 ± 2.1**
C→D	22.0 ± 0.6	31.8 ± 2.7	32.6 ± 0.7	35.2 ± 1.0	27.7 ± 3.7	36.0 ± 4.1	**41.1 ± 1.3**	41.8 ± 2.8
A→C	22.6 ± 0.3	29.5 ± 1.9	35.3 ± 0.5	35.6 ± 0.4	30.1 ± 1.2	32.6 ± 1.3	**37.9 ± 0.4**	35.2 ± 0.8
A→W	23.5 ± 0.6	26.7 ± 1.9	31.0 ± 0.7	34.4 ± 0.9	28.0 ± 2.0	34.7 ± 6.3	35.7 ± 0.9	**38.4 ± 5.4**
W→C	16.1 ± 0.4	24.2 ± 0.9	21.7 ± 0.4	27.2 ± 0.5	26.7 ± 2.3	32.8 ± 1.2	29.3 ± 0.4	**35.5 ± 0.9**
W→A	20.7 ± 0.6	26.7 ± 1.1	27.5 ± 0.5	31.1 ± 0.7	29.0 ± 1.2	38.7 ± 0.7	35.5 ± 0.7	**40.0 ± 1.0**
D→A	27.7 ± 0.4	28.8 ± 1.5	32.0 ± 0.4	32.5 ± 0.5	29.2 ± 0.8	32.5 ± 0.9	**36.1 ± 0.4**	34.9 ± 1.3
D→W	53.1 ± 0.6	71.5 ± 2.1	66.0 ± 0.5	74.9 ± 0.6	69.8 ± 2.0	81.5 ± 1.0	79.1 ± 0.7	**84.2 ± 1.0**
mean	25.8	33.9	35.4	38.5	33.9	41.3	41.9	**44.2**

merchant), the *Caltech-256* image collection [24], *Webcam* (images taken from a webcam) and *DSLR* (images taken from a high resolution digital SLR camera). Those domains are respectively noted in the remainder as A, C, W and D. A feature extraction method is used to preprocess those images; it namely consists in computing SURF descriptors [23], which allows to transform each image into a 800 bins histogram, which are then subsequently normalized and reduced to standard scores. We followed the experimental protocol exposed in [14]: each dataset is considered in turn as the source domain and used to predict the others. Within those datasets, 10 classes of interest are extracted. The source domain are formed by picking 20 elements per class for domains A,C and W, and 8 for D. The training set is then formed by adapting these samples to the target domain. The latter is composed of all the elements in the test domain. The classification is conducted using a 1-Nearest Neighbor classifier, which avoids cross-validation of hyper-parameters. As for the toy example above, we repeat each experiment 20 times and report the overall classification accuracy and the associated standard deviation. We compare the results of the three transport models (**OT-ori, OT-reg** and **OT-reglab**) against both a classification conducted without adaptation (**no adapt.**) and 3 state-of-the-art methods: 1) the surrogate kernel approach (**SuK**), which in [3] was shown to outperform both the Transfer Component Analysis method [10] and the reweighing scheme of [2]; 2) the (**SGF**) method proposed in [13] and 3) the Geodesic Flow Kernel (**GFK**) approach proposed in [14]. Note that this last method can also efficiently incorporate label information: therefore we make a distinctions between methods, which do not incorporate label information (**no adapt, SuK, SGF, GFK, OT-ori** and **OT-reg**) and those that do (**GFK-lab** and **OT-reglab**). For each setting we used the recommended parameters to tune the competing methods. Results are reported in Table. 1.

When no label information is used, (**OT-reg**) usually performs best. In some cases (notably when considering the adaptation from (W→A or D→W), it can even surpass the (**GFK-lab**) method, which uses labels information.**OT-ori** usually enhances the result obtained without adaptation, but remains less efficient than the competing methods (except in the case of W→A where it surpasses **SGF** and **SuK**. Among all the methods, **OT-reglab** usually performs best, and with a significant increase in the classification performances for some cases (W→C or D→W). Yet, our method does not reach state-of-the-art performance in two cases: A→C and D→A. Finally, the overall mean value (last line of the table) shows a consistent increase of the performances with the proposed **OT-reglab**, which outperforms in average **GFK-lab** by 2%. Also note that the regularized unsupervised version **OT-reg** outperforms all the competing methods by at least 3%.

6   Conclusion and Discussion

We have presented in this paper a new method for unsupervised domain adaptation based on the optimal transport of discrete distributions from a source to a target domain. While the classical optimal transport provide satisfying results, it fails in some cases to provide state-of-the-art performances in the tested classification approaches. We proposed to regularize the transport by relaxing some sparsity constraints in the probabilistic coupling of the source and target distributions, and to incorporate the label information by penalizing couplings that mix samples issued from different classes. This was made possible by a Majoration Minimization strategy that exploits a $\ell_p - \ell_1$ norm, which promotes sparsity among the different classes. The corresponding algorithm is fast, and allows to work efficiently with sets of several thousand samples. With this regularization, competitive results were achieved on challenging domain adaptation datasets thanks to the ability of our approach to express both class relationship and non-linear transformations of the domains.

Possible improvements of our work are numerous, and include: *i)* extension to a multi-domain setting, by finding simultaneously the best minimal transport among several domains, *ii)* extension to semi-supervised problems, where several unlabeled samples in the source domain, or labelled samples in the target domain are also available. In this last case, the group sparsity constraint should not only operate over the columns but also the lines of the coupling matrix, which makes the underlying optimization problem challenging. *iii)* Definition of the transport in a RKHS, in order to exploit the manifold structure of the data.

Acknowledgements. We thank the anonymous reviewers for their critics and suggestions. This work has been partially funded by a visiting professor grant from EPFL, by the French ANR under reference ANR-13-JS02-0005-01 (Asterix project). and by the Swiss National Science Foundation (grant 136827, http://p3.snf.ch/project-136827).

References

1. Pan, S.J., Yang, Q.: A survey on transfer learning. IEEE Trans. Knowl. Data Eng. 22(10), 1345–1359 (2010)

2. Sugiyama, M., Nakajima, S., Kashima, H., Buenau, P.V., Kawanabe, M.: Direct importance estimation with model selection and its application to covariate shift adaptation. In: NIPS (2008)
3. Zhang, K., Zheng, V.W., Wang, Q., Kwok, J.T., Yang, Q., Marsic, I.: Covariate shift in Hilbert space: A solution via surrogate kernels. In: ICML (2013)
4. Kantorovich, L.: On the translocation of masses. C.R (Doklady) Acad. Sci. URSS (N.S.) 37, 199–201 (1942)
5. Villani, C.: Optimal transport: old and new. Grundlehren der mathematischen Wissenschaften. Springer (2009)
6. Cuturi, M.: Sinkhorn distances: Lightspeed computation of optimal transportation. In: NIPS, pp. 2292–2300 (2013)
7. Solomon, J., Rustamov, R., Leonidas, G., Butscher, A.: Wasserstein propagation for semi-supervised learning. In: Proceedings of The 31st International Conference on Machine Learning, pp. 306–314 (2014)
8. Bickel, S., Brückner, M., Scheffer, T.: Discriminative learning for differing training and test distributions. In: ICML (2007)
9. Daumé III., H.: Frustratingly easy domain adaptation. In: Ann. Meeting of the Assoc. Computational Linguistics (2007)
10. Pan, S.J., Yang, Q.: Domain adaptation via transfer component analysis. IEEE Trans. Neural Networks 22, 199–210 (2011)
11. Kumar, A., Daumé III, H., Jacobs, D.: Generalized multiview analysis: A discriminative latent space. In: CVPR (2012)
12. Wang, C., Mahadevan, S.: Manifold alignment without correspondence. In: International Joint Conference on Artificial Intelligence, Pasadena, CA (2009)
13. Gopalan, R., Li, R., Chellappa, R.: Domain adaptation for object recognition: An unsupervised approach. In: ICCV, pp. 999–1006. IEEE (2011)
14. Gong, B., Shi, Y., Sha, F., Grauman, K.: Geodesic flow kernel for unsupervised domain adaptation. In: CVPR, pp. 2066–2073. IEEE (2012)
15. Zheng, J., Liu, M.-Y., Chellappa, R., Phillips, P.J.: A grassmann manifold-based domain adaptation approach. In: ICPR, pp. 2095–2099 (November 2012)
16. Rubner, Y., Tomasi, C., Guibas, L.J.: A metric for distributions with applications to image databases. In: ICCV, pp. 59–66 (January 1998)
17. Bonneel, N., van de Panne, M., Paris, S., Heidrich, W.: Displacement interpolation using lagrangian mass transport. ACM Transaction on Graphics 30(6), 158:1–158:12 (2011)
18. Ferradans, S., Papadakis, N., Rabin, J., Peyré, G., Aujol, J.-F.: Regularized discrete optimal transport. In: Scale Space and Variational Methods in Computer Vision, SSVM, pp. 428–439 (2013)
19. Knight, P.: The sinkhorn-knopp algorithm: Convergence and applications. SIAM J. Matrix Anal. Appl. 30(1), 261–275 (2008)
20. Candes, E.J., Wakin, M.B., Boyd, S.P.: Enhancing sparsity by reweighted l1 minimization. Journal of Fourier Analysis and Applications 14(5), 877–905 (2008)
21. Hunter, D.R., Lange, K.: A Tutorial on MM Algorithms. The American Statistician 58(1), 30–38 (2004)
22. Tsybakov, A.: Introduction to Nonparametric Estimation. Springer Publishing Company, Incorporated (2008)
23. Saenko, K., Kulis, B., Fritz, M., Darrell, T.: Adapting visual category models to new domains. In: Daniilidis, K., Maragos, P., Paragios, N. (eds.) ECCV 2010, Part IV. LNCS, vol. 6314, pp. 213–226. Springer, Heidelberg (2010)
24. Griffin, G., Holub, A., Perona, P.: Caltech-256 Object Category Dataset. Technical Report CNS-TR-2007-001, California Institute of Technology (2007)

Discriminative Subnetworks with Regularized Spectral Learning for Global-State Network Data

Xuan Hong Dang, Ambuj K. Singh,
Petko Bogdanov, Hongyuan You, and Bayyuan Hsu

Department of Computer Science, University of California Santa Barbara, USA
{xdang,ambuj,petko,hyou,soulhsu}@cs.ucsb.edu

Abstract. Data mining practitioners are facing challenges from data with network structure. In this paper, we address a specific class of *global-state* networks which comprises of a set of network instances sharing a similar structure yet having different values at local nodes. Each instance is associated with a global state which indicates the occurrence of an event. The objective is to uncover a small set of discriminative subnetworks that can optimally classify global network values. Unlike most existing studies which explore an exponential subnetwork space, we address this difficult problem by adopting a space transformation approach. Specifically, we present an algorithm that optimizes a constrained dual-objective function to learn a low-dimensional subspace that is capable of discriminating networks labelled by different global states, while reconciling with common network topology sharing across instances. Our algorithm takes an appealing approach from spectral graph learning and we show that the globally optimum solution can be achieved via matrix eigen-decomposition.

1 Introduction

With the increasing advances in hardware and software technologies for data collection and management, practitioners in data mining are now confronted with more challenges from the collected datasets: the data are no longer as simple as objects with flattened representation but now embedded with relationships among variables describing the objects. This sort of data is often referred to as *network* or *graph* data. In the literature, there are a large number of techniques developed to mine useful patterns from network databases, ranging from frequent (sub)networks mining [15], network classification/clustering [1,18] to anomaly detection [2]. Often, even for the same data mining task, we may need different algorithms to be developed depending on whether the networks are *directed* or *indirected*, or whether the data resides at nodes, edges or both of them [15].

In this work, the focus is on a specific class of interesting networks in which we have a series of network instances that share a common structure but may have different dynamic values at local nodes and/or edges. In addition, each network

T. Calders et al. (Eds.): ECML PKDD 2014, Part I, LNCS 8724, pp. 290–306, 2014.
© Springer-Verlag Berlin Heidelberg 2014

instance is associated with a global state indicating the occurrence of an event. Such a class of *global-state* network data can be used to model a number of real-world applications ranging from opinion evolution in social networks [21], regulatory networks in biology [22] to brain networks in neuroscience [10]. For example, we possess the same set of genes (nodes) embedded in regulatory networks. Yet, research in systems biology shows that the gene expression levels (node values) may vary across individuals and for some specific genes, their over-expressions may impact those in the neighbors through the regulatory network. These local effects may jointly encode a logical function that determines the occurrence of a disease [22,26]. In analyzing these types of network data, a natural question to be asked is how one can learn a function that can determine the global-state values of the networks based on the dynamic values captured at their local nodes along with the network topology? More specifically, is it possible to identify a small succinct set of influential discriminative subnetworks whose local-node values have the maximum impact on the global states and thus uncover the complex relationships between local entities and the global-state network properties? In searching for an answer, obviously, a naive approach would be to enumerate all possible subnetworks and seek those who have the most discriminative potential. Nonetheless, as the number of subnetworks is *exponentially* proportional to the numbers of nodes and edges, this approach generally is analytically intractable and might not be feasible for large scale networks. A more practical approach is to perform heuristic sampling from the space of subnetworks. Though greatly reducing the number of subnetworks to be visited, the sampling approaches might still suffer from suboptimal solutions and might further lose explanation capability due to the large number of generating subnetworks.

In this paper, we propose a novel algorithm for mining a set of concise subnetworks whose local-state node values discriminate networks of different global-state values. Unlike the existing techniques that directly search through the exponential space of subnetworks, our proposed method is fundamentally different by investigating the discriminative subnetworks in a low dimensional transformed subspace. Toward this goal, we construct on top of the network database three meta-graphs to learn the network neighboring relationships. The first meta-graph is built to capture the network topology sharing across network instances which serves as the network constraint in our subspace learning function, whereas the two subsequent ones are build to essentially capture the relationships between neighboring networks, especially those located close to the potential discriminative boundary. By this setting, our algorithm aims to discover a unique low dimensional subspace to which: i) networks sharing similar global state values are mapped close to each other while those having different global values are mapped far apart; ii) the common network topology is smoothly preserved through constraints on the learning process. In this way, our algorithm helps to attack two challenging issues at the same time. It first avoids searching through the original space of exponential number of subnetworks by learning a single subspace via the optimization of a single dual-objective function. Second, our network topology constraint not only matches properly with our subspace learning function, its quadratic form naturally imposes the L_2-norm shrinkage over the connecting nodes, resulting in an effective selection of relevant and dominated nodes for the subnetworks embedded in the induced subspace. Additionally, the principal technical contributions of our work is the formulation of our learning objective function that is mathematically founded on spectral learning and its advantages therefore not only ensure the stability but also the global optimum of the uncovered solutions.

In summary, we claim the following contributions: (i) *Novelty:* We formulate the problem of mining discriminative subnetworks by transformed subspace learning—an approach that is fundamentally different from most existing techniques that address the problem in the original high-dimensional network space. (ii) *Flexibility:* We propose a novel dual-objective function along with constraints to ensure learning of a single subspace in which different global state networks are well discriminated while smoothly retaining their common topology. (iii) *Optimality:* We develop a mathematically sound solution to solve the constrained optimization problem and show that the optimal solution can be achieved via matrix eigen-decomposition. (iv) *Practical relevance:* We evaluate the performance of the proposed technique on both synthetic and real world datasets and demonstrate its appealing performance against related techniques in the literature.

2   Preliminaries and Problem Setting

In this section, we first introduce some preliminaries related to network data with global state values and then give the definition of our problem on mining discriminative subgraphs to distinguish global state networks.

Definition 1. (Network data instance) *Given $V_i = \{v_1, v_2, \ldots, v_{n_i}\}$ as a set of nodes and $E_i \subseteq V_i \times V_i$ as a set of edges, each connecting two nodes (v_p, v_q) if they are known to relate or influence each other, we define a network instance (or snapshot) N_i as a quadruple $N_i = (V_i, E_i, L_i, S_i)$ in which L_i is a function operating on the local states of nodes $L_i : V_i \to \mathbb{R}$ and S_i encodes the global network state of N_i.*

We consider N_i as an *indirected* network and values at its local nodes are numerical (both continuous and binary) while its global state is a discrete value. Since each N_i is associated with S_i as its state property, N_i is often referred to as a *global-state* network. For example, in the gene expression data, each N_i corresponds to a subject and a local state indicates the gene expression level at node $v_p \in V_i$ whereas the global state encodes the presence or absence of the disease, i.e., $S_i \in \{presence, absence\}$. Likewise in a dynamic social network, a value at each node v_p may encode the political standpoint of an individual whereas the global state indicates the overall political viewpoint of the entire community at some specific time (snapshot). Both local and global states may change across different network snapshots. Note that, for network instances/snapshots with different structures, we may use the null value to denote the state of a missing node and consequently, an edge in a network instance is valid only if it connects two non-null nodes.

Now, let us consider a database consisting m network instances $\mathbb{N} = \{N_1, N_2, \ldots, N_m\}$, we further define the following network over these network instances:

Definition 2. (Generalized network - first meta-graph)
We define the generalized network N as a triple $N = (V, E, K)$ where $V = V_1 \cup V_2 \ldots \cup V_m$ and if $\exists (v_p, v_q) \in E_i$, such an edge also exists in E. For a valid edge $E(p, q) \in E$, we associate a weight $K(p, q)$ as the fraction of network instances having edge $E(p, q)$ in their topology structure,i.e., $K(p, q) = m^{-1} \times \sum_i E_i(p, q)$ with $E_i(p, q) = 1$ if there exists an edge between v_p, v_q in network N_i. As such, $K(p, q)$ is naturally normalized between $(0, 1]$. The value of 1 means the corresponding edge exists in all N_i's while a value close to 0 shows that the edge only exists in a small fraction of network data.

It should be noted here that while we have no edge values at individual networks N_i's, we have non-zero value associated with each existing edge $E(p, q)$ in the generalized network N. Indeed, the corresponding $K(p, q)$ reflects how frequently there is an edge between v_p and v_q or equivalently, how strongly is the mutual influence between two entities v_p and v_q across all networks. As N is defined based on all network instances, we also view N as our first meta-graph with V being its vertices and K capturing its graph topology generalized from the network topology of all network instances. We are now ready to define our problem as follows.

Definition 3. (Mining Discriminative Subnetworks Problem)
Given a database of network data instances/snapshots $\mathbb{N} = \{N_1, N_2, \ldots, N_m\}$, we aim to learn an optimal and succinct set of subnetworks with respect to the topology structure generalized in the first meta-graph that well discriminate network instances with different global state values.

3 Our approach

3.1 Meta-Graphs over Network Instances

As mentioned in the above sections, searching for optimal subnetworks in the fully high dimensional original network space is always challenging and potentially intractable. We adopt an indirect yet more viable approach by transforming the original space into a low dimensional space of which networks with different global-states are well distinguished while concurrently retaining the generalized network topology captured by our first meta-graph. Toward this goal, we develop two neighboring *meta-graphs* based on both the local state values and global state values.

We denote these two meta-graphs respectively by G^+ and G^-. Their vertices correspond to the network instances while a link connecting two vertices represents the neighboring relationship between two corresponding network instances.

For the meta-graph G^+, we denote $\mathbf{A}^+$ as its affinity matrix that captures the similarity of neighboring networks having the same global state values. Likewise, we denote $\mathbf{A}^-$ as the affinity matrix for meta-graph G^- that captures the similarity of neighboring networks yet having different global network states. As such, $\mathbf{A}^+$ and $\mathbf{A}^-$ respectively encode the weights on the vertex-links of two corresponding graphs G^+ and G^-. In computing values for these affinity matrices, with each given network instance N_i, we find its k nearest neighboring networks based on the local state values and divide them into two sets, those sharing similar global state values and those having different global states. More specifically, let $k\mathrm{NN}(N_i)$ be the neighboring set of N_i, then elements of $\mathbf{A}^+$ and $\mathbf{A}^-$ are computed as: $\mathbf{A}_{ij}^+ = \frac{\mathbf{v}_i \cdot \mathbf{v}_j}{\|\mathbf{v}_i\|\|\mathbf{v}_j\|}$ if $S_i = S_j$ and $N_j \in k\mathrm{NN}(N_i)$ or $N_i \in k\mathrm{NN}(N_j)$, otherwise we set $\mathbf{A}_{ij}^+ = 0$. And $\mathbf{A}_{ij}^- = \frac{\mathbf{v}_i \cdot \mathbf{v}_j}{\|\mathbf{v}_i\|\|\mathbf{v}_j\|}$ if $S_i \neq S_j$ and $N_j \in k\mathrm{NN}(N_i)$ or $N_i \in k\mathrm{NN}(N_j)$, otherwise $\mathbf{A}_{ij}^- = 0$. In these equations, we have denoted the boldface letters $\mathbf{v}_i$ and $\mathbf{v}_j$ as the vectors encoding the dynamic local states of N_i's and N_j's nodes, and have used the cosine distance to define the similarity between two network instances. It is worth mentioning that, though existing other measures for network data [28], our using of cosine distance is motivated by the observation that we can view each node as a single feature and thus the network data can be essentially considered as a special case of very high dimensional data. As such, the symmetric cosine measure can be effectively used though obviously the other ones [28] can also be directly applied here.

It is also important to give the intuition behind our above computation. First, notice that both $\mathbf{A}^+$ and $\mathbf{A}^-$ are the affinity matrices having the same size of $m \times m$ since we calculate for every network instance. Second, while $\mathbf{A}^+$ captures the similarity of network instances sharing the same global states and neighboring to each other, $\mathbf{A}^-$ encodes the similarity of different global state networks yet also neighboring to each other. Such networks are likely to locate close to the discriminative boundary function and thus they play essential roles in our subsequent learning function. Third, both $\mathbf{A}^+$ and $\mathbf{A}^-$ are sparse and symmetric matrices since only k neighbors are involved in computing for each network and if N_j is neighboring to N_i, we also consider the inverse relation, i.e., N_i is neighboring to N_j. Moreover, $\mathbf{A}^-$ is generally sparser compared to $\mathbf{A}^+$ as the immediate observation from the second remark.

3.2   Constrained Dual-Objective Function

Let us recall that $\mathbf{v}_i$ is the vector encoding the node states of the corresponding network N_i and let us denote the transformation function that maps $\mathbf{v}_i$ into our novel target subspace by $f(\mathbf{v}_i)$. We first formulate the two objective functions as follows:

$$\arg\min_f \sum_{i=1}^{m} \sum_{j=1}^{m} (f(\mathbf{v}_i) - f(\mathbf{v}_j))^2 \mathbf{A}_{ij}^+ \tag{1}$$

$$\arg\max_f \sum_{i=1}^{m} \sum_{j=1}^{m} (f(\mathbf{v}_i) - f(\mathbf{v}_j))^2 \mathbf{A}_{ij}^- \tag{2}$$

To gain more insights into these setting objectives, let us take a closer look at the first Eq.(1). If two network instances N_i and N_j have similar local states in the original space (i.e., $\mathbf{A}_{ij}^+$ is large), this first objective function will be penalized if the respective points $f(\mathbf{v}_i)$ and $f(\mathbf{v}_j)$ are mapped far part in the transformed space. As such, minimizing this cost function is equivalent to maximizing the similarity amongst instances having the same global network states in the reduced dimensional subspace. On the other hand, looking at Eq.(2) can tell us that the function will incur a high penalty (proportional to $\mathbf{A}_{ij}^-$) if two networks having different global states are mapped close in the induced subspace. Thus, maximizing this function is equivalent to minimizing the similarity among neighboring networks having different global states in the novel reduced subspace. As mentioned earlier, such networks tend to locate close to the discriminative boundary function and hence, maximizing the second objective function leads to the maximal margin among clusters of different global-state networks.

Having the mapping function $f(.)$ to be optimized above, it is crucial to ask which is an appropriate form for it. Either a linear or non-linear function can be selected as long as it effectively optimizes two objectives concurrently. Nonetheless, keeping in mind that our ultimate goal is to derive a set of succinct discriminative subnetworks along with their *explicit* nodes. Optimizing a non-linear function is generally not only more complex but importantly may lose the capability in explaining how the new features have been derived (since they will be the *non-linear* combinations of the original nodes). We therefore would prefer $f(.)$ as in the form of a linear combination function and following this, $f(.)$ can be represented explicitly as a transformation matrix $U_{n \times d}$ that linearly combines n nodes into d novel features $(d \ll n)$ of the induced subspace. For the sake of discussion, we elaborate here for the projection onto 1-dimensional subspace (i.e., $d = 1$). The solution for the general case $d > 1$ will be straightforward once we obtain the solution for this base case. Given this simplification and with little algebra, we recast our first objective function as follows:

$$\operatorname*{arg\,min}_{\mathbf{u}} \sum_{i=1}^{m} \sum_{j=1}^{m} \|\mathbf{u}^T \mathbf{v}_i \quad \mathbf{u}^T \mathbf{v}_j\|^2 \mathbf{A}_{ij}^+ = \sum_{i=1}^{m} \sum_{j=1}^{m} tr\left(\mathbf{u}^T (\mathbf{v}_i - \mathbf{v}_j)(\mathbf{v}_i - \mathbf{v}_j)^T \mathbf{u}\right) \mathbf{A}_{ij}^+$$

$$= tr\left(\sum_{i=1}^{m} \sum_{j=1}^{m} \left(\mathbf{u}^T (\mathbf{v}_i - \mathbf{v}_j)\mathbf{A}_{ij}^+ (\mathbf{v}_i - \mathbf{v}_j)^T\right) \mathbf{u}\right)$$

$$= 2tr\left(\mathbf{u}^T \mathbf{V}\mathbf{D}^+ \mathbf{V}^T \mathbf{u}\right) - 2tr\left(\mathbf{u}^T \mathbf{V}\mathbf{A}^+ \mathbf{V}^T \mathbf{u}\right) = 2tr\left(\mathbf{u}^T \mathbf{V}\mathbf{L}^+ \mathbf{V}^T \mathbf{u}\right) \quad (3)$$

in which we have used $tr(.)$ to denote the trace of a matrix and $\mathbf{V}$ as the matrix whose column ith accommodates the dynamic local states of network instance N_i (i.e., $\mathbf{v}_i$), forming its size of $n \times m$. Also, $\mathbf{D}$ is the diagonal matrix whose $\mathbf{D}_{ii}^+ = \sum_j \mathbf{A}_{ij}^+$ and we have defined $\mathbf{L}^+ = \mathbf{D}^+ - \mathbf{A}^+$, which can be shown to be the Laplacian matrix [12]. For the second objective function in Eq.(2), we can repeat the same computation which yields to the following form:

$$\arg\max_{\mathbf{u}} \sum_{i=1}^{m} \sum_{j=1}^{m} \|\mathbf{u}^T\mathbf{v}_i - \mathbf{u}^T\mathbf{v}_j\|^2 \mathbf{A}_{ij}^-$$
$$= 2tr\left(\mathbf{u}^T\mathbf{V}\mathbf{D}^-\mathbf{V}^T\mathbf{u}\right) - 2tr\left(\mathbf{u}^T\mathbf{V}\mathbf{A}^-\mathbf{V}^T\mathbf{u}\right)$$
$$= 2tr\left(\mathbf{u}^T\mathbf{V}\mathbf{L}^-\mathbf{V}^T\mathbf{u}\right) \tag{4}$$

where again $\mathbf{D}^-$ is the diagonal matrix with $\mathbf{D}_{ii}^- = \sum_j \mathbf{A}_{ij}^-$ and we have defined $\mathbf{L}^- = \mathbf{D}^- - \mathbf{A}^-$.

Notice that while the above formulations aim at discriminating different global state networks in the low dimensional subspace, it has not yet taken into consideration the generalized network structure captured by our first meta-graph. As described previously, the mutual interactions among nodes are also important in determining the global network states. Also according to Definition 2, the larger the value placing on the link between nodes v_p and v_q, the more likely they are being involved in the same process. Therefore, we would expect our mapping vector $\mathbf{u}$ not only separating well different global state networks but also ensuring its smoothness property w.r.t. the generalized network topology characterized by the first meta-graph N.

Toward the above objective, we formulate the network topology as a constraint in our learning objective function, and in order to be consistent with the approach based on spectral graph analysis, we encode the topology captured in N by an $n \times n$ constraint matrix $\mathbf{C}$ whose elements are defined by:

$$\mathbf{C}_{pq} = \mathbf{C}_{qp} = \begin{cases} \sum_q K(p,q) & \text{if } v_p \equiv v_q \\ -K(p,q) & \text{if } v_p \text{ and } v_q \text{ are connected} \\ 0 & \text{otherwise} \end{cases} \tag{5}$$

It is easy to show that, by this definition, $\mathbf{C}$ is also the Laplacian matrix and its quadratic form, taking $\mathbf{u}$ as the vector, is always non-negative:

$$\mathbf{u}^T\mathbf{C}\mathbf{u} = \sum_{p=1}^{n} u_p^2 \sum_{q=1}^{n} K(p,q) - \sum_{p=1}^{n} \sum_{q=1}^{n} u_p u_q K(p,q)$$
$$= \frac{1}{2} \sum_{p=1}^{n} \sum_{q=1}^{n} K(p,q)(u_p - u_q)^2 \geq 0 \tag{6}$$

in which u_p, u_q are components of vector $\mathbf{u}$. It is possible to observe that if $K(p,q)$ is large, indicating nodes v_p and v_q are strongly interacted in large portion of the network instances, the coefficients of u_p and u_q should be similar (i.e., smooth) in order to minimize this equation. From the network-structure perspective, we would say that if v_p is known as a node affecting the global network state, its selection in the transformed space will increase the possibility of being selected of its nearby connected node v_q if $K(p,q)$ is large, leading to the formation of discriminative subnetworks in the induced subspace. Therefore, in combination with the dual-objective function formulated above, we finally claim our constrained optimization problem as follows (the constants can be omitted due to optimization):

$$\mathbf{u}^* = \arg\max_{\mathbf{u}} \left\{ tr\left(\mathbf{u^T V (L^- - L^+) V^T u} \right) \right\}$$

$$\text{subject to } \mathbf{u}^T \mathbf{C} \mathbf{u} \leq t$$

$$\text{and } \mathbf{u}^T \mathbf{V} \mathbf{D}^+ \mathbf{V}^T \mathbf{u} = 1 \tag{7}$$

The first network topology constraint aims to retain the smoothness property of $\mathbf{u}$ whereas the second constraint aims to remove its freedom, meaning that we need $\mathbf{u}$'s direction rather than its magnitude. The network topology constraint is beneficial in two ways. First as presented above, it offers a convenient and natural way to incorporate the network topology into our space transformation learning process. Second, as being formulated in the vector quadratic form, it essentially imposes the features/nodes selection through the coefficients of $\mathbf{u}$ by shrinking those of irrelevant nodes toward zero while crediting large values to those of relevant nodes. Indeed, this quadratic L_2-norm is a kind of regularization which is often referred to as the ridge shrinkage in statistics for regression [13,7]. The parameter t is used to control the amount of shrinkage. The smaller the value of t, the larger the amount of shrinkage.

3.3 Solving the Function

In order to solve our dual objective function associated with constraints, we resort the Lagrange multipliers method and following this, Eq. (7) can be rephrased as follows:

$$\mathcal{L}(\mathbf{u}, \lambda) = \mathbf{u}^T \left(\mathbf{V} \widetilde{\mathbf{L}} \mathbf{V}^T - \alpha \mathbf{C} \right) \mathbf{u} - \lambda \left(\mathbf{u}^T \mathbf{V} \mathbf{D} \mathbf{V}^T \mathbf{u} - 1 \right) \tag{8}$$

of which, to simplify notations, we have denoted $\widetilde{\mathbf{L}} = \mathbf{L}^- - \mathbf{L}^+$, $\mathbf{D} = \mathbf{D}^+$ and α is used in replacement for t as there is a one-to-one correspondence between them [13]. Taking the derivative of $\mathcal{L}(\mathbf{u}, \lambda)$ with respect to vector $\mathbf{u}$ yields:

$$\frac{\partial \mathcal{L}(\mathbf{u}, \lambda)}{\partial \mathbf{u}} = 2 \left(\mathbf{V} \widetilde{\mathbf{L}} \mathbf{V}^T - \alpha \mathbf{C} \right) \mathbf{u} - 2\lambda \mathbf{V} \mathbf{D} \mathbf{V}^T \mathbf{u} \tag{9}$$

And equating it to zero leads to the generalized eigenvalue problem:

$$\left(\mathbf{V} \widetilde{\mathbf{L}} \mathbf{V}^T - \alpha \mathbf{C} \right) \mathbf{u} = \lambda \mathbf{V} \mathbf{D} \mathbf{V}^T \mathbf{u} \tag{10}$$

It is noticed that $\mathbf{V}$ is a singular matrix and its rank is at most $\min(n, m)$, making the combined matrix on the right hand side not directly invertible. We therefore decompose $\mathbf{V} \mathbf{D}^{1/2}$ into $\mathbf{P} \Sigma \mathbf{Q}^T$, where columns in $\mathbf{P}$ and $\mathbf{Q}$ are respectively called the left and right (orthonormal) singular vector of $\mathbf{V} \mathbf{D}^{1/2}$ while Σ

stores its singular values. Note that this decomposition is always possible since $\mathbf{D}$ is a non-negative diagonal matrix of node degrees. Additionally, both $\mathbf{P}$ and $\mathbf{Q}$ can be represented in the square matrices while $\boldsymbol{\Sigma}$ a rectangular one of $n \times m$ size according to the most general decomposition form in [6]. Following this, the combined matrix on the right hand size can be rewritten as:

$$\mathbf{VDV}^T = \mathbf{P}\boldsymbol{\Sigma}^2\mathbf{P}^T \tag{11}$$

And in order to get a stable solution, we keep the top ranked singular values in $\boldsymbol{\Sigma}$ such as their summation explains for no less than 95% of the total singular values[1]. Let us denote $\mathbf{B}^* = \mathbf{P}\boldsymbol{\Sigma}^{-2}\mathbf{P}^T$ as the inversion of the right hand side and before showing our optimal solution, we need the following proposition:

Proposition 1. *Let $\mathbf{P}$ be the matrix of left singular vectors of $\mathbf{VD}^{1/2}$ defined above, then its row vectors are also orthogonal, i.e., $\mathbf{PP}^T = \mathbf{I}$*

Proof. Let $\mathbf{a}$ be an arbitrary vector, we need to show $\mathbf{PP}^T\mathbf{a} = \mathbf{a}$. Due to the orthogonal property of left singular vectors, it is true that $\mathbf{P}^T\mathbf{P} = \mathbf{I}$. The inversion of $\mathbf{P}$ therefore is equal to $\mathbf{P}^T$ and given arbitrary vector $\mathbf{a}$, there is a uniquely determined vector $\mathbf{b}$ such that $\mathbf{Pb} = \mathbf{a}$. Consequently,

$$\mathbf{PP}^T\mathbf{a} = \mathbf{PP}^T\mathbf{Pb} = \mathbf{Pb} = \mathbf{a}$$

It follows that $\mathbf{PP}^T = \mathbf{I}$ since $\mathbf{a}$ is an arbitrary vector.

Theorem 1. *Given $\mathbf{B} = \mathbf{P}\boldsymbol{\Sigma}^2\mathbf{P}^T$, we have $\mathbf{BB}^* = \mathbf{I}$*

Proof. The proof of this theorem is straightforward given Proposition 1.

Now, for simplicity, let us denote $\mathbf{A}$ for the combined matrix $(\mathbf{V}\widetilde{\mathbf{L}}\mathbf{V}^T - \alpha\mathbf{C})$, then it is straightforward to see that $\mathbf{u}$ turns out to be the eigenvector of the equation:

$$\mathbf{B}^*\mathbf{A} = \lambda\mathbf{u} \tag{12}$$

with the maximum value is given by the following theorem.

Theorem 2. *Given matrix $\mathbf{A} = \mathbf{V}\widetilde{\mathbf{L}}\mathbf{V}^T - \alpha\mathbf{C}$ and $\mathbf{B} = \mathbf{VDV}^T$ defined above, the maximum value of $\mathbf{u}^T\mathbf{Au}$ subjected to $\mathbf{u}^T\mathbf{Bu} = 1$ is the largest eigenvalue of $\mathbf{B}^*\mathbf{A}$.*

Proof. Due to Theorem 1, it is straightforward to see that:

$$\mathbf{u}^T\mathbf{Au} = \mathbf{u}^T\mathbf{BB}^*\mathbf{Au}$$

On the other hand, $\mathbf{u}^T\mathbf{BB}^*\mathbf{Au} = \mathbf{u}^T\mathbf{B}\lambda\mathbf{u}$ by equation Eq. (12) and further taking into account our second constraint, it follows that:

$$\max_{\mathbf{u}:\mathbf{u}^T\mathbf{Bu}=1}\{\mathbf{u}^T\mathbf{Au}\} = \max\{\lambda\}$$

---

[1] Note that since $(\mathbf{VD}^{1/2})(\mathbf{VD}^{1/2})^T$ is Hermitian and positive semidefinite, the diagonal entries in $\boldsymbol{\Sigma}$ are always real and nonnegative.

From this theorem, it is safe to say that $\mathbf{u}^* = \mathbf{u}_1$ as the first eigenvector of $\mathbf{B}^*\mathbf{A}$ corresponding to its largest eigenvalue λ_1 is our optimal solution. Since eigenvectors and eigenvalues go in pair, the second optimal solution is the second eigenvector $\mathbf{u}_2$ corresponding to the second largest eigenvalue λ_2 and so on. Consequently, in the general case, if d is the number of unique global network states, our optimal transformed space is the one spanned by the top d eigenvectors. In the next section, we present a method to select optimal features/nodes along with the subnetworks formed by these nodes.

3.4   Subnetwork Selection

In essence, our top d eigenvectors play the role of space transformation which projects network data from the original high dimensional space into the induced subspace of d dimensions. Their coefficients essentially reflect how the original nodes (features) have been combined or more specifically, the degree of node's importance in contributing to the subspace that optimally discriminates network instances. Following the approach adopted in [8] with c as the user parameter, we select top c entries in each $\{\mathbf{u}_i\}_{i=1}^d$ corresponding to the selective nodes. Nonetheless, it is possible that there will be more than c nodes selected by combining from d eigenvectors. Therefore, in practice, we may use a simple approach by first selecting the largest absolute entries across d eigenvectors:

$$\mathbf{v} = \{v_1, \ldots, v_n\} \text{ where } v_p = \max_i |u_{i,p}| \tag{13}$$

where $u_{i,p}$ is the p-th entry of eigenvector $\mathbf{u}_i$, and then selecting nodes according to the top c ranking entries in $\mathbf{v}$. The subnetworks forming from these nodes can be straightforwardly obtained by matching to the nodes in our generalized network N defined in Definition 2, along with their connecting edges stored in E. These subnetworks can be visualized which offers the user an intuitive way to examine the results.

3.5   Computational Complexity

We name our algorithm SNL, an acronym stands for SubNetwork spectral Learning. Its computation complexity is analyzed as follows. We first need to compute edges' weights according to Definition 2 to build our first meta-graph which takes $O(n^2 m)$ since there are at most $n(n-1)/2$ edges in the generalized network N. Second, in building the two subsequent meta-graphs, the cosine distance between any two network instances is computed which amounts to $O(n^2 m)$ or $O(mn \log n)$ in case the multidimensional binary search tree is used [3]. Also, since the size of matrix $\mathbf{VD}^{1/2}$ is $m \times n$, its singular value decomposition takes $O(mn \log n)$ with the Lanczos technique [12]. Likewise, the eigen-decomposition of the matrix $\mathbf{B}^*\mathbf{A}$ takes $O(n^2 \log n)$ since its size is $n \times n$. Therefore, in combination, the overall complexity is at most $O(n^2 m + n^2 \log n)$ assuming that the number of nodes is larger than the number of network instances.

4   Empirical Studies

4.1   Datasets and Experimental Setup

We compare the performance of SNL against MINDS [26] which is among the first approaches formally addressing the global-state network classification problem by a subnetwork sampling. Another algorithm for comparison is the Network Guided Forests (NGF) [11] designed specifically for protein protein interaction (PPI) networks. We use both synthetic and real world datasets for experimentation. Since global states are available in all datasets, we compare average accuracy in 10-fold cross validation for synthetic data, and 5-fold cross validation for real data (due to smaller numbers of network instances). For SNL, the cross validation is further used to select its optimal α parameter (shortly discussed below). Unless otherwise indicated, we set $k = 10$ and use the linear-SVM to perform training and testing in the transformed space (keeping top 50 nodes) in SNL. We set MINDS' parameters as follows: 10000 sampling iterations, 0.8 discriminative potential threshold and $K = 200$ as recommended in the original paper [26]. The Gini index is used for the tree building in NGF and we set its improvement threshold $\epsilon = 0.02$ [11].

4.2   Results on Synthetic Datasets

We use synthetic data to evaluate the performance of our technique in training robust classifiers and selecting relevant subnetworks. We generate scale-free backbone networks by preferential attachment of a predefined size adding 20 edges for each new node. The probabilities of backbone edges are sampled from a truncated Gaussian distributions: $N(0.9, 0.1)$ for edges among *ground truth nodes* (pre-selected nodes of high-correlation with the network state) and $N(0.7, 0.1)$ for the rest of the edges. The weighted backbone serves as our generalized template to generate network instances by independently sampling the existence of every edge based on its probability. The global states are binary $S_i \in \{0, 1\}$ with balanced distribution. We further add noise to both global and local states of ground truth nodes, respectively with levels of 10% and 30%.

Varying $|V_{gt}|$: In the first set of experiments, we aim to test whether the performance of all algorithms is affected by the number of ground truth nodes. To this end, we generate 5 datasets by fixing $m = 1000$ instances, $n = 3000$ nodes and vary the ground truth nodes $|V_{gt}|$ from 10 to 50. In Figure 1, we report the average accuracy (and standard deviation) of all algorithms in 10-fold cross validation. As one may observe, SNL performs stably regardless of the change in the ground truth sizes. Compared to the other techniques, its classification is always consistently higher across all cases. The MINDS technique also performs well on this experimental setting yet the NGF seems to be sensitive to the small ground truth sizes. For small $|V_{gt}|$, the sampling strategy based on density areas employed in NGF has little chance to select the ground truth nodes, making its accuracy close to a random technique. When more ground truth nodes are

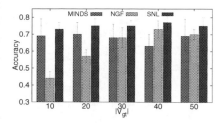

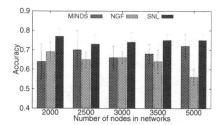

Fig. 1. Accuracy of all algorithms by varying ground truth subnetworks' nodes

Fig. 2. Accuracy of all algorithms by varying network size

introduced, NGF has higher possibility to sample high-utility nodes and in the last two datasets, its performance is on par with that of MINDS. Nonetheless, its accuracy only peaks at 73% in the best case which is lower than 77% in SNL (last column).

Varying Network Size: In the second set of experiments, we evaluate the performance of all algorithms by varying the network sizes. Specifically, we fix $m - 3000$ network instances, $|V_{gt}| = 50$ ground truth nodes and generate 5 datasets having the network size varied from 2000 to 5000 nodes. The classification performance along with the standard deviation is reported in Figure 2. It is possible to see that the performance traits are similar to those in our first set of experiments. SNL's classification accuracy remains high while that of NGF decreases with the increase of network size. This again can be explained by the extension of the searching subnetwork space, leading to the lower likelihood of both NGF and MINDS in identifying relevant subnetworks with potentially discriminative nodes. The slightly better performance of MINDS (compared to NGF) is due to its accuracy thresholding in selecting candidate substructures. The set of MINDS' selected trees are thus qualitatively better. Nonetheless, as compared to SNL, our subspace learning approach show more competitive results. Moreover, since the low-dimensional subspace learnt in SNL is unique and linearly combined from the most discriminative nodes, its performance also shows more stable, indicated by the small standard deviation across all cases.

Effect of Network Topology: In order to provide more insights into the performance of SNL, we further test the network effect. As presented in Section 3, α is the parameter controlling the influence of the network information on the subspace learning process. The higher the α, the more preference putting on the heavily connected nodes. We report in Figures 3(a),4(a) the accuracy of SNL by varying α from 0.1 to 6.5 and in Figures 3(b),4(b) its ability in discovering the ground truth nodes. For the latter case, we validate the performance through the usage of area under the ROC curve (AUC) [13].

As expected, incorporating the network structure in the subspace learning process improves both classification rate and the AUC in uncovering the ground truth nodes. The plots in Figures 3(a),4(a) show that the accuracy initially

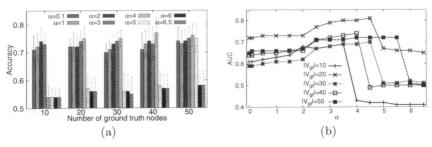

Fig. 3. Network effect on accuracy (a) and AUC performance (b) for different numbers of ground truth nodes

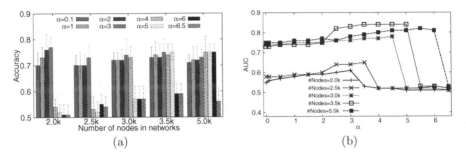

Fig. 4. Network effect on accuracy (a) and AUC (b) for different network sizes

improves for increasing influence of the network ($\alpha \leq 5$) and then decreases as the network component becomes prevalently dominant ($\alpha > 5$). This is because for large α, SNL tends to incorporate irrelevant nodes solely based on their strong connections to the neighbors (yet their local values might not help classifying global state values). Another notable observation is that, in larger instances or ground truth feature sets, the optimal α tends to increase as well. Moreover, the values of α that maximize classification accuracy also result in optimal AUC in identifying the ground truth nodes (Fig. 3(b),4(b)). These experiments clearly show the helpful information provided by the network topology in uncovering the groundtruth features. Also, we exclude NGF and MINDS from these experiments (to save space) and leave the discussion over their AUC performance with the real-world datasets.

4.3   Real-World Datasets

We use 4 real-world datasets to evaluate the performance of SNL and its competing methods. The features in all datasets correspond to micro-array expression measurements of genes; the topology structures relating features correspond to gene interaction networks; and the global network states correspond to phenotypic traits of the subjects/instances. The statistics of our datasets are listed in Table 1. Two of our real-world datasets, breast cancer and embryonic development, were also used for experimentation in the original NGF method [11]. Our other datasets come from a study on maize properties [14] and a human liver metastasis study [19] combined with a functional network [9]. The network

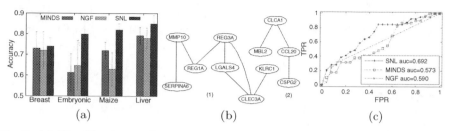

Fig. 5. (a) Classification performance of all algorithms on four real world datasets. (b) Subnetworks identified by the SNL related to the liver metastasis. (c) ROC performance over the liver metastasis (x-axis is false positive rate, y-axis is true positive rate).

Table 1. Real-world dataset statistics and sources

Datasets	Genes	Edges	Instances	Global State
Breast cancer	11203	57235	295	cancer/non-cancer
Embryonic development	1321	5227	35	developmental tissue layer
Maize	8574	298510	344	high/low oil production
Liver metastasis	7383	251916	123	disease/non-disease

samples are used as provided in the original studies, except for maize where we down-sample one of the classes to balance the global state distribution.

Classification Performance: The comparison of classification accuracy for all techniques and datasets is presented in Figure 5(a). We report the average accuracy and standard deviation from the 5-fold stratified cross validation. All techniques perform competitively on the breast cancer data, achieving more than 70% of classification accuracy on average. The accuracy of SNL dominates significantly that of the sampling techniques on the embryonic and maize datasets (at least 15% and 10% improvement respectively) and less so in the liver dataset. The separation is highest in the datasets of small number of instances and big number of feature nodes – the settings in which SNL is particularly effective. Beyond average performance improvement, SNL's accuracy is also more stable across all folds as it considers the global network structure when learning a subspace for classification, while the alternatives perform sampling in the exponential space of substructures.

Subnetwork Discovery: Unlike the synthetic datasets where we can control the ground truth network features, it is generally much harder to obtain ground truth subnetworks for real world datasets. However, as an attempt to look deeper into the results, we choose the Liver metastasis and further investigate the meaningful subnetworks generated by the SNL. For this dataset, out of top 50 nodes of highest coefficient values (ref. Section 3.4), about one third of the nodes are connected into four subnetworks. We depict in Figure 5(b) the two largest ones which respectively contain 7 and 4 connected gene nodes. Among these selected subnetworks, the genes REG1A and REG3A are particularly interesting since they are in agreement with the ones found in [20] which was shown to be involved in the liver metastasis cancer. As a comparison against MINDS and NGF, we notice that both methods generate multiple binary-trees where each node has only a *single* parent. Moreover, while SNL can provide a natural rank of important genes based on their coefficients (from the learnt subspace), it is less trivial to define important genes from NGF and MINDS as they both generate thousands of trees. For the purpose of measuring biological relevance of obtained genes, we define a ranking for these competing techniques based on the frequency of genes appeared in the generated trees. For comparison, we select 46 metastasis-specific genes identified in [20] to serve as a ground truth set (39 intersect with our network and expression data) and plot the ROC performance of all algorithms in Figure 5(c). Note that, this is only a partial ground truth set, since identifying all genes associated with this disease is a subject of ongoing research [20]. It is observed that the ranking produced by SNL includes more ground truth genes than those of NGF and MINDS at increasing false-positive rates. The higher true positive rates of SNL makes it a better method for identifying new genes associated with the phenotype of interest. In practice, this is an important feature of the algorithm since validating even a single gene related to cancer is both time-wise and financially costly. As shown in Figure 5(c), while the ROC performance of NGF and MINDS are only at 0.59 and 0.57 AUC, that value of SNL is 0.69 which clearly demonstrates large gap of better performance.

5   Related Work

Mining discriminative subspaces from global-state networks is a novel and challenging problem. Two lines of work close to this problem are network classification and mining evolving subgraphs from dynamic network data. In the network classification case, most representative algorithms are LEAP [29], graphSig [27], GAIA [17] and COM [16] which generally assume a database consisting of positive and negative networks that need to be classified. These approaches, though diverse in terms of their underlying algorithms, all aim at extracting a set significant subnetworks that are *more frequent* in one class of positive networks and *less frequent* in the negative class. Different from the above problems, we aim to mine subnetworks which are represented in all network instances; yet the node values along with the network structures can discriminate the global states of the networks. Another line of related research focuses on mining dynamic evolving subnetworks [24,4,5]. The problem in this case is to obtain subnetworks over time that evolve significantly (outliers) from other network locations. This setting therefore do not model the problem developed in this paper since the dynamic network snapshots neither contain global-state values nor can remove their temporal property.

Several studies in systems biology have indicated the critical role of the network structure in identifying protein modules related to clinical outcomes, for both regression [23,25,22] and classification [11,26]. In the classification setting which is related to our study, the NGF [11] is an ensemble approach that builds a forest of trees jointly voting for the class of a network instance. Resided at the NGF's core is the CART (classification and Regression tree) technique and in order to build a decision tree within the PPI network, NGF starts with a root node and progressively includes connected nodes as long as the improvement in class separation (measured by Gini index) is no smaller than a given threshold. The study in [26] is the first one to formally introduce the problem of subnetwork mining in global-state networks and further propose the MINDS algorithm to solve it. Similar to NGF, MINDS adopts network-constraint decision trees and is also an ensemble classifier. Nonetheless, it increases the quality of decision trees by developing a novel concept of editing map over the space of potential subnetworks and exploits Monte Carlo Markov Chain sampling over this novel data structure to seek decision trees with maximum classification potential. Unlike the frequency-based and sampling classification discussed above, our approach is fundamentally different as it searches for the most discriminative subnetworks in a single low dimensional subspace through the spectral learning technique, which generally leads to more stable and high-accuracy performance.

6   Conclusion

We proposed a novel algorithm named SNL to address the challenging problem of uncovering the relationship between local state values residing on nodes and the global network events. While most existing studies address this problem by sampling the exponential subnetworks space, we adopt an efficient and effective subspace transformation approach. Specifically, we define three meta-graphs to capture the essential neighboring relationships among network instances and devise a spectral graph theory algorithm to learn an optimal subspace in which networks with different global-states are well separated while the common structure across samples is smoothly respected to enable subnetwork discovery. Through experimental analysis on synthetic data and real-world datasets, we demonstrated its appealing performance in both classification accuracy and the real-world relevance of the discovered discriminative subnetwork features.

Acknowledgements. The research work was supported in part by the NSF (IIS-1219254) and the NIH (R21-GM094649).

References

1. Aggarwal, C.C., Wang, H.: A survey of clustering algorithms for graph data. In: Managing and Mining Graph Data, pp. 275–301 (2010)
2. Akoglu, L., McGlohon, M., Faloutsos, C.: oddball: Spotting anomalies in weighted graphs. In: Zaki, M.J., Yu, J.X., Ravindran, B., Pudi, V. (eds.) PAKDD 2010, Part II. LNCS (LNAI), vol. 6119, pp. 410–421. Springer, Heidelberg (2010)
3. Bentley, J.L.: Multidimensional binary search trees used for associative searching 18(9), 509–517 (1975)
4. Bogdanov, P., Faloutsos, C., Mongiovì, M., Papalexakis, E.E., Ranca, R., Singh, A.K.: Netspot: Spotting significant anomalous regions on dynamic networks. In: SDM, pp. 28–36 (2013)
5. Bogdanov, P., Mongiovì, M., Singh, A.K.: Mining heavy subgraphs in time-evolving networks. In: ICDM, pp. 81–90 (2011)
6. Cline, A.K., Dhillon, I.S.: Computation of the Singular Value Decomposition. In: Handbook of Linear Algebra. CRC Press (2006)

7. Dang, X.H., Assent, I., Ng, R.T., Zimek, A., Schubert, E.: Discriminative features for identifying and interpreting outliers. In: ICDE, pp. 88–99 (2014)
8. Dang, X.H., Micenková, B., Assent, I., Ng, R.T.: Local outlier detection with interpretation. In: Blockeel, H., Kersting, K., Nijssen, S., Železný, F. (eds.) ECML PKDD 2013, Part III. LNCS (LNAI), vol. 8190, pp. 304–320. Springer, Heidelberg (2013)
9. Dannenfelser, R., Clark, N.R., Ma'ayan, A.: Genes2fans: connecting genes through functional association networks. BMC Bioinformatics 13(1), 156 (2012)
10. Davidson, I.N., Gilpin, S., Carmichael, O.T., Walker, P.B.: Network discovery via constrained tensor analysis of fmri data. In: KDD, pp. 194–202 (2013)
11. Dutkowski, J., Ideker, T.: Protein networks as logic functions in development and cancer. PLoS Computational Biology 7(9) (2011)
12. Golub, G.H., Loan, C.F.V.: Matrix Computations, 3rd edn. The Johns Hopkins University Press (1996)
13. Hastie, T., Tibshirani, R., Friedman, J.: The Elements of Statistical Learning. Data Mining, Inference, and Prediction (2001)
14. Hui, L., Zhiyu, P., et al.: Genome-wide association study dissects the genetic architecture of oil biosynthesis in maize kernels. Nature Genetics 45, 43–50 (2013)
15. Jiang, C., Coenen, F., Zito, M.: A survey of frequent subgraph mining algorithms. Knowledge Eng. Review 28(1), 75–105 (2013)
16. Jin, N., Young, C., Wang, W.: Graph classification based on pattern co-occurrence. In: CIKM, pp. 573–582 (2009)
17. Jin, N., Young, C., Wang, W.: Gaia: graph classification using evolutionary computation. In: SIGMOD Conference, pp. 879–890 (2010)
18. Ketkar, N.S., Holder, L.B., Cook, D.J.: Empirical comparison of graph classification algorithms. In: ICDM, pp. 259–266 (2009)
19. Ki, D.H., Jeung, H.-C., Park, C.H., Kang, S.H., Lee, G.Y., Lee, W.S., Kim, N.K., Chung, H.C., Rha, S.Y.: Whole genome analysis for liver metastasis gene signatures in colorectal cancer. International Journal of Cancer 121(9) (2007)
20. Ki, D.H., Jeung, H.-C., Park, C.H., Kang, S.H., Lee, G.Y., Lee, W.S., Kim, N.K., Chung, H.C., Rha, S.Y.: Whole genome analysis for liver metastasis gene signatures in colorectal cancer. Int. J. Cancer 121(9), 2005–2012 (2007)
21. Lee, D., Jeong, O.-R., Lee, S.-G.: Opinion mining of customer feedback data on the web. In: ICUIMC 2008, pp. 230–235. ACM (2008)
22. Li, C., Li, H.: Network-constrained regularization and variable selection for analysis of genomic data. Bioinformatics 24(9), 1175–1182 (2008)
23. Li, C., Li, H.: Variable selection and regression analysis for graph-structured covariates with an application to genomics. The Annals of Applied Statistics 4(3), 1498–1516 (2010)
24. Mongiovì, M., Bogdanov, P., Singh, A.K.: Mining evolving network processes. In: ICDM, pp. 537–546 (2013)
25. Noirel, J., Sanguinetti, G., Wright, P.C.: Identifying differentially expressed subnetworks with mmg. Bioinformatics 24(23), 2792–2793 (2008)
26. Ranu, S., Hoang, M., Singh, A.K.: Mining discriminative subgraphs from global-state networks. In: KDD, pp. 509–517 (2013)
27. Ranu, S., Singh, A.K.: Graphsig: A scalable approach to mining significant subgraphs in large graph databases. In: ICDE, pp. 844–855 (2009)
28. Soundarajan, S., Eliassi-Rad, T., Gallagher, B.: Which network similarity method should you choose? In: Workshop on Information Networks at NYU (2013)
29. Yan, X., Cheng, H., Han, J., Yu, P.S.: Mining significant graph patterns by leap search. In: SIGMOD Conference, pp. 433–444 (2008)

Infinitely Many-Armed Bandits
with Unknown Value Distribution

Yahel David and Nahum Shimkin

Department of Electrical Engineering,
Technion—Israel Institute of Technology,
Haifa 32000, Israel

Abstract. We consider a version of the classical stochastic Multi-Armed bandit problem in which the number of arms is large compared to the time horizon, with the goal of minimizing the cumulative regret. Here, the mean-reward (or *value*) of newly chosen arms is assumed to be i.i.d. We further make the simplifying assumption that the value of an arm is revealed once this arm is chosen. We present a general lower bound on the regret, and learning algorithms that achieve this bound up to a logarithmic factor. Contrary to previous work, we do not assume that the functional form of the tail of the value distribution is known. Furthermore, we also consider a variant of our model where sampled arms are non-retainable, namely are lost if not used continuously, with similar near-optimality results.

1 Introduction

We consider a statistical learning problem where a learning agent is facing a large pool of possible choices, or *arms*, each associated with a distinct numeric *value* which equals the one-stage reward that is obtained by choosing that arm. The goal is to minimize the cumulative n-step regret (relative to the best available arm). The agent has no prior knowledge on the value of unobserved arms, and assumes that the value of each newly observed arm is sampled independently from a common probability distribution. Once an arm is chosen its value is revealed, and the agent may continue to pick a new arm, or return to a previously chosen one. Clearly, this choice represents the essence of the *exploration vs. exploitation* dilemma for this model.

It is assumed that the pool of arms is large enough compared to the time horizon n, so that the agent cannot (or does not find it efficient) to sample them all, hence this pool can be effectively viewed as infinite. A similar model has been considered in [4,5,7,14,15]. In these papers, the observed reward of a given arm is assumed to be stochastic. In contrast, we consider here the simpler case where the reward of each arm is deterministic, so that a single observation is enough to evaluate it precisely[1]. This focuses the problem strictly on the issue of

---

[1] More generally, we may assume that the obtained reward is stochastic, but its mean is revealed once an arm is chosen. This does not affect our results as we consider the expected regret.

T. Calders et al. (Eds.): ECML PKDD 2014, Part I, LNCS 8724, pp. 307–322, 2014.

obtaining new samples, rather than learning the expected value of ones already sampled. On the other hand, the present paper generalizes the models studied in these papers in the following two respects.

- **Prior knowledge:** No prior knowledge is assumed regarding the functional form of the value distribution. Thus, the required sample size need to be estimated from the observed samples.
- **Non-retainable arms:** In addition to the basic model that allows retaining previously observed arms for further use, we consider the case where previously sampled arms are lost if not used again immediately. Discarded arms cannot be used again, but their observed values are useful for learning purposes.

Relaxing the prior knowledge assumption is natural when facing an unknown population for the first time. The non-retainable arms model is motivated by applications where arms are associated with volatile resources such as job offers and positions, apartment rental, business contracts, established routs in an ad-hoc network, and so on. To elaborate on a particular example, consider the problem of video streaming of a movie file to a media client over a wireless channel. After the transmission of each segment of the movie, the provider obtains feedback on the quality of the used channel, and decides whether to use this channel again or try a new one. If a channel is dropped it may be used by another user and hence lost. This scenario may be captured in our model by associating channels with arms, and the perceived channel quality with the obtained rewards.

As mentioned, the infinitely-many arms model has been considered before in [4], [5], [7] [14] and [15]. In [4], the rewards of each arm are assumed to be Bernoulli distributed, while the mean rewards (or *values*) of the different arms are taken to be uniformly distributed. This paper presents algorithms for a fixed horizon n which achieve a cumulative regret of an order of $\sqrt{n}$ for a fixed horizon n, and establishes a lower bound of the same order. Later, in [14] and [5], anytime algorithms were presented for similar reward and value distributions, where [5] also provides a fixed horizon time algorithm which achieves the optimal regret. A more general model was considered in [15], where arm value distribution (or at least its upper tail) is assumed known and to belong to a certain one parameter family. This paper provides a lower bound on the regret, that depends on this parameter, and proposes fixed horizon and anytime algorithms that approaches this bound up to logarithmic factors in n. Motivated by e-commerce applications, a deterministic reward model, similar to ours, is considered in [7]. That paper presents an algorithm which attains the optimal regret bound, under the assumption of known value distribution.

In a broader context, our model may be compared to the continuous multi-armed bandit problem discussed in [11], [2] and [6]. In this model the arm is chosen from a continuous set, and continuity conditions are assumed on the arm values. In contrast, in the model of the present paper no regularity or dependence assumptions are made on the arms; for further discussion and comparison of the two models see [15]. Another similar model is the contextual Multi-armed Bandit with an infinite number of arms or context sets, which is discussed in [12], [13], [10] and [1]. Again, in this model a continuity or another similarity condition is assumed on the arm values.

The non-retainable arm assumption (along with the deterministic reward property) are reminiscent of the celebrated *Secretary problem* of optimal stopping theory. In its basic form, a known number of candidates arrive sequentially for an interview, which reveals their relative merit. The interviewer should decide after each interview whether to stop and hire the last interviewed candidate, with the goal of maximizing the probability of hiring the best one. This problem has been extensively studied and extended, for example see [9] and [3]. An essential difference in our problem is the use of the regret as the performance criterion.

In this paper we present several classes of adaptive sampling algorithms for the infinitely many armed bandit problem. The algorithms are developed gradually, starting with the simpler case of a known tail distribution and generalizing to the unknown distribution case. The presentation proceeds as follows. After presenting the model in Sect. 2, we formulating in Sect. 3 a lower bound that applies to all the cases considered. All our proposed algorithms will be shown to achieve this lower bound up to a logarithmic factor. In Sect. 4 we consider the model with known tail distribution, and in Sect. 5 we address the problem with unknown distribution. Both the retainable arms and non-retainable arms cases are treated in these sections. Section 6 concludes the paper with some directions for further study.

2   Model Formulation

We consider an unlimited pool of possible objects or *arms*. The reward obtained by choosing a particular arm is deterministic, and considered as the *value* of that arm. The value of a newly chosen arm is determined as an independent sample from a fixed probability distribution, with a cumulative distribution function $F(\mu)$, $\mu \in \mathbb{R}$, that represents the empirical value distribution in the population. The obtained value is observed by the learning agent, and remains the same in future choices of that arm.

Let I_F denote the support of the probability measure that corresponds to F. We denote μ^* as the supremal reward, i.e., the maximal value in the support I_F. Our performance measure will be the cumulative regret, which is defined as follows.

Definition 1. *The n-step regret is defined as:*

$$regret(n) = E\left[\sum_{t=1}^{n}(\mu^* - r(t))\right],\tag{1}$$

where $r(t)$ is the reward obtained at time t, namely, the value of the arm chosen at time t.

We assume that all arms values are in the interval $[0, 1]$. We further use the following notations.

- μ stands for a generic random variable with distribution F.
- μ_i is the i-th sampled value from F, i.e., the revealed value of the i-th newly sampled arm.
- For $0 \leq \epsilon \leq 1$, let $\mu_\epsilon^* = \sup\{x \in \mathbb{R} : P(\mu \geq x) \geq \epsilon\}$. Note that $\mu^* = \mu_0^*$. Furthermore, let
$$D(\epsilon) = \mu^* - \mu_\epsilon^*,$$
  Note that $P(\mu \geq \mu^* - D(\epsilon)) \geq \epsilon$, with equality if $\mu_{D(\epsilon)}^*$ is a continuity point of F. We refer to $D(\epsilon)$ as the *tail function* of F.
- Let $\epsilon^*(n)$ be defined as[2]

$$\epsilon^*(n) = \sup\left\{\epsilon \in [0, 1] : nD(\epsilon) \leq \frac{1}{\epsilon}\right\}. \tag{2}$$

Note that $nD(\epsilon_1) \leq \frac{1}{\epsilon^*(n)}$ for $\epsilon_1 < \epsilon^*(n)$, and $nD(\epsilon_2) \geq \frac{1}{\epsilon^*(n)}$ for $\epsilon_2 > \epsilon^*(n)$.

The following property of the distribution F will be needed in Sect. 5.

Assumption 1
$$D(2\epsilon) \geq (C+1) D(\epsilon) \tag{3}$$
for some constant $C > 0$ and every $0 \leq \epsilon \leq \frac{1}{2}$.

Remark 1. We observed that property (3) is satisfied in the following cases, among others:
(a) Suppose that the probability density function (p.d.f.) of μ is strictly positive and bounded, i.e., $0 < c_1 \leq f_\mu(x) \leq c_2$ for some positive constants c_1 and c_2 and for every $x \in I_F$. Then (3) is satisfied for $C = \frac{c_1}{c_2}$.
(b) If $P(\mu \geq \mu^* - \epsilon) = c\epsilon^\beta$ for $\beta > 0$ and for every $0 \leq \epsilon \leq 1$, then $D(\epsilon) = c^{-\frac{1}{\beta}}\epsilon^{\frac{1}{\beta}}$, so that (3) is satisfied for $C = 2^{\frac{1}{\beta}}$ and every $0 \leq \epsilon \leq \frac{1}{2}$.
(c) Suppose that the p.d.f. of μ is non decreasing. Then (3) is satisfied for $C = 1$.

3   Lower Bound and Some Examples

We next present a lower bound on the regret that holds for all our model variations (and, in particular, for the "easiest" case of *known distribution*, retainable arms, and given time horizon).

Theorem 1. *The n-step regret is lower bounded by*

$$regret(n) \geq (1 - \delta_n)\frac{\mu^* - E[\mu]}{16}\frac{1}{\epsilon^*(n)}, \tag{4}$$

---

[2] If the support of μ is a single interval, then $D(\epsilon)$ is continuous. In that case, definition (2) reduced to the equation $nD(\epsilon) = \frac{1}{\epsilon}$ which, by monotonicity, has a unique solution for n large enough. See Sect. 3 for examples.

where $\epsilon^*(n)$ satisfies (2), and

$$\delta_n = 1 - 2\exp\left(-\frac{(\mu^* - E[\mu])^2}{8\epsilon^*(n)}\right).$$

Note that when $\epsilon^*(n) \to 0$ as $n \to \infty$, $\delta_n \to 0$ as $n \to \infty$, so that its effect becomes negligible.

Proof. Let $\{\mu_1, ...\mu_n\}$ denote the values of the first n arms to be drawn from the pool, and assume that these values are revealed beforehand to the learning agent (even if it does not actually draw n new arms in n steps).

For any such sequence $\{\mu_1, ...\mu_n\}$, the smallest possible regret that can be obtained (by any algorithm) is

$$R_n^* = \min_{k \in \{1,...,n\}} \{\Gamma(n, k)\},$$

where

$$\Gamma(n, k) = n\mu^* - \left[\sum_{i=1}^{k} \mu_i + (n - k)\mu_k\right].$$

This is due to the easily varified fact that the optimal policy for given (μ_i) is to continue sampling new arms up to some index k^* and continue pulling the k^*-th arm thereafter.

Define the events

$$A(m, \delta_1) = \left\{\frac{1}{m}\sum_{i=1}^{m} \mu_i < \mu^* - \delta_1\right\}$$

and

$$B(m, \delta_2) = \left\{\max_{i \in \{1,...,m\}} \mu_i < \mu^* - \delta_2\right\}$$

for $m \in \{1, ..., n\}$, $0 \le \delta_1 \le \mu^*$ and $0 \le \delta_2 \le \mu^*$. If these two events are satisfied for some m, δ_1, and δ_2, we obtain that $R_n^* > m\delta_1$, for $m \le k^*$, and $R_n^* > n\delta_2$, for $m \ge k^*$, where

$$\arg\min_{k \in \{1,...,n\}} \{\Gamma(n, k)\} \triangleq k^*.$$

Therefore,

$$R_n^* > \min(m\delta_1, n\delta_2).$$

Also,

$$P(A(m, \delta_1) \cap B(m, \delta_2)) \ge 1 - P(A(m, \delta_1)^c) - P(B(m, \delta_2)^c),$$

where A^c denotes the complement of A. So, for $\delta_1 = \frac{1}{2}(\mu^* - E[\mu])$, by Hoeffding's inequality,

$$P(A(m, \delta_1)^c) \le \exp\left(-\frac{m}{2}(\mu^* - E[\mu])^2\right)$$

and for $\delta_2 = \frac{1}{2}D(2\epsilon^*(n))$,

$$P(B(m,\delta_2)^c) = 1 - \prod_i^m P\left(\boldsymbol{\mu_i} < \mu^* - \delta_2\right) \leq 1 - (1 - 2\epsilon^*(n))^m \leq 2\epsilon^*(n)m \, .$$

Therefore, for $m = \frac{1}{4\epsilon^*(n)}$, and δ_1, δ_2 as above

$$regret(n) \geq (1 - P(A(m,\delta_1)^c) - P(B(m,\delta_2)^c)) \min\left(m\delta_1, n\delta_2\right)$$

$$\geq \left(1 - \exp\left(-\frac{(\mu^* - E[\mu])^2}{8\epsilon^*(n)}\right) - \frac{1}{2}\right) \min\left(\frac{\mu^* - E[\mu]}{8\epsilon^*(n)}, \frac{n}{2}D(2\epsilon^*(n))\right)$$

$$= \left(\frac{1}{2} - \exp\left(-\frac{(\mu^* - E[\mu])^2}{8\epsilon^*(n)}\right)\right) \frac{\mu^* - E[\mu]}{8\epsilon^*(n)} \, ,$$

where the last equality follows by (2), since $\frac{n}{2}D(2\epsilon^*(n)) \geq \frac{1}{2\epsilon^*(n)} \geq \frac{\mu^* - E[\mu]}{8\epsilon^*(n)}$.   □

The main consequence of this bound is that the order of the regret is at least $\frac{1}{\epsilon^*(n)}$. As illustrate in the following examples, the order of $\frac{1}{\epsilon^*(n)}$ is typically polynomial in n. We will show below that all the algorithms presented in this paper attain the lower bound up to a logarithmic factors.

The papers [4] and [15] provide similar lower bounds for specific cases. In [4], a lower bound of $\sqrt{2n}$ is provided for the case where the arms values are uniformly distributed in $[0,1]$ and with Bernoulli rewards. In [15], a lower bound of order $\Omega\left(n^{\frac{\beta}{\beta+1}}\right)$ is provided for the case where $D(\epsilon) = O(\epsilon^\beta)$ with $\beta \geq 0$. Noting Example 1, our bound below is of the same order. Our proof approach is different than that of [15] and applies to more general distribution. Also, we provide a specific coefficient rather than just an order of magnitude.

The following examples serve to illustrate the dependence of $\epsilon^*(n)$ on n. Example 1 is the standard form studied in [15], while the others examples illustrate general cases that are covered by our model.

1. Suppose that for $\epsilon > 0$ (small enough), we have $P\left(\boldsymbol{\mu} \geq \mu^* - \epsilon\right) = \Theta\left(\epsilon^\beta\right)$, where $\beta > 0$. Then $D(\epsilon) = \Theta\left(\epsilon^{\frac{1}{\beta}}\right)$, so that $\epsilon^*(n) = \Theta\left(n^{-\frac{\beta}{\beta+1}}\right)$.
   This is the case considered in [15]. Note that $\beta = 1$ corresponds to a uniform probability distribution.
2. Suppose $\boldsymbol{\mu}$ has the CDF

$$F(\mu) = \begin{cases} (1-a)\frac{\mu}{\mu^*} & 0 \leq \mu < \mu^* \\ 1 & \mu = \mu^* \end{cases} \, ,$$

where $0 \leq a < 1$. This describes a uniform distribution with an added atom of probability a at μ^*. Then $D(\epsilon) = 0$ for $\epsilon \leq a$, and $D(\epsilon) = \frac{\mu^*(\epsilon - a)}{1-a}$ for $\epsilon > a$. Therefore, it follows that $2\epsilon^*(n) = a + \left(a^2 + \frac{4c(1-a)}{n}\right)^{\frac{1}{2}}$.
Note that in this case $\epsilon^*(n) > a$ for all n. Hence, contrary to Example 1, $\epsilon^*(n)$ does not converge to 0 as $n \to \infty$. So, the regret is finite.

3. Suppose we have $P(\mu \geq \mu^* - \epsilon) = -\frac{c}{\ln(\epsilon)}$. We obtain that $D(\epsilon) = e^{-\frac{c}{\epsilon}}$. Therefore, it follows that $\frac{c}{\ln(n)} \leq \epsilon^*(n) \leq \frac{c+1}{\ln(n)}$.

Note that in this case, $\epsilon^*(n)$ decays slower than any polynomial function of n, and the regret grows as $O(\ln(n))$.

4   Known Tail Function

This section discusses the model in which the tail function $D(\epsilon)$ is known (although, of course, the upper value μ^* is unknown). This model specializes the stochastic-arm model presented by Wang et al. [15] to deterministic arms. On the other hand, our model is more general in the sense that it is not restricted to tail functions of the form $D(\epsilon) = \epsilon^\beta$. Furthermore, we consider here both the retainable arms and the non-retainable arms problems, as described in the Introduction.

4.1   Retainable Arms

We propose the following algorithm.

Algorithm 1 (KT&RA – Known Tail and Retainable Arms).

1. *Parameters: Time horizon $n > 1$ and a constant $A > 0$.*
2. *Compute $\epsilon^*(n)$ as defined in (2).*
3. *Pull $N = \lfloor A \ln(n) \frac{1}{\epsilon^*(n)} \rfloor + 1$ arms and keep the best one so far.*
4. *Continue by pulling the saved best arm up to the last stage n.*

The right tradeoff between exploring new arms and pulling the best one so far is obtained by (2). The parameter A allows a further tuning of the algorithm performance. Our regret bound for this algorithm is presented in the following Theorem.

Theorem 2. *For each $n > 1$, the regret of the KT&RA Algorithm with a constant A is upper bounded by*

$$regret(n) \leq (1 + A\ln(n))\frac{1}{\epsilon^*(n)} + n^{1-A} + 1, \tag{5}$$

where $\epsilon^*(n)$ is defined in (2).

By properly choosing A, for example $A = 1$, we obtain an $O\left(\frac{\ln(n)}{\epsilon^*(n)}\right)$ bound on the regret. This bound is of the same order as the lower bound in (4), up to a logarithmic factor. We note that a slightly better choice of A may be obtained by balancing the two terms in the bound (5).

Proof. For $N \geq 1$, let $V_N(1)$ denote the value of the best arm found by sampling N different arms. Clearly,

$$regret(n) \leq N + (n - N)\Delta(N),$$

where $\Delta(N) = E[\mu^* - V_N(1)]$. But for any $0 \le \epsilon \le 1$

$$P\left(\mu^* - V_N(1) > D(\epsilon)\right) \le (1 - \epsilon)^N \qquad (6)$$

(note that equality holds if the distribution function of μ is continuous) so that, since $\mu^* - V_N(1) \le 1$,

$$\Delta(N) \le (1 - \epsilon)^N + D(\epsilon). \qquad (7)$$

Since in step 3 of the algorithm we chose $N = A \ln(n)\frac{1}{\epsilon(n)}$, where $\epsilon(n) < \epsilon^*(n)$, and noting that $(1 - \epsilon)^{\frac{1}{\epsilon}} \le e^{-1}$ for $\epsilon \in (0, 1]$, we obtain that

$$(1 - \epsilon(n))^N \le n^{-A}. \qquad (8)$$

Since, $\epsilon(n) < \epsilon^*(n)$, it follows that $nD\left(\epsilon(n)\right) \le \frac{1}{\epsilon^*(n)}$. Therefore,

$$regret(n) \le \lfloor A \ln(n)\frac{1}{\epsilon^*(n)} \rfloor + 1 + n^{1-A} + nD(\epsilon(n)) \le A \ln(n)\frac{1}{\epsilon^*(n)} + 1 + n^{1-A} + \frac{1}{\epsilon^*(n)}.$$

Hence (5) is obtained. $\qquad\qquad\qquad\qquad\qquad\qquad\qquad\qquad\qquad\qquad\qquad\qquad\square$

4.2   Non-retainable Arms

Here we are not allowed to keep any previously chosen arm except the last one. Therefore, the previous algorithm that keeps the best arm so far while trying out new arms cannot be applied in this case. However, the observed values of discarded arms provide usefull information for the learning agent. We introduce the notation $V_N(m)$ for the m-th largest value obtained after observing N arms.

Algorithm 2 (KT&NA – Known Tail and Non-retainable Arms).

1. *Parameters: Time horizon $n > 1$ and a constant $A \ge 2$.*
2. *Compute $\epsilon^*(n)$ as defined in (2).*
3. *Pull $N = \lfloor 5A \ln(n)\frac{1}{\epsilon^*(n)} \rfloor + 1$ arms and store their values.*
4. *a. Continue pulling new arms until observing a value not smaller than $V_N(m)$, where $m = \lceil 2A \ln(n) \rceil$.*
   b. Once such a value is observed, continue pulling this arm up to the last stage n.

After observing N arms, a threshold which is large on one hand, and on the other hand it is likely enough to find a new arm with a larger value than it is obtained. Then, the algorithm searches for an arm with a larger value than this threshold and keeps pulling this arm. Our regret bound for this algorithm is presented in the following Theorem.

Theorem 3. *For each $n > 1$, the regret of the KT&NA Algorithm with a constant A is upper bounded by*

$$regret(n) \le (5A \ln(n) + 8)\frac{1}{\epsilon^*(n)} + c_A(n), \qquad (9)$$

where $\epsilon^*(n)$ is defined in (2) and for $n \geq 10$ it is obtained that

$$c_A(n) \leq 4 \qquad (10)$$

The exact expression of $c_A(n)$ for $n \geq 10$ is found in (15).

The algorithm starts with a learning period of length N, which allows to assess the values distribution near μ^*. A threshold $V_N(m)$ is then set, and sampling new arms continues until an arm with that value is observed. The threshold $V_N(m)$ is chosen as the m-th largest value in the obtained samples, where m is chosen so that the chances of quickly drawing a new arm with that value or over are high.

By a proper choice of A, for example $A = 2$ we obtain an $O\left(\frac{\ln(n)}{\epsilon^*(n)}\right)$ bound on the regret. This bound is of the same order as the lower bound in (4), up to a logarithmic factor. We note that by considering the exact expression of $c_A(n)$, a slightly better choice of A may be obtained.

The proof of Theorem 3 relies on the following Lemma.

Lemma 1. *Let m and N be positive integers such that $m < N$.*

(a) If $\frac{m}{N} > \epsilon$, then
$$P(V_N(m) > \mu_\epsilon^*) \leq f_0(m, N, \epsilon).$$

(b) If $\frac{m}{N} < \epsilon$, then
$$P(V_N(m) < \mu_\epsilon^*) \leq f_0(m, N, \epsilon),$$

where $f_0(m, N, \epsilon) = \exp\left(-\frac{|m - N\epsilon|^2}{2(N\epsilon + |m - N\epsilon|/3)}\right)$.

For space considerations, the proof of that Lemma is presented in the technical report [8].

Proof of Theorem 3. The regret is bounded by

$$regret(n) \leq N + E[Y(V_N(m))] + nE[\mu^* - V_N(m)], \qquad (11)$$

where N is the number of arms which are sampled in step 3 of the algorithm. The random variable $Y(V)$ is the number of arms which are sampled until an arm with a greater value than V is sampled (or until the end of the time horizon, if such a value is never sampled again). We can find that for any $\epsilon_1 > 0$, the second term of (11) is bounded by

$$E[Y(V_N(m))] \leq P\left(V_N(m) \leq \mu_{\epsilon_1}^*\right) E\left[Y(V_N(m))|V_N(m) \leq \mu_{\epsilon_1}^*\right]$$
$$+ P\left(V_N(m) > \mu_{\epsilon_1}^*\right) E\left[Y(V_N(m))|V_N(m) > \mu_{\epsilon_1}^*\right]$$
$$\leq \frac{1}{\epsilon_1} + nP\left(V_N(m) > \mu_{\epsilon_1}^*\right) \triangleq E_1(\epsilon_1). \qquad (12)$$

By using the fact that $Y(V) \leq n$, the non decreasing of $Y(V)$ in V, and the expected value of a geometric variable. Also, for any $\epsilon_2 > 0$, the third term of (11) is bounded by

$$nE[\mu^* - V_N(m)] \leq nP\left(V_N(m) \geq \mu^*_{\epsilon_2}\right) E\left[\mu^* - V_N(m)|V_N(m) \geq \mu^*_{\epsilon_2}\right]$$
$$+ nP\left(V_N(m) < \mu^*_{\epsilon_2}\right) E\left[\mu^* - V_N(m)|V_N(m) < \mu^*_{\epsilon_2}\right] \qquad (13)$$
$$\leq nD(\epsilon_2) + nP\left(V_N(m) < \mu^*_{\epsilon_2}\right) \triangleq E_2(\epsilon_2).$$

Since it is known that $\mu^* - V_N(m) \leq 1$.

Therefore, by (11), (12), (13) and Lemma 1, for $\epsilon_1 = \frac{\epsilon(n)}{7}$ and $\epsilon_2 = \epsilon(n)$, where $N = 5A\ln(n)\frac{1}{\epsilon(n)}$ for some $\frac{5A\ln(n)\epsilon^*(n)}{5A\ln(n)+\epsilon^*(n)} \leq \epsilon(n) < \epsilon^*(n)$, it is obtained that

$$regret(n) \leq \lfloor \frac{5A\ln(n)}{\epsilon^*(n)} \rfloor + 1 + E_1(\epsilon_1) + E_2(\epsilon_2) \leq \frac{5A\ln(n)}{\epsilon^*(n)} + \frac{8}{\epsilon^*(n)} + c_A(n), \quad (14)$$

where $\epsilon^*(n)$ is defined in (2), and

$$c_A(n) \leq 2n^{1-0.6A} + 2 \qquad (15)$$

for $n \geq 10$. Note that (14) holds since $nD(\epsilon(n)) \leq \frac{1}{\epsilon^*(n)}$ and $\frac{1}{\epsilon(n)} \leq \frac{1}{\epsilon^*(n)} + \frac{1}{5A\ln(n)}$. Hence, (9) is obtained. $\qquad \square$

5   Unknown Tail Function

We now proceed to the harder problem where the tail function $D(\epsilon)$ is unknown. Here, it is impossible to calculate beforehand the optimal number of arms to sample, as done in the algorithms of Sect. 4. To overcome this issue, the algorithms proposed in this section gradually increase the number of sampled arms until a certain condition is satisfied.

The analysis in this section will be carried out under Assumption 1. Note that, the values of the constant C in the assumption is not used in the algorithm, but only in its analysis. Again, we consider here both the retainable arms and the non-retainable arms problems.

5.1   Retainable Arms

Recall that $V_N(m)$ stands for the m-th largest value obtained after observing N arms.

Algorithm 3 (UT&RA – Unknown Tail and Retainable Arms).

1. *Parameters: Time horizon $n > 1$, constants $N \geq 2$, $A \geq 2$.*
   Set $N_0 = \lceil NA_n \rceil$, where $A_n = A\ln(n)$.
2. *Pull $K = N_0$ arms.*
3. *If $\Psi(K, n) < \frac{K}{nA_n}$, where $\Psi(K, n) = V_K(1) - V_K(\lceil 5A_n \rceil)$:*
   a. Pull another K arms.
   b. Continue pulling the best arm so far up to time n.
   Else, if $\Psi(K, n) \geq \frac{K}{nA_n}$:
   a. Pull one more arm, and set $K = K + 1$.
   b. Return to 3.

In this algorithm, the number of sampled arms K is increased until the condition in stage 3 is satisfied. Thereafter, the number of sampled arms is doubled, and then the best arm found is pulled up to time n.

The rational of this algorithm is as follows. Our goal is to ensure that, essentially, the number of samples K is large enough, namely comparable to $\epsilon^*(n)^{-1}$ from (2). This translates to $K > nD(\frac{1}{K})$. The condition in stage 3 indicates that the gap $V_K(1) - V_K(5A_n)$, which is the difference between the largest value and the $5A_n$-th largest value in the first K samples, is small in comparison to K. This gap is related, with high probability, to the difference $D(\frac{2}{K}) - D(\frac{1}{K})$, which, under Assumption 1, upper bounds the size of $D(\frac{1}{K})$. A second sample of K arms is required due to the dependencies between the above stopping condition and values of the the first K samples.

Our regret bound for this algorithm is presented in the following Theorem.

Theorem 4. *Let Assumption 1 hold for some $C > 0$, For each $n > 1$, the regret of the UT&RA Algorithm with a constant A is upper bounded by*

$$regret(n) \leq \left(20A \ln(n) + \frac{1}{\min(1, C)} \right) \frac{1}{\epsilon^*(n)} + c_A(n), \tag{16}$$

where $\epsilon^*(n)$ solves (2) and

$$c_A(n) \leq 2N_0 + 9 \tag{17}$$

for $n \geq 10$. The exact expression of $c_A(n)$ for $n \geq 10$ is given in (30).

Again, by a proper choice of A, for example $A = 2$ we obtain an $O\left(\frac{\ln(n)}{\epsilon^*(n)} \right)$ bound on the regret.

Proof. The regret is bounded by

$$regret(n) \leq E[2\hat{N}] + nE[\mu^* - V_{2\hat{N}}(1)], \tag{18}$$

where $\hat{N}$ is the number of arms sampled by the algorithm until the condition in stage 3 is satisfied.

The first term of (18) is bounded by

$$E[2\hat{N}] \leq 2\left(N_1 + nP(\hat{N} > N_1) \right) \tag{19}$$

for every $N_1 \geq \lceil NA_n \rceil$. To bound the probability $P(\hat{N} > N_1)$, we note that

$$\left\{ \hat{N} > N_1 \right\} \subseteq \left\{ \Psi(N_1, n) \geq \frac{N_1}{nA_n} \right\} \subseteq \left\{ \Psi(N_1, n) > \frac{N_1}{\gamma n A_n} \right\}$$

$$\subseteq \left\{ \Psi(N_1, n) > D\left(\frac{\gamma A_n}{N_1} \right) \right\} \bigcup E_4(\gamma, N_1)$$

$$\subseteq E_3(\gamma, N_1) \bigcup E_4(\gamma, N_1),$$

where $\gamma > 1$,

$$E_3(\gamma, N_1) \triangleq \left\{ V_{N_1}(\lceil 5A_n \rceil) < \mu^*_{\frac{\gamma A_n}{N_1}} \right\}$$

and

$$E_4(\gamma, N_1) \triangleq \left\{ D\left(\frac{\gamma A_n}{N_1}\right) > \frac{N_1}{\gamma n A_n} \right\}.$$

Note that $D(\epsilon) < \frac{1}{n\epsilon}$ for $\epsilon < \epsilon^*(n)$. So, it follows that $E_4(\gamma, N_1)$ is false, when $\frac{\gamma A_n}{N_1} < \epsilon^*(n)$, or $N_1 > \frac{\gamma A_n}{\epsilon^*(n)}$. So, for $N_1 = \max\left(\lceil\frac{\gamma A_n}{\epsilon^*(n)} + 1\rceil, N_0\right)$, it is obtained that

$$\{\hat{N} > N_1\} \subseteq E_3(\gamma, N_1)$$

and by Lemma 1, it follows that

$$P\left(\hat{N} > N_1\right) \le n^{-0.9A} \tag{20}$$

for $\gamma = 10$ and $n \ge 10$. Therefore, by (19),

$$E[2\hat{N}] \le 2\left(\lceil\frac{10A_n}{\epsilon^*(n)} + 1\rceil + N_0 + n^{1-0.9A}\right). \tag{21}$$

For bounding the second term of (18) we note that, for any $N_2 \ge 1$,

$$
\begin{aligned}
nE[\mu^* - V_{2\hat{N}}(1)] &\le nE[\mu^* - V_{\hat{N}}(1)] \\
&\le n\Big(E\left[\mu^* - V_{\hat{N}}(1)|\hat{N} \le N_2\right] P(\hat{N} \le N_2) \\
&\quad + E\left[\mu^* - V_{\hat{N}}(1)|\hat{N} > N_2\right] P(\hat{N} > N_2)\Big) \\
&\le n\left(P(\hat{N} \le N_2) + E\left[\mu^* - V_{N_2+1}(1)\right]\right),
\end{aligned}
\tag{22}
$$

where, starting from the first inequality, we consider only the $\hat{N}$ arms that were sampled after the condition in stage 3 of the algorithm has been satisfied, so that, $\hat{N}$ and the obtained values are independent. In the third inequality we use the fact that $E\left[V_m(1)\right]$ is non decreasing in m.

For bounding $P(\hat{N} \le N_2)$, we note that for every $i \ge N_0$,

$$\left\{\hat{N} \le N_2\right\} = \cup_{N_0 \le i \le N_2}\{A(i)\}, \tag{23}$$

where

$$
\begin{aligned}
A(i) &\triangleq \left\{\Psi(i, n) < \frac{i}{nA_n}\right\} \\
&\subseteq \left\{\Psi(i, n) < D\left(\frac{2A_n}{i}\right) - D\left(\frac{A_n}{i}\right)\right\} \cup \left\{D\left(\frac{2A_n}{i}\right) - D\left(\frac{A_n}{i}\right) < \frac{i}{nA_n}\right\}.
\end{aligned}
$$

Since

$$\Psi(i, n) = V_i(1) - V_i(\lceil 5A_n\rceil)$$

and

$$D\left(\frac{2A_n}{i}\right) - D\left(\frac{A_n}{i}\right) = \mu^*_{\frac{A_n}{i}} - \mu^*_{\frac{2A_n}{i}},$$

it follows that

$$A(i) \subseteq B(i) \bigcup C(i),\tag{24}$$

where

$$B(i) \triangleq \left\{ V_i(\lceil 5A_n \rceil) > \mu^*_{\frac{2A_n}{i}} \right\} \cup \left\{ V_i(1) < \mu^*_{\frac{A_n}{i}} \right\}$$

$$C(i) \triangleq \left\{ \min(1,C)D\left(\frac{A_n}{i}\right) < \frac{i}{nA_n} \right\}$$

and the constant C satisfies that $CD(\epsilon) \leq D(2\epsilon) - D(\epsilon)$ for every $0 \leq \epsilon \leq \frac{1}{2}$. So, by (23) and (24), and since

$$\cup_{N_0 \leq i \leq N_2} \{C(i)\} \subseteq C(N_2),$$

it is obtained that for any $N_2 \geq N_0$ such that $C(N_2)$ is false,

$$\left\{ \hat{N} \leq N_2 \right\} = \cup_{N_0 \leq i \leq N_2} \{A(i)\} \subseteq \cup_{N_0 \leq i \leq N_2} \{B(i)\} .$$

Therefore, by Lemma 1, and similarly to (6) and (8) with $\epsilon = \frac{A_n}{i}$ and $N = i$, it follows that

$$P\left(\hat{N} \leq N_2\right) \leq n(n^{-1.4A} + n^{-A})\tag{25}$$

for $n \geq 10$. Note that for $N_2 < N_0$ it is obtained that $P\left(\hat{N} \leq N_2\right) = 0$.

The remaining issue is to bound the term $E\left[\mu^* - V_{N_2+1}(1)\right]$ from (22) under the same condition that $C(N_2)$ is false. Since $\Delta \triangleq \mu^* - V_{N_2+1}(1) \leq 1$

$$\begin{aligned} E\left[\Delta\right] &\leq D(\frac{A_n}{N_2+1}) + P\left(\Delta > D(\frac{A_n}{N_2+1})\right) \\ &\leq D(\frac{A_n}{N_2+1}) + n^{-A} . \end{aligned}\tag{26}$$

The last inequality follows similarly to (6) and (8) with $\epsilon = \frac{A_n}{N_2+1}$ and $N = N_2+1$. Let $\epsilon(n)$ be defined as

$$\epsilon(n) = \sup\left\{ \epsilon \in [0,1] : n\min(1,C)D(\epsilon) \leq \frac{1}{\epsilon} \right\}.$$

If it is satisfied that

$$E(C) \triangleq \left\{ \min(1,C)D(\epsilon(n)) \geq \frac{1}{n\epsilon(n)} \right\}$$

then, let us choose N_2 as the largest integer for which $\frac{N_2}{A_n} \leq \frac{1}{\epsilon(n)}$. Then, $C(N_2)$ is false, and furthermore $\frac{A_n}{N_2+1} < \epsilon(n)$. So,

$$D(\frac{A_n}{N_2+1}) \leq \frac{1}{n\min(1,C)\epsilon(n)} .$$

On the other hand, if $E(C)$ is not satisfied, then, let us choose N_2 as the largest integer for which $\frac{N_2}{A_n} < \frac{1}{\epsilon(n)}$. Then, again, $C(N_2)$ is false, and furthermore $\frac{A_n}{N_2+1} \le \epsilon(n)$. So,

$$D(\frac{A_n}{N_2+1}) \le D(\epsilon(n)) < \frac{1}{n \min(1,C)\epsilon(n)} \,.$$

Therefore, since $\min(1,C) \le 1$, it is obtained that $\frac{1}{\epsilon(n)} \le \frac{1}{\epsilon^*(n)}$. So,

$$D(\frac{A_n}{N_2+1}) \le \frac{1}{n \min(1,C)\epsilon^*(n)} \,. \tag{27}$$

Therefore, by (22), (25), (26) and (27), it follows that

$$nE[\mu^* - V_{2\hat{N}}(1)] \le n\left(n\left(n^{-1.4A} + n^{-A}\right) + \frac{1}{n \min(1,C)\epsilon^*(n)} + n^{-A}\right) \,. \tag{28}$$

Finally, by (18), (21) and (28), it is obtained that

$$regret(n) \le \left(20A_n + \frac{1}{\min(1,C)}\right)\frac{1}{\epsilon^*(n)} + c_A(n) \,, \tag{29}$$

where

$$c_A(n) = 2n^{1-0.9A} + n^{2-1.4A} + n^{2-A} + n^{1-A} + 2NA_n + 4 \tag{30}$$

for $n \ge 10$. Hence, since $A \ge 2$, it follows that $c_A(n) \le 2N_0 + 9$ for $n \ge 10$, so Theorem 4 is obtained. $\square$

5.2   Non-retainable Arms

Here, as in Sect. 4.2, it is impossible to pull a group of arms and keep the best one of them. So, we combine the UT&RA algorithm from the previous section with the KT&NA algorithm from Sect. 4.2. Recall that (3) is satisfied for a positive constant C and $\epsilon \le \epsilon_0$, where ϵ_0 is known for the learning agent.

Algorithm 4 (UT&NA – Unknown Tail and Non-retainable Arms).

1. *Parameters: Time horizon $n > 1$, constants $N \ge 10$, $A \ge 4$.*
   Set $N_0 = \lceil NA_n \rceil$, where $A_n = A\ln(n)$.
2. *Pull $K = N_0$ arms.*
3. *If $\Psi(K,n) < \frac{K}{nA_n}$, where $\Psi(K,n) = V_K(1) - V_K(\lceil 5A_n \rceil)$:*
   a. Pull another K arms.
   b. Continue pulling new arms until observing a value equal or larger than $V_K(m)$, where $m = \lceil \frac{3A_n}{10} \rceil$.
   c. Continue pulling this arm up to time n.
   Else, if $\Psi(K,n) \ge \frac{K}{nA_n}$:
   a. Pull one more arm, and set $K = K + 1$.
   b. Return to 3.

This algorithm begins, similarly to the UT&RA Algorithm 3, to find a large enough sample size K. Then, since observed arms cannot be retained, it proceeds similarly to KT&NA Algorithm 2, to compute a desired value threshold and sample new arms until such an arm is obtained. Our regret bound for this algorithm is as follows.

Theorem 5. *Let Assumption 1 hold for some $C > 0$. For each $n > 1$, the regret of the UT&NA Algorithm with a constant A is upper bounded by*

$$regret(n) \leq \left(20A\ln(n) + 140 + \frac{1}{\min(1, C)} \right) \frac{1}{\epsilon^*(n)} + c_A(n), \qquad (31)$$

where $\epsilon^*(n)$ solves (2) and

$$c_A(n) \leq 2N_0 + 14N + 13 \qquad (32)$$

for $A \geq 7$ and $n \geq 100$. The full expression of $c_A(n)$ can be found in [8].

Similarly to the UT-LB and the KT-LB Algorithms, by a proper choice of A, for example $A = 7$, we obtain an $O\left(\frac{\ln(n)}{\epsilon^*(n)} \right)$ bound on the regret.

For space considerations, the proof of Theorem 5 is presented in the technical report [8].

6 Conclusion

For the problem of infinitely many armed-bandits with unknown value distribution, we have proposed algorithms that obtain the optimal regret up to a logarithmic factors. Our treatment was focused on the case of deterministic rewards. Further work should naturally consider the extension of our results to the stochastic rewards model, which requires repeated trials of sampled arms (possibly using a UCB-like bandit algorithm similarly to [15]). Another extension of our results, which were presented here for a given time horizon, is to the case of anytime algorithms. This can of course be accomplished using a simple doubling trick, however the development of specific and more effective algorithms for this case should be of interest. Note that in the stochastic rewards problem, it should be of interest to consider the intermediate case, where only a limited number of arms (rather than all or none) can be retained. As mentioned, in the present deterministic rewards model, it is enough to retain only the one arm with the best value so far.

Acknowledgments. This research was partly supported by the Technion-Microsoft Electronic Commerce Research Center.

References

1. Amin, K., Kearns, M., Draief, M., Abernethy, J.D.: Large-scale bandit problems and KWIK learning. In: Proceedings of the 30th International Conference on Machine Learning (ICML 2013), pp. 588–596 (2013)

2. Auer, P., Ortner, R., Szepesvári, C.: Improved rates for the stochastic continuum-armed bandit problem. In: Bshouty, N.H., Gentile, C. (eds.) COLT. LNCS (LNAI), vol. 4539, pp. 454–468. Springer, Heidelberg (2007)
3. Babaioff, M., Immorlica, N., Kempe, D., Kleinberg, R.: Online auctions and generalized secretary problems. ACM SIGecom Exchanges 7(2), 1–11 (2008)
4. Berry, D.A., Chen, R.W., Zame, A., Heath, D.C., Shepp, L.A.: Bandit problems with infinitely many arms. The Annals of Statistics, 2103–2116 (1997)
5. Bonald, T., Proutiere, A.: Two-target algorithms for infinite-armed bandits with Bernoulli rewards. In: Advances in Neural Information Processing Systems 26, pp. 2184–2192. Curran Associates, Inc. (2013)
6. Bubeck, S., Munos, R., Stoltz, G., Szepesvári, C.: X-armed bandits. Journal of Machine Learning Research 12, 1655–1695 (2011)
7. Chakrabarti, D., Kumar, R., Radlinski, F., Upfal, E.: Mortal multi-armed bandits. In: Advances in Neural Information Processing Systems 21, pp. 273–280. Curran Associates, Inc. (2009)
8. David, Y., Shimkin, N.: Infinitely many-armed bandits with unknown value distribution. Technical report, Technion—Israel Institute of Technology (2014), http://webee.technion.ac.il/people/shimkin/PAPERS/ECML14-full.pdf
9. Freeman, P.: The secretary problem and its extensions: A review. International Statistical Review, 189–206 (1983)
10. Kakade, S.M., von Luxburg, U. (eds.): COLT 2011 - The 24th Annual Conference on Learning Theory, Budapest, Hungary, June 9-11. JMLR Proceedings, vol. 19. JMLR.org (2011)
11. Kleinberg, R., Slivkins, A., Upfal, E.: Multi-armed bandits in metric spaces. In: Proceedings of the 40th Annual ACM Symposium on Theory of Computing, pp. 681–690. ACM (2008)
12. Langford, J., Zhang, T.: The epoch-greedy algorithm for multi-armed bandits with side information. In: Advances in Neural Information Processing Systems, pp. 817–824 (2007)
13. Lu, T., Pál, D., Pál, M.: Contextual multi-armed bandits. In: International Conference on Artificial Intelligence and Statistics, pp. 485–492 (2010)
14. Teytaud, O., Gelly, S., Sebag, M.: Anytime many-armed bandits. In: CAP, Grenoble, France (2007)
15. Wang, Y., Audibert, J.-Y., Munos, R.: et al. Infinitely many-armed bandits. In: Advances in Neural Information Processing Systems, vol. 8, pp. 1–8 (2008)

Cautious Ordinal Classification
by Binary Decomposition

Sébastien Destercke and Gen Yang

Université de Technologie de Compiegne U.M.R. C.N.R.S. 7253 Heudiasyc Centre de recherches de Royallieu F-60205 Compiegne Cedex France
{sebastien.destercke,gen.yang}@hds.utc.fr

Abstract. We study the problem of performing cautious inferences for an ordinal classification (a.k.a. ordinal regression) task, that is when the possible classes are totally ordered. By cautious inference, we mean that we may produce partial predictions when available information is insufficient to provide reliable precise ones. We do so by estimating probabilistic bounds instead of precise ones. These bounds induce a (convex) set of possible probabilistic models, from which we perform inferences. As the estimates or predictions for such models are usually computationally harder to obtain than for precise ones, we study the extension of two binary decomposition strategies that remain easy to obtain and computationally efficient to manipulate when shifting from precise to bounded estimates. We demonstrate the possible usefulness of such a cautious attitude on tests performed on benchmark data sets.

Keywords: Ordinal regression, imprecise probabilities, Binary decomposition, Nested dichotomies.

1 Introduction

We are interested in the supervised learning problem known as *ordinal classification* [18] or *regression* [9]. In this problem, the finite set of possible labels are naturally ordered. For instance, the rating of movies can be one of the following labels: Very-Bad, Bad, Average, Good, Very-Good that are ordered from the worst situation to the best. Such problems are different from multi-class classification and regression problems, since in the former there is no ordering between classes and in the latter there exists a metric on the outputs (while in ordinal classification, a 5-star movie should not be considered five times better than a 1-star movie).

A common approach to solve this problem is to associate the labels to their rank, e.g., $\{1,2,3,4,5\}$ in our previous film example, and then to learn a ranking function. In the past years, several algorithms and methods [27] have been proposed to learn such a function, such as SVM techniques [26,22,23,25], monotone functions [28], binary decomposition [20], rule based models [14]. This is not the approach followed in this paper, in which our goal is to estimate the probability of the label conditionally on the observed instance. In this sense, our approach is much closer to the one proposed by Frank et Hall [18].

A common feature of all the previously cited approaches is that, no matter how reliable is the model and the amount of data it is learned from, it will always produce a

T. Calders et al. (Eds.): ECML PKDD 2014, Part I, LNCS 8724, pp. 323–337, 2014.

unique label as prediction. In this paper, we are interested in making partial predictions when information is insufficient to provide a reliable precise one. That is, if we are unsure of the right label, we may abstain to make a precise prediction and instead predict a subset of potentially optimal labels. The goal is similar to the one pursued by the use of a reject option [2,8], and in particular to methods returning subsets of possible classes [1,21]. Yet we will see in the experiments that the two approaches can provide very different results.

Besides the fact that such cautious predictions can prevent bad decisions based on wrong predictions, making such imprecise predictions in an ordinal classification setting can also be instrumental in more complex problems that can be decomposed into sets of ordinal classification problem, such as graded multi-label [7] or label ranking [6]. Indeed, in such problems with structured outputs, obtaining fully reliable precise predictions is much more difficult, hence producing partial but more reliable predictions is even more interesting [5].

To obtain these cautious predictions, we propose to estimate sets of probabilities [10] from the data in the form of probabilistic bounds over specific events, and to then derive the (possibly) partial predictions from it. As computations with generic methods using sets of probabilities (e.g., using imprecise graphical models [10]) can be quite complex, we propose in Section 2 to consider two well-known binary decompositions whose extension to probability sets keep computations tractable, namely Frank & Hall decomposition [18] and nested dichotomies decompositions [17]. In Section 3, we discuss how to perform inferences from such probability sets both with general loss functions and with the classical 0/1 loss function. We end (Section 4) by providing several experiments showing that our cautious approach can help identify hard to predict cases and provides more reliable predictions for those cases.

2   Probability Set Estimation through Binary Decomposition

The goal of ordinal classification is to associate an instance $\mathbf{x} = \mathbf{x}^1 \times \ldots \times \mathbf{x}^p$ coming from an instance space $\mathscr{X} = \mathscr{X}^1 \times \ldots \times \mathscr{X}^p$ to a single label of the space $\mathscr{Y} = \{y_1, \ldots, y_m\}$ of possible classes. Ordinal classification differs from multi-class classification in that labels y_i are ordered, that is $y_i \prec y_{i+1}$ for $i = 1, \ldots, m-1$. An usual task is then to estimate the theoretical conditional probability measure $P_\mathbf{x} : 2^{\mathscr{Y}} \rightarrow [0,1]$ associated to an instance $\mathbf{x}$ from a set of n training samples $(\mathbf{x}_i, \ell_{x_i}) \in \mathscr{X} \times \mathscr{Y}$, $i = 1, \ldots, n$.

In order to derive cautious inferences, we shall explore in this paper the possibility to provide a convex set $\mathscr{P}_\mathbf{x}$ of probabilities as an estimate rather than a precise probability $\hat{P}_\mathbf{x}$, with the idea that the size of $\mathscr{P}_\mathbf{x}$ should decrease as more data (i.e., information) become available, converging to $P_\mathbf{x}$.

Manipulating generic sets $\mathscr{P}_\mathbf{x}$ to compute expectations or make inferences can be tedious, hence it is interesting to focus on collections of assessments that are easy to obtain and induce sets $\mathscr{P}_\mathbf{x}$ that allow for easy computations. Here we focus on the extensions of two particularly attractive binary decomposition techniques already used to estimate a precise $\hat{P}_\mathbf{x}$, namely Frank et Hall [18] technique and nested dichotomies [19].

2.1 Imprecise Cumulative Distributions

In their original paper, Frank et Hall suggest [18] to estimate, for an instance $\mathbf{x}$, the probabilities that its output ℓ_x will be less or equal than y_k, $k = 1, \ldots, m-1$. That is, one should estimate the $m-1$ probabilities $P_{\mathbf{x}}(A_k) := F_{\mathbf{x}}(y_k)$ where $A_k = \{y_1, \ldots, y_k\}$, the mapping $F_{\mathbf{x}} : \mathscr{Y} \to [0,1]$ being equivalent to a discrete cumulative distribution. The probabilities $P_{\mathbf{x}}(\ell_{\mathbf{x}} = y_k)$ can then be deduced through the formula $P_{\mathbf{x}}(\{y_k\}) = F_{\mathbf{x}}(y_k) - F_{\mathbf{x}}(y_{k-1})$.

The same idea can be applied to sets of probabilities, in which case we estimate the bounds

$$\underline{P}_{\mathbf{x}}(A_k) := \underline{F}_{\mathbf{x}}(y_k) \text{ and } \overline{P}_{\mathbf{x}}(A_k) := \overline{F}_{\mathbf{x}}(y_k),$$

where $\underline{F}_{\mathbf{x}}, \overline{F}_{\mathbf{x}} : \mathscr{Y} \to [0,1]$ correspond to lower and upper cumulative distributions. These bounds induce a well-studied [15] probability set $\mathscr{P}_{\mathbf{x}}([\underline{F}, \overline{F}])$. For $\mathscr{P}_{\mathbf{x}}([\underline{F}, \overline{F}])$ to be properly defined, we need the two mappings $\underline{F}_{\mathbf{x}}, \overline{F}_{\mathbf{x}}$ to be increasing with $\underline{F}_{\mathbf{x}}(y_m) = \overline{F}_{\mathbf{x}}(y_m) = 1$ and to satisfy the inequality $\underline{F}_{\mathbf{x}} \leq \overline{F}_{\mathbf{x}}$. In practice, estimates $\underline{F}_{\mathbf{x}}, \overline{F}_{\mathbf{x}}$ obtained from data will always satisfy the latest inequality, however when using binary classifiers on each event A_k, nothing guarantees that they will be increasing, hence the potential need to correct the model. Algorithm 1 provides an easy way to obtain a well-defined probability set. In spirit, it is quite similar to the Frank et Hall estimates $P_{\mathbf{x}}(y_k) = \max\{0, F_{\mathbf{x}}(y_k) - F_{\mathbf{x}}(y_{k-1})\}$, where an implicit correction is performed to obtain well-defined probabilities in case $F_{\mathbf{x}}$ is not increasing.

Algorithm 1. Correction of estimates $\underline{F}_{\mathbf{x}}, \overline{F}_{\mathbf{x}}$ into proper estimates

---

Input: estimates $\underline{F}_{\mathbf{x}}, \overline{F}_{\mathbf{x}}$ obtained from data
Output: corrected estimates $\underline{F}_{\mathbf{x}}, \overline{F}_{\mathbf{x}}$
1 **for** $k=1, \ldots, m-1$ **do**
2   **if** $\overline{F}_{\mathbf{x}}(y_k) > \overline{F}_{\mathbf{x}}(y_{k+1})$ **then** $\overline{F}_{\mathbf{x}}(y_{k+1}) \leftarrow \overline{F}_{\mathbf{x}}(y_k)$;
3   **if** $\underline{F}_{\mathbf{x}}(y_{m-k+1}) < \underline{F}_{\mathbf{x}}(y_{m-k})$ **then** $\underline{F}_{\mathbf{x}}(y_{m-k}) \leftarrow \underline{F}_{\mathbf{x}}(y_{m-k+1})$;

2.2 Nested Dichotomies

The principle of nested dichotomies is to form a tree structure using the class values $y_i \in \mathscr{Y}$. A nested dichotomy consists in recursively partitioning a tree node $C \subseteq \mathscr{Y}$ into two subsets A and B such that $A \cap B = \emptyset$ and $A \cup B = C$, until every leaf-node corresponds to a single class value ($card(C) = 1$). The root node is the whole set of classes $\mathscr{Y}$. To each branch A and B of a node C are associated conditional probabilities $P_{\mathbf{x}}(A|C) = 1 - P_{\mathbf{x}}(B|C)$. In the case of ordinal classifications, events C are of the kind $\{y_i, y_{i+1}, \ldots, y_j\}$ and their splits of the kind $A = \{y_i, y_{i+1}, \ldots, y_k\}$ and $B = \{y_{k+1}, \ldots, y_j\}$.

Generalizing the concept of nested dichotomies is pretty straightforward: it consists in allowing every local conditional probability to be imprecise, that is to each node C can be associated an interval $[\underline{P}_{\mathbf{x}}(A \mid C), \overline{P}_{\mathbf{x}}(A \mid C)]$, precise nested dichotomies being retrieved when $\underline{P}_{\mathbf{x}}(A \mid C) = \overline{P}_{\mathbf{x}}(A \mid C)$ for every node C. By duality of the imprecise probabilities [30, Sec.2.7.4.], we have $\underline{P}_{\mathbf{x}}(A \mid C) = 1 - \overline{P}_{\mathbf{x}}(B \mid C)$ and $\overline{P}_{\mathbf{x}}(A \mid C) = 1 - \underline{P}_{\mathbf{x}}(B \mid C)$. Such an imprecise nested dichotomy is then associated to a set $\mathscr{P}_{\mathbf{x}}$ of joint probabilities, obtained by considering all precise selection $P_x(A \mid C) \in [\underline{P}_x(A \mid C), \overline{P}_x(A \mid C)]$ for each node C. Figure 1 shows examples of a precise and an imprecise nested dichotomy tree when $\mathcal{Y} = \{y_1, y_2, y_3\}$.

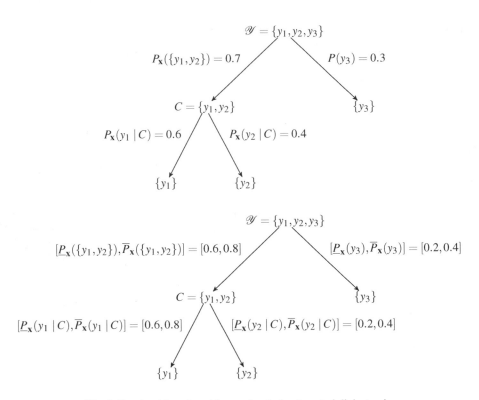

Fig. 1. Precise (above) and imprecise (below) nested dichotomies

3   Inferences

In this section, we expose how inferences (decision making) can be done with our two decompositions, both with general costs and 0/1 costs. While other costs such as the absolute error cost are also natural in an ordinal classification setting [14], we chose to focus on the 0/1 cost, as it is the only one for which a theoretically sound way to compare determinate and indeterminate classifiers, i.e., classifiers returning respectively precise and (potentially) imprecise classification, has been provided [33].

We will first recall the basic of decision making with probabilities and will then present their extensions when considering sets of probabilities. Let us denote by $c_k : \mathcal{Y} \to \mathbb{R}$ the cost (loss) function associated to y_k, that is $c_k(y_j)$ is the cost of predicting y_k when y_j is true. In the case where precise estimates $P_x(y_k)$ are obtained from the learning algorithm, obtaining the optimal prediction is

$$\hat{y} = \arg \min_{y_k \in \mathcal{Y}} \mathbb{E}_x(c_k)$$

with $\mathbb{E}_\mathbf{x}$ the expectation of c_k under $P_\mathbf{x}$, i.e. we predict the value having the minimal expected cost.

In practice, this also comes down to build a preference relation $\succ_P$ on elements of $\mathcal{Y}$, where $y_l \succ_{P_\mathbf{x}} y_k$ iff $\mathbb{E}_\mathbf{x}(c_k) > \mathbb{E}_\mathbf{x}(c_l)$ or equivalently $\mathbb{E}_\mathbf{x}(c_k - c_l) > 0$, that is the expected cost of predicting y_l is lower than the expected cost of predicting y_k. When working with a set $\mathcal{P}_\mathbf{x}$ of probabilities, this can be extended by building a partial order $\succ_{\underline{\mathbb{E}}_\mathbf{x}}$ on elements of $\mathcal{Y}$ such that $y_l \succ_{\underline{\mathbb{E}}_\mathbf{x}} y_k$ iff $\underline{\mathbb{E}}_\mathbf{x}(c_k - c_l) > 0$ with

$$\underline{\mathbb{E}}_\mathbf{x}(c_k - c_l) = \inf_{P_\mathbf{x} \in \mathcal{P}_\mathbf{x}} \mathbb{E}_\mathbf{x}(c_k - c_l).$$

That is, we are sure that the cost of exchanging y_k with y_l will have a positive expectation (hence y_l is preferred to y_k). The final cautious prediction $\hat{Y}$ is then obtained by taking the maximal elements of the partial order $\succ_{\underline{\mathbb{E}}_\mathbf{x}}$, that is

$$\hat{Y} = \{y \in \mathcal{Y} : \nexists y' \neq y \text{ s.t. } y' \succ_{\underline{\mathbb{E}}_\mathbf{x}} y\}$$

and is known under the name maximality criterion [30,29]. In practice, getting $\hat{Y}$ requires at worst a number $m(m-1)/2$ of computations that is quadratic in the number of classes. A conservative approximation (in the sense that the obtained set of non-dominated classes includes $\hat{Y}$) can be obtained by using the notion of interval dominance [29], in which $y_l \succ_{\underline{\mathbb{E}}_\mathbf{x}} y_k$ if $\underline{\mathbb{E}}_\mathbf{x}(c_k) > -\underline{\mathbb{E}}_\mathbf{x}(-c_l)$, thus requiring only $2m$ computations at worst to compare all classes, yet as m is typically low in ordinal classification, we will only consider maximality here.

In particular, 0/1 costs are defined as $c_k(y_j) - 1$ if $j \neq k$ and 0 else. If we note $\mathbf{1}_{(A)}$ the indicator function of A ($\mathbf{1}_{(A)}(x) = 1$ if $x \in A$, 0 else), then $(c_k - c_l) = \mathbf{1}_{(y_l)} - \mathbf{1}_{(y_k)}$ as $c_k(y_j) - c_l(y_j) = -1$ if $j = k$, 1 if $j = l$ and 0 if $j \neq k,l$. Hence we have $y_l \succ_{\underline{\mathbb{E}}_\mathbf{x}} y_k$ iff $\underline{\mathbb{E}}_\mathbf{x}(\mathbf{1}_{(y_l)} - \mathbf{1}_{(y_k)}) > 0$. Table 1 provides an example of the functions over which lower expectations must be computed for 0/1 losses in the case $\mathcal{Y} = \{y_1, \ldots, y_5\}$.

Table 1. 0/1 cost functions comparing y_2 and y_4

	y_1	y_2	y_3	y_4	y_5
c_2	1	0	1	1	1
c_4	1	1	1	0	1
$c_2 - c_4$	0	−1	0	1	0

3.1 Inference with Imprecise Cumulative Distributions

If the probability set $\mathcal{P}_\mathbf{x}([\underline{F}, \overline{F}])$ is induced by the bounding cumulative distributions $[\underline{F}_\mathbf{x}, \overline{F}_\mathbf{x}]$, then it can be shown[1] that the lower expectation of any function f over $\mathcal{Y}$ can be computed through the Choquet Integral: if we denote by $()$ a reordering of elements of $\mathcal{Y}$ such that $f(y_{(1)}) \leq \ldots \leq f(y_{(m)})$, this integral reads

$$\underline{\mathbb{E}}_\mathbf{x}(f) = \sum_{i=1}^{m} (f(y_{(i)}) - f(y_{(i-1)})) \underline{P}_\mathbf{x}(A_{(i)}) \tag{1}$$

---

[1] For details, interested readers are referred to [15]. Shortly speaking, this is due to the supermodularity of the induced lower probability.

with $f(y_{(0)}) = 0$, $A_{(i)} = \{y_{(i)}, \dots, y_{(m)}\}$ and $\underline{P}_{\mathbf{x}}(A_{(i)}) = \inf_{P_{\mathbf{x}} \in \mathscr{P}_{\mathbf{x}}([\underline{F}, \overline{F}])} P(A_{(i)})$ is the lower probability of $A_{(i)}$. In the case of imprecise cumulative distributions, the lower probability of an event A can be easily obtained: let $C = [y_{\underline{j}}, y_{\overline{j}}]$ denote a discrete interval of $\mathscr{Y}$ such that $[y_{\underline{j}}, y_{\overline{j}}] = \{y_i \in \mathscr{Y} : \underline{j} \le i \le \overline{j}\}$, then $\underline{P}_{\mathbf{x}}(C) = \max\{0, \underline{F}_{\mathbf{x}}(y_{\overline{j}}) - \overline{F}_{\mathbf{x}}(y_{\underline{j}-1})\}$ with $\underline{F}_{\mathbf{x}}(y_0) = \overline{F}_{\mathbf{x}}(y_0) = 0$. Any event A can then be expressed as a union of disjoint intervals[2] $A = C_1 \cup \dots \cup C_M$, and we have [15] $\underline{P}_{\mathbf{x}}(A) = \sum_{i=1}^{M} \underline{P}_{\mathbf{x}}(C_i)$.

Table 2. Imprecise cumulative distribution

	y_1	y_2	y_3	y_4	y_5
$\overline{F}_{\mathbf{x}}$	0.15	0.5	0.55	0.95	1
$\underline{F}_{\mathbf{x}}$	0.1	0.4	0.5	0.8	1

Example 1. Consider the imprecise cumulative distributions defined by Table 2 together with a 0/1 loss and the function $c_2 - c_4$ of Table 1. The elements used in the computation of the Choquet integral (1) for this case are summarized in Table 3.

Table 3. Choquet integral components of Example 1

i	$y_{(i)}$	$f_{(i)}$	$A_{(i)}$	$\underline{P}_{\mathbf{x}}(A_{(i)})$
1	y_2	-1	$\mathscr{Y}$	1
2	y_1	0	$\{y_1, y_3, y_4, y_5\}$	0.6
3	y_3	0	$\{y_3, y_4, y_5\}$	0.5
4	y_5	0	$\{y_4, y_5\}$	0.45
5	y_4	1	$\{y_4\}$	0.25

The lower probability of $A_{(2)} = \{y_1, y_3, y_4, y_5\} = \{y_1\} \cup \{y_3, y_4, y_5\}$ is

$$\underline{P}_{\mathbf{x}}(A_{(2)}) = \underline{P}_{\mathbf{x}}(\{y_1\}) + \underline{P}_{\mathbf{x}}(\{y_3, y_4, y_5\})$$
$$= \max\{0, \underline{F}_{\mathbf{x}}(y_1) - \overline{F}_{\mathbf{x}}(y_0)\} + \max\{0, \underline{F}_{\mathbf{x}}(y_5) - \overline{F}_{\mathbf{x}}(y_2)\}$$
$$= 0.1 + 0.5,$$

and the final value of the lower expectation $\underline{\mathbb{E}}_{\mathbf{x}}(c_2 - c_4) = -0.15$, meaning that y_4 is not preferred to y_2 in this case. As we also have $\underline{\mathbb{E}}_{\mathbf{x}}(c_4 - c_2) = -0.2$, y_2 and y_4 are incomparable under a 0/1 loss and given the bounding distributions $\underline{F}_{\mathbf{x}}, \overline{F}_{\mathbf{x}}$. Actually, our cautious prediction would be $\hat{Y} = \{y_2, y_4\}$, as we have $y_i \succ_{\underline{\mathbb{E}}} y_j$ for any $i \in 2, 4$ and $j \in 1, 3, 5$.

---

[2] Two intervals $[y_{\underline{j}}, y_{\overline{j}}], [y_{\underline{k}}, y_{\overline{k}}]$ are said disjoint if $\overline{j} + 1 < \underline{k}$.

3.2   Inference with Nested Dichotomies

In the precise case, computations of expectations with nested dichotomies can be done by backward recursion and local computations (simply applying the law of iterated expectation). That is the global expectation $\mathbb{E}_{\mathbf{x}}(f)$ of a function $f : \mathcal{Y} \to \mathbb{R}$ can be done by computing local expectations for each node, starting from the tree leaves taking values $f(y)$. This provides nested dichotomies with a computationally efficient method to estimate expectations.

It has been shown [13] that the same recursive method can be applied to imprecise nested dichotomies. Assume we have a split $\{A, B\}$ of a node C, and a real-valued (cost) function $f : \{A,B\} \to \mathbb{R}$ defined on $\{A,B\}$. We can compute the (local) lower expectation associated with the node C by :

$$\underline{\mathbb{E}}_{\mathbf{x},C}(f) = \min \left\{ \begin{array}{l} \underline{P}_{\mathbf{x}}(A \,|\, C)f(A) + \overline{P}_{\mathbf{x}}(B \,|\, C)f(B), \\ \overline{P}_{\mathbf{x}}(A \,|\, C)f(A) + \underline{P}_{\mathbf{x}}(B \,|\, C)f(B) \end{array} \right\} \qquad (2)$$

Starting from a function such as the one given in Table 1, we can then go from the leaves to the root of the imprecise nested dichotomy to obtain the associated lower expectation.

Example 2. Consider a problem where we have $\mathcal{Y} = \{y_1, y_2, y_3\}$ and the same imprecise dichotomy as in Figure 1. Figure 2 shows the local computations performed to obtain the lower expectation of $c_1 - c_3$. For instance, using Eq. (2) on node $C = \{y_1, y_2\}$, we get

$$\underline{\mathbb{E}}_{\mathbf{x},\{y_1,y_2\}}(c_1 - c_3) = \min\{-1 \cdot 0.8 \mid 0 \cdot 0.2, -1 \cdot 0.6 + 0 \cdot 0.4\}$$

We finally obtain $\underline{\mathbb{E}}_{\mathbf{x},\mathcal{Y}}(c_1 - c_3) = -0.44$, concluding that y_3 is not preferred to y_1. As the value $\underline{\mathbb{E}}_{\mathbf{x},\mathcal{Y}}(c_3 - c_1) = -0.04$ is also negative, we can conclude that y_1 and y_3 are not comparable. Yet we do have $\underline{\mathbb{E}}_{\mathbf{x},\mathcal{Y}}(c_2 - c_1) > 0$, meaning that y_1 is preferred to y_2, hence $\hat{Y} = \{y_1, y_3\}$.

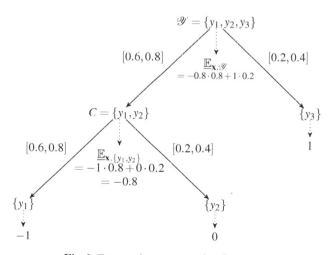

Fig. 2. Expectation computation for $c_1 - c_3$

4   Experimentations

This section presents the experiments we achieved to compare decomposition methods providing determinate predictions and their imprecise counterpart delivering possibly indeterminate predictions.

4.1   Learning Method

In our experiments, we consider a base classifier which can be extended easily to output interval-valued probabilities, so that we can evaluate the impact of allowing for cautiousness in ordinal classification. For this reason, we use the Naive Bayesian Classifier (NBC) which has an extension in imprecise probabilities : the Naive Credal Classifier (NCC) [32].

The NCC preserves the main properties of NBC, such as the assumption of attribute independence conditional on the class. In binary problems where we have to differentiate between two complementary events A and B, NBC reads

$$P(A|x^1,\dots,x^p) = \frac{P(A)\prod_{i=1}^{p} P(x^i\mid A)}{\prod_{i=1}^{p} P(x^i\mid A)P(A) + \prod_{i=1}^{p} P(x^i\mid B)P(B)}, \qquad (3)$$

where $(x_1,\dots,x_p)$ are the feature variables and A,B are the two events whose probability we have to estimate. The NCC consists in using probability bounds in Eq. 3, getting

$$\underline{P}(A|x^1,\dots,x^p) = \min\left\{ \begin{array}{c} \frac{P(A)\prod_{i=1}^{p}\underline{P}(x^i|A)}{\prod_{i=1}^{p}\underline{P}(x^i|A)\underline{P}(A)+\prod_{i=1}^{p}\underline{P}(x^i|B)\overline{P}(B)}, \\[2ex] \frac{\overline{P}(A)\prod_{i=1}^{p}\underline{P}(x^i|A)}{\prod_{i=1}^{p}\underline{P}(x^i|A)\overline{P}(A)+\prod_{i=1}^{p}\underline{P}(x^i|B)\underline{P}(B)} \end{array} \right\} = 1 - \overline{P}(B|x^1,\dots,x^p). \quad (4)$$

and $\underline{P}(B|x^1,\dots,x^p) = 1 - \overline{P}(A|x^1,\dots,x^p)$ can be obtained in the same way. Using the Imprecise Dirichlet Model (IDM) [4], we can compute these probability estimates from the training data by simply counting occurrences :

$$\underline{P}(x^i\mid A) = \frac{occ_{i,A}}{occ_A + s}, \quad \overline{P}(x^i\mid A) = \frac{occ_{i,A}+s}{occ_A + s} \qquad (5)$$

and

$$\underline{P}(A) = \frac{occ_{i,A}}{n_{A,B}+s}, \quad \overline{P}(A) = \frac{occ_{i,A}+s}{n_{A,B}+s} \qquad (6)$$

where $occ_{i,A}$ is the number of instances in the training set where the attribute $\mathscr{X}^i$ is equal to x^i and the class value is in A, occ_A the number of instances in the training set where the class value is in A, $n_{A,B}$ is the number of training sample whose class is either in A or B. The hyper-parameter s that sets the imprecision level of the IDM is usually equal to 1 or 2 [31].

4.2 Evaluation

Comparing classifiers that return cautious (partial) predictions in the form of multiple classes is an hard problem. Indeed, compared to the usual setting, measures of performance have to include the informativeness of the predictions in addition to the accuracy. Zaffalon et al. [33] discuss in details the case of comparing a cautious prediction with a classical one under a 0/1 loss assumption, using a betting interpretation. They show that the discounted accuracy, which rewards a cautious prediction Y class with $1/|Y|$ if the true class is in Y, and zero otherwise, is a measure satisfying a number of appealing properties. However, they also show that discounted accuracy makes no difference between a cautious classifier providing indeterminate predictions and a random classifier: for instance, in a binary setting, a cautious classifier always returning both classes would have the same value as a classifier picking the class at random, yet the determinate classifier displays a lower variance (it always receives $1/2$ as reward, while the random one would receive a reward of 1 half of the time, and 0 the other half).

This is why a decision maker that wants to value cautiousness should consider modifying discounted accuracy by a risk-adverse utility function [33]. Here, we consider the u_{65} function: Let $(\mathbf{x}_i, \ell_i)$, $i = 1, \ldots, n$ be the set of test data and Y_i our (possibly imprecise) predictions, then u_{65} is

$$u_{65} = \frac{1}{n}\sum_{i=1}^{n} -0.6d_i^2 + 1.6d_i,$$

where $d_i = \mathbf{1}_{(Y_i)}(\ell_i)/|Y_i|$ is the discounted accuracy. It has been shown in [33] that this approach is consistent with the use of F_1 measures [12,1] as a way to measure the

Table 4. Data set details

Name	#instances	#features	#classes
autoPrice	159	16	5
bank8FM	8192	9	5
bank32NH	8192	33	5
boston housing	506	14	5
california housing	20640	9	5
cpu small	8192	13	5
delta ailerons	7129	6	5
elevators	16599	19	5
delta elevators	9517	7	5
friedman	40768	11	5
house 8L	22784	9	5
house 16H	22784	17	5
kinematics	8192	9	5
puma8NH	8192	9	5
puma32H	8192	33	5
stock	950	10	5
ERA	1000	5	9
ESL	488	5	9
LEV	1000	5	5

quality of indeterminate classifications. In fact, it is shown in [33] that u_{65} is less in favor of indeterminate classifiers than the use of F_1 measure.

4.3 Results

In this section, our method is tested on 19 datasets of the UCI machine learning repository [16], whose details are given in Table 4. As there is a general lack of benchmark data sets for ordinal classification data, we used regression problems that we turned into ordinal classification by discretizing the output variable, except for the data sets LEV that has 5 ordered classes and ESL, ERA that have 9 ordered classes. The results reported in this section are obtained with a discretization into five classes of equal frequencies. We also performed experiments on the other data sets, using 7 and 9 discretized classes, obtaining the same conclusions.

The results in this section are obtained from a 10-fold cross validation. To build the dichotomy trees, we selected at each node the split $A = \{y_i, y_{i+1}, \ldots, y_k\}$ and $B = \{y_{k+1}, \ldots, y_j\}$ of $C = \{y_i, y_{i+1}, \ldots, y_j\}$ that maximised the u_{65} measure on the binarized data set. We use ordinal logistic regression (logreg) as a base line classifier to compare our results. For each decomposition method, Frank & Hall (F) and nested dichotomies (N), we compared the naive Bayes classifier (B) with its indeterminate counterpart (NCC), picking an hyper-parameter $s = 2$ for the IDM in Eqs. (5)- (6). The naive Bayes classifier was used in a classical way to provide determinate predictions

Table 5. u_{65} Results (and method rank) obtained on the different methods. Log= logistic regression, B = Naive Bayes classifier, C = Naive credal classifier, A= Alonso *et al.* prediction method, F = Frank & Hall, N = Nested Dichotomies.

	Log	B/F	B/F/A	C/F	B/N	B/N/A	C/N
autoPrice	52.2 (5)	58.5 (3)	39.7 (7)	53.8 (4)	59.1 (1)	51.3 (6)	58.6 (2)
bank8FM	68.2 (2)	67.4 (3)	37.3 (7)	68.3 (1)	63.9 (5)	54.9 (6)	64.8 (4)
bank32NH	43.3 (4)	43.6 (3)	30.2 (7)	47.8 (1)	42.9 (5)	40.2 (6)	46.7 (2)
boston hous.	55.6 (4)	55.1 (5)	34.1 (7)	55.8 (3)	56.1 (2)	43.9 (6)	57.4 (1)
california hous.	47.6 (5)	48.2 (4)	32.9 (7)	48.6 (2)	48.3 (3)	43.5 (6)	48.7 (1)
cpu small	58.8 (3)	57 (5)	40.9 (7)	57.1 (4)	60.8 (2)	54.1 (6)	61.1 (1)
delta ail.	50.2 (6)	53.5 (4)	31.8 (7)	53.8 (3)	54.2 (2)	52.1 (5)	54.9 (1)
elevators	42.7 (2)	39.0 (5)	30.5 (7)	39.2 (4)	42.6 (3)	37.9 (6)	42.9 (1)
delta elev.	46.5 (6)	49.9 (5)	34.3 (7)	50.4 (4)	50.8 (3)	53.2 (1)	51.2 (2)
friedman	53.2 (5)	63.8 (2)	32 (7)	64.5 (1)	62.2 (4)	47.3 (6)	63 (3)
house 8L	39.9 (6)	49.6 (2)	34.9 (7)	49.8 (1)	49.4 (4)	43.9 (5)	49.6 (3)
house 16H	41.4 (6)	47.5 (4)	35.3 (7)	47.6 (3)	50.0 (2)	43.9 (5)	50.2 (1)
kinematics	37.7 (5)	44.9 (3)	28.8 (7)	46.2 (1)	44.4 (4)	37.5 (6)	45.4 (2)
puma8NH	30.3 (6)	46.5 (4)	29.7 (7)	47.6 (3)	47.7 (2)	42.9 (5)	48.3 (1)
puma32H	30.5 (6)	48.6 (3)	29.7 (7)	50.9 (1)	47.7 (4)	40.6 (5)	49.9 (2)
stock	61.2 (6)	72.4 (3)	41.7 (7)	71.5 (4)	75.1 (1)	61.2 (5)	74.2 (2)
ERA	23.2 (5)	23.2 (4)	14.1 (7)	28.5 (1)	22.5 (6)	26.8 (2)	26.6 (3)
ESL	12.7 (7)	55.7 (4)	28.1 (6)	53.4 (5)	57 (2)	63 (1)	56.5 (3)
LEV	46.3 (6)	60.5 (2)	44.9 (7)	60.4 (3)	59.8 (4)	61.6 (1)	59.6 (5)
Avg. rank	5	3.6	6.9	2.6	3.1	4.7	2.1

and (B/A) with the F_1 measure of Alonso *et al.* [1] to produce indeterminate predictions (details about this latter method can be found in [1]).

Table 5 show the obtained results in terms of u_{65} (that reduces to classical accuracy for the three determinate methods) as well as the rank of each classifier. Using Demsar's approach by applying the Friedman statistic on the ranks of algorithm performance for each dataset, we obtain a value of 68.16 for the Chi Square, and a 26.8 statistic for the F-distribution. Since the statistic is 1.7 for a p-value of 0.05, we can safely reject the null hypothesis, meaning that the performances of the classifiers are significatively different. This shows that in average the introduced indeterminacy (or cautiousness) in the predictions is not too important and is compensated by more reliable predictions. We use Nemenyi test as a post-hoc test, and obtain that two classifiers are significantly different (with p-value 0.05) if the difference between their mean rank is higher than 2.06.

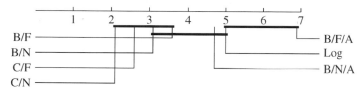

Fig. 3. Post-hoc test results on algorithms. Thick lines links non-significantly different algorithms.

Figure 3 summarises the average ranks of the different methods and shows which one are significatively different from the others. We can see that, although techniques using probability sets (C/N and C/F) have the best average rank, they are not significantly different from their determinate Bayesian counterpart (B/N and B/F) under u_{65} measure. This is not surprising, since the goal of such classifiers is not to outperform Bayesian methods, but to provide more reliable predictions when not enough information is available. It should also be recalled that the u_{65} measure is only slightly favourable to indeterminate classifiers, and that other measures such as F_1 and u_{80} would have given better scores to indeterminate classifiers.

An interesting result is that Alonso *et al.* [1] method, that use a precise probabilistic models and produce indeterminate predictions through the use of specific cost functions (the F_1 measure in our case), performs quite poorly, in particular when applied with the Frank and Hall decomposition (B/F/A). This can be explained by the fact that Alonso *et al.* [1] method will mainly produce indeterminate classifications when the labels having the highest probabilities will have close probability values, i.e., when there will be some ambiguity as to the modal label. However, it is well known that the naive Bayes classifier tends to overestimate model probabilities, therefore acting as a good classifier for 0/1 loss functions, but as a not so good probability density estimator. This latter feature can clearly be counter-productive when using Alonso *et al.* [1] method, that relies on having good probability estimates. On the other hand, indeterminate classification using probability sets can identify situations where information is lacking, even if the underlying estimator is poor. Our results indicate that, while the two methods both produce indeterminate classifications, they do so in very different ways (and therefore present different interests).

Table 6 shows the mean imprecision of indeterminate predictions for all the methods producing such predictions. This sheds additional light on the bad performances of the B/F/A method, which tends to produce rather imprecise predictions without necessarily counterbalancing them with an higher reliability or accuracy. For the other methods, the mean imprecision is comparable.

Table 6. Mean imprecision of predictions (rank)

	B/F/A	C/F	B/N/A	C/N
autoPrice	2.22 (3)	2.25 (4)	1.03 (1)	1.93 (2)
bank8FM	2.06 (4)	1.06 (1)	1.55 (3)	1.08 (2)
bank32NH	2.11 (4)	1.78 (2)	2.01 (3)	1.72 (1)
boston housing	2.23 (4)	1.36 (2)	1.11 (1)	1.51 (3)
california housing	2.17 (4)	1.04 (2)	1.6 (3)	1.04 (1)
cpu small	2.38 (4)	1.03 (1)	1.2 (3)	1.04 (2)
delta ailerons	2.54 (4)	1.03 (1)	1.62 (3)	1.06 (2)
elevators	2.47 (4)	1.03 (1)	1.39 (3)	1.04 (2)
delta elevators	2.47 (4)	1.05 (2)	1.63 (3)	1.04 (1)
friedman	2.06 (3)	1.06 (1)	2.17 (4)	1.06 (2)
house 8L	2.24 (4)	1.01 (1)	1.43 (3)	1.02 (2)
house 16H	2.28 (4)	1.02 (1)	1.25 (3)	1.03 (2)
kinematics	2.12 (3)	1.21 (2)	2.36 (4)	1.2 (1)
puma8NH	2.16 (4)	1.12 (2)	1.89 (3)	1.1 (1)
puma32H	2.47 (4)	1.43 (1)	1.91 (3)	1.5 (2)
stock	2.21 (4)	1.15 (3)	1.04 (1)	1.14 (2)
ERA	4.02 (4)	2.81 (3)	2.24 (1)	2.32 (2)
ESL	3.62 (4)	2.27 (3)	1.39 (1)	1.84 (2)
LEV	2.05 (4)	1.18 (2)	1.52 (3)	1.12 (1)

Figures 4 displays the non-discounted accuracy (that is, we count 1 each time the true class is in the prediction, whether its determinate or not) on those instances where the use of NCC returned an indeterminate classification. On those instances, the accuracy

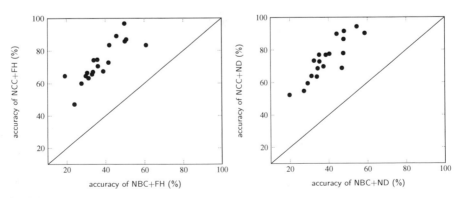

Fig. 4. Non-discounted accuracy of the NBC vs NCC methods for both decompositions on indeterminate instances

of the determinate version (NBC) is on average 10 % lower than the accuracy displayed in Table 5. In contrast, the non-discounted accuracy of the indeterminate version on these instances is much higher, meaning that the indeterminacy actually concerns hard-to-classify instances.

5   Conclusions

In this paper, we have proposed two methods to learn cautious ordinal classifiers, in the sense that they provide indeterminate predictions when information is insufficient to provide a reliable determinate one. More precisely, these methods extend two well-known binary decomposition methods previously used for ordinal classification, namely Frank & Hall decomposition and nested dichotomies. The extension consists in allowing one to provide interval-valued probabilistic estimates rather than precise ones for each binary problem, the width of the interval reflecting our lack of knowledge about the instances.

Our experiments on different data sets show that allowing for cautiousness in ordinal classification methods can increase the reliability of the prediction, while not providing too indeterminate predictions. More specifically, indeterminacy tends to focus on those instances that are hard to classify for determinate classifiers. We could probably improve both the efficiency of inferences, e.g., by studying extensions of labelling trees to imprecise trees [3], or their accuracy by using more complex classifiers, e.g., credal averaging techniques [11]. Yet, as the number m of labels in ordinal classification is usually small, and as the advantages of using binary decompositions are usually lower when using complex estimation methods, the benefits of such extensions would be limited.

In these experiments, we have focused on the 0/1 loss and its extensions to indeterminate classification u_{65}, which is more favourable to determinate classifier than the F_1 measure proposed by Alonso et al. [1].The reason for this is that 0/1 loss is the only one to which the results of Zaffalon et al. [33] that allows to compare determinate and indeterminate classifiers apply. Yet, our approaches can easily handle generic losses (in contrast with the multi-class naive credal classifier [32]), as shows Section 3 and Eqs (1)- (2). Also, there are loss functions such as the absolute error that are at least as natural to use in an ordinal classification problem as the 0/1 loss function. Our future efforts will therefore focus on determining meaningful ways to compare cost-sensitive determinate and indeterminate classifiers. Another drawback of using 0/1 loss function [24], shown by Examples 1 and 2, is that we may obtain indeterminate predictions containing non-consecutive labels. We expect that considering other losses such as L_1 loss could solve this issue.

In addition to that, we plan to apply the methods developed in this paper to more complex problems that can be reduced as a set of ordinal classification problems, such as graded multi-label [7] or label ranking [6].

Acknowledgements. This work was carried out and funded by the French National Research Agency, through the project ANR-13-JS03-0007 RECIF. It was also supported by the French Government, through the LABEX MS2T and the program "Investments for the future" managed by the National Agency for Research (Reference ANR-11-IDEX-0004-02).

References

1. Alonso, J., del Coz, J.J., Díez, J., Luaces, O., Bahamonde, A.: Learning to predict one or more ranks in ordinal regression tasks. In: Daelemans, W., Goethals, B., Morik, K. (eds.) ECML PKDD 2008, Part I. LNCS (LNAI), vol. 5211, pp. 39–54. Springer, Heidelberg (2008)
2. Bartlett, P., Wegkamp, M.: Classification with a reject option using a hinge loss. The Journal of Machine Learning Research 9, 1823–1840 (2008)
3. Bengio, S., Weston, J., Grangier, D.: Label embedding trees for large multi-class tasks. In: NIPS, vol. 23, p. 3 (2010)
4. Bernard, J.: An introduction to the imprecise dirichlet model for multinomial data. International Journal of Approximate Reasoning 39(2), 123–150 (2005)
5. Cheng, W., Hüllermeier, E., Waegeman, W., Welker, V.: Label ranking with partial abstention based on thresholded probabilistic models. In: Advances in Neural Information Processing Systems 25 (NIPS 2012), pp. 2510–2518 (2012)
6. Cheng, W., Hüllermeier, E.: A nearest neighbor approach to label ranking based on generalized labelwise loss minimization
7. Cheng, W., Hüllermeier, E., Dembczynski, K.J.: Graded multilabel classification: The ordinal case. In: Proceedings of the 27th International Conference on Machine Learning (ICML 2010), pp. 223–230 (2010)
8. Chow, C.: On optimum recognition error and reject tradeoff. IEEE Transactions on Information Theory 16(1), 41–46 (1970)
9. Chu, W., Keerthi, S.S.: Support vector ordinal regression. Neural Computation 19(3), 792–815 (2007)
10. Corani, G., Antonucci, A., Zaffalon, M.: Bayesian networks with imprecise probabilities: Theory and application to classification. In: Holmes, D.E., Jain, L.C. (eds.) Data Mining: Foundations and Intelligent Paradigms. ISRL, vol. 23, pp. 49–93. Springer, Heidelberg (2012)
11. Corani, G., Zaffalon, M.: Credal model averaging: an extension of bayesian model averaging to imprecise probabilities. In: Daelemans, W., Goethals, B., Morik, K. (eds.) ECML PKDD 2008, Part I. LNCS (LNAI), vol. 5211, pp. 257–271. Springer, Heidelberg (2008)
12. José del Coz, J., Bahamonde, A.: Learning nondeterministic classifiers. The Journal of Machine Learning Research 10, 2273–2293 (2009)
13. De Cooman, G., Hermans, F.: Imprecise probability trees: Bridging two theories of imprecise probability, vol. 172, pp. 1400–1427
14. Dembczyński, K., Kotłowski, W., Słowiński, R.: Learning rule ensembles for ordinal classification with monotonicity constraints. Fundamenta Informaticae 94(2), 163–178 (2009)
15. Destercke, S., Dubois, D., Chojnacki, E.: Unifying practical uncertainty representations - i: Generalized p-boxes. Int. J. Approx. Reasoning 49(3), 649–663 (2008)
16. Frank, A., Asuncion, A.: UCI machine learning repository (2010), http://archive.ics.uci.edu/ml
17. Frank, E., Kramer, S.: Ensembles of nested dichotomies for multi-class problems. In: ICML 2004, p. 39 (2004)
18. Frank, E., Hall, M.: A simple approach to ordinal classification. In: Flach, P.A., De Raedt, L. (eds.) ECML 2001. LNCS (LNAI), vol. 2167, pp. 145–156. Springer, Heidelberg (2001)
19. Frank, E., Kramer, S.: Ensembles of nested dichotomies for multi-class problems. In: Proceedings of the Twenty-first International Conference on Machine Learning, p. 39. ACM (2004)
20. Fürnkranz, J., Hüllermeier, E., Vanderlooy, S.: Binary decomposition methods for multipartite ranking. In: Buntine, W., Grobelnik, M., Mladenić, D., Shawe-Taylor, J. (eds.) ECML PKDD 2009, Part I. LNCS (LNAI), vol. 5781, pp. 359–374. Springer, Heidelberg (2009)

21. Ha, T.M.: The optimum class-selective rejection rule. IEEE Transactions on Pattern Analysis and Machine Intelligence 19(6), 608–615 (1997)
22. Herbrich, R., Graepel, T., Obermayer, K.: Large margin rank boundaries for ordinal regression. In: Advances in Neural Information Processing Systems, pp. 115–132 (1999)
23. Joachims, T.: Training linear svms in linear time. In: Proceedings of the 12th ACM SIGKDD International Conference on Knowledge Discovery and Data Mining, pp. 217–226. ACM (2006)
24. Kotlowski, W., Slowinski, R.: On nonparametric ordinal classification with monotonicity constraints. IEEE Trans. Knowl. Data Eng. 25(11), 2576–2589 (2013)
25. Li, L., Lin, H.T.: Ordinal regression by extended binary classification. In: Advances in Neural Information Processing Systems, pp. 865–872 (2006)
26. Shashua, A., Levin, A.: Ranking with large margin principle: Two approaches. In: Advances in Neural Information Processing Systems, pp. 937–944 (2002)
27. Sousa, R., Yevseyeva, I., da Costa, J.F.P., Cardoso, J.S.: Multicriteria models for learning ordinal data: A literature review. In: Yang, X.-S. (ed.) Artificial Intelligence, Evolutionary Computing and Metaheuristics. SCI, vol. 427, pp. 109–138. Springer, Heidelberg (2013)
28. Tehrani, A.F., Cheng, W., Hüllermeier, E.: Preference learning using the choquet integral: The case of multipartite ranking. IEEE T. Fuzzy Systems 20(6), 1102–1113 (2012)
29. Troffaes, M.: Decision making under uncertainty using imprecise probabilities. Int. J. of Approximate Reasoning 45, 17–29 (2007)
30. Walley, P.: Statistical reasoning with imprecise Probabilities. Chapman and Hall, New York (1991)
31. Walley, P.: Inferences from multinomial data: learning about a bag of marbles. Journal of the Royal Statistical Society. Series B (Methodological), 3–57 (1996)
32. Zaffalon, M.: The naive credal classifier. J. Probabilistic Planning and Inference 105, 105–122 (2002)
33. Zaffalon, M., Corani, G., Mauá, D.: Evaluating credal classifiers by utility-discounted predictive accuracy. International Journal of Approximate Reasoning (2012)

Error-Bounded Approximations for Infinite-Horizon Discounted Decentralized POMDPs

Jilles S. Dibangoye, Olivier Buffet, and François Charpillet

Inria & Université de Lorraine
Nancy, France
firstname.lastname@inria.fr

Abstract. We address decentralized stochastic control problems represented as decentralized partially observable Markov decision processes (Dec-POMDPs). This formalism provides a general model for decision-making under uncertainty in cooperative, decentralized settings, but the worst-case complexity makes it difficult to solve optimally (NEXP-complete). Recent advances suggest recasting Dec-POMDPs into continuous-state and deterministic MDPs. In this form, however, states and actions are embedded into high-dimensional spaces, making accurate estimate of states and greedy selection of actions intractable for all but trivial-sized problems. The primary contribution of this paper is the first framework for error-monitoring during approximate estimation of states and selection of actions. Such a framework permits us to convert state-of-the-art exact methods into error-bounded algorithms, which results in a scalability increase as demonstrated by experiments over problems of unprecedented sizes.

Keywords: decentralized stochastic control, error-bounded approximations.

Learning and planning algorithms for decentralized stochastic control problems are of importance in a number of practical domains such as network communications and control; rescue, surveillance and exploration tasks; multi-robotics; collaborative games [15]; to cite a few. Decentralized partially observable Markov decision processes (Dec-POMDPs) have emerged as a standard framework for modeling and solving such problems [6]. This formalism involves a set of agents with different, but related, observations about the world, which cooperate to achieve a common long-term goal, but cannot explicitly communicate with one another. While many decentralized stochastic control problems can be formalized as Dec-POMDPs, only a few of them can be solved optimally due to their worst-case complexity: finite horizon problems are in NEXP, and infinite horizon problems are undecidable [6]. This intractability is due to the doubly-exponential growth in required computational resources, making it hard to find an optimal solution for all but the smallest instances [12,5].

A recent scalability increase builds upon two fundamental results [7]. The first result establishes that Dec-POMDPs can be transformed with no loss of optimality into continuous-state and deterministic MDPs, called occupancy MDPs. In this form, the states —called occupancy states— are distributions over the states and action-observation joint histories of the original Dec-POMDPs, and the actions —called decentralized decision rules— are mappings from joint histories to joint actions of the

T. Calders et al. (Eds.): ECML PKDD 2014, Part I, LNCS 8724, pp. 338–353, 2014.

original Dec-POMDPs. Secondly, the optimal value function of a finite-horizon occupancy MDP is a piecewise linear and convex function of the occupancy state. These results allow to combine advances in continuous-state MDP and POMDP algorithms, which (among others) result in the feature-based heuristic search value iteration algorithm (FB-HSVI). This algorithm can produce optimal solutions for medium-sized problems and medium planning horizons, but quickly runs out of time and memory for larger-scale problems and planning horizons. Such limited scalability is mainly because states and actions of occupancy MDPs are embedded into high-dimensional spaces, making accurate estimate of states and greedy selection of actions intractable for all but trivial-sized problems.

A natural question to ask is whether approximate (error-bounded) solutions can be found efficiently for decentralized stochastic control problems. On the one hand, memory-bounded dynamic programming algorithms for solving infinite-horizon discounted Dec-POMDPs are often quite effective at finding good heuristic solutions, while requiring bounded computational resources [20,9,13,10]. However, these methods do not come with rigorous guarantees concerning the quality of the final heuristic solution. On the other hand, error-bounded algorithms for solving infinite-horizon discounted Dec-POMDPs exist. Examples include error-bounded methods for discounted POMDPs that are (or can be) transferred back to discounted Dec-POMDPs: policy iteration (PI) [5]; incremental policy generation (IPG) [2]; point-based value iteration (PBVI) [14,17]; and heuristic search value iteration (HSVI) [7,21]. These algorithms rely either on ϵ-pruning methods[1] (PI and IPG) or/and on exploration strategies that focus on a small subset of the search space (PBVI and HSVI), but they all make use of greedy action-selection and accurate state-estimation operators, which quickly exhausts the available resources before convergence. Furthermore, theoretical analyses of point-based approaches (e.g., PBVI) demonstrate that resulting error bounds are loose and have only a theoretical significance [21,10].

In this paper, we focus on characterizing efficient error-bounded solutions for infinite-horizon discounted decentralized stochastic control problems. The novel approach proceeds by converting infinite-horizon discounted Dec-POMDPs into finite-horizon discounted occupancy MDPs, thereby computing a non-stationary policy over a finite planning horizon. In such a setting, approximations are typically achieved by replacing greedy action-selection and accurate state-estimation operators by approximate counterparts. In addition, we preserve the ability to bound the error with respect to the optimal infinite-horizon value function. Our study differs from previous studies in that it directly bounds the regret, avoiding the max-norm machinery of previous analyses of value and policy iteration algorithms [18,4,17,21,10,19], which may result in tighter bounds. We further extend the state-of-the-art feature-based heuristic search value iteration algorithm to incorporate the error we make in both greedy action-selection and accurate state-estimation. The result is an algorithm that can solve problems of unprecedented sizes from the literature while providing strong theoretical guarantees.

In the remainder of this paper, we will introduce in Section 1 the Dec-POMDP framework and the reformulation into occupancy MDPs. Section 2 extends, from finite-

---

[1] An ϵ-pruning method circumvents regions of the search space that cannot significantly improve the current solution, and the resulting solution is guaranteed to be at ϵ of the optimum.

horizon settings to infinite-horizon ones, recent advances in optimally solving Dec-POMDPs as occupancy MDPs. Then, we describe the novel approximation framework, derive theoretical guarantees and algorithmic extensions in Section 3. Finally, we present in Section 4 experimental results demonstrating the scalability of the resulting error-bounded algorithm.

1   From Dec-POMDPs to OMDPs

This section presents formalisms for the infinite-horizon discounted decentralized partially observable Markov decision process (Dec-POMDP) and its associated occupancy Markov decision process (OMDP).

1.1   Decentralized Stochastic Control Problems as Dec-POMDPs

The Dec-POMDP framework formalizes a discrete stochastic system that evolves under the influence of N agents. A key assumption in this framework is that agents cannot directly observe the true state of the system. In fact, they have different but related observations about the state of the system and cannot explicitly communicate with one another. Nevertheless, they need to cooperate in order to achieve a common long-term goal, i.e., to select actions that maximize the collection of rewards in the long run.

Definition 1 (Dec-POMDP). *An N-agent decentralized partially observable Markov decision process is given as a tuple $M \equiv (S, \{A^i\}, \{Z^i\}, p, r, b_0, \gamma)$, where: S is a finite set of hidden states; A^i is a finite set of private actions of agent $i \in \{1, 2, \ldots, N\}$; Z^i is a finite set of private observations of agent $i \in \{1, 2, \ldots, N\}$; $p^{a,z}(s, s') = Pr(s', z|s, a)$ is a dynamics model of the team of agents as a whole; $r(s, a)$ is a reward model of the team of agents as a whole; b_0 is an initial belief state; and $\gamma \in (0, 1)$ is a discount factor.*

The goal of solving M is to find an N-tuple of private policies $\pi \equiv (\pi^i)_{i \in \{1, 2, \ldots, N\}}$ that yields the highest discounted total reward starting at b_0:

$$V^{\pi}_{M,\gamma,0}(b_0) = E\left\{\sum_{t=0}^{\infty} \gamma^t \cdot r(s_t, a_t) \mid \pi, b_0\right\}. \tag{1}$$

Let decentralized policy π be a N-tuple of private policies $(\pi^i)_{i \in \{1, 2, \ldots, N\}}$. Each private policy π^i is a sequence of private decision rules $(\pi^i_t)_{t \in \{0, 1, \ldots, \infty\}}$. The t-th private decision rule $\pi^i_t \colon \Theta^i_t \mapsto A^i$ of agent i prescribes private actions based on the whole information available to the agent up to time step t, namely its complete history of past actions and observations $\theta^i_t = (a^i_0, z^i_1, \ldots, a^i_{t-1}, z^i_t) \in \Theta^i_t$. We define Θ^i_t to be the set of all length-t private histories of actions and observations agent i may have experienced, $\Theta_t \equiv \times_{i \in \{1, 2, \ldots, N\}} \Theta^i_t$ the set of joint histories and $\Theta = \cup_{t \in \{0, 1, \ldots, T-1\}} \Theta_t$. In addition, we define the t-th decentralized decision rule π_t to be an N-tuple $(\pi^i_t)_{i \in \{1, 2, \ldots, N\}}$ of private decision rules.

Since the history length grows as time goes on, for infinite horizon cases, this would require private decision rules to have infinite memory, which is not possible in practice. Therefore, we shall specify the nature of the decentralized policies we target in more detail. We first notice that the optimal value function over an infinite horizon can be arbitrarily accurately approximated by the optimal value function over a finite horizon.

To this end, we choose finite horizon T so that the regret of operating only over $T = \lceil \log_\gamma ((1 - \gamma)\varepsilon/\|r\|_\infty) \rceil$ steps instead of an infinite number of steps is upper-bounded by any arbitrarily small scalar $\varepsilon > 0$, where $\|r\|_\infty = \max\{|r(s,a)|: \forall s \in S, \forall a \in A\}$. Indeed, the regret is upper-bounded by the cumulated sum of discounted losses from time step T onwards, so that: $\sum_{t=T}^\infty \gamma^t \|r\|_\infty \leq \varepsilon$.

In the remainder of this paper, we restrict the search space to decentralized policies described over planning horizon T. Unlike infinite-horizon decentralized policies, finite-horizon decentralized policies require a finite memory. At the execution phase, agents follow actions their private policies prescribe up to time step T; thereafter they behave randomly. Doing so, we are guaranteed to achieve performance with bounded error as discussed later below. Before proceeding any further, we next consider a reformulation of finite-horizon Dec-POMDPs into occupancy MDPs.

1.2 Occupancy Markov Decision Processes

The decentralized partially observable Markov decision process framework formalizes a decentralized stochastic control problem from a *perspective oriented towards agents*. In such a setting, agents are unaware of which actions the other agents take and which observations they receive; each agent behavior is based only upon its private histories. In this section, however, we formalize decentralized stochastic control problems from a *perspective oriented towards centralized solution methods*. In such a perspective, the system evolves under the control of agents based upon the total information about the state of the system the centralized solution method makes available to all agents prior to the execution phase, namely the information state.

The t-th information state $\zeta_t \equiv (b_0, \pi_0, \ldots, \pi_{t-1})$ is a sequence of decentralized decision rules starting at initial belief state b_0. It satisfies the following recursion: $\zeta_0 \equiv (b_0)$ and $\zeta_t \equiv (\zeta_{t-1}, \pi_{t-1})$, for all $t \in \{1, \ldots, T - 1\}$. Next, it will prove useful to introduce the concept of occupancy states, as a means of maintaining a concise representation of the information state. A t-th occupancy state ξ_t is a distribution $P^{\zeta_t}(s_t, \theta_t)$ over histories and hidden states of M conditional on an information state ζ_t. For the sake of simplicity, we use notation $\Theta(\xi_t)$ to represent histories that are reachable in occupancy state ξ_t. The occupancy state has many important properties. First, it is a sufficient statistic of the information state when estimating the (current and future) reward to be gained by executing a decentralized decision rule: $R(\xi_t, \pi_t) = \sum_{s_t} \sum_{\theta_t} \xi_t(s_t, \theta_t) \cdot r(s_t, \pi_t(\theta_t))$. In addition, it describes a deterministic and Markov decision process, where next occupancy state $\xi_{t+1} = P(\xi_t, \pi_t)$ depends only upon the current occupancy state ξ_t and the next decentralized decision rule π_t:

$$\xi_{t+1}(s', (\theta_t, a_t, z_{t+1})) = \mathbf{1}_{\{a_t\}}(\pi_t(\theta_t)) \sum_{s \in S} \xi_t(s, \theta_t) \cdot p^{a_t, z_{t+1}}(s, s'), \tag{2}$$

for $s' \in S$, $a_t \in A$, $z_{t+1} \in Z$, $\theta_t \in \Theta$ and where $\mathbf{1}_F$ is an indicator function. This process is known as the occupancy Markov decision process.

Definition 2 (OMDP). *Let $\hat{M} \equiv (\triangle, A, R, P, \gamma, \xi_0, T)$ be the T-steps OMDP with respect to Dec-POMDP M, where γ is a discount factor; ξ_0 corresponds to the initial belief in M; $\triangle \equiv \cup_{t \in \{0,1,\ldots,T\}} \triangle_t$ is the set of occupancy states up to time T; $A \equiv \cup_{t \in \{0,1,\ldots,T\}} A_t$*

is the finite set of decentralized decision rules; $R(\xi_t, \pi_t)$ is the reward model; $P(\xi_t, \pi_t)$ is the transition rule; and T is a planning horizon.

It is worth noticing that OMDP $\hat{M}$ can be seen as a generative model for occupancy states $P(\xi_t, \pi_t)$ and rewards $R(\xi_t, \pi_t)$, for all time step $t \in \{0, 1, \ldots, T - 1\}$. A recent result shows that an optimal solution for $\hat{M}$, together with the correct estimation of the occupancy states, will give rise to the optimal solution of the original Dec-POMDP M over finite horizon T [8].

2 Optimally Solving Dec-POMDPs as OMDPs

This section reviews how to optimally solve Dec-POMDPs as OMDPs, a theory originally introduced under the total reward criterion [7,8]. Here, we extend it to deal with the discounted total reward criterion.

2.1 Bellman's Optimality Equations

In this subsection, we extend dynamic programming properties, including Bellman's optimality equations, to OMDPs (respectively Dec-POMDPs). Before proceeding any further, we start with preliminary definitions.

The discounted total reward of a decentralized policy $\pi \equiv (\pi_t)_{t \in \{0,1,\ldots,T-1\}}$ over T time steps and starting at occupancy state ξ_t is

$$V_{\hat{M},\gamma,t}^{\pi}(\xi_t) = \left[\sum_{k=t}^{T-1} \gamma^{k-t} R(\xi_k, \pi_k) \mid \xi_{k+1} = P(\xi_k, \pi_k) \right], \quad (3)$$

where the occupancy state sequence $(\xi_k)_{k \in \{t,t+1,\ldots,T-1\}}$ is generated by the deterministic transition rule P under decentralized policy π: $\xi_{k+1} = P(\xi_k, \pi_k), \forall k \in \{t, t+1, \ldots, T-1\}$ and $\forall t \in \{0, 1, \ldots, T - 1\}$. Therefore, the optimal value function starting at occupancy state ξ_0 is $V_{\hat{M},\gamma,0}^{*}(\xi_0) = \max_{\pi} V_{\hat{M},\gamma,0}^{\pi}(\xi_0)$. Hence, the optimal value function $(V_{\hat{M},\gamma,t}^{*})_{t \in \{0,1,\ldots,T\}}$ is a solution of Bellman's optimality equation for $\hat{M}$, given by:

$$V_{\hat{M},\gamma,t}^{*}(\xi_t) = \max_{\pi_t \in A_t} \left\{ R(\xi_t, \pi_t) + \gamma V_{\hat{M},\gamma,t+1}^{*}(P(\xi_t, \pi_t)) \right\}, \quad \forall \xi_t \in \triangle \quad (4)$$

and for $t = T$, we add a boundary condition $V_{\hat{M},\gamma,T}^{*}(\cdot) = 0$. If it can be solved for $(V_{\hat{M},\gamma,t}^{*})_{t \in \{0,1,\ldots,T\}}$, an optimal decentralized policy $\pi^{*} \equiv (\pi_t^{*})_{t \in \{0,1,\ldots,T-1\}}$ may typically be obtained by maximization of the right-hand side for each ξ_t, i.e.,

$$\pi_t^{*} \in \arg\max_{\pi_t \in A_t} \left\{ R(\xi_t, \pi_t) + \gamma V_{\hat{M},\gamma,t+1}^{*}(P(\xi_t, \pi_t)) \right\}, \quad \forall \xi_t \in \triangle. \quad (5)$$

2.2 Dynamic Programming Update Operators

This subsection formally introduces the dynamic programming update operators involved in solving OMDPs, including: Bayesian state estimation; Bellman's evaluation and backup operators; and greedy action selection. To better understand this, let $\mathcal{V}$ be the set of real-valued functions $f: \triangle_t \mapsto \mathbb{R}$ for all $t \in \{0, 1, \ldots, T\}$.

Definition 3 (Bellman's evaluation operator). *For each decentralized decision rule* $\pi_t \in A_t$, *let* $T_{\pi_t}: \mathcal{V} \mapsto \mathcal{V}$ *be Bellman's evaluation operator, given by:*

$$(T_{\pi_t} V_{\hat{M},\gamma,t+1})(\xi_t) = R(\xi_t, \pi_t) + \gamma V_{\hat{M},\gamma,t+1}(P(\xi_t, \pi_t)), \quad \forall \xi_t \in \triangle, \pi_t \in A_t. \quad (6)$$

Bellman's evaluation operator transforms any arbitrary value function into a new value function based on a specified decentralized decision rule. It is worth noticing that Bellman's optimality equations (Equation 4) and greedy decision rule selections (Equation 5) can be stated in terms of the expression depending on occupancy state, decentralized decision rule and Bellman's evaluation operator. In the following, we formally define greedy selection and Bellman's update operators.

Definition 4 (Greedy action-selection operator). *For each decentralized decision rule* $\pi_t \in A_t$, *let* $G: \mathcal{V} \mapsto (\triangle \mapsto A)$ *be the greedy operator, given by:*

$$(GV_{\hat{M},\gamma,t+1})(\xi_t) = \arg\max_{\pi_t \in A_t} (T_{\pi_t} V_{\hat{M},\gamma,t+1})(\xi_t), \quad \forall \xi_t \in \triangle, V_{\hat{M},\gamma,t+1} \in \mathcal{V}. \quad (7)$$

Together the greedy action-selection and Bellman's evaluation operators permit us to define Bellman's update operator as follows.

Definition 5 (Bellman's update operator). *Let* $T: \mathcal{V} \mapsto \mathcal{V}$ *be Bellman's update operator, given by:*

$$(TV_{\hat{M},\gamma,t+1})(\xi_t) = (T_{(GV_{\hat{M},\gamma,t+1})(\xi_t)} V_{\hat{M},\gamma,t+1})(\xi_t), \quad \forall \xi_t \in \triangle, V_{\hat{M},\gamma,t+1} \in \mathcal{V}. \quad (8)$$

Bellman's update operator maintains the value of a given occupancy state based on the greedy decentralized decision rule for a specified value function. When optimized exactly, the value function, solution of Bellman's optimality equations (Equation 4), is a *piecewise-linear and convex* function of the occupancy states [8]. That is, there exist finite sets of hyperplanes $(\Lambda_t)_{t \in \{0,1,\dots,T-1\}}$, such that: $V^*_{\hat{M},\gamma,t}(\xi_t) = \max_{\lambda_t \in \Lambda_t} \sum_{s_t, \theta_t} \lambda_t(s_t, \theta_t) \cdot \xi_t(s_t, \theta_t)$, where $\lambda_t \in \mathbb{R}^{|S||\Theta_t|}$ for all $t \in \{0, 1, \dots, T - 1\}$.

Mappings G and T serve to define a dynamic programming methodology for the solution of occupancy Markov decision process $\hat{M}$. In particular, the piecewise-linearity and convexity property of the value function, together with mappings G and T, allow to combine advances in continuous-state MDP and POMDP algorithms, which have led to the development of a novel family of exact algorithms, including the *feature-based heuristic search value iteration* [7,8].

2.3 The Feature-Based Heuristic Search Value Iteration

This subsection provides a succinct description of feature-based heuristic search value iteration (FB-HSVI) (Algorithm 1), which was originally introduced under the total reward criterion. Here, we extend it to address the discounted total reward criterion and discuss complexity issues.

The FB-HSVI Algorithm's Description. FB-HSVI extends to decentralized stochastic control problems the *heuristic search value iteration* (HSVI) algorithm, which was originally developed for partially observable Markov decision processes [21]. Similarly to HSVI, it corresponds to a family of trial-based algorithms that searches an optimal solution of an occupancy Markov decision process. FB-HSVI proceeds by generating trajectories of occupancy states, starting at the initial occupancy state. It maintains both upper and lower bounds over the optimal value function. It guides exploration towards occupancy states that are more relevant to the upper bound by greedily selecting decentralized decision rules with respect to the upper bound, and reducing the gap between bounds at visited occupancy states. If the gap between upper and lower bounds at the initial occupancy state is ε, then it terminates. In such a case, we are guaranteed FB-HSVI has converged to an ε-optimal solution, as initially targeted. Though FB-HSVI is already equipped with a mechanism for finding ε-optimal solutions —since it uses greedy action-selection and accurate state-estimation operators— in practice it quickly exhausts the available resources before convergence. To better understand this, we provide a complexity analysis of each operator involved in FB-HSVI.

Algorithm 1. The feature-based heuristic search value iteration for $\hat{M}$ (resp. M)

1 **function** FB-HSVI$(\hat{M}, \varepsilon, (\underline{V}_{\hat{M},\gamma,t})_{t\in\{0,1,\cdots,T\}}, (\bar{V}_{\hat{M},\gamma,t})_{t\in\{0,1,\cdots,T\}})$
2 $\quad$ **while** GAP$(\xi_0) > \varepsilon$ **do** EXPLORE (ξ_0)

3 **function** GAP(ξ_t)
4 $\quad$ **return** $\bar{V}_{\hat{M},\gamma,t}(\xi_t) - \underline{V}_{\hat{M},\gamma,t}(\xi_t)$

5 **function** EXPLORE (ξ_t)
6 $\quad$ **if** GAP$(\xi_t) > \varepsilon/\gamma^t$ **then**
7 $\quad\quad$ $\pi_t^* \leftarrow (G\bar{V}_{\hat{M},\gamma,t+1})(\xi_t)$
8 $\quad\quad$ EXPLORE$(P(\xi_t, \pi_t^*))$
9 $\quad\quad$ $(T\bar{V}_{\hat{M},\gamma,t+1})(\xi_t)$ and $(T\underline{V}_{\hat{M},\gamma,t+1})(\xi_t)$

Complexity of Dynamic Programming Operators. As FB-HSVI proceeds, there are three operations that can significantly affect the overall performance: the greedy action-selection operator G; the accurate state-estimation operator P; and finally, Bellman's update operator T. To better understand the complexity involved in these operations, let $|V|$ be the size of value function V (respectively the upper- or lower-bound value functions). Let $\Theta^i(\xi_t)$ be the set of private histories of agent i involved in occupancy state ξ_t, $|\Theta^*(\xi_t)| = \max_{i\in 1,2,...,N} |\Theta^i(\xi_t)|$ and $|A^*| = \max_{i\in 1,2,...,N} |A^i|$. Algorithm 1 (lines 7 and 9) performs a greedy action-selection operator G, which involves enumerating and evaluating exponentially many decentralized decision rules in the worst case, and requires time complexity $O(|V|^{|\Theta^*(\xi_t)|^{N|A^*|}})$ that grows doubly exponentially with increasing number of private histories involved in the occupancy state ξ_t. In practice, branch-and-bound methods explore only a small portion of this set, which saves considerable time [7,8]. Then, Algorithm 1 (line 8) computes the next occupancy state given the current one and the next decentralized decision rule. Unlike the greedy action-selection operator, this state-estimation rule has complexity $O(|S|^2|\Theta(\xi_t)||Z|)$, that is polynomial in the number of joint histories involved in the occupancy states and the number of joint observations.

However, in the worst case, the number of joint histories increases by a factor of $|Z|$ as time goes on. This may limit ability to perform the greedy action-selection operator later on. Finally, Algorithm 1 (line 9) performs Bellman's update operator T to maintain both upper and lower bounds at a given occupancy state, namely *point-based Bellman's update*. Unlike the full Bellman's update operator, the point-based Bellman's update operator maintains the value function only at a single occupancy state at a time, which makes it significantly more tractable. Nonetheless, the complexity of this operation remains time demanding as it requires performing a greedy action selection.

Given that the complexity of operators G, P and T are prohibitive for a number of realistic decentralized stochastic control problems, the importance of approximate variants is clear.

3   An Error-Bounded Heuristic Search Framework

The primary contribution of this section is an error-bounded heuristic search framework which builds upon approximate variants of greedy action-selection and accurate state-estimation operators. We also provide a provable bound on the error FB-HSVI algorithms would make by using these approximate operators instead of their exact counterparts. The result is a general algorithmic framework that allows for monitoring the divergence between the exact and approximate solutions of infinite-horizon and discounted decentralized stochastic control problems represented as Dec-POMDPs.

3.1   Error-Bounded Action-Selection Operators

This subsection characterizes error-bounded action-selection operators that select decentralized decision rules within α of maximizing the value.

Definition 6. *Let $\alpha \in [0, \infty)^T$ be a real vector. An α-approximate action-selection operator $\widetilde{G} \colon \mathcal{V} \mapsto (\Delta \mapsto A)$ is such that, at each time step $t \in \{0, 1, \ldots, T - 1\}$, the decentralized decision rule found comes within $\alpha(t)$ of maximizing the value:*

$$(T_{(GV_{\hat{M},\gamma,t+1})(\xi_t)} V_{\hat{M},\gamma,t+1})(\xi_t) - (T_{(\widetilde{G}V_{\hat{M},\gamma,t+1})(\xi_t)} V_{\hat{M},\gamma,t+1})(\xi_t) \leq \alpha(t), \quad \forall \xi_t \in \Delta, V_{\hat{M},\gamma,t+1} \in \mathcal{V}.$$

For any positive T-dimensional vector α, a feature-based heuristic search value iteration, together with an α-approximate action-selection operator, terminates with a final estimate $V^{\alpha}_{\hat{M},\gamma,0}(\xi_0)$. The error between this approximate value and the optimal value is bounded and the bound depends only upon parameter α and γ.

Theorem 1. *The error introduced in FB-HSVI by using $\widetilde{G}$ instead of G is bounded by $\sum_{t=0}^{T-1} \gamma^t \alpha(t)$, assuming accurate estimation of the occupancy states during the planning phase. In particular, if $\alpha(t) = \alpha(t + 1) = \ldots = \alpha$ for all $t \in \{0, 1, \ldots, T - 1\}$, then the error is bounded by $\frac{1-\gamma^T}{1-\gamma}\alpha$.*

Proof. Let π^* and π^{α} be decentralized policies that are optimal given that we use (P, G) and $(P, \widetilde{G})$, respectively. Vectors $\xi_1, \ldots, \xi_{T-1}$ being the occupancy states generated from ξ_0 when applying π^*, it follows that:

$$V^{\pi^*}_{\hat{M},\gamma,0}(\xi_0) - V^{\pi^\alpha}_{\hat{M},\gamma,0}(\xi_0)$$
$$= \left(\textstyle\sum_{t=0}^{T-1} \gamma^t R(\xi_t, \pi_t^*)\right) - V^{\pi^\alpha}_{\hat{M},\gamma,0}(\xi_0) \quad \text{(definition of } V^{\pi^*}_{\hat{M},\gamma,0}(\xi_0)),$$
$$= \left(\textstyle\sum_{t=0}^{T-1} \gamma^t R(\xi_t, \pi_t^*)\right) + \textstyle\sum_{t=1}^{T-1} \left(\gamma^t V^{\pi^\alpha_{t:T-1}}_{\hat{M},\gamma,t}(\xi_t) - \gamma^t V^{\pi^\alpha_{t:T-1}}_{\hat{M},\gamma,t}(\xi_t)\right) - V^{\pi^\alpha}_{\hat{M},\gamma,0}(\xi_0) \quad \text{(adding zero).}$$

Next, we use the fact that $V^{\pi^\alpha}_{\hat{M},\gamma,T}(\xi_T) = 0$ to re-arrange terms:

$$= \left(\textstyle\sum_{t=0}^{T-1} \gamma^t R(\xi_t, \pi_t^*)\right) + \left(\gamma^T V^{\pi^\alpha}_{\hat{M},\gamma,T}(\xi_T) + \textstyle\sum_{t=1}^{T-1} \gamma^t V^{\pi^\alpha_{t:T-1}}_{\hat{M},\gamma,t}(\xi_t)\right) - \left(\gamma^0 V^{\pi^\alpha}_{\hat{M},\gamma,0}(\xi_0) + \textstyle\sum_{t=1}^{T-1} \gamma^t V^{\pi^\alpha_{t:T-1}}_{\hat{M},\gamma,t}(\xi_t)\right),$$
$$= \left(\textstyle\sum_{t=0}^{T-1} \gamma^t R(\xi_t, \pi_t^*)\right) + \left(\textstyle\sum_{t=0}^{T-1} \gamma^{t+1} V^{\pi^\alpha_{t+1:T-1}}_{\hat{M},\gamma,t+1}(P(\xi_t, \pi_t^*))\right) - \left(\textstyle\sum_{t=0}^{T-1} \gamma^t V^{\pi^\alpha_{t:T-1}}_{\hat{M},\gamma,t}(\xi_t)\right),$$
$$= \textstyle\sum_{t=0}^{T-1} \gamma^t \left(R(\xi_t, \pi_t^*) + \gamma V^{\pi^\alpha_{t+1:T-1}}_{\hat{M},\gamma,t+1}(P(\xi_t, \pi_t^*)) - V^{\pi^\alpha_{t:T-1}}_{\hat{M},\gamma,t}(\xi_t)\right),$$
$$= \textstyle\sum_{t=0}^{T-1} \gamma^t \left(V^{\pi_t^*, \pi^\alpha_{t+1:T-1}}_{\hat{M},\gamma,t}(\xi_t) - V^{\pi^\alpha_{t:T-1}}_{\hat{M},\gamma,t}(\xi_t)\right),$$
$$= \textstyle\sum_{t=0}^{T-1} \gamma^t \left((T_{\pi_t^*} V^{\pi^\alpha_{t+1:T-1}}_{\hat{M},\gamma,t+1})(\xi_t) - (T_{\pi_t^\alpha} V^{\pi^\alpha_{t+1:T-1}}_{\hat{M},\gamma,t+1})(\xi_t)\right),$$
$$\leq \textstyle\sum_{t=0}^{T-1} \gamma^t \left((T_{(GV^{\pi^\alpha_{t+1:T-1}}_{\hat{M},\gamma,t+1})(\xi_t)}V^{\pi^\alpha_{t+1:T-1}}_{\hat{M},\gamma,t+1})(\xi_t) - (T_{(\widetilde{G}V^{\pi^\alpha_{t+1:T-1}}_{\hat{M},\gamma,t+1})(\xi_t)}V^{\pi^\alpha_{t+1:T-1}}_{\hat{M},\gamma,t+1})(\xi_t)\right),$$
$$= \textstyle\sum_{t=0}^{T-1} \gamma^t \alpha(t).$$

Thus, the error between $V^{\pi^*}_{\hat{M},\gamma,0}(\xi_0)$ and $V^{\pi^\alpha}_{\hat{M},\gamma,0}(\xi_0)$ is bounded by $\sum_{t=0}^{T-1} \gamma^t \alpha(t)$. □

To the best of our knowledge, in decentralized stochastic control theory, this is the first attempt to monitor and bound the error made by using approximate action-selection instead of greedy action-selection. This bound comes with a natural interpretation: *all time steps are not equally relevant to the final error.* Indeed, due to discounted errors, approximate action-selection operators give more credit to errors they make at the earlier stages of the process. In other words, one can tolerate more approximation error at occupancy states that appear later in the process.

The problem of assigning errors to time steps goes beyond the scope of this paper, and will be addressed in the future. However, given the error vector α, another problem consists in finding a practical algorithm for selecting error-bounded actions over time steps. To do so, one can make use of the same branch-and-bound algorithms used for selecting greedy actions [7,8]. Except that, now, these algorithms need to be interrupted whenever the gap between lower and upper bounds is α. In that case, we are guarantee the returned action has value within α of the optimal value, as targeted.

3.2   Error-Bounded State-Estimation Operators: Definition and Example

This subsection discusses the long term behavior of successive applications of an approximate state-estimation operator. Next, we formally define the family of approximate state-estimation operators we target. Then, we exhibit one such operator. And finally, we derive theoretical guarantees.

Since we are interested in quantifying the error between occupancy states, we choose the *total variational distance* as a metric for measuring their distance. The total variational distance between two probability distributions ξ and ξ' on $[0,1]^{\|S\|\|\Theta\|}$ is defined by $\|\xi - \xi'\|_{\text{TV}} = \frac{1}{2} \sum_{s \in S, \theta \in \Theta} |\xi(s,\theta) - \xi'(s,\theta)|, \quad \forall \xi, \xi' \in \Delta$. Informally, the total variational distance $\|\xi - \xi'\|_{\text{TV}}$ defines the minimal probability mass that would have to be re-assigned in order to transform occupancy state ξ into occupancy state ξ'. The following definition of approximate state-estimation operator $\widetilde{P}_{\pi_t}$ guarantees that, for any occupancy state $\xi_t \in \Delta_t$, we have $\|\xi_t P_{\pi_t} - \xi_t \widetilde{P}_{\pi_t}\|_{TV} \leq \delta$.

Definition 7. *Let $\delta \in [0,1]$ be a small scalar. Then, for each decentralized decision rule $\pi_t \in A_t$, transition matrix $\widetilde{P}_{\pi_t}$ is a δ-approximation of P_{π_t} if, for any occupancy state $\xi_t \in \Delta_t$, there exists $\delta' \in [0,\delta]$ and $(\xi'_{t+1}, \xi''_{t+1}, \widetilde{\xi}''_{t+1}) \in \Delta^3_{t+1}$ such that*

$$\xi_t P_{\pi_t} = (1-\delta')\xi'_{t+1} + \delta'\xi''_{t+1} \quad and \quad \xi_t \widetilde{P}_{\pi_t} = (1-\delta')\xi'_{t+1} + \delta'\widetilde{\xi}''_{t+1}.$$

Now, we introduce and describe Algorithm 2 for constructing an artificial occupancy state that is within δ (in terms of variational distance) from the original occupancy state. To ensure the total variational distance between artificial and original occupancy states is upper bounded by δ, the algorithm clusters together private histories of the original occupancy state that are close enough (see Definition 8). Then, it replaces each such cluster with a unique private history in that cluster. Finally, this private history represents the cluster in the artificial occupancy state.

Algorithm 2. The occupancy state approximation algorithm (OSA).

1 **function** OSA(ξ_t, π_t, δ)
2     $\widetilde{\xi}_{t+1} \leftarrow [0,1]^{|S\|\Theta|}$ and $C \leftarrow$ LABELS(ξ_t, π_t, δ)
3   **foreach** $s \in S$ and $c \in C$ **do** $\widetilde{\xi}_{t+1}(s,l) \leftarrow \sum_{\theta \in [c]_{(\xi_t P_{\pi_t},\delta)}} \zeta_t P_{\pi_t}(s,\theta)$ **return** ξ_{t+1}

4 **function** LABELS(ξ_t, π_t, δ)
5   **foreach** $i \in \{1,2,\dots,N\}$ **do**
6        $C^i \leftarrow \emptyset$ and $\Theta^i \leftarrow \Theta^i(\xi_t P_{\pi_t})$
7   **while** $\Theta^i \neq \emptyset$ **do**
8           $c^i \leftarrow \arg\max_{\theta^i \in \Theta^i} |[\theta^i]_{(\xi_t P_{\pi_t},\delta)}|$
9           $C^i \leftarrow C^i \cup \{c^i\}$ and $\Theta^i \leftarrow \Theta^i \backslash [c^i]_{(\xi_t P_{\pi_t},\delta)}$
10   **return** $\otimes_{i \in \{1,2,\dots,N\}} C^i$

Before proceeding any further, we introduce the criterion we use, namely the *approximate probabilistic* measure.

Definition 8. *Let ξ_t be an occupancy state, and θ^i and $\bar{\theta}^i$ be two private histories in set $\Theta^i(\xi_t)$. We say that θ^i and $\bar{\theta}^i$ are δ-probabilistically close if and only if:*

$$\|Pr(X_t, Y_t|\xi_t, \theta^i) - Pr(X_t, Y_t|\xi_t, \bar{\theta}^i)\|_{TV} \leq \delta, \tag{9}$$

where X_t and Y_t denote random variables associated with states and other agent histories, respectively. We also denote $[\theta^i]_{(\xi_t,\delta)}$ the entire set of private histories $\bar{\theta}^i \in \Theta(\xi_t)$ that are δ-probabilistically close to θ^i and with respect to ξ_t.

By clustering private histories that are δ-probabilistically close with a single private history in that cluster, we produce (from Definition 8) an approximate occupancy state

$\tilde{\xi}_t$ with respect to the original occupancy state ξ_t such that: $\|\xi_t - \tilde{\xi}_t\|_{\mathrm{TV}} \le \delta$. Notice that Algorithm 2 is not guarantee to produce an occupancy state with the minimum number of private histories. A more promising goal, which we do not address here, would be to find a clustering method that can identify the minimum number of clusters of private histories so that the total variational distance between original and artificial occupancy states is upper-bounded by δ.

3.3   Error-Bounded State-Estimation Operators: Theoretical Analysis

We are now ready to bound the regret of using an approximate occupancy state instead of the accurate occupancy state. To do so, let $P_{\pi_{0:t-1}} = P_{\pi_0} P_{\pi_1} \cdots P_{\pi_{t-1}}$ for all time steps $t \in \{1, 2, \ldots, T\}$. Our analysis monitors the error we make step by step using approximate occupancy states.

Lemma 1. *The total variational distance between $\xi_0 \widetilde{P}_{\pi_{0:t-1}}$ and $\xi_0 P_{\pi_{0:t-1}}$ is bounded:*

$$\|\xi_0 \widetilde{P}_{\pi_{0:t-1}} - \xi_0 P_{\pi_{0:t-1}}\|_{\mathrm{TV}} \le 1 - (1-\delta)^t, \quad \forall t \in \{1, 2, \ldots, T\}. \tag{10}$$

Proof. The proof holds directly by expanding $\xi_0 \widetilde{P}_{\pi_{0:t}}$ and $\xi_0 P_{\pi_{0:t}}$ using Definition 7.

$$
\begin{aligned}
&\|\xi_0 \widetilde{P}_{\pi_{0:t}} - \xi_0 P_{\pi_{0:t}}\|_{\mathrm{TV}}, \\
&\le \|(1-\delta)\xi_1' P_{\pi_{1:t}} + \delta\tilde{\xi}_1'' \widetilde{P}_{\pi_{1:t}} - (1-\delta)\xi_1' P_{\pi_{1:t}} - \delta\xi_1'' P_{\pi_{1:t}}\|_{\mathrm{TV}}, \\
&\le \|(1-\delta)^2\xi_2' P_{\pi_{2:t}} + \delta(1-\delta)\tilde{\xi}_2' \widetilde{P}_{\pi_{2:t}} + \delta\tilde{\xi}_1'' \widetilde{P}_{\pi_{1:t}} - (1-\delta)^2\xi_2' P_{\pi_{2:t}} - \delta(1-\delta)\xi_2' P_{\pi_{2:t}} - \delta\xi_1'' P_{\pi_{1:t}}\|_{\mathrm{TV}}, \\
&\le \|(1-\delta)^t\xi_t' P_{\pi_{t:t}} + \sum_{k=1}^{t}\delta(1-\delta)^{k-1}\tilde{\xi}_k'' \widetilde{P}_{\pi_{k:t}} - (1-\delta)^t\xi_t' P_{\pi_{t:t}} - \sum_{k=1}^{t}\delta(1-\delta)^{k-1}\xi_k'' P_{\pi_{k:t}}\|_{\mathrm{TV}}, \\
&= \|\sum_{k=1}^{t}\delta(1-\delta)^{k-1}\tilde{\xi}_k'' \widetilde{P}_{\pi_{k:t}} - \sum_{k=1}^{t}\delta(1-\delta)^{k-1}\xi_k'' P_{\pi_{k:t}}\|_{\mathrm{TV}}, \\
&\le \sum_{k=1}^{t}\delta(1-\delta)^{k-1}\|\tilde{\xi}_k'' \widetilde{P}_{\pi_{k:t}} - \xi_k'' P_{\pi_{k:t}}\|_{\mathrm{TV}}, \\
&\le \sum_{k=1}^{t}\delta(1-\delta)^{k-1}, \\
&= 1 - (1-\delta)^t. \qquad \square
\end{aligned}
$$

It is worth noticing that approximation errors tend to increase exponentially as time goes on. The following derives the regret induced by approximating state estimates.

Theorem 2. *Let $\delta \in [0, \infty)^T$ be a scalar vector. The error introduced in FB-HSVI by using a δ-approximate state-estimation operator instead of the exact state-estimation operator is bounded by $2\|r\|_\infty \sum_{t=0}^{T-1} \gamma^t [1 - \prod_{k=1}^{t}(1 - \delta(k))]$, assuming we use G for selecting decentralized decision rules. In particular, if $\delta(t) = \delta$ for all time steps $t \in \{0, 1, \ldots, T-1\}$, then the error is bounded by $2\left(\frac{1-\gamma^T}{1-\gamma} - \frac{1-[\gamma(1-\delta)]^T}{1-\gamma(1-\delta)}\right)\|r\|_\infty$.*

Proof. Let π^* and $\tilde{\pi}$ be decentralized policies that are optimal given that we use accurate or approximate state-estimation operators, respectively. Define $\widetilde{V}_{\hat{M},\gamma,0}^{\pi^*}(\xi_0)$ as follows: $\widetilde{V}_{\hat{M},\gamma,0}^{\pi^*}(\xi_0) = \sum_{t=0}^{T-1}\gamma^t R(\xi_0 \widetilde{P}_{\pi_{0:t}^*}, \pi_t^*)$. Clearly, we have $\widetilde{V}_{\hat{M},\gamma,0}^{\pi^*}(\xi_0) \le V_{\hat{M},\gamma,0}^{\tilde{\pi}}(\xi_0)$ by definition of decentralized policy $\tilde{\pi}$. Using this property, we know that:

$$
\begin{aligned}
V_{\hat{M},\gamma,0}^{\pi^*}(\xi_0) - V_{\hat{M},\gamma,0}^{\tilde{\pi}}(\xi_0) &\le V_{\hat{M},\gamma,0}^{\pi^*}(\xi_0) - \widetilde{V}_{\hat{M},\gamma,0}^{\pi^*}(\xi_0), \\
&= \sum_{t=0}^{T-1}\gamma^t \left(R(\xi_0 P_{\pi_{0:t-1}^*}, \pi_t^*) - R(\xi_0 \widetilde{P}_{\pi_{0:t-1}^*}, \pi_t^*)\right).
\end{aligned}
$$

Since the value function is piecewise-linear and convex, $\boldsymbol{R}^{\pi^*_t} \equiv \boldsymbol{R}(\cdot, \pi^*_t)$ is a linear function of occupancy states. Thus, if we let $\langle \cdot, \cdot \rangle$ be an inner product, then we have

$$
\begin{aligned}
V^{\pi^*}_{\hat{M},\gamma,0}(\xi_0) - \widetilde{V}^{\pi^*}_{\hat{M},\gamma,0}(\xi_0) &= \textstyle\sum_{t=0}^{T-1} \gamma^t \langle \boldsymbol{R}^{\pi^*_t}, \xi_0 \boldsymbol{P}_{\pi^*_{0:t-1}} - \xi_0 \widetilde{\boldsymbol{P}}_{\pi^*_{0:t-1}} \rangle, &&\text{(by linearity of } \boldsymbol{R}^{\pi^*_t}) \\
&\leq \textstyle\sum_{t=0}^{T-1} \gamma^t \|\boldsymbol{R}^{\pi^*_t}\|_\infty \|\xi_0 \boldsymbol{P}_{\pi^*_{0:t-1}} - \xi_0 \widetilde{\boldsymbol{P}}_{\pi^*_{0:t-1}}\|_1, &&\text{(Hölder's inequality)} \\
&\leq 2\textstyle\sum_{t=0}^{T-1} \gamma^t \|\boldsymbol{R}^{\pi^*_t}\|_\infty \|\xi_0 \boldsymbol{P}_{\pi^*_{0:t-1}} - \xi_0 \widetilde{\boldsymbol{P}}_{\pi^*_{0:t-1}}\|_{\mathrm{TV}}, &&\text{(where } \|x\|_1 = 2\|x\|_{\mathrm{TV}}) \\
&\leq 2\|r\|_\infty \textstyle\sum_{t=0}^{T-1} \gamma^t \|\xi_0 \boldsymbol{P}_{\pi^*_{0:t-1}} - \xi_0 \widetilde{\boldsymbol{P}}_{\pi^*_{0:t-1}}\|_{\mathrm{TV}}, &&\text{(where } \|\boldsymbol{R}^{\pi^*_t}\|_\infty \leq \|r\|_\infty) \\
&\leq 2\|r\|_\infty \textstyle\sum_{t=0}^{T-1} \gamma^t \left[1 - \prod_{k=1}^{t}(1 - \delta(k)) \right].
\end{aligned}
$$

This proves the result for any arbitrary $\delta \in [0, \infty)^T$. If we let $\delta(t) = \delta$ for all time step $t \in \{0, 1, \ldots, T-1\}$, then geometric series produce the following bound:

$$
V^{\pi^*}_{\hat{M},\gamma,0}(\xi_0) - V^{\tilde{\pi}}_{\hat{M},\gamma,0}(\xi_0) \leq 2 \left(\frac{1-\gamma^T}{1-\gamma} - \frac{1-[\gamma(1-\delta)]^T}{1-\gamma(1-\delta)} \right) \|r\|_\infty,
$$

which concludes the proof.   □

Once again, in decentralized stochastic control settings, this is the first attempt to monitor and bound the error made by using approximate state-estimation operators. We note that, as time goes on, these operators become more tolerant to approximation errors. But there is no free lunch: approximation errors tend to increase as time goes on. This new bound provides a way to analyze this trade-off.

3.4   Convergence and Error Bounds

Given any arbitrary state-estimation and action-selection operators $\widetilde{P}$ and $\widetilde{G}$, which come with provable guarantees, the feature-based heuristic search value iteration produces an estimate $V_{\hat{M},\gamma,0}(\xi_0)$. The error between $V_{\hat{M},\gamma,0}(\xi_0)$ and the true value function $V^*_{\hat{M},\gamma,0}(\xi_0)$ is bounded. The error depends on quantities ϵ, δ and α, each of which comes from a relaxation of the original problem. First, ϵ results from transforming an infinite horizon problem into a finite horizon one. Second, δ represents the vector of errors the state-estimation operator allows at each time step. Finally, α denotes the vector of errors the action-selection operator produces at each time step.

Theorem 3. *Let $\delta \in [0, \infty)^T$ be the estimation operator parameter and $\alpha \in [0, \infty)^T$ be the greedy operator parameter. The error of the feature-based heuristic search value iteration introduced by using $\widetilde{P}$ and $\widetilde{G}$ instead of P and G is bounded by:*

$$
2\|r\|_\infty \sum_{t=0}^{T-1} \gamma^t \left[1 - \prod_{k-1}^{t}(1 - \delta(k)) \right] + \left(\sum_{t=0}^{T-1} \gamma^t \alpha(t) \right) + \varepsilon, \tag{11}
$$

for any planning horizon $T = \lceil \log_\gamma ((1 - \gamma)\varepsilon / \|r\|_\infty) \rceil$.

Proof. Let π^*, π^α and $\pi^{\alpha,\delta}$ be decentralized policies that are optimal given that we use (P, G), $(P, \widetilde{G})$ and $(\widetilde{P}, \widetilde{G})$, respectively. Then,

$$
\begin{aligned}
V^{\pi^*}_{\hat{M},\gamma,0}(\xi_0) - V^{\pi^{\alpha,\delta}}_{\hat{M},\gamma,0}(\xi_0) &= \left(V^{\pi^*}_{\hat{M},\gamma,0}(\xi_0) - V^{\pi^\alpha}_{\hat{M},\gamma,0}(\xi_0) \right) + \left(V^{\pi^\alpha}_{\hat{M},\gamma,0}(\xi_0) - V^{\pi^{\alpha,\delta}}_{\hat{M},\gamma,0}(\xi_0) \right), \\
&\leq 2\|r\|_\infty \textstyle\sum_{t=0}^{T-1} \gamma^t \left[1 - \prod_{k=1}^{t}(1 - \delta(k)) \right] + \left(\sum_{t=0}^{T-1} \gamma^t \alpha(t) \right).
\end{aligned} \tag{12}
$$

This bound together with the fact that we search only for T-step policies is sufficient to demonstrate that the result holds.   □

This theorem provides the first result quantifying the influence of different approximate operators in the overall performance of an algorithm for solving Dec-POMDPs. To the best of our knowledge, no similar results exist in Dec-POMDPs.

4   Experiments

This section presents experiments on a selection of infinite-horizon γ-discounted Dec-POMDPs including small-sized benchmarks (broadcast channel, multi-agent tiger, recycling robots and meeting in a 3x3 grid) and large-sized benchmarks (box-pushing, mars rover and wireless). For each benchmark, we ran the error-bounded feature-based heuristic search value iteration (EB-FB-HSVI) algorithm using parameters ϵ (pruning criterion), α (action-selection tolerance), and δ (state-estimation tolerance). Notice that, over the selection of benchmarks, action-selection tolerance α has only minor influence on performance results, so we set $\alpha = 0$ for many domains. We selected greedy actions using a constraint programming software, namely toulbar2 [11]. EB-FB-HSVI ran on a Mac OSX machine with 2.4GHz Dual-Core Intel and 2GB of RAM available.

Table 1. Results for infinite-horizon decentralized POMDPs with $\gamma = 0.9$, and by default we set $\epsilon = 0.001$, $\alpha = 0$ and $\delta = 0$. Higher $\underline{V}(\xi_0)$ is better. Results for Mealy NLP, EM, PeriEM, PI, MPBVI and IPG were likely computed on different platforms, an therefore time comparisons may be approximate at best.

Algorithm	$|\Delta|$	Time	$\underline{V}(\xi_0)$
Broadcast ($|S| = 4$, $|A^i| = 2$, $|Z^i| = 2$)			
FB-HSVI	102	19.8s	9.271
FB-HSVI($\delta = 0.01$)	435	7.8s	9.269
MPBVI	36	< 18000s	9.27
NLP	2	1s	9.1
Dec-tiger ($|S| = 2$, $|A^i| = 3$, $|Z^i| = 2$)			
FB-HSVI($\delta = 0.01$)	52	6s	13.448
FB-HSVI	25	157.3s	13.448
MPBVI	231	< 18000s	13.448
Peri	10×30	220s	13.45
PeriEM	7×10	6540s	9.42
Goal-directed	11	75s	5.04
Mealy NLP	4	29s	−1.49
EM	6	142s	−16.3
Recycling robots ($|S| = 4$, $|A^i| = 3$, $|Z^i| = 2$)			
FB-HSVI	109	2.6s	31.929
FB-HSVI($\delta = 0.01$)	108	0s	31.928
MPBVI	37	< 18000s	31.929
Mealy NLP	1	0s	31.928
Peri	6×30	77s	31.84
PeriEM	6×10	272s	31.80
EM	2	13s	31.50
IPG	4759	5918s	28.10
PI	15552	869s	27.20
Meeting in a 3x3 grid ($|S| = 81$, $|A^i| = 5$, $|Z^i| = 9$)			
FB-HSVI	108	67s	5.802
FB-HSVI($\delta = 0.01$)	88	45s	5.794
Peri	20×70	9714s	4.64

Algorithm	$|\Delta|$	Time	$\underline{V}(\xi_0)$
Box-pushing ($|S| = 100$, $|A^i| = 4$, $|Z^i| = 5$)			
FB-HSVI($\delta = 0.01$)	331	1715.1s	224.43
FB-HSVI($\alpha = 1, \delta = 0.05$)	288	1405.7s	224.26
FB-HSVI($\epsilon = 30$)	264	15.24s	199.42
MPBVI	305	> 18000s	224.12
Goal-directed	5	199s	149.85
Peri	15 × 30	5675s	148.65
Mealy NLP	4	774s	143.14
PeriEM	4 × 10	7164s	106.68
Mars rover ($|S| = 256$, $|A^i| = 6$, $|Z^i| = 8$)			
FB-HSVI($\delta = 0.01$)	136	74.31s	26.94
FB-HSVI($\alpha = 0.2$)	149	85.72s	26.92
FB-HSVI($\epsilon = 1$)	155	32.5s	26.77
Peri	10 × 30	6088s	24.13
Goal-directed	6	956s	21.48
Mealy NLP	3	396s	19.67
PeriEM	3 × 10	7132s	18.13
EM	3	5096s	17.75
Wireless ($|S| = 64$, $|A^i| = 2$, $|Z^i| = 6$)			
FB-HSVI($\delta = 0.01$)	897	6309s	−144.24
FB-HSVI($\alpha = 0.1$)	408	6740s	−140.37
FB-HSVI($\epsilon = 20$)	866	6084s	−176.59
MPBVI	374	> 18000s	−167.10
EM	3	6886s	−175.40
Peri	15 × 100	6492s	−181.24
PeriEM	2 × 10	3557s	−218.90
Mealy NLP	1	9s	−294.50

We compare EB-FB-HSVI for infinite-horizon Dec-POMDPs with state-of-the-art approximate and exact algorithms, including: optimal policy iteration (PI) [5]; incremental policy iteration (IPG) [2]; nonlinear programming (NLP and Mealy NLP) [1]; goal-directed algorithm [3]; periodic expectation maximisation algorithm (EM, Peri and PeriEM) [16]; and modified point-based value iteration (MPBVI) [14]. Note that, while PI and IPG are optimal in theory, in practice they do not produce optimal solutions due to resources being exhausted before convergence. Table 1 reports performance results. For each domain and each algorithm, we report the lower-bound value function at the initial occupancy state $\underline{V}(\xi_0)$, the computation time required to achieve that value, and the memory requirement $|\underline{\Lambda}|$, which represents either the number of hyperplanes or the number of nodes in a policy graph.

In all tested benchmarks, EB-FB-HSVI achieves values higher or equal to the highest values that have been recorded so far, while being multiple orders of magnitude faster than state-of-the-art algorithms over many domains. In particular, over small-sized problems, EB-FB-HSVI demonstrates the best trade-off between the quality of the solution and the computation time. In addition, it is the only algorithm to provide provable bounds on the resulting solutions. In the broadcast channel, for example, both EB-FB-HSVI and MPBVI provide the highest value known so far, but EB-FB-HSVI comes with two advantages over MPBVI. First, it guarantees that value 9.271 is within 0.001 of the optimum. Second, it computed this value four orders of magnitude faster than MPBVI. Over large-sized problems, EB-FB-HSVI terminated with the highest values over all benchmarks for parameters $\alpha = 0, \delta = 0.01$ and $c = 0.001$. In the wireless problem, for example, the distance between the previous best value and EB-FB-HSVI's value is about 27, and EB-FB-HSVI was one order of magnitude faster than the previous best solver (MPBVI).

Table 2. Theoretical guarantees of EB-FB-HSVI($\delta = 0.01, \epsilon = 0.001$). We denote $\varepsilon_{apriori}$ the error computed a priori based on parameters δ and ϵ, and $\varepsilon_{aposteriori}$ the error computed a posteriori given approximation errors observed during the planning phase, both using Equation (11). Gap(ξ_0) = $V(\xi_0) - \bar{V}(\xi_0)$, where $\underline{V}(\xi_0)$ is provided by FB-HSVI($\delta = 0.01, \epsilon = 0.001$) and $\bar{V}(\xi_0)$ results from EB-FB-HSVI(ϵ) for some ϵ.

$\varepsilon_{apriori}$	$\varepsilon_{aposteriori}$	Gap(ξ_0)	$\varepsilon_{apriori}$	$\varepsilon_{aposteriori}$	Gap(ξ_0)
Broadcast			*Mars rover*		
1.651	0.018	0.003	16.62	0.773	0.83
Dec-tiger			*Box-pushing*		
166.7	0.727	0.001	16.74	1.8	4.99
Recycling robots			*Wireless*		
8.25	0.052	0.002	456.53	10.85	7.12

We continue the study of the performance of EB-FB-HSVI with respect to tightness of error bounds. For each benchmark, we report in Table 2: a priori and a posteriori errors based on Equation (11) for FB-HSVI($\delta = 0.01, \epsilon = 0.001$); and gap Gap($\xi_0$) $= \underline{V}(\xi_0) - \bar{V}(\xi_0)$ based on FB-HSVI(ϵ). Notice that a posteriori errors were computed based on approximation errors observed during the planning phase. Overall, a posteriori errors are tighter than a priori errors and closer to gaps. The tightness of a posteriori error is mainly because the observed approximation errors were significantly smaller than the targeted ones. In the tiger problem, for example, the a priori error is about 166.7 whereas the a posteriori error and the gap are close: 0.727 and 0.001, respectively. Surprisingly, in some domains such as mars rover and box-pushing, a posteriori errors are even smaller than gaps. This phenomenon occurs when EB-FB-HSVI(ϵ) exhausts the total available resources before convergence, i.e., the gap is larger than targeted error ϵ. The closeness between the gaps and the a posteriori errors demonstrate, at least over all tested domains, the tightness of our error bounds.

5   Discussion, Conclusion and Future Work

This paper presented two relatively interdependent contributions towards error-bounded solutions for infinite-horizon discounted Dec-POMDPs. First, we introduce the first error-bounded algorithmic framework for monitoring and bounding the error we make by using approximate action-selection and state-estimation operators instead of their exact counterparts. Second, we extend the state-of-the-art algorithm for solving finite-horizon Dec-POMDPs, namely the feature-based heuristic search value iteration algorithm, to infinite-horizon discounted Dec-POMDPs. The major difference being that we can now use approximate operators instead of exact operators while still being able to provide theoretical guarantees on the quality of the resulting solution. Experimental results demonstrate that, when compared to state-of-the-art algorithms, the error-bounded feature-based heuristic search value iteration algorithm improves both values and computation times in many domains from the literature.

Though this paper provides the first attempts to monitor and bound the error made by using approximate operators in decentralized stochastic control, similar results exist in simpler settings. Such results can be traced back to max-norm-based analyses of value and policy iteration algorithms for γ-discounted MDPs [18], which prove that for some error α at each iteration there exists a stationary policy within $\frac{2\gamma}{(1-\gamma)^2}\alpha$ of the optimum. This result led to the development of much research on convergence arguments for γ-discounted MDPs and extensions including partially observable cases [17,21] and decentralized stochastic control settings [10]. Closer to our performance guarantees, [19,4] developed variations of value and policy iteration algorithms for computing non-stationary policies in γ-discounted MDPs for which the performance bounds can be significantly improved by a factor of $\frac{1}{1-\gamma}$. Hence, Theorem 1 can be viewed as an extension of [19] to decentralized stochastic control settings. However, Theorem 2 differs from previous performance bounds in many aspects. First, it is not derived from the max-norm analysis; instead we measure state-estimation errors we made steps by steps, which may result in tighter performance bounds. As a consequence, it does not fit within the standard scheme of performance bounds. Nonetheless, it allows us to accurately estimate errors made in practice on all tested benchmarks.

In the future, we plan to extend the feature-based heuristic search value iteration algorithm so as to learn how to dynamically assign approximation errors over time steps in order to minimize the total computation time while providing the targeted error bound. Another avenue we plan to follow, relies on how to automatically find the minimum number of clusters of private histories such that the artificial occupancy state based on clusters is within δ of the original occupancy state.

References

1. Amato, C., Bernstein, D., Zilberstein, S.: Optimizing fixed-size stochastic controllers for POMDPs and decentralized POMDPs. Autonomous Agents and Multi-Agent Systems 21(3), 293–320 (2010)
2. Amato, C., Dibangoye, J., Zilberstein, S.: Incremental policy generation for finite-horizon Dec-POMDPs. In: ICAPS (2009),
   http://aaai.org/ocs/index.php/ICAPS/ICAPS09/paper/view/711/1086
3. Amato, C., Zilberstein, S.: Achieving goals in decentralized POMDPs. In: AAMAS (2009)
4. Bagnell, A., Kakade, S., Ng, A., Schneider, J.: Policy search by dynamic programming. In: NIPS, vol. 16 (2003)
5. Bernstein, D.S., Amato, C., Hansen, E.A., Zilberstein, S.: Policy iteration for decentralized control of Markov decision processes. Journal of Artificial Intelligence Research 34, 89–132 (2009)
6. Bernstein, D.S., Givan, R., Immerman, N., Zilberstein, S.: The complexity of decentralized control of Markov decision processes. Mathematics of Operations Research 27(4), 819–840 (2002)
7. Dibangoye, J.S., Amato, C., Buffet, O., Charpillet, F.: Optimally solving Dec-POMDPs as continuous-state MDPs. In: IJCAI (2013)
8. Dibangoye, J.S., Amato, C., Buffet, O., Charpillet, F.: Optimally solving Dec-POMDPs as continuous-state MDPs: Theory and algorithms. Tech. Rep. RR-8517, Inria (April 2014)
9. Dibangoye, J.S., Mouaddib, A.I., Chai-draa, B.: Point-based incremental pruning heuristic for solving finite-horizon Dec-POMDPs. In: AAMAS (2009)
10. Dibangoye, J.S., Mouaddib, A.I., Chaib-draa, B.: Toward error-bounded algorithms for infinite-horizon Dec-POMDPs. In: AAMAS (2011)
11. de Givry, S., Heras, F., Zytnicki, M., Larrosa, J.: Existential arc consistency: Getting closer to full arc consistency in weighted CSPs. In: IJCAI (2005)
12. Hansen, E.A., Bernstein, D.S., Zilberstein, S.: Dynamic programming for partially observable stochastic games. In: AAAI (2004)
13. Kumar, A., Zilberstein, S.: Point-based backup for decentralized POMDPs: Complexity and new algorithms. In: AAMAS (2010)
14. MacDermed, L.C., Isbell, C.: Point based value iteration with optimal belief compression for Dec-POMDPs. In: NIPS (2013)
15. Oliehoek, F.A., Spaan, M.T.J., Dibangoye, J.S., Amato, C.: Heuristic search for identical payoff Bayesian games. In: AAMAS (2010)
16. Pajarinen, J., Peltonen, J.: Periodic finite state controllers for efficient POMDP and DEC-POMDP planning. In: NIPS (2011)
17. Pineau, J., Gordon, G., Thrun, S.: Point-based value iteration: An anytime algorithm for POMDPs. In: IJCAI (2003)
18. Puterman, M.L.: Markov Decision Processes, Discrete Stochastic Dynamic Programming. Wiley-Interscience, Hoboken (1994)
19. Scherrer, B.: Improved and generalized upper bounds on the complexity of policy iteration. In: NIPS (2013)
20. Seuken, S., Zilberstein, S.: Formal models and algorithms for decentralized decision making under uncertainty. Autonomous Agents and Multi-Agent Systems 17(2), 190–250 (2008)
21. Smith, T., Simmons, R.G.: Point-based POMDP algorithms: Improved analysis and implementation. In: UAI (2005),
   http://dblp.uni-trier.de/db/conf/uai/uai2005.html#SmithS05

Approximate Consistency: Towards Foundations of Approximate Kernel Selection

Lizhong Ding and Shizhong Liao*

School of Computer Science and Technology
Tianjin University, Tianjin 300072, China
szliao@tju.edu.cn

Abstract. Kernel selection is critical to kernel methods. Approximate kernel selection is an emerging approach to alleviating the computational burdens of kernel selection by introducing kernel matrix approximation. Theoretical problems faced by approximate kernel selection are how kernel matrix approximation impacts kernel selection and whether this impact can be ignored for large enough examples. In this paper, we introduce the notion of approximate consistency for kernel matrix approximation algorithm to tackle the theoretical problems and establish the preliminary foundations of approximate kernel selection. By analyzing the approximate consistency of kernel matrix approximation algorithms, we can answer the question that, under what conditions, and how, the approximate kernel selection criterion converges to the accurate one. Taking two kernel selection criteria as examples, we analyze the approximate consistency of Nyström approximation and multilevel circulant matrix approximation. Finally, we empirically verify our theoretical findings.

1 Introduction

Since learning is ill-posed and data by itself is not sufficient to find the solution [1], some extra assumptions should be made to have a unique solution. The set of assumptions we make to have learning possible is called the *inductive bias* [21]. Model selection is the process of choosing the inductive bias, which is fundamental to learning. For kernel based learning, model selection involves the selection of the kernel function, which determines the reproducing kernel Hilbert space (RKHS), and the regularization parameter. The selection of regularization parameter has typically been solved by means of cross validation, generalized cross validation [14], theoretical estimation [8] or regularization path [15]. This paper focuses on the kernel selection problem, which is a challenging and central problem in kernel based learning [20].

Kernel selection is to select the optimal kernel in a prescribed kernel set by minimizing some kernel selection criterion that is usually defined via the estimate of the expected error [3]. The estimate can be empirical or theoretical. The k-fold cross validation (CV) is a commonly used empirical estimate of the expected error and the extreme form of cross validation, leave-one-out (LOO), gives an almost unbiased estimate of the expected error [6]. However, CV and LOO require training the learning algorithm

---

* Corresponding author.

T. Calders et al. (Eds.): ECML PKDD 2014, Part I, LNCS 8724, pp. 354–369, 2014.

for every candidate kernel for several times, unavoidably bringing high computational burdens. For the sake of efficiency, some approximate CV approaches are proposed, such as, generalized cross validation (GCV)[14], generalized approximate cross validation (GACV) [27] and Bouligand influence function cross validation (BIFCV) [18]. Minimizing theoretical estimate bounds of the expected error is an alternative to kernel selection. The commonly used theoretical estimates usually introduce some measures of complexity [3], such as VC dimension [26], Rademacher complexity [4], maximal discrepancy [3], radius-margin bound [6] and compression coefficient [19].

Approximate kernel selection is an emerging approach to alleviating the computational burdens of kernel selection for the large scale application by introducing kernel matrix approximation into the kernel selection domain [9,10]. As pointed out in [6,5], a kernel selection criterion is not required to be an unbiased estimate of the expected error, instead the primary requirement is merely for the minimum of the kernel selection criterion to provide a reliable indication of the minimum of the expected error in kernel parameter space. Therefore, we argue that it is sufficient to calculate an approximate criterion that can discriminate the optimal kernel from the candidates. Although the idea of approximate kernel selection has been successfully applied in model selection of the least squares support vector machine (LSSVM) [9,10], two theoretical problems are still open: how kernel matrix approximation impacts the kernel selection criterion and whether this impact can be ignored for large enough examples.

In this paper, we define the notion of approximate consistency for kernel matrix approximation algorithm to tackle the theoretical problems and establish the preliminary foundations of approximate kernel selection. By analyzing the approximate consistency of different kernel matrix approximation algorithms, we can answer the question that, under what conditions, and how, the approximate kernel selection criterion converges to the accurate one. It is worth noting that the approximate consistency is defined for kernel matrix approximation algorithms and different from the classical "consistency", which is defined for learning algorithms. For two kernel selection criteria, we analyze the approximate consistency of two typical kernel matrix approximation algorithms. The results demonstrate the appositeness of kernel matrix approximation for kernel selection in a hierarchical structure. Empirical studies are also conducted to verify our theoretical findings on benchmark and synthetic data.

The rest of this paper is organized as follows. In Section 2, we introduce related work and contributions of this paper. In Section 3, we define two kernel selection criteria and simply demonstrate the approximate kernel selection scheme. Section 4 gives the definition of approximate consistency and further analyzes the approximate consistency of several kernel matrix approximation algorithms. We empirically study the approximate consistency in Section 5. Finally, we conclude in Section 6.

2   Related Work and Contributions

Kernel matrix approximation is an effective tool for reducing the computational burdens of kernel based learning. In order to achieve linear complexity in l, where l is the number of examples, approximations from subsets of columns are considered: Nyström method [29], modified Nyström method [28], sparse greedy approximations [22] or

incomplete Cholesky decomposition [12]. These methods are all low-rank approximations and have time complexity $O(p^2l)$ for an approximation of rank p. Constructing
multilevel circulant matrix (MCM) to approximate kernel matrix is another effective
strategy [24,23,9], which allows the multi-dimensional fast Fourier transform (FFT) to
be utilized in solving learning algorithms with complexity of $O(l \log(l))$.

Column sampling and MCM approximation have been theoretically analyzed a lot
[17,28,13,24,9]. However, the analysis provides a bound on the matrix approximation
error for an appropriate norm (typically spectral, Frobenius and trace norm), but this is
independent of specific learning problem and can not reveal the influence of kernel matrix approximation on learning. Recent literatures [7,30,16,2,9] measure the influence
of kernel matrix approximation on hypothesis, but none of them reveal the influence
of kernel matrix approximation on kernel selection criterion. Approximate consistency
defined in this paper, for the first time, is to measure the difference between the accurate
criterion calculated by original kernel matrix and the approximate criterion calculated
by approximate kernel matrix, and further show the convergence of the difference for
large enough examples.

The approach of approximate model selection was first proposed in [9], in which
MCM approximation is adopted. Later, an extension to Nyström approximation for
approximate model selection was proposed in [10]. Approximate model selection is
a promising topic, especially for large scale applications. However, two fundamental
questions require answering: how different kernel matrix approximation algorithms influence the kernel selection criterion and whether the approximate criterion converges
to the accurate one. This paper provides answers to these questions.

3   Approximate Kernel Selection

In this section, we first present two kernel selection criteria and then give a brief introduction of approximate kernel selection.

We use $\mathcal{X}$ to denote the input space and $\mathcal{Y}$ the output domain. Usually we will have
$\mathcal{X} \subseteq \mathbb{R}^d$, $\mathcal{Y} = \{-1, 1\}$ for binary classification and $\mathcal{Y} = \mathbb{R}$ for regression. We as-
sume $|y| \leq M$ for any $y \in \mathcal{Y}$, where M is a constant. The training set is denoted
by $\mathcal{S} = \{(\boldsymbol{x}_1, y_1), \ldots, (\boldsymbol{x}_l, y_l)\} \in (\mathcal{X} \times \mathcal{Y})^l$. The kernel κ considered in this paper is a function from $\mathcal{X} \times \mathcal{X}$ to the field $\mathbb{R}$ such that for any finite set of inputs
$\{\boldsymbol{x}_1, \ldots, \boldsymbol{x}_l\} \subseteq \mathcal{X}$, the matrix $\mathbf{K} = [\kappa(\boldsymbol{x}_i, \boldsymbol{x}_j)]_{i,j=1}^l$ is symmetric and positive definite (SPD). $\mathbf{K}$ is the kernel matrix. The reproducing kernel Hilbert space (RKHS) $\mathcal{H}_\kappa$
associated with the kernel κ can be defined as $\mathcal{H}_\kappa = \overline{\operatorname{span}}\{\kappa(\boldsymbol{x}, \cdot) : \boldsymbol{x} \in \mathcal{X}\}$, and
the inner product $\langle \cdot, \cdot \rangle_{\mathcal{H}_\kappa}$ on $\mathcal{H}_\kappa$ is determined by $\langle \kappa(\boldsymbol{x}, \cdot), \kappa(\boldsymbol{x}', \cdot) \rangle_{\mathcal{H}_\kappa} = \kappa(\boldsymbol{x}, \boldsymbol{x}')$
for $\boldsymbol{x}, \boldsymbol{x}' \in \mathcal{X}$. We use $\|\mathbf{K}\|_2$, $\|\mathbf{K}\|_F$ and $\|\mathbf{K}\|_*$ to denote the spectral, Frobenius and
trace norm of $\mathbf{K}$. We use $\lambda_t(\mathbf{K})$ for $t = 1, \ldots, l$ to denote the eigenvalues of $\mathbf{K}$ in the
descending order.

Now we present two kernel selection criteria, which are both derived from the error
estimation. The first one is from the regularized empirical error functional

$$\mathcal{R}(f) = \frac{1}{l} \sum_{i=1}^{l} (f(\boldsymbol{x}_i) - y_i)^2 + \mu \|f\|_{\mathcal{H}_\kappa}^2, \tag{1}$$

where μ is the regularization parameter. We denote the target function as f_κ and $f_\kappa = \arg\min_{f \in \mathcal{H}_\kappa} \mathcal{R}(f)$. Using the representer theorem, f_κ can be represented as $f_\kappa = \sum_{i=1}^{l} \alpha_i \kappa(\boldsymbol{x}_i, \cdot)$ with $\boldsymbol{\alpha} = (\alpha_1, \ldots, \alpha_l)^{\mathrm{T}} = (\mathbf{K} + \mu l \mathbf{I})^{-1} \boldsymbol{y}$, where $\boldsymbol{y} = (y_1 \ldots, y_l)^{\mathrm{T}}$ and $\mathbf{I}$ is the identity matrix. Denoting $\mathbf{K} + \mu l \mathbf{I} = \mathbf{K}_\mu$, we have $\|f_\kappa\|_{\mathcal{H}_\kappa}^2 = \boldsymbol{\alpha}^{\mathrm{T}} \mathbf{K} \boldsymbol{\alpha} = \boldsymbol{y}^{\mathrm{T}} \mathbf{K}_\mu^{-1} \mathbf{K} \mathbf{K}_\mu^{-1} \boldsymbol{y}$. We use bold $\boldsymbol{f}_\kappa$ to denote $(f_\kappa(\boldsymbol{x}_1), \ldots, f_\kappa(\boldsymbol{x}_l))^{\mathrm{T}}$ and hence $\boldsymbol{f}_\kappa = \mathbf{K}\boldsymbol{\alpha} = \mathbf{K}\mathbf{K}_\mu^{-1}\boldsymbol{y}$, which implies $\boldsymbol{f}_\kappa - \boldsymbol{y} = -\mu l \mathbf{K}_\mu^{-1}\boldsymbol{y}$. Therefore,

$$
\begin{aligned}
\mathcal{R}(f_\kappa) &= \frac{1}{l}(\boldsymbol{f}_\kappa - \boldsymbol{y})^{\mathrm{T}}(\boldsymbol{f}_\kappa - \boldsymbol{y}) + \mu\|f_\kappa\|_{\mathcal{H}_\kappa}^2 \\
&= \mu^2 l \boldsymbol{y}^{\mathrm{T}} \mathbf{K}_\mu^{-1} \mathbf{K}_\mu^{-1} \boldsymbol{y} + \mu \boldsymbol{y}^{\mathrm{T}} \mathbf{K}_\mu^{-1} \mathbf{K} \mathbf{K}_\mu^{-1} \boldsymbol{y} \\
&= \mu \boldsymbol{y}^{\mathrm{T}} \mathbf{K}_\mu^{-1} \boldsymbol{y}.
\end{aligned}
\tag{2}
$$

There is a bijection between the set of kernels on $\mathcal{X}$ and that of reproducing kernel Hilbert spaces (RKHSs) on $\mathcal{X}$. For different RKHSs $\mathcal{H}_\kappa$, we may obtain different target functions f_κ. Then from all target functions, we select the one making $\mathcal{R}(f_\kappa)$ the smallest and the corresponding kernel will be the optimal one. We denote

$$
\mathcal{C}_1(\mathbf{K}) = \mathcal{R}(f_\kappa) = \mu \boldsymbol{y}^{\mathrm{T}} \mathbf{K}_\mu^{-1} \boldsymbol{y}.
\tag{3}
$$

Supposing we are given a prescribed set of kernels $\mathcal{K} = \{\kappa_1, \ldots, \kappa_n\}$, we can find the optimal kernel as $\kappa^* = \arg\min_{\kappa \in \mathcal{K}} \mathcal{C}_1(\mathbf{K})$.

We further present another kernel selection criterion, which is derived by the bias-variance decomposition of in-sample prediction error estimation [2]. In most practical cases, the observed output $\boldsymbol{y} = (y_1, \ldots, y_l)^{\mathrm{T}}$ is corrupted by some noises. We assume $y_i = \dot{y}_i + \xi_i, 1 \leq i \leq l$, where $\dot{\boldsymbol{y}} = [\dot{y}_1, \ldots \dot{y}_l]^{\mathrm{T}}$ is the unknown true output and the noise vector $\boldsymbol{\xi} = [\xi_1, \ldots, \xi_i]^{\mathrm{T}}$ is a random vector with mean 0 and finite covariance matrix $\mathbf{C}$. $\boldsymbol{f}_\kappa$ is a linear function of $\boldsymbol{y}$, which is an estimate of $\dot{\boldsymbol{y}}$. The expected prediction error of $\boldsymbol{f}_\kappa$ can be represented as

$$
\begin{aligned}
\frac{1}{l}\mathbb{E}_{\boldsymbol{\xi}}\|\boldsymbol{f}_\kappa - \dot{\boldsymbol{y}}\|^2 &= \frac{1}{l}\|\mathbb{E}_{\boldsymbol{\xi}}\boldsymbol{f}_\kappa - \dot{\boldsymbol{y}}\|^2 + \frac{1}{l}\mathrm{trace}(\mathrm{var}_{\boldsymbol{\xi}}(\boldsymbol{f}_\kappa)) \\
&= \frac{1}{l}\|\mathbf{K}\mathbf{K}_\mu^{-1}\dot{\boldsymbol{y}} - \dot{\boldsymbol{y}}\|^2 + \frac{1}{l}\mathrm{trace}(\mathbf{C}\mathbf{K}^2\mathbf{K}_\mu^{-2}) \\
&= \underbrace{\mu^2 l \dot{\boldsymbol{y}}^{\mathrm{T}} \mathbf{K}_\mu^{-2} \dot{\boldsymbol{y}}}_{\text{bias}(\mathbf{K})} + \underbrace{\frac{1}{l}\mathrm{trace}(\mathbf{C}\mathbf{K}^2\mathbf{K}_\mu^{-2})}_{\text{variance}(\mathbf{K})}.
\end{aligned}
\tag{4}
$$

For $\mathbf{C} = \sigma^2 \mathbf{I}$ with σ^2 as variance, we denote

$$
\mathcal{C}_2(\mathbf{K}) = \text{bias}(\mathbf{K}) + \text{variance}(\mathbf{K}) = \mu^2 l \dot{\boldsymbol{y}}^{\mathrm{T}} \mathbf{K}_\mu^{-2} \dot{\boldsymbol{y}} + \frac{\sigma^2}{l}\mathrm{trace}(\mathbf{K}^2\mathbf{K}_\mu^{-2}).
\tag{5}
$$

Now we simply review approximate kernel selection [9,10]. Supposing we have the training data $\mathcal{S}$, a prescribed kernel set $\mathcal{K}$, a kernel selection criterion $\mathcal{C}(\mathbf{K})$ and a kernel matrix approximation algorithm $\mathcal{A}$, which takes the kernel matrix $\mathbf{K}$ as input and generate the approximate matrix $\tilde{\mathbf{K}}$, we can describe the approximate kernel selection scheme as in Algorithm 1. The optimal kernel is selected by minimizing $\mathcal{C}(\tilde{\mathbf{K}})$.

Algorithm 1. Approximate Kernel Selection Scheme

Input: $\mathcal{S} = \{(\boldsymbol{x}_i, y_i)\}_{i=1}^{l}, \mathcal{K}, \mathcal{C}(\mathbf{K}), \mathcal{A}$;
Output: κ^*;
Initialize: $\mathcal{C}_{\mathrm{opt}} = \infty$;
for *each* $\kappa \in \mathcal{K}$ **do**
    Generate the kernel matrix $\mathbf{K}$ with κ and $\mathcal{S}$;
    Calculate the approximate kernel matrix $\tilde{\mathbf{K}}$ by $\mathcal{A}(\mathbf{K})$;
    Calculate the approximate kernel selection criterion $\mathcal{C}(\tilde{\mathbf{K}})$ with $\tilde{\mathbf{K}}$;
    if $\mathcal{C}(\tilde{\mathbf{K}}) \le \mathcal{C}_{\mathrm{opt}}$ **then**
        $\mathcal{C}_{\mathrm{opt}} = \mathcal{C}(\tilde{\mathbf{K}})$;
        $\kappa^* = \kappa$;
return κ^*;

The computational cost for the criteria $\mathcal{C}_1(\mathbf{K})$ and $\mathcal{C}_2(\mathbf{K})$ is $O(l^3)$, which is prohibitive for large scale data. The computation of $\mathcal{C}(\tilde{\mathbf{K}})$ could be much more efficient than that of $\mathcal{C}(\mathbf{K})$ due to the specific structure of $\tilde{\mathbf{K}}$. For Nyström approximation [10] Woodbury formula could be used for calculating $\mathcal{C}(\tilde{\mathbf{K}})$ and for MCM approximation [9] multi-dimensional fast Fourier transform (FFT) could be used. The computational cost can even be reduced from $O(l^3)$ to $O(l \log(l))$[9].

However, to demonstrate the rationality of approximate kernel selection, we need analyze what the difference between $\mathcal{C}(\mathbf{K})$ and $\mathcal{C}(\tilde{\mathbf{K}})$ is and whether this difference converges for large enough examples. In the next section, we will discuss these problems by defining the approximate consistency. At the end of this section, we introduce two typical kernel matrix approximation algorithms that will be discussed in this paper: Nyström approximation and MCM approximation.

The Nyström approximation generates a low rank approximation of $\mathbf{K}$ using a subset of the columns of $\mathbf{K}$.[1] Suppose we randomly sample c columns of $\mathbf{K}$. Let $\mathbf{C}$ denote the $l \times c$ matrix formed by theses columns. Let $\mathbf{D}$ be the $c \times c$ matrix consisting of the intersection of these c columns with the corresponding c rows of $\mathbf{K}$. The Nyström approximation matrix is $\tilde{\mathbf{K}} = \mathbf{C} \mathbf{D}_k^{\dagger} \mathbf{C}^{\mathrm{T}} \approx \mathbf{K}$, where $\mathbf{D}_k$ is the optimal rank k approximation to $\mathbf{D}$ and $\mathbf{D}_k^{\dagger}$ is the Moore-Penrose generalized inverse of $\mathbf{D}_k$. We further introduce the modified Nyström approximation [28], which shows tighter error bound than the standard Nyström method. The approximation matrix is $\tilde{\mathbf{K}} = \mathbf{C} \left(\mathbf{C}^{\dagger} \mathbf{K} \left(\mathbf{C}^{\dagger} \right)^{\mathrm{T}} \right) \mathbf{C}^{\mathrm{T}}$.

We now present the MCM approximation. We first briefly review the definition of MCM.[2] Let $\mathbb{N}$ denote the set of positive integers. For $m \in \mathbb{N}$, let $[m] = \{0, 1, \ldots, m - 1\}$. For a fixed positive integer p, let $\boldsymbol{m} = (m_0, m_1, \ldots, m_{p-1}) \in \mathbb{N}^p$. We set $\Pi_{\boldsymbol{m}} = m_0 m_1 \ldots m_{p-1}$ and $[\boldsymbol{m}] = [m_0] \times [m_1] \times \cdots \times [m_{p-1}]$. A multilevel circulant matrix [25] is defined recursively. A 1-level circulant matrix is an ordinary circulant matrix. For any positive integer s, an $(s + 1)$-level circulant matrix is a block circulant matrix whose blocks are s-level circulant matrices. According to [25], for $\boldsymbol{m} \in \mathbb{N}^p$, $\mathbf{A}_{\boldsymbol{m}} = [a_{\boldsymbol{i},\boldsymbol{j}} : \boldsymbol{i}, \boldsymbol{j} \in [\boldsymbol{m}]]$ is a p-level circulant matrix if, for any $\boldsymbol{i}, \boldsymbol{j} \in [\boldsymbol{m}]$,

---

[1]   Different sampling distributions have been considered [11,31,17,13].
[2]   Detailed definition can be seen in [25,9].

$a_{i,j} = a_{i_0-j_0 \pmod{m_0}, \ldots, i_{p-1}-j_{p-1} \pmod{m_{p-1}}}$. $\mathbf{A_m}$ is determined by its first column $a_{i,0}$ with $\mathbf{0} = (0, \ldots, 0) \in \mathbb{R}^p$. We write $\mathbf{A_m} = \mathrm{circ}_m[a_i : i \in [m]]$, where $a_i = a_{i,0}$, for $i \in [m]$. For the kernel function κ, we can construct MCM approximation of the kernel matrix $\mathbf{K}$ following the procedure given in Equation (12)-(14) of [9].

4 Approximate Consistency

In this section, we give the definition of approximate consistency and analyze the approximate consistency of Nyström approximation and MCM approximation.

Definition 1. *Suppose we are given a kernel selection criterion $\mathcal{C}(\mathbf{K})$, which is a functional of the kernel matrix $\mathbf{K}$, and a kernel matrix approximation algorithm $\mathcal{A}$, which takes the kernel matrix $\mathbf{K}$ as input and generate the approximate matrix $\tilde{\mathbf{K}}$. We say the kernel matrix approximation algorithm $\mathcal{A}$ is of strong approximate consistency for the kernel selection criterion $\mathcal{C}(\mathbf{K})$, if*

$$|\mathcal{C}(\mathbf{K}) - \mathcal{C}(\tilde{\mathbf{K}})| \leq \varepsilon(l), \tag{6}$$

where $\lim_{l \to \infty} \varepsilon(l) \to 0$. We say $\mathcal{A}$ is of p-order approximate consistency for $\mathcal{C}(\mathbf{K})$ if

$$|\mathcal{C}(\mathbf{K}) - \mathcal{C}(\tilde{\mathbf{K}})| \leq \varepsilon(l), \tag{7}$$

where $\lim_{l \to \infty} \varepsilon(l)/l^p \to 0$. There are two scenarios: if $\mathcal{A}$ is a deterministic algorithm, the approximate consistency is defined deterministically; if $\mathcal{A}$ is a stochastic algorithm, (6) or (7) is established under expectation or with high probability.

The approximate consistency reveals the convergence of the difference between the approximate kernel selection criterion and the accurate one. When kernel matrix approximation was applied in the kernel selection problem, the approximate consistency can be considered as a fundamental property of kernel matrix approximation algorithm to test its appositeness for kernel selection.

4.1 Approximate Consistency of Nyström Approximation

In this section we analyze the approximate consistency of Nyström approximation for the kernel selection criterion $\mathcal{C}_1(\mathbf{K})$.

Although there are many different versions of Nyström approximation, we concentrate on those with $(1 + \epsilon)$ relative-error bounds, where ϵ does not depend on l. Two $(1+\epsilon)$ relative error bounds have been reported for the standard Nyström [13] and modified Nyström method [28]. The bound for the standard Nyström method [13] states that for a failure probability $\delta \in (0, 1]$ and an approximation factor $\epsilon \in (0, 1]$,

$$\|\mathbf{K} - \tilde{\mathbf{K}}\|_* \leq (1 + \epsilon)\|\mathbf{K} - \mathbf{K}_k\|_* \tag{8}$$

holds with probability at least $0.6 - \delta$. The bound for the modified Nyström method [28] states that,

$$\mathbb{E}\left(\|\mathbf{K} - \tilde{\mathbf{K}}\|_F\right) \leq (1 + \epsilon)\|\mathbf{K} - \mathbf{K}_k\|_F. \tag{9}$$

Before stating the main theorem of this section, we introduce two assumptions.

Assumption 1. For $\rho \in (0, 1/2)$ and the rank parameter $k \le c \ll l$, $\lambda_k(\mathbf{K}) = \Omega(l/c^\rho)$ and $\lambda_{k+1}(\mathbf{K}) = O(l/c^{1-\rho})$, where ρ is to characterize the eigengap.

Assumption 2. We always assume that the rank parameter k is a constant and the sampling size c is a small ratio r of l.

Assumption 1 states the large eigengap in the spectrum of kernel matrix $\mathbf{K}$, i.e., the first few eigenvalues of the full kernel matrix are much larger than the remaining eigenvalues. This assumption has been adopted in [30,16] and empirically tested in [30]. Assumption 2 is one of common settings for Nyström approximation.

The following theorem shows the approximate consistency of the standard Nyström approximation using leverage score sampling [13]. The proof is given in Appendix A.

Theorem 1. *For the kernel selection criterion* $\mathcal{C}_1(\mathbf{K})$ *defined in* (3), *if Assumption 1 and 2 hold, we have for* $\delta \in (0, 1]$ *and* $\epsilon \in (0, 1]$,

$$|\mathcal{C}_1(\mathbf{K}) - \mathcal{C}_1(\tilde{\mathbf{K}})| \le \varepsilon(l) \tag{10}$$

holds with probability at least $0.6 - \delta$, *where* $\tilde{\mathbf{K}}$ *is produced by the standard Nyström approximation using leverage score sampling,* $\varepsilon(l) = \frac{\tau M^2 (1+\epsilon)}{\mu r^{1-\rho} l^{1-\rho}}(l - k)$ *for some constant* τ *and* $\lim_{l \to \infty} \varepsilon(l)/l^{\frac{1}{2}} \to 0$.

Theorem 1 demonstrates the $\frac{1}{2}$-order approximate consistency of the standard Nyström approximation for $\mathcal{C}_1(\mathbf{K})$. The strong approximate consistency has not been established. This is because the trace norm bound shown in (8), which is, to the best of our knowledge, the tightest bound for the standard Nyström approximation, is still not tight enough. If $(1 + \epsilon)$ relative-error bound for spectral norm can be proved, we can derive the strong approximate consistency.

The following theorem shows the approximate consistency of the modified Nyström approximation [28]. The proof can be seen in Appendix A.

Theorem 2. *For the kernel selection criterion* $\mathcal{C}_1(\mathbf{K})$ *defined in* (3), *if Assumption 1 and 2 hold, we have*

$$\mathbb{E}\left(|\mathcal{C}_1(\mathbf{K}) - \mathcal{C}_1(\tilde{\mathbf{K}})|\right) \le \varepsilon(l) \tag{11}$$

where $\tilde{\mathbf{K}}$ *is produced by the modified Nyström approximation,* $\varepsilon(l) = \frac{\tau M^2 (1+\epsilon)}{\mu r^{1-\rho} l^{1-\rho}}\sqrt{l - k}$ *for some constant* τ *and* $\lim_{l \to \infty} \varepsilon(l) \to 0$.

Theorem 2 demonstrates the strong approximate consistency of the modified Nyström approximation for $\mathcal{C}_1(\mathbf{K})$.

4.2 Approximate Consistency of MCM Approximation

In this section, we analyze the approximate consistency of MCM approximation for the criteria $\mathcal{C}_1(\mathbf{K})$ and $\mathcal{C}_2(\mathbf{K})$.

We use $\mathbf{U}_m$ to denote the MCM that approximates the kernel matrix $\mathbf{K}$. To facilitate the analysis, we will rewrite the kernel matrix $\mathbf{K}$ in multilevel notation. For a given

$m \in \mathbb{N}^p$, we assume the number of elements in $\mathcal{S}$ is Π_m, that is, $|\mathcal{S}| = l = \Pi_m$. We relabel the elements in $\mathcal{S}$ using multi-index, $\mathcal{S} = \{(x_i, y_i) : i \in [m]\}$. In this notation, we rewrite $\mathbf{K}$ as $\mathbf{K}_m = [K(\|x_i - x_j\|_2) : i, j \in [m]].^3$

The following theorem demonstrates the strong approximate consistency of MCM approximation for the criterion $\mathcal{C}_1(\mathbf{K})$. The proof is provided in Appendix B.

Theorem 3. *If the following assumptions:*

(H1) there exist positive constants c_0 and β such that $|K(s) - K(t)| \leq c_0 |s - t|^\beta$ for $s, t \in \mathbb{R}$;

(H2) there exists a positive constant h such that $h_{m,j} \geq h$ for $m \in \mathbb{N}^p$ and $j \in [p]$;

(H3) there exist positive constants λ_1 and c_1 such that $|K(s)| \leq c_1 e^{-\lambda_1 |s|}$ for $s \in \mathbb{R}$;

(H4) there exist positive constants λ_2 and c_2 such that for any $m \in \mathbb{N}^p$, $i, j \in [m]$, $\left| \|x_i - x_j\|_2 - \|[(i_s - j_s)h_{m,s} : s \in [p]]\|_2 \right| \leq c_2 \sum_{s \in [p]} \left(e^{-\lambda_2 \delta_{m_s}(i_s)} + e^{-\lambda_2 \delta_{m_s}(j_s)} \right)$, where $\delta_m(j) = \frac{m}{2} - |\frac{m}{2} - j|$ for $m \in \mathbb{N}$ and $j \in [m]$;

hold and in addition, there exist positive constants c_3 and r_1 such that for any $m \in \mathbb{N}^p$ and $i \in [m]$, $|y_i| \leq c_3 e^{-r_1 \nu_m(i)}$, where $\nu_m(i) = \left\| \frac{m}{2} - i \right\|_2$, then we have

$$\lim_{m \to \infty} |\mathcal{C}_1(\mathbf{K}_m) - \mathcal{C}_1(\mathbf{U}_m)| = 0, \tag{12}$$

where $m \to \infty$ means all of its components go to infinity.

For the criterion $\mathcal{C}_2(\mathbf{K})$, we first give the following theorem. The detailed proof can be seen in Appendix B.

Theorem 4. *If the assumptions (H1), (H2), (H3) and (H4) in Theorem 3 hold, then there exists a positive constant c such that*

$$|\text{variance}(\mathbf{K}_m) - \text{variance}(\mathbf{U}_m)| \leq c\sigma^2 (m_{\min})^{-1}, \tag{13}$$

where $m_{\min} = \min\{m_s : s \in [p]\}$. If in addition, there exist positive constants c_3 and r_1 such that for any $m \in \mathbb{N}^p$ and $i \in [m]$, $|\dot{y}_i| \leq c_3 e^{-r_1 \nu_m(i)}$, where $\nu_m(i) = \left\| \frac{m}{2} - i \right\|_2$, then there exist positive constants c and r such that for any $m \in \mathbb{N}^p$

$$|\text{bias}(\mathbf{K}_m) - \text{bias}(\mathbf{U}_m)| \leq c\mu^2 \Pi_m^{3/2} e^{-r m_{\min}}. \tag{14}$$

By Theorem 4, we can obtain the following theorem.

Theorem 5. *If the assumptions (H1), (H2), (H3) and (H4) in Theorem 3 hold, we have*

$$\lim_{m \to \infty} |\mathcal{C}_2(\mathbf{K}_m) - \mathcal{C}_2(\mathbf{U}_m)| = 0. \tag{15}$$

Theorem 5 shows the strong approximate consistency of MCM approximation for the criterion $\mathcal{C}_2(\mathbf{K})$.

---

^{3} There exists a real-valued function $K \in L^1(\mathbb{R})$ on $\mathcal{X}$ such that $K(\|x - x'\|_2) = \kappa(x, x')$ for all $x, x' \in \mathcal{X}$.

5 Empirical Studies

In this section, we empirically study the approximate consistency of different kernel matrix approximation algorithms. We compare 6 approximation algorithms, including optimal rank k approximation (OptApp), Nyström approximation with uniform sampling (Uniform) [17], column norm based sampling (ColNorm) [11], and leverage score based sampling (Leverage) [13], modified Nyström approximation (Modified)[4] [28] and MCM approximation (MCM) [9].

We set the rank parameter $k = 20$ and the sampling size $c = 0.2l$. To avoid the randomness, we run all Nyström methods 20 times. We use Gaussian kernels $\kappa(\boldsymbol{x}, \boldsymbol{x}') = \exp\left(-\gamma\|\boldsymbol{x} - \boldsymbol{x}'\|_2^2\right)$ with different width γ as our candidate kernel set $\mathcal{K}$. This paper does not focus on tuning the regularization parameter μ, so we just set $\mu = 0.005$. Since the regularized kernel matrix $\mathbf{K}_\mu = \mathbf{K} + \mu l \mathbf{I}$, $\mu = 0.005$ is not too small.

We conduct experiments on benchmark and synthetic data. The benchmark data sets are 12 public available data sets from UCI repository[5] and LIBSVM Data:[6] 7 data sets for classification and 5 data sets for regression. To evaluate the evolution of the approximate consistency as the number of examples increases, we also generate the synthetic data. The target function is

$$f(\boldsymbol{x}) = \frac{1}{10}\left(\|\boldsymbol{x}\|_2 + 2\mathrm{e}^{-8(\frac{4}{3}\pi - \|\boldsymbol{x}\|_2)^2} - 2\mathrm{e}^{-8(\frac{1}{2}\pi - \|\boldsymbol{x}\|_2)^2} - \mathrm{e}^{-8(\frac{3}{2}\pi - \|\boldsymbol{x}\|_2)^2}\right). \quad (16)$$

The points are $\{(\boldsymbol{x}_j, y_j), j \in [\boldsymbol{m}]\} \in \mathbb{R}^2 \times \mathbb{R}$ for $\boldsymbol{m} = (10, 10), (20, 20), (30, 30)$. The sampled inputs $\boldsymbol{x}_j$ is centered at 0 with fixed difference of any two successive numbers 0.1, and $y_j = f(\boldsymbol{x}_j) + \xi$, where the noise ξ is normally distributed with mean 0 and standard deviation 0.01.

For each γ, we generate the kernel matrix $\mathbf{K}$ and then use different approximation algorithms to produce the approximate kernel matrices $\tilde{\mathbf{K}}$. We compare the values of $\mathcal{C}(\mathbf{K})$ and $\mathcal{C}(\tilde{\mathbf{K}})$. The results for benchmark data[7] are shown in Fig. 1 and Fig. 2. We can find that for most data sets the curves of the accurate and approximate criteria are close, and for the rest of data sets, although the values of the criteria are different, the lowest points of the curves are close, which means that the optimal kernels selected by minimizing the accurate and approximate criteria are close. Modified Nyström approximation shows better approximate consistency than the standard Nyström approximation, which is in accord with the theoretical results. The results for synthetic data are given in Fig. 3 to demonstrate the evolution of the approximate consistency as the number of examples increases. We can find that the more the number of examples is, the closer the curves of accurate and approximate criteria are.

---

[4] We only adopt uniform sampling for modified Nyström approximation.

[5] http://www.ics.uci.edu/~mlearn/MLRepository.html

[6] http://www.csie.ntu.edu.tw/~cjlin/libsvm

[7] To satisfy the assumption (H4) in Theorem 3, we only conduct experiments on synthetic data for MCM approximation.

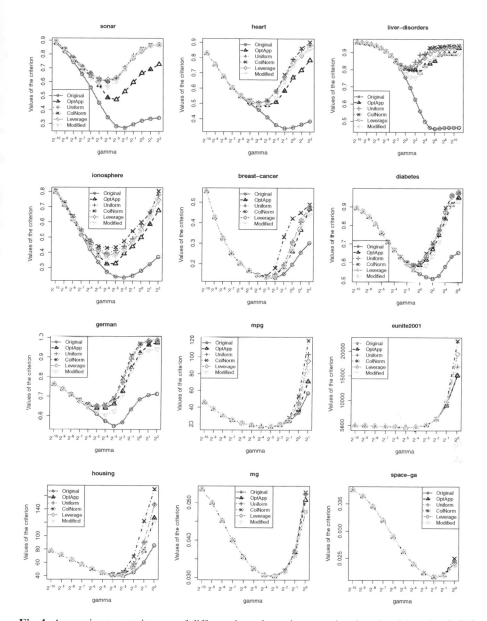

Fig. 1. Approximate consistency of different kernel matrix approximation algorithms for $\mathcal{C}_1(\mathbf{K})$

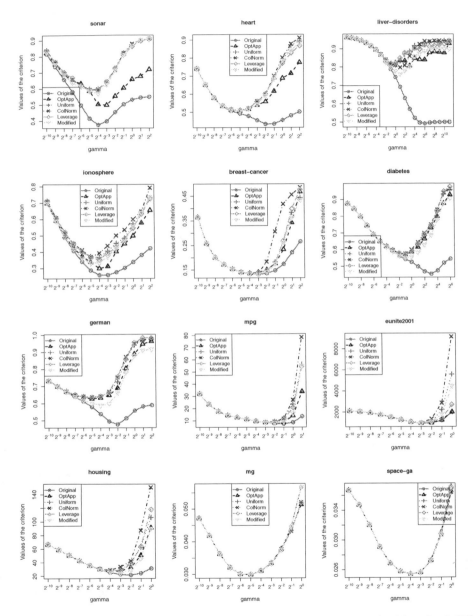

Fig. 2. Approximate consistency of different kernel matrix approximation algorithms for $\mathcal{C}_2(\mathbf{K})$

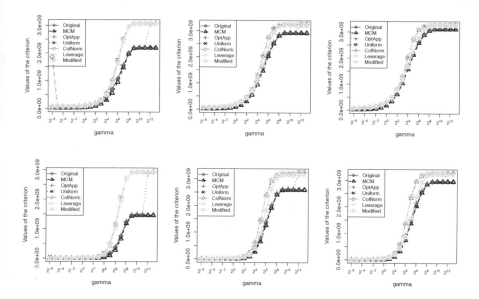

Fig. 3. Evolution of approximate consistency as the number of examples increases. The top 3 subfigures are for $\mathcal{C}_1(\mathbf{K})$ with $m = (10, 10), (20, 20), (30, 30)$ and the bottom 3 are for $\mathcal{C}_2(\mathbf{K})$.

6   Conclusion

In this paper, we defined the notion of approximate consistency for kernel matrix approximation algorithms, which tackles the theoretical problems faced by approximate kernel selection and establishes the preliminary foundations of approximate kernel selection. When kernel matrix approximation was applied in the kernel selection problem, approximate consistency can be considered as a fundamental property of kernel matrix approximation algorithm to test its appositeness for kernel selection. We have theoretically and empirically studied the approximate consistency of different kernel matrix approximation algorithms. To complete the foundations of approximate kernel selection, we will give the notion of hypothesis consistency, that is, the consistency of approximate optimal hypothesis and accurate optimal hypothesis, which are respectively learned using approximate and accurate optimal selected kernels, for future work.

Acknowledgments. The work is supported in part by the National Natural Science Foundation of China under grant No. 61170019.

Appendix A: Proof of Theorem 1 and Theorem 2

The proofs given in this section are partly based on the results in [7].

Let $\tilde{\mathbf{K}} = \mathcal{A}(\mathbf{K})$ be the produced approximate kernel matrix. We need to bound

$$
\begin{aligned}
\mathcal{C}_1(\mathbf{K}) - \mathcal{C}_1(\tilde{\mathbf{K}}) &= \mu \boldsymbol{y}^{\mathrm{T}}(\mathbf{K} + \mu l\mathbf{I})^{-1}\boldsymbol{y} - \mu \boldsymbol{y}^{\mathrm{T}}(\tilde{\mathbf{K}} + \mu l\mathbf{I})^{-1}\boldsymbol{y} \\
&= -\mu \boldsymbol{y}^{\mathrm{T}}[(\mathbf{K} + \mu l\mathbf{I})^{-1}(\mathbf{K} - \tilde{\mathbf{K}})(\tilde{\mathbf{K}} + \mu l\mathbf{I})^{-1}]\boldsymbol{y},
\end{aligned}
\tag{17}
$$

where the second equality follows that, for any invertible matrices $\mathbf{A}$, $\mathbf{B}$, the equality $\mathbf{A}^{-1} - \mathbf{B}^{-1} = -\mathbf{A}^{-1}(\mathbf{A} - \mathbf{B})\mathbf{B}^{-1}$ holds. Then

$$
\begin{aligned}
|\mathcal{C}_1(\mathbf{K}) - \mathcal{C}_1(\tilde{\mathbf{K}})| \leq &\mu \|\boldsymbol{y}^{\mathrm{T}}\| \|(\mathbf{K} + \mu l \mathbf{I})^{-1}\|_2 \|\mathbf{K} - \tilde{\mathbf{K}}\|_2 \|(\tilde{\mathbf{K}} + \mu l \mathbf{I})^{-1}\|_2 \|\boldsymbol{y}\| \\
\leq &\frac{\mu \|\boldsymbol{y}^{\mathrm{T}}\| \|\mathbf{K} - \tilde{\mathbf{K}}\|_2 \|\boldsymbol{y}\|}{\lambda_{\min}(\mathbf{K} + \mu l \mathbf{I}) \lambda_{\min}(\tilde{\mathbf{K}} + \mu l \mathbf{I})} \\
\leq &\frac{1}{\mu l^2} \|\boldsymbol{y}^{\mathrm{T}}\| \|\mathbf{K} - \tilde{\mathbf{K}}\|_2 \|\boldsymbol{y}\|,
\end{aligned}
\tag{18}
$$

where $\lambda_{\min}(\mathbf{M})$ denotes the smallest eigenvalue of $\mathbf{M}$. Since $\|\boldsymbol{y}\| \leq \sqrt{l}M$, we have

$$
|\mathcal{C}_1(\mathbf{K}) - \mathcal{C}_1(\tilde{\mathbf{K}})| \leq \frac{M^2}{\mu l} \|\mathbf{K} - \tilde{\mathbf{K}}\|_2.
\tag{19}
$$

Now we give the proofs of Theorem 1 and Theorem 2.

Proof (Theorem 1). Since $\|\mathbf{K} - \tilde{\mathbf{K}}\|_2 \leq \|\mathbf{K} - \tilde{\mathbf{K}}\|_*$, substituting (8) into (19), we have for $\delta \in (0, 1]$ and $\epsilon \in (0, 1]$,

$$
|\mathcal{C}_1(\mathbf{K}) - \mathcal{C}_1(\tilde{\mathbf{K}})| \leq \frac{M^2}{\mu l}(1 + \epsilon) \|\mathbf{K} - \mathbf{K}_k\|_*
\tag{20}
$$

holds with probability at least $0.6 - \delta$. We know that $\|\mathbf{K} - \mathbf{K}_k\|_* = \sum_{t=k+1}^{l} \lambda_t(\mathbf{K})$. Therefore $\|\mathbf{K} - \mathbf{K}_k\|_* \leq (l - k)\lambda_{k+1}(\mathbf{K})$. Combining Assumption 1 and 2, we can obtain

$$
|\mathcal{C}_1(\mathbf{K}) - \mathcal{C}_1(\tilde{\mathbf{K}})| \leq \varepsilon(l)
\tag{21}
$$

with $\varepsilon(l) = O\left(\frac{M^2(1+\epsilon)}{\mu r^{1-\rho} l^{1-\rho}}(l - k)\right)$ and $\lim_{l \to \infty} \varepsilon(l)/l^{\frac{1}{2}} \to 0$, since $\rho < \frac{1}{2}$.

Proof (Theorem 2). Following the similar procedure as the proof of Theorem 1, we can obtain

$$
\mathbb{E}\left(|\mathcal{C}_1(\mathbf{K}) - \mathcal{C}_1(\tilde{\mathbf{K}})|\right) \leq \varepsilon(l)
\tag{22}
$$

with $\varepsilon(l) = O\left(\frac{M^2(1+\epsilon)}{\mu r^{1-\rho} l^{1-\rho}} \sqrt{l - k}\right)$ and $\lim_{l \to \infty} \varepsilon(l) = 0$, since $\rho < \frac{1}{2}$.

Appendix B: Proof of Theorem 3 and Theorem 4

The proofs given in this section are mainly based on the results in [24,23].

We first introduce the "distances" of an entry in a multilevel matrix to its diagonal, to its upper right corner and to its lower left corner at each level [24]. For $t \in \{0, 1\}$, $m \in \mathbb{N}$, $i, j \in [m]$, we set $d_m(t, i, j) := t|i - j| + (1 - t)(m - |i - j| - 1)$ and for $t \in \{0, 1\}^p$, $\boldsymbol{m} \in \mathbb{N}^p$, $\boldsymbol{i}, \boldsymbol{j} \in [\boldsymbol{m}]$, let $\boldsymbol{d_m}(\boldsymbol{t}, \boldsymbol{i}, \boldsymbol{j}) := [d_{m_s}(t_s, i_s, j_s) : s \in [p]]$. For any $s \in [p]$, $d_{m_s}(1, i_s, j_s)$ is the distance of the entry at the position (i_s, j_s) to the diagonal at level s and $d_{m_s}(0, i_s, j_s)$ is the distance to the upper right and lower left corners at level s. For any $\boldsymbol{m} \in \mathbb{N}^p$, $\boldsymbol{i}, \boldsymbol{j} \in \mathbb{N}_{\boldsymbol{m}}$ and $r > 0$, let $E_{r,p,\boldsymbol{m}}(\boldsymbol{i}, \boldsymbol{j}) := \sum_{\boldsymbol{t} \in \{0,1\}^p} \mathrm{e}^{-r\|\boldsymbol{d_m}(\boldsymbol{t}, \boldsymbol{i}, \boldsymbol{j})\|_2}$. In what follows, we write $\mathcal{A} = \{\mathbf{A}_{\boldsymbol{m}} : \boldsymbol{m} \in \mathbb{N}^p\}$, $\mathcal{A}^{-1} = \{\mathbf{A}_{\boldsymbol{m}}^{-1} : \boldsymbol{m} \in \mathbb{N}^p\}$, $\mathcal{B} = \{\mathbf{B}_{\boldsymbol{m}} : \boldsymbol{m} \in \mathbb{N}^p\}$ and $\mathcal{AB} = \{\mathbf{A}_{\boldsymbol{m}}\mathbf{B}_{\boldsymbol{m}} : \boldsymbol{m} \in \mathbb{N}^p\}$.

We introduce two definitions about a class of matrices whose entries have an exponential decay property [24].

Definition 2. *A sequence of positive definite matrices* $\mathcal{A}$ *belongs to* $\mathcal{E}_r$ *for a positive constant* r *if it satisfies the following conditions: (i) there exists a positive constant* κ *such that* $\|\mathbf{A}_m^{-1}\|_2 \leq \kappa$ *for any* $m \in \mathbb{N}^p$; *(ii) there exists a positive constant* c *such that for any* $m \in \mathbb{N}^p$ *and* $i, j \in [m]$, $|a_{i,j}| \leq cE_{r,p,m}(i,j)$.

Definition 3. *We say* $\mathcal{A}$ *and* $\mathcal{B}$ *are asymptotically equivalent in* $\mathcal{E}_r$ *if* $\mathcal{A}, \mathcal{B} \in \mathcal{E}_r$ *and there exists a positive constant* c *such that for any* $m \in \mathbb{N}^p$ *and* $i, j \in [m]$, $|a_{i,j} - b_{i,j}| \leq c \sum_{s \in [p]} (e^{-r\delta_{m_s}(i_s)} + e^{-r\delta_{m_s}(j_s)})$, *where* $\delta_m(j) = \frac{m}{2} - |\frac{m}{2} - j|$, *for* $m \in \mathbb{N}$, $j \in [m]$. *We use the notation* $\mathcal{A} \sim_{\mathcal{E}_r} \mathcal{B}$.

We further introduce two propositions and a lemma [23].

Proposition 1. *If* $\mathcal{A}, \mathcal{B} \in \mathcal{E}_r$, *then* $\mathcal{AB} \in \mathcal{E}_{r_0}$ *for any* $r_0 < r$.

Proposition 2. *If* $\mathcal{A} \sim_{\mathcal{E}_r} \mathcal{B}$, *then* $\mathcal{A}^{-1} \sim_{\mathcal{E}_{r'}} \mathcal{B}^{-1}$ *for any* $r' < r$.

Lemma 1. *If* $\mathcal{A} \sim_{\mathcal{E}_r} \mathcal{B}$ *and* $\mathcal{D} \in \mathcal{E}_r$, *then* $\mathcal{DA} \sim_{\mathcal{E}_{r_1}} \mathcal{DB}$ *and* $\mathcal{AD} \sim_{\mathcal{E}_{r_1}} \mathcal{BD}$ *for some* $r_1 < r$.

Using above results [23], we prove the following three propositions.

Proposition 3. *If* $\mathcal{A} \sim_{\mathcal{E}_r} \mathcal{B}$, *then* $\mathcal{A}^2 \sim_{\mathcal{E}_{r'}} \mathcal{B}^2$, *for some* $r' < r$.

Proof. From Proposition 1, we have $\mathcal{A}^2 = \mathcal{AA}$, $\mathcal{B}^2 = \mathcal{BB}$, and $\mathcal{AB}$ are in $\mathcal{E}_{r_0}$ for any $r_0 < r$. From Lemma 1, we can obtain $\mathcal{A}^2 \sim_{\mathcal{E}_{r_1}} \mathcal{AB}$ for some $r_1 < r$ and $\mathcal{AB} \sim_{\mathcal{E}_{r_2}} \mathcal{B}^2$ for some $r_2 < r$. We take $r' = \min\{r_1, r_2\}$. According to Definition 3 and triangular inequality, we have $\mathcal{A}^2 \sim_{\mathcal{E}_{r'}} \mathcal{B}^2$.

Proposition 4. *If* $\mathcal{A} \sim_{\mathcal{E}_{r_1}} \mathcal{B}$ *and* $\mathcal{C} \sim_{\mathcal{E}_{r_2}} \mathcal{D}$, *then* $\mathcal{AC} \sim_{\mathcal{E}_{r_0}} \mathcal{BD}$, *for some* $r_0 < \min\{r_1, r_2\}$.

Proposition 5. *If* $\mathcal{A} \in \mathcal{E}_{r_1}$ *and* $\mathcal{B} \in \mathcal{E}_{r_2}$ *then* $\mathcal{AB} \in \mathcal{E}_{r_0}$ *for any* $r_0 < \min\{r_1, r_2\}$.

The proof of Proposition 5 is similar to Proposition 1. Now we prove Proposition 4.

Proof (Proposition 4). Following Lemma 1 and Proposition 5, we have $\mathcal{AC} \sim_{\mathcal{E}_{r_3}} \mathcal{BC}$ for some $r_3 < \min\{r_1, r_2\}$ and $\mathcal{BC} \sim_{\mathcal{E}_{r_4}} \mathcal{BD}$ for some $r_4 < \min\{r_1, r_2\}$. Therefore, $\mathcal{AC} \sim_{\mathcal{E}_{r_0}} \mathcal{BD}$ for some $r_0 = \min\{r_3, r_4\} < \min\{r_1, r_2\}$.

Now we prove Theorem 3 and Theorem 4.

Proof (Theorem 3). We write $\mathcal{K}_\mu = \{\mathbf{K}_m + \mu l \mathbf{I}_m : m \in \mathbb{N}^p\}$ and $\mathcal{U}_\mu = \{\mathbf{U}_m + \mu l \mathbf{I}_m : m \in \mathbb{N}^p\}$. Theorem 5.2.3 in [23] states that if the assumptions (H1), (H2), (H3) and (H4) in Theorem 3 hold, $\mathcal{K}_\mu \sim_{\mathcal{E}_r} \mathcal{U}_\mu$, where $r = \min\{\lambda_1 h, \lambda_2 \beta\}$. Combining Theorem 5.2.3 in [23] with Theorem 4.2.2 in [23], we can obtain Theorem 3.

Proof (Theorem 4). We denote $\mathcal{K}_\mu^{-q} = \{(\mathbf{K}_m + \mu l \mathbf{I}_m)^{-q} : m \in \mathbb{N}^p\}$ and $\mathcal{U}_\mu^{-q} = \{(\mathbf{U}_m + \mu l \mathbf{I}_m)^{-q} : m \in \mathbb{N}^p\}$ for $q = 1, 2, \ldots$. According to Proposition 3 and Theorem 5.2.3 in [23], we have $\mathcal{K}_\mu^{-2} \sim_{\mathcal{E}_{r'}} \mathcal{U}_\mu^{-2}$, for some $r' < r$. Based on Proposition 4.2.1 in [23], we have $|\text{bias}(\mathbf{K}_m) - \text{bias}(\mathbf{U}_m)| \leq c\mu^2 \Pi_m^{3/2} e^{-r' m_{\min}}$, which is the bound for bias term in Theorem 2. Since the kernel matrix considered in this paper is assumed to be SPD, the corresponding multilevel circulant matrix is also positive definite. We have the matrix sets $\mathcal{K} = \{\mathbf{K}_m : m \in \mathbb{N}^p\}$ and $\mathcal{U} = \{\mathbf{U}_m : m \in \mathbb{N}^p\}$ are asymptotically equivalent. Therefore, according to Proposition 3 and 4, we have $\mathcal{K}^2 \mathcal{K}_\mu^{-2} \sim_{\mathcal{E}_{r_0}} \mathcal{U}^2 \mathcal{U}_\mu^{-2}$ for some $r_0 < r'$. By Proposition 4.3.2 of [23], we can obtain $|\mathrm{trace}(\mathbf{K}_m^2(\mathbf{K}_m + \mu l \mathbf{I}_m)^{-2}) - \mathrm{trace}(\mathbf{U}_m^2(\mathbf{U}_m + \mu l \mathbf{I}_m)^{-2})| \leq c \Pi_m (m_{\min})^{-1}$. Therefore, we have $|\mathrm{variance}(\mathbf{K}_m) - \mathrm{variance}(\mathbf{U}_m)| \leq c\sigma^2 (m_{\min})^{-1}$, which is the bound for variance term in Theorem 4.

References

1. Alpaydin, E.: Introduction to Machine Learning. MIT Press, Cambridge (2004)
2. Bach, F.: Sharp analysis of low-rank kernel matrix approximations. In: Proceedings of the 26th Annual Conference on Learning Theory (COLT), pp. 185–209 (2013)
3. Bartlett, P.L., Boucheron, S., Lugosi, G.: Model selection and error estimation. Machine Learning 48(1–3), 85–113 (2002)
4. Bartlett, P., Mendelson, S.: Rademacher and Gaussian complexities: Risk bounds and structural results. Journal of Machine Learning Research 3, 463–482 (2002)
5. Cawley, G., Talbot, N.: On over-fitting in model selection and subsequent selection bias in performance evaluation. Journal of Machine Learning Research 11, 2079–2107 (2010)
6. Chapelle, O., Vapnik, V., Bousquet, O., Mukherjee, S.: Choosing multiple parameters for support vector machines. Machine Learning 46(1-3), 131–159 (2002)
7. Cortes, C., Mohri, M., Talwalkar, A.: On the impact of kernel approximation on learning accuracy. In: Proceedings of the 13th International Conference on Artificial Intelligence and Statistics (AISTATS), pp. 113–120 (2010)
8. De Vito, E., Caponnetto, A., Rosasco, L.: Model selection for regularized least-squares algorithm in learning theory. Foundations of Computational Mathematics 5(1), 59–85 (2005)
9. Ding, L.Z., Liao, S.Z.: Approximate model selection for large scale LSSVM. Journal of Machine Learning Research - Proceedings Track 20, 165–180 (2011)
10. Ding, L., Liao, S.: Nyström approximate model selection for LSSVM. In: Tan, P.-N., Chawla, S., Ho, C.K., Bailey, J. (eds.) PAKDD 2012, Part I. LNCS (LNAI), vol. 7301, pp. 282–293. Springer, Heidelberg (2012)
11. Drineas, P., Mahoney, M.W.: On the Nyström method for approximating a Gram matrix for improved kernel-based learning. Journal of Machine Learning Research 6, 2153–2175 (2005)
12. Fine, S., Scheinberg, K.: Efficient SVM training using low-rank kernel representations. Journal of Machine Learning Research 2, 243–264 (2002)
13. Gittens, A., Mahoney, M.W.: Revisiting the Nyström method for improved large-scale machine learning. In: Proceedings of the 30th International Conference on Machine Learning (ICML), pp. 567–575 (2013)
14. Golub, G.H., Heath, M., Wahba, G.: Generalized cross-validation as a method for choosing a good ridge parameter. Technometrics 21(2), 215–223 (1979)
15. Hastie, T., Rosset, S., Tibshirani, R., Zhu, J.: The entire regularization path for the support vector machine. Journal of Machine Learning Research 5, 1391–1415 (2004)
16. Jin, R., Yang, T.B., Mahdavi, M., Li, Y.F., Zhou, Z.H.: Improved bounds for the Nyström method with application to kernel classification. IEEE Transactions on Information Theory 5(10), 6939–6949 (2013)
17. Kumar, S., Mohri, M., Talwalkar, A.: Sampling methods for the Nyström method. Journal of Machine Learning Research 13, 981–1006 (2012)

18. Liu, Y., Jiang, S., Liao, S.: Efficient approximation of cross-validation for kernel methods using Bouligand influence function. In: Proceedings of the 31st International Conference on Machine Learning (ICML), pp. 324–332 (2014)
19. Luxburg, U.V., Bousquet, O., Schölkopf, B.: A compression approach to support vector model selection. Journal of Machine Learning Research 5, 293–323 (2004)
20. Micchelli, C.A., Pontil, M.: Learning the kernel function via regularization. Journal of Machine Learning Research 6, 1099–1125 (2005)
21. Mitchell, T.M.: Machine Learning. McGraw Hill, New York (1997)
22. Smola, A.J., Schölkopf, B.: Sparse greedy matrix approximation for machine learning. In: Proceedings of the 17th International Conference on Machine Learning (ICML), pp. 911–918 (2000)
23. Song, G.H.: Approximation of kernel matrices in machine learning. Ph.D. thesis, Syracuse University, Syracuse, NY, USA (2010)
24. Song, G.H., Xu, Y.S.: Approximation of high-dimensional kernel matrices by multilevel circulant matrices. Journal of Complexity 26(4), 375–405 (2010)
25. Tyrtyshnikov, E.E.: A unifying approach to some old and new theorems on distribution and clustering. Linear Algebra and its Applications 232, 1–43 (1996)
26. Vapnik, V.: The Nature of Statistical Learning Theory. Springer, New York (1995)
27. Wahba, G., Lin, Y., Zhang, H.: GACV for support vector machines. In: Advances in Large Margin Classifiers. MIT Press, Cambridge (1999)
28. Wang, S.S., Zhang, Z.H.: Improving CUR matrix decomposition and the Nyström approximation via adaptive sampling. Journal of Machine Learning Research 14, 2729–2769 (2013)
29. Williams, C.K.I., Seeger, M.: Using the Nyström method to speed up kernel machines. In: Advances in Neural Information Processing Systems 13, pp. 682–688 (2001)
30. Yang, T.B., Li, Y.F., Mahdavi, M., Jin, R., Zhou, Z.H.: Nyström method vs random Fourier features: A theoretical and empirical comparison. In: Advances in Neural Information Processing Systems 24, pp. 1060–1068 (2012)
31. Zhang, K., Kwok, J.T.: Clustered Nyström method for large scale manifold learning and dimension reduction. IEEE Transactions on Neural Networks 21(10), 1576–1587 (2010)

Nowcasting with Numerous Candidate Predictors

Brendan Duncan and Charles Elkan

Department of Computer Science and Engineering,
University of California, San Diego,
La Jolla, CA 92093-0404, USA
{baduncan,elkan}@cs.ucsd.edu

Abstract. The goal of nowcasting, or "predicting the present," is to estimate up-to-date values for a time series whose actual observations are available only with a delay. Methods for this task leverage observations of correlated time series to estimate values of the target series. This paper introduces a nowcasting technique called FDR (false discovery reduction) that combines tractable variable selection with a time series model trained using a Kalman filter. The FDR method guarantees that all variables selected have statistically significant predictive power. We apply the method to sales figures provided by the United States census bureau, and to a consumer sentiment index. As side data, the experiments use time series from Google Trends of the volumes of search queries. In total, there are 39,059 potential correlated time series. We compare results from the FDR method to those from several baseline methods. The new method outperforms the baselines and achieves comparable performance to a state-of-the-art nowcasting technique on the consumer sentiment time series, while allowing variable selection from over 250 times as many side data series.

Keywords: Nowcasting, time series analysis, Kalman filter, feature selection, economic data, supervised learning, forecasting.

1 Introduction

Many important measurements are published on a periodic basis. For example, the United States government releases GDP figures every quarter, and unemployment figures every month. These data are published with a lag; the employment rate for March of 2014 was released in April of 2014. Even once published, many of these time series are still subject to later revisions as more information becomes known.

Because of these issues, such data do not provide an up-to-date estimate of the statistic they are tracking. The goal of nowcasting, which is also called "predicting the present," is to provide an up-to-date estimate of the current value of a time series. Nowcasting methods employ correlated data that are real-time, or more frequently updated than the desired statistic. The experiments in this paper use Google Trends data, which track the daily volume of search queries by geographic region. These data are described further in Section 2. We use a

T. Calders et al. (Eds.): ECML PKDD 2014, Part I, LNCS 8724, pp. 370–385, 2014.

forward selection algorithm to select relevant Google Trends queries. We then
combine the selected auxiliary time series with a random walk model of the target
series. The Kalman filter is used both for training and for making predictions (or
more correctly, "nowcasts") of the target series. We call the proposed method
FDR, because the variable selection process aims to reduce the false discovery
rate and, consequently, to reduce overfitting. The computational tractability
of the FDR method allows variable selection from a large number of potential
side variables, which reduces the need for choosing a small set of potentially
predictive auxiliary time series by hand, and thereby also allows for the discovery
of unexpected correlations.

2   Google Trends Data

Google Trends data have been shown to be effective when used as predictors for
other time series, such as financial time series [4] and disease outbreaks [3]. For
example, searches about flu remedies and immunization have been shown to be
predictive of the number of people who currently have the flu [6], and Google has
has a website called Flu Trends which gives nowcasts for flu activity in different
regions based on Google query volumes. In recent years, however, Google Flu
Trends has come under criticism for overestimating peak flu levels [2].

Google Trends tracks daily query volume, but only makes data from the past
90 days available for download by outsiders. Weekly data are available starting
in January 2004. Volume is calculated as a percentage of all queries from a given
region that match the given query description during the time period. The entire
time series is then normalized to fall between 0 and 100. Although Google Trends
allows comparing the relative volumes of individual searches, it does not provide
a way to download a collection of data series that preserves the relative volumes
of queries. We therefore use only individually normalized series in this paper.

In addition to providing volumes of individual queries, Google Trends orga-
nizes queries into a hierarchical structure of categories. For example, Arts &
Entertainment is a top-level category, which includes subcategories Movies,
Comics & Animation, and Music & Audio, each of which has its own subcategories. There are 25 categories at the highest level and 278 second-level subcategories, 120 of which are leaf nodes. The longest path in the hierarchical tree is
of length seven. Google Trends also supports filtering query volume by region.
For example, if we were interested in the number of people who have flu in the
US, we can request the number of flu searches that originated in the US.

3   Time Series Model

We model the target time series using a random walk model. Let y_t be a value
in the time series we would like to nowcast, and let $\hat{y}_t$ be the estimate of y_t
according to the model. The estimate is based on a combination of a hidden
variable μ_t and multiple auxiliary data series, $\mathbf{x}_t$, which are chosen by variable

selection as explained in Section 6. The regression coefficients at time t are the vector $\boldsymbol{\beta}_t$. The model is

$$\hat{y}_t = \mu_t + \boldsymbol{\beta}_t^T \mathbf{x}_t$$
$$\begin{pmatrix} \mu_t \\ \boldsymbol{\beta}_t \end{pmatrix} = \begin{pmatrix} \mu_{t-1} \\ \boldsymbol{\beta}_{t-1} \end{pmatrix} + \begin{pmatrix} v_t \\ \boldsymbol{w}_t \end{pmatrix} \qquad (1)$$
$$v_t \sim \mathcal{N}(0, v)$$
$$\boldsymbol{w}_t \sim \mathcal{N}(0, W).$$

The hidden variable μ_t is called the level; it is allowed to change over time based on a random process. The changes v_t in its value are called *innovation steps* rather than error terms, because they are considered part of the model. The initial regression coefficients $\boldsymbol{\beta}_0$ are learned as described in Section 5. These coefficients change over time via *innovation steps* with covariance W. They need not be thought of as hidden time series variables, but could instead be fixed. However, we find that updating $\boldsymbol{\beta}_t$ yields slightly better results.

We choose this time series model because it incorporates auxiliary data observations and works well with the Kalman filter. Vector autoregression methods such as that in [5] could also be used to model a nowcasting problem. This would involve combining a time series and the side data as a single vector. However, this would mean we would have to predict the side data as well, which is not the goal of nowcasting. Online learning methods are another candidate for nowcasting time series models. The algorithm described in [1] requires fewer constraints on the innovation step behavior and loss functions used. However, our model performs well even with the stricter constraints, and the online learning method cited is a vector autoregressive model with the added complexity of modeling the auxiliary data.

If the time series has a trend, our model can have a hidden trend variable in addition to the level μ_t, as in [13]. Although the experiments in this paper use series with a noticeable trend component, we have found that the simpler random walk model works just as well. Other time series models exist to deal with seasonal data, but for our experiments we deseasonalize the data in preprocessing, as described in Section 7.

4   The Kalman Filter for Nowcasting

The Kalman filter, originally described in [9], is used to find the maximum-likelihood sequence of values of a set of hidden continuous time series. It assumes that hidden variables $\boldsymbol{\nu}_t$ are updated according to a recursive process governed by a state transition matrix F_t and process noise $\boldsymbol{\omega}_t$, as given by the equation $\boldsymbol{\nu}_t = F_t \boldsymbol{\nu}_{t-1} + \boldsymbol{\omega}_t$. The hidden variables determine the observable variables $\boldsymbol{y}_t$ based on the linear model $\boldsymbol{y}_t = H_t \boldsymbol{\nu}_t + \boldsymbol{e}_t$, where H_t is the observation model matrix and $\boldsymbol{e}_t$ is the observation noise. The Kalman filter assumes that the process and observation noise are both normally distributed: $\boldsymbol{\omega}_t \sim \mathcal{N}(0, \Omega_t)$ and $\boldsymbol{e}_t \sim \mathcal{N}(0, E_t)$.

Given the assumptions about the noise distributions, the Kalman filter finds the maximum-likelihood estimate of the hidden variables $\boldsymbol{\nu}_t$ based on the observations $\boldsymbol{y}_t$. The Kalman filter is an inference algorithm moving forward in time that can be divided into two steps, a predict step and and update step:

Predict step:
$$\hat{\boldsymbol{\nu}}_t = F_t \boldsymbol{\nu}_{t-1}$$
$$\hat{\boldsymbol{y}}_t = H_t \hat{\boldsymbol{\nu}}_t$$
$$\hat{P}_t = F_t P_{t-1} F_t^T + \Omega_t$$

Update step:
$$\boldsymbol{r}_t = \boldsymbol{y}_t - \hat{\boldsymbol{y}}_t$$
$$S_t = H_t \hat{P}_t H_t^T + E_t$$
$$K_t = \hat{P}_t H_t^T S_t^{-1}$$
$$\boldsymbol{\nu}_t = \hat{\boldsymbol{\nu}}_t + K_t \boldsymbol{r}_t$$
$$P_t = (I - K_t H_t)\hat{P}_t.$$

The P_t matrices represent the error covariance and K_t is the optimal Kalman gain matrix. Symbols with hats represent predictions before the actual observation $\boldsymbol{y}_t$ is available. Symbols without hats represent updated values after $\boldsymbol{y}_t$ is available. This is a slight abuse of notation, since the true values of variables are never really known, except for the observations $\boldsymbol{y}$. The hyperparameters of the Kalman filter are the noise covariance matrices Ω_t and E_t, and base recursion values $\boldsymbol{\nu}_0$ and P_0. The matrices F_t and H_t are usually fixed based on knowledge of the time series process.

Kalman filters have been applied before to economic time series, for example in [11], where the authors apply a Kalman filter to determine the "true price" of an asset given its price series. In their model, the true price of an asset follows a random walk model while the reported or actual price is subject to observational noise. In this case, the values $\boldsymbol{\omega}_t$ represent the *innovation steps* taken by the random walk, and are thus not considered process noise but part of the actual process update. This perspective is consistent with our time series models discussed in Section 3.

In order to apply the Kalman filter model from above to the random walk time series model of Equation (1), we make the following variable definitions:
$$\boldsymbol{\nu}_t = [\mu_t \ \boldsymbol{\beta}_t]^T$$
$$\mathbf{y}_t = y_t$$
$$F_t = I$$
$$H_t = [1 \ \mathbf{x}_t^T]$$
$$\Omega_t = \begin{pmatrix} v & \mathbf{0}^T \\ \mathbf{0} & W \end{pmatrix} = \text{diag}([v, v_\beta, \ldots, v_\beta]^T)$$
$$E_t = \sigma^2$$
$$\boldsymbol{e}_t = \epsilon_t.$$

Note that y_t, σ^2, ϵ_t, and v_β are scalars, and that the covariance matrix Ω_t is diagonal. The reason for using this Ω_t is described in more detail in Section 5. These substitutions result in the following predict and update steps:

Predict step:

$$[\hat{\mu}_t \; \hat{\boldsymbol{\beta}}_t^T] = [\mu_{t-1} \; \boldsymbol{\beta}_{t-1}^T]$$

$$\hat{y}_t = [1 \; \mathbf{x}_t^T][\hat{\mu}_t \; \hat{\boldsymbol{\beta}}_t^T]^T = \hat{\mu}_t + \hat{\boldsymbol{\beta}}_t^T \mathbf{x}_t$$

$$\hat{P}_t = P_{t-1} + \mathrm{diag}([v, v_\beta, ..., v_\beta]^T)$$

Update step:

$$r_t = y_t - \hat{y}_t \tag{2}$$

$$s_t = [1 \; \mathbf{x}_t^T]\hat{P}_t[1 \; \mathbf{x}_t^T]^T + \sigma^2$$

$$\boldsymbol{k_t} = \frac{1}{s_t}\hat{P}_t[1 \; \mathbf{x}_t^T]^T$$

$$[\mu_t \; \boldsymbol{\beta}_t^T]^T = [\hat{\mu}_t \; \hat{\boldsymbol{\beta}}_t^T]^T + r_t\boldsymbol{k_t}$$

$$P_t = (I - \boldsymbol{k_t}[1 \; \mathbf{x}_t^T])\hat{P}_t.$$

In addition to learning the hidden time series level, the Kalman filter also gives a natural means of performing a nowcast: the value $\hat{y}_t$ in the predict step is the estimated current value of the time series.

5   Training the Model

The method introduced here for training a time series model with auxiliary data involves two uses of a Kalman filter: one to compute initial regression coefficients $\boldsymbol{\beta}_0$, and then one to train the full time series model in Equation (2). To perform training, we split the time series into three periods: a training period, a validation period, and a testing period. More detail is in Section 7. The training period is used to determine $\boldsymbol{\beta}_0$, and the validation period is used to determine hyperparameters σ^2, v and v_β. The training in this section occurs after side data variables have been selected. Variable selection is described in detail in Section 6.

First, we train a random walk model *without* side data using the Kalman filter. The model is

$$\tilde{y}_t = \mu_t$$

$$\mu_t = \mu_{t-1} + v_t \tag{3}$$

$$v_t \sim \mathcal{N}(0, v).$$

In order to train this model, we simplify the predict and update steps of Equation (2) by removing the side data $\mathbf{x}_t$ and regression coefficients $\boldsymbol{\beta}_t$. The $\tilde{y}_t$ notation indicates an estimate from the model that does not include side data. We apply the Kalman filter to obtain a series of estimates $\tilde{y}_t$ for all t in the training period. Then we compute the corresponding residual $\tilde{r}_t = y_t - \tilde{y}_t$. Equation (1) gives $\hat{y}_t = \mu_t + \boldsymbol{\beta}_t^T \mathbf{x}_t$ and the Kalman filter gives $y_t = \hat{y}_t + \epsilon_t$. If we constrain μ_t to be the same as in Equation (1) we can solve for $\tilde{r}_t$ as

$$\tilde{r}_t = \boldsymbol{\beta}_t^T \mathbf{x}_t + \epsilon_t \tag{4}$$

where $\epsilon_t = y_t - \hat{y}_t$ is the error of the complete model. We assume constant regression coefficients over the training period, $\boldsymbol{\beta}_t = \boldsymbol{\beta}$ for all t in the training period, and estimate $\boldsymbol{\beta}$ by performing a linear regression of the residuals $\tilde{r}_t$ on the side data after variable selection.[1]

After determining $\boldsymbol{\beta}_0$, we run the Kalman filter to produce a set of predictions $\hat{y}_t$ across the training and validation periods. We then compute the mean absolute error of logarithms, described in Section 7, over the validation period to obtain an estimate of the nowcasting performance. The process of training the parameters $\boldsymbol{\beta}_0$ and computing the validation error can be repeated with different assignments to hyperparameters σ^2, v, and $v_{\boldsymbol{\beta}}$. We choose the hyperparameters and resulting initial coefficients $\boldsymbol{\beta}_0$ that yield the lowest error on the validation period.

In order to simplify the search for hyperparameter values, we assume that the error variance σ^2 remains fixed whether performing a simple random walk or training the full model with side data. That is, we assume $\tilde{r}_t \sim \mathcal{N}(0, \sigma^2)$ and $\epsilon_t \sim \mathcal{N}(0, \sigma^2)$. We find that this does not greatly affect the final results. Again, for simplicity, we choose $P_0 - I$. It is important to note that the innovation steps v_t and $\mathbf{w}_t$ are outputs of the Kalman filter, not parameters.

The form of the innovation step covariance matrix,

$$\Omega_t = \begin{pmatrix} v & \mathbf{0}^T \\ \mathbf{0} & W \end{pmatrix} = \mathrm{diag}([v, v_{\boldsymbol{\beta}}, ..., v_{\boldsymbol{\beta}}]^T), \tag{5}$$

also simplifies the search for hyperparameters. The time series model in Equation (1) already assumes that the innovation steps for the level are independent of the innovation steps for the regression coefficients. We additionally assume that the innovation step for each regression coefficient is independent of the others, and that each has the same variance.

The variable selection procedure reduces the chance that auxiliary variables are highly correlated because it sweeps out each chosen predictor from the other variables as they are chosen (Section 6). Therefore a diagonal covariance matrix is a reasonable approximation. Each side data variable is measured on a similar normalized scale, as described in Section 2, so assuming equal variance is also reasonable. Experiments give good results even with these constraints.

6   FDR Variable Selection

Before we can train the time series model (Section 5), we use false discovery reduction (FDR) variable selection to choose a subset of the possible side data

---

[1] If one instead solved for linear regression coefficients of $\mathbf{y}$ on X, the coefficients would constitute a standard linear model, and would not correspond to the $\boldsymbol{\beta}_t$ in Equation (1).

variables to use in the model. FDR variable selection is a modified version of the method of [7]. That paper performs variable selection over a variety of different types of potential side data predictors, including binary and sparse variables, to predict bankruptcy. Because we are nowcasting a time series using other time series as side data, calculations for p-values and variances are different.

Detailed pseudocode of the FDR variable selection is Algorithm 1. The bar notation $\bar{X}$ and $\bar{\mathbf{x}}_t$ indicates the full set of potential side data variables, before variable selection. First, the algorithm iterates through all variables, and creates a set V_q of those variables that have a p-value $\leq \alpha/p$ where α is a significance parameter and p represents the number of potential predictors. Of the significant variables, the algorithm adds the variable $\mathbf{z}$ that maximizes $SS(\mathbf{z})$, where $SS(\mathbf{z})$ can be thought of as a guaranteed reduction in residual sum of squares based on a confidence interval. This idea is similar to that used in upper confidence bound search [10].

After a variable is added, we repeat the process to select additional variables. Each time a new predictor is added to the side data X, we sweep all the current predictors from $\mathbf{y}$ and from each remaining potential predictor $\mathbf{z}$ using the projection matrix H. In addition, each time the model grows, we increase the significance threshold to $\alpha q/p$, where q is the cardinality of the current model.

For the experimental results reported in this paper, we choose $\alpha = 0.005$, which performs slightly better than $\alpha = 0.05$ and $\alpha = 0.2$ in experiments. Although this threshold appears strict, we find that the number of variables selected, and the specific variables selected do not change much for these different values of α. The reason is that the cardinality of V_q decreases quickly as variables are added to the model, so once selected variables are swept away, few correlated variables remain. The similarity of selected variables for different values of α suggests that the highly predictive variables tend to be the same ones that minimize $SS(\mathbf{z})$.

In Algorithm 1, the variables that are chosen are those that are predictive for linear regression. Linear regression can be performed over different dependent-independent variable combinations. The simple case is using raw side data $\bar{X}$ as a predictor for time series $\mathbf{y}$, but more complicated models can be used. In particular, one can use the changes $\bar{\mathbf{x}}_t - \bar{\mathbf{x}}_{t-1}$ in predictors to predict changes $\dot{\mathbf{y}}$ in the target time series. Or, one can use the side data $\bar{X}$ to predict the residuals $\tilde{r}_t$ after applying the Kalman filter without side data. This latter method has the theoretical advantage of optimizing the same loss function as that used to compute regression coefficients (Equation 4). However in experiments, we find that the simplest approach gives similar results.

Instead of separating variable selection and training, a sparse prior on $\hat{\boldsymbol{\beta}}$ could be used to perform variable selection and to train $\hat{\boldsymbol{\beta}}$ simultaneously. One such method is given in [14]. These methods are more computationally expensive, and it is more difficult to determine how the sparsity parameter corresponds to statistical significance. The same comments apply to general-purpose feature selection methods such as QPFS [12]. Another greedy variable selection algorithm that may be promising for nowcasting is orthogonal matching pursuit [8].

Algorithm 1. FDR variable selection

Input: Training period time steps $T_{\text{train}} = \{1, \ldots, n\}$
Input: Complete side data, $\bar{\mathbf{x}}_t \in \mathbb{R}^p$, $\forall t \in T_{\text{train}}$
Input: Dependent variable time series $\mathbf{y} \in \mathbb{R}^n$
Input: Significance parameter α
Set $X = (1, \ldots, 1)^T$, the set of current predictors, initially a $n \times 1$ matrix
Set $M = \{\}$
for $q = 1, \ldots, p$ **do**
   $H = X(X^T X)^{-1} X^T$. H projects $\mathbf{y}$ to least-squares prediction $\hat{\mathbf{y}}$ for current X.
   $\mathbf{y}_H = (I - H)\mathbf{y}$, the residuals between $\mathbf{y}$ and current least-squares prediction $H\mathbf{y}$
   $df = n - q$, the degrees of freedom
   Set $V_q = \{\}$
   $p^* = \frac{\alpha q}{p}$, the maximum two-sided p-value considered significant
   Set t^* such that $1 - F(df, t^*) = p^*/2$, where $F(df, t)$ is the CDF of the t distribution
   for each side variable $\mathbf{z}_i$ **do**
      $\mathbf{z}_{iH} = (I - H)\mathbf{z}_i$, the variable with chosen predictors removed
      $\hat{\beta}_i = \mathbf{z}_{iH}^T \mathbf{y}_H / \mathbf{z}_{iH}^T \mathbf{z}_{iH}$, the least-squares regression slope between $\mathbf{z}_{iH}$ and $\mathbf{y}_H$
      $t_i = \hat{\beta}_i / se(\hat{\beta}_i)$, where $se(\hat{\beta}_i)$ is the standard error of the slope
      $p_i = 2(1 - F(df, t_i))$, the corresponding two-sided p-value
      if $p_i \leq p^*$ **then** Set $V_q = V_q \cup \{i\}$
   end for
   if $V_q = \{\}$ **then break**, search found no additional significant predictors
   Choose $i^* = \underset{i \in V_q}{\text{argmax}}\ SS(\mathbf{z}_i)$, where $SS(\mathbf{z}_i) = (\mathbf{z}_{iH}^T \mathbf{z}_{iH})(|\hat{\beta}_i| - t^* se(\hat{\beta}_i))^2$
   Set $X = \left(X\ \mathbf{z}_{i^*} \right)$
   Set $M = M \cup \{i^*\}$
end for
return $X = \left(\mathbf{x}_1\ \mathbf{x}_2 \ldots \mathbf{x}_n \right)^T$ where the $\mathbf{x}_t$ are the new side data vectors

---

Variable selection may also be repeated after a certain time interval in order to deal with time series whose predictors change over time. Our experiments, however, achieve good results with a single round of variable selection.

7   Design of Experiments

After using the Kalman filter to compute a series of predictions $\hat{y}_t$ for all t in the test period, we can compute any error measure between these predictions and the corresponding observed y_t. Although other measures such as mean squared error are also reasonable, we focus on mean absolute error of logarithms (MAEL). This measure is approximately equal to the mean absolute proportional error, which gives more interpretable results than MSE, since an average error of 8% is more meaningful than a MSE of 1000. MAEL is calculated as

$$\sum_{t \in T_{\text{test}}} |\log \hat{y}_t - \log y_t|$$

where T_{test} is the testing period. We multiply MAEL values by 100 and present them as percentages.

Google Trends provides weekly data starting in January 2004. We choose to use a training period of three years and a validation period of three years. This allows for a test period of almost four years, depending on the experiment. The precise dates are given in the following table. The end of each test period is given in the corresponding experiment in Section 8.

Period	Time range
Training period	February 2004 - January 2007
Validation period	February 2007 - January 2010
Testing period	February 2010 -

7.1 Preprocessing Auxiliary Time Series

The times series available from Google Trends are query volumes by week. The experiments are to nowcast monthly time series, so first we create monthly auxiliary data series by taking the mean of each query category over the corresponding month, ensuring that a week included in this monthly average never overlaps with the following month. Next, we remove seasonality from each of the 278 second-level Google Trends categories using a stable filter based on code from MathWorks.[2] This filter first subtracts the time series average over a one year moving window, and then computes a seasonal component on the remaining time series. The seasonal component is a zero-mean time series that repeats every year, representing the effect of the month on the time series. For example, searches for "Christmas card" are highest in December, while searches for "sunblock" are highest in June. Overall, the original data can be thought of as consisting of a seasonal component, a smoothed yearly average component, and a residual component. We subtract only the seasonal component as preprocessing.

Next, we compute interactions and squares. We shift each time series so that it consists of only nonnegative values. Then an element-wise multiplication is performed between two time series to compute an interaction, or between a time series and itself to compute a square. This results in a total of 39,059 potential side variables: the 278 original series, 278 squares, and $\binom{278}{2}$ interactions.

7.2 Baseline Methods

We compare the results of the FDR method to several baseline methods: a lagged time series model, a pure regression model, an autoregressive nowcasting model, and a simple random walk model. In addition, we compare our consumer sentiment nowcasting results to the state-of-the-art nowcasting method from [13]. This subsection describes the baseline methods. The symbol β represents a vector of regression coefficients, while the variance symbols v and σ^2 are reused. Although these symbols are shared by different methods, we train each method separately, so the values are not necessarily equal across methods. The vector $\mathbf{x}_t$ is the same side variables as those chosen by FDR.

---

[2] http://www.mathworks.com/help/econ/seasonal-adjustment.html

Lagged Time Series. The lagged time series is the simple model that assumes that the current value will be equal to the previous value, i.e. $\hat{y}_t = y_{t-1}$. It is well-known that for many time series, this model works remarkably well. Surprisingly, it is often not included as a baseline method in nowcasting research.

Simple Random Walk. This method assumes that there is a hidden variable μ_t that changes according to a random innovation step v, as in the FDR model. In this model, however, there is no contribution from the side data. This model is shown as equation 3. Its hyperparameters are the innovation step variance v and the error variance σ^2. We determine hyperparameters using grid search, as described in Section 5.

Simple Regression. This model assumes that the change in y can be estimated by looking at the change in $\mathbf{x}$. That is, $\dot{\hat{y}}_t = \boldsymbol{\beta}^T \dot{\mathbf{x}}_t$, where, $\dot{\hat{y}}_t$ is an estimate for $y_t - y_{t-1}$ and $\dot{\mathbf{x}}_t = \mathbf{x}_t - \mathbf{x}_{t-1}$. We choose $\boldsymbol{\beta}$ by performing least squares regression over the training and validation periods.

AR-1 Plus Side Data. An AR-n model is an autoregressive model that looks at the previous n values y_{t-1} to y_{t-n}. It is a linear model over y_{t-1} and $\mathbf{x}_t$. In [4] this model is used for nowcasting, and in [13] it is a baseline method. Specifically, $\hat{y}_t = b_1 y_{t-1} + \boldsymbol{\beta}^T \mathbf{x}_t$. We find b_1 and $\boldsymbol{\beta}$ by performing least squares regression over the training and validation periods.

7.3   How Results Are Reported

For each nowcasting experiment in Section 8, we provide a figure showing the time series to be nowcast ($\mathbf{y}$) along with the FDR predictions over the training, validation, and test periods. In addition, we provide a table which compares the FDR method to the baseline methods. Error statistics are mean absolute logarithmic error, represented as percentages as mentioned above.

Each table also includes g_{test}, which is how often the prediction of $\text{sign}(y_t - y_{t-1})$ is correct. That is, if we use a nowcasting method to estimate the direction of the change in y over the test period, g_{test} is the percentage of correct estimates:

$$g_{\text{test}} = \frac{100}{|T_{test}|} \sum_{t \in T_{test}} 1\{\text{sign}(\hat{y}_t - y_{t-1}) = \text{sign}(y_t - y_{t-1})\}. \qquad (6)$$

Each table also reports a p-value, which is the probability that flipping a fair coin to guess $\text{sign}(y_t - y_{t-1})$ would perform better than the method. For methods that require grid search to choose hyperparameters (as described in section 5), we report the learned hyperparameter values.

8   Results of Experiments

In order to verify that FDR variable selection produces sensible outcomes, we first confirm that the method chooses intuitive variables for simple variable selection tasks. For this experiment, we perform "trendcasting:" nowcasting a single Google Trends query category using the other query categories. We find that

Table 1. Variables selected for trendcasting

Time series	Variables chosen
Engineering & Technology	Technical Reference, Technology News x Software
Fantasy Sports	Sport News
Medical Literature & Resources	Mental Health x Health Conditions
Outdoors	Water Activities x Campers & RVs
Ticket Sales	Events & Listings x Events & Listings

Table 2. Trendcasting results. δ_{lag} and δ_{rw} are % improvements in MAEL of the FDR method over lagged and random walk baseline methods. g_{test} is from Equation (6).

Time series	δ_{lag}	δ_{rw}	g_{test}
Engineering & Technology	85.6	74.3	93.2
Fantasy Sports	63.1	34.8	86.4
Medical Literature & Resources	65.7	52.3	90.9
Outdoors	43.3	37.4	77.3
Ticket Sales	30.1	17.5	77.3

applying FDR significantly improves accuracy over baseline methods for most query categories. In addition, we find variables that are intuitively correlated to the categories being predicted. Table 1 shows several intuitive correlations found by FDR. Under "variables chosen" we list all the queries chosen by FDR. Separate variables are delimited using commas, and x denotes a variable interaction. For each of the 278 original trend series, one to three variables are chosen, except for two series without significant predictors. Table 2 gives the percentage improvement over the lagged and random walk models (δ_{lag} and δ_{rw}). In all cases these were the two most competitive baseline methods.

Because of the results of the trendcasting experiment, we can be confident that the variables selected in the following experiments are new and unexpected correlations. Many of the variables chosen may not have a clear intuitive interpretation, but the FDR method guarantees that they are statistically significant.

8.1 Nowcasting Results

We examine the performance of the FDR method on four time series provided by the US census bureau: Auto and Other Motor Vehicles, Electronics and Appliance Stores, Paper and Paper Products, and Chemicals and Allied Products. We also nowcast a consumer sentiment index published by the University of Michigan. We choose this last series in order to compare the results of FDR with the nowcasting method from [13].

Motor Vehicle Sales. This experiment nowcasts the advance monthly sales figures for `Auto and Other Motor Vehicles: U.S. Total` (NAICS codes 4411 and 4412) as reported by the US census bureau. We use data from January 2004 to November 2013. The FDR method selects one side data variable, namely the interaction `Classifieds x Movies`. Performance is shown in Figure 1, and a comparison to baseline methods is given in Table 3. The FDR method yields a test error improvement of 6.9% over the next-best method, the lagged model.

Electronics and Appliance Stores. This experiment looks at another time series from the US census bureau: the advance monthly sales figures for `Electronics and Appliance Stores: U.S. Total` (NAICS code 443). The data range from January 2004 to November 2013. FDR variable selection finds two side data variables, the interactions `Auctions x Energy & Utilities` and `Mass Merchants & Department Stores x Bicycles & Accessories`. Figure 1 shows the performance of FDR, and Table 4 compares with baseline methods. The improvement of FDR over the lagged time series is 9.4%, and 5.8% over a simple random walk.

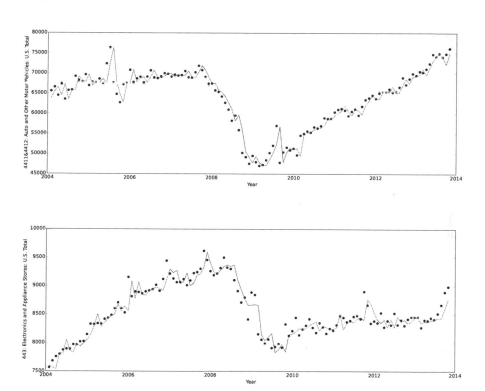

Fig. 1. Nowcasts from the FDR method (line) and original time series (dots). Green: training period, blue: validation period, red: testing period. Top: motor vehicles, bottom: electronics and appliances. Next page top: paper and paper products, middle: chemicals, bottom: consumer sentiment.

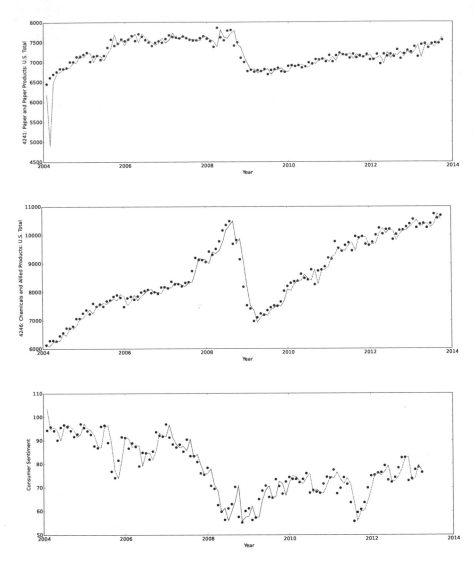

Fig. 1. (*continued*)

Paper and Paper Products. This experiment nowcasts the advance monthly sales figures for `Paper and Paper Products: U.S. Total` (NAICS code 4241) as reported by the US census bureau. We use data from January 2004 to October 2013. The FDR method selects one time series, `Retirement & Pension x Magazines`. Figure 1 and Table 5 show performance. The FDR method yields a test error improvement of 6.8% over the lagged model, and a 2.5% improvement over the simple random walk model.

Table 3. Comparison of accuracy between methods for nowcasting motor vehicle sales. Table entries are explained in Section 7.3. Lower is better for ϵ and higher is better for g_{test}. The lagged model always predicts $\hat{y}_t - y_{t-1} = 0$, so g_{test} and its corresponding p-value are not applicable.

Method	ϵ_{train}	ϵ_{val}	ϵ_{test}	g_{test}	p-value	(σ^2, v, v_β)
FDR method	2.83	3.04	1.53	65.22	3.90	$(10^{-5}, 10^4, 1.0)$
Lagged time series[2]	2.83	3.08	1.64	-	-	
Random Walk	2.83	3.08	1.64	60.87	14.04	$(10^{-5}, 10^4, -)$
AR-1 baseline	3.33	3.06	1.86	33.33	2.53	
Pure regression	2.81	3.02	2.04	66.67	2.53	

Table 4. Electronics and appliance stores method comparison

Method	ϵ_{train}	ϵ_{val}	ϵ_{test}	g_{test}	p-value	(σ^2, v, v_β)
FDR method	1.24	1.60	1.29	63.04	7.68	$(10^{-4}, 10^{-4}, 10^{-5})$
Lagged time series[2]	1.11	1.69	1.42	-	-	
Random Walk	1.10	1.69	1.37	65.22	3.90	$(10^{-5}, 10^{-4}, -)$
AR-1 baseline	1.48	1.75	1.38	55.56	45.61	
Pure regression	1.09	1.59	1.42	60.00	17.97	

Table 5. Paper and paper products method comparison

Method	ϵ_{train}	ϵ_{val}	ϵ_{test}	g_{test}	p-value	(σ^2, v, v_β)
FDR method	2.21	1.30	1.18	68.89	1.13	$(10^{-5}, 10^{-4}, 10^{-4})$
Lagged time series[2]	1.31	1.40	1.26	-	-	
Random Walk	1.33	1.38	1.21	82.22	0.00	$(10^{-5}, 10^{-4}, -)$
AR-1 baseline	1.81	1.38	1.28	55.56	45.61	
Pure regression	1.29	1.30	1.36	42.22	29.67	

Table 6. Chemicals and allied products method comparison

Method	ϵ_{train}	ϵ_{val}	ϵ_{test}	g_{test}	p-value	(σ^2, v, v_β)
FDR method	1.72	2.71	1.83	64.44	5.26	$(10^{-5}, 0.01, 10^{-5})$
Lagged time series[2]	1.57	2.71	1.93	-	-	
Random Walk	1.57	2.71	1.93	53.33	65.47	$(10^{-5}, 10^4, -)$
AR-1 baseline	3.09	2.67	2.03	51.11	88.15	
Pure regression	1.67	2.67	1.88	60.00	17.97	

Table 7. Consumer sentiment method comparison

Method	ϵ_{train}	ϵ_{val}	ϵ_{test}	g_{test}	p-value	(σ^2, v, v_β)
FDR method	4.27	5.43	4.33	56.41	42.33	$(10^{-5}, 10^4, 10.0)$
Lagged time series[2]	4.30	5.57	4.36	-	-	
Random Walk	4.30	5.57	4.36	46.15	63.10	$(10^{-5}, 10^4, -)$
AR-1 baseline	4.89	5.57	4.50	35.00	5.78	
Pure regression	4.10	5.45	4.24	55.00	52.71	

Chemicals and Allied Products. In this experiment we nowcast advance monthly sales figures for NAICS code 4246, Chemicals and Allied Products. The data are from January 2004 to October 2013. FDR variable selection selects the variables Shopping Portals & Search Engines x Energy & Utilities and Bus & Rail x E-Books. Figure 1 and Table 6 show results. The improvement of FDR over the next-best method, the lagged model, is 5.1%.

Consumer Sentiment. This experiment uses data from the University of Michigan monthly survey of consumer sentiment from January 2004 to April 2013. FDR variable selection selects a single interaction variable, Classifieds x Energy & Utilities. Figure 1 plots the FDR predictions along with the original time series, and Table 7 compares FDR and baseline methods. The FDR method achieves a mean absolute logarithmic error of 4.33% on the test set. An experiment in [13] using this same dataset reports a mean absolute logarithmic error of 4.5%, although the paper does not specify the training and testing periods used. Its results were obtained using a hand-selected set of 151 potential side data variables. The results here suggest that FDR is competitive with current state-of-the-art nowcasting methods, but can deal with a much larger set of potential variables.

9   Discussion

This paper introduces a novel method for nowcasting, called FDR for false discovery reduction. The method combines a time series model with tractable variable selection, which allows for nowcasting with a large number of potential side predictors. Variable selection is a particularly important issue in the context of nowcasting because nowcasting relies on observations of correlated side data to make an up-to-date estimate, and in many cases the number of potential auxiliary time series to choose among is large. We demonstrate the performance of FDR variable selection from 39,059 potential predictors. Reducing the rate of discovery of false predictors reduces overfitting, and leads to simple final predictive models, although the statistically significant predictors chosen can be unintuitive. The FDR method outperforms baseline methods when nowcasting sales data from the United States census bureau and consumer sentiment, and has performance that is comparable with the state-of-the-art nowcasting method in [13], while allowing selection from over 250 times as many auxiliary time series.

References

1. Anava, O., Hazan, E., Mannor, S., Shamir, O.: Online learning for time series prediction. In: Shalev-Shwartz, S., Steinwart, I. (eds.) COLT. JMLR Proceedings, vol. 30, pp. 172–184. JMLR.org (2013)
2. Butler, D.: When Google got flu wrong. Nature 494(7436), 155 (2013)
3. Carneiro, H.A., Mylonakis, E.: Google Trends: A web-based tool for real-time surveillance of disease outbreaks. Clinical Infectious Diseases 49(10), 1557–1564 (2009)

4. Choi, H., Varian, H.: Predicting the present with Google Trends. Economic Record 88(s1), 2–9 (2012)

5. Davis, R.A., Zang, P., Zheng, T.: Sparse vector autoregressive modeling. arXiv preprint arXiv:1207.0520 (2012)

6. Doornik, J.A.: Improving the timeliness of data on influenza-like illnesses using Google search data. Working paper (2009),
http://www.doornik.com/flu/Doornik2009_Flu.pdf

7. Foster, D.P., Stine, R.A.: Variable selection in data mining: Building a predictive model for bankruptcy. Journal of the American Statistical Association 99(466), 303–313 (2004)

8. Joseph, A.: Variable selection in high dimension with random designs and orthogonal matching pursuit. Journal of Machine Learning Research 14(1), 1771–1800 (2013)

9. Kalman, R.E.: A new approach to linear filtering and prediction problems. Journal of Basic Engineering 82(1), 35–45 (1960)

10. Kocsis, L., Szepesvári, C.: Bandit based Monte-Carlo planning. In: Fürnkranz, J., Scheffer, T., Spiliopoulou, M. (eds.) ECML 2006. LNCS (LNAI), vol. 4212, pp. 282–293. Springer, Heidelberg (2006)

11. Moody, J., Wu, L.: High frequency foreign exchange rates: Price behavior analysis and "true price" models. In: Dunis, C., Zhaou, B. (eds.) Nonlinear Modeling of High Frequency Financial Time Series. John Wiley & Sons (1998)

12. Rodriguez-Lujan, I., Huerta, R., Elkan, C., Cruz, C.S.: Quadratic programming feature selection. Journal of Machine Learning Research 11, 1491–1516 (2010)

13. Scott, S.L., Varian, H.R.: Bayesian variable selection for nowcasting economic time series. Tech. rep., National Bureau of Economic Research (2013),
http://www.nber.org/papers/w19567.pdf

14. Wainwright, M.J.: Sharp thresholds for high-dimensional and noisy sparsity recovery using constrained quadratic programming (Lasso). IEEE Transactions on Information Theory 55(5), 2183–2202 (2009)

Revisit Behavior in Social Media:
The Phoenix-R Model and Discoveries

Flavio Figueiredo[1], Jussara M. Almeida[1],
Yasuko Matsubara[2,3], Bruno Ribeiro[3], and Christos Faloutsos[3]

[1] Department of Computer Science, Universidade Federal de Minas Gerais
[2] Department of Computer Science, Kumamoto University
[3] Department of Computer Science, Carnegie Mellon University

Abstract. How many listens will an artist receive on a online radio? How about plays on a YouTube video? How many of these visits are new or returning users? Modeling and mining popularity dynamics of social activity has important implications for researchers, content creators and providers. We here investigate the effect of revisits (successive visits from a single user) on content popularity. Using four datasets of social activity, with up to tens of millions media objects (e.g., YouTube videos, Twitter hashtags or LastFM artists), we show the effect of revisits in the popularity evolution of such objects. Secondly, we propose the PHOENIX-R model which captures the popularity dynamics of individual objects. PHOENIX-R has the desired properties of being: (1) parsimonious, being based on the minimum description length principle, and achieving lower root mean squared error than state-of-the-art baselines; (2) applicable, the model is effective for predicting future popularity values of objects.

1 Introduction

How do we quantify the popularity of a piece of content in social media applications? Should we consider only the audience (unique visitors) or include revisits as well? Can the revisit activity be explored to create more realistic popularity evolution models? These are important questions in the study of social media popularity. In this paper, we take the first step towards answering them based on four large traces of user activity collected from different social media applications: Twitter, LastFM, and YouTube[1].

Understanding the popularity dynamics of online content is both a challenging task, due to the vast amount and variability of content available, as it can also provide invaluable insights into the behaviors of human consumption [6] and into more effective engineering strategies for online services. A large body of previous work has investigated the popularity dynamics of social media content, focusing mostly on modeling and predicting the *total number of accesses* a piece of content receives [5, 6, 9, 17, 21].

However, a key aspect that has not been explored by most previous work is the effect of revisits on content. The distinction between audience (unique users), revisits (returning users), and popularity (the sum of the previous two) can have large implications for different stakeholders of these applications - from content providers to content

---

[1] http://twitter.com, http://lastfm.com, http://youtube.com

T. Calders et al. (Eds.): ECML PKDD 2014, Part I, LNCS 8724, pp. 386–401, 2014.
© Springer-Verlag Berlin Heidelberg 2014

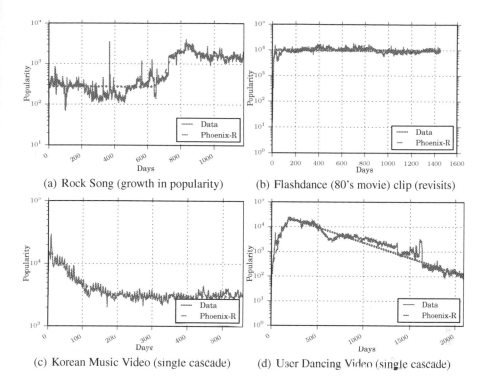

(a) Rock Song (growth in popularity)   (b) Flashdance (80's movie) clip (revisits)

(c) Korean Music Video (single cascade)   (d) User Dancing Video (single cascade)

Fig. 1. Different YouTube videos as captured by the PHOENIX-R model

producers - as well as for internal and external services that rely on social activity data. For example, marketing services should care most about the audience of a particular content, as opposed to its total popularity, as each access does not necessarily represent a new exposed individual. Even system level services, such as geographical sharding [8, 23], can be affected by such distinction, as a smaller audience served by one data center does not necessarily imply that a smaller volume of activity (and thus lower load) should be expected. As prior studies of content popularity in social media do not clearly distinguish between unique and returning visits, the literature still lacks fundamental knowledge about content popularity dynamics in this environment.

Goals: We here aim at investigating and modeling the effect of revisits on popularity, thus complementing prior efforts on the field of social media popularity. Our goals are: (1) Characterizing the revisits phenomenon and show how it affects the evolution of popularity of different objects (videos, artists or hashtags) on social media applications; (2) Introducing the PHOENIX-R model that captures the evolution of popularity of individual objects, while explicitly accounting for revisits. Also, we develop the model so that it can capture multiple cascades, or outbreaks, of interest in a given object.

Discoveries: Among other findings, we show that when analyzing total popularity values, revisits account from 40% to 96% of the popularity of an object (on median), depending on the application. Moreover, when looking at small time windows (e.g., hourly) revisits can be up to 14x more common than new users accessing the object.

PHOENIX-R Results: The PHOENIX-R model explicitly addresses revisits in social media behavior and is able to automatically identify multiple cascades [13] using only popularity time series data. This is in contrast to previous methods such as the SpikeM [18] approach, which models single cascades only, and the TemporalDynamics [21] models, are linear in nature. Figure 1 shows the different behaviors which can be captured by the Phoenix-R model. Notice how the model captures a growth in the popularity of video (a), videos which have a plateau like popularity after the upload (b), and two different single cascade dynamics (c-d). The PHOENIX-R model is also scalable. Fitting is done in linear time and no parameters are required.

Outline: Section 2 presents an overview of definitions and background. This is followed by Section 3 which presents our characterization. PHOENIX-R is described in Section 4, whereas it's applicability is presented in Section 5. Related work is discussed on Section 6. Finally, we conclude the paper in Section 7.

2   Definitions and Background

In this section we present the definitions used throughout the paper (Section 2.1). Next, we discuss existing models of popularity dynamics of individual objects (Section 2.2).

2.1   Definitions

We define an **object** as a piece of media content stored on an application. Specifically, an object on YouTube is a video, whereas, on an online radio like LastFM, we consider (the webpage of) an artist as an object. We also define an object on Twitter as a hashtag or a *musictag*[2]. A **social activity** is the act of accessing - posting, re-posting, viewing or listening to - an object on a social media application. The **popularity** of an object is the aggregate behavior of social activities on that object. We here study popularity in terms of the most general activities in each application: number of views for YouTube videos, number of plays for LastFM artists, and number of tweets with a hashtag. The popularity of an object is the sum of **audience** (user's first visit) and, **revisits**, (or returning users), and the evolution of the popularity of an object over time defines a **time series**.

2.2   Existing Models of Object Popularity Dynamics

Epidemic Models: Previous work on information propagation on online social networks (OSNs) has exploited epidemic models [12] to explain the dynamics of the propagation process. An epidemic model describes the transmission of a "disease" through individuals. The simplest epidemic model is the Susceptible-Infected (SI) model. The SI model considers a fixed population divided into S susceptible individuals and I infected individuals. Starting with $S(0)$ susceptible individuals and $I(0)$ infected individuals, at each time step $\beta S(t-1)I(t-1)$ individuals get infected, and transition from the S state to the I state. The product $S(t-1)I(t-1)$ accounts for all the possible connections between individuals. The parameter β is the strength of the infectivity, or virus.

---

[2] Users informing their followers which artists they are listening to.

Table 1. Comparison of PHOENIX-R with other approaches

	Revisits	Non-Linear	Forecasting	Multi Cascade
SI [12]		✓		
SpikeM [18]		✓	✓	
TemporalDynamics [21]			✓	
PHOENIX-R	✓	✓	✓	✓

Cha *et. al* used an SI model to study how information (i.e., the "disease") disseminates through social links on Flickr [4], whereas Matsubara *et. al* [18] proposed an alternative model called SpikeM. SpikeM builds on an SI model by adding, among other things, a decaying power law infectivity per newly infected individual, which produces a behavior that is similar to the model proposed in [6]. The SpikeM model was used to captured the time series popularity for a single cascade. One of the reasons why the SI model is useful to represent online cascades of information propagation is that individuals usually do not delete their posts, tweets or favorite markings [4, 18]. Thus, once an individual is infected he/she remains infected forever (as captured by the SI model).

Temporal Dynamics Models: Other models that can be explored in the study of content popularity dynamics are auto-regressive models and state space models, such as the Holt-Winters model and its extensions [21]. However, these models are linear in nature, and thus cannot account for more complex temporal dynamics observed in online content [18]. Although, these models have been successful in predicting *normalized* query behavior in search engines [21], the descriptive power of such models is less attractive. For example, Holt-Winters based models are very general, that is, they are used to predict time series behavior, but will not take into account cascades, revisits or information dissemination. From a descriptive point of view, these models are of little help to understand the actual process that drives popularity evolution.

Multiple Cascades: Very recently, the work of Hu *et. al* focused on the defining longevity of social impulses, or multiple cascades [13]. However, unlike our approach, the authors are not focused on modeling the long term popularity of objects.

Table 1 summarizes the key properties of the aforementioned models as well as of our new PHOENIX-R model. In comparison these approaches, PHOENIX-R explicitly captures both revisits and multiple cascades, allows for non-linear solutions, and can be used for accurate forecasting. The next section presents the effect of revisits in both long and short term content popularity evolution for real world datasets. This is followed by the definition of the PHOENIX-R model.

3   Content Revisit Behavior in Social Media

We now analyze the revisit behavior in various social media applications. We first describe the datasets used in this analysis as well as in the evaluation of our model, and then discuss our main characterization findings.

3.1   Datasets

Our study is performed on four large social activity datasets:

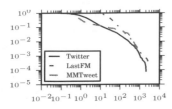

Fig. 2. Distributions of $\frac{\#Revisits}{Audience}$

Table 2. Relationships between revisits, audience and popularity

Dataset	Median $\frac{\#Revisits}{Audience}$	Median $\frac{\#Revisits}{Popularity}$	% objects with $\frac{\#Revisits}{Audience} > 1$
Twitter	1.70	0.62	66%
MMTweet	0.68	0.40	33%
LastFM	25.39	0.96	100%

- The Million Musical Tweets Dataset (MMTweet): consists of 1,086,808 tweets of users about artists they are listening to at the time [11]. We focus on the artist of each tweet as an object. 25,060 artists were mentioned in tweets.
- The 2010 LastFM listening habits dataset (LastFM): consists of the whole listening habits (until May 5th 2009) of nearly 1,000 users, with over 19 million activities on 107,428 objects (artists) [3].
- The 476 million Twitter tweets corpus (Twitter): accounts for roughly 20% to 30% of the tweets from June 1 2009 to December 31 2009 [24], and includes over 50 million objects (hashtags) tweeted by 17 million users.
- The YouTube dataset: Recently, YouTube began to provide the full daily time series (known as insight data) of visits for videos in the page of each video. We crawled the time series of over 3 million YouTube videos similar to as done in [9].

3.2   Main Findings

Our goal is to analyze how the popularity acquired by different objects, in the long and short runs, is divided into audience and revisits. In particular, we aim at assessing to which extent the number of revisits may be *larger* than the size of the audience, in which case popularity is largely a sum of repeated user activities. Since this property may vary depending on the type of content, we perform our characterization on the LastFM, MMTweet, and Twitter datasets. We leave the YouTube dataset out of this analysis since, unlike the other datasets, it does not contain individual social activities, but only popularity time series. We will make use of the YouTube dataset to fit and evaluate our PHOENIX-R model, in the next section.

We first analyze the distribution of the final values[3] of popularity, audience, and revisits across objects in each dataset. For illustration purposes, Figure 2 shows the complementary cumulative distribution function of the ratio of the number of revisits to the audience size for all datasets, computed for objects with popularity greater than 500. We filtered out very unpopular objects, which attract very little attention during the periods of our datasets (over 6 months each). Note that the probability of an object having more revisits than audience (ratio greater than 1) is large. Indeed, though rare, the ratio of revisits to audience size reaches 10^2 and even 10^3.

In order to better understand these findings across all datasets, Table 2 shows, for each dataset: (1) the median of the ratio of number of revisits to audience size, (2) the median of the ratio of number of revisits to total popularity; and (3) the percentage

---

[3] Values computed at the time the data was collected.

Table 3. Quartiles of the ratio $\frac{\#Revisits}{Audience}$ for various time windows w

Dataset	Time window (w)	25th percentile	Median	75th percentile
Twitter	1 hour	1.08	3.93	12
	1 day	1	2.5	6.28
	1 week	0.66	1.69	4.28
	1 month	0.56	1.44	3.75
MMTweet	1 hour	0.25	0.66	12.5
	1 day	0.55	0.83	1.26
	1 week	0.41	0.73	1.41
	1 month	0.31	0.56	1.17
LastFM	1 hour	20	21	25
	1 day	21	28	41
	1 week	20	30.5	55.25
	1 month	14	25	48

of objects where the revisits dominate the popularity (i.e., ratio of number of revisits to the audience size greater than 1). Note that revisits dominate popularity in 66% of the Twitter objects. Moreover, on median, 62% of the total popularity of these objects is composed of revisits, which account for 1.7 times more activities than the visits by new users (audience size). Again, for LastFM artists, revisits are over 25 times more frequent than the visits by new users (on median), and the revisits dominate popularity in *all* objects. In contrast, the ratios of number of revisits to audience size and to total popularity are smaller for MMTweet objects, most likely because users do not tweet about artists they are listening to all the time, but rather only when they wish to share this activity with their followers. Yet, the revisits dominate popularity in 33% of the objects. These results provide evidence that, at least in the long run, revisits are much more common than new users for many objects in different applications. For microblogs, though less intense, this behavior is still non-negligible.

We further analyze the effect of revisits on popularity, focusing now in the short term, by zooming into smaller time windows w. Specifically, we analyze the distributions of the ratios of number of revisits to audience size computed for window sizes w equal to one hour, one day, one week, and one month. Table 3 shows the three distribution quartiles for the various window sizes and datasets considered. These quartiles were computed considering only window sizes during which the popularity acquired by the object exceeds 20. We adopted this threshold to avoid biases in time windows with very low popularity, focusing on the periods where the objects had a minimal attention (note that 20 is still small considering that each trace has millions of activities).

Focusing first on the LastFM dataset, we note that, regardless of the time window size, the number of revisits is at least one order of magnitude (14x) larger than the audience size for at least 75% of the analyzed windows (25th percentile). In fact, the ratio between the two measures exceeds 55 for 25% of the windows (75th percentile) on the weekly case. In contrast, in the MMTweet dataset, once again, the ratios are much smaller. Nevertheless, at least 25% of the of the windows we observe a burst of revisits in very short time, with the ratio exceeding 12 for the hourly cases. Once again, we suspect that these lower ratios may simply reflect that users do not tweet about every artist they list to. Thus, in general, we have strong evidence that, for music-related content, popularity is mostly governed by revisits, as opposed to new users (audience).

The same is observed, though with less intensity, in the Twitter dataset. Revisits are more common than new users in 50% of the time windows, for all sizes considered. Indeed, considering hourly time windows, popularity is dominated by revisits for 75% of the cases. While large ratios, such as those observed for LastFM, do not occur, the number of revisits can still be 12 times larger than the audience size during a single hour in 25% of the Twitter hourly windows.

Summary of Findings: Our main conclusions so far are: (1) for most objects in the Twitter and LastFM datasets, popularity, measured both in the short (as short as 1 hour periods) and long runs, is mostly due to revisits than to audience size; and (2) revisits are less common on the MMTweet dataset, which we believe is due to data sparsity, but are still a significant component of the popularity acquired by a large fraction of the objects (in both long and short runs). These findings motivate the need for models that explicitly account for revisits in the popularity dynamics, which we discuss next.

4   The PHOENIX-R Model

In this section we introduce the proposed PHOENIX-R model(Section 4.1), show how we fit the model to a given popularity time series (Section 4.2). In the next section we present results on the efficacy of the model on our datasets when compared to state-of-the-art alternatives, and the applicability of the PHOENIX-R model.

Notation: We present vectors ($\mathbf{x}$) in bold. Sets are shown in non-bold calligraphy letters ($\mathcal{X}$), and variables are represented by lower case letters or Greek symbols (x, β). Moreover, $\mathbf{x}(i)$ means data index i (from 1), and $\mathbf{x}(: i)$ means sub-vector up to i.

4.1   Deriving the Model

The PHOENIX-R model is built based on the 'Susceptible-Infected-Recovery' (SIR) compartments, extending for revisits and multiple cascades. Specifically, it captures the following behavior *for each individual object*:

- We assume a fixed population of individuals, where each individual can be in one of three states: susceptible, infected and recovered.
- At any given time s_i, an external shock i causes initial interest in the object. The shock can be any event that draws attention to the object, such as a video being uploaded to YouTube, a news event about a certain subject, or even a search engine indexing a certain subject for the same time (thus making an object easier to be found). We assume that the initial shock s_1 is always caused by one individual.
- New individuals discover the object by being infected by the first one. Moreover, after discovery, these "newly infected" can also infect other individuals, thus contributing to the propagation.
- Infected individuals may access (watch, play or tweet) the object. It is important to note that being infected does not necessarily imply in an access. For example, people may talk about a trending video before actually watching it. Each infected individual accesses the object following a Poisson process with rate ω ($\omega > 0$)[4].

---

[4] Both [1, 14] show the poissonian behavior of mutiple visits from the same user.

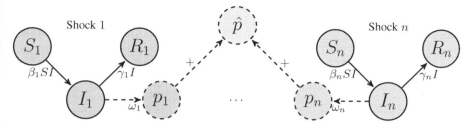

Fig. 3. Individual shocks that when added up account for the PHOENIX-R model

- After some time, individuals lose interest in the object, which, in the model, is captured by a recovery rate γ.
- Multiple external shocks may occur for a single object.

Figure 3 presents the PHOENIX-R model. In the figure, three compartments are shown for each shock S_i, I_i, and R_i, which represent the number of susceptible, infected and recovered individuals for shock i, respectively. Variable p_i, associated with shock i, measures the popularity acquired by the object due this shock. The total popularity of the object, i.e., the sum of the values of p_i for all shocks, is denoted by $\hat{p}$. We first present the model for a single shock, and then generalize the solution for multiple shocks. Also, we drop the subscripts while discussing a single shock. We present the model assuming discrete time, referring to each time tick as a time window.

Each shock begins with a given susceptible population ($S(0)$) and one infected individual ($I(0) = 1$). The total population is fixed and given by ($N = S(0) + 1$). The R compartment captures the individuals that lost interest in the object. Similarly the SI model, βSI susceptible individuals become infected in each time window. Moreover, γI individuals loose interest in (i.e., recover from) the object in each window. Revisits to the object are captured by the rate ω. Thus ω is the expected number of accesses of an individual up to time t, the probability of the individual accessing the object k times during a time interval of τ windows is given by $P(v(t+\tau) - v(t) = k) = \frac{(\omega\tau)^k e^{-\omega\tau}}{k!}$.

We assume that the shock starts at time zero, thus focusing the dynamics *after* the shock. Under this assumption, the equations that govern a single shock are:

$$S(t) = S(t-1) - \beta S(t-1)I(t-1) \tag{1}$$
$$I(t) = I(t-1) + \beta S(t-1)I(t-1) - \gamma I(t-1) \tag{2}$$
$$R(t) = R(t-1) + \gamma I(t-1) \tag{3}$$
$$p(t) = \omega I(t). \tag{4}$$

The equation $p(t) = \omega I(t)$ accounts for the expected number of times infected individuals access the object, thus capturing the popularity of the object at time t due to the shock. We can also define the expected audience size of the object at time t due to the shock, $a(t)$, as: $a(t) = (1 - e^{-\frac{\omega}{\gamma}})\beta S(t-1)I(t-1)$. Each newly infected individual ($\beta S(t-1)I(t-1)$) will stay infected for γ^{-1} windows (see [12]). The probability of generating at least one access while the individual is infected is: $1 - P(v(t+\gamma^{-1}) - v(t) = 0) = 1 - e^{-\frac{\omega}{\gamma}}$. Thus, we here capture the individuals which where infected at some time and generated at least one access.

The PHOENIX-R model is thus defined as the sum of the popularity values due to multiple shocks. We discuss how to determine the number of shocks in the next section. Given a set of shocks $\mathcal{S}$, where shock i starts at given time s_i, the popularity $\hat{p}$ is:

$$\hat{p}(t) = \sum_{i, s_i \in \mathcal{S}} p_i(t - s_i) \mathbb{1}[t > s_i] \qquad (5)$$

where $\mathbb{1}[t > s_i]$ is an indicator function that takes value of 1 when $t > s_i$, and 0 otherwise. Audience, size $\hat{a}(t)$ can be similarly defined. Also, both in the single shock and in the PHOENIX-R models, the number of revisits at time t, $\hat{r}(t)$, can be computed as $\hat{r}(t) = \hat{p}(t) - \hat{a}(t)$. The overall population that can be infected is defined by $N = \sum_i N_i = \sum_i S(0)_i + 1$.

Note that we assume that the population of different shocks do not interact, that is, an infected individual from shock s_i does not interact with a susceptible one from shock s_j, where $i \neq j$. While this may not hold for some objects (e.g., people may hear about the same content from two different populations), it may be a good approximation for objects that become popular in large scale (e.g., objects that are propagated world wide). It also allows us to have different β_i, γ_i, and ω_i values for each population. Intuitively, the use of different parameters for each shock captures the notion that some objects may be more (or less) interesting for different populations. For example, samba songs may attract more interest from people in Brazil.

Adding a Period: In some cases, the popularity of an object may be affected by periodical factors. For example, songs may get more plays on weekends. We add a period to the PHOENIX-R model by making ω fluctuate in a periodic manner. That is:

$$\omega_i(t) = \omega_i * (1 - \frac{m}{2} * (sin(\frac{2\pi(t + h)}{e}) + 1)). \qquad (6)$$

e is the period, and sin is a sine function. For example, for daily series we may set $e = 7$ if more interest is expected on weekends. Since an object may have been uploaded on a Wednesday, we use the shift h parameter to correct the sine wave to peak on weekends. The amplitude m captures oscillation in visits. The same period parameters are applied to every shock model. This approach is similar to the one adopted in [18].

The final PHOENIX-R model will have 5 parameters to be estimated from the data *for each* shock, namely, $S(0)_i$, β_i, γ_i, ω_i, s_i; plus the m and h period parameters. The last two do not change for individual shocks. We decided to fix e in our experiments to 7 days, when using daily time windows, and $e = 24$ hours when using hourly series.

4.2   Fitting the Model

We now discuss how to fit the PHOENIX-R parameters to real world data. Our goal is to produce a model that delivers a good trade-off between parsimony (i.e., small number of parameters) and accuracy. To that end, three issues must be addressed: (1) the identification of the start time of each individual shock; (2) an estimation of the cost of the model associated with multiple shocks; and, (3) the fitting algorithm itself. Note that one key component of the fitting algorithm is model selection: it is responsible for

---

Algorithm 1. Fitting the PHOENIX-R model. Only the time series is required as input.

```
 1: function FITPHOENIXR(t)
 2:     ε = 0.05
 3:     s ← {}
 4:     p, s′ ← FindPeaks(t)
 5:     s[1] = 0
 6:     s ← append(s′)
 7:     𝒫 ← {}
 8:     min_cost ← ∞
 9:     for i ← 1  to  |s| do
10:         ℱ ← LM(t, s(: i))
11:         m ← PhoenixR(ℱ)
12:         mdl_cost ← Cost(m, t, ℱ)
13:         if mdl_cost < min_cost then
14:             min_cost ← mdl_cost
15:             𝒫 ← ℱ
16:         end if
17:         if mdl_cost > min_cost * (1 + ε) then
18:             break
19:         end if
20:     end for
21:     return 𝒫
22: end function
```

---

determining the number of shocks that will compose the PHOENIX-R model, choosing a value based on the cost estimate and model accuracy.

Finding the Start Times s_i of the Shocks: Intuitively, we expect each shock to correspond to a peak in the time series. Indeed, previous work has looked at the dynamics of single shock cascades, finding a single prominent peak in each cascade [2, 18]. With this in mind, instead of searching for s_i directly, we initially attempt to find peaks. We can achieve both tasks using a continuous wavelet transform based peak finding algorithm [7]. We chose this algorithm since it has the following key desirable properties. Firstly, it can find peaks regardless of the "volume" (or popularity in the present context) in the time windows surrounding the peaks. It does so by only considering peaks with a high signal to noise ratio in the series, that is, peaks that can be distinguished the time series signal around the candidate peak. Secondly, the algorithm is fast, with complexity in the order of the length, n, of the time series ($O(n)$). Lastly and more importantly, using the algorithm we can estimate both the peaks and the start times of the shocks that caused each peak. We shall refer to the algorithm as $FindPeaks$.

As stated $FindPeaks$ makes use of a continuous wavelet transform to find the peaks of the time series. Specifically, we apply the Mexican Hat Wavelet[5] for this task. The Mexican Hat Wavelet is parametrized by a half-width l. We use half-widths (l) of values $\{1, 2, 4, 8, 16, 32, 64, 128, 256\}$ to find the peaks. Thus, for the peak identified at position k_i, with wavelet determined by the parameter l_i, we define the start point of the shock s_i as: $s_i = k_i - l_i$. We found that using the algorithm with the default parameters presented in [7], combined with our MDL fitting approach (see below), proved accurate in modeling the popularity of objects[6].

Estimating the Cost of the Model with Multiple Shocks: we estimate the cost of a model with $|\mathcal{S}|$ shocks based on the minimum description length (MDL) principle [10,

---

[5] https://en.wikipedia.org/wiki/Mexican_hat_wavelet

[6] We used the open source implementation available with SciPy (http://scipy.org).

19], which is largely used for problems of model selection. To apply the MDL principle, we need a coding scheme that can be used to compress both the model parameters and the likelihood of the data given the model. We here provide a new intuitive coding scheme, based on the MDL principle, for describing the PHOENIX-R model with $|\mathcal{S}|$ shocks, assuming a popularity time series of n elements (time windows). As a general approach, we code natural numbers using the $\log^*$ function (universal code length for integers)[7] [10], and fix the cost of floating point numbers at $c_f = 64$ bits.

For each shock i, the complexity of the description of the set of parameters associated with i consists of the following terms: $\log^*(n)$ for the s_i parameter (since the start time of i can be at any point in the time series); $\log^*(S_i(0))$ for the initial susceptible population; and $3*c_f$ for β_i, γ_i, and ω_i. We note that an additional cost of $\log^*(7) + 2*c_f$ is incurred if a period is added to the model. However, we ignore this component here since it is fixed for all models. Therefore, it does not affect model selection. The cost associated with the set of parameters $\mathcal{P}$ of all $|\mathcal{S}|$ shocks is:

$$Cost(\mathcal{P}) = |\mathcal{S}| \times (\log^*(n) + \log^*(S_i(0)) + 3*c_f) + \log^*|\mathcal{S}|. \qquad (7)$$

Given the full parameter set $\mathcal{P}$, we can encode the data using Huffman coding, i.e., a number of bits is assigned to each value which is the logarithm of the inverse of the probability of the values (here, we use a Gaussian distribution as suggested in [19] for the cases when not using probabilistic models.).

Thus, the cost associated with coding of the time series given the parameters is:

$$Cost(\mathbf{t} \mid \mathcal{P}) = -\sum_{i=1}^{n} \log(p_{gaussian}(\mathbf{t}(i) - \mathbf{m}(i); \mu, \sigma)). \qquad (8)$$

where $\mathbf{t}$ is the time series data and $\mathbf{m}$ is the time series produced by the model (i.e., $\mathbf{t}(i) - \mathbf{m}(i)$ is the error of the model at time window i.) Here, $p_{gaussian}$ is the probability density function of a Gaussian distribution with mean μ and standard deviation σ estimated from the model errors. We do not include the costs of encoding μ and σ because, once again, they are constant for all models. The total cost is:

$$Cost(\mathbf{t}; \mathcal{P}) = \log^* n + Cost(\mathcal{P}) + Cost(\mathbf{t} \mid \mathcal{P}). \qquad (9)$$

This accounts for the parameters cost, the likelihood cost, and the cost of the data size.

Fitting Algorithm: The model fitting approach is summarized in Algorithm 1. The algorithm receives as input a popularity time series $\mathbf{t}$. It first identifies candidate shocks using the $FindPeaks$ method, which returns the peaks $\mathbf{p}$ and the start times $\mathbf{s}'$ of the corresponding shocks in decreasing order of peak volume (line 3). To account for the upload of the object, we include one other shock starting at time $s_1 = 0$, in case a shock was not identified in this position. Each s_i is stored in vector $\mathbf{s}$, ordered by the volume of the each identified peak (with the exception of $s_1 = 0$ which is always in the first position) (lines 4 and 5). We then fit the PHOENIX-R model using the Levenberg-Marquardt (LM) algorithm adding one shock at a time, in the order they appear in $\mathbf{s}$ (loop in line 9), that is, in decreasing order of peak volume (after the initial shock).

---

[7] $\log^*(x) = 1 + \log^*(\log x)$ if $x > 1$. $\log^*(x) = 1$ otherwise. We use base-2 logarithms.

Intuitively, shocks that lead to larger peaks account for more variance in the data. For each new shock added, we evaluate the MDL cost (line 12). We keep adding new shocks as long as the MDL cost decreases (line 13) or provided that an increase of at most ϵ over the best model is observed[8] (line 17). We set the Levenberg-Marquardt algorithm to evaluate the mean squared errors of the model and adopt a threshold ϵ equal to 5%. We also note that we initialize each parameter randomly (uniform from 0 to 1), except for $S_i(0)$ values. For the first shock we do test multiple initial values: $S_1(0) = 10^3, 10^4, 10^5$, and 10^6. The other $S_i(0)$ values are initialized to the corresponding peak volume.

5   Experiments

In this section we discuss the experimental evaluation of the PHOENIX-R model. Initially, we present results on the efficacy of the model on our datasets when compared to state-of-the-art alternatives (Section 5.1) Next, we show results for the applicability of the model for popularity prediction (Section 5.2)[9].

5.1   Is PHOENIX-R Better than Alternatives?

We compare PHOENIX-R with two state-of-the-art alternatives: the TemporalDynamics [21], used to model query popularity; and the SpikeM model [18], which captures single cascades. We compare these models in terms of time complexity, accuracy, estimated by the root mean squared errors (RMSE), and cost-benefit. For the latter, we use the Bayesian Information Criterion (BIC) [21], which captures the tradeoff between cost (number of parameters) and accuracy of the model.

In terms of time complexity, we note that the PHOENIX-R model scales linearly with the length of the time series n. This is shown in Figure 4, which presents the number of seconds (y-axis) required to fit a time series with a given number of time windows (x-axis). TemporalDynamics also has linear time complexity [21]. In contrast, the equations that govern the SpikeM model requires quadratic ($O(n^2)$) runtime on the time series length, making it much less scalable to large datasets.

In terms of accuracy, we make an effort to compare PHOENIX-R with the alternatives in fair settings, with datasets with similar characteristics from those used in the original papers. In particular, when comparing with TemporalDynamics, we run the models proposed in [21] selecting the best one (i.e., the one with smallest root mean squared error) for each time series. Moreover, we use long term daily time series (over 30 days), with a total popularity of at least 1,000[10]. We compare PHOENIX-R and TemporalDynamics under these settings in our four datasets, including YouTube.

When comparing with SpikeM, we use Twitter hourly time series trimmed to 128 time windows around the largest peak (most popular hour). We focus on the 500 most popular of these times series for comparison. We chose this approach since this is the same dataset explored by the authors. Moreover, we focus on a smaller time scale because the SpikeM was proposed for single cascades only.

---

[8] MDL based costs will decrease with some variance and then increase again. The ϵ threshold is a guard against local minima due to small fluctuations.

[9] All of our source code is provided at: http://github.com/flaviovdf/phoenix

[10] Similar results were achieved using other thresholds.

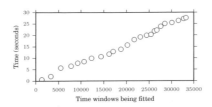

Fig. 4. Scalability of PHOENIX-R

Table 4. Comparison of PHOENIX-R with TemporalDynamics [21] and SpikeM [18]: Average RMSE values (with 95% confidence intervals in parentheses). Statistically significant results (including ties) are shown in bold.

| | PHOENIX-R vs. TemporalDynamics (daily series) | | PHOENIX-R vs. SpikeM (hourly series) | |
	RMSE PHOENIX-R	RMSE TemporalDynamics	RMSE PHOENIX-R	RMSE SpikeM
MMTweet	**2.93** (± 0.23)	4.18 (± 0.49)	-	-
LastFM	**7.09** (± 0.23)	8.31 (± 0.32)	-	-
Twitter	**72.05** (± 6.08)	194.79 (± 20.49)	**10.83** (± 1.61)	**9.77** (± 2.24)
YouTube	**280.58** (± 29.29)	3429.19 (± 577.76)	-	-

Table 4 shows the average RMSE (along with corresponding 95% confidence intervals) computed over the considered time series for all models. Best results of each comparison (including statistical ties) are shown in bold. Note that PHOENIX-R has statistically lower RMSE than TemporalDynamics in all datasets. These improvements come particularly from the non-linear nature of PHOENIX-R , which better fits the long term popularity dynamics of most objects. The difference between the models is more striking for the YouTube dataset, where most time series cover long periods (over 4 years in some cases). The linear nature of TemporalDynamics largely affects its performance in those cases, as many objects do not experience a linear popularity evolution over such longer periods of time. As result, PHOENIX-R produces reductions on average RMSE of over one order of magnitude. In contrast, the gap between both models is smaller in the LastFM dataset, where the fraction of objects (artists) for which a linear fit is reasonable is larger. Yet, PHOENIX-R produces results that are still statistically better, with a reduction on average RMSE of 15%.

When comparing with SpikeM, the PHOENIX-R model produces results that are statistically tied. We consider this result very positive, given that this comparison favors SpikeM: the time series cover only 128 hours, and thus there is no much room for improvements from capturing multiple cascades, one key feature of PHOENIX-R . Yet, we note that our model is more general and suitable to modeling popularity dynamics in the longer run, besides being much more scalable, as discussed above.

As a final comparison, we evaluate the cost-benefit of the models using BIC, as suggested by [21]. We found that we beat TemporalDynamics in terms of BIC on at least 80% of the objects in all datasets but LastFM. For LastFM objects, the reasonable linear evolution of popularity of many objects, makes the cost-benefit of TemporalDynamics superior. Yet, PHOENIX-R is still the preferred option in 30% of the objects in this dataset. Compared to SpikeM we also find that again, statistically equal BIC scores are achieved for both models.

Table 5. Comparing Phoenix-R with TemporalDynamics [21] for prediction. The values on the table are RMSE. Statistically significant results are in bold.

		5%			25%			50%		
		1	7	30	1	7	30	1	7	30
MMTweet	PhoenixR	**11.61**	**12.78**	15.15	**8.67**	**6.74**	**8.82**	**4.08**	**6.87**	**13.58**
	TempDynamics	17.07	17.41	**16.52**	9.63	10.78	14.46	25.19	23.08	30.39
Twitter	PhoenixR	**53.68**	**60.78**	**215.76**	**132.21**	**135.15**	**210.30**	**75.58**	**229.59**	**254.93**
	TempDynamics	104.45	129.36	255.69	643.39	643.83	786.50	420.74	587.86	598.75
LastFM	PhoenixR	**2.37**	**3.97**	**5.71**	**8.60**	**12.06**	**14.66**	**11.34**	**15.03**	**15.43**
	TempDynamics	6.47	7.03	8.00	11.15	14.62	17.86	14.91	18.15	18.80
YouTube	PhoenixR	**91.62**	**106.38**	**138.88**	**83.76**	**113.14**	**147.04**	**127.53**	**97.97**	**115.97**
	TempDynamics	3560.65	3631.09	3661.81	5091.82	5107.82	5143.70	4136.14	4139.73	4169.26

5.2 Predicting Popularity with PHOENIX-R

We here assess the efficacy of PHOENIX-R for predicting the popularity of objects a few time windows into the future, comparing it against TemporalDynamics[11]. To that end, we train the PHOENIX-R and TemporalDynamics models for each time series using 5%, 25%, and 50% of the initial daily time windows. We then use the δ time windows following the training period as validation set to learn model parameters. In each setting, we train 10 models for each time series, selecting the best one on the validation period. We then use the selected model to estimate the popularity of the object δ windows after the validation (test period). We experiment with δ equal to 1, 7 and 30 windows.

Table 5 shows the average RMSE of both models on the test period. Confidence intervals are omitted for the sake of clarity, but the best results (and statistical ties) in each setting are shown in bold. PHOENIX-R produces more accurate predictions than TemporalDynamics in practically all scenarios and datasets. Again, the improvements are quite striking for the YouTube dataset, mainly because the time series cover long periods (over 4 years in some cases). While the linear TemporalDynamics model fits reasonably well the popularity dynamics of some objects, it performs very poorly on others, thus leading to high variability in the results. In contrast, PHOENIX-R is much more robust, producing more accurate predictions for most objects, and thus being more suitable for modeling and predicting long periods of social activity.

6   Related Work

Popularity prediction of social media has gained a lot of attention recently, with many efforts focused on linear methods to achieve this task [20–22]. However, not all of these methods are useful for modeling *individual* time series. For example, linear regression based methods [20, 22] can be used for prediction but are not explanatory of individual time series. Moreover, as we showed in our experiments, there is strong evidence that linear methods are less suitable for modeling popularity dynamics than non-linear ones,

---

[11] We do not use SpikeM for this task, as it is suitable for tail forecasting only (i.e., predicting after the peak).

particularly for long term dynamics. This comes from the non-linear behavior of social cascades [18]. Li *et. al.* [17] proposed a non-linear popularity prediction model. However they focused on modeling the video propagation through links on a single online social network, and not on general time series data, as we do here.

Recent work has also focused on modeling the dynamics of news evolution [18], or posts on news aggregators [2, 15, 16]. These prior efforts do not explicitly account for revisits nor multiple cascades, as we do. For example, the authors either assume unique visits only [18], or focus on applications that do not allow revisits (e.g., once a user likes a news posted on a application, she/he cannot like it a second time) [15]. In other cases, the models do not distinguish between a first visit by a user and a revisit [2].

Very recently, Anderson *et. al.* [1] analyzed revisits in social media applications. However, unlike we do here, the authors were not focused on modeling the evolution of popularity of individual objects, but rather the aggregate and user behavior.

7   Conclusions

In this paper we presented the PHOENIX-R model for social media popularity time series. Before introducing the model, we showed the effect of revisits on the popularity of objects on large social activity datasets. Our main findings are: ·

- **Discoveries:** We explicitly show the effect of revisits in social media popularity.
- **Explanatory model:** We define the PHOENIX-R , which explicitly accounts for revisits and multiple cascades. Factors not captured by state-of-the art alternatives.
- **Scalable and Parsimonious:** Our fitting approach make's use of the MDL principle to achieve a parsimonious description of the data. We also show that fitting the model is scalable (linear time).
- **Effectiveness** of model: We showed the effectiveness of the model not only when describing popularity time series, but also when predicting future popularity values for individual objects. Gains can be up to one order of magnitude larger than baseline approaches, depending on the dataset.

As future work we intend on extending the PHOENIX-R model to deal with: (1) interacting populations between shocks; (2) multiple cascades from a single population; and, (3) fitting on multiple time series at once (e.g., audience and revisits).

Acknowledgments. This research is in part funded by the Brazilian National Institute of Science and Technology for Web Research (MCT/CNPq/INCT Web Grant Number 573871/2008-6), and by the authors' individual grants from Google's Brazilian Focused Research Award, CNPq, CAPES and Fapemig. It was also supported by NSF grants CNS-1065133 and CNS-1314632, as well as ARL Cooperative Agreements W911NF-09-2-0053 and W911NF-11-C-0088. The views and conclusions contained in this document are those of the authors and should not be interpreted as representing the official policies, either expressed or implied of the NSF, ARL, or other funding parties. The U.S. Government is authorized to reproduce and distribute reprints for Government purposes notwithstanding any copyright notation here on.

References

1. Anderson, A., Kumar, R., Tomkins, A., Vassilvitski, S.: Dynamics of Repeat Consumption. In: Proc. WWW (2014)
2. Bauckhage, C., Kersting, K., Hadiji, F.: Mathematical Models of Fads Explain the Temporal Dynamics of Internet Memes. In: Proc. ICWSM (2013)
3. Celma, O.: Music Recommendation and Discovery in the Long Tail, 1st edn. Springer (2010)
4. Cha, M., Mislove, A., Adams, B., Gummadi, K.P.: Characterizing social cascades in flickr. In: Proc. WOSN (2008)
5. Cha, M., Mislove, A., Gummadi, K.P.: A Measurement-Driven Analysis of Information Propagation in the Flickr Social Network. In: Proc. WWW (2009)
6. Crane, R., Sornette, D.: Robust Dynamic Classes Revealed by Measuring the Response Function of a Social System. Proceedings of the National Academy of Sciences 105(41), 15649–15653 (2008)
7. Du, P., Kibbe, W.A., Lin, S.M.: Improved peak detection in mass spectrum by incorporating continuous wavelet transform-based pattern matching. Bioinformatics 22(17), 2059–2065 (2006)
8. Duong, Q., Goel, S., Hofman, J., Vassilvitskii, S.: Sharding social networks. In: Proc. WSDM (2013)
9. Figueiredo, F., Benevenuto, F., Almeida, J.: The Tube Over Time: Characterizing Popularity Growth of YouTube Videos. In: Proc. WSDM (2011)
10. Hansen, M.H., Yu, B.: Model Selection and the Principle of Minimum Description Length. Journal of the American Statistical Association 96(454), 746–774 (2001)
11. Hauger, D., Schedl, M., Kosir, A., Tkalci, M.: The Million Musical Tweets Dataset: What Can we Learn from Microblogs. In: Proc. ISMIR (2013)
12. Hethcote, H.W.: The Mathematics of Infectious Diseases. SIAM Review 42(4), 599–653 (2000)
13. Hu, Q., Wang, G., Yu, P.S.: Deriving Latent Social Impulses to Determine Longevous Videos. In: Proc. WWW (2014)
14. Huang, C., Li, J., Ross, K.W.: Can Internet Video-on-Demand be Protable. In: Proc. SIGCOMM (2007)
15. Lakkaraju, H., McAuley, J., Leskovec, J.: What's in a Name? Understanding the Interplay between Titles, Content, and Communities in Social Media. In: Proc. ICWSM (2013)
16. Lerman, K., Hogg, T.: Using a Model of Social Dynamics to Predict Popularity of News. In: Proc. WWW (2010)
17. Li, H., Ma, X., Wang, F., Liu, J., Xu, K.: On popularity prediction of videos shared in online social networks. In: Proc. CIKM (2013)
18. Matsubara, Y., Sakurai, Y., Prakash, B.A., Li, L., Faloutsos, C.: Rise and Fall Patterns of Information Diffusion. In: Proc. KDD (2012)
19. Nannen, V.: A Short Introduction to Model Selection, Kolmogorov Complexity and Minimum Description Length (MDL). Complexity (Mdl), 20 (2010)
20. Pinto, H., Almeida, J., Gonçalves, M.: Using Early View Patterns to Predict the Popularity of YouTube Videos. In: Proc. WSDM (2013)
21. Radinsky, K., Svore, K., Dumais, S., Teevan, J., Bocharov, A., Horvitz, E.: Behavioral Dynamics on the Web: Learning, Modeling, and Prediction. ACM Transactions on Information Systems 32(3), 1–37 (2013)
22. Szabo, G., Huberman, B.A.: Predicting the Popularity of Online Content. Communications of the ACM 53(8), 80–88 (2010)
23. Vakali, A., Giatsoglou, M., Antaris, S.: Social networking trends and dynamics detection via a cloud-based framework design. In: Proc. WWW (2012)
24. Yang, J., Leskovec, J.: Patterns of temporal variation in online media. In: Proc. WSDM (2011)

FASOLE: Fast Algorithm for Structured Output LEarning

Vojtech Franc

Czech Technical University in Prague
Faculty of Electrical Engineering
Technicka 2, 166 27 Prague 6, Czech Republic
xfrancv@cmp.felk.cvut.cz

Abstract. This paper proposes a novel Fast Algorithm for Structured Ouput LEarning (FASOLE). FASOLE implements the sequential dual ascent (SDA) algorithm for solving the dual problem of the Structured Output Support Vector Machines (SO-SVM). Unlike existing instances of SDA algorithm applied for SO-SVM, the proposed FASOLE uses a different working set selection strategy which provides nearly maximal improvement of the objective function in each update. FASOLE processes examples in an on-line fashion and it provides certificate of optimality. FASOLE is guaranteed to find the ε-optimal solution in $\mathcal{O}(\frac{1}{\varepsilon^2})$ time in the worst case. In the empirical comparison FASOLE consistently outperforms the existing state-of-the-art solvers, like the Cutting Plane Algorithm or the Block-Coordinate Frank-Wolfe algorithm, achieving up to an order of magnitude speedups while obtaining the same precise solution.

1 Introduction

The Structured Output Support Vector Machines SO-SVM [17,19] is a supervised algorithm for learning parameters of a linear classifiers with possibly exponentially large number of classes. SO-SVM translate learning into a convex optimization problem size of which scales with the number of classes which rules out application of common off-the-shelf solvers. The specialized solvers can be roughly split to batch methods and on-line solvers. The batch methods, like variants of the Cutting Plane Algorithm (CPA) [18,8] or the column generation algorithm [19], approximate the SO-SVM objective by an iteratively built global under-estimator called the cutting plane model (the column generation algorithm instead approximates the feasible set). Optimizing the cutting plane model is cheaper than the original problem and, in addition, it provides a certificate of optimality. The bottle-neck of the CPA is the expensive per-iteration computational complexity. Namely, computation of a single element (the cutting plane) of the cutting plane model requires calling the *classification oracle* on all training examples. Note that in structured setting the classification is often time consuming. Moreover, many iterations are typically needed before the cutting plane model becomes tight and the approximate solution sufficiently precise.

The on-line methods, like the Stochastic Gradient Descent (SGD) [13,15], process the training examples one by one with a cheap update requiring a single call of the classification oracle. A disadvantage of the SGD is its sensitivity to setting of the step size

T. Calders et al. (Eds.): ECML PKDD 2014, Part I, LNCS 8724, pp. 402–417, 2014.
© Springer-Verlag Berlin Heidelberg 2014

and a missing clear stopping condition as the method does not provide a certificate of optimality. Unlike the SGD which optimizes the primal objective, the recently proposed Sequential Dual Method for Structural SVMs (SDM) [1] and the Block-Coordinate Frank-Wolfe (BCFW) [9] are on-line solvers maximizing the Lagrange dual of the SO-SVM problem. SDM and BCFW are instances of the same optimization framework in the sequel denoted as the Sequential Dual Ascent (SDA) method. The SDA methods iteratively update blocks of dual variables each of them being associated with a single training example. Particular instances of SDA method, like SDM or BCFW, differ in the strategy used to select the working set containing the dual variables to be updated. The SDA methods if compared to the SGD algorithm have two main advantages. First, for fixed working set the optimal step size can be computed analytically. Second, the optimized dual objective provides a certificate of optimality useful for defining a rigorous stopping condition.

The convergence speed of the SDA methods is largely dependent on the working set selection strategy. In this paper, we propose a novel SDA algorithm for solving the SO-SVM dual using a working set selection strategy which yields nearly maximal improvement of the dual objective in each iteration. We named the proposed solver as the Fast Algorithm for Structured Ouput LEarning (FASOLE). The same idea has been previously applied for optimization of simpler QP tasks emerging in learning of two-class SVM classifiers with L2-hinge loss [4,5] and L1-hinge loss [3,6]. The SDA solver using a similar working set selection strategy is implemented for example in popular LibSVM [2]. Our paper extends these solvers to the structured output setting. The structured output setting imposes several difficulties, namely, the SO-SVM dual problem has exponentially large number of variables and m linear equality constraints in contrast to the two-class SVM dual having only m variables and single equality constraint. The extreme size of the SO-SVM does not permit operations feasible in two-class case like maintaining all dual variables and buffering the columns of the Hessian matrix. The proposed method thus introduces a sparse representation of the SO-SVM dual and a set of heuristics to reflect the mentioned difficulties. In addition, we provide a novel convergence analysis which guarantees that the proposed SDA solver finds ε-optimal solution in $\mathcal{O}(\frac{1}{\varepsilon^2})$ time. We experimentally compare the proposed FASOLE against BCFW, SDM and CPA showing that FASOLE consistently outperforms all competing methods achieving up to an order of magnitude speedup. We remark that recently proposed BCFW and SDM have not been compared so far hence their empirical study is an additional contribution of this paper.

The paper is organized as follows. The problem to be solved is formulated in Section 2. Section 3 describes the proposed solver. Relation to existing methods is discussed in Section 4. Section 5 presents experiments and Section 6 concludes the paper.

2   Formulation of the Problem

Let us consider a linear classifier $h\colon \mathcal{X} \times \mathbb{R}^n \to \mathcal{Y}$ defined as

$$h(x; \boldsymbol{w}) \in \operatorname*{Argmax}_{y \in \mathcal{Y}} \langle \boldsymbol{w}, \boldsymbol{\psi}(x, y) \rangle$$

which assigns label $y \in \mathcal{Y}$ to observations $x \in \mathcal{X}$ according to a linear scoring function given by a scalar product between a feature vector $\boldsymbol{\psi} \colon \mathcal{X} \times \mathcal{Y} \to \mathbb{R}^n$ and a parameter vector $\boldsymbol{w} \in \mathbb{R}^n$ to be learned from data. In the structured output setting the label set $\mathcal{Y}$ is finite but typically very large, e.g. $\mathcal{Y}$ contains all possible image segmentations. Given a set of examples $\{(x_1, y_1), \ldots, (x_m, y_m)\} \in (\mathcal{X} \times \mathcal{Y})^m$, the SO-SVM algorithm translates learning of the parameters $\boldsymbol{w} \in \mathbb{R}^n$ into the following convex problem

$$\boldsymbol{w}^* = \underset{\boldsymbol{w} \in \mathbb{R}^n}{\operatorname{argmin}} P(\boldsymbol{w}) := \frac{\lambda}{2} \|\boldsymbol{w}\|^2 + \frac{1}{m} \sum_{i \in \mathcal{I}} \max_{y \in \mathcal{Y}} \left(\Delta_i(y) + \langle \boldsymbol{w}, \boldsymbol{\psi}_i(y) \rangle \right) \qquad (1)$$

where $\mathcal{I} = \{1, \ldots, m\}$, $\boldsymbol{\psi}_i(y) = \boldsymbol{\psi}(x_i, y) - \boldsymbol{\psi}(x_i, y_i)$, $\Delta_i(y) = \Delta(y_i, y)$ is a loss function and $\lambda > 0$ is a regularization constant. For a convenience of notation we assume that the loss function satisfies $\Delta(y, y) = 0$, $\forall y \in \mathcal{Y}$, which implies that $\Delta_i(y_i) + \langle \boldsymbol{w}, \boldsymbol{\psi}_i(y_i) \rangle = 0$. Note that all common loss functions like e.g. Hamming loss satisfy this assumption. The problem (1) can be equivalently expressed as a quadratic program whose Lagrange dual, denoted as SO-SVM dual, reads

$$\boldsymbol{\alpha}^* = \underset{\boldsymbol{\alpha} \in \mathcal{A}}{\operatorname{argmax}} D(\boldsymbol{\alpha}) := \langle \boldsymbol{b}, \boldsymbol{\alpha} \rangle - \frac{1}{2} \|\boldsymbol{A}\boldsymbol{\alpha}\|^2 , \qquad (2)$$

where $\boldsymbol{\alpha} = (\alpha_i(y) \mid i \in \mathcal{I}, y \in \mathcal{Y}) \in \mathbb{R}^d$ is vector of $d = m|\mathcal{Y}|$ dual variables, $\boldsymbol{b} = (\Delta_i(y) \mid i \in \mathcal{I}, y \in \mathcal{Y}) \in \mathbb{R}^d$ is a vector containing losses on training examples, $\boldsymbol{A} = (\boldsymbol{\psi}_i(y)/\sqrt{\lambda} \mid i \in \mathcal{I}, y \in \mathcal{Y}) \in \mathbb{R}^{n \times d}$ is a matrix of feature vectors, and $\mathcal{A} = \{\boldsymbol{\alpha} \in \mathbb{R}^d \mid \boldsymbol{\alpha} \geq 0 \land \sum_{y \in \mathcal{Y}} \alpha_i(y) = \frac{1}{m}, i \in \mathcal{I}\}$ denotes a feasible set. Since the primal problem (1) is convex and non-degenerate, the duality gap at the optimum is zero, i.e. $P(\boldsymbol{w}^*) = D(\boldsymbol{\alpha}^*)$. The optimal primal variables can be computed from the optimal dual variables by $\boldsymbol{w}^* = -\frac{1}{\sqrt{\lambda}} \boldsymbol{A} \boldsymbol{\alpha}^*$.

In this paper we propose a solver which iteratively maximizes the SO-SVM dual (2). Although maintaining the dual problem in computer memory is not feasible, an approximate solution can be found thanks to the problem sparsity. For any prescribed $\varepsilon > 0$, our solver finds an ε-optimal solution $\hat{\boldsymbol{w}}$ satisfying $P(\hat{\boldsymbol{w}}) \leq P(\boldsymbol{w}^*) + \varepsilon$ in time not bigger than $\mathcal{O}(\frac{1}{\varepsilon^2})$. Our solver is modular: it accesses the problem only via a classification oracle solving so called loss-augmented inference task, i.e. for given $i, \boldsymbol{w}$ the classification oracle returns an optimal solution and the optimal value of

$$\max_{y \in \mathcal{Y}} \left(\Delta_i(y) + \langle \boldsymbol{w}, \boldsymbol{\psi}_i(y) \rangle \right) . \qquad (3)$$

3   Proposed Algorithm Solving SO-SVM Dual

In this section we describe the proposed solver. For the sake of space we put derivations and proofs to a supplementary material available online[1].

---

[1] ftp://cmp.felk.cvut.cz/pub/cmp/articles/franc/
Franc-FasoleSupplementary-ECML2014.pdf

3.1   Generic SDA

In this section we first outline the idea of a generic SDA algorithm for solving SO-SVM dual (2). In the next section we then describe the proposed instance of the generic SDA.

A generic SDA algorithm converts optimization of the SO-SVM dual (2) into a series of simpler auxiliary QP tasks solvable analytically. Starting from a feasible point $\alpha \in \mathcal{A}$, an SDA algorithm iteratively applies the update rule

$$\alpha^{\text{new}} := \underset{\alpha \in \mathcal{A}_L}{\text{argmax}}\, D(\alpha)\,, \tag{4}$$

where $\mathcal{A}_L \subset \mathcal{A}$ is a line between the current solution α and a point β selected from $\mathcal{A}$,

$$\mathcal{A}_L = \{\alpha' \mid \alpha' = (1 - \tau)\alpha + \tau\beta\,, \tau \in [0, 1]\}\,. \tag{5}$$

The update rule (4) is an auxiliary QP task having the same objective as the original SO-SVM dual but the feasible set is reduced to a line inside $\mathcal{A}$. This implies that the new solution α^{new} is also feasible. Let us define a single-variable quadratic function

$$D_L(\tau) = D\big((1 - \tau)\alpha + \tau\beta\big)$$

corresponding to the dual objective $D(\alpha)$ restricted to the line $\mathcal{A}_L$. If the point β is selected such that the derivative of $D(\alpha)$ along the line $\mathcal{A}_L$ evaluated at α is positive, i.e. $D_L(0)' > 0$, then the update (4) strictly increases the objective function, i.e. $D(\alpha^{\text{new}}) - D(\alpha) > 0$ holds. Moreover, the update (4) has a simple analytical solution (supplementary material, Sec 1.1)

$$\alpha^{\text{new}} := (1 - \tau)\alpha + \tau\beta$$

where

$$\tau := \underset{\tau' \in [0,1]}{\text{argmax}}\, D_L(\tau') = \min\left(1, \frac{\langle \beta - \alpha, b - A^T A \alpha \rangle}{\langle \alpha - \beta, A^T A(\alpha - \beta) \rangle}\right)\,. \tag{6}$$

Algorithm 1. Generic SDA for solving SO-SVM dual (2)

Initialization: select a feasible $\alpha \in \mathcal{A}$
repeat
    Select $\beta \in \mathcal{A}$ such that $D_L(0)' > 0$.
    Compute τ by (6).
    Compute the new solution $\alpha := (1 - \tau)\alpha + \tau\beta$.
until *until convergence*;

Algorithm (1) outlines the generic SDA algorithm. The recently proposed SDM [1] and the BCFW Algorithm [9] are instances of the generic SDA which differ in the way how they construct the point β. Note, that the point β determins which variables will be modified (the working set) by the update rule. For example, the BCFW simultaneously updates all $|\mathcal{Y}|$ dual variables associated with one training example, while, the method

SDM updates just two variables at a time. In Section 4 we describe the related instances of the generic SDA in more details.

Our algorithm is also instance of the generic SDA which adopts a different selection strategy originally proposed for two-class SVM solvers independently proposed by [3] and [4]. This method, in [3] coined the Working Set Selection strategy using second order information (WSS2), selects two variables ensuring nearly maximal improvement of the objective. In the next section we describe the adaptation of the WSS2 to solving the SO-SVM dual.

3.2 Proposed SDA Solver with WSS2 Selection Strategy

The proposed solver constructs the point β as follows

$$\beta_j(y) := \begin{cases} \alpha_i(u) + \alpha_i(v) & \text{if } j = i \wedge y = u \\ 0 & \text{if } j = i \wedge y = v \\ \alpha_j(y) & \text{otherwise} \end{cases} \tag{7}$$

where $(i, u, v) \in \mathcal{I} \times \mathcal{Y} \times \mathcal{Y}$ is a triplet such that $u \neq v$ (the way how (i, u, v) is selected will be described below). We denote the SDA update rule (4) using β constructed by (7) as the Sequential Minimal Optimization (SMO) rule [7]. The SMO rule changes the minimal number of dual variables, in our case two, without escaping the feasible set $\mathcal{A}$. There are $m|\mathcal{Y}|(|\mathcal{Y}| - 1)$ SMO rules in total out of which we need to select one in each iteration. A particular SMO rule determined by the choice of (i, u, v) changes the variables $\alpha_i(u)$ and $\alpha_i(v)$ associated with labels u and v and the i-th example. We now describe our strategy to select (i, u, v).

Selection of i. Recall that the ultimate goal is to find an ε-optimal solution of the SO-SVM dual (2). Specifically, we aim to find a primal-dual pair $(w, \alpha) \in \mathbb{R}^n \times \mathcal{A}$ such that the duality gap is at most ε, i.e. $G(w, \alpha) = P(w) - D(\alpha) \leq \varepsilon$ holds. Let us define shorthands

$$w = -\frac{1}{\sqrt{\lambda}} A\alpha \quad \text{and} \quad s_i(y, \alpha) = \Delta_i(y) + \langle w, \psi_i(y) \rangle$$

for the primal solution w constructed from the dual solution α and the score function $s_i(y, w)$ of the classification oracle (3) with parameters w, respectively. Using this notation it is not difficult to show (supplementary material, Sec. 1.3) that the duality gap can be computed by the following formula

$$G(w, \alpha) = \frac{1}{m} \sum_{i \in \mathcal{I}} G_i(w, \alpha)$$

where

$$G_i(w, \alpha) = \max_{y \in \mathcal{Y}} s_i(y, w) - m \sum_{y \in \mathcal{Y}} \alpha_i(y) s_i(y, w) .$$

The proposed solver goes through the examples and it uses the value of $G_i(w, \alpha)$ to decide whether the block of variables $\alpha_i = (\alpha_i(y) \mid y \in \mathcal{Y})$, associated with the i-th example, should be updated. In particular, if $G_i(\boldsymbol{w}, \boldsymbol{\alpha}) > \varepsilon$ holds then $\boldsymbol{\alpha}_i$ is updated otherwise they are skipped and the next block of variables is checked. It is clear that if $G_i(\boldsymbol{w}, \boldsymbol{\alpha}) \leq \varepsilon$ holds for all $i \in \mathcal{I}$, i.e. no variables are selected for update, the target duality gap $G(\boldsymbol{w}, \boldsymbol{\alpha}) \leq \varepsilon$ has been already achieved and thus no update is needed.

Selection of u and v. Here we employ the WSS2 strategy of [3,4]. Given block of variables $\boldsymbol{\alpha}_i$, our goal is to select among them two, $\alpha_i(u)$ and $\alpha_i(v)$, which if used in the SMO rule will cause a large increment, $\delta(i, u, v) = D(\boldsymbol{\alpha}^{\text{new}}) - D(\boldsymbol{\alpha})$, of the dual objective. First, we need to identify those pairs (u, v) which guarantee positive change of dual objective, i.e. those for which $D'_L(0) > 0$ holds. Using (7) the condition $D'_L(0) > 0$ can be written as (supplementary material, Sec. 1.2)

$$\alpha_i(v)\big(s_i(u, \boldsymbol{w}) - s_i(v, \boldsymbol{w})\big) > 0 . \tag{8}$$

Provided the condition (8) holds, application of the SMO rule given by a triplet (i, u, v) increases the dual objective exactly by the quantity (supplementary material, Sec. 1.2)

$$\delta(i, u, v) = \begin{cases} \dfrac{\lambda(s_i(u, \boldsymbol{w}) - s_i(v, \boldsymbol{w}))}{\|\boldsymbol{\psi}_i(u) - \boldsymbol{\psi}_i(v)\|^2} & \text{if } \tau < 1 \\[2ex] \alpha_i(v)(s_i(u, \boldsymbol{w}) - s_i(v, \boldsymbol{w})) - \dfrac{\alpha_i(v)^2}{2\lambda}\|\boldsymbol{\psi}_i(u) - \boldsymbol{\psi}_i(v)\|^2 & \text{if } \tau = 1 \end{cases} \tag{9}$$

The optimal strategy would be to find the pair (u, v) maximizing $\delta(i, u, v)$, however, this is not feasible due to a large number of candidate pairs $(u, v) \in \mathcal{Y} \times \mathcal{Y}$. Instead we use a cheaper WSS2 strategy which has been shown to yield nearly the same improvement as trying all pairs [4]. We first find $\hat{u}$ by maximizing the dominant term $s_i(u, \boldsymbol{w})$ appearing in the improvement formulas. This corresponds to fining the most violated primal constraint associated with i-th example by solving

$$\hat{u} \in \underset{y \in \mathcal{Y}}{\text{argmax}}\, s_i(y, \boldsymbol{w}) \tag{10}$$

via using the classification oracle. When $\hat{u}$ is fixed, we find $\hat{v}$ which brings the maximal improvement by

$$\hat{v} \in \underset{y \in \mathcal{Y}_i}{\text{argmax}}\, \delta(i, \hat{u}, y) \tag{11}$$

where $\mathcal{Y}_i = \{y \in \mathcal{Y} \mid \alpha_i(y) > 0\}$ is a set of labels corresponding to non-zero dual variables associated with the i-th example. Note that the maximization tasks (10) and (11) do not need to have a unique solution in which case we take any maximizer.

The SDA solver using WSS2 is summarized in Algorithm 2. Note that the SDA-WSS2 algorithm is an on-line method passing thought the examples and updating those blocks of variables found to be sub-optimal. The algorithm stops if $G_i(\boldsymbol{w}, \boldsymbol{\alpha}) \leq \varepsilon$, $\forall i \in \mathcal{I}$, implying that the duality gap $G(\boldsymbol{w}, \boldsymbol{\alpha})$ is not larger than ε. Besides primal variables $\boldsymbol{w} \in \mathbb{R}^n$ the algorithm maintains the non-zero dual variables, $\alpha_i(y) > 0$, $y \in \mathcal{Y}_i$, their number is upper bounded by the number of updates. Although the primal variables are at any time related to the dual variables by $\boldsymbol{w} = -\frac{1}{\sqrt{\lambda}}A\boldsymbol{\alpha}$ it is beneficial to maintain the vector $\boldsymbol{w}$ explicitly as it speeds up the computation of the score

$$s_i(y, \boldsymbol{w}) = \Delta_i(y) + \langle \boldsymbol{\psi}_i(y), \boldsymbol{w} \rangle = \Delta_i(y) - \frac{1}{\lambda} \sum_{j \in \mathcal{Y}} \sum_{y' \in \mathcal{Y}_i} \alpha_j(y') \langle \boldsymbol{\psi}_i(y), \boldsymbol{\psi}_j(y') \rangle .$$

Algorithm 2. SDA-WSS2 algorithm for solving SO-SVM dual (2)

Input: precision parameter $\varepsilon > 0$, regularization constant $\lambda > 0$
Output: ε-optimal primal-dual pair $(\boldsymbol{w}, \boldsymbol{\alpha})$
Initialization:

$$\boldsymbol{w} = \boldsymbol{0}, \mathcal{Y}_i = \{y_i\}, \alpha_i(y) = \begin{cases} \frac{1}{m} & \text{if } y = y_i \\ 0 & \text{otherwise} \end{cases}, i \in \mathcal{I}$$

repeat
> num_updates := 0
> **forall the** $i \in \mathcal{I}$ **do**
>> $u_1 := \mathrm{argmax}_{y \in \mathcal{Y}} s_i(y, \boldsymbol{w})$
>> $u_2 := \mathrm{argmax}_{y \in \mathcal{Y}_i} s_i(y, \boldsymbol{w})$
>> **if** $s_i(u_1, \boldsymbol{w}) > s_i(u_2, \boldsymbol{w})$ **then**
>>> $\hat{u} := u_1$
>>
>> **else**
>>> $\hat{u} := u_2$
>>
>> **if** $s_i(\hat{u}, \boldsymbol{w}) - m \sum_{y \in \mathcal{Y}_i} \alpha_i(y) s_i(y, \boldsymbol{w}) > \varepsilon$ **then**
>>> num_updates := num_updates + 1
>>> $\hat{v} := \underset{y \in \{y' \in \mathcal{Y}_i | \alpha_i(y') > 0\}}{\mathrm{argmax}} \delta(i, \hat{u}, y)$
>>> $\tau := \min\left\{1, \frac{\lambda(s_i(\hat{u}, \boldsymbol{w}) - s_i(\hat{v}, \boldsymbol{w}))}{\alpha_i(\hat{v}) \|\boldsymbol{\psi}_i(\hat{u}) - \boldsymbol{\psi}_i(\hat{v})\|^2}\right\}$
>>> $\boldsymbol{w} := \boldsymbol{w} + (\boldsymbol{\psi}_i(\hat{v}) - \boldsymbol{\psi}_i(\hat{u})) \frac{\tau \alpha_i(\hat{v})}{\lambda}$
>>> $\alpha_i(\hat{u}) := \alpha_i(\hat{u}) + \tau \alpha_i(\hat{v})$
>>> $\alpha_i(\hat{v}) := \alpha_i(\hat{v}) - \tau \alpha_i(\hat{v})$
>>> **if** $\hat{u} = u_1$ **then**
>>>> $\mathcal{Y}_i := \mathcal{Y}_i \cup \{u_1\}$

until num_updates = 0;

Note, however, that all computations can be carried out in terms of the dual variables $\boldsymbol{\alpha}$. This property allows to kernelize the algorithm by replacing $\langle \boldsymbol{\psi}_i(y), \boldsymbol{\psi}_j(y') \rangle$ with a selected kernel function.

The computational bottle neck of the SDA-WSS2 is calling the classification oracle to solve $u_1 := \mathrm{argmax}_{y \in \mathcal{Y}} s_i(y, \boldsymbol{w})$. The other maximization problems over $\mathcal{Y}_i$, which are required to select u_2 and v_2, can be solved exhaustively since the set $\mathcal{Y}_i$ contains only those $\boldsymbol{\psi}_i(y)$ corresponding to at least once updated dual variable $\alpha_i(y)$.

The convergence of the SDA-WSS2 is ensured by the following theorem:

Theorem 1. *For any $\varepsilon > 0$ and $\lambda > 0$, Algorithm 2 terminates after*

$$T = \frac{8LD^2}{\varepsilon^2 \lambda}$$

updates at most where $L = \max_{y \in \mathcal{Y}, y' \in \mathcal{Y}} \Delta(y, y')$ *and* $D = \max_{i \in \mathcal{I}, y \in \mathcal{Y}} \|\boldsymbol{\psi}(x_i, y)\|$.

Proof is given in the supplementary material, Section 2.

We point out without giving a formal proof that the convergence Theorem 1 is valid not only for the SDA-WSS2 algorithm but also for a broader class of SDA solvers using the SMO update rule. In particular, the idea behind the proof of Theorem 1 applies to the DCM of [1] for which no such bound has been published so far.

The competing methods like the BCFW [9] and the CPA [18] are known to converge to the ε-optimal solution in $\mathcal{O}(\frac{1}{\varepsilon})$ time which is order of magnitude better compared to our bound $\mathcal{O}(\frac{1}{\varepsilon^2})$ for the SDA-WSS2 algorithm. However, all the bounds are obtained by the worst case analysis and little is known about their tightness. The empirical simulations provided in Section 5 show that the actual number of updates required by the proposed algorithm is consistently much smaller (up to order of magnitude) compared to the competing methods.

3.3   FASOLE: Fast Algorithm for Structured Output LEarning

In this section we describe a set of heuristics significantly decreasing the absolute computational time of SDA-WSS2 without affecting its convergence guarantees. We denote SDA-WSS2 algorithm with the implemented heuristics as the Fast Algorithm for Structured Output LEarning (FASOLE). A pseudo-code of the FASOLE is summarized by Algorithm 3. The implemented heuristics aim at i) further reducing the number of oracle calls and ii) using a tighter estimate of the duality gap which is used as a certificate of the ε-optimality. In particular, FASOLE uses the following set of heuristics:

Reduced problem. FASOLE maintains vector $\hat{b}$ and matrix $\hat{A}$ containing a subset of coefficients (b, A) of the SO-SVM dual (2). The coefficients $(\hat{b}, \hat{A})$ correspond to the dual variables $\hat{\alpha} = (\alpha_i(y) \mid y \in \mathcal{Y}_i, i \in \mathcal{I})$ which has been updated in the course of algorithm. The remaining variables are zero hence the corresponding coefficients need not be maintained. At the end of each pass through the examples we use $\hat{b}$ and $\hat{A}$ to find the optimal setting of $\hat{\alpha}$ by solving a reduced dual problem

$$\hat{\alpha} := \operatorname*{argmax}_{\alpha' \in \mathcal{A}} \langle \hat{b}, \alpha' \rangle - \frac{1}{2} \|\hat{A}\alpha'\|^2 \tag{12}$$

We use the SDA-WSS2 Algorithm 2, "worm"-started from the current solution $\hat{\alpha}$, to find ε-optimal solution of (12), i.e. FASOLE uses one loop optimizing over all variables and second loop for optimizing those variables which have been selected by the first loop. This strategy reduces the number oracle calls and is cheap due to the warm start.

Variable shrinking. SDA-WSS2 Algorithm 2 checks optimality of dual variables α_i in each iteration irrespectively if they have been found optimal in the previous pass. FA-SOLE instead introduces binary flag, satisfied(i), which is set true if the corresponding variables α_i already satisfied the partial duality gap constraint $G_i(w, \alpha) \leq \varepsilon$. Once all variables have been found optimal, i.e. satisfied(i) = true, $\forall i \in \mathcal{I}$, the flags are reset to false and the process starts again. This strategy allows to concentrate on non-optimal variables without wasting oracle calls on already optimal ones.

On-line and batch regime. The SDA-WSS2 Algorithm 2 stops if $G_i(w, \alpha) \leq \varepsilon$, $\forall i \in \mathcal{I}$, holds which implies $G(w, \alpha) \leq \varepsilon$. This stopping conditions is cheap and can be evaluated in an on-line manner, however, it is overly stringent because the actual duality gap $G(w, \alpha)$ is usually a way below ε. In fact $G(w, \alpha) = \varepsilon$ holds only

Algorithm 3. FASOLE: Fast Algorithm for Structured Output LEarning

---

Input: precision parameter $\varepsilon > 0$, regularization constant $\lambda > 0$
Output: ε-optimal primal-dual pair $(\boldsymbol{w}, \boldsymbol{\alpha})$
Initialization:

$$\boldsymbol{w} = \boldsymbol{0}, \; \big(\mathcal{Y}_i = \{y_i\}, i \in \mathcal{I}\big), \; \left(\alpha_i(y) = \begin{cases} \frac{1}{m} & \text{if } \; y = y_i \\ 0 & \text{otherwise} \end{cases}, i \in \mathcal{I}\right)$$

$\hat{\boldsymbol{A}} := (\boldsymbol{\psi}_i(y_i) \mid i \in \mathcal{I}), \hat{\boldsymbol{b}} := (\Delta_i(y_i) \mid i \in \mathcal{I})$
converged := false, regime := online, $\big(\text{satisfied}(i) := \text{false}, i \in \mathcal{I}\big)$
repeat

    G := 0, num_satisfied := 0, num_checked := 0

    forall the $i \in \mathcal{I}$ **do**

        if satisfied(i) = false **then**

            num_checked := num_checked + 1

            $u_1 := \text{argmax}_{y \in \mathcal{Y}}\, s_i(y, \boldsymbol{w})$

            $u_2 := \text{argmax}_{y \in \mathcal{Y}_i}\, s_i(y, \boldsymbol{w})$

            if $s_i(u_1, \boldsymbol{w}) > s_i(u_2, \boldsymbol{w})$ **then**
                $\hat{u} := u_1$

            else
                $\hat{u} := u_2$

            $G_i := s_i(\hat{u}, \boldsymbol{w}) - m \sum_{y \in \mathcal{Y}_i} \alpha_i(y) s_i(y, \boldsymbol{w})$

            $G := G + G_i$

            if $G_i \le \varepsilon$ **then**

                satisfied(i) := true

                num_satisfied := num_satisfied + 1

            else

                if $\hat{u} = u_1$ **then**

                    $\mathcal{Y}_i := \mathcal{Y}_i \cup u_1$

                    $\hat{\boldsymbol{A}} := \hat{\boldsymbol{A}} \cup \boldsymbol{\psi}_i(u_1)$

                    $\hat{\boldsymbol{b}} = \hat{\boldsymbol{b}} \cup \Delta_i(u_1)$

                if regime = online **then**

                    $\hat{v} := \underset{y \in \{y' \in \mathcal{Y}_i \mid \alpha_i(y') > 0\}}{\text{argmax}} \delta(i, \hat{u}, y)$

                    $\tau := \min\left\{1, \frac{\lambda(s_i(\hat{u}, \boldsymbol{w}) - s_i(\hat{v}, \boldsymbol{w}))}{\alpha_i(\hat{v}) \| \boldsymbol{\psi}_i(\hat{u}) - \boldsymbol{\psi}_i(\hat{v}) \|^2}\right\}$

                    $\boldsymbol{w} := \boldsymbol{w} + (\boldsymbol{\psi}_i(\hat{v}) - \boldsymbol{\psi}_i(\hat{u})) \frac{\tau \alpha_i(\hat{v})}{\lambda}$

                    $\alpha_i(\hat{u}) := \alpha_i(\hat{u}) + \tau \alpha_i(\hat{v})$

                    $\alpha_i(\hat{v}) := \alpha_i(\hat{v}) - \tau \alpha_i(\hat{v})$

    if num_checked = m **then**

        if regime = batch $\wedge$ $G \le \varepsilon$ **then**
            converged := true

        if $m \cdot K \le$ num_satisfied **then**
            regime := batch

    if $\forall i \in \mathcal{I}$, satisfied(i) = true **then**
        satisfied(i) := false, $i \in \mathcal{I}$

    if converged = false **then**
        Update $\hat{\boldsymbol{\alpha}} = (\alpha_i(y) \mid y \in \mathcal{Y}_i, i \in \mathcal{I})$ by solving the reduced problem (12)

until converged = true;

---

in a rare case when $G(\boldsymbol{w}, \boldsymbol{\alpha}) = \varepsilon, \forall i \in \mathcal{I}$. We resolve the problem by introducing two optimization regimes: on-line and batch. FASOLE is started in the on-line regime, regime = online, during which the "for" loop instantly updates the dual variables identified as non-optimal. As soon as it gets close to the optimum it switches from online to the batch regime. In the batch regime, FASOLE uses the "for" loop only to select new non-optimal variables and to simultaneously evaluate the actual duality gap $G(\boldsymbol{w}, \boldsymbol{\alpha})$. The variable update in the batch regime is done solely by solving the reduced problem (12). FASOLE switches from on-line to batch when a large portion of dual variables are found optimal, in particular, if $G_i(\boldsymbol{w}, \boldsymbol{\alpha}) \leq \varepsilon$ holds for $m \cdot K$ variables at least. We used the value $K = 0.9$ in all our experiments.

4   Relation to Existing Methods

In this section we describe relation between the proposed SDA-WSS2 algorithm (and FASOLE, respectively) and other two instances of the generic SDA algorithm 1 that have been recently proposed for solving the SO-SVM dual. We also mention a relation to two-class SVM solvers which use a similar optimization strategy.

Sequential Dual Method for Structural SVMs (SDM) [1]   SDM is among the existing solvers the most similar to our approach. SDM is an instance of the generic SDA using the SMO updated rule (7) similarly to the proposed SDA-WSS2. The main difference lies in the strategy for selecting the variables for update. SDM uses so called maximal violating pair (MVP) strategy which returns the variables most violating Karush-Kuhn-Tucker (KKT) conditions of the SO-SVM dual (2). Specifically, it finds $\hat{u}$ by (10), similarly to our approach, however $\hat{v}$ is set to

$$\hat{v} = \underset{y \in \{y' \in \mathcal{Y} | \alpha_i(y') > 0\}}{\operatorname{argmin}} s_i(v, \boldsymbol{w}) \,.$$

The MVP strategy can be seen as a cruel approximation of WSS2 strategy. Indeed, MVP maximizes the improvement $\delta(i, u, v)$ if we neglect the terms containing λ and $\|\boldsymbol{\psi}_i(u) - \boldsymbol{\psi}_i(v)\|$ in the formula (9). Note that WSS2 strategy introduces only a negligible computational overhead if compared to the MVP. We show experimentally that the proposed the SDA with WSS2 strategy consistently outperforms the SDM using the MVP strategy as it requires consistently less number of oracle calls. Similar behavior showing that the WSS2 strategy outperforms the MVP strategy has been observed for the two-class SVM solvers in [3,4].

Block-Coordinate Frank-Wolfe (BCFW) [9]   BCFW is an instance of the generic SDA constructing the point β as

$$\beta_j(y) = \begin{cases} \frac{1}{m} & \text{if } y = u \wedge i = j \\ 0 & \text{if } y \neq u \wedge i = j \\ \alpha_j^{(t)}(y) & \text{if } j \neq i \end{cases} \tag{13}$$

where $\hat{u}$ is selected by (10); we call this variable selection strategy as the BCFW update rule. Unlike the SMO update rule used by SDA-WSS2 and SDM, the BCFW rule changes the whole block of $|\mathcal{Y}|$ variables α_i at once. It can be shown that the BCFW rule selects among the admissible points the one in which direction the derivative of the SO-SVM dual objective is maximal. Hence, the resulting SDA algorithm using the BCFW rule can be seen as the steepest feasible ascent method. Empirically it also behaves similarly to the steepest ascent methods, i.e. it exhibits fast convergence at the first iterations but stalls as it approaches optimum. The slow convergence is compensated by simplicity of the method. Specifically, it can be shown that the BCFW rule admits to express the update rule, and consequently the whole algorithm, without explicitly maintaining the dual variables α. That is, the BCFW algorithm operates only with the primal variables though it maximizes the dual SO-SVM objective. The empirical evaluation shows that the BCFW converges significantly slower compared to the SDA-WSS2, as well as SDM, both using the SMO update rule. Similar behavior have been observed when the BCFW update rule is applied to two-class SVM problem [4].

Two-class SVM solvers using the SDA with WSS2 [3][4][5] The SDA methods with WSS2 have been first applied for solving the two-class SVM with L2-hinge loss [4][5] and with L1-hinge loss in [3]. A similar method was also proposed in [6]. The SDA with WSS2 is the core solver of LibSVM [2] being currently the most popular SVM implementation. The main difference to the proposed SDA-WSS2 lies in the form and the size of the quadratic programs these methods optimize. In particular, the two-class SVM dual has only a single linear constraint and m variables. In contrast, the SO-SVM dual has m linear constraints and $m|\mathcal{Y}|$ variables. The extreme size of the SO-SVM does not admit operations used in two-class SVM solvers like maintaining all dual variables and buffering the columns of the Hessian matrix. In addition, selection of the most violated constraint via the classification oracle is expensive in the case of SO-SVM and must be reduced. The proposed method thus introduces a sparse representation of the SO-SVM dual and a set of heuristics to reflect the mentioned difficulties.

In addition, our convergence Theorem 1 provides an upper bound on the number of updates to achieve the ε-optimal solution. To our best knowledge no similar result is known for the two-class SVM solvers. In particular, only asymptotical convergence of the SMO type algorithms have been proved so far [16].

5   Experiments

Compared methods. We compared the proposed solver FASOLE (Algorithm 3) against the SDM [1] and BCFW [9] which are all instances of the generic SDA algorithm 1. In addition, we compare against the the Cutting Plane Algorithm (CPA) [8,18] being the current gold-standard for SO-SVM learning (e.g. implemented in popular StructSVM library). We also refer to [9] which provides a thorough comparison showing that the BCFW consistently outperforms approximate on-line methods including the exponentiated gradient [11] and the stochastic sub-gradient method [14] hence these methods are excluded from our comparison.

Datasets. We used three public benchmarks which fit to the SO-SVM setting. First, we learn a linear multi-class classifier of isolated handwritten digits from the USPS dataset [10]. Second, we learn OCR for a sequence of segmented handwritten letters modeled by the Markov Chain classifier [17]. Third, we learn a detector of landmarks in facial images based on a deformable part models [12]. For USPS and OCR classifiers we use normalized image intensities as dense features. For the LANDMARK detector we use high-dimensional sparse feature descriptors based on the Local Binary Patterns as suggested in [12]. The classification oracle can be solved by enumeration in the case of USPS and by Viterbi algorithm in the case of OCR and LANDMARK. The three applications require different loss function $\Delta(y, y')$ to measure the performance of the structured classifier. Specifically, we used the 0/1-loss (classification loss) for USPS data, Hamming loss normalized to the number of characters the OCR problem, and the loss for LANDMARK data was the absolute average deviation between the estimated and the ground-truth landmark positions measured in percents of the face size. The datasets are summarized in Table 1.

Implementation. The competing methods are implemented in Matlab. CPA, SDM and FASOLE use the same inner loop quadratic programming solver written in C. SDM and FASOLE implement the same framework described by Algorithm 3 but SDM uses the SMO update with MVP selection strategy of [1]. We do not implement different heuristics proposed in [1] in order to measure the effect of different variable selection strategy being the main contribution of our paper. All methods use the same classification oracles. The oracles for OCR and USPS are implemented in Matlab. The oracle for LANDMARK is implemented in C. All methods use the same stopping condition based on monitoring the duality gap. In contrast to FASOLE and SDM, the authors of BCFW do not provide an efficient way to compute the duality gap. Hence we simply evaluate the gap every iteration and stop BCFW when the goal precision is achieved but we DO NOT count the duality gap computation to the convergence speed and thus the wall clock times for BCFW are biased to lower values. The experiments were run on the AMD Opteron CPU 2600 MHz/256GB RAM.

Evaluation. We measure convergence speed in terms of the effective iterations. One effective iteration equals to m oracle calls where m is the number of examples. Note that CPA algorithm requires one effective iteration to compute a single cutting plane. In contrast, one effective iteration of SDM, BCFW and FASOLE corresponds to m updates. The effective iteration is an implementation independent measure of the convergence time which is correlated with the the real CPU time when the oracle calls dominate the other computations. This is the case e.g. for the OCR and LANDMARK where the oracle calls are expensive. We also record the wall clock time because the competing methods have different overheads, e.g. CPA, SDM and FASOLE call an inner QP solver. We run all methods for a range of regularization constants, specifically, $\lambda \in \{10, 1, 0.1, 0.01\}$. We stopped each method when the ε-optimal solution has been achieved. We set the precision parameter to $\varepsilon = 0.001$ for USPS and OCR, and $\varepsilon = 0.1$ for the LANDMARK problem. Note that the target precision ε is given in terms of the risk function which has different units for different application. Specifically, for USPS and OCR the units is the probability (hence $\varepsilon = 0.001$ seems sufficient) while for LANDMARK the units is the percentage (hence $\varepsilon = 0.1$).

Table 1. Parameters of the benchmark datasets used in comparison

dataset	#training examples	#testing examples	#parameters	structure
USPS	7,291	2,007	2,570	flat
OCR	5,512	1,365	4,004	chain
LANDMARK	5,062	3,512	232,476	tree

Table 2. The number of effective iterations and the wall clock time needed to converge to ε-optimal solution for different setting of the regularization constant λ. The time is measured in seconds for USPS and in hours for OCR and LANDMARK. The last two rows correspond to accumulated time/iterations and the speedup achieved by the proposed solver FASOLE (speedup={CPA,BCFW,SDM}/FASOLE). The BCFW method had problems to converge for low values of λ hence we stopped BCFW when it used the same number of effective iterations as CPA (the slowest method which converged). These cases are marked with brackets. The best results, i.e. minimal number of iterations and the shortest time are printed in bold.

USPS

λ	CPA iter	CPA time	BCFW iter	BCFW time	SDM iter	SDM time	FASOLE iter	FASOLE time [s]
1.000	62	**3.6**	18	61.6	**5**	9.5	**5**	9.8
0.100	101	**6.0**	70	214.2	6	11.6	**5**	9.9
0.010	197	**10.9**	(197)	(538.0)	13	39.12	**5**	14.0
0.001	380	**26.7**	(380)	(1,018.8)	24	399.9	7	30.5
total	740	**47.3**	665	1,832.6	48	460.2	**22**	64.2
speedup	33.6	0.73	30.2	28.6	2.2	7.1	1	1

OCR

λ	CPA iter	CPA time	BCFW iter	BCFW time	SDM iter	SDM time	FASOLE iter	FASOLE time [h]
1.000	63	0.23	26	0.41	**8**	0.09	9	**0.04**
0.100	111	0.39	89	1.28	**10**	0.16	13	**0.07**
0.010	257	0.91	(257)	(3.55)	20	0.60	**16**	**0.15**
0.001	655	2.31	(655)	(9.47)	49	6.04	**20**	**0.43**
total	1086	3.83	1027	14.70	87	6.89	**58**	**0.70**
speedup	18.7	5.5	17.7	21.0	1.5	9.8	1	1

LANDMARK

λ	CPA iter	CPA time	BCFW iter	BCFW time	SDM iter	SDM time	FASOLE iter	FASOLE time [h]
10.00	93	2.43	4	0.18	8	0.32	**6**	**0.21**
1.00	165	4.71	20	0.78	11	0.40	**8**	**0.28**
0.10	261	7.82	(261)	(8.52)	30	1.42	**15**	**0.55**
0.01	446	12.20	(446)	(12.14)	131	12.30	**39**	**1.79**
total	965	27.25	731	21.62	180	14.42	**68**	**2.83**
speedup	14.2	9.6	10.8	7.6	2.6	5.1	1	1

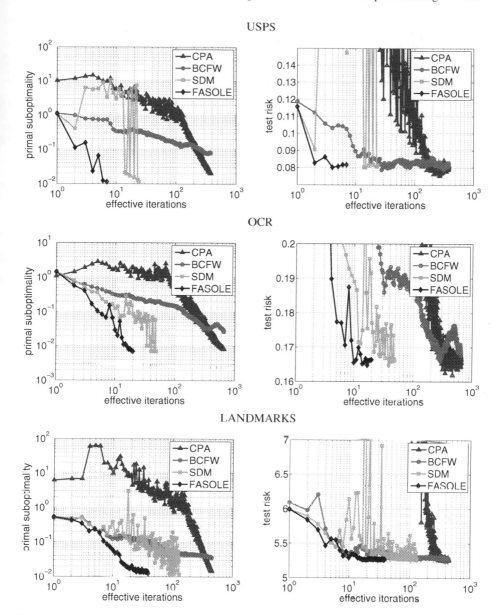

Fig. 1. Convergence curves for the regularization parameter λ which produces the minimal test risk, in particular, $\lambda = 0.01$ for USPS and LANDMARK and $\lambda = 0.001$ for OCR. The left column shows convergence in terms of the primal sub-optimality and the right column convergence of the test risk.

Discussion of the results. Table 2 summarizes the numbers of effective iteration and the wall clock time required by competing methods to achieve the target ε-precision. In sake of space we do not included the final objective because they are almost the same (must not differ more than ε).

The results show that if compared to CPA and BCFW, the proposed FASOLE requires order of magnitude less number of effective iterations on all three datasets. This leads to the speedup in terms of the wall clock time ranging from 5.5 to 21.0 on structured problems OCR and LANDMARKS. The CPA algorithm requires less time on the non-structured USPS benchmark because the risk function of multi-class SVM required by CPA can be evaluated by a single matrix multiplication (very effective in Matlab) unlike the SDA solvers (BCFW, SDM, FASOLE) which compute the loss for individual examples separately (calling function in a loop not effective in Matlab).

Comparison to SDM, the closest method to FASOLE, shows that SDM requires approximately only two times more effective iterations than FASOLE. However, SDM requires much more time in the inner loop, optimizing over buffered features, especially for low values of λ. This results to significantly slower convergence in terms of the wall clock time, specifically, the speedup achieved by FASOLE relatively to SDM ranges from 5.1 to 9.8.

These results show that FASOLE converges to the ε-optimal solution consistently faster on all problems which translate to significant speedup in terms of wall-clock time for the cases where the oracle calls are expensive. The advantage of the FASOLE is especially significant for small values of λ when the speed up is often an order of magnitude. Figure 1 (column 1) shows convergence of the competing methods in terms of the primal sub-optimality $(P(w) - P(w^*))/P(w^*)$, i.e. relative deviation of the primal objective from the optimal value, where $P(w^*)$ was replaced by the maximal achieved dual value. The figures show that FASOLE converges consistently faster from the beginning to the end. This implies that it beats the competing methods for the whole range of the precision parameter ε and not only the particular setting the results of which are reported in Table 2.

Some authors advocate that the optimality in terms of the objective function is not the primal goal and instead they propose to stop the algorithm based on monitoring the test risk. Therefore we also record convergence of the test risk which is presented in the second column of Figure 1. We see that the convergence of the test risk closely resembles the convergence of the objective function (compare the first and the second column).

6   Conclusions

In this paper we have proposed a variant of the sequential dual ascent algorithm for optimization of the SO-SVM dual. The proposed algorithm, called FASOLE, uses working set selection strategy which has been previously used for optimization of simpler QP tasks emerging in learning the two-class SVM. We provide a novel convergence analysis which guarantees that FASOLE finds the ε-optimal solution in $\mathcal{O}(\frac{1}{\varepsilon^2})$ time. The empirical comparison indicates that FASOLE consistently outperforms the existing state-of-the-art solvers for the SO-SVM achieving up to an order of magnitude speedups while obtaining the same precise solution.

Acknowledgment. The author was supported by the project ERC-CZ LL1303.

References

1. Balamurugan, P., Shevade, S.K., Sundararajan, S., Sathiya Keerthi, S.: A sequential dual method for structural SVMs. In: SIAM Conference on Data Mining (2011)
2. Chang, C.-C., Lin, C.-J.: LIBSVM: A library for support vector machines. ACM Transactions on Intelligent Systems and Technology 2, 27:1–27:27 (2011)
3. Fan, R., Chen, P.H., Lin, C.J.: Working set selection using second order information for training support vector machines. JMLR 6, 1889–1918 (2005)
4. Franc, V.: Optimization Algorithms for Kernel Methods. PhD thesis, Center for Machine Perception, K13133 FEE Czech Technical University, Prague, Czech Republic (2005)
5. Franc, V., Hlaváč, V.: Simple solvers for large quadratic programming tasks. In: Kropatsch, W.G., Sablatnig, R., Hanbury, A. (eds.) DAGM 2005. LNCS, vol. 3663, pp. 75–84. Springer, Heidelberg (2005)
6. Glasmachers, T., Igel, C.: Maximum-gain working set selection for SVMs. Journal of Machine Learning Research (2006)
7. Platt, J.C.: Fast training of support vector machines using sequential minimal optimization. In: Advances in Kernel Methods: Support Vector Machines. MIT Press (1998)
8. Joachims, T., Finley, T., Yu, C.: Cutting-plane training of structural SVMs. Machine Learning 77(1), 27–59 (2009)
9. Lacoste-Julien, S., Jaggi, M., Schmidt, M., Pletscher, P.: Block-coordinate frank-wolfe optimization for structural SVMs. In: ICML (2013)
10. LeCun, Y., Boser, B., Denker, J.S., Henderson, D., Howard, R.E., Hubbard, W., Jackel, L.D.: Handwritten digit recognition with a back-propagation network. In: Lisboa, P.G.J. (ed.) Neural Networks, Current Applications. Chappman and Hall (1992)
11. Collins, M., Globerson, A., Koo, T., Carreras, X., Bartlett, P.L.: Exponentiated gradient algorithms for conditional random fileds and max-margin markov networks. JMLR 9, 1775–1822 (2008)
12. Uricar, M., Franc, V., Hlavac, V.: Detector of facial landmarks learned by the structured output SVM. In: VISAPP (2012)
13. Ratliff, N., Bagnell, J.A., Zinkevich, M.: (online)subgradient methods for structured predications. In: AISTATS (2007)
14. Shalev-Shwartz, S., Singer, Y., Sebro, N., Cotter, A.: Pegasos: primal estimated sub-gradient solver for svm. In: ICML (2007)
15. Shalev-Shwartz, S., Zhang, T.: Trading accuracy for sparsity in optimization problems with sparsity constraints. SIAM Journal on Optimization 20(6), 2807–2832 (2010)
16. Takahashi, N., Nishi, T.: Rigorous proof of termination of smo algorithm for support vector machines. IEEE Transactions on Neural Networks (2005)
17. Taskar, B., Guestrin, C., Koller, D.: Max-margin markov networks. In: NIPS, vol. 16. MIT Press (2003)
18. Teo, C.H., Vishwanathan, S.V.N., Smola, A.J., Le., Q.V.: Bundle methods for regularized risk minimization. JMLR (2010)
19. Tsochantaridis, I., Joachims, T., Hoffman, T., Altun, Y.: Large margin methods for structured and interdependent output variables. JMLR 6, 1453–1484 (2005)

Neutralized Empirical Risk Minimization with Generalization Neutrality Bound

Kazuto Fukuchi and Jun Sakuma

University of Tsukuba, 1-1-1 Tennodai, Tsukuba, Ibaraki, 305-8577 Japan
kazuto@mdl.cs.tsukuba.ac.jp, jun@cs.tsukuba.ac.jp

Abstract. Currently, machine learning plays an important role in the lives and individual activities of numerous people. Accordingly, it has become necessary to design machine learning algorithms to ensure that discrimination, biased views, or unfair treatment do not result from decision making or predictions made via machine learning. In this work, we introduce a novel empirical risk minimization (ERM) framework for supervised learning, neutralized ERM (NERM) that ensures that any classifiers obtained can be guaranteed to be neutral with respect to a viewpoint hypothesis. More specifically, given a viewpoint hypothesis, NERM works to find a target hypothesis that minimizes the empirical risk while simultaneously identifying a target hypothesis that is neutral to the viewpoint hypothesis. Within the NERM framework, we derive a theoretical bound on empirical and generalization neutrality risks. Furthermore, as a realization of NERM with linear classification, we derive a max-margin algorithm, neutral support vector machine (SVM). Experimental results show that our neutral SVM shows improved classification performance in real datasets without sacrificing the neutrality guarantee.

Keywords: neutrality, discrimination, fairness, classification, empirical risk minimization, support vector machine.

1 Introduction

Within the framework of empirical risk minimization (ERM), a supervised learning algorithm seeks to identify a hypothesis f that minimizes empirical risk with respect to given pairs of *input* x and *target* y. Given an input x without the target value, hypothesis f provides a prediction for the target of x as $y = f(x)$. In this study, we add a new element, *viewpoint hypothesis* g, to the ERM framework. Similar to hypothesis f, which is given an input x without the viewpoint value, viewpoint hypothesis g provides a prediction for the viewpoint of the x as $v = g(x)$. In order to distinguish between the two different hypotheses, f and g, f will be referred to as the *target hypothesis*. Examples of the viewpoint hypothesis are given with the following specific applications.

With this setup in mind, we introduce our novel framework for supervised learning, *neutralized ERM* (NERM). Intuitively, we say that a target hypothesis is neutral to a given viewpoint hypothesis if there is low correlation between

T. Calders et al. (Eds.): ECML PKDD 2014, Part I, LNCS 8724, pp. 418–433, 2014.

the target $f(x)$ and viewpoint $g(x)$. The objective of NERM is to find a target hypothesis f that minimizes empirical risks while simultaneously remaining neutral to the viewpoint hypothesis g. The following two application scenarios motivate NERM.

Application 1 (Filter bubble). Suppose an article recommendation service provides personalized article distribution. In this situation, by taking a user's access logs and profile as input x, the service then predicts that user's preference with respect to articles using supervised learning as $y = f(x)$ (target hypothesis). Now, suppose a user strongly supports a policy that polarizes public opinion (such as nuclear power generation or public medical insurance). Furthermore, suppose the user's opinion either for or against the particular policy can be precisely predicted by $v = g(x)$ (viewpoint hypothesis). Such a viewpoint hypothesis can be readily learned by means of supervised learning, given users' access logs and profiles labeled with the parties that the users support. In such situations, if predictions by the target hypothesis f and viewpoint hypothesis g are closely correlated, recommended articles are mostly dominated by articles supportive of the policy, which may motivate the user to adopt a biased view of the policy [12]. This problem is referred to as the *filter bubble* [10]. Bias of this nature can be avoided by training the target hypothesis so that the predicted target is independent of the predicted viewpoint.

Application 2 (Anti-discrimination). Now, suppose a company wants to make hiring decisions using information collected from job applicants, such as age, place of residence, and work experience. While taking such information as input x toward the hiring decision, the company also wishes to predict the potential work performance of job applicants via supervised learning, as $y = f(x)$ (target hypothesis). Now, although the company does not collect applicant information on sensitive attributes such as race, ethnicity, or gender, suppose such sensitive attributes can be sufficiently precisely predicted from an analysis of the non-sensitive applicant attributes, such as place of residence or work experience, as $v = g(x)$ (viewpoint hypothesis). Again, such a viewpoint hypothesis can be readily learned by means of supervised learning by collecting moderate number of labeled examples. In such situations, if hiring decisions are made by the target hypothesis f and if there is a high correlation with the sensitive attribute predictions $v = g(x)$, those decisions might be deemed discriminatory [11]. In order to avoid this, the target hypothesis should be trained so that the decisions made by f are not highly dependent on the sensitive attributes predicted by g. Thus, this problem can also be interpreted as an instance of NERM.

The neutrality of a target hypothesis should not only be guaranteed for given samples, but also for unseen samples. In the article recommendation example, the recommendation system is trained using the user's past article preferences, whereas recommendation neutralization is needed for unread articles. In the hiring decision example, the target hypothesis is trained with information collected from the past histories of job applicants, but the removal of discrimination from hiring decisions is the desired objective.

Given a viewpoint hypothesis, we evaluate the degree of neutrality of a target hypothesis with respect to given and unseen samples as *empirical neutrality risk* and *generalization neutrality risk*, respectively. The goal of NERM is to show that the generalization risk is theoretically bounded in the same manner as the standard ERM [2,1,6], and, simultaneously, to show that the generalization neutrality risk is also bounded with respect to given viewpoint hypothesis.

Our Contribution. The contribution of this study is three-fold. First, we introduce our novel NERM framework in which, assuming the target hypothesis and viewpoint hypothesis output binary predictions, it is possible to learn a target hypothesis that minimizes empirical and empirically neutral risks. Given samples and a viewpoint hypothesis, NERM is formulated as a convex optimization problem where the objective function is the linear combination of two terms, the empirical risk term penalizing the target hypothesis prediction error and the neutralization term penalizing correlation between the target and the viewpoint. The predictive performance and neutralization can be balanced by adjusting a parameter, referred to as the neutralization parameter. Because of its convexity, the optimality of the resultant target hypothesis is guaranteed (in Section 4).

Second, we derive a bound on empirical and generalization neutrality risks for NERM. We also show that the bound on the generalization neutrality risk can be controlled by the neutralization parameter (in Section 5). As discussed in Section 2, a number of diverse algorithms targeting the neutralization of supervised classifications have been presented. However, none of these have given theoretical guarantees on generalization neutrality risk. To the best of our knowledge, this is the first study that gives a bound on generalization neutrality risk.

Third, we present a specific NERM learning algorithm for neutralized linear classification. The derived learning algorithm is interpreted as a *support vector machine* (SVM) [14] variant with a neutralization guarantee. The kernelized version of the neutralization SVM is also derived from the dual problem (in Section 6).

2   Related Works

Within the context of removing discrimination from classifiers, the need for a neutralization guarantee has already been extensively studied. Calders & Verwer [4] pointed out that elimination of sensitive attributes from training samples does not help to remove discrimination from the resultant classifiers. In the hiring decision example, even if we assume that a target hypothesis is trained with samples that have no race or ethnicity attributes, hiring decisions may indirectly correlate with race or ethnicity through addresses if there is a high correlation between an individual's address and his or her race or ethnicity. This indirect effect is referred to as a *red-lining effect* [3].

Calders & Verwer [4] proposed the Calders–Verwer 2 Naïve Bayes method (CV2NB) to remove the red-lining effect from the Naïve Bayes classifier. The CV2NB method is used to evaluate the Calders–Verwer (CV) score, which is a measure that evaluates discrimination of naïve Bayes classifiers. The CV2NB method learns the naïve Bayes classifier in a way that ensures the CV score is made as small as possible. Based on this idea, various situations where discrimination can occur have been discussed in other studies [16,7]. Since a CV score is empirically measured with the given samples, naïve Bayes classifiers with low CV scores result in less discrimination for those samples. However, less discrimination is not necessarily guaranteed for unseen samples. Furthermore, the CV2NB method is designed specifically for the naïve Bayes model and does not provide a general framework for anti-discrimination learning.

Zemel et al. [15] introduced the learning fair representations (LFR) model for preserving classification fairness. LFR is designed to provide a map, from inputs to prototypes, that guarantees the classifiers that are learned with the prototypes will be fair from the standpoint of statistical parity. Kamishima et al. [8] presented a prejudice remover regularizer (PR) for fairness-aware classification that is formulated as an optimization problem in which the objective function contains the loss term and the regularization term that penalizes mutual information between the classification output and the given sensitive attributes. The classifiers learned with LFR or PR are empirically neutral (i.e., fair or less discriminatory) in the sense of statistical parity or mutual information, respectively. However, no theoretical guarantees related to neutrality for unseen samples have been established for these methods.

Fukuchi et al. [5] introduced η-neutrality, a framework for neutralization of probability models with respect to a given viewpoint random variable. Their framework is based on maximum likelihood estimation and neutralization is achieved by maximizing likelihood estimation while setting constraints to enforce η-neutrality. Since η-neutrality is measured using the probability model of the viewpoint random variable, the classifier satisfying η-neutrality is expected to preserve neutrality for unseen samples. However, this method also fails to provide a theoretical guarantee for generalization neutrality.

LFR, PR, and η-neutrality incorporate a hypothesis neutrality measure into the objective function in the form of a regularization term or constraint; however, these are all non-convex. One of the reasons why generalization neutrality is not theoretically guaranteed for these methods is the non-convexity of the objective functions. In this study, we introduce a convex surrogate for a neutrality measure in order to provide a theoretical analysis of generalization neutrality.

3   Empirical Risk Minimization

Let X and Y be an input space and a target space, respectively. We assume $D_n = \{(x_i, y_i)\}_{i=1}^n \in Z^n$ ($Z = X \times Y$) to be a set of i.i.d. samples drawn from an unknown probability measure ρ over $(Z, \mathcal{Z})$. We restrict our attention to binary classification, $Y = \{-1, 1\}$, but our method can be expanded to handle multi-valued classification via a straightforward modification. Given the i.i.d. samples, the supervised learning objective is to construct a target hypothesis $f : X \to \mathbb{R}$ where the hypothesis is chosen from a class of measurable functions $f \in \mathcal{F}$. We assume that classification results are given by sgn $\circ$ $f(x)$, that is,

$y = 1$ if $f(x) > 0$; otherwise $y = -1$. Given a loss function $\ell : Y \times \mathbb{R} \to \mathbb{R}^+$, the generalization risk is defined by

$$R(f) = \int \ell(y, f(x)) d\rho.$$

Our goal is to find $f^* \in \mathcal{F}$ that minimizes the generalization risk $R(f)$. In general, ρ is unknown and the generalization risk cannot be directly evaluated. Instead, we minimize the empirical loss with respect to sample set D_n

$$R_n(f) = \frac{1}{n} \sum_{i=1}^{n} \ell(y_i, f(x_i)).$$

This is referred to as *empirical risk minimization (ERM)*.

In order to avoid overfitting, a regularization term $\Omega : \mathcal{F} \to \mathbb{R}^+$ is added to the empirical loss by penalizing complex hypotheses. Minimization of the empirical loss with a regularization term is referred to as *regularized ERM (RERM)*.

3.1   Generalization Risk Bound

Rademacher Complexity measures the complexity of a hypothesis class with respect to a probability measure that generates samples. The Rademacher Complexity of class $\mathcal{F}$ is defined as

$$\mathcal{R}_n(\mathcal{F}) = \mathrm{E}_{D_n, \boldsymbol{\sigma}} \left[\sup_{f \in \mathcal{F}} \frac{1}{n} \sum_{i=1}^{n} \sigma_i f(x_i) \right]$$

where $\boldsymbol{\sigma} = (\sigma_1, ..., \sigma_n)^T$ are independent random variables such that $\Pr(\sigma_i = 1) = \Pr(\sigma_i = -1) = 1/2$. Bartlett & Mendelson [2] derived a generalization loss bound using the Rademacher complexity as follows:

Theorem 1 (Bartlett & Mendelson [2]). *Let ρ be a probability measure on $(Z, \mathcal{Z})$ and let $\mathcal{F}$ be a set of real-value functions defined on X, with $\sup\{|f(x)| : f \in \mathcal{F}\}$ finite for all $x \in X$. Suppose that $\phi : \mathbb{R} \to [0, c]$ satisfies and is Lipschitz continuous with constant L_ϕ. Then, with probability at least $1 - \delta$, every function in $\mathcal{F}$ satisfies*

$$R(f) \leq R_n(f) + 2 L_\phi \mathcal{R}_n(\mathcal{F}) + c \sqrt{\frac{\ln(2/\delta)}{2n}}.$$

4   Generalization Neutrality Risk and Empirical Neutrality Risk

In this section, we introduce the viewpoint hypothesis into the ERM framework and define a new principle of supervised learning, neutralized ERM (NERM), with the notion of *generalization neutrality risk*. Convex relaxation of the neutralization measure is also discussed in this section.

4.1 +1/−1 Generalization Neutrality Risk

Suppose a measurable function $g : X \to \mathbb{R}$ is given. The prediction of g is referred to as the *viewpoint* and g is referred to as the *viewpoint hypothesis*. We say the target hypothesis f is neutral to the viewpoint hypothesis g if the target predicted by the learned target hypothesis f and the viewpoint predicted by the viewpoint hypothesis g are not mutually correlating. In our setting, we assume the target hypothesis f and viewpoint hypothesis g to give binary predictions by sgn $\circ f$ and sgn $\circ g$, respectively. Given a probability measure ρ and a viewpoint hypothesis g, the neutrality of the target hypothesis f is defined by the correlation between sgn $\circ f$ and sgn $\circ g$ over ρ. If $f(x)g(x) > 0$ holds for multiple samples, then the classification sgn $\circ f$ closely correlates to the viewpoint sgn $\circ g$. On the other hand, if $f(x)g(x) \leq 0$ holds for multiple samples, then the classification sgn $\circ f$ and the viewpoint sgn $\circ g$ are inversely correlating. Since we want to suppress both correlations, our neutrality measure is defined as follows:

Definition 1 (+1/-1 Generalization Neutrality Risk). *Let $f \in \mathcal{F}$ and $g \in \mathcal{G}$ be a target hypothesis and viewpoint hypothesis, respectively. Let ρ be a probability measure over $(Z, \mathcal{Z})$. Then, the $+1/-1$ generalization neutrality risk of target hypothesis f with respect to viewpoint hypothesis g over ρ is defined by*

$$C_{\text{sgn}}(f, g) = \left| \int \text{sgn}(f(x)g(x))d\rho \right|.$$

When the probability measure ρ cannot be obtained, a $+1/-1$ generalization neutrality risk $C_{\text{sgn}}(f, g)$ can be empirically evaluated with respect to the given samples D_n.

Definition 2 (+1/−1 Empirical Neutrality Risk). *Suppose that $D_n = \{(x_i, y_i)\}_{i=1}^n \in Z^n$ is a given sample set. Let $f \in \mathcal{F}$ and $g \in \mathcal{G}$ be the target hypothesis and the viewpoint hypothesis, respectively. Then, the $+1/-1$ empirical neutrality risk of target hypothesis f with respect to viewpoint hypothesis g is defined by*

$$C_{n,\text{sgn}}(f, g) = \frac{1}{n} \left| \sum_{i=1}^n \text{sgn}(f(x_i)g(x_i)) \right|. \tag{1}$$

4.2 Neutralized Empirical Risk Minimization (NERM)

With the definition of neutrality risk, a novel framework, the *Neutralized Empirical Risk Minimization* (NERM) is introduced. NERM is formulated as minimization of the empirical risk and empirical $+1/−1$ neutrality risk:

$$\min_{f \in \mathcal{F}} R_n(f) + \Omega(f) + \eta C_{n,\text{sgn}}(f, g). \tag{2}$$

where $\eta > 0$ is the neutralization parameter which determines the trade-off ratio between the empirical risk and the empirical neutrality risk.

4.3   Convex Relaxation of $+1/-1$ Neutrality Risk

Unfortunately, the optimization problem defined by Eq (2) cannot be efficiently solved due to the nonconvexity of Eq (1). Therefore, we must first relax the absolute value function of $C_{\mathrm{sgn}}(f, g)$ into the max function. Then, we introduce a convex surrogate of the sign function, yielding a convex relaxation of the $+1/-1$ neutrality risk.

By letting I be the indicator function, the $+1/-1$ generalization neutrality risk can be decomposed into two terms:

$$C_{\mathrm{sgn}}(f, g) = \left| \underbrace{\int I(\mathrm{sgn}(g(x)) = \mathrm{sgn}(f(x))) d\rho}_{\text{prob. that } f \text{ agrees with } g} - \underbrace{\int I(\mathrm{sgn}(g(x)) \neq \mathrm{sgn}(f(x))) d\rho}_{\text{prob. that } f \text{ disagrees with } g} \right|$$
$$:= |C_{\mathrm{sgn}}^{+}(f, g) - C_{\mathrm{sgn}}^{-}(f, g)| \tag{3}$$

The upper bound of the $+1/-1$ generalization neutrality risk $C_{\mathrm{sgn}}(f, g)$ is tight if $C_{\mathrm{sgn}}^{+}(f, g)$ and $C_{\mathrm{sgn}}^{-}(f, g)$ are close. Thus, the following property is derived.

Proposition 1. *Let $C_{\mathrm{sgn}}^{+}(f, g)$ and $C_{\mathrm{sgn}}^{-}(f, g)$ be functions defined in Eq (3). For any $\eta \in [0.5, 1]$, if*

$$C_{\mathrm{sgn}}^{\max}(f, g) := \max(C_{\mathrm{sgn}}^{+}(f, g), C_{\mathrm{sgn}}^{-}(f, g)) \leq \eta,$$

then

$$C_{\mathrm{sgn}}(f, g) = |C_{\mathrm{sgn}}^{+}(f, g) - C_{\mathrm{sgn}}^{-}(f, g)| \leq 2\eta - 1.$$

Proposition 1 shows that $C_{\mathrm{sgn}}^{\max}(f, g)$ can be used as the generalization neutrality risk instead of $C_{\mathrm{sgn}}(f, g)$. Next, we relax the indicator function contained in $C_{\mathrm{sgn}}^{\pm}(f, g)$.

Definition 3 (Relaxed Convex Generalization Neutrality Risk). *Let $f \in \mathcal{F}$ and $g \in \mathcal{G}$ be a classification hypothesis and viewpoint hypothesis, respectively. Let ρ be a probability measure over $(Z, \mathcal{Z})$. Let $\psi : \mathbb{R} \to \mathbb{R}^{+}$ be a convex function and*

$$C_{\psi}^{\pm}(f, g) = \int \psi(\pm g(x) f(x)) d\rho.$$

Then, the relaxed convex generalization neutrality risk of f with respect to g is defined by

$$C_{\psi}(f, g) = \max(C_{\psi}^{+}(f, g), C_{\psi}^{-}(f, g)).$$

The empirical evaluation of relaxed convex generalization neutrality risk is defined in a straightforward manner.

Definition 4 (Convex Relaxed Empirical Neutrality Risk). *Suppose $D_n = \{(x_i, y_i)\}_{i=1}^{n} \in Z^n$ to be a given sample set. Let $f \in \mathcal{F}$ and $g \in \mathcal{G}$ be the target hypothesis and the viewpoint hypothesis, respectively. Let $\psi : \mathbb{R} \to \mathbb{R}^{+}$ be a* *convex function and*

$$C_{n,\psi}^{\pm}(f,g) = \frac{1}{n} \sum_{i=1}^{n} \psi(\pm g(x_i)f(x_i)).$$

Then, relaxed convex empirical neutrality risk *of f with respect to g is defined by*

$$C_{n,\psi}(f,g) = \max(C_{n,\psi}^{+}(f,g), C_{n,\psi}^{-}(f,g)).$$

$C_{n,\psi}^{\pm}(f,g)$ is convex because it is a summation of the convex function ψ. Noting that $\max(f_1(x), f_2(x))$ is convex if f_1 and f_2 are convex, $C_{n,\psi}(f,g)$ is convex as well.

4.4   NERM with Relaxed Convex Empirical Neutrality Risk

Finally, we derive the convex formulation of NERM with the relaxed convex empirical neutrality risk as follows:

$$\min_{f \in \mathcal{F}} R_n(f) + \Omega(f) + \eta C_{n,\psi}(f,g). \tag{4}$$

If the regularized empirical risk is convex, then this is a convex optimization problem.

The neutralization term resembles the regularizer term in the formulation sense. Indeed, the neutralization term is different from the regularizer in that it is dependent on samples. We can interpret the regularizer as a prior structural information of the model parameters, but we cannot interpret the neutralization term in the same way due to its dependency on samples. PR and LFR have similar neutralization terms in the sense of adding the neutrality risk to objective function, and neither can be interpreted as a prior structural information. Instead, the neutralization term can be interpreted as a prior information of *data*. The notion of a prior data information is relevant to *transfer learning* [9], which aims to achieve learning dataset information from other datasets. However, further research on the relationships between the neutralization and transfer learning will be left as an area of future work.

5   Generalization Neutrality Risk Bound

In this section, we show theoretical analyses of NERM generalization neutrality risk and generalization risk. First, we derive a probabilistic uniform bound of the generalization neutrality risk for any $f \in \mathcal{F}$ with respect to the empirical neutrality risk $C_{n,\psi}(f,g)$ and the Rademacher complexity of $\mathcal{F}$. Then, we derive a bound on the generalization neutrality risk of the optimal hypothesis.

For convenience, we introduce the following notation. For a hypothesis class $\mathcal{F}$ and constant $c \in \mathbb{R}$, we denote $-\mathcal{F} = \{-f : f \in \mathcal{F}\}$ and $c\mathcal{F} = \{cf : f \in \mathcal{F}\}$. For any function $\phi : \mathbb{R} \to \mathbb{R}$, let $\phi \circ \mathcal{F} = \{\phi \circ f : f \in \mathcal{F}\}$. Similarly, for any function $g : X \to \mathbb{R}$, let $g\mathcal{F} = \{h : f \in \mathcal{F}, h(x) = g(x)f(x) \ \forall x \in X\}$.

5.1   Uniform Bound of Generalization Neutrality Risk

A probabilistic uniform bound on $C_\psi(f,g)$ for any hypothesis $f \in \mathcal{F}$ is derived as follows.

Theorem 2. *Let $C_\psi(f,g)$ and $C_{n,\psi}(f,g)$ be the relaxed convex generalization neutrality risk and the relaxed convex empirical neutrality risk of $f \in \mathcal{F}$ w.r.t. $g \in \mathcal{G}$. Suppose that $\psi : \mathbb{R} \to [0,c]$ satisfies and is Lipschitz continuous with constant L_ψ. Then, with probability at least $1 - \delta$, every function in $\mathcal{F}$ satisfies*

$$C_\psi(f,g) \leq C_{n,\psi}(f,g) + 2L_\psi \mathcal{R}_n(g\mathcal{F}) + c\sqrt{\frac{\ln(2/\delta)}{2n}}.$$

As proved by Theorem 2, $C_\psi(f,g) - C_{n,\psi}(f,g)$, the approximation error of the generalization neutrality risk is uniformly upper-bounded by the Rademacher complexity of hypothesis classes $g\mathcal{F}$ and $O(\sqrt{\ln(1/\delta)/n})$, where δ is the confidence probability and n is the sample size.

5.2   Generalization Neutrality Risk Bound for NERM Optimal Hypothesis

Let $\hat{f} \in \mathcal{F}$ be the optimal hypothesis of NERM. We derive the bounds on the empirical and generalization neutrality risks achieved by $\hat{f}$ under the following conditions:

1. Hypothesis class $\mathcal{F}$ includes a hypothesis f_0 s.t. $f_0(x) = 0$ for $\forall x$, and   (A)
2. the regularization term of f_0 is $\Omega(f_0) = 0$.

The conditions are relatively moderate. For example, consider the linear hypothesis $f(\boldsymbol{x}) = \boldsymbol{w}^T\boldsymbol{x}$ and $\Omega(f) = \|\boldsymbol{w}\|_2^2$ (ℓ_2^2 norm) and let $W \subseteq \mathbb{R}^D$ be a class of the linear hypothesis. If $\boldsymbol{0} \in W$, the two conditions above are satisfied. Assuming that $\mathcal{F}$ satisfies these conditions, the following theorem provides the bound on the generalization neutrality risk.

Theorem 3. *Let $\hat{f}$ be the optimal target hypothesis of NERM, where the viewpoint hypothesis is $g \in \mathcal{G}$ and the neutralization parameter is η. Suppose that $\psi : \mathbb{R} \to [0,c]$ satisfies and is Lipschitz continuous with constant L_ψ. If conditions (A) are satisfied, then with probability at least $1 - \delta$,*

$$C_\psi(\hat{f},g) \leq \psi(0) + \phi(0)\frac{1}{\eta} + 2L_\psi \mathcal{R}_n(g\mathcal{F}) + c\sqrt{\frac{\ln(2/\delta)}{2n}}.$$

For the proof of Theorem 3, we first derive the upper bound of the empirical neutrality risk of $\hat{f}$.

Corollary 1. *If the conditions (A) are satisfied, then the empirical relaxed convex neutrality risk of $\hat{f}$ is bounded by*

$$C_{n,\psi}(\hat{f},g) \leq \psi(0) + \phi(0)\frac{1}{\eta}.$$

Theorem 3 is immediately obtained from Theorem 2 and Corollary 1.

5.3   Generalization Risk Bound for NERM

In this section, we compare the generalization risk bound of NERM with that of a regular ERM. Theorem 1 denotes a uniform bound of the generalization risk. This theorem holds with the hypotheses which are optimal in terms of NERM and ERM. However, the hypotheses which are optimal in terms of NERM and ERM have different empirical risk values. The empirical risk of NERM is greater than that of ERM since NERM has a term that penalizes less neutrality. More precisely, if we let $\bar{f}$ be the optimal hypothesis in term of ERM, we have

$$R_n(\hat{f}) - R_n(\bar{f}) \geq 0. \tag{5}$$

The reason for this is that empirical risk of any other hypothesis is greater than one of $\bar{f}$ since $\bar{f}$ minimizes empirical risk. Furthermore, due to $\hat{f}$ is a minimizer of $R_n(f) + \eta C_{n,\phi}(f, g)$, we have

$$R_n(\hat{f}) + \eta C_{n,\phi}(\hat{f}, g) - R_n(\bar{f}) - \eta C_{n,\phi}(\bar{f}, g) \leq 0$$
$$R_n(\hat{f}) - R_n(\bar{f}) \leq \eta(C_{n,\phi}(\bar{f}, g) - C_{n,\phi}(\hat{f}, g)). \tag{6}$$

Since the left term of this inequality is greater than zero due to Eq (5), the empirical risk becomes greater if the empirical neutrality risk becomes lower.

6   Neutral SVM

6.1   Primal Problem

SVMs [14] are a margin-based supervised learning method for binary classification. The algorithm of SVMs can be interpreted as minimization of the empirical risk with regularization term, which follows the RERM principle. In this section, we introduce a SVM variant that follows the NERM principle.

The soft-margin SVM employs the linear classifier $f(\boldsymbol{x}) = \boldsymbol{w}^T\boldsymbol{x} + b$ as the target hypothesis. In the objective function, the hinge loss is used for the loss function, as $\phi(yf(x)) = \max(0, 1 - yf(x))$, and the ℓ_2 norm is used for the regularization term, $\Omega(f) = \lambda\|f\|_2^2/2n$, where $\lambda > 0$ denotes the regularization parameter. In our SVM in NERM, referred to as the neutral SVM, the loss function and regularization term are the same as in the soft-margin SVM. For a surrogate function of the neutralization term, the hinge loss $\psi(\pm g(x)f(x)) = \max(0, 1 \mp g(x)f(x))$ was employed. Any hypothesis can be used for the viewpoint hypothesis. Accordingly, following the NERM principle defined in Eq (4), the neutral SVM is formulated by

$$\min_{\boldsymbol{w},b} \sum_{i=1}^{n} \max(0, 1 - y_i(\boldsymbol{w}^T\boldsymbol{x}_i + b)) + \frac{\lambda}{2}\|\boldsymbol{w}\|_2^2 + \eta C_{n,\psi}(\boldsymbol{w}, b, g), \tag{7}$$

where

$$C_{n,\psi}(\boldsymbol{w}, b, g) = \max(C_{n,\psi}^+(\boldsymbol{w}, b, g), C_{n,\psi}^-(\boldsymbol{w}, b, g)),$$

$$C_{n,\psi}^{\pm}(\boldsymbol{w}, b, g) = \sum_{i=1}^{n} \max(0, 1 \mp g(\boldsymbol{x}_i)(\boldsymbol{w}^T\boldsymbol{x}_i + b)).$$

Since the risk, regularization, and neutralization terms are all convex, the objective function of the neutral SVM is convex. The primal form can be solved by applying the subgradient method [13] to Eq (7).

6.2   Dual Problem and Kernelization

Next, we derive the dual problems of the problem of Eq (7), from which the neutral SVM kernelization is naturally derived. First, we introduce slack variables $\xi, \xi^{\pm}$, and ζ into Eq (7) to represent the primal problem:

$$\min_{\substack{\boldsymbol{w},b, \\ \boldsymbol{\xi},\boldsymbol{\xi}^{\pm},\zeta}} \sum_{i=1}^{n} \xi_i + \frac{\lambda}{2}\|\boldsymbol{w}\|_2^2 + \eta\zeta \tag{8}$$

$$\text{sub to } \sum_{i=1}^{n} \xi_i^+ \leq \zeta, \sum_{i=1}^{n} \xi_i^- \leq \zeta, 1 - y_i(\boldsymbol{w}^T\boldsymbol{x}_i + b) \leq \xi_i,$$

$$1 - v_i(\boldsymbol{w}^T\boldsymbol{x}_i + b) \leq \xi_i^+, 1 + v_i(\boldsymbol{w}^T\boldsymbol{x}_i + b) \leq \xi_i^-,$$

$$\xi_i \geq 0, \xi_i^+ \geq 0, \xi_i^- \geq 0, \zeta \geq 0$$

where slack variables ξ_i, ξ_i^+, and ξ_i^- denote measures of the degree of misclassification, correlation, and inverse correlation, respectively. The slack variable ζ, derived from max function in $C_{n,\psi}(\boldsymbol{w}, b, g)$, measures the imbalance of the degree of correlation and inverse correlation. From the Lagrange relaxation of the primal problem Eq (8), the dual problem is derived as

$$\max_{\boldsymbol{\alpha},\boldsymbol{\beta}^{\pm}} \lambda \sum_{i=1}^{n} b_i - \frac{1}{2} \sum_{i}^{n} \sum_{j}^{n} a_i a_i k(x_i, x_j) \tag{9}$$

$$\text{sub to } \sum_{i}^{n} a_i = 0, 0 \leq \alpha_i \leq 1, 0 \leq \beta_i^{\pm}, \beta_i^+ + \beta_i^- \leq \eta$$

where $b_i = \alpha_i + \beta_i^+ + \beta_i^-, a_i = \alpha_i y_i + \beta_i^+ v_i - \beta_i^- v_i$. As seen in the dual problem, the neutral SVM is naturally kernelized with kernel function $\boldsymbol{x}_i^T\boldsymbol{x}_j = k(x_i, x_j)$. The derivation of the dual problem and kernelization thereof is described in the supplemental document in detail. The optimization of Eq (9) is an instance of *quadratic programming (QP)* that can be solved by general QP solvers, although it does not scale well with large samples due to its large memory consumption. In the supplemental documentation, we also show the applicability of the well-known *sequential minimal optimization* technique to our neutral SVM.

7   Experiments

In this section, we present experimental evaluation of our neutral SVM for synthetic and real datasets. In the experiments with synthetic data, we experimentally evaluate the change of generalization risk and generalization neutrality risk according to the number of samples, in which their relations are described in Theorem 2. In the experiments for real datasets, we compare our method with CV2NB [4], PR [8] and η-neutral logistic regression (ηLR for short) [5] in terms of risk and neutrality risk. The CV2NB method learns a naïve Bayes model, and then modifies the model parameters so that the resultant CV score approaches zero. The PR and ηLR are based on maximum likelihood estimation of a logistic regression (LR) model. These methods have two parameters, the regularizer parameter λ, and the neutralization parameter η. The PR penalizes the objective function of the LR model with mutual information. The ηLR performs maximum likelihood estimation of the LR model while enforcing η-neutrality as constraints. The neutralization parameter of neutral SVM and PR balances risk minimization and neutrality maximization. Thus, it can be tuned in the same manner used to determine the regularizer parameter. The neutralization parameter of ηLR determines the region of the hypothesis in which the hypotheses are regarded as neutral. The tuning strategy of the regularizer parameter and neutralization parameter are different in all these methods. We determined the neutralization parameter tuning range of these methods via preliminary experiments.

7.1 Synthetic Dataset

In order to investigate the change of generalization neutrality risk with sample size n, we performed our neutral SVM experiments for a synthetic dataset. First, we constructed the input $\boldsymbol{x}_i \subset \mathbb{R}^{10}$ with the vector being sampled from the uniform distribution over $[-1,1]^{10}$. The target y_i corresponding to the input $\boldsymbol{x}_i$ is generated as $y_i = \mathrm{sgn}(\boldsymbol{w}_y^T \boldsymbol{x}_i)$ where $\boldsymbol{w}_y \subset \mathbb{R}^{10}$ is a random vector drawn from the uniform distribution over $[-1,1]^{10}$. Noises are added to labels by inverting the label with probability $1/(1 + \exp(-100|\boldsymbol{w}_y^T \boldsymbol{x}_i|))$. The inverting label probability is small if the input $\boldsymbol{x}_i$ is distant from a plane $\boldsymbol{w}_y^T \boldsymbol{x} = 0$. The viewpoint v_i corresponding to the input $\boldsymbol{x}_i$ is generated as $v_i = \mathrm{sgn}(\boldsymbol{w}_v^T \boldsymbol{x}_i)$, where the first element of $\boldsymbol{w}_v$ is set as $w_{v,1} = w_{y,1}$ and the rest of elements are drawn from the uniform distribution over $[-1,1]^9$. Noises are added in the same manner as the target. The equality of the first element of $\boldsymbol{w}_y$ and $\boldsymbol{w}_v$ leads to correlation between y_i and v_i. Set the regularizer parameter as $\lambda = 0.05n$. The neutralization parameter was varied as $\eta \in \{0.1, 1.0, 10.0\}$. In this situation, we evaluate the approximation error of the generalization risk and the generalization neutrality risk by varying sample size n. The approximation error of generalization risk is the difference of the empirical risk between training and test samples, while that of the generalization neutrality risk is the difference of the empirical neutrality risk between training and test samples. Five fold cross-validation was used for evaluation of the approximation error of the empirical risk and empirical neutrality; the average of ten different folds are shown as the results.

Results. Fig 1 shows the change of the approximation error of generalization risk (the difference of the empirical risks w.r.t. test samples and training samples), and the approximation error of generalization neutrality risk (the difference of the empirical neutrality risks w.r.t. test samples and training samples)

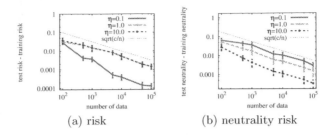

(a) risk   (b) neutrality risk

Fig. 1. Change of approximation error of generalization risk (left) and approximation error of generalization neutrality risk (right) by neutral SVM (our proposal) according to varying the number of samples n. The horizontal axis shows the number of samples n, and the error bar shows the standard deviation across the change of five-fold division. The line "sqrt(c/n)" denotes the convergence rate of the approximation error of the generalization risk (in Theorem 1) or the generalization neutrality risk (in Theorem 2). Each line indicates the results with the neutralization parameter $\eta \in \{0.1, 1.0, 10.0\}$. The regularizer parameter was set as $\lambda = 0.05n$.

Table 1. Specification of Datasets

dataset	#Inst.	#Attr.	Viewpoint	Target
Adult	16281	13	gender	income
Dutch	60420	10	gender	income
Bank	45211	17	loan	term deposit
German	1000	20	foreign worker	credit risk

with changing sample size n. The plots in Fig 1 left and right show the approximation error of generalization risk and the approximation error of generalization neutrality risk, respectively.

Recall that the discussions in Section 5.3 showed that the approximation error of generalization risk decreases with $O(\sqrt{\ln(1/\delta)/n})$ rate. As indicated by the Theorem 1, Fig 1 (left) clearly shows that the approximation error of the generalization risk decreases as sample size n increases. Similarly, discussions in Section 5.1 revealed that the approximation error of generalization neutrality risk also decreases with $O(\sqrt{\ln(1/\delta)/n})$ rate, which can be experimentally confirmed in Fig 1 (right). The plot clearly shows that the approximation error of the generalization neutrality risk decreases as the sample size n increases.

7.2   Real Datasets

We compare the classification performance and neutralization performance of neutral SVM with CV2NB, PR, and ηLR for a number of real datasets specified in Table 1. In Table 1, #Inst. and #Attr. denote the sample size and the number of attributes, respectively; "Viewpoint" and "Target" denote the attributes used as the target and the viewpoint, respectively. All dataset attributes were discretized by the same procedure described in [4] and coded by 1-of-K representation for PR, ηLR, and neutral SVM. We used the primal problem of neutral

Table 2. Range of neutralization parameter

method	range of neutralization parameter
PR	0, 0.01, 0.05, 0.1, ..., 100
ηLR	0, 5×10^{-5}, 1×10^{-4}, 5×10^{-4}, ..., 0.5
neutral SVM	0, 0.01, 0.05, 0.1, ..., 100

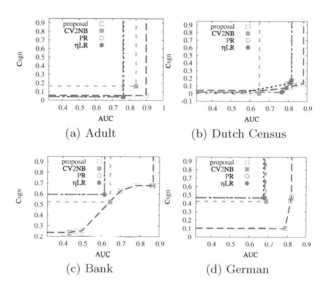

(a) Adult   (b) Dutch Census

(c) Bank   (d) German

Fig. 2. Performance of CV2NB, PR, ηLR, and neutral SVM (our proposal). The vertical axis shows the AUC, and horizontal axis shows $C_{n,sgn}(f, g)$. The points in these plots are omitted if they are dominated by others. The bottommost line shows limitations of neutralization performance, and the rightmost line shows limitations of classification performance, which are shown only as guidelines.

SVM (non-kernelized version) to compare our method with the other methods in the same representation. For PR, ηLR, and neutral SVM, the regularizer parameter was tuned in advance for each dataset in the non-neutralized setting by means of five-fold cross validation, and the tuned parameter was used for the neutralization setting. CV2NB has no regularization parameter to be tuned. Table 2 shows the range of the neutralization parameter used for each method.

The classification performance and neutralization performance was evaluated with *Area Under the receiver operating characteristic Curve* (AUC) and $+1/-1$ empirical neutrality risk $C_{n,sgn}(f, g)$, respectively. Both measures were evaluated with five-fold cross-validation and the average of ten different folds are shown in the plots.

Results. Fig 2 shows the classification performance (AUC) and neutralization performance ($C_{n,sgn}(f, g)$) at different setting of neutralization parameter η. In the graph, the best result is shown at the right bottom. Since the classification performance and neutralization performance are in a trade-off relationship, as indicated by Theorem Eq (6), the results dominated by the other parameter settings are omitted in the plot for each method.

CV2NB achieves the best neutrality in Dutch Census, but is less neutral compared to the other methods in the rest of the datasets. In general, the classification performance of CV2NB is lower than those of the other methods due to the poor classification performance of naíve Bayes. PR and ηLR achieve competitive performance to neutral SVM in Adult and Dutch Census in term of the neutrality risk, but the results are dominated in term of AUC. Furthermore, the results of PR and ηLR in Bank and German are dominated. The results of neutral SVM are dominant compared to the other methods in Bank and German dataset, and it is noteworthy that the neutral SVM achieves the best AUC in almost all datasets. This presumably reflects the superiority of SVM in the classification performance, compared to the naíve Bayes and logistic regression.

8   Conclusion

We proposed a novel framework, NERM. NERM provides a framework that learns a target hypothesis that minimizes the empirical risk and that is empirically neutral in terms of risk to a given viewpoint hypothesis. Our contributions are as follows: (1) We define NERM as a framework for guaranteeing the neutrality of classification problems. In contrast to existing methods, the NERM can be formulated as a convex optimization problem by using convex relaxation. (2) We provide theoretical analysis of the generalization neutrality risk of NERM. The theoretical results show the approximation error of the generalization neutrality risk of NERM is uniformly upper-bounded by the Rademacher complexity of hypothesis class $g\mathcal{F}$ and $O(\sqrt{\ln(1/\delta)/n})$. Moreover, we derive a bound on the generalization neutrality risk for the optimal hypothesis corresponding to the neutralization parameter η. (3) We present a specific learning algorithms for NERM, neutral SVM. We also extend the neutral SVM to the kernelized version.

Suppose the viewpoint is set to some private information. Then, noting that neutralization reduces correlation between the target and viewpoint values, outputs obtained from the neutralized target hypothesis do not help to predict the viewpoint values. Thus, neutralization realizes a certain type of privacy preservation. In addition, as already mentioned, NERM can be interpreted as a variant of transfer learning by regarding the neutralization term as data-dependent prior knowledge. Clarifying connection to privacy-preservation and transfer learning is remained as an area of future work.

Acknowledgments. The authors would like to thank Toshihiro Kamishima and the anonymous reviewers for their valuable suggestion. The work is supported by JST CREST program of Advanced Core Technologies for Big Data Integration, JSPS KAKENHI 12913388, and 25540094.

References

1. Bartlett, P.L., Bousquet, O., Mendelson, S.: Local rademacher complexities. The Annals of Statistics 33(4), 1497–1537 (2005)
2. Bartlett, P.L., Mendelson, S.: Rademacher and gaussian complexities: Risk bounds and structural results. Journal of Machine Learning Research 3, 463–482 (2002)
3. Calders, T., Kamiran, F., Pechenizkiy, M.: Building classifiers with independency constraints. In: Saygin, Y., Yu, J.X., Kargupta, H., Wang, W., Ranka, S., Yu, P.S., Wu, X. (eds.) ICDM Workshops, pp. 13–18. IEEE Computer Society (2009)
4. Calders, T., Verwer, S.: Three naive bayes approaches for discrimination-free classification. Data Mining and Knowledge Discovery 21(2), 277–292 (2010)
5. Fukuchi, K., Sakuma, J., Kamishima, T.: Prediction with model-based neutrality. In: Blockeel, H., Kersting, K., Nijssen, S., Železný, F. (eds.) ECML PKDD 2013, Part II. LNCS (LNAI), vol. 8189, pp. 499–514. Springer, Heidelberg (2013)
6. Kakade, S.M., Sridharan, K., Tewari, A.: On the complexity of linear prediction: Risk bounds, margin bounds, and regularization. In: Koller, D., Schuurmans, D., Bengio, Y., Bottou, L. (eds.) NIPS, pp. 793–800. Curran Associates, Inc. (2008)
7. Kamiran, F., Calders, T., Pechenizkiy, M.: Discrimination aware decision tree learning. In: 2010 IEEE 10th International Conference on Data Mining (ICDM), pp. 869–874. IEEE (2010)
8. Kamishima, T., Akaho, S., Asoh, H., Sakuma, J.: Fairness-aware classifier with prejudice remover regularizer. In: Flach, P.A., De Bie, T., Cristianini, N. (eds.) ECML PKDD 2012, Part II. LNCS (LNAI), vol. 7524, pp. 35–50. Springer, Heidelberg (2012)
9. Pan, S.J., Yang, Q.: A survey on transfer learning. IEEE Trans. Knowl. Data Eng. 22(10), 1345–1359 (2010)
10. Pariser, E.: The Filter Bubble: What The Internet Is Hiding From You. Viking, London (2011)
11. Pedreschi, D., Ruggieri, S., Turini, F.: Measuring discrimination in socially-sensitive decision records. In: Proceedings of the SIAM Int'l Conf. on Data Mining, pp. 499–514. Citeseer (2009)
12. Resnick, P., Konstan, J., Jameson, A.: Measuring discrimination in socially-sensitive decision records. In: Proceedings of the 5th ACM Conference on Recommender Systems: Panel on The Filter Bubble, pp. 499–514 (2011)
13. Shor, N.Z., Kiwiel, K.C., Ruszcayński, A.: Minimization Methods for Non-differentiable Functions. Springer-Verlag New York, Inc., New York (1985)
14. Vapnik, V.N.: Statistical learning theory (1998)
15. Zemel, R.S., Wu, Y., Swersky, K., Pitassi, T., Dwork, C.: Learning fair representations. In: ICML (3). JMLR Proceedings, vol. 28, pp. 325–333. JMLR.org (2013)
16. Zliobaite, I., Kamiran, F., Calders, T.: Handling conditional discrimination. In: 2011 IEEE 11th International Conference on Data Mining (ICDM), pp. 992–1001. IEEE (2011)

Effective Blending of Two and Three-way Interactions for Modeling Multi-relational Data

Alberto García-Durán, Antoine Bordes, and Nicolas Usunier

Université de Technologie de Compiègne - CNRS
Heudiasyc UMR 7253
Compiègne, France
{agarciad,bordesan,nusunier}@utc.fr

Abstract. Much work has been recently proposed to model relational data, especially in the multi-relational case, where different kinds of relationships are used to connect the various data entities. Previous attempts either consist of powerful systems with high capacity to model complex connectivity patterns, which unfortunately usually end up overfitting on rare relationships, or in approaches that trade capacity for simplicity in order to fairly model all relationships, frequent or not. In this paper, we propose a happy medium obtained by complementing a high-capacity model with a simpler one, both pre-trained separately and jointly fine-tuned. We show that our approach outperforms existing models on different types of relationships, and achieves state-of-the-art results on two benchmarks of the literature.

Keywords: Representation learning, Multi-relational data.

1 Introduction

Predicting new links in multi-relational data plays a key role in many areas and hence triggers a growing body of work. Multi-relational data are defined as directed graphs whose nodes correspond to *entities* and *edges* are in the form of triples (*head*, *label*, *tail*) (denoted (h, ℓ, t)), each of which indicates that there exists a relationship of name *label* between the entities *head* and *tail*. Figure 1 displays an example of such data with six entities (*Jane*, *Patti*, *John*, *Mom*, *Miami* and *Austin*) and two relationships (born_in and child_of). Link prediction in this context consists in attempting to create new connections between entities and to determine their type; this is crucial in social networks, knowledge management or recommender systems to name a few.

Performing predictions in multi-relational data is complex because of their heterogeneous nature. Any such data can equivalently be seen as a set of directed graphs that share the same nodes but that usually present drastically different properties in terms of sparsity or connectivity. As illustration, we can look at some statistics of a subset of the knowledge base FREEBASE, named FB15K in the following, which we use for our experiments. This data set contains ~15k

T. Calders et al. (Eds.): ECML PKDD 2014, Part I, LNCS 8724, pp. 434–449, 2014.

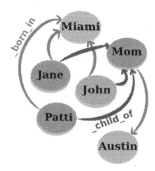

Fig. 1. Example of multi-relational data with 6 entities and 2 relationships

entities connected by ∼1.5k relationships to form a graph of ∼500k triples. Even if FB15K is just a small sample, it is likely that its characteristics are shared with most real-world multi-relational data, but at a different scale. The relationships of FB15K have a mean number of triples of ∼400 and a median of 21: a vast number of relationships appear in very few triples, while others provide a large majority of the connections. Besides, roughly 25% of the relationships are of type 1-to-1, that is a *head* is connected to at most one *tail* (think of a spouse_of link for instance), but on the opposite 25% of the relationships are of type Many-to-Many, that is, multiple *head* can be linked to a *tail* and vice-versa (for instance, a like_product link). Creating new connections for Many-to-Many relationships can be possible by relying on several existing links of the same kind, whereas in the 1-to-1 case, the only way to be able to generalize is to count on the other relationships, especially if the relationship of interest happens to be rare.

In contrast to (pseudo-) symbolic approaches for link prediction based on Markov logic networks [11] or random walks [13], most recent effort towards solving this problem concern latent factor models (e.g. [19,10,22,16,1,17,9]) because they tend to scale better and to be more robust w.r.t. the heterogeneity of multi-relational data. These models represent entities with latent factors (usually low-dimensional vectors or *embeddings*) and relationships as operators destined to combine those factors. Operators and latent factors are trained to fit the data using reconstruction [17], clustering [22] or ranking costs [1]. The multi-relational quality of the data is modeled through the sharing of the latent factors across relationships which grants a transfer of information from one relationship to another; operators are normally specific to each relationships, except in [9] where some parameters are shared among relationships.

Learning these latent factor models can be seen as a kind of multitask training, with one task per relationship: one model is fitted for each task and some parameters are shared across tasks. All existing latent factor approaches define the same formulation for each task. This is natural since hand-crafting a particular model for each relationship seems daunting since there can be several thousands of them. However, as all relationships have very different properties, this also induces important drawbacks.

The standard modeling assumption is to consider 3-way interactions between *head, label* and *tail*, i.e. to consider that they are all interdependent: the validity of a triple (h, ℓ, t) depends jointly on h, ℓ and t. This generally results in models where entities are represented by vectors and relationships by matrices. An exception is Parafac [8,5] that models multi-relational data as a binary tensor and factorizes it as a sum of rank one tensors. Other tensor factorization methods derived from Tucker decompositions [23] or Dedicom [7] like RESCAL [17,18] end up with vectorial latent factors for entities and low rank matrices for relationships. This formulation is also shared by many non-parametric Bayesian approaches such as extensions of the Stochastic Block Models [10,16,24], or joint entities and relationships clustering approaches [22], which end up modeling the data with similar architectures but with drastically different training and inference procedures. Linear Relational Embeddings [19] were proposed as a 3-way model trained using a regression loss: the vector representing t is learned so that it can be reconstructed using the embedding of h and the matrix encoding ℓ, if (h, ℓ, t) is valid. This work was later followed by the Structured Embeddings model (SE) [1] where the regression loss was replaced by a ranking loss for learning embeddings of entities.

Three-way models are appropriate for multi-relational data since they can potentially represent any kind of interaction. However, this comes at a price since they have to allocate the same large capacity to model each relationship. While this is beneficial for frequent ones, this can be problematic for rare relationships and cause major overfitting. Hence, one must control the capacity either by regularizing, which is not straightforward since different penalties might need to be used for each relationship, or by reducing the expressivity of the model. The second option is implemented in two recent embedding models, SME [2] and TransE [3], that choose to represent multi-relational data as combination of 2-way interactions. The idea is to assume that the validity of a triple (h, ℓ, t) is governed by binary interaction terms (h, t), (t, ℓ) and (h, ℓ), which allows to represent a relationship as a vector as with the other entities. Such a model, TransE [3], outperforms most 3-way approaches on various data sets, which indicates that less expressivity can be beneficial overall for a whole database, especially for relationships where the number of training triples is reduced. However, by design, methods based on 2-way interactions are limited and can not hope to represent all kinds of relations between entities.

In this paper, we introduce Tatec (for *Two And Three-way Embeddings Combination*), a latent factor model which successfully combines well-controlled 2-way interactions with high-capacity 3-way ones. We demonstrate in the following that our proposal is a generalization of many previous methods. Unlike recent work like the Latent Factor Model (LFM) of [9] or the Neural Tensor Model (NTN) of [21] that proposed similar joint formulations mixing several interaction terms, we deliberately choose not to share parameters between the 2- and 3-way interaction components of our model. Previous work use the same embeddings for entities in all terms of their models, which seems to be a sound and natural idea to obtain the possible latent representations. However, we discovered that

2- and 3-way models do not respond to the same data patterns, and that they do not necessarily encode the same kind of information in the embeddings: sharing them among interaction terms can hence be detrimental because it can make some features to be missed by the embeddings or to be destroyed. On the contrary, we explain in the following that using different embeddings for both terms allows to detect distinct kinds of patterns in the data. To ensure that Tatec satisfactorily collects both kinds of patterns, we pre-train separately a 2-way and a 3-way model, which are then combined and jointly fine-tuned in a second stage. We show in various experiments that this combination process is powerful since it allows to jointly enjoy both nice properties of 2 and of 3-way interactions. As a result, Tatec is more robust than previous work w.r.t. the number of training samples and the type of relationships. It consistently outperforms most models in all conditions and achieves state-of-the-art results on two benchmarks from the literature, FB15K [3] and SVO [9].

This paper is organized as follows. Section 2 introduces our formulation and our training procedure, divided in a pre-training phase followed by a fine-tuning step both conducted via stochastic gradient descent. We justify our particular modeling choices in Section 3. Finally, we display and discuss our experimental results on FB15K, SVO and a synthetic dataset in Section 4.

2   Model

We now describe our model, and the training algorithm associated to it. The motivation underlying our parameterization is given in the next section.

2.1   Scoring Function

The data S is a set of relations between entities in a fixed set of entities in $\mathcal{E} = \{c^1, ..., e^E\}$. Relations are represented as triples (h, ℓ, t) where the head h and the tail t are indexes of entities (i.e. $h, t \in [\![E]\!] = \{1, ..., E\}$), and the label ℓ is the index of a relationship in $\mathcal{L} = \{l^1, ..., l^L\}$, which defines the type of the relation between the entities e^h and e^t. Our goal is to learn a discriminant scoring function on the set of all possible triples $\mathcal{E} \times \mathcal{L} \times \mathcal{E}$ so that the triples which represent likely relations receive higher scores than triples that represent unlikely relations. Our proposed model, Tatec, learns embeddings of entities in low dimensional vector spaces, and parameters of operators on $\mathbb{R}^d \times \mathbb{R}^d$, most of them being associated to single relationships. More precisely, the score given by Tatec to a triple (h, ℓ, t), denoted by $s(h, \ell, t)$, is defined as:

$$s(h, \ell, t) = s_1(h, \ell, t) + s_2(h, \ell, t) \tag{1}$$

where s_1 and s_2 have the following form:

(B)   Bigrams or the 2-way interactions terms:

$$s_1(h, \ell, t) = \langle \mathbf{r}_1^\ell | \mathbf{e}_1^h \rangle + \langle \mathbf{r}_2^\ell | \mathbf{e}_1^t \rangle + \langle \mathbf{e}_1^h | \mathbf{D} | \mathbf{e}_1^t \rangle,$$

where $\mathbf{e}_1^h, \mathbf{e}_1^t$ are embeddings in $\mathbb{R}^{d_1}$ of the head and tail entities of (h, ℓ, t) respectively, $\mathbf{r}_1^\ell$ and $\mathbf{r}_2^\ell$ are vectors in $\mathbb{R}^{d_1}$ that depend on the relationship l^ℓ, and $\mathbf{D}$ is a diagonal matrix that does not depend on the input triple.

As a general notation throughout this section, $\langle . | . \rangle$ is the canonical dot product, and $\langle \mathbf{x} | \mathbf{A} | \mathbf{y} \rangle = \langle \mathbf{x} | \mathbf{A}\mathbf{y} \rangle$ where $\mathbf{x}$ and $\mathbf{y}$ are two vectors in the same space and $\mathbf{A}$ is a square matrix of appropriate dimensions.

(T)   Trigram or the 3-way interactions term:

$$s_2(h, \ell, t) = \left\langle \mathbf{e}_2^h \middle| \mathbf{R}^\ell \middle| \mathbf{e}_2^t \right\rangle,$$

where $\mathbf{R}^\ell$ is a matrix of dimensions (d_2, d_2), and $\mathbf{e}_2^h$ and $\mathbf{e}_2^t$ are embeddings in $\mathbb{R}^{d_2}$ of the head and tail entities respectively. The embeddings of the entities for this term are not the same as for the 2-way term; they can have different dimensions for instance.

The embedding dimensions d_1 and d_2 are hyperparameters of our model. All other vectors and matrices are learned without any additional parameter sharing.

The 2-way interactions term of the model is similar to that of [2], but slightly more general because it does not contain any constraint between the relation-dependent vectors $\mathbf{r}_1^\ell$ and $\mathbf{r}_2^\ell$. It can also be seen as a relaxation of the translation model of [3], which is the special case where $\mathbf{r}_1^\ell = -\mathbf{r}_2^\ell$, $\mathbf{D}$ is the identity matrix, and the 2-norm of the entities embeddings are constrained to equal 1.

The 3-way term corresponds exactly to the model used by the collective factorization method RESCAL [17], and we chose it for its high expressivity on complex relationships. Indeed, as we said earlier in the introduction, 3-way models can basically represent any kind of interaction among entities. The usage of combinations of 2-way and 3-way terms has already been used in [9,21], but, besides a different parameterization, Tatec contrasts with them by the choice of not sharing the embeddings between the two models. In LFM [9], constraints were imposed on the relation-dependent matrix of the 3-way terms (low rank in a limited basis of rank-one matrices), the relation vectors $\mathbf{r}_1^\ell$ and $\mathbf{r}_2^\ell$ were constrained to be a constant linear function of the matrix ($\mathbf{D} = \mathbf{0}$ in their work). These global constraints severely limited the expressivity of the 3-way model, and act as powerful regularization in that respect. However, their global constraints also reduces the expressivity of the 2-way model, which, as we explain in Section 3, should be left with maximum degrees of freedom. The fact that we do not share any parameter between relations is similar to NTN [21]. Our overall scoring function is similar to this model with a single layer, with the fundamental difference that we use different embedding spaces and do not use any non-linear transfer function, which results in a facilitated training (the gradients have a larger magnitude, for instance).

2.2   Training

Training is carried out using gradient descent, with a ranking criterion as training objective. The optimization approach is similar to the one used for TransE [3], but the models are very different. Our loss function takes training triples, and tries to give them higher scores than to corrupted versions, where the corruption consists in either replacing the head or the tail of each triple by a random entity. Since we are learning with positive examples only, this kind of criterion implements the prior knowledge that unobserved triples are likely to be invalid. Such corruption approaches are widely used when learning embeddings of knowledge bases [1,3] or words in the context of language models [4,14].

Given a training triple (h, ℓ, t), the set of possible corrupted triples, denoted by $\mathcal{C}(h, \ell, t)$, is defined as:

$$\mathcal{C}(h, \ell, t) = \{(h', \ell, t') \in [\![E]\!] \times \{\ell\} \times [\![E]\!] | h' = h \text{ or } t' = t\}.$$

The loss function we optimize is then:

$$\sum_{(h,\ell,t)\in\mathcal{S}} \sum_{(h',\ell,t')\in\mathcal{C}(h,\ell,t)} \max(0, 1 - s(h, \ell, t) + s(h', \ell, t')) \qquad (2)$$

Stochastic gradient is performed in a minibatch setting. The dataset $\mathcal{S}$ is shuffled at each epoch, minibatches of $m << |\mathcal{S}|$ training triples are selected, and, for each one of them, a corresponding mini-batch of corrupted triples is sampled at random to create ranking pairs. We only create a single corrupted triple per training sample. The learning rate of the stochastic gradient is kept constant, and optimization is stopped using early stopping on a validation set.

Several regularization schemes were tried during training: either by forcing the entity embeddings to have, at most, a 2-norm of r (for radius), or by adding 2-norm regularization inside the sum of (2) of the form $\lambda \| \mathbf{x} \|_2^2$ for each parameter $\mathbf{x}$ (relation vectors and diagonal matrix in the 2-way term or relation matrix in the 3-way term) that appears in $\max(0, 1 - s(h, \ell, t) + s(h', \ell, t'))$. The first kind of regularization is carried out after each minibatch by projecting the entities into the 2-norm ball of radius r.

A random initialization of the scoring function (1) can lead to a poor local minimum, but can also prevent the different embeddings used in the 2- and 3-way terms to capture different patterns as we expect (see next section). Hence, following many previous work on deep architecture, we decided to first pre-train separately the bigrams and trigram terms on the training set. When their pre-training is over (i.e. stopped using early stopping on a validation set), we initialize the parameter of the full score (1) using these learned weights and fine-tuned it by running stochastic gradient descent on the training set with the full model.

3   Interpretation and Motivation of the Model

This section discusses the motivations underlying the parameterization of Tatec, and in particular our choice of 2-way model to complement the 3-way term.

3.1   2-Way Interactions as One Fiber Biases

It is common in regression, classification or collaborative filtering to add biases (also called offsets or intercepts) to the model. For instance, a critical step of the best-performing techniques of the Netflix prize was to add user and item biases, i.e. to approximate a user-rating R_{ui} according to (see e.g. [12]):

$$R_{ui} \approx \langle \mathbf{P}_u | \mathbf{Q}_i \rangle + \alpha_u + \beta_i + \mu \tag{3}$$

where $\mathbf{P} \in \mathbb{R}^{U \times k}$, with each row $\mathbf{P}_u$ containing the k-dimensional embedding of the user (U is the number of users), $\mathbf{Q} \in \mathbb{R}^{I \times k}$ containing the embeddings of the I items, $\alpha_u \in \mathbb{R}$ a bias only depending on a user and $\beta_i \in \mathbb{R}$ a bias only depending on an item (μ is a constant that we do not consider further on).

The 2-way + 3-way interactions model we propose can be seen as the 3-mode tensor version of this "biased" version of matrix factorization: the trigram term $(\mathbf{T})$ is the collective matrix factorization parametrization of the RESCAL algorithm [17] and plays a role analogous to the term $\langle \mathbf{P}_u | \mathbf{Q}_i \rangle$ of the matrix factorization model for collaborative filtering (3). The bigram term $(\mathbf{B})$ then plays the role of biases for each fiber of the tensor,[1] i.e.

$$s_1(h, \ell, t) \approx B_{\ell,h}^1 + B_{\ell,t}^2 + B_{h,t}^3 \tag{4}$$

and thus is the analogous for tensors to the term $\alpha_u + \beta_i$ in the matrix factorization model (3). The exact form of $s_1(h, \ell, t)$ given in $(\mathbf{B})$ corresponds to a specific form of collective factorization of the fiber-wise bias matrices $\mathbf{B}^1 = \left[B_{\ell,h}^1 \right]_{\ell \in [\![L]\!], h \in [\![E]\!]}$, $\mathbf{B}^2$ and $\mathbf{B}^3$ of Equation 4. We do not exactly learn one bias by fiber because many such fibers have very little data, while, as we argue in the following, the specific form of collective factorization we propose in $(\mathbf{B})$ should allow to share relevant information between different biases.

3.2 The Need for Multiple Embeddings

A key feature of Tatec is to use different embedding spaces for the 2-way and 3-way terms, while existing approaches that have both types of interactions use the same embedding space [9,21]. We motivate this choice in this section.

It is important to notice that biases in the matrix factorization model (3), or the bigram term in the overall scoring function (1) do not affect the model expressiveness, and in particular do not affect the main modeling assumptions that embeddings should have low rank. The user/item-biases in (3) only boil down to adding two rank-1 matrices $\alpha \mathbf{1}^T$ and $\mathbf{1}\beta^T$ to the factorization model. Since the rank of the matrix is a hyperparameter, one may simply add 2 to this hyperparameter and get a slightly larger expressiveness than before, with reasonably little impact since the increase in rank would remain small w.r.t. its original value (which is usually 50 or 100 for large collaborative filtering data sets). The critical feature of these biases in collaborative filtering is how they interfere with capacity control terms other than the rank, namely the 2-norm regularization: in [12] for instance, all terms of (3) are trained using a squared error as a measure

---

[1] Fibers are the higher order analogue of matrix rows and columns for tensors and are defnied by fixing every index but one.

of approximation and regularized by $\lambda \left(\parallel \mathbf{P}_u \parallel_2^2 + \parallel \mathbf{Q}_i \parallel_2^2 + \alpha_u^2 + \beta_i^2 \right)$, where $\lambda > 0$ is the regularization factor. This kind of regularization is a weighted trace norm regularization [20] on $\mathbf{PQ}^T$. Leaving aside the "weighted" part, the idea is that at convergence, the quantity $\lambda \left(\sum_u \parallel \mathbf{P}_u \parallel_2^2 + \sum_i \parallel \mathbf{Q}_i \parallel_2^2 \right)$ is equal to 2λ times the sum of the singular values of the matrix $\mathbf{PQ}^T$. However, $\lambda \parallel \boldsymbol{\alpha} \parallel_2^2$, which is the regularization applied to user biases, is *not* 2λ times the singular value of the rank-one matrix $\boldsymbol{\alpha} \mathbf{1}^T$, which is equal to $\sqrt{I} \parallel \boldsymbol{\alpha} \parallel_2$, and can be much larger than $\parallel \boldsymbol{\alpha} \parallel_2^2$. Thus, if the pattern user+item biases exists in the data, but very weakly because it is hidden by stronger factors, it will be less regularized than others and the model should be able to capture it. Biases, which are allowed to fit the data more than other factors, offer the opportunity of relaxing the control of capacity on some parts of the model but this translates into gains if the patterns that they capture are indeed useful patterns for generalization. Otherwise, this ends up relaxing the capacity to lead to more overfitting.

Our bigram terms are closely related to the trigram term: the terms $\langle \mathbf{r}_1^\ell | \mathbf{e}_1^h \rangle$ and $\langle \mathbf{r}_2^\ell | \mathbf{e}_1^t \rangle$ can be added to the trigram term by adding constant features in the entities' embeddings, and $\langle \mathbf{e}_1^h | \mathbf{D} | \mathbf{e}_1^t \rangle$ is directly in an appropriate quadratic form. Thus, the only way to gain from the addition of bigram terms is to ensure that they can capture useful patterns, but also that capacity control on these terms is less strict than on the trigram terms. In tensor factorization models, and especially 3-way interaction models with parameterizations such as $(\mathbf{T})$, capacity control through the regularization of individual parameters is still not well understood, and as it turns out in experiments is more detrimental than effective. The only effective parameter is the admissible rank of the embeddings, which leads to the conclusion that the bigram term can be really useful in addition to the trigram term if higher-dimensional embeddings are used. Hence, in absence of clear and concrete way of effectively controlling the capacity of the trigram term, we believe that different embedding spaces should be used.

3.3   2-Way Interactions as Entity Types+Similarity

Having a part of the model that is less expressive, but less regularized than the other part is only useful if the patterns it can learn are meaningful for the prediction task at hand. In this section, we give the motivation for our 2-way interactions term for the task of modeling multi-relational data.

Most relationships in multi-relational data, and in knowledge bases like FB15K in particular, are strongly typed, in the sense that only well-defined and specific subsets of entities can be either heads or tails of selected relationships. For instance, a relationship like capital_of expects a (big) city as head and a country as tail for any valid relation. Large knowledge bases have huge amounts of entities, but those belong to many different types. Identifying the expected types of head and tail entities of relationships, with an appropriate granularity of types (e.g. person or artist or writer), is likely to filter out 95% of the entity set during prediction. The exact form of the first two terms $\langle \mathbf{r}_1^\ell | \mathbf{e}_1^h \rangle + \langle \mathbf{r}_2^\ell | \mathbf{e}_1^t \rangle$ of the

Table 1. Statistics of the data sets used in this paper and extracted from an artificial database, FAMILY, and from two knowledge bases: FB15K and SVO

DATA SET	FAMILY	FB15K [3]	SVO [9]
ENTITIES	721	14,951	30,605
RELATIONSHIPS	5	1,345	4,547
TRAINING EXAMPLES	5,748	483,142	1,000,000
VALIDATION EXAMPLES	1,935	50,000	50,000
TEST EXAMPLES	1,955	59,071	250,000

2-way interaction model (**B**), which corresponds to a low-rank factorization of the per bias matrices (*head, label*) and (*tail, label*) in which *head* and *tail* entities have the same embeddings, is based on the assumption that the types of entities can be predicted based on few (learned) features, and these features are the same for predicting *head*-types as for predicting *tail*-types. As such, it is natural to share the entities embeddings in the first two terms of (**B**).

The last term, $\langle e_1^h | \mathbf{D} | e_1^t \rangle$, is intended to account for a global similarity between entities. For instance, predicting the capital of France can easily be performed correctly by saying that we search for the city with strongest overall connections with France in the knowledge base. A country and a city may be strongly linked through their geographical positions, independent of their respective types. The diagonal matrix **D** allows to re-weight features of the embedding space to account for the fact that the features used to describe types may not be the same as those that can describe the similarity between objects of different types. The use of a diagonal matrix is strictly equivalent to using a general symmetric matrix in place of **D**.[2] The reason for using a symmetric matrix comes from the intuition that the direction of many relationships is arbitrary (i.e. the choice between having triples "Paris is capital of France" rather than "France has capital Paris"), and the model should be invariant under arbitrary inversions of the directions of the relationships (in the case of an inversion of direction, the relations vectors $\mathbf{r}_1^\ell$ and $\mathbf{r}_2^\ell$ are swapped, but all other parameters are unaffected). For tasks in which such invariance is not desirable, the diagonal matrix could be replaced by an arbitrary matrix.

4   Experiments

This section presents a series of experiments that we conducted to compare Tatec to previous models from the literature on two benchmarks, FB15K, a subset of FREEBASE [3], and SVO, a database of nouns connected by verbs and introduced

---

[2] We can see the equivalence by taking the eigenvalue decomposition of a symmetric **D**: apply the change of basis to the embeddings to keep only the diagonal part of **D** in the term $\langle e_1^h | \mathbf{D} | e_1^t \rangle$, and apply the reverse transformation to the vectors $\mathbf{r}_1^\ell$ and $\mathbf{r}_2^\ell$. Note that since rotations preserve euclidian distances, the equivalence still holds under 2-norm regularization of the embeddings.

Table 2. Baselines. Rules used by our symbolic baselines, an upper and a lower one, to predict a *tail* given a *head* and a *label* on the FAMILY data set. Similar symmetric rules have been used to predict a *head* given a *tail* and a *label* for these relations.

RELATIONSHIP	BASELINE	RULES FOR PREDICTING A *tail* GIVEN A *head*
cousin_of	Lower	Any entity of the same layer as the *head* of the families where the *head* has a cousin.
	Upper	Any entity whose parent is a sibling(-in-law) of the parents of the *head*.
sibling_of	Lower	Any entitiy of the same layer as the *head* of the families where the *head* has a sibling.
	Upper	Any entity whose parent is the same as the parents of *head*.
married_to	Lower	Any entity of the same layer as the *head* whose marriage would not be forbidden with.
	Upper	Entity who has children in common with the *head*.
parent_of	Lower	Any entity of the lower layer of *head* of the families where *head* has a child.
	Upper	Any entity being children of the spouse of *head*.
uncle_of	Lower	Any entity of the lower layer of *head* of the families where *head* has a niece/nephew.
	Upper	Any entity being child of a sibling/sibling a law of *head* or the spouse of *head*.

in [9], as well an artificial data set that we created (FAMILY). The statistics of these data sets are given in Table 1.

For evaluation, we use a ranking procedure as in [1,3]. For each test triplet, the head is removed and replaced by each of the entities of the dictionary in turn. Scores of those corrupted triplets are computed by the models and sorted by descending order and the rank of the correct entity is stored. This whole procedure is repeated when removing the tail instead or the head. We report the *mean* of those predicted ranks and the *hits@10*, i.e. the proportion of correct entities ranked in the top 10.

4.1 Synthetic Data

"Family" Data Set. This database contains triples expressing family relationships among the members of 5 families along 6 generations, each family being organized in a layered tree structure where each layer refers to a generation. These 5 families are first created independent of each other by recursively sampling the number of children of each node of a layer using the normal distribution $\mathcal{N}(3, 1.5)$ to create a new generation. Then, families are connected by marriage links between two members. We use pre defined rules to avoid non-typical situations, like marriages between cousins and brothers, as well as marriages between two members of different generations. To control the number of connections between families, only $i - 1$ marriages are allowed in the generation i.

After all families are created, we build a multi-relational data set by collecting the pairs of entities connected using the following relationships: cousin_of, married_to, parent_of, sibling_of and uncle_of. We end up with a data set with 721 entities, 5 relationships and ~9k triples which is later split into training, validation and test sets. There is a large variation in the numbers of triples: there are only 30 examples with married_to but 5,060 with cousin_of. FAMILY is a realistic and challenging study case, but for which we know the underlying semantics.

Baselines. We created two symbolic baselines since we know the underlying rules used to generate the data. These baselines can be used to assess what kind of pattern is caught by our model. Our first baseline, Lower, uses simplistic rules

Table 3. Synthetic data set. We compare our Bigrams, Trigram and Tatec models in terms of mean rank, with both baselines on the FAMILY dataset.

RELATION	cousin_of	married_to	parents_of	sibling_of	uncle_of
Lower	76	4	25	35	34
Upper	4	3	7	4	3
Bigrams	10	102	9	8	13
Trigram	7	162	7	5	8
Tatec	6	69	5	4	6

in order to find a set of potential candidates among all entities given a *label* and either a *head* or a *tail* as input. Our second baseline, Upper, returns candidate answers using a much more refined knowledge about the underlying semantics of the relationships, such as, if *John* and *Mary* have children together then they are likely to be married. We made sure than, for any triple from the data set, the correct missing element given a *label* and either a *head* or a *tail* as input, is always contained in the sets returned by Lower and Upper (the second being included in the first one). Then, for each test triple, we computed the mean rank of the answers given by the baselines, by sorting the elements of the returned candidate sets by the number of occurrences of each entity in the training set. It is worth noting that the entities of the Upper set always form triples expressing true knowledge. The rules used to defined both baselines are given in Table 2.

Implementation. To pre-train our Bigrams and Trigram models we selected the learning rate for the stochastic gradient descent among $\{0.1, 0.01, 0.001, 0.0001\}$, and the embedding dimension among $\{5, 10, 20\}$. The margin was fixed to 1, and the radius determining the maximim 2-norm of the entity embeddings was validated among $\{1, 10, 100\}$. For fine-tuning Tatec, the learning rate was selected among the same values as above, independent of the values chosen for pre-training. For all three models, training was limited to a maximum of 500 epochs, and we used the mean rank on the validation set (computed every 50 epochs) as stopping criterion.

Results. Table 3 presents our results. Tatec outperforms the best performance of Bigrams and Trigram counterparts for each relationship, indicating that the biases brought to the 3-way model by the bigram terms are indeed beneficial to detect complementary patterns in the data. As a result, Tatec gets a really close performance to that of the upper baseline, indicating that it can perform relatively sophisticated inference, as long as the amount of data is sufficient. Indeed, all embedding-based models fail completely on the married_to relationship: with very few training samples, those models cannot learn any meaningful information that would grant non-trivial predictions for this complex relationship.

4.2 Encoding Freebase

Freebase Data Set (FB15k). FREEBASE is a huge and growing database of general facts; there are currently around 1.2 billion triples and more than 80

Table 4. Link prediction results. We compare Bigrams, Trigram and several version of Tatec with various methods from the literature on the FB15K dataset. Results are displayed in the filtered setting, see the text for more details.

Interaction	Model	Mean Rank	Hits@10
2-way	SME(LINEAR) [2]	154	40.8
	TransE [3]	125	47.1
3-way	SE [1]	162	39.8
	SME(BILINEAR) [2]	158	41.3
	RESCAL [17]	683	44.1
2 + 3-way	NTN [21]	332	27.0
	LFM [9]	164	33.1
2-way	Bigram	133	44.7
3-way	Trigram	156	42.7
2 + 3-way	Tatec-NO-PRETRAIN	133	44.7
	Tatec-SHARED-EMBS	137	45.0
	Tatec-LINEAR-COMB	115	51.7
	Tatec	**111**	**52.6**

million entities. We used FB15K, a data set based on FREEBASE introduced in [3], This small data set is based on a subset of entities that are also present in the WIKILINKS database³ and that also have at least 100 mentions in FREEBASE (for both entities and relationships). This results in $592,213$ triples with $14,951$ entities and $1,345$ relationship which were randomized and split.

Baselines. We compare Tatec with various models from previous work: RESCAL [17], LFM [9] , SE [1] , SME [2] and TransE [3]. Results were extracted from [3] since we follow the same experimental protocol here. We also include comparisons with NTN [21]. For this method, we ran experiments with the code provided by the authors. The embedding dimension was selected between $\{25, 50\}$ and the number of slices of the tensor layer was fixed to 2 for computational considerations. We chose the regularization hyperparameter among $\{0, 0.1, 0.01, 0.001\}$ and tanh as nonlinear element-wise function. The negative triplets were generated as before in a proportion of 10 negative to 1 positive triple. The model ran for 700 iterations and was validated every 50 iterations.

Besides Bigrams, Trigram and Tatec, we also propose the performance of 3 other versions of Tatec:

- Tatec-NO-PRETRAIN: Tatec without pre-training $s_1(h, \ell, t)$ and $s_2(h, \ell, t)$.
- Tatec-SHARED: Tatec but sharing the embeddings between $s_1(h, \ell, t)$ and $s_2(h, \ell, t)$ and without pre-training.
- Tatec-LINEAR-COMB: this version simply combines the bigram and trigram terms using a linear combination, without jointly fine-tuning their parameters. The score is hence defined as follows:

$$s(h, \ell, t) = \delta_1^\ell \langle \mathbf{r}_1^\ell | \mathbf{e}_1^h \rangle + \delta_2^\ell \langle \mathbf{r}_2^\ell | \mathbf{e}_1^t \rangle + \delta_3^\ell \langle \mathbf{e}_1^h | \mathbf{D} | \mathbf{e}_1^t \rangle + \delta_4^\ell \langle \mathbf{e}_2^h | \mathbf{R}^\ell | \mathbf{e}_2^t \rangle$$

---

³ code.google.com/p/wiki-links

Table 5. Detailed results by category of relationship. We compare our Bigrams, Trigram and Tatec models in terms of Hits@10 (in %) on FB15K in the filtered setting against other models of the literature. (M. stands for MANY).

TASK	PREDICTING *head*				PREDICTING *tail*			
REL. CATEGORY	1-TO-1	1-TO-M.	M.-TO-1	M.-TO-M.	1-TO-1	1-TO-M.	M.-TO-1	M.-TO-M.
SE [1]	35.6	62.6	17.2	37.5	34.9	14.6	68.3	41.3
SME(LINEAR) [2]	35.1	53.7	19.0	40.3	32.7	14.9	61.6	43.3
SME(BILINEAR) [2]	30.9	69.6	19.9	38.6	28.2	13.1	76.0	41.8
TransE [3]	43.7	65.7	18.2	47.2	43.7	**19.7**	66.7	**50.0**
Bigrams	55.4	73.2	25.5	49.3	51.3	11.4	78.5	37.4
Trigram	44.3	69.6	29.0	48.0	41.2	8.3	72.6	35.1
Tatec	**65.8**	**84.8**	**40.0**	**58.9**	**62.3**	15.1	**87.0**	42.3

The combination weights δ_i^ℓ depend on the relationship and are learned by optimizing the ranking loss defined in (2) using L-BFGS, with an additional quadratic penalization term, $\sum_k \frac{||\delta^k||_2^2}{\sigma_k + \epsilon}$, subject to $\sum_k \sigma_k = \lambda$. Training is carried out in an iterative way, by alternating optimization of δ parameters via L-BFGS, and update of σ parameters using $\sigma_i^* = \frac{\lambda ||\delta_i||_2}{\sum_i ||\delta_i||_2}$, until some stopping criterion is reached. The δ parameters are initialized to 1 and the λ value is validated among $\{0.1, 1, 10, 100, 250, 500, 1000\}$. The intuition behind this particular penalization for the δs is that it is equivalent to a LASSO penalization [6] and our initial idea was to enforce sparsity among δ parameters. However we found experimentally that the best performance was obtained with a λ of 250, which does not yield a sparse solution.

Implementation. Tatec, and its 3 alternative versions have been trained and validated in the same setting that was used for the FAMILY experiments, except that we chose the embedding dimensions among $\{25, 50\}$.

Results. Table 4 displays the mean rank and hits@10 for all the aforementioned methods. These results have been computed in a *filtered setting* as defined in [3]: to reduce the error introduced by true triples that might be ranked above the target triple in test, all the entities forming existing triples in the train, validation and test sets but the target one are removed from the candidate set of entities to be ranked. This grants a clearer view on ranking performance.

First of all, we can notice that our plain 2- and 3– way models (Bigrams and Trigram respectively) are performing comparably as other similarly expressive models: Bigrams is better than SME(linear) but worse than TransE, and Trigram performs roughly like SME(bilinear). RESCAL is interesting since it achieves a very poor mean rank but almost the best hits@10 value: we believe that this is due to overfitting. To make it scale on large data sets, RESCAL has to be ran without regularization, this causes a major overfitting on rare relationships and hence a poor mean rank. But, it seems that one can reach a very decent hits@10 nonetheless. Interestingly, Tatec is able to significantly outperform both its constituents Bigrams

Table 6. Verb prediction results. We compare our Bigrams, Trigram and Tatec models with baselines from the literature on the SVO dataset.

	Median/Mean Rank	Hits@5%	Hits@20%
Counts [9]	48/517	72	83
LFM [9]	50/195	78	95
SME [2]	56/199	77	95
Bigrams	52/210	78	98
Trigram	**44**/188	79	95
Tatec	**44/183**	**80**	**99**

and Trigram, which indicates that they can encode complementary information. This is confirmed by the comparison with the baseline 2 + 3-way models, LFM and NTN.[4] By sharing their embeddings between their 2- and 3-way terms, they constrain their model too much. We can see a similar behavior if we look at the results of Tatec-SHARED-EMBS, which are much worse than those of Tatec. The pre-training is very useful: without pre-training, Tatec only achieves the same performance as the 2-way term alone. Tatec-LINEAR-COMB performs only slightly worse than Tatec, which indicates that, with proper regularization, a simpler combination of 2- and 3-way terms can be efficient. Overall, Tatec outperforms all previous models by a wide margin, especially in hits@10.

We also broke down the results by type of relation, classifying each relationship according to the cardinality of their *head* and *tail* arguments. A relationship is considered as 1-to-1, 1-to-M, M-to-1 or M-M regarding the variety of arguments *head* given a *tail* and vice versa. If the average number of different *heads* for the whole set of unique pairs (*label, tail*) given a relationship is below 1.5 we have considered it as 1, and the same in the other way around. The number of relations classified as 1-to-1, 1-to-M, M-to-1 and M-M is 353, 305, 380 and 307, respectively. The results are displayed in the Table 5. Most results point out that Tatec consistently outperforms all models we compared it with, except for the relations 1-to-M and M-to-M when predicting the *tail*.

4.3   Predicting Verbs

In this last experimental section, we present results of ranking *label* given *head* and *tail*. We do so by working on a verb prediction task, where one has to assign the correct verb given two noun phrases acting subject and direct object.

Subject-Verb-Object Data Set (SVO). This data set was generated by extracting sentences from Wikipedia articles whose syntactic structure is (subject, verb, direct object) and where the verb appears in the WordNet lexicon [15] and where the subject and direct object are noun phrases from WordNet as well. Due to the high number of relations in this data set, this is an interesting benchmark for *label* prediction.

---

[4] Results of NTN are worse than expected. As we said earlier, we tried a large number of hyperparameter values but NTN might require to cover an even wider range.

Baselines. We compare Tatec with 3 different approaches: LFM, Counts and SME(linear). Counts is based on the direct estimation of probabilities of triples (*head, label, tail*) by using the number of occurrences of pairs (*head, label*) and (*label, tail*) in the training set. The results for these models have been extracted from [9], and we followed the same experimental setting.

Implementation. Due to the different nature of the application, the negative triples have been generated here by replacing the verb of a given positive triple by a random verb. The rest of the experimental setting is identical to the one used for FAMILY and FB15K, but running only 100 epochs and validating every 10 epochs in the pre-training phase, since we found that the models were converging much faster. For Tatec, we even validated every epoch.

Results. Table 6 shows the results for this database. The measure hits@z% indicates the proportion of predictions for which the correct verb is ranked in the top z% of the verb list. The performance of Tatec is also excellent in this case since it outperforms all previous methods on all metrics, including LFM, another model combining 2- and 3-way interactions.

5 Conclusion

This paper introduced Tatec, a new method for performing link prediction in multi-relational data, which is made of the combination a 2- and 3-way interactions terms. Both terms do not share their embeddings and this, along with a two-phase training (pre-training and fine-tuning), allows for the model to encode complementary information into its parameters. As a result, Tatec outperforms by a wide margin many methods from the literature, some based on 2-way, on 3-way interactions or on both.

Acknowledgments. This work was carried out in the framework of the Labex MS2T (ANR-11-IDEX-0004-02), and was funded by the French National Agency for Research (EVEREST-12-JS02-005-01).

References

1. Bordes, A., Weston, J., Collobert, R., Bengio, Y.: Learning structured embeddings of knowledge bases. In: Proc. of the 25th Conf. on Artif. Intel. (AAAI) (2011)
2. Bordes, A., Glorot, X., Weston, J., Bengio, Y.: A semantic matching energy function for learning with multi-relational data. Machine Learning (2013)
3. Bordes, A., Usunier, N., Garcia-Duran, A., Weston, J., Yakhnenko, O.: Translating embeddings for modeling multi-relational data. In: Advances in Neural Information Processing Systems, pp. 2787–2795 (2013)
4. Collobert, R., Weston, J., Bottou, L., Karlen, M., Kavukcuoglu, K., Kuksa, P.: Natural language processing (almost) from scratch. Journal of Machine Learning Research 12, 2493–2537 (2011)

5. Franz, T., Schultz, A., Sizov, S., Staab, S.: Triplerank: Ranking semantic web data by tensor decomposition. In: Bernstein, A., Karger, D.R., Heath, T., Feigenbaum, L., Maynard, D., Motta, E., Thirunarayan, K. (eds.) ISWC 2009. LNCS, vol. 5823, pp. 213–228. Springer, Heidelberg (2009)
6. Grandvalet, Y.: Least absolute shrinkage is equivalent to quadratic penalization. In: The International Conference on Artificial Neural Networks (1998)
7. Harshman, R., Lundy, M.: Three-way dedicom: Analyzing multiple matrices of asymmetric relationships. An. Meeting of the N. Amer. Psych. Society (1992)
8. Harshman, R.A., Lundy, M.E.: Parafac: parallel factor analysis. Comput. Stat. Data Anal. 18(1), 39–72 (1994)
9. Jenatton, R., Le Roux, N., Bordes, A., Obozinski, G., et al.: A latent factor model for highly multi-relational data. In: NIPS 25 (2012)
10. Kemp, C., Tenenbaum, J.B., Griffiths, T.L., Yamada, T., Ueda, N.: Learning systems of concepts with an infinite relational model. In: Proc. of the 21st National Conf. on Artif. Intel. (AAAI), pp. 381–388 (2006)
11. Kok, S., Domingos, P.: Statistical predicate invention. In: Proceedings of the 24th International Conference on Machine Learning, ICML 2007 pp. 433–440 (2007)
12. Koren, Y., Bell, R., Volinsky, C.: Matrix factorization techniques for recommender systems. Computer 42(8), 30–37 (2009)
13. Lao, N., Mitchell, T., Cohen, W.W.: Random walk inference and learning in a large scale knowledge base. In: Proceedings of the Conference on Empirical Methods in Natural Language Processing, pp. 529–539. Association for Computational Linguistics (2011)
14. Mikolov, T., Sutskever, I., Chen, K., Corrado, G.S., Dean, J.: Distributed representations of words and phrases and their compositionality. In: Advances in Neural Information Processing Systems, pp. 3111–3119 (2013)
15. Miller, G.: WordNet: a Lexical Database for English. Communications of the ACM 38(11), 39–41 (1995)
16. Miller, K., Griffiths, T., Jordan, M.: Nonparametric latent feature models for link prediction. In: Bengio, Y., Schuurmans, D., Lafferty, J., Williams, C.K.I., Culotta, A. (eds.) Advances in Neural Information Processing Systems 22 (2009)
17. Nickel, M., Tresp, V., Kriegel, H.P.: A three-way model for collective learning on multi-relational data. In: Proceedings of the 28th International Conference on Machine Learning (ICML 2011), pp. 809–816 (2011)
18. Nickel, M., Tresp, V., Kriegel, H.P.: Factorizing yago: scalable machine learning for linked data. In: Proceedings of the 21st International Conference on World Wide Web, WWW 2012, pp. 271–280 (2012)
19. Paccanaro, A., Hinton, G.: Learning distributed representations of concepts using linear relational embedding. IEEE Trans. on Knowl. and Data Eng. 13 (2001)
20. Salakhutdinov, R., Srebro, N.: Collaborative filtering in a non-uniform world: Learning with the weighted trace norm. tc (X) 10, 2 (2010)
21. Socher, R., Chen, D., Manning, C.D., Ng, A.Y.: Reasoning With Neural Tensor Networks For Knowledge Base Completion. In: Advances in Neural Information Processing Systems 26 (2013)
22. Sutskever, I., Salakhutdinov, R., Tenenbaum, J.: Modelling relational data using bayesian clustered tensor factorization. In: Adv. in Neur. Inf. Proc. Syst. 22 (2009)
23. Tucker, L.R.: Some mathematical notes on three-mode factor analysis. Psychometrika 31, 279–311 (1966)
24. Zhu, J.: Max-margin nonparametric latent feature models for link prediction. In: Proceedings of the 29th Intl Conference on Machine Learning (2012)

Automatic Design of Neuromarkers
for OCD Characterization

Oscar García Hinde[1], Emilio Parrado-Hernández[1], Vanessa Gómez-Verdejo[1],
Manel Martínez-Ramón[1], and Carles Soriano-Mas[2,3]

[1] Signal Processing and Communications, Universidad Carlos III de Madrid,
Leganés, 28911, Spain
{oghinde,emipar,vanessa,manel}@tsc.uc3m.es
[2] Bellvitge Biomedical Research Institute-IDIBELL,
Psychiatry Service, Bellvitge University Hospital, Barcelona, Spain
csoriano@idibell.cat
[3] Carlos III Health Institute, Centro de Investigación Biomédica en Red de Salud
Mental (CIBERSAM), Spain

Abstract. This paper proposes a new paradigm to discover biomarkers
capable of characterizing obsessive-compulsive disorder (OCD). These
biomarkers, named neuromarkers, will be obtained through the analysis
of sets of magnetic resonance images (MRI) of OCD patients and control
subjects.

The design of the neuromarkers stems from a method for the automatic discovery of clusters of voxels relevant to OCD recently published
by the authors. With these clusters as starting point, we will define the
neuromarkers as a set of measurements describing features of these individual regions. The principal goal of the project is to come up with a
set of about 50 neuromarkers for OCD characterization that are easy to
interpret and handle by the psychiatric community.

1 Introduction

In some areas of medicine it is quite common to find punctuation systems that
allow for state evaluation and patient diagnosis. For instance APACHE II (Acute
Physiology and Chronic Health Evaluation) [17] is one of the most widely used
score systems to quantify the seriousness of critical patient's state by means of
12 factors or routine physiological measures (blood pressure, body temperature,
heart rate, etc.). Among these punctuation systems we can also find the Ranson
criterion, which predicts the severity of acute pancreatitis [26], the Glasgow scale
[15], used to measure a person's conscience level, or the SAPS II index (Simplified
Acute Physiology Score) [19] which, as the APACHE II index, estimates the
severity of a patient's state. It has been shown that the adequate use of these
scores provides a better characterization of the illness and helps researchers
analyse the success of new therapies and compare their effectiveness in different
hospitals.

However, psychiatry lacks direct and objective indicators of a subject's physiological state for the diagnosis of a certain pathology or its evolution analysis

T. Calders et al. (Eds.): ECML PKDD 2014, Part I, LNCS 8724, pp. 450–465, 2014.

[23]. To this end, psychiatrists usually use the Diagnostic and Statistical Manual of Mental Disorders, which provides a classification of mental illnesses along with descriptions of the diagnostic categories based on the patient's medical history and the disorders they may show. Over the past few years, neuroanatomical and neurofunctional analysis have become common practise in the evaluation of certain mental conditions by means of Magnetic Resonance Imaging (MRI), either structural (sMRI) or functional (fMRI), aimed at the study of pathologies and the detection of the structural brain anomalies that cause them [4] [31]. For this purpose, different techniques have been proposed in the literature, such as voxel based morphometry" (VBM) [2], enabling the analysis of structural abnormalities in the brain; or the "General Linear Model" [1], which establishes a mathematical model to either analyse sMRI data or obtain the functional response of the brain in fMRI studies. These research lines have laid the basis for the re-evaluation of previous neuroanatomical hypotheses that were considered to be associated with certain disorders, and the proposal of new models with a sound biological foundation, although in some occasions these results have not been correctly translated to the clinical practise [23]. As a result, there has been a growing interest in the application of other analysis strategies, such as machine learning (ML) methods, since they are able to describe differences between patient and control groups and to obtain mathematical models that allow discerning between them [20].

ML techniques have positioned themselves as some of the most promising options to extract relevant information from the neuroimaging data through statistical learning methods. These approaches have the main characteristic of being able to automatically learn a model of data from a collection of examples, which in many occasions can enable the detection of information that would otherwise be hidden from the eyes of an expert. For this reason, ML methods are being successfully used in data based diagnosis in many fields of medicine. For instance, they are being used in the classification of tissue-cells, the segmentation of retinopathy, the detection breast-cancer or auricular arrhythmia, just to name a few.

Furthermore, the multivariate nature of these techniques, as well as their ability to extract the greatest amount of available information possible when the number of data is limited (a very common situation with MRI data) has favoured the widespread use of ML tools in neuroimaging analysis [25] and the diagnosis from this type of data [16]. So far, scientific production in relation to neuroimaging and ML methods has followed a path in which the psychiatric community provides MRI data from an experiment designed to study the brain, and the ML community directly applies standard techniques. Because of this, we can find many examples of the application of ML approaches to magnetic resonance experiments, such as brain mapping from fMRI data sequences [37], temporal fMRI series analysis [18] or brain state decoding [14] [21]. Clinical applications can also be found, in which the goal is to detect a particular mental illness, such as Alzheimer's disease [34], schizophrenia [5] or obsessive compulsive disorder (OCD) [29] [24].

Due to the small sample problem presented by MRI data bases (the number of dimensions is several orders of magnitude greater than the number of training examples), it is common to find ML approximations that apply an intermediate feature selection or extraction step, thus reducing the problem's dimensionality and making the outcome of the processing step easier to interpret. By this process, the final machine or classifier that detects the illness will only use a subset of the original voxels (three dimensional pixels) or some transformation thereof. Among these approximations, ideas such as those proposed in [33] or [36] stand out. Their approach is to directly define voxel groups and represent each one by the mean value of the voxels they comprise. The methods presented in [6] apply a t-Test as a step prior to the classification process to eliminate irrelevant voxels. The approximations introduced in [8] and [9] employ a Recursive Feature Elimination method (RFE) [11] to select the voxel subset that is most relevant to the classification phase. Other distinguishable lines of work include those proposed in [32], [27] and [7] in which sparsity inducing regularizations are directly included in the classifier to obtain the relevant voxel subset during the design of the classifier.

The vast majority of methods proposed in the literature using ML with MRI data focus on analysing differences between patient and control groups. These methods provide a decision on the class to which each MRI belongs in the form of a probability value or a binary value (patient/control), further proving that the images contain relevant information for the diagnosis. In the best cases these studies also provide a subset of voxels or regions that characterize the pathology, which can indicate the psychiatrist or neurologist that a particular region of the brain presents structural or functional differences between healthy and ill subjects. However, given the isolated analysis of these regions in an MRI scan from a single patient, the psychiatrist or neurologist is unable to determine whether the subject is ill or not: the discrimination pattern provided by the classifier comprises, together with these regions and groups, a series of mathematical relations between them that are not directly manageable and are practically impossible to interpret in most cases.

The goal of this paper is to propose a new paradigm to discover biomarkers capable of characterizing OCD. These biomarkers, which we will call neuromarkers, will be obtained through the automatic analysis of sets of MRIs of OCD patients and control subject differences. In order for these neuromarkers to have penetration in clinical psychiatry, they will have to be interpretable and manageable.

The design of these neuromarkers stems from a method for the automatic discovery of clusters of voxels relevant to OCD recently proposed in [24]. With these regions as a starting point, we will define several candidates to become neuromarkers, that is, a set of measurements describing features of these individual regions. In order to obtain a reduced subset of neuromarkers for OCD characterization, we will apply different selection strategies to remove irrelevant features. This will result in a small set of neuromarkers that is easy to interpret and handle by the psychiatric community. Experiments will analyse the suitability of each subset of neuromarker candidates, as well as the different selection strategies, showing that we can handle a subset of no more than 50 neuromarkers maintaining the original performance in terms of classification error.

The rest of paper is organized as follows: Section 2 reviews the method presented in [24], that will be used as starting point to define the relevant regions for OCD characterization; Section 3 introduces the different kinds of neuromarker candidates as well as the different strategies considered for their selection; experimental results will be presented in Section 4; finally, Section 5 summarizes the main conclusions and proposes some future research lines.

2 Related Work: A Review on Discovering Brain Regions

2.1 Bagged Support Vector Machines

An MRI brain scan provides a vector in which each element is a voxel associated with the probability of it being gray matter. Therefore, linear classifiers in such an input space admit a pretty straightforward interpretation of the role of each voxel in the discriminant function. A linear classifier assigns each brain scan of D voxels, $\mathbf{x} = [x_1, \ldots, x_D]^T$, to a possible output class, $\hat{y}(\mathbf{x})$, using

$$\hat{y}(\mathbf{x}) = \text{sign}\left(\mathbf{w}^T \mathbf{x} + b\right) = \text{sign}\left(\sum_{d=1}^{D} w_d x_d + b\right) \qquad (1)$$

where $\mathbf{w} = [w_1, \ldots, w_D]^T$ and b are the weight vector and the bias term of the classifier, respectively.

Borrowing some ideas from the starplots method of [3], the bagging procedure applied here trains a linear SVM with a subset of M instances of the training data and repeats this procedure a significant number of times, R. It then counts the number of times that each weight w_d takes positive and negative values, classifying w_d in one of these two groups:

- Those w_d that take positive (or negative) values in at least r iterations.
- Those w_d that are not sign consistent in at least r iterations.

The contribution to the final classification of those voxels belonging to the second group depends on the particular selection of the training set, thus they can be considered as non critical for the discriminative task and can be discarded. Therefore, the selection method consists in picking the features from the first group, since they are consistently relevant for the classification.

2.2 Refined Voxel Selection with Conformal Analysis

The above voxel selection method still suffers from the small sample problem. This can be further alleviated with some ideas from Conformal Analysis (CA) [35]. In a nutshell, CA is based on assessing the likelihood of every test sample being assigned to each possible output class, and choosing the class that is most likely.

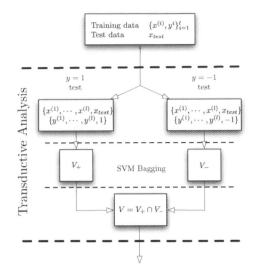

Fig. 1. Workflow of the CBS algorithm

The CA philosophy completes the voxel selection in the following manner. For every test sample $\mathbf{x}_t$, we carry out the voxel selection twice. We first obtain a subset of voxels, V_+, considering $y_t = +1$. Then we obtain a second subset, V_-, considering $y_t = -1$. The final set of voxels used in the training of the classifier, that we will label V, is the intersection of V_+ and V_-. The intuition behind this is that the voxels present in one of the subsets but not in the intersection depend strongly on the specific labelling of $\mathbf{x}_t$, therefore they are depicting specific subject traits, instead of characterizing the pathology.

Figure 1 shows the workflow employed to apply the bagging SVM approach in combination with the CA refinement.

2.3 Discovering Brain Regions Relevant to OCD

The application of the above process to the OCD problem (details of this dataset are given in Section 4) has allowed us to discard 92% of the voxels, finding a subset of approximately 40.000 voxels which are clustered in regions. We have then applied a post-processing algorithm targeted at discovering regions by simply including connected voxels in the same cluster. On average, across the different LOO (Leave One Out) iterations, the clustering finds 718 ± 40 groups of connected voxels.

The relevance of these groups of voxels towards the characterization of the OCD patology is made clear when they are used to classify patients and controls, since the use of this subset of voxels provides a classification error of around 26.2%, whereas state-of-art approaches get performances around 35%-40%. Further details of this procedure, as well as the results obtained over the OCD dataset, can be found in [24].

3   Discovering Neuromarkers

3.1   Defining and Selecting Neuromarkers

Despite the usefulness of the aforementioned process, the obtained subset of voxels presents an unfriendly characterization of the OCD pathology, since its huge size (40.000 voxels) makes it unmanageable and difficult to understand for the clinical community. It is almost impossible to relate the value of each voxel with the brain deformity or dystrophy which may characterize the disorder.

However, we can exploit the grouped distribution of these voxels to define a set of measurements, each of them associated to a single brain region, which are able to represent the relevant information of these brain regions in a friendlier way. Due to the fact that these measurements must be useful for disease characterization, we will denote them as neural biomarkers or simply neuromarkers.

Taking into account that each voxel of an MRI scan is characterized by its grey matter probability with the variable x_d, $d = 1, \ldots, D$, and that we have grouped the relevant voxels in a subset of brain regions, each of them indexed by S_g, $g = 1, \ldots, G$, we can now characterize the gray matter probability of these regions with some measurements that we will exploit as neuromarkers:

1. **Averaged (AV) grey matter probability**
   This first measurement directly obtains a single parameter over each brain region by averaging the gray matter probability values of the voxels that belong to it. That is,

   $$\text{AV}_g = \frac{1}{\mid S_q \mid} \sum_{i \subset S_g} x_i$$

   where $\mid S_g \mid$ is the number of voxels in a brain region g.
2. **Variance (VAR) of the grey matter probability**
   We consider the variance of the voxel gray matter probability to represent each brain region. Thus, each VAR marker is computed as:

   $$\text{VAR}_g = \frac{1}{\mid S_g \mid} \sum_{i \in S_g} (x_i - \text{AV}_g)^2$$

3. **Accumulated (AC) grey matter probability**
   Another interesting parameter can be obtained by summing all the x_i probabilities belonging to the same region. In this way, the AC parameter is given by:

   $$\text{AC}_g = \sum_{i \in S_g} x_i$$

   Note that, unlike AV markers, this marker is not dividing by the brain region size. Therefore, it is indirectly including the brain region volume.
4. **SVM weighted (WE) grey matter probability**
   Finally, we can use the information provided by the linear SVM classifier to extract the relevant information of each brain region. A linear SVM classifier

applied over the overall set of selected voxels S computes the output for a sample $\mathbf{x}$ as:

$$f(\mathbf{x}) = \sum_{i \in S} w_i x_i + b \tag{2}$$

If we split the index set S into the different brain regions ($S = S_1 \cup S_2 \ldots \cup S_G$), (2) can be rewritten as:

$$f(\mathbf{x}) = \sum_{g=1}^{G} \sum_{i \in S_g} w_i x_i + b \tag{3}$$

and each term of the inner summation would be summarizing the information of each region. So, we can define the WE neuromarker as:

$$WE = \sum_{i \in S_g} w_i x_i$$

This gives us four kinds of neromarkers for OCD characterisation, providing a single parameter for each brain region. However, with the intention of characterizing the pathology with even fewer markers, we will also introduce some feature selection strategies to be applied over these sets of neuromarkers.

In particular, we will consider the following feature selection approaches:

1. **Ranking based on variance**
   A quick glance over the neuromarker values reveals that some of them are constant over all subjects, regardless of whether they are patients or controls. Therefore, a simple criterion to remove this redundancy is to rank the neuromarkers according to their variance.
2. **T-test**
   The second criterion applies a standard t-test [12] to analyze the statistical differences of the neuromarkers belonging to patient and control populations. The resulting p-values of the test allows us to rank the neuromarkers according to its significance.
3. **Recursive Feature Elimination (RFE)**
   This method [11] aims at finding the subset on N features that are able to provide the largest classification margin in a SVM classifier. For this purpose, the RFE approach carries out a backward feature elimination by removing at each iteration the data feature that least decreases the SVM margin. When a linear SVM classifier is applied, this process is simplified by iteratively training a linear SVM and removing the feature with the smallest value $|w_i|$. As in previous criteria, this recursive elimination process will provide a sorted list of neuromarkers in order of relevance.
4. **Ranking based on correlation**
   This criterion supposes that good neuromarkers should be highly correlated with the classification task. Thus, another straightforward selection procedure is to rank the neuromarkers according to their correlation with the patient/control labels.

5. **Ranking based on HSIC**
   The previous criterion analyses the linear relationships. This strategy extends this idea by measuring non linear relationships by means of the Hilbert-Schmidt independence criterion [10], [28].

4   Experimental Work

4.1   Dataset Description

Eighty-six subjects with OCD (44 males; mean±SD age, 34.23±9.25 years) were recruited from the outpatient service of the Department of Psychiatry of the Bellvitge University Hospital, Barcelona (Spain), paired with a group of 86 healthy control subjects, with the same age and gender distribution (43 males; 33.47±9.94 years old).

Images were acquired with a 1.5-T Signa Excite system (General Electric, Milwaukee, Wisconsin) equipped with an 8-channel phased-array head coil. A high-resolution T1-weighted anatomical image was obtained for each subject using a 3-dimensional fast spoiled gradient inversion-recovery prepared sequence with 130 contiguous slices (TR, 11.8 milliseconds; TE, 4.2 milliseconds; flip angle, 15°; field of view, 30 cm; 256×256 pixel matrix; slice thickness, 1.2 mm). Imaging data were transferred and processed on a MS Windows platform using MATLAB 7.8 and Statistical Parametric Mapping (SPM8). Following the inspection of image artifacts, image preprocessing was performed. Briefly, native-space MRIs were segmented into the three tissue types (gray and white matter, and cerebrospinal fluid, although only gray matter segment were used in the present study) and normalized to the SPM-T1 template by means of a DARTEL approach. Additionally, the Jacobian determinants derived from the spatial normalization were used to modulate image voxel values to restore volumetric information. Finally, images were smoothed with a 4 mm full-width at half maximum (FWHM) Gaussian kernel.

4.2   Analysis of Neuromarker Performance

In this section we analyse the usefulness of each neuromarker and the extent to which we can reduce their number by means of the aforementioned feature selection strategies. To this end we shall consider a neuromarker to be useful if we can maintain a classification error that is similar to the 26.2% obtained before its construction. Furthermore, to select the optimum number of neuromarkers to be used, a nested LOO process, which provides a validation error for each number of selected neuromarkers, has been applied.

Table 1 illustrates the effectiveness of our neuromarker types, paired with each selection strategy, at characterising OCD by means of the classification error and the number of neuromarkers that yield said error. For comparison, the first row shows the error rate obtained with the full set of neuromarkers.

Overall, the most capable neuromaker is by far the WE type. Most selection criteria converge at errors of around 30% when applied with it. Moreover, the

Table 1. Analysis of the LOO classification errors (CE) and number of neuromarkers (# NM) obtained by selection criteria and neuromarker type

		AV	VAR	ACC	WE
All neuromarkers	CE (%)	35.47	49.42	33.14	28.49
	# NM	718	718	718	718
Variance ranking	CE (%)	37.21	50.58	36.63	28.49
	# NM	80.30	144.81	40.02	38.84
t-Test selection	CE (%)	34.88	52.91	38.37	32.56
	# NM	258.25	198.75	286.65	525.83
RFE	CE (%)	36.05	51.16	32.56	30.23
	# NM	132.88	245.03	224.45	60.26
HSIC-Test ranking	CE (%)	38.37	47.09	40.12	31.98
	# NM	96.81	148.20	54.76	48.41
Correlation ranking	CE (%)	40.12	50.00	40.70	32.56
	# NM	575.40	435.61	513.93	527.51

number of relevant features needed to characterise the pathology using this neuromarker is well under 100 with the variance ranking, HSIC ranking and RFE methods.

Specifically, the most effective criterion is the WE neuromarker type paired with a variance ranking selection strategy. This combination produces an error of 28.49%, which is only slightly greater than the error obtained with no feature selection, while the number of voxels it employs is one order of magnitude smaller (around 718 versus an average of 38.84).

4.3   Neuromarker Interpretation

Given the results of the previous section, we will now analyse the relevance and neuroanatomical position of each WE neuromarker. In particular we shall focus on the subset of neuromarkers selected by the variance ranking criterion.

Due to the fact that we are managing 172 iterations of a LOO procedure, we have obtained 172 different subsets of neuromarkers with an average size of 38.84. In order to obtain a subset that is easy to interpret, we have merged all these subsets into a single one comprising 59 neuromarkers. Note that these neuromarkers present varying consistencies in the voxels they contain over the 172 iterations, meaning that some of these voxels appear in every iteration while others in just a few.

To analyse the relevance of each neuromarker, we have studied the classification error variations provided by the classification system when one neuromarker is removed from the training set. Table 2 shows these CE rate deviations for the most important neuromarkers, that is, those which cause a significant CE increment when they are not used during training. To complete this analysis, Table 2 also includes the most significant MNI neuroanatomical [30] regions (those whose consistency is greater than 50%) and Figures 2-6 show the localization of the five

Table 2. Neuroanatomical analysis for the most important neuromarkers

NM ranking	Δ CE (%)	MNI ROIs
1	6.39	Temporal Sup-Mid L
2	5.81	Frontal Inf Tri-Orb R; Insula R
3	4.65	Insula L; Putamen L; Pallidum L
4	4.65	Parietal Inf L; SupraMarginal L
5	3.49	Frontal Sup-Mid R
6	2.9	Calcarine L-R; Lingual L; Precuneus L-R; Cerebelum 6 L; Vermis 4-5-6
7	2.9	Olfactory L-R ; Frontal Med Orb L; Rectus L; Cingulum Ant R; Lingual L-R; Occipital Inf R; Fusiform L-R; Precuneus L Caudate L; Pallidum L; Thalamus L-R; Temporal Inf R Cerebelum Crus-1-3-4-5-6-7b-9-10 L-R; Vermis 1-2-3-4-5-7-10
8	2.9	Temporal Sup-Mid R
9	2.9	Frontal Inf Oper-Tri R; Insula R; Putamen R; Pallidum R; Heschl R; Temporal Sup-Pole Sup R
10	2.9	Precentral R; Frontal Mid R; Postcentral R; Parietal Inf R; SupraMarginal R
11	2.9	Parietal Inf R; SupraMarginal R; Angular R; Temporal Sup R
12	2.9	Frontal Mid L
13	2.9	Cerebelum Crus2-7b-8 R
14	2.32	Lingual R
15	2.32	Fusiform L; Temporal Inf L
16	2.32	Frontal Sup-Mid L
17	2.32	Frontal Med Orb L R; Rectus L-R
18	2.32	Frontal Sup Orb L; Rectus L
19	1.74	Cuneus L; Parietal Sup L; Precuneus L
20	1.16	Occipital Inf L; Parietal Sup-Inf L; SupraMarginal L; Angular L; Temporal Sup-Mid L
21	1.16	Temporal Sup-Mid-Inf R
22	1.16	Temporal Mid L
23	0.58	Precentral; Frontal Sup-Mid L
24	0.58	Frontal Sup Medial L-R; Cingulum Ant L-R
25	0.58	Precentral L; Frontal Mid L
26	0.58	Hippocampus L-R
27	0.58	Occipital Sup-Mid R

most important neuromarkers, with color intensity indicating the consistency of each voxel (white means a voxel was selected in all the iterations).

The five most relevant neuromarkers to OCD are located in the frontal, temporal and parietal lobes. Three of them appear in regions traditionally associated with the disorder, such as the orbitofrontal cortex (right inferior frontal and middle frontal gyri) and the striatum (putamen and globus pallidum, extending to the adjacent insular cortex). Such regions are part of the distributed cortico-striatal circuits know to be involved in OCD pathophysiology [13]. Specifically, while striatal regions seem to be hyperactive (and volume increased), prefrontal areas seem to be hypoactive (and volume decreased) and inefficient in regulating

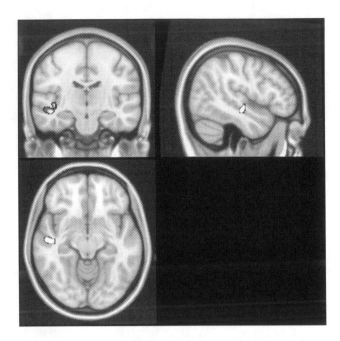

Fig. 2. Position of the first ranked neuromarker, which is mainly located in the ROIs: Temporal Sup-Mid L. Each voxel's color intensity indicates its consistency.

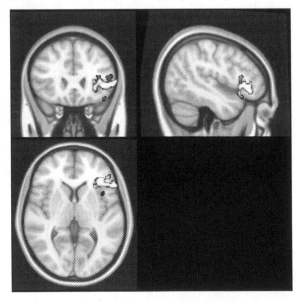

Fig. 3. Position of the second ranked neuromarker, which is mainly located in the ROIs: Frontal Inf Tri-Orb R and Insula R. Each voxel's color intensity indicates its consistency.

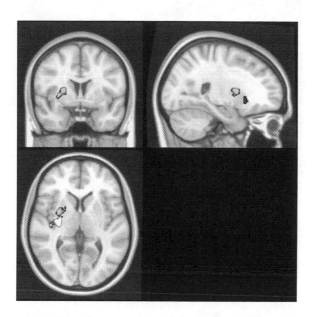

Fig. 4. Position of the third ranked neuromarker, which is mainly located in the ROIs: Insula L, Putamen L and Pallidum L. Each voxel's color intensity indicates its consistency.

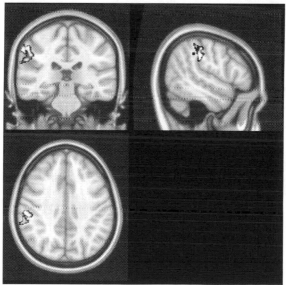

Fig. 5. Position of the fourth ranked neuromarker, which is mainly located in the ROIs: Parietal Inf L; SupraMarginal L. Each voxel's color intensity indicates its consistency.

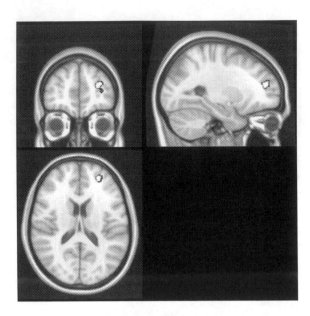

Fig. 6. Position of the fiveth ranked neuromarker, which is mainly located in the ROIs: Frontal Sup-Mid R. Each voxel's color intensity indicates its consistency.

enhanced striatal activity, which leads to the development of the repetitive and ritualized behaviors characteristic of the disorder.

The other two regions (superior temporal and supramarginal gyri) have less frequently been associated with the disorder, although they are also connected to subcortical striatal regions and thus may also be considered as part of the extended cortico-striatal circuitry. Indeed, the role of the parietal cortex (i.e., supramariginal gyrus) in striatal regulation and the importance of such parieto-striatal connectivity for OCD has already been incorporated in more recent neurobiological models of the disease [22].

5   Conclusions

This paper establishes a framework to automatically obtain a set of neuromarkers capable of characterizing obsessive-compulsive disorder (OCD). The presented work analyses different kinds of candidates for neuromarkers, as well as different feature selection criteria to reduce their number as much as possible.

Experimental results reveal that the definition of neuromakers from the weights of a linear SVM classifier in combination with a selection process based on their variance is able to provide a subset of no more than 50 values that are easy to interpret and handle by the psychiatric community.

Further work will be focused on studying whether these neuromarkers can also be used to analyse the patient's evolution, detect a pathology's subtype or even to aid in the prescription process. Furthermore, we also intend to extend this framework to other pathologies that could benefit from being characterized by neuromarkers.

Acknowledgments. Funding from the Spanish MINECO grants TEC2011-22480 and TIN2011-24533, the Carlos III Health Institute (PI09/01331, CP10/00604, PI13/01958, CIBER-CB06/03/0034) and the Agencia de Gestio d'Ajuts Universitaris i de Recerca (AGAUR; 2009SGR1554). Dr. Soriano-Mas is funded by a Miguel Servet contract from the Carlos III Health Institute (CP10/00604). We thank Janice Kay Hinde for revising the paper.

References

1. Friston, K.J., Ashburner, J., Kiebel, S.J., Nichols, T.E., Penny, W.D.: Statistical Parametric Mapping: The Analysis of Functional Brain Images. Academic Press (2007)
2. Ashburner, J., Friston, K.J.: Voxel-based morphometrythe methods. Neuroimage 11(6), 805–821 (2000)
3. Bi, J., Bennett, K., Embrechts, M., Breneman, C., Song, M.: Dimensionality reduction via sparse support vector machines. JMLR 3, 1229–1243 (2003)
4. Carter, C.S., MacDonald, A.W., Ross, L.L., Stenger, V.A.: Anterior cingulate cortex activity and impaired self-monitoring of performance in patients with schizophrenia: an event-related fmri study. American Journal of Psychiatry 158(9), 1423–1428 (2001)
5. Castro, E., Martínez-Ramón, M., Pearlson, G., Sui, J., Calhoun, V.D.: Characterization of groups using composite kernels and multi-source fMRI analysis data: application to schizophrenia. Neuroimage 58(2), 526–536 (2011)
6. Costafreda, S.G., Chu, C., Ashburner, J., Fu, C.H.: Prognostic and diagnostic potential of the structural neuroanatomy of depression. PLoS One 4(7), e6353 (2009)
7. Cuingnet, R., Gerardin, E., Tessieras, J., Auzias, G., Lchéricy, S., Habert, M.O., Chupin, M., Benali, H., Colliot, O.: Automatic classification of patients with Alzheimer's disease from structural MRI: a comparison of ten methods using the ADNI database. Neuroimage 56(2), 766–781 (2011)
8. De Martino, F., Valente, G., Staeren, N., Ashburner, J., Goebel, R., Formisano, E.: Combining multivariate voxel selection and support vector machines for mapping and classification of fMRI spatial patterns. Neuroimage 43(1), 44–58 (2008)
9. Ecker, C., Rocha-Rego, V., Johnston, P., Mourao-Miranda, J., Marquand, A., Daly, E.M., Brammer, M.J., Murphy, C., Murphy, D.G.: Investigating the predictive value of whole-brain structural mr scans in autism: a pattern classification approach. Neuroimage 49(1), 44–56 (2010)
10. Gretton, A., Bousquet, O., Smola, A.J., Schölkopf, B.: Measuring statistical dependence with Hilbert-Schmidt norms. In: Jain, S., Simon, H.U., Tomita, E. (eds.) ALT 2005. LNCS (LNAI), vol. 3734, pp. 63–77. Springer, Heidelberg (2005)
11. Guyon, I., Weston, J., Barnhill, S., Vapnik, V.: Gene selection for cancer classification using support vector machines. Machine Learning 46, 389–422 (2002)
12. Guyon, I., Elisseeff, A.: An introduction to variable and feature selection. The Journal of Machine Learning Research 3, 1157–1182 (2003)

13. Harrison, B.J., Soriano-Mas, C., Pujol, J., Ortiz, H., López-Solà, M., Hernández-Ribas, R., Deus, J., Alonso, P., Yücel, M., Pantelis, C., et al.: Altered corticostriatal functional connectivity in obsessive-compulsive disorder. Archives of General Psychiatry 66(11), 1189–1200 (2009)
14. Haynes, J.D., Rees, G.: Predicting the orientation of invisible stimuli from activity in human primary visual cortex. Nature Neuroscience 8(5), 686–691 (2005)
15. Iankova, A.: The Glasgow coma scale clinical application in emergency departments: ANDRIANA IANKOVA discusses the use of the Glasgow Coma Scale assessment tool in adult patients with head injury whose levels of consciousness be compromised by alcohol. Emergency Nurse 14(8), 30–35 (2006)
16. Klöppel, S., Abdulkadir, A., Jack Jr., C.R., Koutsouleris, N., Mourão-Miranda, J., Vemuri, P.: Diagnostic neuroimaging across diseases. Neuroimage 61(2), 457–463 (2012)
17. Knaus, W.A., Draper, E.A., Wagner, D.P., Zimmerman, J.E.: Apache II: a severity of disease classification system. Critical Care Medicine 13(10), 818–829 (1985)
18. LaConte, S., Strother, S., Cherkassky, V., Anderson, J., Hu, X.: Support vector machines for temporal classification of block design fMRI data. NeuroImage 26(2), 317–329 (2005)
19. Le Gall, J.R., Lemeshow, S., Saulnier, F.: A new simplified acute physiology score (SAPS II) based on a European/North American multicenter study. Jama 270(24), 2957–2963 (1993)
20. Lemm, S., Blankertz, B., Dickhaus, T., Müller, K.R.: Introduction to machine learning for brain imaging. Neuroimage 56(2), 387–399 (2011)
21. Martínez-Ramón, M., Koltchinskii, V., Heileman, G.L., Posse, S.: fMRI pattern classification using neuroanatomically constrained boosting. Neuroimage 31(3), 1129–1141 (2006)
22. Menzies, L., Chamberlain, S.R., Laird, A.R., Thelen, S.M., Sahakian, B.J., Bullmore, E.T.: Integrating evidence from neuroimaging and neuropsychological studies of obsessive-compulsive disorder: the orbitofronto-striatal model revisited. Neuroscience & Biobehavioral Reviews 32(3), 525–549 (2008)
23. Orrù, G., Pettersson-Yeo, W., Marquand, A.F., Sartori, G., Mechelli, A.: Using support vector machine to identify imaging biomarkers of neurological and psychiatric disease: a critical review. Neuroscience & Biobehavioral Reviews 36(4), 1140–1152 (2012)
24. Parrado-Hernández, E., Gómez-Verdejo, V., Martínez-Ramón, M., Shawe-Taylor, J., Alonso, P., Pujol, J., Menchón, J.M., Cardoner, N., Soriano-Mas, C.: Discovering brain regions relevant to obsessive–compulsive disorder identification through bagging and transduction. Medical Image Analysis 18(3), 435–448 (2014)
25. Pereira, F., Mitchell, T., Botvinick, M.: Machine learning classifiers and fMRI: a tutorial overview. Neuroimage 45(1), S199–S209 (2009)
26. Ranson, J., Rifkind, K., Roses, D., Fink, S., Eng, K., Spencer, F.: Prognostic signs and the role of operative management in acute pancreatitis. Surgery, Gynecology & Obstetrics 139(1), 69–81 (1974)
27. Ryali, S., Supekar, K., Abrams, D.A., Menon, V.: Sparse logistic regression for whole-brain classification of fMRI data. NeuroImage 51(2), 752–764 (2010)
28. Song, L., Smola, A., Gretton, A., Borgwardt, K.M., Bedo, J.: Supervised feature selection via dependence estimation. In: Proceedings of the 24th International Conference on Machine Learning, pp. 823–830. ACM (2007)
29. Soriano-Mas, C., Pujol, J., Alonso, P., Cardoner, N., Menchón, J.M., Harrison, B.J., Deus, J., Vallejo, J., Gaser, C.: Identifying patients with obsessive–compulsive disorder using whole-brain anatomy. Neuroimage 35(3), 1028–1037 (2007)

30. Tzourio-Mazoyer, N., Landeau, B., Papathanassiou, D., Crivello, F., Etard, O., Delcroix, N., Mazoyer, B., Joliot, M.: Automated anatomical labeling of activations in SPM using a macroscopic anatomical parcellation of the MNI MRI single-subject brain. Neuroimage 15(1), 273–289 (2002)
31. Ungar, L., Nestor, P.G., Niznikiewicz, M.A., Wible, C.G., Kubicki, M.: Color stroop and negative priming in schizophrenia: an fMRI study. Psychiatry Research: Neuroimaging 181(1), 24–29 (2010)
32. Van Gerven, M.A., Cseke, B., De Lange, F.P., Heskes, T.: Efficient bayesian multivariate fMRI analysis using a sparsifying spatio-temporal prior. NeuroImage 50(1), 150–161 (2010)
33. Varoquaux, G., Gramfort, A., Thirion, B.: Small-sample brain mapping: sparse recovery on spatially correlated designs with randomization and clustering. arXiv preprint arXiv:1206.6447 (2012)
34. Vemuri, P., Gunter, J.L., Senjem, M.L., Whitwell, J.L., Kantarci, K., Knopman, D.S., Boeve, B.F., Petersen, R.C., Jack Jr., C.R.: Alzheimer's disease diagnosis in individual subjects using structural MR images: validation studies. Neuroimage 39(3), 1186–1197 (2008)
35. Vovk, V., Gammerman, A., Shafer, G.: Algorithmic learning in a random world. Springer, New York (2005)
36. Wang, Y., Fan, Y., Bhatt, P., Davatzikos, C.: High-dimensional pattern regression using machine learning: from medical images to continuous clinical variables. Neuroimage 50(4), 1519–1535 (2010)
37. Wang, Z.: A hybrid SVM–GLM approach for fMRI data analysis. Neuroimage 46(3), 608–615 (2009)

Importance Weighted
Inductive Transfer Learning for Regression

Jochen Garcke and Thomas Vanck

Institut für Numerische Simulation,
Wegelerstr. 6, 53115 Bonn, Germany
garcke@ins.uni-bonn.de, vanck@math.tu-berlin.de

Abstract. We consider inductive transfer learning for dataset shift, a situation in which the distributions of two sampled, but closely related, datasets differ. When the target data to be predicted is scarce, one would like to improve its prediction by employing data from the other, secondary, dataset. Transfer learning tries to address this task by suitably compensating such a dataset shift. In this work we assume that the distributions of the covariates and the dependent variables can differ arbitrarily between the datasets. We propose two methods for regression based on importance weighting. Here to each instance of the secondary data a weight is assigned such that the data contributes positively to the prediction of the target data. Experiments show that our method yields good results on benchmark and real world datasets.

Keywords: inductive transfer learning, importance weighting, dataset shift.

1 Introduction

In a standard machine learning setting one has given data $\mathcal{X} \subset \mathbb{R}^{N \times D}$ and corresponding labels $\mathcal{Y} \subset \mathbb{R}^N$. It is assumed that the data is distributed according to $p(x, y)$ and that this distribution never changes; in particular it remains the same for new data. According to this assumption, a good model learned with the training data will also perform well when predicting for such new data. However, this assumption might not always be true, and there are quite often situations in which the underlying distribution changes. In general these situations are called dataset shift. Mathematically speaking, a dataset shift is given if two datasets are samples from two different distributions [5,12]. For instance, suppose one had given the dataset $(\mathcal{X}^P, \mathcal{Y}^P)$, which is distributed according to $p^P(x, y)$, and additionally $(\mathcal{X}^S, \mathcal{Y}^S)$, which was sampled according to $p^S(x, y)$, called primal data and secondary (or supplementary) data, respectively. A dataset shift is given if $p^P(x, y) \neq p^S(x, y)$. An example for such a dataset shift is the so-called covariate shift where the functional relationship between the dependent variable y remains the same, i.e. $p(y|x^P) = p(y|x^S)$ if $x^P = x^S$, but the distribution of the covariates are not the same, i.e. $p(x^P) \neq p(x^S)$ [5,12,16]. Another example is a situation where the distribution of the dependent variable y changes but the distribution of the covariates remains the same. This is referred to as prior probability shift [5].

In this work we will investigate situations where the distribution of the primal (or P) data differs from the distribution of the secondary (or S) data in both dependent variable

T. Calders et al. (Eds.): ECML PKDD 2014, Part I, LNCS 8724, pp. 466–481, 2014.

y and covariate x, i.e. $p^P(x,y) \neq p^S(x,y)$. For example, consider earthquake data that has been measured in California. A model learned with this data is suitable for making predictions for California, but might not be appropriate for making predictions for earthquakes in Japan due to a shift in the data caused by a change of location. However, if the data provided for Japan is very small a separate model learned solely on the Japan data might not provide a good prediction quality. Although the distributions for the California data and Japan data differ in general, it is reasonable to assume that in some aspects the distributions are very similar or almost equal. Therefore, it might be helpful to augment the Japan data with the California data to improve the prediction quality. This augmentation, also called knowledge borrowing, is commonly known as inductive transfer learning (ITL). Here, the data for California is the supplementary data and the Japan data the primal data. Other such data shift situations occur when the distribution drifts in time. A situation like this occurs, for example, in data that describe the causes of delays of aircrafts. This shift might be due to new airports that have been opened recently or a new aircraft model that is more reliable. Therefore, the data can shift from year to year. Other examples arise in the case of classification of text data where one would like to transfer knowledge obtained on texts about one topic to texts about a different topic.

Formally speaking, inductive transfer learning (ITL) refers to a situation of at least two datasets, which are sampled from the distributions $p^P(x,y)$ and $p^S(x,y)$ with, in general, $p^P(x,y) \neq p^S(x,y)$. Furthermore, the number of the P data is typically much smaller then that of the S data. Additionally, due to the small number of data, a model learned solely on the P data will usually not provide a good prediction quality. However, it is assumed that the distribution p^P and p^S are similar to some degree, which even could result in some connected sets of (x,y) with $p^P(x,y) \approx p^S(x,y)$. In ITL one tries to achieve a good prediction quality of a model for the P data by employing the S data.

In this work we will, motivated by the concept of importance sampling, investigate two new approaches for improving regression in the ITL setting by assigning each instance in the S data a weight. The first one is a supervised and the second an unsupervised method. Although both methods employ labels from the S and P data, we consider the one approach unsupervised since it does not directly employ a cost function for estimating an error between actual and predicted labels. The resulting weights are then used in a modified ridge regression in combination with the S and P data in order to improve the prediction quality on the P data. Experiments show that both approaches yield good results.

This work is structured in the following way: section 2 presents an overview of related work on the topic of ITL while section 3 gives a brief description of ITL in general. Section 4, 5 and 6 explain our idea and state practical instructions. Finally section 7 demonstrates the performance of our algorithm on several datasets.

2   Related Work

The task of inductive transfer learning has been tackled in the past by various approaches. One is the so-called instance based transfer, where each instance in the S domain gets some weight for indicating how much influence it will get for predicting the target data. TrAdaBoost [8] and an extension [1] are methods that assign a weight to each datapoint such that some S data have an influence on the prediction quality for the P data. [13] states a similar boosting approach for regression. Other recent work based on instance transfer has been put forward by [18], [20] and [22] in which they use multiple input sources to improve the prediction quality of classifiers.

Other existing work that implement instance based reweighting methods focus primarily on the covariate shift setting. Important work on this topic has been put forward by [7,10,17], see also [5,16]. It is possible to apply these methods to inductive transfer learning setting. However, the major drawback of these methods in the ITL setting is that they do not take the information about the target labels from the P data into account. This can lead to situations where a S datapoint still gets a high weight assigned due to the similarity to the covariates of the P data although the label, which eventually is what one wants, is fundamentally different from the ones in the P data. To compensate this shortcoming our second approach (explained in section 6.3) is inspired by [17] such that it takes also the labels of the P data into account.

Kernel based ideas have been presented by [15,6], where a special kernel matrix is learned that reflects the similarities between the S and P data. A further method is given in [14], in which an informative prior is constructed from the S data in order to improve a model on the P data. An additional advance is feature representation transfer [3]. This method learns a projection of the S and P data onto a lower dimensional subspace such that the common or shared information of both data can be used for the model on the P data. Learning feature representation is in particular common in the domain of natural language processing (NLP). Since due to differences in vocabulary and writing style learning approaches tend to perform worse in different domains. In this area, [9] proposed a simple, but often well performing, kernel-mapping function for NLP problems, which maps the data from both source and target domains to a high-dimensional feature space, where standard discriminative learning methods are used.

Model transfer or hypothesis transfer learning comprise another class of approaches for treating ITL. In the model transfer setting a model parameter θ_S is learned on the S data. Assuming that the models should be similar, the idea is to regularize the model parameter for the P data θ_P with the help of the parameter θ_S. Recent work on this topic is given by [11] and [19].

Note that, in contrast to multi-task learning [3], inductive transfer learning is not concerned with the prediction quality on both the S and P data, but concentrates only on the prediction of the P data; the S data is exclusively used as data that helps to improve the prediction quality for the P data.

3   Problem Formulation

For inductive transfer learning we now assume a situation where the two datasets $(\mathcal{X}^S, \mathcal{Y}^S)$, the S data, and $(\mathcal{X}^P, \mathcal{Y}^P)$, the P data, are given by:

$$(\mathcal{X}^S, \mathcal{Y}^S) \sim p^S(x, y) \quad \text{and} \quad (\mathcal{X}^P, \mathcal{Y}^P) \sim p^P(x, y).$$

Further, the number M of P data is assumed to be much smaller than the number N of S data, i.e. $|\mathcal{X}^P| \ll |\mathcal{X}^S|$, and the two distributions from which the data was sampled are not equal, i.e. $p^P(x, y) \neq p^S(x, y)$. Nevertheless, it is assumed that the two datasets are somehow related to each other, so that in some parts of the domain the distributions are similar (or even equal), i.e.:

$$p^S(\tilde{x}, \tilde{y}) \approx p^P(\tilde{x}, \tilde{y}) \text{ for some } (\tilde{x}, \tilde{y}).$$

Therefore, one can employ the S data to improve the prediction on the P data. By assumption, we have $p^P(x, y) \neq p^S(x, y)$, and consequently we cannot simply combine the S and P data. The crucial part is to determine points from the S data that contribute positively to the P data prediction and neglect points that have a negative influence. A solution to this problem is based on a measure of similarity between the two distributions. A common way to achieve this is importance sampling, a technique that reweights a given distribution p such that the reweighted p equals another distribution q. Defining the importance weight function as $w(x, y) := \frac{p^P(x,y)}{p^S(x,y)}$ one could reweight the S data distribution by:

$$p^P(x, y) = w(x, y)p^S(x, y) = \frac{p^P(x, y)}{p^S(x, y)}p^S(x, y). \tag{1}$$

With the help of the function $w(x, y)$ it becomes possible to assign each S datapoint (x^S, y^S) an individual and appropriate weight. A weight close to one indicates a preferable point, while a weight far from one indicates the opposite. Hence this approach seems suitable for tackling the induction transfer learning setting. However, this definition of the importance function requires knowledge of both distributions, which is not available. Therefore, an approximation of the importance function $w(x, y)$ is needed instead. By employing an appropriate approximation, the idea of importance sampling offers a guideline for solving the task of ITL.

4   New Instance Based Approach

4.1   Reweighting of the Prediction Function

We start by assuming that the given data $(\mathcal{X}, \mathcal{Y})$ is distributed according to an (unknown) distribution $p(x, y)$. This distribution can be expressed by:

$$p(x, y) = p(y|x)p(x) \quad \text{or} \quad p(x, y) = p(x|y)p(y).$$

Although our suggested method can be applied to both cases, the discriminative and the generative one, we will concentrate in the following on the first equation for the discriminative approach. Predictions are obtained by:

$$\hat{y}^* = \text{argmax}_y \left(p(y|x^*)p(x^*) \right). \tag{2}$$

By assumption, the new data x^* and its corresponding (unknown) label y^* is distributed according to $p(x, y)$, and therefore one can make a prediction by applying (2).

However, in the setting of inductive transfer learning we have two different distributions, which gives the following two expressions for the prediction of y^P:

$$y_P^P = \text{argmax}_y \left(p^P(y|x^P)p^P(x^P) \right)$$
$$y_S^P = \text{argmax}_y \left(p^S(y|x^P)p^S(x^P) \right).$$

In general the prediction of y^P based on p^S for the S data, namely y_S^P, can differ arbitrarily from the prediction y_P^P based on the P distribution. Therefore, in order to make better predictions for the P data using the distribution of the S data, we will now reweight the S distribution as suggested in (1):

$$
\begin{aligned}
y^P &= \operatorname{argmax}_y \left(p^P(y|x^P) p^P(x^P) \right) \\
&= \operatorname{argmax}_y \left(\frac{p^P(y|x^P) p^P(x^P)}{p^S(x^P, y)} p^S(y|x^P) p^S(x^P) \right) \\
&= \operatorname{argmax}_y \left(w(x^P, y) p^S(y|x^P) p^S(x^P) \right) .
\end{aligned}
\tag{3}
$$

From this derivation one can see that this also is an unbiased estimator for the P data.

Due to the lack of knowledge about the true distributions p^P and p^S one cannot obtain the correct importance function. Instead we will aim for an approximation $\hat{w}(x, y)$. To determine suitable weights $\hat{w}$ we will now introduce two approaches for their estimation.

4.2 Model Based Estimation of the Weight Function

The first approach will be referred to as the direct method or DITL (Direct ITL) because it will directly rely on the prediction performance of a model learned on the S data. The goal of our model is to minimize the prediction error, i.e.

$$
\min \| Y^P - \hat{Y}^P \|^2
$$

where Y^P is the vector of the real labels $\{y_i\}_{i=1,\dots,M}$ and $\hat{Y}^P$ the vector of the model predictions. Therefore, by following this approach, and with the help of expression (3), an optimization problem for the estimation of a weight function can be stated as:

$$
\min_{\hat{w}} \sum_{i=1}^{M} \left(y_i^P - \operatorname{argmax}_y \left(\hat{w}(x_i^P, y) p^S(y|x_i^P) p^S(x_i^P) \right) \right)^2 .
$$

The idea behind this approach is that the computation of the weights $\hat{w}$ is performed with respect to the known labels Y^P. Therefore this approach provides a supervised method for adjusting the weights $\hat{w}$. Since for a given point x^P the argmax does not depend on $p^S(x^P)$ that term can be omitted, which leads to:

$$
\min_{\hat{w}} \sum_{i=1}^{M} \left(y_i^P - \operatorname{argmax}_y \left(\hat{w}(x_i^P, y) p^S(y|x_i^P) \right) \right)^2 .
\tag{4}
$$

4.3 Distribution Based Estimation of the Weight Function

Additionally, we propose a method which does not depend directly on prediction models and can be regarded as an unsupervised approach. Following the idea of [17] we will minimize the Kullback-Leibler divergence between two distributions and straight-forwardly extend the approach [17] for covariate shift by also taking the labels into account:

$$\text{argmin}_{\hat{w}} \text{KL}(p^P(x,y)||\hat{w}(x,y)p^S(x,y))$$

$$= \text{argmin}_{\hat{w}} \left(\int p^P(x,y) \log \left(\frac{p^P(x,y)}{\hat{w}(x,y)p^S(x,y)} \right) dxdy \right)$$

$$= \text{argmin}_{\hat{w}} \left(- \int p^P(x,y) \log \left(\hat{w}(x,y) \right) dxdy \right).$$

The last expression can be approximated by the empirical mean:

$$\Rightarrow \min_{\hat{w}} \sum_{i=1}^{M} - \log \left(\hat{w}(x_i^P, y_i^P) \right). \tag{5}$$

Additionally, one obtains the following constraint for normalization [17]:

$$p^P(x,y) = w(x,y)p^S(x,y)$$

$$\Rightarrow 1 = \int p^P(x,y)dxdy = \int w(x,y)p^S(x,y)dxdy$$

$$\Rightarrow N = \sum_{j=1}^{N} \hat{w}(x_j^S, y_j^S), \tag{6}$$

it enforces that the reweighted distribution $w \cdot p^S$ still has measure one. We will refer to this approach as the indirect method or KLITL (Kullback-Leibler ITL).

5   Weighted Kernel Ridge Regression for ITL

Assuming one has obtained suitable weights, their application in regression requires adjusted models for prediction. We will propose a weighted kernel ridge regression model, which we will call ITL-KRR. The modified ridge regression model is given by:

$$J_W(\theta) = \frac{1}{2} \left(\sum_{i=1}^{M} (\theta^t \phi(x_i^P) - y_i^P)^2 + \sum_{j=1}^{N} w_j(\theta^t \phi(x_j^S) - y_j^S)^2 \right) + \frac{\lambda}{2} \sum_{d=1}^{D} \theta_d^2 \tag{7}$$

where $\theta \in \mathbb{R}^D$ denotes the model parameter, λ the regularization parameter, ϕ the feature map that maps the input x into the feature space (see e.g. [4]), and $w_j :=$ $\hat{w}(x_j^S, y_j^S)$ denotes the weight for each supplementary datapoint from $(\mathcal{X}^S, \mathcal{Y}^S)$. Some what surprisingly, such a natural extension of a regression approach for applying impor-tance weights has, to our knowledge, not been stated and used in the context of ITL so far. Dualization is given straightforwardly by defining the diagonal matrix $W \in \mathbb{R}^{(M+N)\times(M+N)}$:

$$W := \begin{bmatrix} I_M & 0 \\ 0 & \text{diag}\left(w(x_1^S, y_1^S), \ldots, w(x_N^S, y_N^S)\right) \end{bmatrix}$$

with I_M the identity matrix for M dimensions and appending the S data to the P data ('|' denotes vertical concatenation):

$$X^{PS} = \left(X^P | X^S\right) \text{ and } Y^{PS} = \left(Y^P | Y^S\right)$$
$$K = \phi(X^{PS})^t \phi(X^{PS}) \text{ the kernel matrix,}$$

with the data matrices $X^P \in \mathbb{R}^{M \times D}, X^S \in \mathbb{R}^{N \times D}$ and label vectors $Y^P \in \mathbb{R}^M, Y^S \in \mathbb{R}^N$. As the dual optimization problem one obtains:

$$\frac{1}{2} a^t KWKa - a^t KWY^{PS} + \frac{1}{2} Y^{PS} WY^{PS} + \frac{\lambda}{2} a^t Ka.$$

We apply Gaussian kernels, with the bandwidth denoted by σ, in our experiments.

6   Determination of Individual Weights

We will now specify how the weights can be obtained computationally for both approaches.

6.1   Weight Function

Until now we have not been specific in the concrete representation of the weight function $\hat{w}(x, y)$. We employ in this work the common approach of linear combination of Gaussian kernels for an approximation of the importance function, i.e.:

$$\hat{w}^\alpha(x, y) = \sum_{l=1}^{N} \alpha_l \exp\left(-\frac{||(x, y) - (x_l', y_l')||^2}{2\eta^2}\right).$$

The centerpoints $(x_l', y_l')_{l=1}^N$ will be set to the S datapoints. We use the S data instead of the P data since in (14) we optimize over the P data; using the P data as centerpoints would exhibit a higher risk of overfitting. Hence each $\hat{w}_l$ in (13) becomes

$$\hat{w}_j^\alpha(x^*, y) = \alpha_j \exp\left(-\frac{||(x^*, y) - (x_j^S, y_j^S)||^2}{2\eta^2}\right). \tag{8}$$

Other function representations are possible as well, but out of the scope of this work.

6.2   Direct Approach (DITL)

To derive the direct approach, let us remind the abstract modeling of a prediction function in a standard machine learning setting for the discriminative case:

$$\hat{y}^* = \text{argmax}_y p(y|x^*). \tag{9}$$

Here, x^* denotes a data point to be predicted on, and $\hat{y}^*$ the prediction. As a concrete model $f(x)$ for the S data following (9) we again employ kernel ridge regression, which can be stated as:

$$\text{argmax}_y p(y|x^*) \approx f(x^*) = a^t \underline{k}(x^*), \tag{10}$$

where $\underline{k}(x^*) := (k(x_1, x^*), \ldots, k(x_N, x^*))^t$ is the kernel map of the new datapoint x^* and the data $\mathcal{X} \subset \mathbb{R}^{N \times D}$ on which the model has been learned, with $k(x_l, x^*) := \phi(x_l)^t \phi(x^*)$, and a is the vector of coefficients for the linear combination in the feature space. Hence for (4) one needs a different mathematical approximation:

$$\text{argmax}_y \left(\hat{w}(x^*, y) p(y|x^*) \right) \approx f_{\hat{w}(x^*, y)}(x^*) \tag{11}$$

where the model f now also depends on the weight function $\hat{w}$.

We now suggest a weighted prediction model derived from the kernel ridge regression approximation and consider a weighted formulation:

$$J_W(\theta) = \frac{1}{2} \sum_{l=1}^{N} \hat{w}_l \left(y_l - \theta^t \phi(x_l) \right)^2 + \frac{\lambda}{2} ||\theta||^2 \tag{12}$$

where θ again denotes the model parameter, ϕ is the feature map and $\hat{w}_l$ is a weight coefficient for each datapoint x_l. By the process of dualization of the ridge regression [4], one gets the weighted prediction function as:

$$0 = \nabla J_W(\theta) \Leftrightarrow \theta = \sum_{l=1}^{N} \hat{w}_l \underbrace{\left(-\frac{1}{\lambda}(y_l - \theta^t \phi(x_l)) \right)}_{=:\hat{a}_l} \phi(x_l).$$

Here, $\hat{a}_l = a_l \hat{w}_l$ are the coefficients for the linear combination in the feature space. Analogously to (10), this prediction function can be taken as an approximation for the weighted prediction, i.e.:

$$\text{argmax}_y \left(\hat{w}(x^*, y) p(y|x^*) \right) \approx f_{\hat{w}(x^*, y)}(x^*) = a^t \hat{W}(x^*, y) \underline{k}(x^*) \tag{13}$$

where $\hat{W}$ denotes a $N \times N$ diagonal matrix where each entry is a weight $\hat{w}_l$ that corresponds to the kernel function $k_l(\cdot) := k(x_l, \cdot)$ and coefficient a_l of the lth-data point of S. Obviously this prediction function contains the label that is to be predicted. Therefore, label prediction for new data points is not possible with (13). However, we are not actually interested in making predictions using this model; rather we would like to estimate appropriate weights $\hat{w}$ for the subsequent step, in which we apply the weights to learn a model on the P data combined with the weighted S data. (4) provides a framework for getting the best possible weights by conditioning the expression to the labels of the P data. Inserting (13) into (4) we get:

$$\min_{\hat{w}} \sum_{i=1}^{M} \left(y_i^P - a^t \hat{W}(x_i^P, y_i^P) \underline{k}(x_i^P) \right)^2. \tag{14}$$

By making the approximation (8) we get a weight function that is defined by a given set of αs, which can now be estimated by (14).

Note that in early experiments we saw that a direct application of (14) sometimes returns αs where only one or very few elements dominate. In order to avoid such an overfitting we additionally add a regularization term to (14) which penalizes large coefficients:

$$\min_{\alpha \geq 0} \sum_{i=1}^{M} \left(y_i^P - a^t \hat{W}^\alpha(x_i^P, y_i^P) \underline{k}(x_i^P) \right)^2 + \gamma ||\alpha||^2. \tag{15}$$

The estimated αs define the weight function $\hat{w}$ which will then be subsequently used during the actual ITL-KRR.

Learning the weights and a better model jointly from the P data and weighted S data requires a three step procedure for the direct approach. Problem (15) depends on a model of the S data for adjusting the αs. Therefore the first step requires the inference of a kernel ridge regression model solely on the S data, which returns the coefficients a for the prediction function (13). With these a a solution to (15) has to be found which yields proper αs. These αs are then used in (7) for calculating the weight for each S datapoint. The procedure can be stated as:

1. Learn a model a for the normal kernel ridge regression using solely the S data and ignore any P data.
2. Use the coefficients vector a from step 1 to determine appropriate αs for the weight function (8) by using the weighted prediction model (13) and solve (15).
3. After having determined the αs in step 2, use these to calculate the weight for the application of the ITL-KRR (7). Use the resulting model to make predictions for new P data.

The optimization in step 2 is convex and therefore guarantees a single optimal solution. Good parameters in each step are estimated by performing standard cross-validation on the P data. We employ Gaussian kernels in the kernel ridge regression, therefore we need to estimate σ and λ in step 1 and 3 similarly to the two parameters γ and η in step 2.

6.3   Indirect Approach (KLITL)

In addition to the direct approach we state a procedure for the indirect approach. Following the derivation in section 4.3, using expression (5) as the objective and expression (6) as the constraint, we proceed as follows:

1. Optimize the following with a standard solver for constrained problems:

$$\max_{\alpha} \frac{1}{M} \sum_{i=1}^{M} \log \left(\hat{w}^\alpha(x_i^P, y_i^P) \right) \quad \text{s.t. } N = \sum_{j=1}^{N} \hat{w}^\alpha(x_j^S, y_j^S) \text{ and } \alpha \geq 0. \tag{16}$$

2. Use the αs from step 1 to compute the weights $\hat{w}$ of each S datapoint for the optimization of the ITL-KRR (7). Use the resulting model to make predictions for new P data.

We employ here the same representation of the weight function (8) as for the direct approach. For the estimation of a good η in (16) we will apply a modified version of cross-validation that is explained in the experimental section 7.1 of this work.

6.4   Comparison of the Direct and Indirect Approach

Comparing the two approaches, an advantage for the indirect approach is that it does not require the estimation of a model on the S data. This might be advantageous in case when a lot of S data is available. Additionally, the method requires the estimation of just one parameter η for the kernel width used in the weight function. However, on the downside is the fact that this is an unsupervised method. By this we mean a method that does not consider an objective cost function for the parameter inference. Therefore it is less likely to obtain robust or reliable estimations for α. On the other hand DITL applies a supervised optimization problem that takes a subset of the target labels in order to assess the quality of parameter inference. As mentioned further in section 6.2 the additional regularization term allows a higher control of the fitting process. As a consequence the DITL method is much more robust in compensating the dataset shift. The experimental section shows the conditions under which this becomes advantageous. The disadvantage is a higher calculation cost since it requires the calculation of an additional model on the S data and the parameters η and γ.

7   Experiments

In the experimental section we will compare the performance of the direct (DITL) and indirect (KLITL) approaches versus the boosting for transfer learning (TLB) method, another instance-weighted approach, described in [13]. Further we applied the "Frustratingly Easy Domain Adaptation" by [9], a simple, but often well performing feature learning approach, in combination with kernel ridge regression (in the following referred to as FS-KRR). As the final approach for dataset shift, we compare with ATL [6], which is based on Gaussian process (GP) regression and calculates a special correlation matrix for the GP. As a natural baseline, we provide the performance of a normal kernel ridge regression for regression problems learned from the three dataset combinations: P data, S data, and P & S data. As a weighted baseline we also take KLIEP [17] in a normal covariate shift setting for determining instance weights into account, i.e. this approach does not see the labels of the data during weight estimation, only their distribution in x. As an alternative we also employ Kernel Mean Matching (KMM) [10].

7.1   Parameter Selection

DITL applies a kernel ridge regression (KRR), a weight estimation procedure and the ITL-KRR. In each of the three steps we will perform 5-fold standard cross validation for the parameter estimation. For the KRR and the ITL-KRR we used RBF kernel functions for the calculation of the kernel matrix K. Denoting the bandwidth parameter of the RBF kernels with σ we have to estimate two parameters σ and the regularization parameter λ in step 1 (KRR) on the S data, and step 3 (ITL-KRR) on the S and P data. In the second step DITL requires the estimation of the parameters η (the bandwidth for the importance function approximation) and γ (the regularization parameter for the α vector). Since all problems are quadratic, one can use standard algorithms for quadratic programming.

    KLITL is different in the parameter estimation from the DITL method. KLITL requires just two steps. In the first step we solve problem (16); i.e. we simply maximize the

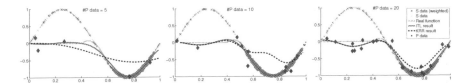

Fig. 1. Illustrative toy example for DITL. From left to right the number of P data is: 5, 10 and 20 datapoints. The location of an S datapoint is marked by a red cross '×'. The round purple points indicate how much weight an S datapoint gets assigned. The thicker the point the more weight it has. As can be seen from the example, in one dimension 20 datapoints are already dense enough to learn a reliable kernel ridge regression.

sum under the normalization constraint. In order to get a good estimate for η we propose a selection criteria that will choose the η from all the proposed η values that maximizes (16). Since KLITL in the first step is unsupervised, we use a similar method to cross-validation to get a more stable selection result. Given the original S dataset, $(\mathcal{X}^S, \mathcal{Y}^S)$, we split the dataset into five disjoint parts, $(\mathcal{X}^S, \mathcal{Y}^S)_{k=1}^{5}$. Each split $(\mathcal{X}^S, \mathcal{Y}^S)_k$ should contain enough samples of the S data but due to our assumption this is not a problem. Now for a fixed parameter η we will maximize expressions (16) for each dataset combination $\{(\mathcal{X}^S, \mathcal{Y}^S)_k, (\mathcal{X}^P, \mathcal{Y}^P)\}$. We pick the parameter with the highest mean of these five maximas. Therefore we obtain a more robust method for estimating an adequate parameter.

7.2 Datasets

First, for illustration purposes, we show by using a toy example how the proposed DITL algorithm learns weights, and how these weights influence the model prediction. The performance of our methods is then verified on some standard benchmark datasets that have been slightly modified. Finally, we apply our methods to three real world datasets, a dataset describing earthquakes, a second describing delays of aircrafts and a third describing radio signal strengths from WiFi access points for indoor location estimation.

Toy Examples. The toy example mainly serves as an illustrative demonstration of how and where the DITL algorithm learns weights for the S data, and shows the consequences for the prediction of the P data when taking additional S data into account. Similar results can be obtained by applying KLITL, which we omit for space reasons.

The dataset is generated by sampling datapoints from two functions that are - as we assume for our methods - partially almost identical. The S data is sampled as:

$$f_s(x) = sin(2\pi x) + \sigma_S \mathcal{N}(0,1), \tag{17}$$

where σ_S is a factor for controlling the influence of the variance (in our experiments we used $\sigma_S = 0.1$). The P data is sampled according to:

$$f_p(x) = \begin{cases} 0 + \sigma_P \mathcal{N}(0,1) & 0 \le x \le 1/2 \\ sin(2\pi x) + \sigma_P \mathcal{N}(0,1) & 1/2 < x \le 1 \end{cases} \tag{18}$$

where σ_P, as in the case for the S data, is the sample variance (in our experiments $\sigma_P = 0.4$). Parameter selection is performed as described in the previous section 7.1. The experiments in Fig. 1 only apply a very small number of P datapoints (just 5,10 and 20). The reason for this is that, for our example, the performance of a standard KRR is already very good at 20 datapoints. This is due to the fact that in one dimension we get a non-sparse dataset very quickly. Since we want to illustrate that the lack of datapoints (as by our assumption), and hence sparseness of data, leads to models that perform poorly on predicting new data, this setting for our toy example is reasonable. However, in high dimensions the situation is different and the number of P datapoints can be much larger, in parts due to the empty space phenomenon.

Benchmark Datasets. We now apply DITL, KLITL, ATL, FS-KRR, KMM, KLIEP and TLB to standard benchmark datasets. The experimental setup is as follows: We took the following standard benchmark datasets for evaluation: abalone, elevators[1], and the kin family datasets[2]. From the kin dataset we took the n datasets (n for nonlinear) with 8 dimensions. We used the nm (non linear medium variance) as the S data and nh (non linear high variance) data as the P data. Since abalone and elevators do not necessarily comprise a dataset shift we will determine the S and P data according to a special selection criteria. The selection process is performed up front and independently of the ITL method. In the first step we normalized the covariates $\mathcal{X}$ to $[0, 1]$ for each dimension. Then the following three values are calculated randomly; First, a dimension $d \in \{1, \ldots, D\}$ is selected randomly. In the same way we choose a threshold value $\vartheta \in [0, 1]$ randomly and finally we sample a selection probability $p_{select} \in [0, 1]$. All values are selected according to a uniform distribution on the corresponding domain. After that we fix these three values for the actual data generation process. For the dataset generation we select a datapoint (x, y) from the set $(\mathcal{X}, \mathcal{Y})$, take the $x \in \mathcal{X}$ and then consider the value for dimension d, i.e. x^d. If x^d is larger than the threshold ϑ we will add this (x, y) combination with probability p_{select} to the S data $(\mathcal{X}^S, \mathcal{Y}^S)$, and to the P dataset $(\mathcal{X}^P, \mathcal{Y}^P)$ otherwise. That way we randomly generate 50 instances of the data sets for each individual experiment with a drift, i.e. a covariate shift, in the distribution. In order to get also a shift in the labels we apply the function

$$f(y) = y + \nu \sin(2\pi y), \quad \nu \in [0, 1] \tag{19}$$

to the labels of the P data only. For instance $\nu = 0$ means no shift in the labels. This way we generate datasets that account for the ITL setting and, due to the ν parameter, gives control about the strength of the shift such that S and P data still have something in common.

Table 1 shows the results for each method for a different number of P data. For illustration we give one result with $\nu = 0$, i.e. with only a covariate shift. As one would expect, a standard KRR using both S and P data performs best, since for $\nu = 0$ the datasets only contain a covariate shift. Nevertheless, this experiment verifies that the introduced ITL methods learn proper weights in order to employ the right S datapoints for improving prediction of the P data. Their prediction performance is best over all

---

[1] Abalone and elevators can be found on mldata.org.

[2] Kin datasets are part of the delve dataset repository.

Table 1. Results on different benchmark datasets for mean square error. Sampling of the S and P data is explained in the text. Each experiment has been performed 50 times and the results have been normalized by the error on the P data. Therefore, each number in the other columns denotes the proportion in percent. Further comments on the results can be found in the text. Error calculation has been performed on a randomly sampled P_{eval} for each trial. (KRR* = KRR on $S \cup P$). Best results are marked as bold text.

	KRR (on P)	KRR (S)	KRR*	FS-KRR	KMM	ATL	TLB	KLIEP	KLITL	DITL
Abalone $\nu = 0$ (no additional label shift), error on $|P_{eval}| = 1000$ and $|S| = 1000$										
#P 50	0.0017 / 1.00	0.88	**0.72**	0.87	0.91	0.85	0.83	0.90	0.84	0.78
#P 100	0.0016 / 1.00	0.94	**0.77**	0.94	0.95	0.92	0.90	0.94	0.89	0.85
#P 200	0.0014 / 1.00	0.98	**0.85**	0.97	0.96	0.96	0.98	0.98	0.96	0.89
#P 300	**0.0012 / 1.00**	1.03	**0.99**	1.02	**1.00**	**1.00**	**1.00**	**1.00**	1.01	**0.99**
Abalone ($\nu = 1/2$), error on $|P_{eval}| = 1000$ and $|S| = 1000$										
#P 50	0.0024 / 1.00	1.53	1.46	0.92	0.93	0.89	0.81	1.48	0.80	**0.76**
#P 100	0.0019 / 1.00	1.41	1.38	0.93	0.96	0.91	0.85	1.38	0.87	**0.80**
#P 200	0.0016 / 1.00	1.42	1.27	0.96	1.01	0.94	0.93	1.25	0.92	**0.89**
#P 300	0.0013 / 1.00	1.45	1.20	1.01	1.00	**0.99**	**0.99**	1.21	1.00	**0.97**
Elevators ($\nu = 1.0$), error on $|P_{eval}| = 1000$ and $|S| = 2000$										
#P 50	6.5e-6 / 1.00	1.61	1.51	0.91	0.89	0.88	0.74	1.53	0.76	**0.68**
#P 100	5.7e-6 / 1.00	1.51	1.40	0.97	0.95	0.91	0.78	1.45	0.79	**0.71**
#P 200	4.1e-6 / 1.00	1.42	1.38	0.99	0.98	0.97	0.94	1.35	**0.90**	0.89
#P 300	**3.6e-6 / 1.00**	1.49	1.29	1.01	1.02	1.01	**1.00**	1.30	1.01	**0.99**
kin dataset ($\nu = 1/4$), error on $|P_{eval}| = 1000$ and $|S| = 2000$										
#P 50	0.065 / 1.00	1.30	1.28	0.88	0.90	0.87	0.83	1.27	0.84	**0.79**
#P 100	0.056 / 1.00	1.34	1.23	0.91	0.93	0.89	0.88	1.24	0.88	**0.84**
#P 150	0.050 / 1.00	1.32	1.19	0.95	0.95	0.94	**0.92**	1.18	0.91	0.91
#P 200	**0.044 / 1.00**	1.30	1.15	1.03	**1.00**	**1.00**	**1.00**	1.12	**1.00**	**1.00**

approaches which aim to take a shift into account, both the covariate shift procedures and the full dataset shift procedures. Experimental results are qualitatively the same for the other datasets, therefore we omit them.

When adding a shift to the labels with $\nu > 0$ to have a full dataset shift setting the situation is as expected differently. The KRR learned exclusively on the S data does not show any performance gain by adding P data. This is to be expected since the P data has no influence on the learning procedure but only serves as an evaluation dataset. On the other hand, if learned on $P \cup S$ the results improve slightly but they are still biased by the S data. Over all approaches, as the proportion of the P data grows the error gets reduced. FS-KRR and ATL show comparable errors, this can be explained by the similarity in these approaches, by construction both do not use weights for each instance but one weight for the correlation of P and S data. Consequently, each S datapoint has an equal influence. For KMM we considered the $\frac{p^P(x,y)}{p^S(x,y)}$ for the ratio calculation since that better fits the ITL setting. Note that KMM does not provide a method for parameter selection, and it is unsupervised since it does not use a subset of the target labels to adjust

Table 2. Results for the mean square error on the real world datasets. Since these datasets exhibit real dataset shifts the advantage of applying weighted S data becomes obvious. Further the robustness of the methods become apparent when the shift is artificially intensified (i.e. (19) with $\nu > 0$). Best results are marked as bold text.

	KRR (on P)	KRR (S)	KRR*	FS-KRR	KMM	ATL	TLB	KLIEP	KLITL	DITL
	Earthquake $\nu = 0$, error on $|P_{eval}| = 1000$ and $|S| = 841$									
#P 20	0.0138 / 1.00	1.13	1.12	1.13	0.91	1.04	0.64	0.97	0.60	**0.51**
#P 30	0.0106 / 1.00	1.15	1.07	1.14	0.95	1.09	0.88	0.99	0.83	**0.78**
#P 50	0.0076 / 1.00	1.20	1.04	1.19	0.98	1.13	0.96	1.00	0.95	**0.93**
#P 70	**0.0064 / 1.00**	1.23	1.04	1.24	1.02	1.16	**0.99**	1.01	1.02	**1.00**
	Flight Data $\nu = 0$, error on $|P_{eval}| = 1000$ and $|S| = 2000$									
#P 50	898.01 / 1.00	0.96	0.95	1.01	0.88	0.92	0.53	0.92	0.55	**0.51**
#P 200	611.39 / 1.00	1.02	0.99	1.04	0.92	0.99	0.78	0.96	0.76	**0.71**
#P 400	265.97 / 1.00	1.36	1.23	1.35	0.99	1.10	0.89	0.97	0.90	**0.86**
#P 800	**211.12 / 1.00**	1.41	1.36	1.42	1.01	1.14	1.01	1.02	**1.00**	0.99
	Wireless $\nu = 0$, error on $|P_{eval}| = 1000$ and $|S| = 2000$									
#P 50	256.83 / 1.00	1.02	0.96	0.99	0.91	0.93	0.74	0.95	0.71	**0.69**
#P 100	230.74 / 1.00	0.98	1.00	1.01	0.95	0.97	0.82	0.97	**0.78**	0.79
#P 200	197.23 / 1.00	1.10	1.12	1.05	0.97	1.02	0.93	0.99	0.89	**0.87**
#P 400	153.21 / 1.00	1.13	1.15	1.10	1.03	1.08	0.99	1.01	0.98	**0.96**
	Wireless $\nu - 1$ (with additional label shift), error on $|P_{eval}| = 1000$ and $|S| = 2000$									
#P 50	431.23 / 1.00	1.78	1.17	1.20	1.10	1.18	0.86	1.34	0.84	**0.74**
#P 100	398.19 / 1.00	1.65	1.13	1.14	1.05	1.12	0.90	1.38	0.88	**0.83**
#P 200	354.21 / 1.00	1.77	1.10	1.09	1.07	1.10	0.97	1.40	0.94	**0.92**
#P 400	**299.85 / 1.00**	1.59	1.07	1.04	1.02	1.05	1.01	1.45	**0.99**	**1.00**

the parameters, which overall makes it less robust and shows moderate performance. KLIEP used as a baseline covariate shift approach shows a poor performance, which is reasonable since it is not adapted to the ITL setting. The performance differences to the other methods show that it makes sense to treat ITL and covariate shift as two separate problem classes. Note that we also considered other (related) methods for covariate shift [16] in our experiments, their performance was similar to KLIEP and we therefore do not report their detailed results. TLB and KLITL show a similar performance. DITL performs best, we assume that this is due to the supervised way for estimating the weights. Nearly all methods eventually converge to a value of 1.00 because, as demonstrated by the toy example in section 7.2, with some data set size the P data provides enough information about its structure to allow a good prediction performance.

Real World Datasets. We now investigate the more interesting situation of real data that very likely contains a distribution shift. The first dataset [2] decribes measurements taken during earthquakes in Japan and California. The features describe values such as magnitude or distance to the center. A categorical feature describes the type of the earthquake. We augmented the dataset and assigned a separate dimension for each category, which turns one dimension into three. It seems natural to assume that the shift within this data is due to the different locations. The label to predict is the so-called PGA (Peak Ground Acceleration) value.

The second real world dataset describes the flight arrival and departure details for all commercial flights within the USA[3]. The complete dataset contains records from October 1987 to April 2008. We took the data from 2007 as the S data and the 2008 data as the P data. Also here one can argue that the measurement taken in 2008 are different to 2007 due to a shift in time. The predicted value is the delay of a particular flight.

The third dataset [21] comprises data for indoor location estimation from radio signal strengths received by a user device (like a PDA) from various WiFi Access Points. The measurements are taken at different locations and therefore contain a dataset shift.

The results are shown in table 2. Besides FS-KRR and ATL all approaches which take a shift into account consistently improve the result in comparison to the baseline approach of KRR on P (and/or S). Adjusting for a covariate shift with KLIEP only slightly improves the result, whereas approaches which also adjust with weights stemming from a dataset shift view achieve much better performance. The supervised approach DITL consistently performs best, with KLITL and TLB as second.

In a final experiment we added additional distortion to the labels with (19) and thereby increased the shift in the labels artificially. The purpose of this additional shift is to investigate the robustness of the methods, assuming that with a stronger shift the methods become more sensitive in the weight calculation, which might lead to a higher error rate. The results confirm this expectation, but also show that it is reasonable to assume that DITL provides a better robustness to stronger shifts than other methods. We only give results for one dataset, the results for other datasets are qualitatively similar.

8 Conclusions

In this paper we suggest two new approaches for tackling the problem of inductive transfer learning. The first one DITL, a supervised method, is motivated by a reweighted and unbiased prediction function of the S data. The second method uses an approximation of the Kullback-Leibler divergence to measure the difference in the distributions of the S and P data. The results indicate that both methods are suitable to account for dataset shifts while the supervised method performs better.

Due to its unsupervised nature, future work on the robustness of KLITL will be an interesting topic. Furthermore, we will investigate the application of the methods in a classification setting. Here, the direct method will need a different optimization than the current formulation (4), which is not suited for classification. Of interest would also be the case of a small number of labeled P data, but large number of unlabeled P data, here one might want to combine covariate shift adaptation with inductive transfer learning.

References

1. Al-Stouhi, S., Reddy, C.K.: Adaptive boosting for transfer learning using dynamic updates. In: Gunopulos, D., Hofmann, T., Malerba, D., Vazirgiannis, M. (eds.) ECML PKDD 2011, Part I. LNCS (LNAI), vol. 6911, pp. 60–75. Springer, Heidelberg (2011)

---

[3] Flight dataset available at http://stat-computing.org/dataexpo/2009/

2. Allen, T., Wald, D.: Evaluation of ground-motion modeling techniques for use in global shakemap–a critique of instrumental ground-motion prediction equations, peak ground motion to macroseismic intensity conversions, and macroseismic intensity predictions in different tectonic settings. U.S. Geological Survey Open-File Report 2009-1047 (2009)
3. Argyriou, A., Evgeniou, T., Pontil, M.: Convex multi-task feature learning. Machine Learning 73(3), 243–272 (2008)
4. Bishop, C.M.: Pattern recognition and machine learning. Springer (2006)
5. Quiñonero Candela, J.N., Sugiyama, M., Schwaighofer, A., Lawrence, N.D. (eds.): Dataset Shift in Machine Learning. MIT Press (2009)
6. Cao, B., Pan, S.J., Zhang, Y., Yeung, D.Y., Yang, Q.: Adaptive transfer learning. In: AAAI. AAAI Press (2010)
7. Cortes, C., Mohri, M.: Domain adaptation and sample bias correction theory and algorithm for regression. Theoretical Computer Science 519, 103–126 (2014)
8. Dai, W., Yang, Q., Rong Xue, G., Yu, Y.: Boosting for transfer learning. In: International Conference on Machine Learning (ICML) (2007)
9. Daumé III, H.: Frustratingly easy domain adaptation. In: ACL, pp. 256–263 (2007)
10. Gretton, A., Smola, A., Huang, J., Schmittfull, M., Borgwardt, K., Schölkopf, B.: Covariate Shift by Kernel Mean Matching, pp. 131–160. MIT Press (2009)
11. Kuzborskij, I., Orabona, F.: Stability and hypothesis transfer learning. In: International Conference on Machine Learning (ICML) (2013)
12. Pan, S.J., Yang, Q.: A Survey on Transfer Learning. IEEE Transactions on Knowledge and Data Engineering 22(10), 1345–1359 (2010)
13. Pardoe, D., Stone, P.: Boosting for regression transfer. In: International Conference on Machine Learning (ICML) (2010)
14. Raina, R., Ng, A.Y., Koller, D.: Constructing informative priors using transfer learning. In: International Conference on Machine Learning (ICML), pp. 713–720 (2006)
15. Rückert, U., Kramer, S.: Kernel-based inductive transfer. In: Daelemans, W., Goethals, B., Morik, K. (eds.) ECML PKDD 2008, Part II. LNCS (LNAI), vol. 5212, pp. 220–233. Springer, Heidelberg (2008)
16. Sugiyama, M., Kawanabe, M.: Machine learning in non-stationary environments: Introduction to covariate shift adaptation. MIT Press, Cambridge (2012)
17. Sugiyama, M., Nakajima, S., Kashima, H., Bünau, P., Kawanabe, M.: Direct importance estimation with model selection and its application to covariate shift adaptation. In: NIPS 20, pp. 1433–1440 (2008)
18. Tan, B., Zhong, E., Wei, E., Yang, X.Q.: Multi-transfer: Transfer learning with multiple views and multiple sources. In: SDM (2013)
19. Tommasi, T., Orabona, F., Caputo, B.: Safety in numbers: Learning categories from few examples with multi model knowledge transfer. In: CVPR, pp. 3081–3088. IEEE (2010)
20. Yang, P., Gao, W.: Multi-view discriminant transfer learning. In: Rossi, F. (ed.) IJCAI. IJCAI/AAAI (2013)
21. Yang, Q., Pan, S.J., Zheng, V.W.: Estimating location using wi-fi. IEEE Intelligent Systems 23(1), 8–13 (2008)
22. Zhang, D., He, J., Liu, Y., Si, L., Lawrence, R.D.: Multi-view transfer learning with a large margin approach. In: Apté, C., Ghosh, J., Smyth, P. (eds.) KDD, pp. 1208–1216. ACM (2011)

Policy Search for Path Integral Control

Vicenç Gómez[1,2], Hilbert J. Kappen[2], Jan Peters[3,4], and Gerhard Neumann[3]

[1] Universitat Pompeu Fabra, Barcelona
Department of Information and Communication Technologies,
E-08018 Barcelona, Spain
vicen.gomez@upf.edu
[2] Radboud University, Nijmegen,
Donders Institute for Brain, Cognition and Behaviour
6525 AJ Nijmegen, The Netherlands
b.kappen@science.ru.nl
[3] Technische Universität Darmstadt,
Intelligent Autonomous Systems,
Hochschulstr. 10, 64289 Darmstadt, Germany
{neumann,peters}@ias.tu-darmstadt.de
[4] Max Planck Institute for Intelligent Systems,
Spemannstr. 38, 72076 Tübingen, Germany

Abstract. Path integral (PI) control defines a general class of control problems for which the optimal control computation is equivalent to an inference problem that can be solved by evaluation of a path integral over state trajectories. However, this potential is mostly unused in real-world problems because of two main limitations: first, current approaches can typically only be applied to learn open-loop controllers and second, current sampling procedures are inefficient and not scalable to high dimensional systems. We introduce the efficient Path Integral Relative-Entropy Policy Search (PI-REPS) algorithm for learning feedback policies with PI control. Our algorithm is inspired by information theoretic policy updates that are often used in policy search. We use these updates to approximate the state trajectory distribution that is known to be optimal from the PI control theory. Our approach allows for a principled treatment of different sampling distributions and can be used to estimate many types of parametric or non-parametric feedback controllers. We show that PI-REPS significantly outperforms current methods and is able to solve tasks that are out of reach for current methods.

Keywords: Path Integrals, Stochastic Optimal Control, Policy Search.

1 Introduction

Stochastic Optimal Control is a powerful framework for computing optimal controllers in noisy systems with continuous states and actions. Optimal control computation usually involves estimation of the value function (or optimal cost-to-go) which, except for the simplest case of a linear system with quadratic rewards and Gaussian noise, is hard to perform exactly. In all other cases, we either have to rely on approximations of the

T. Calders et al. (Eds.): ECML PKDD 2014, Part I, LNCS 8724, pp. 482–497, 2014.

system dynamics, e.g. by linearizations [22] or the value function [12][1]. However, such approximations can significantly degenerate the quality of the estimated controls and hinder the application for complex, non-linear tasks.

Path integral (PI) control theory [7,20] defines a general class of stochastic optimal control problems for which the optimal cost-to-go (and the optimal control) is given explicitly in terms of a path integral. Its computation only involves the path costs of sample roll-outs or (state) trajectories, which are given by the reward along the state trajectory plus the log-probability of the trajectory under the uncontrolled dynamics. The optimal trajectory distribution of the system corresponds to a soft-max probability distribution that uses the path costs in its exponent. This fact allows for using probabilistic inference methods for the computation of the optimal controls, which is one of the main reasons why PI control theory has recently gained a lot of popularity.

However, PI control theory suffers from limitations that reduce its direct application in real-world problems. First, to compute the optimal control, one has to sample many trajectories starting from a certain (initial) state x_0. Such procedure is clearly infeasible for real stochastic environments, as the re-generation of a large number of sample trajectories would be required for each time-step. Hence, current algorithms based on PI control theory are so far limited to optimize state-independent controllers, such as open-loop torque control [19] or parametrized movement primitives such as Dynamic Movement Primitives [18,5].

Second, PI control theory requires sampling from the uncontrolled process. Such procedure requires a huge amount of samples in order to reach areas with low path costs. While open-loop iterative approaches [19] address this problem by importance sampling using a mean control trajectory, they do not provide a principled treatment for adjusting also the variance of the sampling policy. As the uncontrolled process might have small variance, such procedure still takes a large amount of samples to converge to the optimal policy. While some approaches that are used in practice relax these theoretical conditions and also change the sampling variance heuristically [16], they disregard the theoretical basis of PI control and are also restricted to open-loop controllers.

In this paper we introduce Path Integral Relative-Entropy Policy Search (PI-REPS), a new policy search approach that learns to sample from the optimal state trajectory distribution. We reuse insights from the policy search community and require that the information loss of the trajectory distribution update is bounded [11]. Such strategy ensures a stable and smooth learning process. However, instead of explicitly maximizing the expected reward as it is typically done in policy search, our aim is now to approximate the optimal state trajectory distribution obtained by PI control. This computation involves minimizing the Kullback-Leibler (KL) divergence between the trajectory distribution obtained after the policy update and the desired distribution under additional constraints. PI-REPS includes the probability distribution of the initial state x_0 in the KL optimization. This allows direct applicability of the method for learning state feedback controllers and leads to an improvement in terms of sampling efficiency.

In the next section we review current control methods based on path integral theory. In section 3, we describe in detail PI-REPS. In section 4, we show empirically that

---

[1] In [12], the function that is approximated is called desirability function which corresponds to the exp-transformed value function.

PI-REPS outperforms current PI-based and policy search methods on a double-link swing-up as well as on a quad-link swing-up problem.

2   Path Integral Control

We now briefly review the concepts of PI control that are relevant for the present paper. We consider the following stochastic dynamics of the state vector $\mathbf{x}_t \in \mathbb{R}^n$ under controls $\mathbf{u}_t \in \mathbb{R}^m$

$$d\mathbf{x}_t = \mathbf{f}(\mathbf{x}_t)dt + \mathbf{G}(\mathbf{x}_t)(\mathbf{u}_t dt + d\boldsymbol{\xi}_t), \qquad (1)$$

where $\boldsymbol{\xi}_t$ is $m-$dimensional Wiener noise with covariance $\boldsymbol{\Sigma}_\mathbf{u} \in \mathbb{R}^{m \times m}$ and $\mathbf{f}$ and $\mathbf{G}$ are arbitrary functions. For zero control, the system is driven uniquely by the deterministic drift $\mathbf{f}(\mathbf{x}_t)dt = \mathbf{f}_t dt$ and the local diffusion $\mathbf{G}(\mathbf{x}_t)d\boldsymbol{\xi}_t = \mathbf{G}_t d\boldsymbol{\xi}_t$. The cost-to-go is defined as an expectation over all trajectories starting at $\mathbf{x}_0$ with control path $\mathbf{u}_{0:T-dt}$

$$J(\mathbf{x}_0, \mathbf{u}_{0:T-dt}) = \left\langle r_T(\mathbf{x}_T) + \sum_{t=0}^{T-dt} C_t(\mathbf{x}_t, \mathbf{u}_t)dt \right\rangle. \qquad (2)$$

The terms $r_T(\mathbf{x}_T)$ and $C_t(\mathbf{x}_t, \mathbf{u}_t)$ denote the cost at end-time T and the immediate (running) cost respectively. $C_t(\mathbf{x}_t, \mathbf{u}_t)$ is expressed as a sum of an arbitrary state-dependent term $r_t(\mathbf{x}_t)$ and a quadratic control term $\mathbf{u}_t^\mathsf{T} \mathbf{R} \mathbf{u}_t$, i.e.,

$$C_t(\mathbf{x}_t, \mathbf{u}_t) = r_t(\mathbf{x}_t) + \frac{1}{2}\mathbf{u}_t^\mathsf{T} \mathbf{R} \mathbf{u}_t.$$

Minimization of (2) leads to the Hamilton-Jacobi-Bellman (HJB) equations, which in the general case are non-linear, second order partial differential equations. However, if the cost matrix and noise covariance are such that $\mathbf{R} = \lambda \boldsymbol{\Sigma}_\mathbf{u}^{-1}$ the resulting equation is *linear* in the exponentially transformed cost-to-go function $\Psi(\mathbf{x}_0)$, where $J(\mathbf{x}_0) = -\lambda \log \Psi(\mathbf{x}_0)$. The function $\Psi(\mathbf{x}_0)$ is called desirability function. The solution for $\Psi(\mathbf{x}_0)$ using the optimal controls is given by the Feynman-Kac formula as a path integral [7]

$$\Psi(\mathbf{x}_0) = \int p_{\mathrm{uc}}(\boldsymbol{\tau}|\mathbf{x}_0) \exp\left(-\frac{\sum_{t=0}^{T} r_t(\mathbf{x}_t)}{\lambda}\right) d\boldsymbol{\tau}, \qquad (3)$$

where $p_{\mathrm{uc}}(\boldsymbol{\tau}|\mathbf{x}_0)$ is the conditional probability of a state trajectory $\boldsymbol{\tau} = \mathbf{x}_{dt:T}$ starting at $\mathbf{x}_0$ and following the uncontrolled process.

The relation $\mathbf{R} = \lambda \boldsymbol{\Sigma}_\mathbf{u}^{-1}$ forces control and noise to act in the same dimensions, but in an inverse relation. Thus, for fixed λ, the larger the noise, the cheaper the control and vice-versa. Parameter λ can be seen as a temperature: higher values of λ result in optimal solutions that are closer to the uncontrolled process.

Define the path value of a trajectory $\boldsymbol{\tau}$ as

$$S(\boldsymbol{\tau}|\mathbf{x}_0) = \sum_{t=0}^{T} r_t(\mathbf{x}_t) - \lambda \log p_{\mathrm{uc}}(\boldsymbol{\tau}|\mathbf{x}_0). \qquad (4)$$

The optimal path distribution can be obtained from (3) and is given by

$$p^*(\boldsymbol{\tau}|\mathbf{x}_0) = \frac{\exp(-S(\boldsymbol{\tau}|\mathbf{x}_0)/\lambda)}{\int \exp(-S(\boldsymbol{\tau}|\mathbf{x}_0)/\lambda)d\boldsymbol{\tau}}. \tag{5}$$

The optimal control is given as an expectation of the first direction of the noise $d\boldsymbol{\xi}_0$ over the optimal trajectory distribution (5). This is an inference problem that can be solved using Monte Carlo methods, e.g, by forward sampling from the uncontrolled process, as proposed in [7]. However, as the optimal trajectory distribution depends on the initial state $\mathbf{x}_0$, the sampling process has to be repeated at each state which limits the application of PI control in practice. This restriction can be ignored as in [3], at the cost of losing theoretical guarantees of optimality.

2.1 Iterative Path Integral Control

Sampling from uncontrolled process will often result in a poor estimate of the optimal trajectory distribution as the uncontrolled process typically leads to areas of high state costs, i.e., most generated trajectories will have very small probability under the optimal trajectory distribution. Formally, the main problem is the evaluation of the integral in the normalization of equation (5), as this integral is performed over the whole trajectory space. To alleviate this problem, importance sampling schemes that use a (baseline) controlled process to improve the sampling efficiency has been proposed [7]. In this case, the path cost (4) has to be corrected for the extra drift term introduced by the baseline control. An iterative version of this approach was formally derived in [19] and has resulted in several applications [14,2].

There are two main problems with this approach: first, it only considers the mean control trajectory. Since it neglects the state-dependence of the control beyond the initial state, the result is an open-loop controller that may perform poorly when applied to a stochastic system. Second, this approach does not provide a principled treatment for adapting the sampling variance, and hence, might need a large amount of samples if the variance of the uncontrolled process is low.

2.2 Policy Improvement with Path Integrals (PI^2)

Inspired by the PI theory, [18] introduced the PI^2 algorithm in the reinforcement learning community, which has been successfully applied to a variety of robotic systems for tasks such as planning, gain scheduling and variable stiffness control [3,15,17].

PI^2 uses parametrized policies to represent trajectories in the state space. Typically, PI^2 uses open-loop controllers such as Dynamic Movement Primitives (DMPs) [6]. PI^2 identifies the parameters $\boldsymbol{\theta}_t$ of the DMP with the control commands $\mathbf{u}_t$ in eq. (1). Such strategy, however, renders the constraint $R = \lambda\Sigma_{\mathbf{u}}^{-1}$ meaningless. This constraint is also often neglected which might even lead to better performance [16]. The method is model-free in the sense that no model needs to be learned. However, it is implicitly assumed that all the noise of the system is generated by the exploration in the DMP parameters, which is an unrealistic assumption. The noise $\boldsymbol{\xi}_t$ in PI^2 is interpreted as user controlled exploration noise that acts on $\boldsymbol{\theta}_t$.

2.3 Kullback Leibler Divergence Minimization

The PI class of control problems is included in a larger class of (discrete) problems, also known as linearly solvable Markov Decision Processes (MDP) or KL-control [20,21,8] for which the control cost can be expressed as a KL divergence between a controlled process $p(\tau|\mathbf{x}_0)$ and $p_{\text{uc}}(\tau|\mathbf{x}_0)$.

Unlike the continuous case where the controls act as a drift on the uncontrolled process (1), the controls in the discrete case can fully reshape the state-transition probabilities $p(\mathbf{x}_{t+1}|\mathbf{x}_t)$, with the only restriction of being compatible with the uncontrolled process, i.e. $p(\mathbf{x}_{t+1}|\mathbf{x}_t) = 0, \forall \mathbf{x}_{t+1}$ such that $p_{\text{uc}}(\mathbf{x}_{t+1}|\mathbf{x}_t) = 0$. Policy iteration algorithms for that broader class of problems also consider KL minimization have been proposed recently in [1,12]. However, in continuous state spaces, these approaches typically rely on an iterative approximation of the desirability function. Similar to value function approximation, the errors of such approximation can accumulate and damage the policy update. Moreover, these methods do not provide a principled treatment for setting the variance of the sampling policy. Another extension of the PI control theory can be found in [13], where the path integrals are embedded in a reproducing Kernel Hilbert Space (RKHS). While this is also a promising approach, it again relies on approximation of the desirability function $\Psi(\mathbf{x})$ which we typically want to avoid.

In the area of policy search, a common approach is to bound the KL-divergence between the old and the new policy. A bound on the KL-divergence can be more efficient as penalizing the KL for determining the policy update as we obtain a pre-specified step-size of the policy update in the space of probability distributions. This step-size enables to control the exploration that is performed by the policy update in a more principled way. This insight has led to the development of several successful policy search algorithms, such as the relative entropy policy search (REPS) algorithm [11], contextual REPS [10] or a hierarchical extension of REPS [4]. Bounding the KL-divergence between the old and the new policy is almost equivalent to penalizing it, however, it qualifies us to let the temperature of the soft-max distribution be set by the KL-bound instead of hand-tuning the temperature or using heuristics.

REPS divides the policy updates into two steps. It first computes a probability for each observed state action pair by solving an optimization problem with the objective of maximizing the expected rewards while bounding the KL-divergence between the new and the old distributions. This probability corresponds to the desired probability that this sample is used by the new policy. Subsequently, these probabilities are used as weights to estimate a new parametric policy by performing a weighted maximum likelihood update. While our approach is inspired by REPS-based algorithms, there are significant differences: REPS is used to either directly learn in the parameter space of low-level controllers [4,10], which is restricted to controllers with a small number of parameters, such as DMPs [5] or it used to estimate the probability of state action samples [11].

3   Path Integral - Relative Entropy Policy Search (PI-REPS)

PI-REPS considers problems of the PI class but uses an explicit representation of a time-dependent stochastic policy $\pi_t(\mathbf{u}_t|\mathbf{x}_t), \forall t < T$, that maps a state $\mathbf{x}_t$ into a probability distribution over control actions $\mathbf{u}_t$. The objective of PI-REPS is to find the optimal policy π^* that generates the optimal distribution of state trajectories $p^*(\tau|\mathbf{x}_0)$ given by the PI theory, Eq. (5), under some additional constraints.

For that, it alternates two steps. In the first step, the target distribution (5) is approximated from samples generated by the current policy. We can evaluate the target distribution up to a normalization constant from the samples. At the same time, the information loss to the old policy is bounded to avoid overly greedy policy updates [11,4]. The result of this optimization problem is specified by a weight for each of the seen sample trajectories.

This weight is used in the second step, where the current policy $\tilde{\pi}_t(\mathbf{u}_t|\mathbf{x}_t)$ is updated in a way that can reproduce the desired weighted trajectory distribution. This policy update is computed in a (weighted) maximum likelihood sense. These two steps are iterated until convergence. We describe the details of PI-REPS in the following subsections.

3.1 Learning the Optimal Trajectory Distribution

In the first step of PI-REPS, the current control policy is used to generate data in the form of sample trajectories $\mathcal{D} = \{\mathbf{x}_{0:T}^{[i]}\}_{i=1...N}$. Based on these data, we obtain a new trajectory distribution that minimizes the expected KL-divergence to the optimal distribution $p^*(\tau|\mathbf{x}_0)$, i.e.,

$$\text{argmin}_p \int \mu(\mathbf{x}_0)\text{KL}\left(p(\tau|\mathbf{x}_0) \| p^*(\tau|\mathbf{x}_0)\right) d\mathbf{x}_0. \tag{6}$$

As we want to learn a trajectory distribution, we can not directly use the average reward as optimization criterion as this is done in REPS. REPS would choose a trajectory distribution that might be infeasible, while for PI-REPS we know that the target distribution $p^*(\tau|\mathbf{x}_0)$ is optimal and, hence, feasible.

In addition to this objective, we also want to stay close to the old trajectory distribution $q(\tau|\mathbf{x}_0)$, i.e., we bound

$$\int \mu(\mathbf{x}_0)\text{KL}\left(p(\tau|\mathbf{x}_0) \| q(\tau|\mathbf{x}_0)\right) d\mathbf{x}_0 \leq \epsilon. \tag{7}$$

As in REPS, the parameter ϵ can be used as trade-off between exploration and exploitation. As additional constraint, we require that $p(\tau|\mathbf{x}_0)$ defines a proper probability distribution, i.e.,

$$\forall \mathbf{x}_0 : \int p(\tau|\mathbf{x}_0)d\tau = 1.$$

However, this optimization problem requires that we obtain many trajectory samples for each initial state, which is infeasible in many situations. We want to be able to deal with situations where only one trajectory per initial state $\mathbf{x}_0$ can be obtained. For this reason, we extend our objective to optimize also over the initial state distribution, i.e., we optimize over the joint distribution $p(\tau, \mathbf{x}_0)$. The resulting objective is given by

$$\text{argmin}_p \int p(\boldsymbol{\tau}, \mathbf{x}_0) \log \frac{p(\boldsymbol{\tau}, \mathbf{x}_0)}{p^*(\boldsymbol{\tau}, \mathbf{x}_0)} d\boldsymbol{\tau} d\mathbf{x}_0$$

$$= \text{argmax}_p \int p(\boldsymbol{\tau}, \mathbf{x}_0) \left(\frac{1}{\lambda} S(\boldsymbol{\tau}|\mathbf{x}_0) + \log \mu(\mathbf{x}_0) - \log p(\boldsymbol{\tau}, \mathbf{x}_0) \right) d\boldsymbol{\tau} d\mathbf{x}_0, \quad (8)$$

where $p^*(\boldsymbol{\tau}, \mathbf{x}_0) = p^*(\boldsymbol{\tau}|\mathbf{x}_0)\mu(\mathbf{x}_0)$. However, the initial state distribution $\mu(\mathbf{x}_0)$ can not be freely chosen as it is given by the task. Hence, we need to ensure that the marginal distribution $p(\mathbf{x}_0) = \int p(\boldsymbol{\tau}, \mathbf{x}_0) d\boldsymbol{\tau}$ matches the given state distribution $\mu(\mathbf{x}_0)$ for all states $\mathbf{x}_0$. Note that by introducing these constraints we would end up in the original optimization problem (6), but with an infinite number of constraints. However, a common approach to relax this condition is to only match state-feature averages of the marginals [10,4], i.e.,

$$\int p(\mathbf{x}_0)\phi(\mathbf{x}_0)d\mathbf{x}_0 = \int \mu(\mathbf{x}_0)\phi(\mathbf{x}_0)d\mathbf{x}_0 = \hat{\phi}_0,$$

where $\hat{\phi}_0$ is the mean feature vector of the samples corresponding to the initial state. The feature vector $\phi(\cdot)$ can be, for example, all linear and quadratic terms of the initial state. In this case, we would match mean and covariance of both distributions. The complete optimization problem reads[2]

$$\text{argmax}_p \int p(\boldsymbol{\tau}, \mathbf{x}_0) \left(\frac{S(\boldsymbol{\tau}|\mathbf{x}_0)}{\lambda} - \log p(\boldsymbol{\tau}, \mathbf{x}_0) \right) d\boldsymbol{\tau} d\mathbf{x}_0,$$

$$\text{s.t.:} \int p(\mathbf{x}_0)\phi(\mathbf{x}_0)d\mathbf{x}_0 = \hat{\phi}_0,$$

$$\int p(\boldsymbol{\tau}, \mathbf{x}_0) \log \frac{p(\boldsymbol{\tau}, \mathbf{x}_0)}{q(\boldsymbol{\tau}, \mathbf{x}_0)} d\boldsymbol{\tau} d\mathbf{x}_0 \leq \epsilon,$$

$$\int p(\boldsymbol{\tau}, \mathbf{x}_0) d\boldsymbol{\tau} d\mathbf{x}_0 = 1. \quad (9)$$

We solve the above optimization problem using the method of Lagrange multipliers. The solution for $p(\boldsymbol{\tau}, \mathbf{x}_0)$ can be obtained in closed form (see supplement for the derivation)

$$p(\boldsymbol{\tau}, \mathbf{x}_0) \propto q(\boldsymbol{\tau}, \mathbf{x}_0)^{\frac{\eta}{\eta+1}} \exp\left(\frac{S(\boldsymbol{\tau}|\mathbf{x}_0) - \phi(\mathbf{x}_0)^{\mathsf{T}}\boldsymbol{\theta}}{\eta+1} \right), \quad (10)$$

where η and $\boldsymbol{\theta}$ are the Lagrange multipliers corresponding to the KL-divergence and the feature constraints respectively. Their optimal values can be found by optimizing the corresponding dual function $g(\boldsymbol{\theta}, \eta)$

$$[\boldsymbol{\theta}^*, \eta^*] = \text{argmin}_{\boldsymbol{\theta}, \eta} g(\boldsymbol{\theta}, \eta), \quad \text{s.t:} \ \eta > 0, \quad (11)$$

which is also given in the supplement. Note that the solution $p(\boldsymbol{\tau}, \mathbf{x}_0)$ represents a geometric average between the old distribution and the optimal distribution. The parameter η, which specifies how much we want to interpolate, is chosen by the optimization.

---

[2] Note that the log $\mu(\mathbf{x}_0)$ term can be neglected. Due to the initial state constraints, the path-cost component which is only dependent on the initial state has no influence.

3.2   Weighted Maximum Likelihood Policy Updates

From estimates of the probability $p(\tau, \mathbf{x}_0)$ of the sample trajectories, we can fit a parametrized policy $\hat{\pi}_t(\mathbf{u}_t | \mathbf{x}_t; \boldsymbol{\omega}_t)$ for each time-step $t < T$ that can reproduce the trajectory distribution $p(\tau, \mathbf{x}_0)$. For each time-step, we want to find the policy $\hat{\pi}_t$ such that the resulting transition probabilities $p^{\hat{\pi}}(\mathbf{x}_{t+1} | \mathbf{x}_t) = \int P(\mathbf{x}_{t+1} | \mathbf{x}_t, \mathbf{u}_t) \hat{\pi}_t(\mathbf{u}_t | \mathbf{x}_t; \boldsymbol{\omega}_t) d\mathbf{u}_t$ match the estimated transition distribution from $p(\tau, \mathbf{x}_0)$, where $P(\mathbf{x}_{t+1} | \mathbf{x}_t, \mathbf{u}_t)$ corresponds to the model dynamics, assumed to be known. This is an inference problem with latent variables $\mathbf{u}_t$. To solve it, we first compute, for each transition, the action $\mathbf{u}_t^*$ that is most likely to have generated the transition. This controls $\mathbf{u}_t^*$ can be computed from the given control affine system dynamics with $\mathbf{u}_t^* = (\mathbf{G}_t^T \mathbf{G}_t)^{-1} \mathbf{G}_t^T (d\mathbf{x}_t - \mathbf{f}(\mathbf{x}_t) dt)/dt^3$. Subsequently we extract a parametric policy out of the trajectory distribution $p(\tau, \mathbf{x}_0)$ computed from PI-REPS by minimizing

$$
\begin{aligned}
\boldsymbol{\omega}_t^* &= \operatorname{argmin}_{\boldsymbol{\omega}_t} \operatorname{KL}\left(p(\tau, \mathbf{x}_0) \,\|\, p^{\hat{\pi}}(\tau, \mathbf{x}_0)\right) \\
&= \operatorname{argmin}_{\boldsymbol{\omega}_t} \int p(\tau, \mathbf{x}_0) \log\left(\frac{p(\mathbf{x}_{t+1} | \mathbf{x}_t)}{p^{\hat{\pi}}(\mathbf{x}_{t+1} | \mathbf{x}_t)}\right) d\tau d\mathbf{x}_0 + \text{const} \\
&\approx \operatorname{argmax}_{\boldsymbol{\omega}_t} \int p(\tau, \mathbf{x}_0) \log \hat{\pi}_t(\mathbf{u}_t^* | \mathbf{x}_t; \boldsymbol{\omega}_t) d\tau d\mathbf{x}_0 + \text{const} \\
&= \operatorname{argmax}_{\boldsymbol{\omega}_t} \sum_i \frac{p(\tau^{[i]}, \mathbf{x}_0^{[i]})}{q(\tau^{[i]}, \mathbf{x}_0^{[i]})} \log \hat{\pi}_t(\mathbf{u}_t^{*[i]} | \mathbf{x}_t^{[i]}; \boldsymbol{\omega}_t).
\end{aligned}
$$

The division by $q(\tau^{[i]}, \mathbf{x}_0^{[i]})$ in the fourth row of the equation results from using samples from $q(\tau^{[i]}, \mathbf{x}_0^{[i]})$ to approximate the integral. This minimization problem can be seen as a weighted maximum likelihood problem with weights $d_i, i = 1 \ldots N$ given by

$$
d_i = q(\tau^{[i]}, \mathbf{x}_0^{[i]})^{\frac{-1}{\eta+1}} \exp\left(\frac{S(\tau^{[i]} | \mathbf{x}_0^{[i]}) - \boldsymbol{\theta}^T \boldsymbol{\phi}_{\mathbf{x}_0}^{[i]}}{\eta + 1}\right).
$$

In the presented approach we use time-dependent Gaussian policies that are linear in the states, i.e. $\hat{\pi}_t(\mathbf{u}_t | \mathbf{x}_t) \sim \mathcal{N}(\mathbf{u}_t | \mathbf{k}_t + \mathbf{K}_t \mathbf{x}_t, \boldsymbol{\Sigma}_t)$. The resulting update equations for $\mathbf{k}_t$, $\mathbf{K}_t$ and $\boldsymbol{\Sigma}_t$ are given by a weighted linear regression and the weighted sample-covariance matrix, respectively [10]. The estimate of $\boldsymbol{\Sigma}_t$ will also contain the variance of the control noise. As this noise is automatically added by the system dynamics, we do not need to add this noise as exploration noise of the policy. Hence, we subtract the control noise from the estimated variance of the policy while ensuring that $\boldsymbol{\Sigma}_t$ stays positive (semi-)definite.

3.3   Step-Based versus Episode-Based Weight Computation

So far, we computed a single weight per trajectory and used this weight to update the policy for all time-steps. However, we can use the simple observation that, if a trajectory distribution is optimal for the time-steps $t = 1 \ldots T$, it also has to be optimal

---

[3] If the controls $\mathbf{u}_t$ and the noise ϵ_t can be observed, $\mathbf{u}_t^*$ can be computed by $\mathbf{u}_t^* = \mathbf{u}_t + \epsilon_t$.

for all time segments that also end in T but start at $t' > 1$. Hence we can perform the optimization for each time-step separately and use the obtained weights to fit the policy for the corresponding time step. We path value to come $S_t(\tau'|\mathbf{x}_t) = \sum_{h=t}^{T} r_h(\mathbf{x}_h) + \lambda \log p_{uc}(\tau'|\mathbf{x}_t)$, where $\tau' = \tau_{t+dt:T}$ in the exponent of the optimal trajectory distribution. For the initial feature constraints, we use the observed average state features from the old policy at this time step. Such an approach has the advantage that it can considerably reduce the variance of the weights computed for later time steps, and, hence, render PI-REPS more sample efficient. However, as the optimization has now to be done for each time step, the step-based variant is also computationally more demanding.

3.4   Relation to Existing Approaches

To point out the contributions of this paper, we summarize the novel aspects of PI-REPS with respect to the previously mentioned approaches. In comparison to the REPS algorithm, we use our algorithm to generate a weighting of trajectories, not state-action pairs. As we learn trajectory distributions, we can not freely choose the desired trajectory distribution as certain distributions might not be feasible. In PI-REPS we circumvent this problem by minimizing the Kullback-Leibler divergence to the optimal trajectory distribution instead of maximizing the reward. Due to the optimization over trajectory distributions, the weighted maximum likelihood update is different as we need to obtain $\mathbf{u}_t^*$ from the system dynamics instead of using the executed action $\mathbf{u}_t$.

The constrained optimization problem also leads to a very different solution: while PI-REPS interpolates between the old (initial) and the optimal trajectory distribution (eq. 13), REPS is always affected by the influence of the initial distribution. PI-REPS also considers the initial state in the optimization. Although a similar constraint has been used in REPS for contextual policy search [10], our use is novel since it allows a time step version of the algorithm that, as we show in section 3.3, improves significantly the sample efficiency.

4   Experiments

We evaluated PI-REPS on two simulated benchmark tasks, a double-link swing-up and a quad-link swing-up task. We compared it against variants of previous approaches such as iterative PI control (open loop) [14] and a closed loop extension by fitting a policy with weighted maximum likelihood as performed by our approach. Moreover, we compare the episode-based and the step-based version of PI-REPS and also present the first experiments for model-based reinforcement learning, where in addition to learning a controller, we also learn the forward model of the robot. Finally, we evaluated the influence of the control noise, the KL-bound ϵ as well as the influence of the initial policy and the number of samples used for the policy update. Our experiments show that PI-REPS is a promising approach for stochastic optimal control and model-based reinforcement learning that can find high-quality policies for tasks that are beyond the reach of current methods.

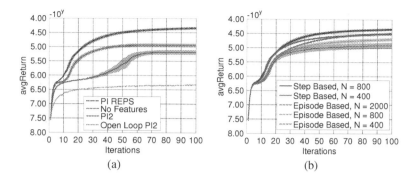

Fig. 1. (a) Comparison of iterative PI with open loop control with our extension of learning closed-loop controller by weighted maximum likelihood, PI-REPS without the feature constraints and the step-based PI-REPS algorithm. PI-REPS outperforms all other methods. (b) Comparison of the step-based variant of PI-REPS with the episode-based variant.

4.1 Double Link Swing-Up

In this task, we used a two link pendulum that needs to swing-up from the bottom position and balance at the top position. Each link had a length of 1m and a weight of 1kg. The torque of each motor was limited to $|u_i| < 10$Nm. We used a viscous friction model to damp the joint velocities. One episode was composed of 70 time steps with $dt = 66$ms. The state rewards were $r_t(\mathbf{q}, \dot{\mathbf{q}}) = -10^4 \mathbf{q}^T \mathbf{q}$, which punishes the squared distance of the joint angles to the upright position. The reward was given for the last 20 time steps only. The default standard deviation of the control noise was set to $\mathbf{\Sigma}_u = 0.5/dt\mathbf{I}$. We used the double link swing-up task for exhaustive parameter evaluation and comparison for being a challenging task but still feasible for running a large number of experiments.

Comparison of different path integral algorithms. We compared our method to different versions of current PI algorithms. We used the step-based variants of all algorithms in this comparison. The episode variants basically show the same results with a slower convergence rate. In the first approach, we applied the iterative path integral method with open loop control to our problem with control noise, as described in section 2.1. Here we simply used a constant action for each time step. In order to estimate this action from samples we used the weighting $d_i = \exp(S(\boldsymbol{\tau}^{[i]}|\mathbf{x}_0^{[i]})/\lambda)$ for each sample. As in the original PI^2 approach [18], we scaled the λ parameter by the range of the path integral values $S(\boldsymbol{\tau}^{[i]}|\mathbf{x}_0^{[i]})$, i.e. $\lambda = \lambda_{\text{PI}^2}/(\max_i S(\boldsymbol{\tau}^{[i]}|\mathbf{x}_0^{[i]}) - \min_i S(\boldsymbol{\tau}^{[i]}|\mathbf{x}_0^{[i]}))$. The value for λ_{PI^2} was empirically optimized. Subsequently, we extended this approach to use the time-dependent linear feedback policy and maximum likelihood updates for the policy as introduced by our approach. Note that this approach is also equivalent to the state of the art policy search approach PoWER [9]. However, in contrast to our method, PoWER as well as PI^2 do not use a principled approach to set the temperature of the soft-max distribution. Moreover, they do not account for the state-dependent part of the path integral as it is done by the use of our baseline. We also evaluated the effect of the

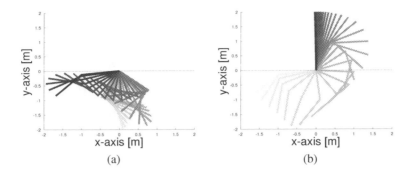

Fig. 2. Illustration of the estimated swing-up movement with the double link. (a) time steps 1 to 25. (b) time steps 36 to 70. Lighter colors indicate an earlier time step.

baseline by using PI-REPS without state features. In each iteration of the algorithm, we sampled 800 new trajectories. Although this number of trajectories is too large for a real robot application, PI-REPS is model-based and therefore we can first learn a model of the robot using the real robot interactions and, subsequently, use the model to generate an arbitrary number of trajectories. The results of such a model-based reinforcement learning approach are presented at the end of this subsection.

Fig. 1(a) shows a global comparison of the methods. As expected, it can be seen that the open-loop control policy, as used in our version of PI^2, can not deal with the stochastic setup. If we extend PI^2 to learn a linear feedback controller, we can learn successfully the task, but convergence is very slow. As a next step, we introduce the information theoretic policy update to obtain a more principled treatment of the temperature parameter, but we still disable the features used for the baseline in our approach. This method is denoted as *No Features* in Fig. 1(a). While the convergence rate is now significantly improved, ignoring the initial state-distribution constraint results in a bias of the resulting solution and the resulting policy can not reach the quality of the proposed approach with a state-dependent base line. We also compared our approach to state of the art optimal control methods that are based on linearization of the system dynamics, such as the AICO approach [22], but we were not able to find good policies due to the high non-linearities in the task. Clearly, we can see that PI-REPS with a state-dependent base line produces policies of the highest quality. An illustration of the learned movement can be seen in Fig. 2.

Step-based versus Episode-based Weighting Computation. We now compare the step-based and the episode-based versions of PI-REPS. From Fig. 1(b), we observe that the step-based version is clearly more sample-efficient, as it reduces the variance of the estimates of the weights for later time steps. However, it is also computationally more demanding, since we need to compute the weights for every time step. The episode-based version with 2000 samples reaches the performance of the step-based version with 400 samples. Hence, if generating samples from the model is cheap, the episode-based version can also be used efficiently.

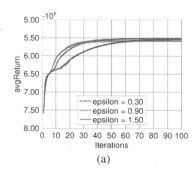
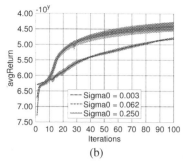

(a)   (b)

Fig. 3. Exploration in PI-REPS. (a) The value of ϵ determines the convergence and the quality of the obtained policy. For large ϵ, changes in the policy are large, resulting in faster convergence, but too little exploration. For too small ϵ, convergence is slow. (b) Evaluation of the initial exploration rate. If we just use the variance of the uncontrolled process for exploration from the beginning, we get very slow convergence. However, PI-REPS allows for using different sampling policies which are updated by policy search.

Exploration in PI-REPS. Exploration is determined by two parameters: the exploration rate Σ_0 of the initial policy and the KL-bound ϵ. For large values of ϵ, PI-REPS converges quickly to the target distribution and stops exploring too soon. In contrast, too small values of ϵ result in too conservative policy updates. This behavior can be seen in Fig. 3(a). We identified an optimal value of $\epsilon = 0.9$ and used it in all other experiments. A second factor that determines exploration is Σ_0. If we would fully rely on the noise of the uncontrolled process for exploration, the policy search procedure would take a long time. Therefore, we start with a highly stochastic policy and slowly move to the target distribution by the information theoretic policy updates. From Fig. 3(b), we can clearly see that only using the noise of the system is very inefficient, but higher values of the initial variance lead to a compelling performance.

Influence of the Control Noise. In this experiment we evaluated our approach with respect to the control noise Σ_u of the system. Note that, by changing the control noise, we also inherently change the reward function in the path integral framework. Fig. 4 (a) shows the performance for different control noise values. As we can see, good policies can be found for all noise levels, while the costs are decreased with higher noise levels due to the smaller control costs.

Model-Based Reinforcement Learning. While the focus on this paper is to derive an efficient stochastic optimal control method that is based on path integral, we can also directly apply our method to model-based reinforcement learning if we combine the PI-REPS policy updates with a probabilistic model learning technique. In this case, the trajectories generated by the real robot are only used to update the model. From the model, a large number of virtual samples are generated to perform the policy updates. As a proof of concept, we used a simple time-dependent linear model with Gaussian noise, i.e., $P_t(\mathbf{x}_{t+1}|\mathbf{x}_t, \mathbf{u}_t) = \mathcal{N}(\mathbf{x}_{t+1}|\mathbf{A}_t\mathbf{x}_t + \mathbf{B}_t\mathbf{u}_t + \mathbf{a}_t, \Sigma_t)$.

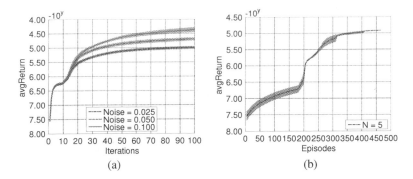

(a)   (b)

Fig. 4. (a) Evaluation for different values of the control noise. PI-REPS could learn high-quality policies even in the existence of a large amount of noise. The difference in the obtained reward is because the reward depends on the noise variance. (b) Experiment with model-based reinforcement learning. We learned time-varying linear models at each time step. A good swing-up policy could be learned already after 300 episodes.

We estimated such a model for each time step by performing maximum likelihood on the transition samples at each time step. As the model is time-varying, it can also capture non-linear dynamics. We started the algorithm with 25 initial trajectories and subsequently collected 5 trajectories in each iteration. From the learned models, we generated 1000 trajectories for the policy updates. Fig. 4(b) shows these results. We observe that a high-quality policy can be learned after 300 episodes, which is remarkable if compared to state of the art policy search approaches [4,18]. Yet, the performance of the final policy is affected by the simplicity of the learned model in comparison to the policy found on the real model of the robot.

4.2   Quad-Link Swing-Up

To conclude this experiment section, we used a quad-link pendulum swing-up task. We used the same physical properties, i.e., link length of 1m and a mass of 1kg, the same reward function as well as the same number of time steps as in the double link experiment. Given the increased complexity and weight of the whole robot, we increased the maximum torques to 20Nm. We evaluated the episode-based version of our algorithm with a different number of samples.

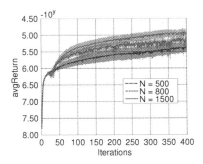

Fig. 5. Learning curve for the quad-link. An increasing number of samples always increases the performance.

The results can be seen in Figure 5. We observe that, due to the increased dimensionality of the problem, more samples are needed to solve the task. However, in contrast to competing methods, PI-REPS is still able to learn high quality policies for this complex task. An illustration of the swing-up movement can be seen in Fig. 6.

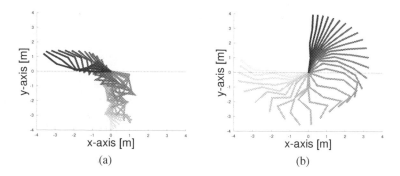

Fig. 6. Illustration of the estimated swing-up movement with the quad link. (a) time steps 1 to 35. (b) time steps 36 to 70. Lighter colors indicate an earlier time step.

5 Conclusions

In this paper we presented PI-REPS, the first approach for PI control that can be used to learn state-feedback policies in an efficient manner. PI-REPS has several benefits to previous PI methods. It allows for a principled treatment of the adaptation of the sampling policy by the information theoretic policy updates. This type of update specifies the temperature of the soft-max distribution. In previous approaches, this temperature had to be chosen heuristically, resulting, as our experiments show, in a poor quality of the estimated policy.

The PI-REPS policy update is based on a weighted maximum likelihood estimate. This is a general approach, not limited to the time varying linear policies that we considered in this paper. Using more complex models such as mixture models, Gaussian processes or neural networks seems to be a promising research direction. We will also investigate the use of more sophisticated model-learning techniques to improve the sample-efficiency in terms of real robot interactions.

Acknowledgements. This work was supported by the European Community Seventh Framework Programme (FP7/2007-2013) under grant agreement 270327 (CompLACS).

References

1. Azar, M.G., Gómez, V., Kappen, H.J.: Dynamic policy programming. Journal of Machine Learning Research 13, 3207–3245 (2012)
2. Berret, B., Yung, I., Nori, F.: Open-loop stochastic optimal control of a passive noise-rejection variable stiffness actuator: Application to unstable tasks. In: Intelligent Robots and Systems, pp. 3029–3034. IEEE (2013)
3. Buchli, J., Stulp, F., Theodorou, E., Schaal, S.: Learning variable impedance control. The International Journal of Robotics Research 30(7), 820–833 (2011)
4. Daniel, C., Neumann, G., Peters, J.: Hierarchical Relative Entropy Policy Search. In: International Conference on Artificial Intelligence and Statistics (2012)

5. Ijspeert, A., Schaal, S.: Learning Attractor Landscapes for Learning Motor Primitives. In: Advances in Neural Information Processing Systems 15, MIT Press, Cambridge (2003)
6. Ijspeert, A.J., Nakanishi, J., Schaal, S.: Learning attractor landscapes for learning motor primitives. In: Advances in Neural Information Processing Systems, pp. 1523–1530 (2002)
7. Kappen, H.J.: Linear theory for control of nonlinear stochastic systems. Physical Review Letters 95(20), 200201 (2005)
8. Kappen, H.J., Gómez, V., Opper, M.: Optimal control as a graphical model inference problem. Machine Learning 87, 159–182 (2012)
9. Kober, J., Peters, J.: Policy search for motor primitives in robotics. Mach. Learn. 84(1-2), 171–203 (2011)
10. Kupcsik, A., Deisenroth, M.P., Peters, J., Neumann, G.: Data-Efficient Contextual Policy Search for Robot Movement Skills. In: Proceedings of the National Conference on Artificial Intelligence (2013)
11. Peters, J., Mülling, K., Altün, Y.: Relative entropy policy search. In: Proceedings of the 24th AAAI Conference on Artificial Intelligence, pp. 1607–1612 (2010)
12. Rawlik, K., Toussaint, M., Vijayakumar, S.: On stochastic optimal control and reinforcement learning by approximate inference. In: International Conference on Robotics Science and Systems (2012)
13. Rawlik, K., Toussaint, M., Vijayakumar, S.: Path integral control by reproducing kernel Hilbert space embedding. In: Proceedings of the Twenty-Third International Joint Conference on Artificial Intelligence, pp. 1628–1634. AAAI Press (2013)
14. Rombokas, E., Theodorou, E., Malhotra, M., Todorov, E., Matsuoka, Y.: Tendon-driven control of biomechanical and robotic systems: A path integral reinforcement learning approach. In: International Conference on Robotics and Automation, pp. 208–214 (2012)
15. Stulp, F., Schaal, S.: Hierarchical reinforcement learning with movement primitives. In: 11th IEEE-RAS International Conference on Humanoid Robots, pp. 231–238 (2011)
16. Stulp, F., Sigaud, O.: Path Integral Policy Improvement with Covariance Matrix Adaptation. In: International Conference Machine Learning (2012)
17. Stulp, F., Theodorou, E., Buchli, J., Schaal, S.: Learning to grasp under uncertainty. In: International Conference on Robotics and Automation, pp. 5703–5708. IEEE (2011)
18. Theodorou, E., Buchli, J., Schaal, S.: A generalized path integral control approach to reinforcement learning. Journal of Machine Learning Research 11, 3137–3181 (2010)
19. Theodorou, E., Todorov, E.: Relative entropy and free energy dualities: connections to path integral and KL control. In: IEEE 51st Annual Conference on Decision and Control, pp. 1466–1473 (2012)
20. Todorov, E.: Linearly-solvable Markov decision problems. In: Advances in Neural Information Processing Systems 19, pp. 1369–1376. MIT Press, Cambridge (2006)
21. Todorov, E.: Policy gradients in linearly-solvable MDPs. In: Advances in Neural Information Processing Systems, pp. 2298–2306 (2010)
22. Toussaint, M.: Robot Trajectory Optimization using Approximate Inference. In: Proceedings of the 26th International Conference on Machine Learning (2009)

Appendix: Dual Function for PI-REPS

We derive the dual function. For notation simplicity, we use $p_{\tau\mathbf{x}_0}$ for $p(\tau, \mathbf{x}_0)$, $\phi_{\mathbf{x}_0}$ for $\phi(\mathbf{x}_0)$, $q_{\tau\mathbf{x}_0}$ for $q(\tau, \mathbf{x}_0)$ and S_τ for $S(\tau, \mathbf{x}_0)$. The Lagrangian is

$$
\begin{aligned}
\mathcal{L} &= \int_{\tau,\mathbf{x}_0} p_{\tau\mathbf{x}_0} \left(S_\tau - \log p_{\tau\mathbf{x}_0} - \lambda - \phi_{\mathbf{x}_0}^\mathsf{T} \theta - \eta \log \frac{p_{\tau\mathbf{x}_0}}{q_{\tau\mathbf{x}_0}} \right) d\tau d\mathbf{x}_0 + \lambda + \hat{\phi}_0^\mathsf{T}\theta + \eta\epsilon \\
&= \int_{\tau,\mathbf{x}_0} p_{\tau\mathbf{x}_0} \left(S_\tau - (\eta + 1) \log p_{\tau\mathbf{x}_0} + \eta \log q_{\tau\mathbf{x}_0} - \lambda - \phi_{\mathbf{x}_0}^\mathsf{T}\theta \right) d\tau d\mathbf{x}_0 \\
&\quad + \lambda + \hat{\phi}_0^\mathsf{T}\theta + \eta\epsilon\,,
\end{aligned}
\tag{12}
$$

where θ, η and λ appear due to the constraints of the features, the KL-bound and the normalization, respectively. Taking derivative and solving for $p_{\tau\mathbf{x}_0}$ gives

$$
\frac{\partial\mathcal{L}}{\partial p_{\tau\mathbf{x}_0}} = S_\tau - (\eta + 1)(\log p_{\tau\mathbf{x}_0} + 1) - \lambda - \phi_{\mathbf{x}_0}^\mathsf{T}\theta + \eta\log q_{\tau\mathbf{x}_0} = 0
$$

$$
\log p_{\tau\mathbf{x}_0} = \frac{\eta\log q_{\tau\mathbf{x}_0}}{\eta + 1} + \frac{S_\tau - \phi_{\mathbf{x}_0}^\mathsf{T}\theta}{\eta + 1} + \frac{-\lambda - (\eta + 1)}{\eta + 1}
$$

$$
p_{\tau\mathbf{x}_0} = Z^{-1} q_{\tau\mathbf{x}_0}^{\frac{\eta}{\eta+1}} \exp\left(\frac{S_\tau - \phi_{\mathbf{x}_0}^\mathsf{T}\theta}{\eta + 1} \right), \qquad Z = \exp\left(\frac{\lambda + (\eta + 1)}{\eta + 1} \right). \tag{13}
$$

From the normalization constraint

$$
\exp\left(\frac{\lambda + (\eta + 1)}{\eta + 1} \right) = \int_{\tau,\mathbf{x}_0} q_{\tau\mathbf{x}_0}^{\frac{\eta}{\eta+1}} \exp\left(\frac{S_\tau - \phi_{\mathbf{x}_0}^\mathsf{T}\theta}{\eta + 1} \right) d\tau d\mathbf{x}_0. \tag{14}
$$

Plugging (13) into (12) and simplifying we arrive to the following dual function

$$
g(\theta, \eta) = (\eta + 1) + \lambda + \hat{\phi}_0^\mathsf{T}\theta + \eta\epsilon. \tag{15}
$$

Reinserting (14) in (15)

$$
\begin{aligned}
g(\theta, \eta) &= (\eta + 1) + \lambda + \hat{\phi}_0^\mathsf{T}\theta + \eta\epsilon = (\eta + 1) \left[\frac{(\eta + 1) + \lambda}{\eta + 1} \right] + \hat{\phi}_0^\mathsf{T}\theta + \eta\epsilon \\
&= (\eta + 1) \log\left(\int_{\tau,\mathbf{x}_0} q_{\tau\mathbf{x}_0}^{\frac{\eta}{\eta+1}} \exp\left(\frac{S_\tau - \phi_{\mathbf{x}_0}^\mathsf{T}\theta}{\eta + 1} \right) d\tau d\mathbf{x}_0 \right) + \hat{\phi}_0^\mathsf{T}\theta + \eta\epsilon.
\end{aligned}
$$

Replacing the integral by a sum over sample trajectories generated by q yields

$$
g(\theta, \eta) = (\eta + 1) \log\left(\frac{1}{N} \sum_i q_{\tau\mathbf{x}_0}^{[i]\,\frac{-1}{\eta+1}} \exp\left(\frac{S_\tau^{[i]} - \theta^\mathsf{T}\phi_{\mathbf{x}_0}^{[i]}}{\eta + 1} \right) \right) + \hat{\phi}_0^\mathsf{T}\theta + \eta\epsilon.
$$

The dual function can be evaluated from the state trajectory samples. The distribution $q_{\tau\mathbf{x}_0}^{[i]}$ can be computed using the current policy and the model, i.e. [4],

$$
q_{\tau\mathbf{x}_0}^{[i]} = \mu(\mathbf{x}_0) \prod_{t=0}^{T-1} \int_{\mathbf{u}_t} P_t(\mathbf{x}_{t+1}|\mathbf{x}_t, \mathbf{u}_t) \pi_t(\mathbf{u}_t|\mathbf{x}_t) d\mathbf{u}_t. \tag{16}
$$

---

[4] In the case of Gaussian policies such as we consider here, the integral in (16) can be computed analytically.

How Many Topics?
Stability Analysis for Topic Models

Derek Greene, Derek O'Callaghan, and Pádraig Cunningham

School of Computer Science & Informatics, University College Dublin
{derek.greene,derek.ocallaghan,padraig.cunningham}@ucd.ie

Abstract. Topic modeling refers to the task of discovering the underlying thematic structure in a text corpus, where the output is commonly presented as a report of the top terms appearing in each topic. Despite the diversity of topic modeling algorithms that have been proposed, a common challenge in successfully applying these techniques is the selection of an appropriate number of topics for a given corpus. Choosing too few topics will produce results that are overly broad, while choosing too many will result in the"over-clustering" of a corpus into many small, highly-similar topics. In this paper, we propose a term-centric stability analysis strategy to address this issue, the idea being that a model with an appropriate number of topics will be more robust to perturbations in the data. Using a topic modeling approach based on matrix factorization, evaluations performed on a range of corpora show that this strategy can successfully guide the model selection process.

1 Introduction

From a general text mining perspective, a *topic* in a text corpus can be viewed as either a probability distribution over the terms present in the corpus or a cluster that defines weights for those terms [26]. Considerable research on topic modeling has focused on the use of probabilistic methods such as variants of Latent Dirichlet Allocation (LDA) [5] and Probabilistic Latent Semantic Analysis (PLSA) [11]. Non-probabilistic algorithms, such as Non-negative Matrix Factorization (NMF) [20], have also been applied to this task [26,1]. Regardless of the choice of algorithm, a key consideration in successfully applying topic modeling is the selection of an appropriate number of topics k for the corpus under consideration. Choosing a value of k that is too low will generate topics that are overly broad, while choosing a value that is too high will result in "over-clustering" of the data. For some corpora, coherent topics will exist at several different resolutions, from coarse to fine-grained, reflected by multiple appropriate k values.

When a clustering result is generated using an algorithm that contains a stochastic element or requires the selection of one or more key parameter values, it is important to consider whether the solution represents a "definitive" solution that may easily be replicated. Cluster validation techniques based on this concept have been shown to be effective in helping to choose a suitable number of clusters in data [17,21]. The *stability* of a clustering model refers to its ability to

T. Calders et al. (Eds.): ECML PKDD 2014, Part I, LNCS 8724, pp. 498–513, 2014.

consistently replicate similar solutions on data originating from the same source. In practice, this involves repeatedly clustering using different initial conditions and/or applying the algorithm to different samples of the complete data set. A high level of agreement between the resulting clusterings indicates high stability, in turn suggesting that the current model is appropriate for the data. In contrast, a low level of agreement indicates that the model is a poor fit for the data. Stability analysis has most frequently been applied in bioinformatics [7,4], where the focus has been on model selection for classical clustering approaches, such as k-means [17,3] and agglomerative hierarchical clustering [21,4].

In the literature, the output of topic modeling procedures is often presented in the form of lists of top-ranked terms suitable for human interpretation. Motivated by this, we propose a term-centric stability approach for selecting the number of topics in a corpus, based on the agreement between term rankings generated over multiple runs of the same algorithm. We employ a "top-weighted" ranking measure, where higher-ranked terms have a greater degree of influence when calculating agreement scores. To ensure that a given model is robust against perturbations, we use both sampling of documents from a corpora and random matrix initialization to produce diverse collections of topics on which stability is calculated. Unlike previous applications of the concept of stability in NMF [7] or LDA [25,8], our approach is generic in the sense that it does not rely on directly comparing probability distributions or topic-term matrices. So although we highlight the use of this method in conjunction with NMF, it could be applied in conjunction with other topic modeling and document clustering techniques.

This paper is organized as follows. Section 2 provides a brief overview of existing work in the areas of matrix factorization, stability analysis, and rank agreement. In Section 3 we discuss the problem of measuring the similarity between sets of term rankings, and describe a solution that can be used to quantify topic stability. Using a topic modeling approach based on matrix factorization, in Section 4 we present an empirical evaluation of the proposed solution on a range of text corpora. The paper finishes with some conclusions and suggestions for future work in Section 5.

2 Related Work

2.1 Matrix Factorization

While work on topic models has largely focused on the use of LDA [5,25], Non-negative Matrix Factorization (NMF) can also be applied to textual data to reveal topical structures [26]. NMF seeks to decompose a data matrix into factors that are constrained so that they will not contain negative values. Given a document-term matrix $\mathbf{A} \in \mathbb{R}^{m \times n}$ representing m unique terms present in a corpus of n documents, NMF generates a reduced rank-k approximation in the form of the product of two non-negative factors $\mathbf{A} \approx \mathbf{WH}$, where the objective is to minimize the reconstruction error between $\mathbf{A}$ and the low-dimensional approximation. The columns or *basis vectors* of $\mathbf{W} \in \mathbb{R}^{m \times k}$ can be interpreted as topics, defined with non-negative weights relative to the m terms. The entries in the matrix $\mathbf{H} \in \mathbb{R}^{k \times n}$ provide document memberships with respect to the k topics. Note that, unlike LDA which operates on raw frequency counts, NMF can be applied to a non-negative matrix $\mathbf{A}$ that has been previously normalized using common pre-processing procedures such as TF-IDF term weighting and document length normalization. As with LDA, document-topic assignments are not discrete, allowing a single document to be associated with multiple topics.

For NMF, the key model selection challenge is the selection of the user-defined parameter k. Although no definitive approach for choosing k has been identified, a number of heuristics exist in the literature. A simple technique is to calculate the Residual Sum of Squares (RSS) between the approximation given by a pair of NMF factors and the original matrix [12], which indicates the degree of variation in the dependent variables the NMF model did not explain. The authors suggest that, by examining the RSS curve for a range of candidate values of k, an inflection point might be identified to provide a robust estimate of the optimal reduced rank.

2.2   Stability Analysis

A range of methods based on the concept of *stability analysis* have been proposed for the task of model selection. The *stability* of a clustering algorithm refers to its ability to consistently produce similar solutions on data originating from the same source [17,3]. Since only a single set of data items will be generally available in unsupervised learning tasks, clusterings are generated on perturbations of the original data. The primary application of stability analysis has been as a robust approach for selecting key algorithm parameters [18], specifically when estimating the optimal number of clusters for a given data set. These methods are motivated by the observation that, if the number of clusters in a model is too large, repeated clusterings will lead to arbitrary partitions of the data, resulting in unstable solutions. On the other hand, if the number of clusters is too small, the clustering algorithm will be constrained to merge subsets of objects which should remain separated, also leading to unstable solutions. In contrast, repeated clusterings generated using some optimal number of clusters will generally be consistent, even when the data is perturbed or distorted.

The most common approach to stability analysis involves perturbing the data by randomly sampling the original objects to produce a collection of subsamples for clustering using values of k from a pre-defined range [21]. The stability of the clustering model for each candidate value of k is evaluated using an agreement measure evaluated on all pairs of clusterings generated on different subsamples. One or more values of k are then recommended, selected based on the highest mean agreement scores.

Brunet *et al.* proposed an initial stability-based approach for NMF model selection based on discretized cluster assignments of items (rather than features) across multiple runs of the same algorithm using different random initializations [7]. Specifically, for each NMF run applied to the same data set of n items, a $n \times n$ *connectivity matrix* is constructed, where an entry $(i, j) = 1$ if items i and j are assigned to the same discrete cluster, and $(i, j) = 0$ otherwise. By repeating this process over τ runs, a *consensus matrix* can be calculated as the average of all τ connectivity matrices. Each entry in this matrix indicates the fraction of times two items were clustered together. To measure the stability of a particular value of k, a cophenetic correlation coefficient is calculated on a hierarchical clustering of the connectivity matrix. The authors suggest a heuristic for selecting one or more values of k, based on a sudden drop in the correlation score as k increases.

In their work on LDA, Steyvers and Griffiths noted the importance of identifying those topics that will appear repeatedly across multiple samples of related data [25], which closely resembles the more general concept of stability analysis [21]. The authors suggested comparing two runs of LDA by examining a topic-topic matrix constructed from the symmetric Kullback Liebler (KL) distance between topic distributions from the two runs. Alternative work on measuring the stability of LDA topic models was described in [8]. The authors proposed a document-centric approach, where topics from two different LDA runs are matched together based on correlations between rows of the two corresponding document-topic matrices. The output was represented as a document-document correlation matrix, where block diagonal structured induced by the correlation values are indicative of higher stability. In this respect, the approach is similar to the Brunet *et al.* approach for NMF.

Other evaluation measures used for LDA have included those based on the semantic coherence of the top terms derived from a single set of topics, with respect to term co occurrence within the same corpus or an external background corpus. For example, Newman *et al.* calculated correlations between human judgements and a set of proposed measures, and found that a Pointwise Mutual Information (PMI) measure achieved best or near-best out of all those considered [23]. However, such measures have not focused on model selection and do not consider the robustness of topics over multiple runs of an algorithm.

2.3 Ranking Comparison

A variety of well-known simple metrics exist for measuring the distance or similarity between pairs of ranked lists of the same set of items, notably Spearman's footrule distance and Kendall's tau function [14]. However, Webber *et al.* [27] note that many problems will involve comparing *indefinite rankings*, where items appear in one list but not in another list, but standard metrics do not consider such cases. For other applications, it will be desirable to employ a *top-weighted* ranking agreement measure, such that changing the rank of a highly-relevant item at the top of a list results in a higher penalty than changing the rank of an irrelevant item appearing at the tail of a list. This consideration is important in the case of comparing query results from different search engines, though, as we demonstrate later, it is also a key consideration when comparing rankings of terms arising in topic modeling.

Motivated by basic set overlap, Fagin *et al.* [9] proposed a top-weighted distance metric between indefinite rankings, also referred to as Average Overlap (AO) [27], which calculates the mean intersection size between every pair of subsets of d top-ranked items in two lists, for $d = [1, t]$. This naturally accords a higher positional weight to items at the top of the lists. More recently, Kumar and Vassilvitskii proposed a generic framework for measuring the distance between a pair of rankings [16], supporting both positional weights and item relevance weights. Based on this framework, generalized versions of Kendall's tau and Spearman's footrule metric were derived. However, the authors did not focus on the case of indefinite rankings.

3   Methods

In this section we describe a general stability-based method for selecting the number of topics for topic modeling. Unlike previous unsupervised stability analysis methods, we focus on the use of features or terms to evaluate the suitability of a model. This is motivated by the term-centric approach generally taken in topic modeling, where precedence is generally given to the term-topic output and topics are summarized using a truncated set of top terms. Also, unlike the approach proposed in [7] for genetic data, our method does not assume that topic clusters are entirely disjoint and does not require the calculation of a dense connectivity matrix or the application of a subsequent clustering algorithm.

Firstly, in Section 3.1 we describe a similarity metric for comparing two ranked lists of terms. Using this measure, in Section 3.2 we propose a measure of the agreement between two topic models when represented as ranked term lists. Subsequently, in Section 3.3 we propose a stability analysis method for selecting the number of topics in a text corpus.

3.1   Term Ranking Similarity

A general way to represent the output of a topic modeling algorithm is in the form of a *ranking set* containing k ranked lists, denoted $S = \{R_1, \ldots, R_k\}$. The i-th topic produced by the algorithm is represented by the list R_i, containing the top t terms which are most characteristic of that topic according to some criterion. In the case of NMF, this will correspond to the highest ranked values in each column of the k basis vectors, while for LDA this will consist of the terms with the highest probabilities in the term distribution for each topic. For partitional or hierarchical document clustering algorithms, this might consist of the highest ranked terms in each cluster centroid.

A variety of symmetric measures could be used to assess the similarity between a pair of ranked lists (R_i, R_j). A naïve approach would be to employ a simple set overlap method, such as the Jaccard index [13]. However, such measures do not take into account positional information. Terms occurring at the top of a ranked list generated by an algorithm such as NMF will naturally be more relevant to a topic than those occurring at the tail of the list, which correspond to zero or near-zero values in the original basis vectors. Also, in practice, rather than considering all m terms in a corpus, the results of topic modeling are presented using the top $t << m$ terms. Similarly, when measuring the similarity between ranked lists, it may be preferable to consider truncated lists with only t terms, for economy

Table 1. Example of Average Jaccard (AJ) term ranking similarity, for two ranked lists of terms up to depth $d = 5$. The value Jac_d indicates the Jaccard score at depth d only, while AJ indicates the current AJ similarity at that depth.

d	$R_{1,d}$	$R_{2,d}$	Jac_d	AJ
1	album	sport	0.000	0.000
2	album, music	sport, best	0.000	0.000
3	album, music, best	sport, best, win	0.200	0.067
4	album, music, best, award	sport, best, win, medal	0.143	0.086
5	album, music, best, award, win	sport, best, win, medal, award	0.429	0.154

of representation and to reduce the computational cost of applying multiple similarity operations. However, this will often lead to indefinite rankings, where different subsets of terms are being compared.

Therefore, following the ranking distance measure proposed by Fagin *et al.* [9], we propose the use of a top-weighted version of the Jaccard index, suitable for calculating the similarity between pairs of indefinite rankings. Specifically, we define the *Average Jaccard* (AJ) measure as follows. We calculate the average of the Jaccard scores between every pair of subsets of d top-ranked terms in two lists, for depth $d \in [1, t]$. That is:

$$AJ(R_i, R_j) = \frac{1}{t} \sum_{d=1}^{t} \gamma_d(R_i, R_j) \tag{1}$$

where

$$\gamma_d(R_i, R_j) = \frac{|R_{i,d} \cap R_{j,d}|}{|R_{i,d} \cup R_{j,d}|} \tag{2}$$

such that $R_{i,d}$ is the head of list R_i up to depth d. This is a symmetric measure producing values in the range $[0, 1]$, where the terms through a ranked list are weighted according to a decreasing linear scale. To demonstrate this, a simple illustrative example is shown in Table 1. Note that, although the Jaccard score at depth $d = 5$ is comparatively high (0.429), the mean score is much lower (0.154), as the similarity between terms occurs towards the tails of the lists – these terms carry less weight than those at the head of the lists, such as "album" and "sport".

3.2 Topic Model Agreement

We now consider the problem of measuring the agreement between two different k-way topic models, represented as two ranking sets $S_x = \{R_{x1}, \ldots, R_{xk}\}$ and $S_y = \{R_{y1}, \ldots, R_{yk}\}$, both containing k ranked lists. We construct a $k \times k$ similarity matrix $\mathbf{M}$, such that the entry M_{ij} indicates the agreement between R_{xi} and R_{yj} (*i.e.* the i-th topic in the first model and the j-th topic in the second model), as calculated using the Average Jaccard score (Eqn. 1). We then find the best match between the rows and columns of $\mathbf{M}$ (*i.e.* the ranked lists in S_x

Ranking set $\mathcal{S}_1$:

$R_{11} = \{\text{sport, win, award}\}$
$R_{12} = \{\text{bank, finance, money}\}$
$R_{13} = \{\text{music, album, band}\}$

	R_{21}	R_{22}	R_{23}
R_{11}	0.00	0.07	0.50
R_{12}	0.50	0.00	0.07
R_{13}	0.00	0.61	0.00

Ranking set $\mathcal{S}_2$:

$R_{21} = \{\text{finance, bank, economy}\}$
$R_{22} = \{\text{music, band, award}\}$
$R_{23} = \{\text{win, sport, money}\}$

$\pi = (R_{11}, R_{23}), (R_{12}, R_{21}), (R_{13}, R_{22})$

$agree(\mathcal{S}_1, \mathcal{S}_2) = \dfrac{0.50 + 0.50 + 0.61}{3} = 0.54$

Fig. 1. A simple example of measuring the agreement between two different topic models, each containing $k = 3$ topics, represented by a pair of ranking sets. Term ranking similarity values are calculated using Average Jaccard, up to depth $d = 3$.

and $\mathcal{S}_y$). The optimal permutation π may be found in $O(k^3)$ time by solving the minimal weight bipartite matching problem using the Hungarian method [15]. From this, we can produce an agreement score:

$$agree(\mathcal{S}_x, \mathcal{S}_y) = \frac{1}{k} \sum_{i=1}^{k} AJ(R_{xi}, \pi(R_{xi})) \tag{3}$$

where $\pi(R_{xi})$ denotes the ranked list in $\mathcal{S}_y$ matched to R_{xi} by the permutation π. Values for the above take the range $[0, 1]$, where a comparison between two identical k-way topic models will result in a score of 1. A simple example illustrating the agreement process is shown in Fig. 1.

3.3 Selecting the Number of Topics

Building on the agreement measure defined in Section 3.2, we now propose a model selection approach for topic modeling. For each value of k in a broad predefined range $[k_{min}, k_{max}]$, we proceed as follows. We firstly generate an initial topic model on the complete data set using an appropriate algorithm (ideally this should be deterministic in nature), which provides a reference point for analyzing the stability afforded by using k topics. We represent this as a *reference ranking set* $\mathcal{S}_0$, where each topic is represented by the ranked list of its top t terms. Subsequently, τ samples of the data set are constructed by randomly selecting a subset of $\beta \times n$ documents without replacement, where $0 \leq \beta \leq 1$ denotes the sampling ratio controlling the number of documents in each sample. We then generate τ k-way topic models by applying the topic modeling algorithm to each of the samples, resulting in alternative ranking sets $\{\mathcal{S}_1, \dots, \mathcal{S}_\tau\}$, where all topics are also represented using t top terms. To measure the overall stability at k, we calculate the mean agreement between the reference ranking set and all other ranking sets using Eqn. 3:

$$stability(k) = \frac{1}{\tau} \sum_{i=1}^{\tau} agree(\mathcal{S}_0, \mathcal{S}_i) \tag{4}$$

1. Randomly generate τ samples of the data set, each containing $\beta \times n$ documents.
2. For each value of $k \in [k_{min}, k_{max}]$:
    1. Apply the topic modeling algorithm to the complete data set of n documents to generate k topics, and represent the output as the reference ranking set S_0.
    2. For each sample $\mathbf{X}_i$:
        (a) Apply the topic modeling algorithm to $\mathbf{X}_i$ to generate k topics, and represent the output as the ranking set S_i.
        (b) Calculate the agreement score $agree(S_0, S_i)$.
    3. Compute the mean agreement score for k over all τ samples (Eqn. 4).
3. Select one or more values for k based upon the highest mean agreement scores.

Fig. 2. Summary of the proposed stability analysis method for topic models

This process is repeated for each candidate $k \in [k_{min}, k_{max}]$. A summary of the entire process is given in Fig. 2. Note that the proposed approach is similar to the strategy for item stability analysis proposed in [21], in that a single reference point is used for each value of k, involving τ comparisons between solutions. This contrasts with the approach used by other authors in the literature (e.g. [18]) which involves comparing all unique pairs of results, requiring $\frac{\tau \times (\tau - 1)}{2}$ agreement comparisons.

By examining a plot of the stability scores produced with Eqn. 4, a final value k may be identified based on peaks in the plot. The presence of more than one peak indicates that multiple appropriate topic schemes exist for the corpus under consideration, which is analogous to the existence of multiple alternative solutions in many general cluster analysis problems [2]. An example of this case is shown in Fig. 3(a) for the *guardian-2013* corpus. This data set has six annotated category labels, but we also see a peak at $k = 3$ in the stability plots, suggesting that thematic structure exists at a more coarse level too. On the other hand, a flat curve with no peaks, combined with low stability values, strongly suggests that no coherent topics exist in the data set. This is analogous to the general problem of identifying "clustering tendency" [21]. The example in Fig. 3(b) shows plots generated for a synthetic data set of 500 randomly generated documents. As one might expect, no strong peak appears in the stability plots.

4   Evaluation

4.1   Data

We now evaluate the stability analysis method proposed in Section 3 to assess its usefulness in guiding the selection of the number of topics for NMF. The evaluation is performed on a number of text corpora, each of which has annotated "ground truth" document labels, such that each document is assigned a single label. When pre-processing the data, terms occurring in < 20 documents

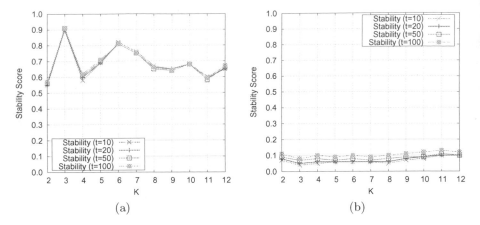

Fig. 3. Stability analysis plots generated using $t = 10/20/50/100$ top terms for (a) the *guardian-2013* corpus of news articles, (b) a synthetic dataset of 500 documents generated randomly from 1,500 terms

were removed, along with English language stop words, but no stemming was performed. Standard log TF-IDF and L2 document length normalization procedures were then applied to the term-document matrix. Descriptions of the corpora are provided in Table 2, and pre-processed versions are made available online for further research[1].

4.2   Experimental Setup

In our experiments we compare the proposed stability analysis method with a popular existing approach for selecting the reduced rank for NMF based on the cophenetic correlation of a consensus matrix [7]. The experimental process involved applying both schemes to each corpus across a reasonable range of values for k, and comparing plots of their output. Here we use $k \in [2, 12]$, based on the fact that the numbers of ground truth labels in the corpora listed in Table 2 are within this range.

To provide a fair comparison, both schemes use information coming from the same collection of matrix factorizations. These were generated using the fast alternating least squares variant of NMF introduced by [22], with random initialization to samples of the data. In all cases we allowed the factorization process to run for a maximum of 50 iterations. We use a sampling ratio of $\beta = 0.8$ (*i.e.* 80% of documents are randomly chosen for each run), with a total of $\tau = 100$ runs to minimize any variance introduced by sampling. For our stability analysis method, we also generate reference ranking sets for each candidate value of k by applying NMF to the complete data set with Nonnegative Double Singular Value Decomposition (NNDSVD) initialization to ensure a deterministic solution [6].

---

[1] http://mlg.ucd.ie/howmanytopics/

Table 2. Details of the corpora used in our experiments, including the total number of documents n, terms m, and number of labels $\hat{k}$ in the associated "ground truth"

Corpus	n	m	$\hat{k}$	Description
bbc	2,225	3,121	5	General news articles from the BBC [10].
bbc-sport	737	969	5	Sports news articles from the BBC [10].
guardian-2013	6,520	10,801	6	New corpus of news articles published by The Guardian during 2013.
irishtimes-2013	3,246	4,832	7	New corpus of news articles published by The Irish Times during 2013.
nytimes-1999	9,551	12,987	4	A subset of the New York Times Annotated Corpus from 1999 [24].
nytimes-2003	11,527	15,001	7	As above, with articles from 2003.
wikipedia-high	5,738	17,311	6	Subset of a Wikipedia dump from January 2014, where articles are assigned labels based on their high level WikiProject.
wikipedia-low	4,986	15,441	10	Another Wikipedia subset. Articles are labeled with fine-grained WikiProject sub-groups.

4.3   Model Selection

Initially, for stability analysis we examined a range of values $t = 10/20/50/100$ for the number of top terms used to represent each topic when measuring agreement between ranked lists. However, the resulting stability scores generated for each value of t were highly correlated across all corpora considered in our evaluation (see Table 3 for average correlations). A typical example of this behavior is shown in Fig. 3(a) for the *guardian-2013* corpus, where the plots almost perfectly overlap. This behavior is perhaps unsurprising as, given the definition of the Average Jaccard measure in Eqn. 1, terms occurring further down ranked lists will naturally carry less weight. Therefore, the difference between scores generated with, say $t = 50$ and $t = 100$ will be minimal. For the remainder of this section we report stability scores for $t = 20$, which provided the highest pairwise mean correlation (0.977) with the results from other values of t examined, while also providing economy of representation for topics.

Figures 4 and 5 show plots generated on the eight corpora for $k \in [2, 12]$, comparing the proposed stability method with the consensus method from [7]. Although both measures can produce values in the range $[0, 1]$, in all experiments the

Table 3. Pearson correlation coefficient scores between stability scores for different numbers of top terms t, as averaged across all corpora in our evaluations

# Terms	$t = 10$	$t = 20$	$t = 50$	$t = 100$	Mean
$t = 10$	-	0.964	0.929	0.926	0.940
$t = 20$	0.964	-	0.985	0.982	**0.977**
$t = 50$	0.929	0.985	-	0.997	0.970
$t = 100$	0.926	0.982	0.997	-	0.968

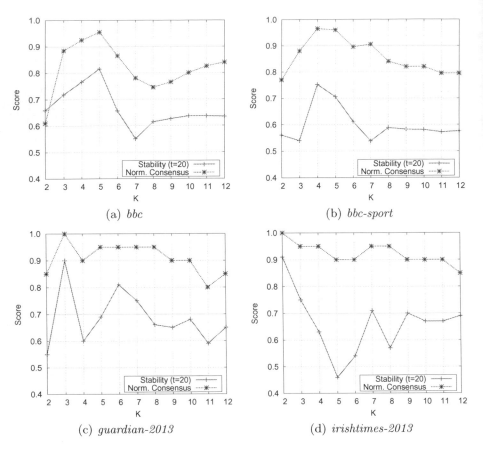

Fig. 4. Comparison of plots generated for stability analysis ($t = 20$) and consensus matrix analysis for values of $k \in [2, 12]$. In both cases we attempt to identify one or more suitable values for k based on peaks in the plots.

observed consensus scores were > 0.8 and often close to 1. Therefore, for the purpose of plotting the results, we apply min-max normalization to the consensus scores (with minimum value 0.8) to rescale the values to a more interpretable range.

We now summarize the results for each of the corpora in detail. The *bbc* corpus contains five well-separated annotated categories for news articles, such as "business" and "entertainment". Therefore it is unsurprising that in Fig. 4(a) we find a strong peak for both methods at $k = 5$, with a sharp fall-off for the stability method after this point. This reflects the fact that the five categories are accurately recovered by NMF. For the *bbcsport* corpus, which also has five ground truth news categories, we see a peak at $k = 4$, followed by a lower peak at $k = 5$ – see Fig. 4(b). The consensus method also exhibits a peak at this point. Examining the top terms for the reference ranking set indicates that the two smallest categories, "athletics" and "tennis" have been assigned to a single larger topic, while the other three categories are clearly represented as topics.

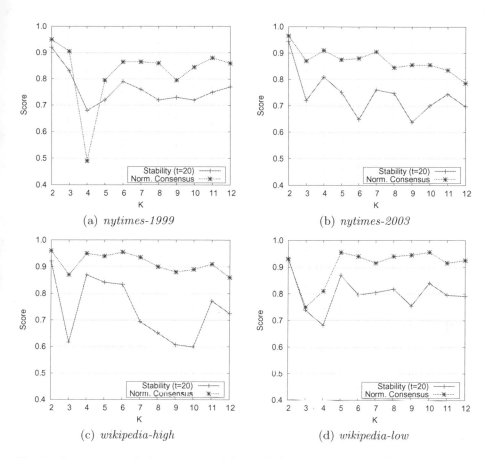

Fig. 5. Comparison of plots generated for stability analysis ($t = 20$) and consensus matrix analysis for values of $k \in [2, 12]$

In the ground truth for the *guardian-2013* corpus, each article is labeled based upon the section in which it appeared on the guardian.co.uk website. From Fig. 4(c) we see that the stability method correctly identifies a peak at $k = 6$ corresponding to the six sections in the corpus, which is not found by the consensus method. However, both methods also suggest a more coarse clustering at $k = 3$. Inspecting the reference ranking set (see Table 4(a)) suggests an intuitive explanation – "books", "fashion" and "music" sections were merged in a single culture-related topic, documents labeled as "politics" and "business" were clustered together, while "football" remains as a distinct topic.

Articles in the *irishtimes-2013* corpus also have annotated labels based on their publication section on irishtimes.com. In Fig. 4(d) we see high scores at $k = 2$ for both methods, and a subsequent peak identified by the stability method at $k = 7$, corresponding to the seven publication sections. In the former case, the top ranked reference set terms indicate a topic related to sports and a catch-all news topic – see Table 4(b).

Next we consider the two corpora of news articles coming from the New York Times Annotated Corpus. Interestingly, for *nytimes-1999*, in Fig. 5(a) both methods exhibit a trough for $k = 4$ topics, even though the ground truth for this corpus contains four news article categories. Inspecting the term rankings shown inTable 4(c) provide a potential explanation of this instability: across the 100 factorization results, the ground truth "sports" category is often but not always split into two topics relating to baseball and basketball. For the *nytimes-2003* corpus, which contains seven article categories, both methods produce high scores at $k = 2$, with subsequent peaks at $k = 4$ and $k = 7$ – see Fig. 5(b). As with the *irishtimes-2013* corpus, the highest-level structures indicate a simple separation between sports articles and other news. The reference topics at $k = 4$ indicates that smaller categories among the New York Times articles, such as "automobiles" and "dining & wine" do not appear as strong themes in the data.

Finally, we consider the two collections of Wikipedia pages, where pages are given labels based on their assignment to WikiProjects[2] at varying levels of granularity. For *wikipedia-high*, from Fig. 5(c) we see that both methods achieve high scores for $k = 2$ and $k = 4$ topics. In the case of the former, the top terms in the reference ranking set indicate a split between Wikipedia pages related to music and all other pages (Table 4(d)). While at $k = 4$ (Table 4(e)), we see coherent topics covering "music", "sports", "space", and a combination of the "military" & "transportation" WikiProject labels. The "medicine" WikiProject is not clearly represented as a topic at this level. In the case of *wikipedia-low*, which contains ten low-level page categories, both methods show spikes at $k = 5$, and $k = 10$. At $k = 5$, NMF recovers topics related to "ice hockey", "cricket", "World War I", a topic covering a mixture of musical genres, and a seemingly incoherent group that includes all other pages. The relatively high stability score achieved at this level (0.87) suggests that this configuration regularly appeared across the 100 NMF runs.

4.4   Discussion

Overall, it is interesting to observe that, for a number of data sets, both methods evaluated here exhibited peaks at $k = 2$, where one might expect far more fine-grained topics in these types of data sets. This results from high agreement between the term ranking sets generated at this level of granularity. A closer inspection of document membership weights for these cases shows that this phenomenon generally arises from the repeated appearance of one small "outlier" topic and one large "merged" topic encompassing the rest of the documents in the corpus (*e.g.* the examples shown in Table 4(b,d)). In a few cases we also see that the ground truth does not always correspond well to the actual data (*e.g.* for the sports-related articles in *nytimes-1999*). This problem arises from time to time when meta-data is used to provide a ground truth in machine learning benchmarking experiments [19].

---

[2] See http://en.wikipedia.org/wiki/Wikipedia:WikiProject

Table 4. Examples of top 10 terms for reference ranking sets generated by NMF on a number of text corpora for different values of k

(a) *guardian-2013* ($k = 3$)

Rank	Topic 1	Topic 2	Topic 3
1	book	league	bank
2	music	club	government
3	fashion	season	labour
4	people	team	growth
5	life	players	uk
6	album	united	economy
7	time	manager	tax
8	novel	game	company
9	love	football	party
10	world	goal	market

(b) *irishtimes-2013* ($k = 2$)

Rank	Topic 1	Topic 2
1	game	cent
2	against	government
3	team	court
4	ireland	health
5	players	ireland
6	time	minister
7	cup	people
8	back	tax
9	violates	dublin
10	win	irish

(c) *nytimes-1999* ($k = 4$)

Rank	Topic 1	Topic 2	Topic 3	Topic 4
1	game	company	yr	mets
2	knicks	stock	bills	yankees
3	team	market	bond	game
4	season	business	rate	inning
5	coach	companies	infl	valentine
6	points	shares	bds	season
7	play	stocks	bd	torre
8	league	york	month	baseball
9	players	investors	municipal	run
10	sprewell	bank	buyer	clemens

(d) *wikipedia-high* ($k = 2$)

Rank	Topic 1	Topic 2
1	album	team
2	band	war
3	song	star
4	music	air
5	released	season
6	songs	aircraft
7	chart	ship
8	video	army
9	rock	line
10	albums	world

(e) *wikipedia-high* ($k = 4$)

Rank	Topic 1	Topic 2	Topic 3	Topic 4
1	album	war	team	star
2	band	air	season	planet
3	song	ship	race	sun
4	music	aircraft	league	earth
5	released	army	game	stars
6	songs	ships	championships	orbit
7	chart	squadron	games	mass
8	video	battle	cup	planets
9	rock	station	world	system
10	albums	british	championship	solar

(f) *wikipedia-low* ($k = 5$)

Rank	Topic 1	Topic 2	Topic 3	Topic 4	Topic 5
1	season	album	cricket	division	opera
2	league	band	test	infantry	stakes
3	team	released	match	battalion	race
4	nhl	metal	innings	war	car
5	hockey	music	runs	battle	racing
6	games	song	wickets	brigade	engine
7	cup	tour	against	army	old
8	game	jazz	australia	regiment	horse
9	goals	songs	england	german	stud
10	club	albums	wicket	squadron	derby

In relation to computational time, the requirement to run a complete hierarchical clustering on the document-document consensus matrix before calculating cophenetic correlations leads to substantially longer running times on all corpora, when compared to the stability analysis method using a reference ranking set.

In addition, the latter can be readily parallelized, as agreement scores can be calculated independently for each of the factorization results.

5   Conclusion

A key challenge when applying topic modeling is the selection of an appropriate number of topics k. We have proposed a new method for choosing this parameter using a term-centric stability analysis strategy, where a higher level of agreement between the top-ranked terms for topics generated across different samples of the same corpus indicates a more suitable choice. Evaluations on a range of text corpora have suggested that this method can provide a useful guide for selecting one or more values for k.

While our experiments have focused on the application of the proposed method in conjunction with NMF, the use of term rankings rather than raw factor values or probabilities means that it can potentially generalize to any topic modeling approach that can represent topics as ranked lists of terms. This includes probabilistic techniques such as LDA, together with more conventional partitional algorithms for document clustering such as k-means and its variants. In further work, we plan to examine the usefulness of stability analysis in conjunction with alternative algorithms.

Acknowledgements. This publication has emanated from research conducted with the financial support of Science Foundation Ireland (SFI) under Grant Number SFI/12/RC/2289.

References

1. Arora, S., Ge, R., Moitra, A.: Learning topic models – Going beyond SVD. In: Proc. 53rd Symp. Foundations of Computer Science, pp. 1–10. IEEE (2012)
2. Bae, E., Bailey, J.: Coala: A novel approach for the extraction of an alternate clustering of high quality and high dissimilarity. In: Proc. 6th International Conference on Data Mining, pp. 53–62. IEEE (2006)
3. Ben-David, S., Pál, D., Simon, H.U.: Stability of k-means clustering. In: Bshouty, N.H., Gentile, C. (eds.) COLT. LNCS (LNAI), vol. 4539, pp. 20–34. Springer, Heidelberg (2007)
4. Bertoni, A., Valentini, G.: Random projections for assessing gene expression cluster stability. In: Proc. IEEE International Joint Conference on Neural Networks (IJCNN 2005)., vol. 1, pp. 149–154 (2005)
5. Blei, D.M., Ng, A.Y., Jordan, M.I.: Latent dirichlet allocation. Journal of Machine Learning Research 3, 993–1022 (2003)
6. Boutsidis, C., Gallopoulos, E.: SVD based initialization: A head start for nonnegative matrix factorization. Pattern Recognition (2008)
7. Brunet, J.P., Tamayo, P., Golub, T.R., Mesirov, J.P.: Metagenes and molecular pattern discovery using matrix factorization. Proc. National Academy of Sciences 101(12), 4164–4169 (2004)
8. De Waal, A., Barnard, E.: Evaluating topic models with stability. In: 19th Annual Symposium of the Pattern Recognition Association of South Africa (2008)

9. Fagin, R., Kumar, R., Sivakumar, D.: Comparing top k lists. SIAM Journal on Discrete Mathematics 17(1), 134–160 (2003)
10. Greene, D., Cunningham, P.: Producing accurate interpretable clusters from high-dimensional data. In: Jorge, A.M., Torgo, L., Brazdil, P.B., Camacho, R., Gama, J. (eds.) PKDD 2005. LNCS (LNAI), vol. 3721, pp. 486–494. Springer, Heidelberg (2005)
11. Hofmann, T.: Probabilistic latent semantic analysis. In: Proc. 15th Conference on Uncertainty in Artificial Intelligence, pp. 289–296. Morgan Kaufmann (1999)
12. Hutchins, L.N., Murphy, S.M., Singh, P., Graber, J.H.: Position-dependent motif characterization using non-negative matrix factorization. Bioinformatics 24(23), 2684–2690 (2008)
13. Jaccard, P.: The distribution of flora in the alpine zone. New Phytologist 11(2), 37–50 (1912)
14. Kendall, M., Gibbons, J.D.: Rank Correlation Methods. Edward Arnold, London (1990)
15. Kuhn, H.W.: The hungarian method for the assignment problem. Naval Research Logistics Quaterly 2, 83–97 (1955)
16. Kumar, R., Vassilvitskii, S.: Generalized distances between rankings. In: Proc. 19th International Conference on World Wide Web, pp. 571–580. ACM (2010)
17. Lange, T., Roth, V., Braun, M.L., Buhmann, J.M.: Stability-based validation of clustering solutions. Neural Computation 16(6), 1299–1323 (2004)
18. Law, M., Jain, A.K.: Cluster validity by bootstrapping partitions. Tech. Rep. MSU-CSE-03-5, University of Washington (February 2003)
19. Lee, C., Cunningham, P.: Community detection: effective evaluation on large social networks. Journal of Complex Networks (2013)
20. Lee, D.D., Seung, H.S.: Learning the parts of objects by non-negative matrix factorization. Nature 401, 788–791 (1999)
21. Levine, E., Domany, E.: Resampling method for unsupervised estimation of cluster validity. Neural Computation 13(11), 2573–2593 (2001)
22. Lin, C.: Projected gradient methods for non-negative matrix factorization. Neural Computation 19(10), 2756–2779 (2007)
23. Newman, D., Lau, J.H., Grieser, K., Baldwin, T.: Automatic Evaluation of Topic Coherence. In: Proc. Conf. North American Chapter of the Association for Computational Linguistics (HLT 2010), pp. 100–108 (2010)
24. Sandhaus, E.: The New York Times Annotated Corpus. Linguistic Data Consortium 6(12), e26752 (2008)
25. Steyvers, M., Griffiths, T.: Probabilistic topic models. In: Handbook of latent semantic analysis, vol. 427(7), pp. 424–440 (2007)
26. Wang, Q., Cao, Z., Xu, J., Li, H.: Group matrix factorization for scalable topic modeling. In: Proc. 35th SIGIR Conference on Research and Development in Information Retrieval, pp. 375–384. ACM (2012)
27. Webber, W., Moffat, A., Zobel, J.: A similarity measure for indefinite rankings. ACM Transactions on Information Systems (TOIS) 28(4), 20 (2010)

Joint Prediction of Topics in a URL Hierarchy

Michael Großhans[1], Christoph Sawade[2], Tobias Scheffer[1], and Niels Landwehr[1]

[1] University of Potsdam, Department of Computer Science, August-Bebel-Straße 89,
14482 Potsdam, Germany
{grosshan,landwehr,scheffer}@cs.uni-potsdam.de
[2] SoundCloud Ltd., Greifswalderstraße 212, 10405 Berlin, Germany
christoph@soundcloud.com

Abstract. We study the problem of jointly predicting topics for all web pages within URL hierarchies. We employ a graphical model in which latent variables represent the predominant topic within a subtree of the URL hierarchy. The model is built around a generative process that infers how web site administrators hierarchically structure web site according to topic, and how web page content is generated depending on the page topic. The resulting predictive model is linear in a joint feature map of content, topic labels, and the latent variables. Inference reduces to message passing in a tree-structured graph; parameter estimation is carried out using concave-convex optimization. We present a case study on web page classification for a targeted advertising application.

1 Introduction

Web page classification of entire web domains has numerous applications. For instance, topic labels can be used to match individual web pages to related advertisements; topic labels can be aggregated over pages that a user has visited in order to create a profile of the user's interests. Classifying the child suitability of web pages is another typical use case.

There is a rich body of research on topic classification of web pages based on page content [8]. Classification does not have to rely on page content alone. For instance, collective classification schemes that exploit the hyperlink structure within the world wide web have also been widely studied [6,11]. Collective classification approaches define probabilistic models over web page content and the observed hyperlink structure; inference in the models yields the most likely joint configuration of topic labels. Typically, discriminative models over topics given page content and link structure are studied, in order to avoid having to estimate high-dimensional distributions over page content. For example, maximum margin Markov networks have been shown to give excellent results in hypertext classification domains [11,2].

In this paper, we study models that exploit the information inherent in the URL hierarchy of a web domain, rather than the information contained in the hyperlink structure. Many web domains organize their individual pages in a meaningful hierarchy in which subtrees tend to contain web pages of similar

T. Calders et al. (Eds.): ECML PKDD 2014, Part I, LNCS 8724, pp. 514–529, 2014.

topics. The predominant topic within a particular subtree constitutes a latent variable from the learner's perspective; correctly inferring these latent variables and propagating the topic information to pages within the subtree has the potential to boost predictive accuracy if topic correlation within subtrees is strong.

The presence of latent variables constitutes a key difference of our problem setting in comparison to collective classification based on hyperlinks; models developed for collective hypertext classification are typically not applicable as they cannot deal with latent variables during learning. The problem could instead be modeled using latent variable structured output models, such as latent variable SVMstruct [14]. The challenge when using this approach is to correctly model the interaction of latent and observed variables using a joint feature map and specifying appropriate loss functions while ensuring that decoding remains tractable.

We instead follow an approach in which we formulate a generative model of how URL trees are populated with topic labels and content is generated for web pages within that URL tree. The model can be formulated conveniently using topic-correlation models and standard exponential-family distributions for page content given a topic. The model is then trained discriminatively by maximizing the conditional distribution of topic labels given page content and the URL tree. This conditional distribution has the form of a linear model with a joint feature map of the URL hierarchy, page content, and the (observable and latent) topic labels. Efficient decoding is possible by message passing in a tree-structured graph. In this formulation, the feature map, decoding algorithm, and optimization criterion directly result from the probabilistic modeling assumptions

The rest of this paper is organized as follows. Section 2 states the problem setting and introduces notation. Section 3 presents the probabilistic model, and Section 4 discusses parameter learning and inference. Section 5 reports on an empirical study on web classification for a targeted advertising application. Section 6 discusses related work, Section 7 concludes.

2   Notation and Problem Setting

A URL tree $\mathcal{T} = (\mathcal{V}, \mathcal{E})$ consists of vertices and edges. The vertices $\mathcal{V}$ include n leaves that we will write as $v_1, \ldots, v_n$, and k inner nodes, written as $v_{n+1}, \ldots, v_{n+k}$. The leaves correspond to URLs of individual web pages (such as *washingtonpost.com/local/crime/murder.htm*); inner nodes corresponds to prefixes of these URLs that end in a separator—typically, the slash symbol. An inner node, such as *washingtonpost.com/local/*, thus represents a subtree in the URL hierarchy. Figure 1 (left) shows an imaginary URL tree for a web domain.

The content of each of the leaf nodes is encoded as a vector $\mathbf{x}_i \in \mathbb{R}^m$; we denote the content of the entire web portal as matrix $\mathbf{X} = (\mathbf{x}_1 \ldots \mathbf{x}_n) \in \mathbb{R}^{m \times n}$. The topic space is denoted by $\mathcal{Y}$; the vector $\mathbf{y} = (y_1, ..., y_n)^{\mathsf{T}} \in \mathcal{Y}^n$ denotes topic labels for the n leaf nodes that correspond to web pages.

Web domains are generated by an unknown distribution $p(\mathbf{y}, \mathbf{X}, \mathcal{T})$; we will make specific modeling assumptions about this distribution in the next section. Distinct web domains are drawn independently and from an identical distribution. Note, however, that the random variables $(\mathbf{x}_1, y_1), ..., (\mathbf{x}_n, y_n)$ that represent

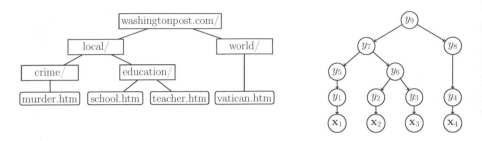

Fig. 1. URL tree (left) and graphical model (right) of an exemplary news domain

content and topic information for any single domain are not assumed to be independent. Typically, the variables will correlate as a function of their position in the URL hierarchy $\mathcal{T}$. A training sample of several labeled web domains is drawn according to this distribution.

Finally, the URL tree $\mathcal{T}^*$ and the content matrix $\mathbf{X}^*$ of a target web domain are drawn. In addition, topic labels y_i^* for a limited number (possibly zero) of leaf nodes are disclosed. The goal is to infer the most likely complete vector of topic labels $\mathbf{y}^*$ for this target domain.

3   Graphical Model

In order to define an appropriate probabilistic model for the problem stated in Section 2, we first define a model of a generative process that we assume to have generated the observable data in Section 3.1. In Section 3.2, we express this model as a member of the exponential family. Deriving the conditional distribution $p(\mathbf{y}|\mathbf{X}; \mathcal{T})$ in this model results in a linear structured-output model. In our application, inference can be carried out efficiently using message passing inference in the tree-structured model (Section 4.1). Parameters can be estimated according to maximal conditional likelihood using CCCP (Section 4.2).

3.1   A Generative Process for Web Domains

In this section, we define a generative process for populating a given URL tree $\mathcal{T}$ with topic labels and word count information (we will make no modeling assumptions about the distribution $p(\mathcal{T})$ over URL trees).

The general assumption that underlies our model is that web site administrators hierarchically structure web site content according to topic, such that topics of pages within one subtree of the URL hierarchy correlate. To represent the predominant topic within specific subtrees, we associate a vector of latent topic variables $\bar{\mathbf{y}} = (y_{n+1}, ..., y_{n+k}) \in \mathcal{Y}^k$ to the k inner nodes $v_{n+1}, \ldots, v_{n+k}$ of the URL tree $\mathcal{T}$. These latent topic variables will couple the observable topic variables $\mathbf{y} = (y_1, ..., y_n)$ through the URL hierarchy. Throughout the paper, we denote by $v_{\to i} \in \mathcal{V}$ the unique parent of the node $v_i \in \mathcal{V}$ specified by the edge set $\mathcal{E}$ of URL tree $\mathcal{T}$. We extend this definition to topic labels as follows: for a topic variable $y_i \in \{y_1, ..., y_{n+k}\}$, we write $y_{\to i}$ to denote the latent topic variable associated with the node $v_{\to i}$.

We assume a top-down generative process for topic variables by modeling the dependency between a topic y_i and the topic $y_{\to i}$ of the parent node as a distribution $p(y_i|y_{\to i}; \lambda)$. The distribution is modeled as a *normalized exponential*

$$p(y_i|y_{\to i}; \lambda) = \frac{\exp\left(-\lambda\Delta(y_i, y_{\to i})\right)}{\sum\limits_{y' \in \mathcal{Y}} \exp\left(-\lambda\Delta(y', y_{\to i})\right)} \tag{1}$$

where the function $\Delta(y_i, y_{\to i})$ measures topic distance in $\mathcal{Y}$. A simple choice for topic distance would be $\Delta(y, y') = 0$ if $y = y'$ and $\Delta(y, y') = 1$ if $y \neq y'$; other distance functions may be employed to reflect a specific structure on the topic space. The parameter λ controls the degree of correlation expected between the topic variables y_i and $y_{\to i}$. This generative process corresponds to the assumption that when administrators add novel material that covers topic $y \in \mathcal{Y}$ to a web domain, they insert a corresponding URL subtree under a parent node that is associated with a topic $y_{\to i}$ close to y. We assume this process is carried out recursively up to and including the leaf nodes, that is, novel URL subtrees are again populated with subtrees and eventually web pages with topics that are close within the topic space $\mathcal{Y}$. The prior distribution $p(y_{n+k}|\tau)$ over the topics of the root is a categorical distribution over topics, parametrized by τ.

In order to complete the specification of the data-generating process we have to assume a distribution $p(\mathbf{x}|y)$ over word-count information $\mathbf{x}$ given the web page topic y. At this point, we only assume that this distribution is a member of the exponential family and follows the general form

$$p\left(\mathbf{x}|y; \boldsymbol{\eta}\right) = h(\mathbf{x})\exp\left(\boldsymbol{\eta}^{\mathsf{T}}\phi(\mathbf{x}, y) - g_{\eta}(\boldsymbol{\eta}, y)\right). \tag{2}$$

In Equation 2, $h(\mathbf{x})$ is called the base measure, $g_{\eta}(\boldsymbol{\eta}, y)$ is the log-partition function that ensures correct normalization of the distribution, $\phi(\mathbf{x}, y)$ is a joint feature map of the web page $\mathbf{x}$ and topic y, and $\boldsymbol{\eta}$ is a parameter vector.

By defining a joint feature map of $\mathbf{x}$ and y, we subsume the case of modeling topic-specific parameter vectors for $\phi(\mathbf{x}, y) = \Lambda(y) \otimes \phi(\mathbf{x})$, where operator $\otimes$ denotes the Kronecker product and $\Lambda(y) = (\llbracket y = \bar{y} \rrbracket)_{\bar{y} \in \mathcal{Y}}$ denotes the one-of-k encoding of y, but can also encode structural prior knowledge, for instance about a structured topic space.

By combining the generative process for topic variables based on Equation 1 and the categorical distribution $p(y_{n+k}|\tau)$ with the conditional distribution defined by Equation 2, we obtain a generative model $p(\mathbf{y}, \bar{\mathbf{y}}, \mathbf{X}|\mathcal{T}; \lambda, \boldsymbol{\eta}, \tau)$ of all topic variables given web page texts and the URL tree:

$$p(\mathbf{y}, \bar{\mathbf{y}}, \mathbf{X}|\mathcal{T}; \lambda, \boldsymbol{\eta}, \tau) = p(y_{n+k}|\tau)\left(\prod_{i=1}^{n+k-1} p\left(y_i|y_{\to i}; \lambda\right)\right)\prod_{i=1}^{n} p\left(\mathbf{x}_i|y_i; \boldsymbol{\eta}\right). \tag{3}$$

Figures 1 (right) shows the graphical model representation of this model for the example URL tree shown in Figure 1 (left).

3.2   A Discriminative Joint Topic Model

Starting from the generative process defined in Section 3.1, we now derive a discriminative model for page topics based on URL hierarchy $\mathcal{T}$ and page texts $\mathbf{X}$.

We begin the derivation by casting the generative process $p(\mathbf{y}, \bar{\mathbf{y}}, \mathbf{X} | \mathcal{T}; \lambda, \boldsymbol{\eta}, \boldsymbol{\tau})$ defined in Section 3.1 into an exponential-family model using a joint feature map $\Phi(\mathbf{X}, \mathbf{y}, \bar{\mathbf{y}})$. Note that $p(y_i | y_{\to i}; \lambda)$ can be written in the canonical form of an exponential family by $p(y_i | y_{\to i}; \lambda) = \exp\left(-\lambda \Delta(y_i, y_{\to i}) - g_\lambda(\lambda, y_{\to i})\right)$ where $g_\lambda(\lambda, y_{\to i}) = \log \sum_{y' \in \mathcal{Y}} \exp\left(-\lambda \Delta(y', y_{\to i})\right)$, and $p(y_{n+k} | \boldsymbol{\tau})$ can be written in exponential family form by $p(y_{n+k} | \boldsymbol{\tau}) = \exp\left(\boldsymbol{\tau}^\mathsf{T} \Lambda(y) - g_{\boldsymbol{\tau}}(\boldsymbol{\tau})\right)$ where $g_{\boldsymbol{\tau}}(\boldsymbol{\tau}) = \log(\mathbf{1}^\mathsf{T} \exp(\boldsymbol{\tau}))$. Then, Equation 3 can be written as

$$p(\mathbf{y}, \bar{\mathbf{y}}, \mathbf{X} | \mathcal{T}; \boldsymbol{\theta})$$

$$= \exp\left(\boldsymbol{\tau}^\mathsf{T} \Lambda(y) - g_{\boldsymbol{\tau}}(\boldsymbol{\tau})\right) \left(\prod_{i=1}^{n+k-1} \exp\left(-\lambda \Delta(y_i, y_{\to i}) - g_\lambda(\lambda, y_{\to i})\right)\right)$$

$$\prod_{i=1}^{n} h(\mathbf{x}_i) \exp\left(\boldsymbol{\eta}^\mathsf{T} \phi(\mathbf{x}_i, y_i) - g_{\boldsymbol{\eta}}(\boldsymbol{\eta}, y_i)\right)$$

$$= \exp\left(\boldsymbol{\tau}^\mathsf{T} \Lambda(y) - g_{\boldsymbol{\tau}}(\boldsymbol{\tau})\right) \exp\left(-\lambda \sum_{i=1}^{n+k-1} \Delta(y_i, y_{\to i}) - \sum_{i=1}^{n+k-1} g_\lambda(\lambda, y_{\to i})\right)$$

$$\left(\prod_{i=1}^{n} h(\mathbf{x}_i)\right) \exp\left(\boldsymbol{\eta}^\mathsf{T} \sum_{i=1}^{n} \phi(\mathbf{x}_i, y_i) - \sum_{i=1}^{n} g_{\boldsymbol{\eta}}(\boldsymbol{\eta}, y_i)\right)$$

$$= h(\mathbf{X}) \exp\left(\boldsymbol{\theta}^\mathsf{T} \Phi(\mathbf{X}, \mathbf{y}, \bar{\mathbf{y}}) - g_{\boldsymbol{\tau}}(\boldsymbol{\tau})\right), \tag{4}$$

where we define a joint parameter vector $\boldsymbol{\theta} = (\boldsymbol{\eta}^\mathsf{T}, \boldsymbol{\tau}^\mathsf{T}, \lambda, \boldsymbol{\gamma}^\mathsf{T})^\mathsf{T}$, a feature map

$$\Phi(\mathbf{X}, \mathbf{y}, \bar{\mathbf{y}}) = \begin{pmatrix} \sum_{i=1}^{n} \phi(\mathbf{x}_i, y_i) \\ \Lambda(y_{n+k}) \\ \sum_{i=1}^{n+k-1} \Delta(y_i, y_{\to i}) \\ \sum_{i=1}^{n+k-1} \Lambda(y_{\to i}) \end{pmatrix}, \tag{5}$$

and base measure $h(\mathbf{X}) = \prod_{i=1}^{n} h(\mathbf{x}_i)$. Note that in Equation 4 we subsumed the sum of log-partition functions $\sum_{i=1}^{n} g_{\boldsymbol{\eta}}(\boldsymbol{\eta}, y_i)$ into the model parameter $\boldsymbol{\eta}$ by adding a constant feature to the feature map $\phi(\mathbf{x}, y)$ for each $y \in \mathcal{Y}$. Additionally we subsumed the sum of log-partition functions $\sum_{i=1}^{n+k-1} g_\lambda(\lambda, y_{\to i})$ into an additional model parameter $\boldsymbol{\gamma}$.

The conditional distribution $p(\mathbf{y}, \bar{\mathbf{y}} | \mathbf{X}, \mathcal{T}; \boldsymbol{\theta})$ over observable and latent topic variables given web page content and the URL tree is now given by

$$
\begin{aligned}
p(\mathbf{y}, \bar{\mathbf{y}} | \mathbf{X}, \mathcal{T}; \boldsymbol{\theta}) &= \frac{p(\mathbf{y}, \bar{\mathbf{y}}, \mathbf{X} | \mathcal{T}; \boldsymbol{\theta})}{\sum_{\mathbf{y}', \bar{\mathbf{y}}'} p(\mathbf{y}', \bar{\mathbf{y}}', \mathbf{X} | \mathcal{T}; \boldsymbol{\theta})} \\
&= \frac{h(\mathbf{X}) \exp(g_{\boldsymbol{\tau}}(\boldsymbol{\tau})) \exp\left(\boldsymbol{\theta}^{\mathsf{T}} \Phi(\mathbf{X}, \mathbf{y}, \bar{\mathbf{y}})\right)}{h(\mathbf{X}) \exp(g_{\boldsymbol{\tau}}(\boldsymbol{\tau})) \sum_{\mathbf{y}', \bar{\mathbf{y}}'} \exp\left(\boldsymbol{\theta}^{\mathsf{T}} \Phi(\mathbf{X}, \mathbf{y}', \bar{\mathbf{y}}')\right)} \\
&= \frac{\exp\left(\boldsymbol{\theta}^{\mathsf{T}} \Phi(\mathbf{X}, \mathbf{y}, \bar{\mathbf{y}})\right)}{\sum_{\mathbf{y}', \bar{\mathbf{y}}'} \exp\left(\boldsymbol{\theta}^{\mathsf{T}} \Phi(\mathbf{X}, \mathbf{y}', \bar{\mathbf{y}}')\right)}.
\end{aligned} \tag{6}
$$

Note that Equation 6 defines a linear structured-output model in the joint feature map $\Phi(\mathbf{X}, \mathbf{y}, \bar{\mathbf{y}})$ because $\arg\max_{\mathbf{y}, \bar{\mathbf{y}}} p(\mathbf{y}, \bar{\mathbf{y}} | \mathbf{X}, \mathcal{T}; \boldsymbol{\theta}) = \arg\max_{\mathbf{y}, \bar{\mathbf{y}}} \boldsymbol{\theta}^{\mathsf{T}} \Phi(\mathbf{X}, \mathbf{y}, \bar{\mathbf{y}})$.

4 Inference and Parameter Estimation

We now turn toward the problem of inferring topic variables and obtaining maximum-a-posteriori estimates of model parameters from data.

4.1 Inferring Topics for New Web Portals

For a given new web domain, inference might target the most likely joint assignment of topics to web pages by summing out the latent variables $\bar{\mathbf{y}}$,

$$
\mathbf{y}^* = \arg\max_{\mathbf{y}} \sum_{\bar{\mathbf{y}}} p(\mathbf{y}, \bar{\mathbf{y}} | \mathbf{X}, \mathcal{T}; \boldsymbol{\theta}). \tag{7}
$$

Unfortunately, this problem is NP-hard even for tree-structured graphs [4]. Instead, we are able to infer the most likely topic assignment y_i of the i-th page by summing out latent variables $\bar{\mathbf{y}}$ and topic variables of all other web pages $\mathbf{y}_{\bar{i}}$,

$$
y_i^* = \arg\max_{y} \sum_{\mathbf{y}_{\bar{i}}, \bar{\mathbf{y}}} p(\mathbf{y}, \bar{\mathbf{y}} | \mathbf{X}, \mathcal{T}; \boldsymbol{\theta}). \tag{8}
$$

Alternatively, we can infer the most likely joint state of all topic variables,

$$
(\mathbf{y}^*, \bar{\mathbf{y}}^*) = \arg\max_{\mathbf{y}, \bar{\mathbf{y}}} p(\mathbf{y}, \bar{\mathbf{y}} | \mathbf{X}, \mathcal{T}; \boldsymbol{\theta}), \tag{9}
$$

thereby also inferring topics for inner nodes in the URL hierarchy $\mathcal{T}$. For the application motivating this paper, the latter approach is advantageous if web sites are very dynamic: if novel pages are added to the URL tree and there is insufficient time to carry out a full inference, the topic assigned to the parent node of the added page can be used to label the novel page quickly. This is often the only feasible approach for real-time systems, and is in fact implemented in the targeted-advertisement company that we collaborate with.

Moreover, if topics $\mathbf{y}_{\bar{S}}$ of a subset of web pages $\bar{S} \subseteq \{1, \ldots, n\}$ are already observed, the most likely conditional joint assignment for the unobserved labels $\mathbf{y}_S$ where $S = \{1, \ldots, n\} \backslash \bar{S}$ and the latent variables $\bar{\mathbf{y}}$ given the observed labels is

$$(\mathbf{y}_S^*, \bar{\mathbf{y}}^*) = \arg\max_{\mathbf{y}_S, \bar{\mathbf{y}}} p(\mathbf{y}_S, \bar{\mathbf{y}} | \mathbf{X}, \mathbf{y}_{\bar{S}}, \mathcal{T}_U; \boldsymbol{\theta}). \tag{10}$$

Due to the tree-structured form of the model given by Equation 3, the optimization problems given by Equation 8, 9, and 10 can be solved efficiently using standard message passing algorithms [7].

4.2   Parameter Estimation

To estimate model parameters, we minimize the regularized discriminative negative log-likelihood over all URL trees:

$$\boldsymbol{\theta}^* = \arg\min_{\boldsymbol{\theta}} \; \Omega(\boldsymbol{\theta}) - \log \prod_{j=1}^{u} p(\mathbf{y}^j | \mathbf{X}^j, \mathcal{T}_j; \boldsymbol{\theta})$$

$$= \arg\min_{\boldsymbol{\theta}} \; \Omega(\boldsymbol{\theta}) + \sum_{j=1}^{u} \ell_{log}(\boldsymbol{\theta}, \mathbf{X}^j, \mathbf{y}^j), \tag{11}$$

where the loss function is given by

$$\ell_{log}(\boldsymbol{\theta}, \mathbf{X}, \mathbf{y}) = \log \sum_{\mathbf{y}', \bar{\mathbf{y}}'} \exp\left(\boldsymbol{\theta}^{\mathsf{T}} \Phi(\mathbf{X}, \mathbf{y}', \bar{\mathbf{y}}')\right) - \log \sum_{\bar{\mathbf{y}}} \exp\left(\boldsymbol{\theta}^{\mathsf{T}} \Phi(\mathbf{X}, \mathbf{y}, \bar{\mathbf{y}})\right).$$

In order to specify the regularizer $\Omega(\boldsymbol{\theta}) = \Omega_{\boldsymbol{\eta}, \boldsymbol{\gamma}}(\boldsymbol{\eta}, \boldsymbol{\gamma}) + \Omega_{\boldsymbol{\tau}}(\boldsymbol{\tau}) + \Omega_{\lambda}(\lambda_j)$, we assume a zero-mean Gaussian prior $(\boldsymbol{\eta}, \boldsymbol{\gamma}) \sim \mathcal{N}(\mathbf{0}, \sigma_{\boldsymbol{\eta}, \boldsymbol{\gamma}}^2 \mathbf{I})$ with variance $\sigma_{\boldsymbol{\eta}, \boldsymbol{\gamma}}^2$ over $\boldsymbol{\eta}$ and $\boldsymbol{\gamma}$, a Dirichlet prior over the topic distribution $p(y_{n+k} | \boldsymbol{\tau})$ at the root node, and an inverse gamma prior $\lambda \sim \text{InvGam}(1, \sigma_\lambda^2)$ over the coupling parameter λ, where the inverse gamma distribution is parameterized using mean and variance. Given these prior distributions the regularizing terms are defined by:

$$\Omega_{\boldsymbol{\eta}, \boldsymbol{\gamma}}(\boldsymbol{\eta}, \boldsymbol{\gamma}) = \frac{1}{2\sigma_{\boldsymbol{\eta}, \boldsymbol{\gamma}}^2} \left(\|\boldsymbol{\eta}\|_2^2 + \|\boldsymbol{\gamma}\|_2^2\right), \quad \Omega_{\lambda}(\lambda) = \frac{\sigma^{-2} + 1}{\lambda} + (\sigma_\lambda^{-2} + 3)\log(\lambda),$$

$$\Omega_{\boldsymbol{\tau}}(\boldsymbol{\tau}) = \log(\mathbf{1}^{\mathsf{T}} \exp(\boldsymbol{\tau})) \mathbf{1}^{\mathsf{T}} (\boldsymbol{\alpha} - \mathbf{1}) - (\boldsymbol{\alpha} - \mathbf{1})^{\mathsf{T}} \boldsymbol{\tau}.$$

In Equation 11 we determine the minimizing argument of a sum of a convex and a concave function $\boldsymbol{\theta}^* = \arg\min_{\boldsymbol{\theta}} f_{\cup}(\boldsymbol{\theta}) + f_{\cap}(\boldsymbol{\theta})$ where

$$f_{\cup}(\boldsymbol{\theta}) = \Omega_{\cup}(\boldsymbol{\theta}) + \sum_{j=1}^{u} \log \sum_{\mathbf{y}', \bar{\mathbf{y}}'} \exp\left(\boldsymbol{\theta}^{\mathsf{T}} \Phi(\mathbf{X}^j, \mathbf{y}', \bar{\mathbf{y}}')\right)$$

$$f_{\cap}(\boldsymbol{\theta}) = \Omega_{\cap}(\boldsymbol{\theta}) - \sum_{j=1}^{u} \log \sum_{\bar{\mathbf{y}}} \exp\left(\boldsymbol{\theta}^{\mathsf{T}} \Phi(\mathbf{X}^j, \mathbf{y}^j, \bar{\mathbf{y}})\right),$$

Algorithm 1. Concave-Convex Procedure

1: Initialize $\boldsymbol{\theta}^*$
2: **repeat**
3:    Construct upper bound on $f_\cap(\boldsymbol{\theta})$ for some c: $f_/(\boldsymbol{\theta}) \leftarrow \boldsymbol{\theta}^\mathsf{T} [\nabla f_\cap(\boldsymbol{\theta})]_{\boldsymbol{\theta}=\boldsymbol{\theta}^*} + c$
4:    $\boldsymbol{\theta}^* \leftarrow \arg\min_{\boldsymbol{\theta}} f_\cup(\boldsymbol{\theta}) + h(\boldsymbol{\theta})$
5: **until** $\boldsymbol{\theta}^*$ converges
6: return $\boldsymbol{\theta}^*$

---

and where

$$\Omega_\cup(\boldsymbol{\theta}) = \Omega_{\boldsymbol{\eta},\boldsymbol{\gamma}}(\boldsymbol{\eta},\boldsymbol{\gamma}) + \Omega_\tau(\boldsymbol{\tau}) + \frac{\sigma^{-2}+1}{\lambda}, \qquad \Omega_\cap(\boldsymbol{\theta}) = (\sigma_\lambda^{-2}+3)\log(\lambda)$$

subsume the convex and the concave part of the regularization term. Thus, the optimization problem given by Equation 11 is in general not convex—although it is convex in $\boldsymbol{\eta}$, since $f_\cap(\boldsymbol{\theta})$ is a linear function in $\boldsymbol{\eta}$. In order to solve the optimization problem, we use the Concave-Convex Procedure, which is guaranteed to converge to a local optimum [15]. Algorithm 1 iteratively upper-bounds the concave part $f_\cap(\boldsymbol{\theta})$ by the linear function $f_/(\boldsymbol{\theta})$ (see Line 4) and solves the resulting convex optimization problem in Line 5 using standard gradient descend methods. The gradients with respect to $\boldsymbol{\theta}$ are given by:

$$\nabla f_\cup(\boldsymbol{\theta}) = \nabla\Omega_\cup(\boldsymbol{\theta}) + \sum_{j=1}^{u} \mathbb{E}_{\mathbf{y}',\bar{\mathbf{y}}}^j \left[\Phi(\mathbf{X}^j, \mathbf{y}', \bar{\mathbf{y}})\right],$$

$$\nabla f_\cap(\boldsymbol{\theta}) = \nabla\Omega_\cap(\boldsymbol{\theta}) + \sum_{j=1}^{u} \mathbb{F}_{\bar{\mathbf{y}}}^j \left[\Phi(\mathbf{X}^j, \mathbf{y}^j, \bar{\mathbf{y}})\right],$$

where expectation $\mathbb{E}_{\mathbf{y}',\bar{\mathbf{y}}}^j$ bases on distribution $p(\mathbf{y}', \bar{\mathbf{y}}|\mathbf{X}^j, \mathcal{T}_j; \boldsymbol{\theta})$ and expectation $\mathbb{E}_{\bar{\mathbf{y}}}^j$ bases on $p(\bar{\mathbf{y}}|\mathbf{y}^j, \mathbf{X}^j, \mathcal{T}_j; \boldsymbol{\theta}) = p(\mathbf{y}^j, \bar{\mathbf{y}}|\mathbf{X}^j, \mathcal{T}_j; \boldsymbol{\theta})/\sum_{\bar{\mathbf{y}}'} p(\mathbf{y}^j, \bar{\mathbf{y}}'|\mathbf{X}^j, \mathcal{T}_j; \boldsymbol{\theta})$. The gradients of the regularization parts are given by

$$\nabla\Omega_\cup(\boldsymbol{\theta}) = \sigma_{\boldsymbol{\eta},\boldsymbol{\gamma}}^{-2}\boldsymbol{\eta} + \sigma_{\boldsymbol{\eta},\boldsymbol{\gamma}}^{-2}\boldsymbol{\gamma} + \frac{\mathbf{1}^\mathsf{T}(\boldsymbol{\alpha}-\mathbf{1})}{\mathbf{1}^\mathsf{T}\exp(\boldsymbol{\tau})}\exp(\boldsymbol{\tau}) - (\boldsymbol{\alpha}-\mathbf{1}) - \frac{\sigma_\lambda^2+1}{\lambda^2}$$

$$\nabla\Omega_\cap(\boldsymbol{\theta}) = \frac{\sigma_\lambda^{-2}+3}{\lambda}.$$

The following proposition states that the gradients can be evaluated efficiently using message passing.

Proposition 1. *Let* $\phi(\mathbf{x}, y) = \Lambda(y) \otimes \mathbf{x}$. *Then the expectations* $\mathbb{E}_{\mathbf{y},\bar{\mathbf{y}}}[\Phi(\mathbf{X},\mathbf{y},\bar{\mathbf{y}})]$ *and* $\mathbb{E}_{\bar{\mathbf{y}}}[\Phi(\mathbf{X},\mathbf{y},\bar{\mathbf{y}})]$ *can be computed in time*

$$\mathcal{O}(|\mathcal{Y}|^3(n+k)^2 + |\mathcal{Y}|nm),$$

where m *denotes the number of features,* n *the number of leaf nodes, and* k *the number of inner nodes of the URL tree.*

A proof of Proposition 1 can be found in the appendix. Computations are based on variations of standard message passing; additional computational savings are realized by reusing specific messages within the overall computation of the gradient. These savings depend on the URL structure under study and thus do not influence asymptotic complexity but significantly influence empirical execution time for the web domains that we have studied.

5 Empirical Study

We empirically investigate the predictive performance of the proposed joint topic model using data from a large targeted advertisement company. The data set contains 36,579 web pages within nine web domains that have been manually annotated with topic labels by human labelers employed by the targeted advertising company. We use a total of $|\mathcal{Y}| = 30$ labels. Out of the nine web domains, five are general news portals run by large newspaper publishers, two are more topic-specific web portals run by family magazines, and two are web portals run by TV stations. Web page content is represented using a binary bag-of-words encoding. Words that occur fewer than ten times in the training data are removed from the dictionary, this results in 94,624 distinct bag-of-words features.

We study the model proposed in Sections 3 and 4 (denoted $JointInf_{Text}$), where we choose a linear feature map $\phi(\mathbf{x}, y) = \Lambda(y) \otimes \mathbf{x}$. The topic distance $\Lambda(y, y')$ in Equation 1 is one if $y = y'$ and zero otherwise. As a baseline, we obtain predictions from a logistic regression model that independently predicts topics for individual web pages (denoted $LogReg_{Text}$). As a further baseline, we study a logistic-regression model based on an augmented feature representation that concatenates the text features with a binary bag-of-words representation of the page's URL where numbers or special characters are used as separator between words (denoted $LogReg_{Text+URL}$). We also study our model using this augmented feature representation (denoted $JointInf_{Text+URL}$). Such URL features have been shown to be predictive for web page classification; for example, Baykan et al. have studied web page classification based on URL features only [1].

5.1 Topic Classification Performance

We study topic prediction in each of the nine web domains (the *target domain*). Training data includes instances from the remaining eight web domains (the *training domains*) as well as a varying number of instances from the target domain. Specifically, a training set is obtained by sampling 100 labeled web pages from each of the training domains and between $n = 0$ and $n = 100$ labeled web pages from the target domain. At $n = 0$ this corresponds to a setting in which predictions for novel web domains have to be obtained given a training set of existing web domains. At $n > 0$ this corresponds to a setting in which a small set of manually labeled seed pages is available from the target domain, and topic labels for the remaining pages need to be predicted. For $JointInf_{Text}$ and $JointInf_{Text+URL}$, the labeled pages from the target domain constitute observed

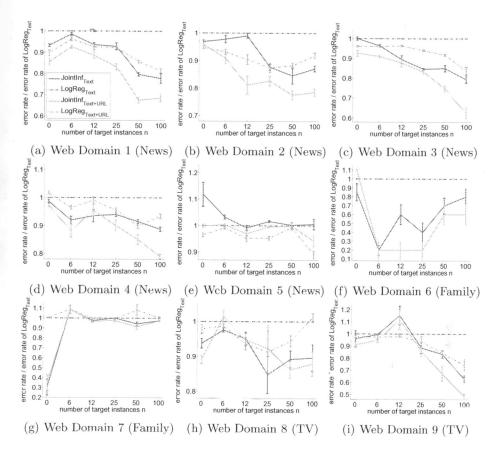

Fig. 2. Average error ratio for different target portals and different number of labeled web sites from target portal

variables, inference is carried out conditioned on these observations (see Equation 10). Predictive performance is evaluated on a sample of 500 pages from the remaining web pages of the target domain.

Hyperparameters of all models are tuned using grid search and a leave-one-domain-out cross-validation on the training data. News domains generally exhibit more diverse topic labels than the topic-specific web portals run by TV stations and family magazines; for tuning we therefore evaluate the models only on domains from the same of these two groups as the target domain. For $JointInf_{Text}$ and $JointInf_{Text+URL}$ hyperparameters are the coupling parameter σ_λ^2 and the regularization parameter $\sigma_{\eta,\gamma}^2$; for $LogReg_{Text}$ and $LogReg_{Text+URL}$, the standard regularization parameter. The hyperparameter of the Dirichlet prior is set to $\boldsymbol{\alpha} = (2,...,2)^\mathsf{T}$ (Laplace smoothing).

Figure 2 shows predictive performance for all web domains as a function of n, averaged over five resampling iterations of web pages from the training and target domains. Specifically, we report the ratio of the mean zero-one error rate of each

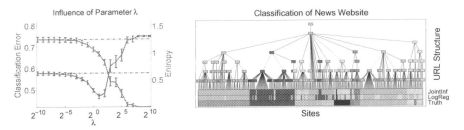

Fig. 3. Classification error and empirical entropy for different choices of parameter λ on *News Portal 1* (left). Labeled URL structure (right) for sample of *News Portal 1* for $\lambda = 1$ and zero labeled web sites from target domain

method to the mean zero-one error of the baseline $LogReg_{Text}$. If both methods incur an error rate of zero, the quotient is defined to be one. In one experiment (Web Domain 6, $LogReg_{Text+URL}$) the quotient was undefined because the error of $LogReg_{Text}$ was zero while the error of $LogReg_{Text+URL}$ was nonzero. Thus the curve for $LogReg_{Text+URL}$ is missing from Figure 2(f).

From Figure 2 we observe that, on average, the methods $JointInf_{Text}$ and $JointInf_{Text+URL}$ predict topic labels more accurately than their corresponding baseline $LogReg_{Text}$ or $LogReg_{Text+URL}$. Additionally we observe that the inclusion of URL features in the feature representation on average improves predictive accuracy. Performance varies for different web domains and values of n, with the best case being a reduction in error rate of approximately 80% and the worst case an increase in error rate by approximately 20% compared to the baseline model.

5.2   Effect of the Model Parameter λ

We also study the influence of λ—which controls the structural homogeneity of the classifier prediction (Equation 1)—for *News Portal 1* and $n = 0$. Figure 3 (left) shows the error rate of $JointInf_{Text}$ (blue solid line) and the $LogReg_{Text}$ (blue dashed line) as a function of λ. In these experiments, the regularization parameter $\sigma_{\eta,\gamma}^2 = 1$ is fixed and only model parameters η, γ and τ are optimized; results are averaged over 20 resampling iterations. Figure 3 (left) also shows the corresponding empirical entropy of the predicted labels (red curves). If λ converges to zero, the model assumes no correlation between topics of nodes and their parents in the URL tree; in this case, $JointInf_{Text}$ reduces to $LogReg_{Text}$. High values of λ couple topic labels strongly within the URL tree, therefore more uniform topic labels are assigned and the empirical entropy of the predicted labels is reduced. Predictive accuracy is maximal for intermediate values of λ.

Figure 3 (right) shows the label tree inferred by the joint topic model, predictions of the logistic regression baseline, and the ground truth for *News Portal 1*, $\lambda = 1$, and $n = 0$ labeled web pages of the target domain. We used a color scheme that maps related topics to related colors. It shows that topic labels are more uniform when using $JointInf_{Text}$ instead of $LogReg_{Text}$.

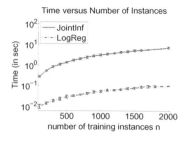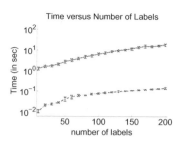

Fig. 4. Execution time for computation of expectation $\mathbb{E}_{\mathbf{y},\bar{\mathbf{y}}}\left[\Phi(\mathbf{X},\mathbf{y},\bar{\mathbf{y}})\right]$ (*JointInf$_{Text}$*) and $\mathbb{E}_{\mathbf{y}}\left[\Phi(\mathbf{X},\mathbf{y})\right]$ (*LogReg$_{Text}$*) for different number of instances (left) and different number of labels (right). Error bars show standard errors.

5.3   Execution Time

In our experiments, both the logistic regression model and the structured model are optimized using a gradient descent approach. Thus, the main computational difference is caused by the gradients of their loss functions: The computational time for the gradient of the structured loss $\ell_{\log}$ is dominated by evaluating the quantities $\mathbb{E}_{\mathbf{y},\bar{\mathbf{y}}}\left[\Phi(\mathbf{X},\mathbf{y},\bar{\mathbf{y}})\right]$ and $\mathbb{E}_{\bar{\mathbf{y}}}\left[\Phi(\mathbf{X},\mathbf{y},\bar{\mathbf{y}})\right]$ (see Proposition 1). The gradient of the logistic loss function can be written as $\mathbb{E}_{\mathbf{y}}\left[\Phi(\mathbf{X},\mathbf{y})\right]-\Phi(\mathbf{X},\mathbf{y})$, where

$$\mathbb{E}_{\mathbf{y}}\left[\Phi(\mathbf{X},\mathbf{y})\right]=\sum_{i=1}^{n}\frac{\exp(\boldsymbol{\eta}^{\mathsf{T}}\phi(\mathbf{x}_i,y_i))\phi(\mathbf{x}_i,y_i)}{\sum_{y'}\exp(\boldsymbol{\eta}^{\mathsf{T}}\phi(\mathbf{x}_i,y_i'))}. \qquad (12)$$

We compare the execution time for computation of expectation $\mathbb{E}_{\mathbf{y},\bar{\mathbf{y}}}\left[\Phi(\mathbf{X},\mathbf{y},\bar{\mathbf{y}})\right]$ for the joint topic model and the corresponding quantity given by Equation 12 for the logistic regression model. Figure 4 (left) shows the execution time for different number of training instances and a fixed number of labels $|\mathcal{Y}|=30$. Figure 4 (right) shows the execution time for different number of labels—randomly assigned to instances—and a fixed number of instances $n=500$. We found that the noticeable difference in time complexities—$\mathcal{O}(|\mathcal{Y}|^3(n+k)^2+|\mathcal{Y}|nm)$ for joint topic model and $\mathcal{O}(|\mathcal{Y}|nm)$ for logistic regression—reduces approximately to a constant factor, when we reuse messages over different variations of the message passing scheme.

6   Related Work

There is a rich body of work on general web page classification [8]. In addition to textual information on pages, hyperlink structure is often used to improve classification accuracy [9,6].

Some earlier work has studied using URL tree information for web page classification. Kan and Thi [3] and Baykan et al. [1] use models over features of URLs to classify web pages based on URL information only. Shih and Karger [10] use URL trees and page layout information encoded in an HTML tree for ad blocking and predicting links that are of interest to a particular user. They employ a generative probabilistic model similar to the coupling model defined by Equation 1 to represent correlations within URL trees. In contrast to our approach, their model does not include web page text or other page content.

Tian et al. [12] study models based on URL tree information with the goal of assigning topics to entire web sites rather than individual web pages. Kumar et al. [5] study the problem of segmenting a web site into topically uniform regions based on the URL tree structure and predictions of a node-level topic classification algorithm. Their central result is that segmentations that are optimal according to certain cost measures can be computed efficiently using dynamic programming.

The prediction problem we study can be phrased as a structured output problem involving latent variables; such problems have been studied, for example, by Wang et al. [13] and Yu and Joachims [14]. The latter model, latent variable structured SVM, is also trained using CCCP. Its margin-based objective leads to a learning algorithm alternating between performing point estimates of latent variables and model parameters, while in our maximum conditional likelihood formulation latent variables are summed out during learning. In the application-specific model that we present, these summations as well as decoding for structured prediction can be carried out efficiently because both problems reduce to message passing in a tree-structured factor graph.

7   Conclusions

We have studied the problem of jointly predicting topic labels for all web pages within a URL hierarchy. Section 3.1 defines a generative process for web page content that captures our intuition about how web site administrators hierarchically structure web sites according to content; latent variables in this process reflect the predominant topic within a URL subtree. Section 3.2 shows that deriving the conditional distribution over topic labels given page content in this model results in a structured output model that is linear in a joint feature map of page content, topic labels, and latent topic variables. Parameter estimation can be carried out using a concave-convex procedure. Proposition 1 shows that parameter estimation and decoding in the model are efficient. An empirical study in a targeted advertisement domain shows that joint inference of topic labels with the proposed model is more accurate than inferring topic labels independently based on features derived from page content and the URL.

Acknowledgment. This work was carried out in collaboration with nugg.ad AG.

References

1. Baykan, E., Henzinger, M., Weber, I.: Web page language identification based on URLs. Proceedings of the VLDB Endowment 1(1), 176–187 (2008)
2. Huynh, T., Mooney, R.: Max-margin weight learning for Markov logic networks. In: Proceedings of the European Conference on Machine Learning (2009)

3. Kan, M., Thi, H.: Fast webpage classification using URL features. In: Proceedings of the 14th ACM International Conference on Information and Knowledge Management (2005)
4. Koller, D., Friedman, N.: Probabilistic Graphical Models, vol. 1. MIT Press (2009)
5. Kumar, R., Punera, K., Tomkins, A.: Hierarchical topic segmentation of websites. In: Proceedings of the 12th ACM SIGKDD International Conference on Knowledge Discovery and Data Mining (2006)
6. McDowell, L., Gupta, K., Aha, D.: Cautious inference in collective classification. In: Proceedings of the 22nd AAAI Conference on Artificial Intelligence (2007)
7. Pearl, J.: Fusion, propagation, and structuring in belief networks. Artificial Intelligence 29(3), 241–288 (1986)
8. Qi, X., Davidson, B.: Web page classification: Features and algorithms. ACM Computing Surveys 41(2), 12:1–12:31 (2009)
9. Shen, D., Sun, J., Yang, Q., Chen, Z.: A comparison of implicit and explicit links for web page classification. In: Proceedings of the 15th International World Wide Web Conference (2006)
10. Shih, L., Karger, D.: Using URLs and table layout for web classification tasks. In: Proceedings of the 13th World Wide Web Conference (2004)
11. Taskar, B., Guestrin, C., Koller, D.: Max-margin Markov models. In: Proceedings of the 17th Annual Conference on Neural Information Processing Systems (2004)
12. Tian, Y., Huang, T., Gao, W.: Two-phase web site classification based on hidden Markov tree models. Web Intelligence and Agent Systems 2(4), 249–264 (2004)
13. Wang, S., Quattoni, A., Morency, L., Demirdjian, D., Darrell, T.: Hidden conditional random fields for gesture recognition. In: Proceedings of the 2006 IEEE Computer Society Conference on Computer Vision and Pattern Recognition (2006)
14. Yu, C.N.J., Joachims, T.: Learning structural SVMs with latent variables. In: Proceedings of the 26th Annual International Conference on Machine Learning, pp. 1169–1176. ACM (2009)
15. Yuille, A.L., Rangarajan, A.: The concave-convex procedure. Neural Computation 15(4), 915–936 (2003)

Appendix

Proof of Proposition 1

We first turn toward the quantity

$$\mathbb{E}_{\mathbf{y},\bar{\mathbf{y}}}\left[\Phi(\mathbf{X},\mathbf{y},\bar{\mathbf{y}})\right] = \frac{\sum_{\mathbf{y},\bar{\mathbf{y}}} \exp\left(\boldsymbol{\theta}^{\mathsf{T}}\Phi(\mathbf{X},\mathbf{y},\bar{\mathbf{y}})\right)\Phi(\mathbf{X},\mathbf{y},\bar{\mathbf{y}})}{\sum_{\mathbf{y},\bar{\mathbf{y}}} \exp\left(\boldsymbol{\theta}^{\mathsf{T}}\Phi(\mathbf{X},\mathbf{y},\bar{\mathbf{y}})\right)}. \tag{13}$$

Let the unnormalized probabilities—the normalizing quantities $h(\mathbf{X})$ and $q_{\boldsymbol{\tau}}(\boldsymbol{\tau})$ can be canceled out in nominator and denominator—be denoted by

$$\psi_{\boldsymbol{\tau}}(y) = \exp(\tau_y) \propto p(y|\boldsymbol{\tau}) \qquad \psi_{\boldsymbol{\eta}}(y,\mathbf{x}) = \exp(\boldsymbol{\eta}^{\mathsf{T}}\phi(\mathbf{x},y)) \propto p(\mathbf{x}|y;\boldsymbol{\eta})$$
$$\psi_{\lambda,\boldsymbol{\gamma}}(y,y') = \exp(-\lambda\Delta(y,y') - \gamma_{y'}) = p(y|y';\lambda).$$

Since $\Delta(y,y')$ is a given problem-specific loss function, $\psi_{\lambda,\boldsymbol{\gamma}}(y,y')$ can be computed in time $\mathcal{O}(|\mathcal{Y}|^2)$ for all $y,y' \in \mathcal{Y}$. Furthermore, under the assumption that $\phi(\mathbf{x}, y) = \Lambda(y) \otimes \mathbf{x}$, we can compute $\psi_{\boldsymbol{\eta}}(y, \mathbf{x}_i)$ for all $i = 1, \ldots, n$ and all $y \in \mathcal{Y}$ in time $\mathcal{O}(|\mathcal{Y}|nm)$. Given these quantities, the denominator in Equation 13 can be computed efficiently using standard message passing [7]. We therefore evaluate

$$\sum_{\mathbf{y}, \bar{\mathbf{y}}} \exp\left(\boldsymbol{\theta}^{\mathsf{T}} \Phi(\mathbf{X}, \mathbf{y}, \bar{\mathbf{y}})\right) = \sum_{y_{n+k}} \psi_{\boldsymbol{\tau}}(y_{n+k}) \prod_{v \to i = v_{n+k}} \mu_i(y_{n+k}) \tag{14}$$

recursively, where the messages $\mu_i(y_{\to i})$ have the form

$$\sum_{y_i} \psi_{\lambda, \gamma}(y_i, y_{\to i}) \begin{cases} \psi_{\boldsymbol{\eta}}(y_i, \mathbf{x}_i) & , \text{if } i \leq n \\ \prod_{v \to j = v_i} \mu_j(y_i) & , \text{otherwise.} \end{cases} \tag{15}$$

In order to evaluate Equation 15 for a given $i \in \{1, \ldots, n+k\}$ and a given $y_{\to i} \in \mathcal{Y}$, we have to compute all $|\mathcal{Y}|$ summands. Each summand contains at most $\max\{c_i + 1, 2\}$ factors, where c_i is the number of children of node v_i. Hence one message can be computed in time $\mathcal{O}(|\mathcal{Y}|c_i)$. Due to the tree structure $\mathcal{T}$, each node has a unique parent node and therefore $\sum_{i=1}^{n+k} c_i = n + k - 1$ holds. Hence the computation for all $i \in \{1, \ldots, n+k\}$ and all $y_{\to i} \in \mathcal{Y}$ can be done in time $\sum_{i=1}^{n+k} \sum_{y_{\to i} \in \mathcal{Y}} \mathcal{O}(|\mathcal{Y}|c_i) = \mathcal{O}(|\mathcal{Y}|^2(n+k))$ using dynamic programming.

We now consider the numerator in Equation 13 and show that the parts of the joint feature map (see Equation 5) that refer to the parameters $\boldsymbol{\eta}$, $\boldsymbol{\gamma}$, $\boldsymbol{\tau}$ and λ can be computed in time $\mathcal{O}(|\mathcal{Y}|^3 n(n+k) + |\mathcal{Y}|nm)$, in time $\mathcal{O}(|\mathcal{Y}|^3(n+k)^2)$, in time $\mathcal{O}(|\mathcal{Y}|^2(n+k))$, and in time $\mathcal{O}(|\mathcal{Y}|^2(n+k)^2)$, respectively. For $\boldsymbol{\eta}$, the numerator can be expressed as

$$\sum_{\mathbf{y}, \bar{\mathbf{y}}} \exp\left(\boldsymbol{\theta}^{\mathsf{T}} \Phi(\mathbf{X}, \mathbf{y}, \bar{\mathbf{y}})\right) \sum_{l=1}^{n} \phi(\mathbf{x}_l, y_l)$$

$$= \sum_{l=1}^{n} \left(\sum_{\mathbf{y}, \bar{\mathbf{y}}} \exp\left(\boldsymbol{\theta}^{\mathsf{T}} \Phi(\mathbf{X}, \mathbf{y}, \bar{\mathbf{y}})\right) \Lambda(y_l)\right) \otimes \mathbf{x}_l. \tag{16}$$

In Equation 16, we reorder the sums and make use of $\phi(\mathbf{x}, y) = \Lambda(y) \otimes \mathbf{x}$. Additionally, we exploit the associativity of the Kronecker product. Note that $\sum_{\mathbf{y}, \bar{\mathbf{y}}} \exp\left(\boldsymbol{\theta}^{\mathsf{T}} \Phi(\mathbf{X}, \mathbf{y}, \bar{\mathbf{y}})\right) \Lambda(y_l)$ is a vector of length $|\mathcal{Y}|$; each component is associated with the case that the l-th web site has a certain label y'. This quantity can be computed by applying the message passing for each label $y' \in \mathcal{Y}$, where the message $\mu_l(y_{\to l})$ is substituted by $\psi_{\lambda, \gamma}(y_l, y_{\to l})\psi_{\boldsymbol{\eta}}(y_l, \mathbf{x}_l)$ if $y_l = y'$ and zero otherwise. In order to evaluate Equation 16, we need to apply standard message passing for all $y' \in \mathcal{Y}$ and for all $l = 1, \ldots, n$, which can be done in time $\mathcal{O}(|\mathcal{Y}|^3 n(n+k))$. The computation of the Kronecker product can be done in time $\mathcal{O}(|\mathcal{Y}|nm)$. Thus, the overall computational time is $\mathcal{O}(|\mathcal{Y}|^3 n(n+k) + |\mathcal{Y}|nm)$.

By reordering the sums, the numerator for γ can be expressed as

$$\sum_{\mathbf{y}, \bar{\mathbf{y}}} \exp\left(\boldsymbol{\theta}^{\mathsf{T}} \Phi(\mathbf{X}, \mathbf{y}, \bar{\mathbf{y}})\right) \sum_{l=1}^{n+k-1} \Lambda(y_{\to l}) = \sum_{l=1}^{n+k-1} \sum_{\mathbf{y}, \bar{\mathbf{y}}} \exp\left(\boldsymbol{\theta}^{\mathsf{T}} \Phi(\mathbf{X}, \mathbf{y}, \bar{\mathbf{y}})\right) \Lambda(y_{\to l}). \tag{17}$$

The term $\sum_{\mathbf{y},\bar{\mathbf{y}}} \exp\left(\boldsymbol{\theta}^{\mathsf{T}} \Phi(\mathbf{X}, \mathbf{y}, \bar{\mathbf{y}})\right) \Lambda(y_{\rightarrow l})$ is a vector of length $|\mathcal{Y}|$, where each component is associated with the case that the parent of the l-th node has certain label y'. For each label $y' \in \mathcal{Y}$, this quantity can be computed by standard message passing, where the message $\mu_l(y_{\rightarrow l})$ is substituted by

$$\psi_{\lambda,\gamma}(y_l, y_{\rightarrow l}) \begin{cases} \psi_{\boldsymbol{\eta}}(y_l, \mathbf{x}_l) & \text{, if } l \leq n \text{ and } y_{\rightarrow l} = y' \\ \prod_{v_{\rightarrow j}=v_l} \mu_j(y_l) & \text{, if } l > n \text{ and } y_{\rightarrow l} = y' \\ 0 & \text{, otherwise.} \end{cases}$$

In order to evaluate Equation 17, we need to apply message passing for all $l = 1, \ldots, n+k-1$ and $y' \in \mathcal{Y}$, which can be done in time $\mathcal{O}(|\mathcal{Y}|^3 (n+k)^2)$.

For τ, the numerator can be evaluated by using message passing

$$\sum_{\mathbf{y},\bar{\mathbf{y}}} \exp\left(\boldsymbol{\theta}^{\mathsf{T}} \Phi(\mathbf{X}, \mathbf{y}, \bar{\mathbf{y}})\right) \Lambda(y_{n+k}) = \sum_{y_{n+k}} \psi_{\tau}(y_{n+k}) \Lambda(y_{n+k}) \prod_{v_{\rightarrow i}=v_{n+k}} \mu_i(y_{n+k}), \quad (18)$$

where $\mu_i(y_{\rightarrow i})$ is defined by Equation 15. Standard message passing as described in Equation 14 requires a summation over $|\mathcal{Y}|$ summands. Instead, in Equation 18 we save each of the $|\mathcal{Y}|$ summands. Hence, the computational time for Equation 18 is the same as for Equation 14, which is $\mathcal{O}(|\mathcal{Y}|^2 (n+k))$.

For λ, the numerator can be expressed as

$$\sum_{\mathbf{y},\bar{\mathbf{y}}} \exp\left(\boldsymbol{\theta}^{\mathsf{T}} \Phi(\mathbf{X}, \mathbf{y}, \bar{\mathbf{y}})\right) \sum_{l=1}^{n+k-1} \Delta(y_l, y_{\rightarrow l})$$

$$= \sum_{l=1}^{n+k-1} \left(\sum_{\mathbf{y},\bar{\mathbf{y}}} \exp\left(\boldsymbol{\theta}^{\mathsf{T}} \Phi(\mathbf{X}, \mathbf{y}, \mathbf{y})\right) \Delta(y_l, y_{\rightarrow l}) \right) \quad (19)$$

by reordering the sums. Again, we use the message passing algorithm in order to evaluate the quantity $\sum_{\mathbf{y},\bar{\mathbf{y}}} \exp\left(\boldsymbol{\theta}^{\mathsf{T}} \Phi(\mathbf{X}, \mathbf{y}, \bar{\mathbf{y}})\right) \Delta(y_l, y_{\rightarrow l})$ for $l = 1, \ldots, n+k-1$. Therefore, we substitute the message $\mu_l(y_{\rightarrow l})$ by

$$\sum_{y_l} \Delta(y_l, y_{\rightarrow l}) \psi_{\lambda,\gamma}(y_l, y_{\rightarrow l}) \begin{cases} \psi_{\boldsymbol{\eta}}(y_l, \mathbf{x}_l) & \text{, if } l \leq n \\ \prod_{v_{\rightarrow j}=v_l} \mu_j(y_l) & \text{, otherwise.} \end{cases}$$

Hence, Equation 19 can be evaluated by applying standard message passing $n+k$ times which can be done in time $\mathcal{O}(|\mathcal{Y}|^2 (n+k)^2)$. This completes the proof for the computational time of the expectation $\mathbb{E}_{\mathbf{y},\bar{\mathbf{y}}} [\Phi(\mathbf{X}, \mathbf{y}, \bar{\mathbf{y}})]$.

The proof for the expectation $\mathbb{E}_{\bar{\mathbf{y}}} [\Phi(\mathbf{X}, \mathbf{y}, \bar{\mathbf{y}})]$ can be done analogously by replacing the sum over $\mathbf{y}$ with the true labels. $\qquad\square$

Learned-Norm Pooling for Deep Feedforward and Recurrent Neural Networks

Caglar Gulcehre, Kyunghyun Cho, Razvan Pascanu, and Yoshua Bengio[*]

Département d'Informatique et de Recherche Opérationelle
Université de Montréal
($\star$) CIFAR Fellow

Abstract. In this paper we propose and investigate a novel nonlinear unit, called L_p unit, for deep neural networks. The proposed L_p unit receives signals from several projections of a subset of units in the layer below and computes a normalized L_p norm. We notice two interesting interpretations of the L_p unit. First, the proposed unit can be understood as a generalization of a number of conventional pooling operators such as average, root-mean-square and max pooling widely used in, for instance, convolutional neural networks (CNN), HMAX models and neocognitrons. Furthermore, the L_p unit is, to a certain degree, similar to the recently proposed maxout unit [13] which achieved the state-of-the-art object recognition results on a number of benchmark datasets. Secondly, we provide a geometrical interpretation of the activation function based on which we argue that the L_p unit is more efficient at representing complex, nonlinear separating boundaries. Each L_p unit defines a superelliptic boundary, with its exact shape defined by the order p. We claim that this makes it possible to model arbitrarily shaped, curved boundaries more efficiently by combining a few L_p units of different orders. This insight justifies the need for learning different orders for each unit in the model. We empirically evaluate the proposed L_p units on a number of datasets and show that multilayer perceptrons (MLP) consisting of the L_p units achieve the state-of-the-art results on a number of benchmark datasets. Furthermore, we evaluate the proposed L_p unit on the recently proposed deep recurrent neural networks (RNN).

Keywords: deep learning, L_p unit, multilayer perceptron.

1 Introduction

The importance of well-designed nonlinear activation functions when building a deep neural network has become more apparent recently. Novel nonlinear activation functions that are unbounded and often piecewise linear but not continuous such as rectified linear units (ReLU) [22,11], or rectifier, and maxout units [13] have been found to be particularly well suited for deep neural networks on many object recognition tasks.

A pooling operator, an idea which dates back to the work in [17], has been adopted in many object recognizers. Convolutional neural networks which often

T. Calders et al. (Eds.): ECML PKDD 2014, Part I, LNCS 8724, pp. 530–546, 2014.

employ max pooling have achieved state-of-the-art recognition performances on various benchmark datasets [20,9]. Also, biologically inspired models such as HMAX have employed max pooling [27]. A pooling operator, in this context, is understood as a way to summarize a high-dimensional collection of neural responses and produce features that are invariant to some variations in the input (across the filter outputs that are being pooled).

Recently, the authors of [13] proposed to understand a pooling operator itself as a nonlinear activation function. The proposed maxout unit *pools* a group of linear responses, or outputs, of neurons, which overall acts as a piecewise linear activation function. This approach has achieved many state-of-the-art results on various benchmark datasets.

In this paper, we attempt to generalize this approach by noticing that most pooling operators including max pooling as well as maxout units can be understood as special cases of computing a normalized L_p norm over the outputs of a set of filter outputs. Unlike those conventional pooling operators, however, we claim here that it is beneficial to estimate the order p of the L_p norm instead of fixing it to a certain predefined value such as ∞, as in max pooling.

The benefit of learning the order p, and thereby a neural network with L_p units of different orders, can be understood from geometrical perspective. As each L_p unit defines a spherical shape in a non-Euclidean space whose metric is defined by the L_p norm, the combination of multiple such units leads to a non-trivial separating boundary in the input space. In particular, an MLP may learn a highly curved boundary efficiently by taking advantage of different values of p. In contrast, using a more conventional nonlinear activation function, such as the rectifier, results in boundaries that are piece-wise linear. Approximating a curved separation of classes would be more expensive in this case, in terms of the number of hidden units or piece-wise linear segments.

In Sec. 2 a basic description of a multi-layer perceptron (MLP) is given followed by an explanation of how a pooling operator may be considered a nonlinear activation function in an MLP. We propose a novel L_p unit for an MLP by generalizing pooling operators as L_p norms in Sec. 3. In Sec. 4 the proposed L_p unit is further analyzed from the geometrical perspective. We describe how the proposed L_p unit may be used by recurrent neural networks in Sec. 5. Sec. 6 provides empirical evaluation of the L_p unit on a number of object recognition tasks.

2   Background

2.1   Multi-layer Perceptron

A multi-layer perceptron (MLP) is a feedforward neural network consisting of multiple layers of nonlinear neurons [29]. Each neuron u_j of an MLP typically receives a weighted sum of the incoming signals $\{a_1, \ldots, a_N\}$ and applies a nonlinear activation function ϕ to generate a scalar output such that

$$u_j\left(\{a_1,\ldots,a_N\}\right) = \phi\left(\sum_{i=1}^{N} w_{ij} a_i\right). \tag{1}$$

With this definition of each neuron[1], we define the output of an MLP having L hidden layers and q output neurons given an input $\mathbf{x}$ by

$$\mathbf{u}(\mathbf{x} \mid \boldsymbol{\theta}) = \phi\left(\mathbf{U}^\top \phi_{[L]}\left(\mathbf{W}_{[L]}^\top \cdots \phi_{[1]}\left(\mathbf{W}_{[1]}^\top \mathbf{x}\right)\cdots\right)\right), \tag{2}$$

where $\mathbf{W}_{[l]}$ and $\phi_{[l]}$ are the weights and the nonlinear activation function of the l-th hidden layer, and $\mathbf{W}_{[1]}$ and $\mathbf{U}$ are the weights associated with the input and output, respectively.

2.2   Pooling as a Nonlinear Unit in MLP

Pooling operators have been widely used in convolutional neural networks (CNN) [21,10,27] to reduce the dimensionality of a high-dimensional output of a convolutional layer. When used to group spatially neighboring neurons, this operator which summarizes a group of neurons in a lower layer is able to achieve the property of (local) translation invariance. Various types of pooling operator have been proposed and used successfully, such as average pooling, root-of-mean-squared (RMS) pooling and max pooling [19,33].

A pooling operator may be viewed instead as a nonlinear activation function. It receives input signal from the layer below, and it returns a scalar value. The output is the result of applying some nonlinear function such as max (max pooling). The difference from traditional nonlinearities is that the pooling operator is not applied element-wise on the lower layer, but rather on groups of hidden units. A *maxout* nonlinear activation function proposed recently in [13] is a representative example of *max* pooling in this respect.

3 L_p Unit

The recent success of maxout has motivated us to consider a more general nonlinear activation function that is rooted in a pooling operator. In this section, we propose and discuss a new nonlinear activation function called an L_p unit which replaces the *max* operator in a *maxout* unit by an L_p norm.

3.1   Normalized L_p-Norm

Given a finite vector/set of input signals $[a_1,\ldots,a_N]$ a normalized L_p norm is defined as

$$u_j\left([a_1,\ldots,a_N]\right) = \left(\frac{1}{N}\sum_{i=1}^{N}|a_i - c_i|^{p_j}\right)^{\frac{1}{p_j}}, \tag{3}$$

---

[1] We omit a bias to make equations less cluttered.

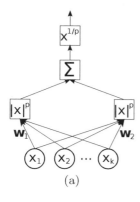

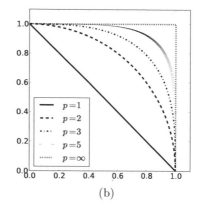

(a)

(b)

Fig. 1. (a) An illustration of a single L_p unit with two sets of incoming signals. For clarity, biases and the division by the number of filters are omitted. The symbol x in each block (square) represents an input signal to that specific block (square) only. (b) An illustration of the effect of p on the shape of an ellipsoid. Only the first quadrant is shown.

where p_j indicates that the order of the norm may differ for each neuron. It should be noticed that when $0 < p_j < 1$ this definition is not a norm anymore due to the violation of triangle inequality. In practice, we re-parameterize p_j by $1 + \log(1 + e^{\rho_j})$ to satisfy this constraint.

The input signals (also called filter outputs) a_i are defined by

$$a_i = \mathbf{w}_i^\top \mathbf{x},$$

where $\mathbf{x}$ is a vector of activations from the lower layer. c_i is a center, or bias, of the i-th input signal a_i. Both p_j and c_i are model parameters that are learned.

We call a neuron with this nonlinear activation function an L_p unit. An illustration of a single L_p unit is presented in Fig. 1 (a).

Each L_p unit in a single layer receives input signal from a subset of linear projections of the activations of the layer immediately below. In other words, we project the activations of the layer immediately below linearly to $A = \{a_1, \ldots, a_N\}$. We then divide A into equal-sized, non-overlapping groups of which each is fed into a single L_p unit. Equivalently, each L_p unit has its private set of filters.

The parameters of an MLP having one or more layers of L_p units can be estimated by using backpropagation [30], and in particular we adapt the order p of the norm.[2] In our experiments, we use Theano [5] to compute these partial derivatives and update the orders p_j (through the parametrization of p_j in terms of ρ_j), as usual with any other parameters.

---

[2] The activation function is continuous everywhere except a finite set of points, namely when $a_i - c_i$ is 0 and the absolute value function becomes discontinuous. We ignore these discontinuities, as it is done, for instance, in maxouts and rectifiers.

3.2   Related Approaches

Thanks to the definition of the proposed L_p unit based on the L_p norm, it is straightforward to see that many previously described nonlinear activation functions or pooling operators are closely related to or special cases of the L_p unit. Here we discuss some of them.

When $p_j = 1$, Eq. (3) becomes

$$u_j\left([a_1, \ldots, a_N]\right) = \frac{1}{N}\sum_{i=1}^{N}|a_i|.$$

If we further assume $a_i \geq 0$, for instance, by using a logistic sigmoid activation function on the projection of the lower layer, the activation is reduced to computing the average of these projections. This is a form of average pooling, where the non-linear projections represent the pooled layer. With a single filter, this is equivalent to the absolute value rectification proposed in [19]. If p_j is 2 instead of 1, the root-of-mean-squared pooling from [33] is recovered.

As p_j grows and ultimately approaches ∞, the L_p norm becomes

$$\lim_{p_j \to \infty} u_j\left([a_1, \ldots, a_N]\right) = \max\left\{|a_1|, \ldots, |a_N|\right\}.$$

When $N = 2$, this is a generalization of a rectified linear unit (ReLU) as well as the absolute value unit [19]. If each a_i is constrained to be non-negative, this corresponds exactly to the maxout unit.

In short, the proposed L_p unit interpolates among different pooling operators by the choice of its order p_j. This was noticed earlier in [8] as well as [33]. However, both of them stopped at analyzing the L_p norm as a pooling operator with a fixed order and comparing those conventional pooling operators against each other. The authors of [4] investigated a similar nonlinear activation function that was inspired by the cells in the primary visual cortex. In [18], the possibility of learning p has been investigated in a probabilistic setting in computer vision.

On the other hand, in this paper, we claim that the order p_j needs to, and can be *learned*, just like all other parameters of a deep neural network. Furthermore, we conjecture that (1) an optimal distribution of the orders of L_p units differs from one dataset to another, and (2) each L_p unit in a MLP requires a different order from the other L_p. These properties also distinguish the proposed L_p unit from the conventional radial-basis function network (see, e.g., [15])

4   Geometrical Interpretation

We analyze the proposed L_p unit from a geometrical perspective in order to motivate our conjecture regarding the order of the L_p units. Let the value of an L_p unit u be given by:

$$u(\mathbf{x}) = \left(\frac{1}{N}\sum_{i=1}^{N}\left|\mathbf{w}_i^\top \mathbf{x} - c_i\right|^p\right)^{\frac{1}{p}}, \tag{4}$$

where $\mathbf{w}_i$ represents the i-th column of the matrix $\mathbf{W}$. The equation above effectively says that the L_p unit computes the p-th norm of the projection of the input $\mathbf{x}$ on the subspace spanned by N vectors $\{\mathbf{w}_1, \ldots, \mathbf{w}_N\}$. Let us further assume that $\mathbf{x} \in \mathbb{R}^d$ is a vector in an Euclidean space.

The space onto which $\mathbf{x}$ is projected may be spanned by linearly dependent vectors $\mathbf{w}_i$'s. Due to the possible lack of the linear independence among these vectors, they span a subspace $\mathcal{S}$ of dimensionality $k \leq N$. The subspace $\mathcal{S}$ has its origin at $\mathbf{c} = [c_1, \ldots, c_N]$.

We impose a non-Euclidean geometry on this subspace by defining a norm in the space to be L_p with p potentially not 2, as in Eq. (4). The geometrical object to which a particular value of the L_p unit corresponds forms a superellipse when projected back into the original input space. [3] The superellipse is centered at the inverse projection of $\mathbf{c}$ in the Euclidean input space. Its shape varies according to the order p of the L_p unit and due to the potentially linearly-dependent bases. As long as $p \geq 1$ the shape remains convex. Fig. 1 (b) draws some of the superellipses one can get with different orders of p, as a function of a_1 (with a single filter).

In this way each L_p unit partitions the input space into two regions – inside and outside the superellipse. Each L_p unit uses a curved boundary of learned curvature to divide the space. This is in contrast to, for instance, a maxout unit which uses piecewise linear hyperplanes and might require more linear pieces to approximate the same curved segment.

4.1   Qualitative Analysis in Low Dimension

When the dimensionality of the input space is 2 and each L_p receives 2 input signals, we can visualize the partitions of the input space obtained using L_p units as well as conventional nonlinear activation functions. Here, we examine some artificially generated cases in a 2-D space.

Two Classes, Single L_p Unit. Fig. 2 shows a case of having two classes (● and ●) of which each corresponds to a Gaussian distribution. We trained MLPs having a single hidden neuron. When the MLPs had an L_p unit, we fixed p to either 2 or ∞. We can see in Fig. 2 (a) that the MLP with the L_2 unit divides the input space into two regions – inside and outside a rotated superellipse.[4] The superellipse correctly identified one of the classes (red).

In the case of $p = \infty$, what we see is a degenerate rectangle which is an extreme form of a superellipse. The superellipse again spotted one of the classes and appropriately draws a separating curve between the two classes.

---

[3] Since $k \leq N$, the superellipse may be degenerate in the sense that in some of the $N - k$ axes the width may become infinitely large. However, as this does not invalidate our further argument, we continue to refer this kind of (degenerate) superellipse simply by an superellipse.

[4] Even though we use $p = 2$, which means an Euclidean space, we get a superellipse instead of a circle because of the linearly-dependent bases $\{\mathbf{w}_1, \ldots, \mathbf{w}_N\}$.

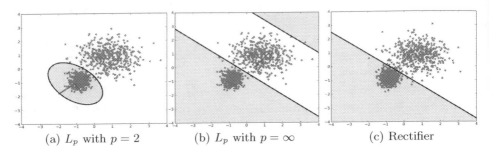

(a) L_p with $p = 2$ (b) L_p with $p = \infty$ (c) Rectifier

Fig. 2. Visualization of separating curves obtained using different activation functions. The underlying data distribution is a mixture of two Gaussian distributions. The red and green dots are the samples from the two classes, respectively, and the black curves are separating curves found by the MLPs. The purple lines indicate the axes of the subspace learned by each L_p unit. Best viewed in color.

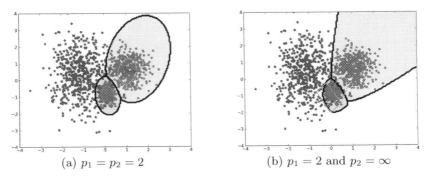

(a) $p_1 = p_2 = 2$ (b) $p_1 = 2$ and $p_2 = \infty$

Fig. 3. Visualization of separating curves obtained using different orders of L_p units. The underlying data distribution is a mixture of three Gaussian distributions. The blue curves show the shape of the superellipse learned by each L_p unit. The red, green and blue dots are the samples. Otherwise, the same color convention as in Fig. 2 has been used.

In the case of rectifier units it could find a correct separating curve, but it is clear that a single rectifier unit can only partition the input space *linearly* unlike L_p units. A combination of several rectifier units can result in a nonlinear boundary, specifically a piecewise-linear one, though our claim is that you need more such rectifier units to get an arbitrarily shaped curve whose curvature changes in a highly nonlinear way.

Three Classes, Two L_p Units. Similarly to the previous experiment, we trained two MLPs having two L_p units on data generated from a mixture of three Gaussian distribution. Again, each mixture component corresponds to each class.

For one MLP we fixed the orders of the two L_p units to 2. In this case, see Fig. 3 (a), the separating curves are constructed by combining two translated superellipses represented by the L_p units. These units were able to locate the two classes, which is sufficient for classifying the three classes (✳, ✳ and ●).

The other MLP had two L_p units with p fixed to 2 and ∞, respectively. The L_2 unit defines, as usual, a superellipse, while the L_∞ unit defines a rectangle. The separating curves are constructed as a combination of the translated superellipse and rectangle and may have more non-trivial curvature as in Fig. 3 (b).

Furthermore, it is clear from the two plots in Fig. 3 that the curvature of the separating curves may change over the input space. It will be easier to model this non-stationary curvature using multiple L_p units with different p's.

Decision Boundary with Non-stationary Curvature: Representational Efficiency. In order to test the potential efficiency of the proposed L_p unit from its ability to learn the order p, we have designed a binary classification task that has a decision boundary with a non-stationary curvature. We use 5000 data points of which a subset is shown in Fig. 4 (a), where two classes are marked with blue dots (●) and red crosses (+), respectively.

On this dataset, we have trained MLPs with either L_p units, L_2 units (L_p units with fixed $p = 2$), maxout units, rectifiers or logistic sigmoid units. We varied the number of parameters, which correspond to the number of units in the case of rectifiers and logistic sigmoid units and to the number of inputs signals to the hidden layer in the case of L_p units, L_2 units and maxout units, from 2 to 16. For each setting, we trained ten randomly initialized MLPs. In order to reduce effects due to optimization difficulties, we used in all cases natural conjugate gradient [23].

From Fig. 4 (c), it is clear that the MLPs with L_p units outperform all others in terms of representing this specific curve. They were able to achieve the zero training error with only three units (i.e., 6 filters) on all ten random runs and achieved the lowest average training error even with less units. Importantly, the comparison to the performance of the MLPs with L_2 units shows that it is beneficial to learn the orders p of L_p units. For example, with only two L_2 units none of the ten random runs succeed while at least one succeeds with two L_p units. All the other MLPs, especially ones with rectifiers and maxout units which can only model the decision boundary with piecewise linear functions, were not able to achieve the similar efficiency of the MLPs with L_p units (see Fig 4 (b)).

Fig. 4 (a) also shows the decision boundary found by the MLP with two L_p units after training. As can be observed from the shapes of the L_p units (purple and cyan dashed curves), each L_p unit learned an appropriate order p that enables them to model the non-stationary decision boundary. Fig. 4 (b) shows the boundary obtained by a rectifier model with four units. We can see that it has to use linear segments to compose the boundary, resulting in not perfectly solving the task. The rectifier model represented here has 64 mistakes, versus 0 obtained by the L_p model.

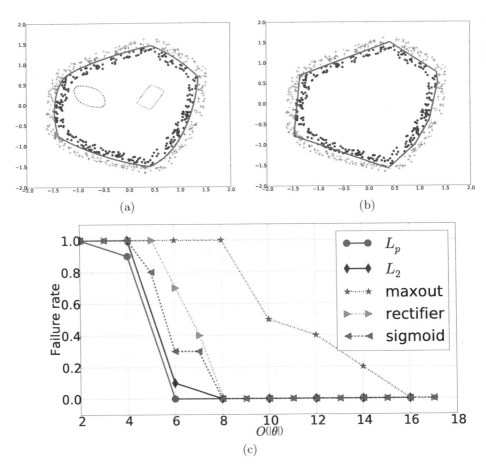

Fig. 4. (a) Visualization of data (two classes, + and •), a decision boundary learned by an MLP with two L_p units (green curve) and the shapes corresponding to the orders p's learned by the L_p units (purple and cyan dashed curves). (b) The same visualization done using four rectifiers. (c) The failure rates computed with MLPs using different numbers of different nonlinear activation functions (L_p: red solid curve with red •, L_2: blue solid curve with blue ◆, maxout: green dashed curve with green ⋆, rectifier: cyan dash-dot curve with cyan ▶ and sigmoid: purple dashed curve with purple ◀). The curves show the proportion of the failed attempts over ten random trials (y-axis) against either the number of units for sigmoid and rectifier model or the total number of linear projection going into the maxout units or L_p units (x-axis).

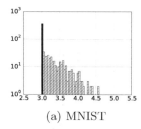
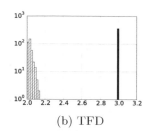
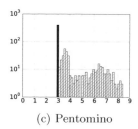

(a) MNIST   (b) TFD   (c) Pentomino

Fig. 5. Distributions of the initial (black bars ■) and learned (shaded bars ╱) orders on MNIST, TFD and Pentomino. x-axis and y-axis show the order and the number of L_p units with the corresponding order. Note the difference in the scales of the x-axes and that the y-axes are in logarithmic scale.

Although this is a low-dimensional, artificially generated example, it demonstrates that the proposed L_p units are efficient at representing decision boundaries which have non-stationary curvatures.

5   Application to Recurrent Neural Networks

A conventional recurrent neural network (RNN) mostly uses saturating nonlinear activation functions such as tanh to compute the hidden state at each time step. This prevents the possible explosion of the activations of hidden states over time and in general results in more stable learning dynamics. However, at the same time, this does not allow us to build an RNN with recently proposed non-saturating activation functions such as rectifiers and maxout as well as the proposed L_p units.

The authors of [24] recently proposed three ways to extend the conventional, *shallow* RNN into a deep RNN. Among those three proposals, we notice that it is possible to use non-saturating activations functions for a deep RNN with deep transition without causing the instability of the model, because a saturating non-linearity (tanh) is applied in sandwich between the L_p MLP associated with each step.

The deep transition RNN (DT-RNN) has one or more intermediate layers between a pair of consecutive hidden states. The transition from a hidden state $\mathbf{h}_{t-1}$ at time $t-1$ to the next hidden state $\mathbf{h}_t$ is

$$\mathbf{h}_t = g\left(\mathbf{W}^\top f\left(\mathbf{U}^\top \mathbf{h}_{t-1} + \mathbf{V}^\top \mathbf{x}_t\right)\right),$$

not showing biases, as previously.

When a usual saturating nonlinear activation function is used for g, the activations of the hidden state $\mathbf{h}_t$ are bounded. This allows us to use any, potentially non-saturating nonlinear function for f. We can simply use a layer of the proposed L_p unit in the place of f.

As argued in [24], if the procedure of constructing a new summary which corresponds to the new hidden state $\mathbf{h}_t$ from the combination of the current input $\mathbf{x}_t$ and the previous summary $\mathbf{h}_{t-1}$ is highly nonlinear, any benefit of the proposed L_p unit over the existing, conventional activation functions in feedforward neural networks should naturally translate to these deep RNNs as well. We show this effect empirically later by training a deep output, deep transition RNN (DOT-RNN) with the proposed L_p units.

6 Experiments

In this section, we provide empirical evidences showing the advantages of utilizing the L_p units. In order to clearly distinguish the effect of employing L_p units from introducing data-specific model architectures, all the experiments in this section are performed by neural networks having densely connected hidden layers.

6.1 Claims to Verify

Let us first list our claims about the proposed L_p units that need to be verified through a set of experiments. We expect the following from adopting L_p units in an MLP:

1. The optimal orders of L_p units vary across datasets
2. An optimal distribution of the orders of L_p units is not close to a (shifted) Dirac delta distribution

The first claim states that there is no universally optimal order p_j. We train MLPs on a number of benchmark datasets to see the resulting distribution of p_j's. If the distributions had been similar between tasks, claim 1 would be rejected.

This naturally connects to the second claim. As the orders are estimated via learning, it is unlikely that the orders of all L_p units will convergence to a single value such as ∞ (maxout or max pooling), 1 (average pooling) or 2 (RMS pooling). We expect that the response of each L_p unit will specialize by using a distinct order. The inspection of the trained MLPs to confirm the first claim will validate this claim as well.

On top of these claims, we expect that an MLP having L_p units, when the parameters including the orders of the L_p units are well estimated, will achieve highly competitive classification performance. In addition to classification tasks using feedforward neural networks, we anticipate that a recurrent neural network benefits from having L_p units in the intermediate layer between the consecutive hidden states, as well.

6.2 Datasets

For feedforward neural networks or MLPs, we have used four datasets; MNIST [21], Pentomino [14], the Toronto Face Database (TFD) [31] and Forest Covertype[5] (data split DS2-581) [32]. MNIST, TFD and Forest Covertype are three

---

[5] We use the first 16 principal components only.

representative benchmark datasets, and Pentomino is a relatively recently proposed dataset that is known to induce a difficult optimization challenge for a deep neural network. We have used three music datasets from [7] for evaluating the effect of L_p units on deep recurrent neural networks.

6.3 Distributions of the Orders of L_p Units

To understand how the estimated orders p of the proposed L_p unit are distributed we trained MLPs with a single L_p layer on MNIST, TFD and Pentomino. We measured validation error to search for good hyperparameters, including the number of L_p units and number of filters (input signals) per L_p unit. However, for Pentomino, we simply fixed the size of the L_p layer to 400, and each L_p unit received signals from six hidden units below.

Table 1. The means and standard deviations of the estimated orders of L_p units

Dataset	Mean	Std. Dev.
MNIST	3.44	0.38
TFD	2.04	0.22
Pentomino	5.81	1.56

In Table 1, the averages and standard deviations of the estimated orders of the L_p units in the single-layer MLPs are listed for MNIST, TFD and Pentomino. It is clear that the distribution of the orders depend heavily on the dataset, which confirms our first claim described earlier. From Fig. 5 we can clearly see that even in a single model the estimated orders vary quite a lot, which confirms our second claim. Interestingly, in the case of Pentomino, the distribution of the orders consists of two distinct modes.

The plots in Fig. 5 clearly show that the orders of the L_p units change significantly from their initial values over training. Although we initialized the orders of the L_p units around 3 for all the datasets, the resulting distributions of the orders are significantly different among those three datasets. This further confirms both of our claims. As a simple empirical confirmation we tried the same experiment with the fixed $p = 2$ on TFD and achieved a worse test error of 0.21.

6.4 Generalization Performance

The ultimate goal of any novel nonlinear activation function for an MLP is to achieve better generalization performance. We conjectured that by learning the orders of L_p units an MLP with L_p layers will achieve highly competitive classification performance.

For MNIST we trained an MLP having two L_p layers followed by a softmax output layer. We used a recently introduced regularization technique called dropout [16]. With this MLP we were able to achieve 99.03% accuracy on the test set, which is comparable to the state-of-the-art accuracy of 99.06% obtained by the MLP with maxout units [13].

On TFD we used the same MLP from the previous experiment to evaluate generalization performance. We achieved a recognition rate of 79.25%. Although we use neither pretraining nor unlabeled samples, our result is close to the current

Table 2. The generalization errors on three datasets obtained by MLPs using the proposed L_p units. The previous state-of-the-art results obtained by others are also presented for comparison.

Data	MNIST	TFD	Pentomino	Forest Covertype
L_p	0.97 %	20.75 %	31.85 %	2.83 %
Previous	0.94 %[1]	21.29 %[2]	44.6 %[3]	2.78 %[4]

state-of-the-art rate of 82.4% on the permutation-invariant version of the task reported by [26] who pretrained their models with a large amount of unlabeled samples.

As we have used the five-fold cross validation to find the optimal hyperparameters, we were able to use this to investigate the variance of the estimations of the p values. Table 3 shows the averages and standard deviations of the estimated orders for MLPs trained on the five folds using the best hyperparameters. It is clear that in all the cases the orders ended up in a similar region near two without too much difference in the variance.

Similarly, we have trained five randomly initialized MLPs on MNIST and observed the similar phenomenon of all the resulting MLPs having similar distributions of the orders. The standard deviation of the averages of the learned orders was only 0.028, while its mean is 2.16.

The MLP having a single L_p layer was able to classify the test samples of Pentomino with 31.38% error rate. This is the best result reported so far on Pentomino dataset [14] without using any kind of prior information about the task (the best previous result was 44.6% error).

On Forest Covertype an MLP having three L_p layers was trained. The MLP was able to classify the test samples with only 2.83% error. The improvement is large compared to the previous state-of-the-art rate of 3.13% achieved by the manifold tangent classifier having four hidden layers of logistic sigmoid units [28]. The result obtained with the L_p is comparable to that obtained with the MLP having maxout units.

These results as well as previous best results for all datasets are summarized in Table 2.

In all experiments, we optimized hyperparameters such as an initial learning rate and its scheduling to minimize validation error, using random search [3], which is generally more efficient than grid search when the number of hyperparameters is not tiny. Each MLP was trained by stochastic gradient descent. All the experiments in this paper were done using the Pylearn2 library [12].

---

[1] Reported in [13].

[2] This result was obtained by training an MLP with rectified linear units which outperformed an MLP with maxout units.

[3] Reported in [14].

[4] This result was obtained by training an MLP with maxout units which outperformed an MLP with rectified linear units.

6.5   Deep Recurrent Neural Networks

We tried the polyphonic music prediction tasks with three music datasets; Nottingam, JSB and MuseData [7]. The DOT-RNNs we trained had deep transition with L_p units and tanh units and deep output function with maxout in the intermediate layer (see Fig. 6 for the illustration). We coarsely optimized the size of the models and the initial leaning rate as well as its schedule to maximize the performance on validation sets. Also, we chose whether to threshold the norm of the gradient based on the validation performance [25]. All the models were trained with dropout [16].

Table 3. The means and standard deviations of the estimated orders of L_p units obtained during the hyperparameter search using the 5-fold cross-validation.

Fold	Mean	Std. Dev.
1	2.00	0.24 $\times 10^{-4}$
2	2.00	0.24 $\times 10^{-4}$
3	2.01	0.77 $\times 10^{-4}$
4	2.02	1.50 $\times 10^{-4}$
5	2.00	0.24 $\times 10^{-4}$

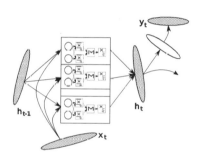

Fig. 6. The illustration of the DOT-RNN using L_p units

As shown in Table 4, we were able to achieve the state-of-the-art results (RNN-only case) on all the three datasets. These results are much better than those achieved by the same DOT-RNNs using logistic sigmoid units in both deep transition and deep output, which suggests the superiority of the proposed L_p units over the conventional saturating activation functions. This suggests that the proposed L_p units are well suited not only to feedforward neural networks, but also to recurrent neural networks. However, we acknowledge that more investigation into applying L_p units is needed in the future to draw more concrete conclusion on the benefits of the L_p units in recurrent neural networks.

Dataset	DOT-RNN		RNN
	L_p	sigmoid*	*
Nottingam	2.95	3.22	3.09
JSB	7.92	8.44	8.01
Muse	6.59	6.97	6.75

Table 4. The negative log-probability of the test sets computed by the trained DOT-RNNs. (⋆) These are the results achieved using DOT-RNNs having logistic sigmoid units, which we reported in [24]. (*) These are the previous best results achieved using conventional RNNs obtained in [2].

7   Conclusion

In this paper, we have proposed a novel nonlinear activation function based on the generalization of widely used pooling operators. The proposed nonlinear activation function computes the L_p norm of several projections of the lower layer. Max-, average- and root-of-mean-squared pooling operators are special cases of the proposed activation function, and naturally the recently proposed maxout unit is closely related under an assumption of non-negative input signals.

An important difference of the L_p unit from conventional pooling operators is that the order of the unit is *learned* rather than pre-defined. We claimed that this estimation of the orders is important and that the optimal model should have L_p units with various orders.

Our analysis has shown that an L_p unit defines a non-Euclidean subspace whose metric is defined by the L_p norm. When projected back into the input space, the L_p unit defines an ellipsoidal boundary. We conjectured and showed in a small scale experiment that the combination of these curved boundaries may more efficiently model separating curves of data with non-stationary curvature.

These claims were empirically verified via training both deep feedforward neural networks and deep recurrent neural networks. We tested the feedforward neural network on on four benchmark datasets; MNIST, Toronto Face Database, Pentomino and Forest Covertype, and tested the recurrent neural networks on the task of polyphonic music prediction. The experiments revealed that the distribution of the estimated orders of L_p units indeed depends highly on dataset and is far away from a Dirac delta distribution. Additionally, our conjecture that deep neural networks with L_p units will be able to achieve competitive generalization performance was empirically confirmed.

Acknowledgments. We would like to thank the developers of Pylearn2 [12] and Theano [6,1]. We would also like to thank CIFAR, and Canada Research Chairs for funding, and Compute Canada, and Calcul Québec for providing computational resources.

References

1. Bastien, F., Lamblin, P., Pascanu, R., Bergstra, J., Goodfellow, I.J., Bergeron, A., Bouchard, N., Bengio, Y.: Theano: new features and speed improvements. In: Deep Learning and Unsupervised Feature Learning NIPS 2012 Workshop (2012)
2. Bayer, J., Osendorfer, C., Korhammer, D., Chen, N., Urban, S., van der Smagt, P.: On fast dropout and its applicability to recurrent networks. arXiv:1311.0701 (cs.NE) (2013)
3. Bergstra, J., Bengio, Y.: Random search for hyper-parameter optimization. J. Machine Learning Res. 13, 281–305 (2012)
4. Bergstra, J., Bengio, Y., Louradour, J.: Suitability of V1 energy models for object classification. Neural Computation 23(3), 774–790 (2011)
5. Bergstra, J., Breuleux, O., Bastien, F., Lamblin, P., Pascanu, R., Desjardins, G., Turian, J., Warde-Farley, D., Bengio, Y.: Theano: a CPU and GPU math expression compiler. In: Proceedings of the Python for Scientific Computing Conference (SciPy) (2010)
6. Bergstra, J., Breuleux, O., Bastien, F., Lamblin, P., Pascanu, R., Desjardins, G., Turian, J., Warde-Farley, D., Bengio, Y.: Theano: a CPU and GPU math expression compiler. In: Proceedings of the Python for Scientific Computing Conference (SciPy). Oral Presentation (June 2010)

7. Boulanger-Lewandowski, N., Bengio, Y., Vincent, P.: Modeling temporal dependencies in high-dimensional sequences: Application to polyphonic music generation and transcription. In: ICML 2012 (2012)
8. Boureau, Y., Ponce, J., LeCun, Y.: A theoretical analysis of feature pooling in vision algorithms. In: Proc. International Conference on Machine learning, ICML 2010 (2010)
9. Ciresan, D., Meier, U., Masci, J., Schmidhuber, J.: Multi column deep neural network for traffic sign classification. Neural Networks 32, 333–338 (2012)
10. Fukushima, K.: Neocognitron: A self-organizing neural network model for a mechanism of pattern recognition unaffected by shift in position. Biological Cybernetics 36, 193–202 (1980)
11. Glorot, X., Bordes, A., Bengio, Y.: Deep sparse rectifier neural networks. In: AISTATS 2011 (2011)
12. Goodfellow, I.J., Warde-Farley, D., Lamblin, P., Dumoulin, V., Mirza, M., Pascanu, R., Bergstra, J., Bastien, F., Bengio, Y.: Pylearn2: a machine learning research library. arXiv preprint arXiv:1308.4214 (2013)
13. Goodfellow, I.J., Warde-Farley, D., Mirza, M., Courville, A., Bengio, Y.: Maxout networks. In: ICML 2013 (2013)
14. Gulcehre, C., Bengio, Y.: Knowledge matters: Importance of prior information for optimization. In: International Conference on Learning Representations, ICLR 2013 (2013)
15. Haykin, S.: Neural Networks and Learning Machines, 3rd edn. Prentice Hall (November 2008)
16. Hinton, G.E., Srivastava, N., Krizhevsky, A., Sutskever, I., Salakhutdinov, R.: Improving neural networks by preventing co adaptation of feature detectors. Technical report, arXiv:1207.0580 (2012)
17. Hubel, D., Wiesel, T.: Receptive fields and functional architecture of monkey striate cortex. Journal of Physiology (London) 195, 215–243 (1968)
18. Hyvärinen, A., Köster, U.: Complex cell pooling and the statistics of natural images. Network: Computation in Neural Systems 18(2), 81–100 (2007)
19. Jarrett, K., Kavukcuoglu, K., Ranzato, M., LeCun, Y.: What is the best multistage architecture for object recognition? In: Proc. International Conference on Computer Vision (ICCV 2009), pp. 2146–2153. IEEE (2009)
20. Krizhevsky, A., Sutskever, I., Hinton, G.: ImageNet classification with deep convolutional neural networks. In: Advances in Neural Information Processing Systems 25, NIPS 2012 (2012)
21. LeCun, Y., Bottou, L., Bengio, Y., Haffner, P.: Gradient-based learning applied to document recognition. Proceedings of the IEEE 86(11), 2278–2324 (1998)
22. Nair, V., Hinton, G.E.: Rectified linear units improve restricted Boltzmann machines. In: Bottou, L., Littman, M. (eds.) Proceedings of the Twenty-seventh International Conference on Machine Learning (ICML 2010), pp. 807–814. ACM (2010)
23. Pascanu, R., Bengio, Y.: Revisiting natural gradient for deep networks. Technical report, arXiv:1301.3584 (2013)
24. Pascanu, R., Gulcehre, C., Cho, K., Bengio, Y.: How to construct deep recurrent neural networks. arXiv:1312.6026 (cs.NE) (December 2013)
25. Pascanu, R., Mikolov, T., Bengio, Y.: On the difficulty of training recurrent neural networks. In: ICML 2013 (2013)
26. Ranzato, M., Mnih, V., Susskind, J.M., Hinton, G.E.: Modeling natural images using gated mrfs. IEEE Transactions on Pattern Analysis and Machine Intelligence 35(9), 2206–2222 (2013)

27. Riesenhuber, M., Poggio, T.: Hierarchical models of object recognition in cortex. Nature Neuroscience (1999)
28. Rifai, S., Dauphin, Y., Vincent, P., Bengio, Y., Muller, X.: The manifold tangent classifier. In: NIPS 2011 (2011)
29. Rosenblatt, F.: Principles of neurodynamics: perceptrons and the theory of brain mechanisms. Report (Cornell Aeronautical Laboratory). Spartan Books (1962)
30. Rumelhart, D.E., Hinton, G.E., Williams, R.J.: Learning representations by back-propagating errors. Nature 323, 533–536 (1986)
31. Susskind, J., Anderson, A., Hinton, G.E.: The Toronto face dataset. Technical Report UTML TR 2010-001, U. Toronto (2010)
32. Trebar, M., Steele, N.: Application of distributed svm architectures in classifying forest data cover types. Computers and Electronics in Agriculture 63(2), 119–130 (2008)
33. Yang, J., Yu, K., Gong, Y., Huang, T.: Linear spatial pyramid matching using sparse coding for image classification. In: Proc. Conference on Computer Vision and Pattern Recognition (CVPR 2010) (2010)

Combination of One-Class Support Vector Machines for Classification with Reject Option

Blaise Hanczar[1,3] and Michèle Sebag[2,3]

[1] CNRS, LRI UMR 8623, Universite Paris-Sud, Orsay, France
[2] TAO Project-team, INRIA Saclay & LRI, Universite Paris-Sud, Orsay, France
[3] LIPADE, Universite Paris Descartes, Paris, France
blaise.hanczar@parisdescartes.fr

Abstract. This paper focuses on binary classification with reject option, enabling the classifier to detect and abstain hazardous decisions. While reject classification produces in more reliable decisions, there is a tradeoff between accuracy and rejection rate. Two type of rejection are considered: ambiguity and outlier rejection. The state of the art mostly handles ambiguity rejection and ignored outlier rejection. The proposed approach, referred as CONSUM, handles both ambiguity and outliers detection. Our method is based on a quadratic constrained optimization formulation, combining one-class support vector machines. An adaptation of the sequential minimal optimization algorithm is proposed to solve the minimization problem. The experimental study on both artificial and real world datasets exams the sensitivity of the CONSUM with respect to the hyper-parameters and demonstrates the superiority of our approach.

Keywords: Supervised classification, Rejection option, Abstaining classifier, Support vector machines.

1 Introduction

One of the most interesting exploitation of the data is the construction of predictive classifiers [9]. For example, in genetic and molecular medicine, gene expression profiles can be used to differentiate different types of tumors with different outcomes and thus assist MD in the selection of an adapted therapeutic treatment if appropriate [20]. A huge number of methods from pattern recognition and machine learning have been developed and deployed on various domains. However, even though these methods produce classifiers with a good accuracy, these are often still insufficiently accurate to be used routinely. For example, a diagnostic or a choice of therapeutic strategy must be based on a very high confidence classifier; an error of the predictive model may lead to tragic consequences. A way of improving the reliability of such classifier is to use *abstaining classifiers* [12] also called *reject classifier* [19] or *selective classifier* [4]. Unlike standard classifiers that associate a predicted label to each example, only a subset of the examples are assigned to a class. Reject classifiers define a rejection

T. Calders et al. (Eds.): ECML PKDD 2014, Part I, LNCS 8724, pp. 547–562, 2014.
© Springer-Verlag Berlin Heidelberg 2014

region including the examples of which confidence is low [1,18,13,7,11]. While
reject classifiers have a higher accuracy than the standard classifiers, there is
a trade-off between accuracy and rejection rate [8]. The higher the classifier
accuracy, the higher the rejection rate.

A contribution of this paper is to investigate and handle two types of rejection: the ambiguity rejection and the outlier rejection. The ambiguity rejection
corresponds to the cases where an example is close to several classes, we cannot
decide between these classes with a high confidence. The ambiguity rejection
region is generally a small region containing the class boundaries. The outlier
rejection corresponds to the cases where an example is far from all classes. The
outlier rejection region is a large region surrounding all classes. Let us to illustrate the difference between these two types of rejection. Assumes that a hospital
use a classifier to identify the lymphoblastic from the myelogenous leukemia of
patients suffering from acute leukemia. A classifier gives a probability of 0.49
for lymphoblastic and 0.51 for myelogenous for a given patient. Although, the
probability of myelogenous is the highest, this class cannot be assigned to the
patient, the difference of probability is too low. This patient must be rejected by
the classifier because no reliable diagnosis can be done. It is an ambiguity rejection. Let another patient file be far from the distribution of both lymphoblastic
and myelogenous. The patient is considered as an outlier and must be also rejected. It is likely that this patient has not an acute leukemia and should pass
tests for other types of leukemia. It is a outlier rejection.

In this paper we propose a new approach of classification with reject option
that defines both the outlier and ambiguity rejection regions. Section 2 introduces the formal background and the state of the art. Section 3 is an overview of
CONSUM, together with the appropriate optimization algorithm for scalability
on large datasets. Experiments are reported and discussed in section 4 shows.
Conclusion and perspectives for future researches are given in section 5.

2   Classification with Reject Option

Let a binary classification problem with a training set $T = \{(x_1, y_1), ..., (x_N, y_N)\}$
where $x_i \in \mathbb{R}^d$ and $y_i \in \{-1, +1\}$. A reject classifier is a function that returns
a class for each example $\Psi : \mathbb{R}^d \to \{-1, +1, R\}$, R represents the reject class
including two subclasses R_a and R_d for ambiguity and outlier rejection. An example can be positive, negative or an outlier (belongs to none of the two classes).
When we use a reject classifier on test examples, 12 different classification results
are possible (three actual classes × four predicted classes). To each of these cases
a cost is associated (table 1). The cost of a good classification is zero. λ_{FP} and
λ_{FN} stand for the cost for false positive and false negative. λ_{ON} and λ_{OP} are
the costs for assigning respectively the positive or negative class to an outlier,
usually we set $\lambda_{ON} = \lambda_{FN}$ and $\lambda_{OP} = \lambda_{FP}$. Finally λ_{Ra} is the cost of ambiguity
rejection. λ_{Rd} is the cost for rejecting a positive or negative example as an out-
lier. Usually the costs of ambiguity and outlier rejection are equal $\lambda_{Ra} = \lambda_{Rd}$.
Note that, in principle, classifying the ambiguity rejection class to an outlier is

Table 1. Cost matrix of a two classes classification problem with ambiguity and outlier reject option

actual class

	+1	-1	O
+1	0	λ_{FP}	λ_{OP}
-1	λ_{FN}	0	λ_{ON}
Ra	λ_{Ra}	λ_{Ra}	0
Rd	λ_{Rd}	λ_{Rd}	0

an error, but it has no impact on the classifier performance, this cost is therefore set to zero. Most of reject classification methods only consider the ambiguity rejection. In this case the cost matrix in table 1 is reduced to its first three rows and first two columns.

Classifiers with rejection involve two main approaches: plug-in and embedded methods. The most popular approach consists to add a rejection rule to a classifier without rejection. These methods are called plug-in methods [1]. Two thresholds t_N and t_P are defined on the output of the classifier $f(x)$:

$$\Psi(x) = \begin{cases} -1 & \text{if} \quad f(x) \leq t_N \\ +1 & \text{if} \quad f(x) \geq t_P \\ R & \text{if} \quad t_N < f(x) < t_P \end{cases} \qquad (1)$$

Chow has introduced the notion of abstaining classifier and his definition of the rejection region is based on the exact posterior probabilities [1]. The thresholds defining the optimal abstaining region are computed directly from the cost matrix. In practice, the exact posterior probabilities are not available since the class distribution is unknown. The Chow's rule must thus be used with an estimation of the posterior probabilities. To overcome the need for the exact posterior probabilities, Tortorella has proposed a method where the abstaining region is computed in selecting two points on the Receiver operating characteristic (ROC) curve describing the performance of the classifier[18]. The two points are identified by their tangent on the ROC curve computed from the cost of rejection and type of error. Note that, Santos-Pereira proved that both the Chow's rule and ROC rule are actually equivalent [15].

Unlike the plug-in methods, the embedded methods compute the rejection region during the learning process. It has been proved that in theory the embedded methods give better classifiers than plug'in rule [3]. Fumera and Roli replace the Hinge loss function of the SVM by a kind of step function where the steps represent the cost of good classification, rejection and miss-classification [5]. The main drawback of this approach is the difficulty of the optimization problem because of the non-convexity of the loss function. Since the natural loss function of reject classification is non-convex, Yuan and Wegkamp proposed to use a surrogate convex loss [22]. They show that these functions are infinite sample consistent (they share the same minimizer) with the original loss function in some conditions.

Grandvalet et al. have proposed to use a double Hinge loss function in a SVM that has the advantage to be convex and linear piece-wise [6].

Dubuisson and Masson introduced the notion of outlier rejection [2]. They defined in a parametric case a threshold on the likelihood in order to reject examples far from the centers of all classes. Landgrebe et al. studied the interaction between error rate, ambiguity and outlier rejection. They optimize the thresholds of the classifiers in plotting the 3D ROC surface and computing the volume under the surface [10].

3   Combination of Two One-Class Support Vector Machines

3.1   Formal Description

Our method involve two interdependent one-class support vector machines, one for each class. The aim of the one-class SVM is to construct the smallest region capturing most of the training examples [16] and is defined by the following optimization problem:

$$min_{w,\rho,\xi_i} \frac{1}{2}||w||^2 - \rho + \frac{1}{VN} \sum_{i=1}^{N} \xi_i$$

$$subject\ to \qquad \langle w, x_i \rangle \geq \rho - \xi_i$$

The hyperplan, defined by w and ρ, separates the training examples from the origin with maximum margin into the feature space. All examples x such that $f(x) = \langle w, \phi(x) \rangle - \rho \geq 0$ are in the one-class SVM, if $f(x) < 0$ then x is out from the one-class SVM. The slack variables ξ_i's represent the empirical loss associated with x_i. $V \in [0,1]$ represents the rate of outliers and controls the trade-off between the size of the one-class SVM and the number of training example outside from the one-class SVM. Without any loss of generality, we consider that the training set is labeled as follow: $y_i = +1$ for $i \in [1,p]$ and $y_i = -1$ for $i \in [p+1, N]$. Moreover we assume that the cost of false positive and false negative are equal $\lambda_{FP} = \lambda_{FN} = \lambda_E$, $\lambda_{ON} = \lambda_{OP} = \lambda_E$ and $\lambda_{Ra} = \lambda_{Rd} = \lambda_R$.

Our method, called CONSUM, is based on the combination of two interdependent one-class SVM. The coupling of the one-class SVM allows a robust estimation of the rejection regions. The intuition behind this model is illustrated by the figure 1. It minimizes the following function:

$$L = \frac{1}{2}||w_+||^2 - \rho_+ + \frac{1}{V_+ N_+} \sum_{i=0}^{p} \xi_i + \frac{1}{2}||w_-||^2 - \rho_- + \frac{1}{V_- N_-} \sum_{i=p+1}^{N} \eta_i$$

$$+ \frac{C}{N} \left(C_r \sum_{i=1}^{N} \theta_i + C_e \sum_{i=1}^{N} \varepsilon_i \right) \qquad (2)$$

$$subject\ to \begin{cases} \langle w_+, \phi(x_i) \rangle \geq \rho_+ - \xi_i & \forall i \in [1, p] \\ \langle w_-, \phi(x_i) \rangle \geq \rho_- - \eta_i & \forall i \in [p+1, N] \\ y_i \langle w_+ - w_-, \phi(x_i) \rangle \geq \rho - \theta_i & \forall i \in [1, N] \\ y_i \langle w_+ - w_-, \phi(x_i) \rangle \geq -\rho - \varepsilon_i & \forall i \in [1, N] \\ \xi_i \geq 0,\ \eta_i \geq 0,\ \theta_i \geq 0,\ \varepsilon_i \geq 0 \end{cases} \quad (3)$$

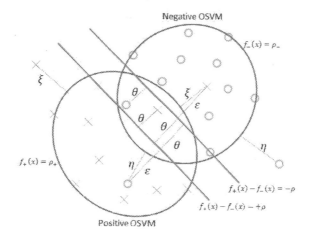

Fig. 1. The two one-class support vector machines classifier CONSUM

Terms w_+, and ρ_+ define the one-class SVM of the positive class, the ξ_i are the slack variables for the violation of the first constraint related to positive examples that are not in the positive one-class SVM. Terms w_-, and ρ_- define the one-class SVM of the negative class, the η_i are the slack variables for the violation of the second constraint related to negative examples that are not in the negative one-class SVM. The θ_i are the slack variables of the third constraint related to the examples in the ambiguity rejection region. The ε_i are the slack variables of the third constraint related to miss-classifications. The interaction between the two one-class SVM is gouverned by the third and fourth constraints, they define a region around the separator $\langle w_+ - w_-, x_i \rangle = 0$. It is similar to the margin region in a standard SVM, excepted that the size of margin (equal to $\frac{2\rho}{||w_+ - w_-||}$) is not maximized in our model. The margin size is independently optimized by the parameter ρ. This region is an approximation of the ambiguity rejection region, i.e. the intersection of the two one-class SVM.

The optimization problem is solved in dual form in introducing the Lagrangian multipliers α_i, γ_i, μ_i, α_i', γ_i', μ_i' for $i = 1..N$. The Lagrangian function can be write as:

$$L = \frac{1}{2}||w_+||^2 - \rho_1 + \frac{1}{V_+ N_+} \sum_{i=1}^{p} \xi_i + \frac{1}{2}||w_-||^2 - \rho_0 + \frac{1}{V_- N_-} \sum_{i=p+1}^{N} \eta_i$$

$$+ \frac{C}{N} \left(C_r \sum_{i=1}^{N} \theta_i + C_e \sum_{i=1}^{N} \epsilon_i \right)$$

$$- \sum_{i=1}^{p} \alpha_i(\langle w_+, \phi(x_i) \rangle - \rho_+ + \xi_i) - \sum_{i=p+1}^{N} \alpha_i(\langle w_-, \phi(x_i) \rangle - \rho_- + \eta_i)$$

$$- \sum_{i=1}^{N} \gamma_i(\langle w_+ - w_-, \phi(x_i) \rangle y_i - \rho + \theta_i) - \sum_{i=1}^{N} \mu_i(\langle w_+ - w_-, \phi(x_i) \rangle y_i - \rho + \varepsilon_i)$$

$$- \sum_{i=1}^{p} \alpha_i' \xi_i - \sum_{i=p+1}^{N} \alpha_i' \eta_i - \sum_{i=1}^{N} \gamma_i' \theta_i - \sum_{i=1}^{N} \mu_i' \varepsilon_i$$

This function is maximized with respect to the Lagrange multiplier and minimized with the respect to primal variables: $max_{\alpha_i, \gamma_i, \mu_i} min_{w_+, w_-, \rho_+, \rho_-, \rho, \xi, \eta, \theta, \varepsilon} L$. Setting the derivatives of L with the respect to the primal variables equal to zero, one yields:

$$w_+ = \sum_{i=1}^{p} \alpha_i x_i + \sum_{i=1}^{N} \gamma_i x_i y_i + \sum_{i=1}^{N} \mu_i x_i y_i \qquad (4)$$

$$w_- = \sum_{i=p+1}^{N} \alpha_i x_i - \sum_{i=1}^{N} \gamma_i x_i y_i - \sum_{i=1}^{N} \mu_i x_i y_i \qquad (5)$$

$$\sum_{i=1}^{p} \alpha_i = 1; \qquad \sum_{i=p+1}^{N} \alpha_i = 1; \qquad \sum_{i=1}^{N} \gamma_i - \sum_{i=1}^{N} \mu_i = 0 \qquad (6)$$

$$0 \le \alpha_i \le \frac{1}{V_+ N_+} \ \forall i \in [1, p] \ ; \qquad 0 \le \alpha_i \le \frac{1}{V_- N_-} \ \forall i \in [p+1, N] \qquad (7)$$

$$0 \le \gamma_i \le \frac{CC_r}{N} \ ; \qquad 0 \le \mu_i \le \frac{CC_e}{N} \qquad (8)$$

Substituting (4)-(8) into L and denoting $K_{ij}' = y_i y_j K_{ij}$ leads to the dual problem that is a quadratic programming problem:

$$min_{\alpha_i, \gamma_i, \mu_i} \quad \frac{1}{2} \sum_{i,j=1}^{p} \alpha_i \alpha_j K_{ij}' + \frac{1}{2} \sum_{i,j=p+1}^{N} \alpha_i \alpha_j K_{ij}' + \sum_{i,j=0}^{N} \gamma_i \gamma_j K_{ij}' + \sum_{i,j=0}^{N} \mu_i \mu_j K_{ij}'$$

$$+ \sum_{i,j=0}^{N} \alpha_i \gamma_j K_{ij}' + \sum_{i,j=0}^{N} \alpha_i \mu_j K_{ij}' + 2 \sum_{i,j=0}^{N} \gamma_i \mu_j K_{ij}' \qquad (9)$$

subject to the constraints (6) (7) and (8)

The classifier is defined by the following decision functions:

$$f_+(x_j) = \langle w_+, x_j \rangle - \rho_+ = \sum_{i=1}^{p} \alpha_i K_{ij} + \sum_{i=0}^{N} \gamma_i K_{ij} y_i + \sum_{i=0}^{N} \mu_i K_{ij} y_i - \rho_+ \quad (10)$$

$$f_-(x_j) = \langle w_-, x_j \rangle - \rho_- = \sum_{i=p+1}^{N} \alpha_i K_{ij} - \sum_{i=0}^{N} \gamma_i K_{ij} y_i - \sum_{i=0}^{N} \mu_i K_{ij} y_i - \rho_- \quad (11)$$

$f_+(x_j)$ returns a positive value if the example x_j is contained in the positive one-class SVM and negative value otherwise. If $f_+(x_j) = 0$, then x_j is on the one-class SVM boundary and we have $0 < \alpha_j < \frac{1}{V_+ N_+}$. This kind of example is exploited in order to recover ρ_+, since x_j satisfies $\rho_+ = \langle w_+, x_j \rangle$. ρ_- is computed in the same way. The final decision of CONSUM is based on the two decision functions.

$$\Psi(x) = \begin{cases} +1 & \text{if} \quad f_+(x) > 0 \land f_-(x) \leq 0 \\ -1 & \text{if} \quad f_+(x) \leq 0 \land f_-(x) > 0 \\ R_a & \text{if} \quad f_+(x) > 0 \land f_-(x) > 0 \\ R_d & \text{if} \quad f_+(x) \leq 0 \land f_-(x) \leq 0 \end{cases} \quad (12)$$

Our model critically depends on the choice of the hyper-parameters V_+, V_-, C, C_r and C_e. V_+ and V_- control the rate of outliers in the training set. By default we consider that the training set contains no outliers. We want these parameters low, we set them to $V_+ = V_- = 0.05$. V_+ and V_- are not set to zero because it would overconstraints the general formulation (2-3). C_r and C_e control the trade-off between the ambiguity rejection and the error rate. The optimal trade-off is given by λ_E and λ_R. In our model the loss of an ambiguity rejected example is $CC_r\theta_i$ and the loss of an error is $C(C_r\theta_i + C_e\varepsilon_i)$. We want that C_e, C_r respect the ratio between the error and rejection costs $\frac{C_r + C_e}{C_r} = \frac{\lambda_E}{\lambda_R}$ and be normalized such that $C_r + C_e = 1$. That leads to $C_r = \frac{\lambda_R}{\lambda_E}$ and $C_e = \frac{\lambda_E - \lambda_R}{\lambda_E}$. The parameters C controls the importance of the error and ambiguity rejection loss, it plays the same role as the C in the usual SVM model. This parameter will be optimized during the model fitting in an inner cross-validation procedure.

3.2   Optimization Algorithm

Our model is formulated as a quadratic programming problem with linear constraints (9). Several approaches are available to solve this type of problems. We propose a modified version of the sequential minimal optimization (SMO) algorithm that was originally developed for SVM training [14], a version for one-class SVM has also been proposed [16]. It has the advantage to have a lower complexity than the other methods and does not need to keep the whole Gram matrix in memory. It is therefore adapted to large datasets. The principle of SMO is to divide the original optimization problem into several optimization tasks of the smallest size. In our case the smallest size is two, we have to optimize over pairs of multipliers. According to the contraints (6), (7) and (8), there are three different

types of multiplier pairs $\{(\alpha_u, \alpha_v)/u, v \in [1, p]]\}$, $\{(\alpha_u, \alpha_v)/u, v \in [p + 1, N]\}$
and all pairs form from the γ and μ. Note that this last types of pairs can
divided into three subtypes $\{(\gamma_u, \gamma_v)\}$, $\{(\mu_u, \mu_v)\}$ and $\{(\gamma_u, \mu_v)\}$. The optimization algorithm consists to select a pair of multipliers and optimize only these
two multipliers; this process is iterated until a stopping criterion.

Initialization. The multipliers can be initialized with any values, the only
conditions is that they respect the constraints (7-8). At the beginning of our
algorithm, the α_i's for $i \in [1, p]$ are initialized to $\frac{1}{N_+}$, the α_i's for $i \in [p + 1, N]$
are initialized to $\frac{1}{N_-}$ and the γ_i's and μ_i's are initialized to $\frac{1}{N}$.

Stopping Criterion. The optimization algorithm is stopped when the gain
of loss is null. However, in our algorithm we have to use a nonzero accuracy
tolerance such that two values are considered equal if they differ by less than e.
In practice the procedure is stopped at the iteration t when $L^t - L^{t-1} < e$. In
our experiments $e = 0.0001$

Multipliers Pair Optimization Step. Our algorithm is a succession of multiplier pair optimization tasks. Suppose that at a given iteration, the pair of
multipliers γ_u and γ_v is selected to be optimized. All others multipliers are con-
sidered as constants during this task. From the constraint (6) we have $\gamma_u + \gamma_v = \sum_{i \neq u, v}^{N} \gamma_i - \sum_{i=0}^{N} \mu_i = D$. The quadratic problem (9) is written in function on
γ_u and γ_v :

$$L = \gamma_i^2 K'_{uu} + \gamma_j^2 K'_{vv} + 2\gamma_i\gamma_j K'_{uv} + 2\gamma_u G_u + 2\gamma_v G_v + G + \gamma_u A_u^+ + \gamma_v A_v^+ + A^+$$
$$+ \gamma_u A_u^- + \gamma_v A_v^- + A^- + 2\gamma_u M_u + 2\gamma_v M_v + 2M - \gamma_u - \gamma_v + X$$

$$A_x^+ = \sum_{i=1}^{p} \alpha_i K'_{ix} \quad A^+ = \sum_{\substack{i=1 \\ j \neq u,v}}^{p} \alpha_i\gamma_j K'_{ij} \quad A_x^- = \sum_{i=p+1}^{N} \alpha_i K'_{ix} \quad A^+ = \sum_{\substack{i=p+1 \\ j \neq u,v}}^{N} \alpha_i\gamma_j K'_{ij}$$

$$G_x = \sum_{i \neq x} \gamma_i K'_{ix} \quad G = \sum_{i,j \neq u,v} \gamma_u\gamma_v K'_{i,j} \quad M_x = \sum_{i \neq x} \mu_i K'_{ix} \quad M = \sum_{i,j \neq u,v} \mu_u\mu_v K'_{i,j}$$

$$X = \frac{1}{2}\sum_{i,j=1}^{p} \alpha_i\alpha_j K'_{ij} + \frac{1}{2}\sum_{i,j=p+1}^{N} \alpha_i\alpha_j K'_{ij} + \sum_{i,j=1}^{N} \alpha_i\mu_j K'_{ij} + \sum_{i,j=1}^{N} \mu_i\mu_j K'_{ij} + \sum_{i=1}^{N} \mu_i$$

Note that A_x^+, A^+, A_x^-, A_-, G_x, G, M_x, M, X are constants in this step.
In using $\gamma_u = D - \gamma_v$, L in expressed only in function on γ_v and compute its
derivative:

$$\frac{\partial L}{\partial \gamma_v} = 2(\gamma_v - D)K'_{uu} + 2(D - 2\gamma_v)K'_{uv} + 2\gamma_v K'_{vv} + A_v^+ y_v - A_u^+ y_u$$

$$+ A_u^- y_u - A_v^- y_v + 2G_v y_v - 2G_u y_u + 2M_v y_v - 2M_u y_u$$

Setting the derivative to zero leads the optimal value of γ_v:

$$\gamma_v^{new} = \frac{2D(K_{uu} - K_{uv}) + A_u^+ - A_v^+ + A_u^- - A_v^- + 2G_u - 2G_v + 2M_u - 2M_v}{2(K_{uu} + K_{vv} - 2K_{uv}Y_{uv})}$$

It is simpler to rewrite this equation in introducing the decision functions f_+ and f_- :

$$\gamma_u^{new} = \gamma_u - \Delta , \qquad \gamma_v^{new} = \gamma_v + \Delta$$

$$\Delta = \frac{y_v(f_-(x_v) - f_+(x_v)) + y_u(f_+(x_u) - f_+(x_u))}{2(K_{uu} + K_{vv} - 2K_{uv}Y_{uv})}$$

Let recall that the multipliers γ_i are still subject to the constraint (8). If γ_u^{new} or γ_v^{new} is outside from the interval $[0, \frac{CC_r}{N}]$, the constraint optimum is found by projecting it into the bound of the allowed interval, then Δ, γ_u^{new} and γ_v^{new} are recomputed. One the multiplier pair has been optimized, the decision functions should be recomputed. However it is not necessary to use the formulas (10-11), for saving computing resources we can updated them by:

$$f_+^{new}(x_i) = f_+(x_i) + \delta_+ \text{ with } \delta_+ = \Delta(K_{ui}y_u - K_{vi}y_v)$$
$$f_-^{new}(x_i) = f_+(x_i) + \delta_- \text{ with } \delta_- = \Delta(K_{vi}y_v - K_{ui}y_u)$$

This procedure is repeated at each iteration, however the formulas differs slightly in function on the type of multiplier pairs. The table 2 gives the formulas of Δ, δ_+ and δ_- for each type of pair.

Table 2. Δ, δ_+ and δ_- used in the different cases of pair optimization tasks

pairs	Δ	δ_+	δ_-
α_u, α_v u,v∈[1,p]	$\frac{f_+(x_u) - f_+(x_v)}{K_{uu} + K_{vv} - 2K_{uv}Y_{uv}}$	$\Delta(K_{vi} - K_{ui})$	0
α_u, α_v u,v∈[p+1,N]	$\frac{f_-(x_u) - f_-(x_v)}{K_{uu} + K_{vv} - 2K_{uv}Y_{uv}}$	0	$\Delta(K_{vi} - K_{ui})$
γ_u, γ_v	$\frac{y_v(f_-(x_v) - f_+(x_v)) + y_u(f_+(x_u) - f_-(x_u))}{2(K_{uu} + K_{vv} - 2K_{uv}Y_{uv})}$	$\Delta(K_{vi}y_v - K_{ui}y_u)$	$\Delta(K_{ui}y_u - K_{vi}y_v)$
μ_u, μ_v	$\frac{y_v(f_-(x_v) - f_+(x_v)) + y_u(f_+(x_u) - f_-(x_u))}{2(K_{uu} + K_{vv} - 2K_{uv}Y_{uv})}$	$\Delta(K_{vi}y_v - K_{ui}y_u)$	$\Delta(K_{ui}y_u - K_{vi}y_v)$
γ_u, μ_v	$\frac{y_u(f_-(x_u) - f_+(x_u)) + y_v(f_-(x_v) - f_+(x_v))}{2(K_{uu} + K_{vv} + 2K_{uv}Y_{uv})}$	$\Delta(K_{ui}y_u - K_{vi}y_v)$	$-\Delta(K_{ui}y_u + K_{vi}y_v)$

Selection of the Multiplier Pairs. At each iteration a multiplier pair is selected to be optimized. The pair selection could be random but an intelligent selection procedure allows to speed up the optimization process. There are three different types of pair, each of these pair types will be handled successively. Let's focus on the optimization of the first type of pair, i.e. $\{(\alpha_u, \alpha_v)/u, v \in [1, p])\}$. We randomly select a positive example x_u that does not respect the following condition :

$$\begin{cases} \text{if } \alpha_u = 0 \text{ then } f_+(x_u) > \rho_+ \\ \text{if } 0 < \alpha_u < \frac{1}{V_+ N_+} \text{ then } f_+(x_u) = \rho_+ \\ \text{if } \alpha_u = \frac{1}{V_+ N_+} \text{ then } f_+(x_u) < \rho_+ \end{cases}$$

If all positive examples respect this condition, x_u is randomly selected among the positive examples. This condition is checked in using the nonzero accuracy tolerance introduced in the section 4.2. The second multipliers α_v is the one maximizing the numerator of Δ i.e. $v = argmax_{i \in \oplus}|f_+(x_u) - f_+(x_i)|$. Then the values of α_u, α_v, ρ_+ and $f_+(x_i)$ are updated. The idea is to select a pair that will produce a large change (large Δ) in the values of α_u and α_v. Only the numerator of Δ is computed for the pair selection because it is very fast since we already have the value $f_+(x_u)$ and $f_+(x_v)$. If we use the real value of Δ, the denominator have to be computed which would be much more slower. Other pairs of the same type $\{(\alpha_u, \alpha_v)/u, v \in \oplus)\}$ are selected and optimized until the gain of loss becomes small i.e. $L^t - L^{t-1} < 10e$. Then the same optimization procedure is run on the two other types of pair. The whole procedure is iterated until the stopping criterion is reached.

4   Experimental Validation

4.1   Experiment Settings

The goal of the experiments is to investigate the sensitivity of CONSUM with respect to the hyper-pameters and to compare its performance to the state of the art. There exist very few methods that construct classifiers with both ambiguity and outlier rejection. Since our method is based on the combination of one-class SVM, we tested the association of two one-class SVM (2OSVM), this approach has been proposed in [17]. A one-class SVM is independently constructed for each class and the decision rule is the same than in Eq.(12). The second method tested is the support vector machine with the ambiguity rejection computed by the Chow rule. The construction of the outlier rejection is more problematic since it has been very few studied in the literature. The best way to have a fair comparison with CONSUM is to base the outlier rejection on a one-class SVM. A one-class SVM is computed on all training examples (both positive and negative class), each example outside from the one-class SVM will considered as an outlier[1]. In our experiments the different methods use a Gaussian kernel whose variance σ is determined in an inner cross-validation loop. The costs of miss-classification and rejection are set to $\lambda_E = 4$ and $\lambda_R = 1$

We did experiments on artificial datasets in order to show the behavior of the different methods and analyze the impact of parameters of our approach. The artificial data are generated from Gaussian distributions with independent

---

[1] Note that since the Chow rule computes two thresholds on the classifier output to define the ambiguity rejection (Eq.(1)), some researchers have proposed to add two other thresholds to define the outlier rejection. These thresholds T_N, T_P are at the extremes of the classifier output, all examples that are not in the interval $[T_N, T_P]$ are considered as outliers. We think that this approach is not good and specially with Gaussian kernel. Indeed, the higher values will be obtained by examples that are in the center of the distribution of positive class and have the highest probabilities to belong to the positive class. It is therefore not correct to consider these examples as outlier rejections.

SVM   2OSVM   CONSUM

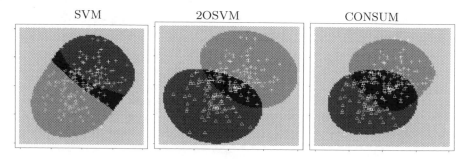

Fig. 2. Classifiers constructed on the artificial dataset with the different methods. The green, blue, black and gray region represents respectively the positive, negative, ambiguity rejection and outlier rejection region. The triangles and crosses represent respectively the positive and negatives examples of the training set.

variables. To simulate the presence of outliers, we add to the test set examples of a third class whose the distribution is uniform on the input space. Some experiments are based on 2-dimension data in order to support visualization of the classifiers.

4.2   Sensitivity Analysis

The figure 2 shows the classifiers obtained with the different methods on the artificial problem. The green, blue, black and gray region represents respectively the positive, negative, ambiguity and outlier rejection region. We see that CONSUM fits the ambiguity and outlier rejection regions better than the other methods. The difference between the methods can be mainly seen in the shape of the ambiguity rejection. The ambiguity rejection region of SVM is spread around the boundary of the non-reject SVM. The ambiguity region of 2OSVM and CONSUM contains only region where positive and negative examples are mixed. The difference between 2OSVM and CONSUM is the size of the ambiguity rejection. The ambiguity region of CONSUM is larger and less regular. In 2OSVM, the two one-class SVM are independent, there is no way to control the trade-off between error and ambiguity rejection. In CONSUM, this trade-off is controlled by the parameters C_E, C_R and the constraints (3), it has more degree of fredoom than 2OSVM. The fact that the size of ambiguity region is large in CONSUM comes from the missclassification cost that is much higher than the rejection cost in our simulation.

One of the crucial points of classification with reject option is the control of the trade-off between the error rate and the rejection rate. In our model, this is done by ρ_+ and ρ_-. We investigate the behavior of the error rate, ambiguity rejection rate and outlier rejection rate in function on these thresholds. A CONSUM classifier has been constructed on artificial data, we vary the value of ρ_+ and ρ_- from 0 to $max\{f_+(x_i), f_-(x_i) | i = 1..N\}$ and observe the impact of the error rate, ambiguity and outlier rejection in the figure 3. The ambiguity rejection is

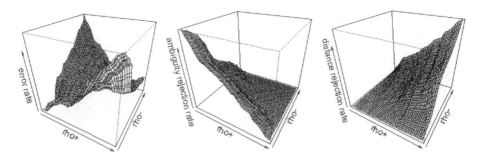

Fig. 3. Evolution of the rate of error, ambiguity rejection and outlier rejection in function of the thresholds ρ_+ and ρ_- of the CONSUM classifier

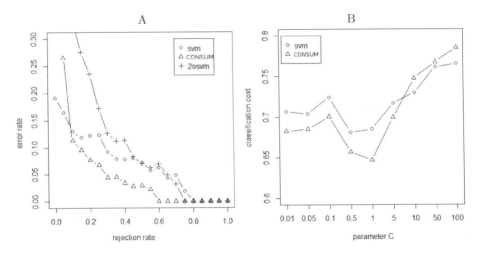

Fig. 4. A: Trade-off between error and rejection rate of the different methods. B: Classification cost in function on the parameter C of SVM and CONSUM (There is no parameter C in 2OSVM).

decreasing with ρ_+ and ρ_-, on the opposite the outlier rejection is increasing with ρ_+ and ρ_-. The error rate is null when $\rho_+ = \rho_- = 0$, this corresponds to the trivial case where all points are outlier rejected, there is therefore no missclassification. On the opposite when both ρ_+ and ρ_- reach their maximum, the two classes greatly overlap, there are few missclassifications. When ρ_+ is maximum and ρ_- is null, the positive class dominates the negative class. All non-rejected negative examples are false positive, the error rate is therefore very high. We have the same thing when ρ_- is maximum and ρ_+ is null. This simulation illustrates the relation between the error rate and the different types of rejection of our model.

When the cost matrix of the classification problem is not known, the most convenient method to compare several classifiers with reject option is to plot their error-rejection curve (ERC). The ERC gives the error of a classifier in function the rejection rate (both ambiguity and outlier). When we consider both ambiguity and outlier rejection, several different classifiers and rejection regions may give the same rejection rate. For a given rejection rate, there are generally several error rates. If we test all values of ρ_+ and ρ_-, the performances of the classifier are illustrated by a scatter plot in the error-rejection space. If we keep only the lowest error for each rejection rate value, the performance of the classifier is represented by a curve. It is the Pareto front of the scatter plot. These curves(ERC) can be used to compare the performance of different classifiers. The figure 4A gives the ERC of the SVM, 2OSVM and CONSUM classifiers on artificial data. For reject.rate$\geq$0.1 CONSUM is the best classifier and reaches error.rate=0 for reject.rate=0.6. 2OSVM is never competitive, it reaches the performance for SVM for reject.rate$\geq$0.45 and the performance of CONSUM for reject.rate$\geq$0.75. Notes that CONSUM has no points for reject.rate=0 because the model does not return empty rejection regions whatever the values of ρ_+ and ρ_-. In theory it is possible to reach reject.rate=0 if both all test points are in one of the one-class SVM and there is no overlap between the two one-class SVM. In practice this case is very rare.

In the next experiments we compare the performances of the classifiers in computing their classification cost on a test set of size N_{ts}. Let's $I(x) = 1$ if x is true, 0 otherwise; the classifier cost is defined by:

$$cost(\Psi) = \frac{1}{N_{ts}} \sum_{i=1}^{N_{ts}} \lambda_E I \left(\Psi(x_i) \neq R_a \vee R_d \right) I \left(\Psi(x_i) \neq y_i \right) + \lambda_R I \left(\Psi(x_i) = R_a \vee R_d \right)$$

The figure 4B gives the classification cost of SVM and CONSUM in function on their hyper-parameter C. The role of C is similar in the SVM and CONSUM classifier. In SVM this parameter controls the trade-off between the maximization of the margin and the good classification of the examples. In CONSUM, it controls the trade-off between the optimization of the two one-class SVM and the constraints on the miss-classifications and rejections of the training examples. We see that the curves of SVM and CONSUM has a similar shape, they reach their minimum around $0.5 < C < 1$. At their optimal parameter, CONSUM gives better results than SVM. In our next experiments, the parameters C is determined by an inner cross-validation procedure.

4.3   Comparative Study

We made some experiments on real data with three different rejection scenarios. We used six datasets from the UCI repository. The table 3 shows the classification cost of the different methods on the six datasets. The first scenario is the classification with no rejection. We use the usual SVM, for 2OSVM and CONSUM a class is assigned to an example according to $\langle w_+ - w_- ; x_i \rangle$. Results are in the first three columns of the table 3. The best results are obtains mainly by the SVM. However CONSUM obtains good results, its performances are close to the SVM. In the second scenario, only ambiguity rejection is considered. The Chow rule is added to the SVM, for 2OSVM and CONSUM the outlier rejection

Table 3. Classification cost on artificial and real data experiments. The best results are in bold.

datasets	dimension	no reject			ambiguity reject			amb.& outlier reject		
		SVM	2OSVM	CONSUM	SVM	2OSVM	CONSUM	SVM	2OSVM	CONSUM
Artif.	2000×50	**1.209**	1.454	1.411	**0.997**	1.169	1.093	0.939	0.933	**0.884**
Artif.	2000×100	0.930	1.092	**0.919**	**0.744**	0.826	0.756	0.778	0.843	**0.750**
wdbc	569×30	0.325	0.410	0.312	0.307	0.279	**0.194**	0.325	0.508	**0.194**
spam	4601×57	**0.262**	0.268	0.309	0.115	0.212	**0.061**	0.248	0.294	**0.152**
madelon	2000×500	1.302	0.1430	1.332	1.100	1.283	**1.017**	0.967	0.953	**0.919**
pop	540×18	**0.261**	0.284	0.272	0.250	0.250	**0.238**	0.355	0.403	**0.351**
transfusion	748×4	0.946	1.002	**0.926**	**0.801**	0.964	0.902	**0.839**	0.959	0.851
bank	1374×4	**0.066**	0.080	0.092	**0.051**	0.074	0.061	0.208	**0.179**	0.182

region is not used. Results are given in the three middle columns. CONSUM has the best performance for four datasets and SVM for two datasets. In the last scenario, we consider both ambiguity and outlier rejection, outliers are added to the test set as in the artificial dataset. A one-class SVM is added to the SVM, 2OSVM and CONSUM are used normally. Results are in the three last columns. These last results are the most important since the scenario represents the practical cases. We see that CONSUM obtains the best performances. The results show that our method gives both a reliable representation of the two classes by one-class SVM and a good trade-off between the rejection and the error rate. It is interesting to note that when there is no rejection, SVM is much better than the other classifiers. When the rejection rate is significant CONSUM becomes better than SVM. These results confirms the conclusion of [3] the best classifier with rejection option is not the best classifier without rejection on which a rejection rule is added. Note also when we compare the "no reject" scenario to the two others, we conclude that the use of a rejection option improve greatly the performance of the classifier whatever the method used.

5   Conclusion

We have introduced a new approach for classification with rejection option CONSUM that constructs simultaneously both the outlier and ambiguity rejection regions. The outlier rejection is generally ignored in the state of the art, but it is essential when the test set contains outliers, that is common in real applications. We showed that CONSUM can be viewed as a quadratic programming problem and we proposed an optimization algorithm adapted for large datasets. The results showed that our method improves the performance of classification.

In this paper we assumed that the cost of false positive and false negative are equal ($\lambda_{FN} = \lambda FP$) and the cost of positive rejection and negative rejection are equal. If the classification problem is cost sensitive or unbalanced, we may want to assigned different costs to the two classes. In this case we introduce different costs for the ambiguity rejection ($\lambda_{RNa} \neq \lambda RPa$) and for the outlier rejection ($\lambda_{RNd} \neq \lambda RPd$). The rejection constraints of the minimization problem (3) are split into constraints for positive examples and constraints for negative examples. Different penalties are assigned to the loss of each classes. The minimization problem can still be solved by the optimization algorithm presented in section

4. The only difference is that the multipliers γ and μ of positive and negative examples have to be optimized separately as with the α.

Our future works should focus on the multi-class problem. We can easily define a one-class SVM for each class but the constraints for the miss-classifications and ambiguity rejection are defined only for binary problems. The number of hyper-parameters is the main challenges since it increase quadratically with the number of classes. Several methods has been proposed to solve the problem of multi-class SVM [21]. They could be a source of inspiration in order to propose a multi-class version of CONSUM.

References

1. Chow, C.: On optimum recognition error and reject tradeoff. IEEE Transactions on Information Theory 16(1), 41–46 (1970)
2. Dubuisson, B., Masson, M.: A statistical decision rule with incomplete knowledge about classes. Pattern Recognition 26(1), 155–165 (1993)
3. El-yaniv, R., Wiener, Y.: Agnostic selective classification. In: Neural Information Processing Systems NIPS (2011)
4. El-Yaniv, R., Wiener, Y.: On the foundations of noise-free selective classification. Journal of Machine Learning Research 99, 1605–1641 (2010)
5. Fumera, G., Roli, F., Giacinto, G.: Multiple reject thresholds for improving classification reliability. In: Amin, A., Pudil, P., Ferri, F., Iñesta, J.M. (eds.) SPR/SSPR 2000. LNCS, vol. 1876, p. 863. Springer, Heidelberg (2000)
6. Grandvalet, Y., Rakotomamonjy, A., Keshet, J., Canu, S.: Support vector machines with a reject option. In: NIPS pp. 537–544 (2008)
7. Hanczar, B., Dougherty, E.: Classification with reject option in gene expression data. Bioinformatics 24(17), 1889–1895 (2008)
8. Hansen, L.K., Liisberg, C., Salamon, P.: The error-reject tradeoff. Open Systems & Information Dynamics 4, 159–184 (1997)
9. Kelley, L., Scott, M.: The evolution of biology. A shift towards the engineering of prediction-generating tools and away from traditional research practice. EMBO Reports 9(12), 1163–1167 (2008)
10. Landgrebe, T., Tax, D., D.M.J., P., R.P.W., D.: The interaction between classification and reject performance for distance-based reject-option classifiers. Pattern Recognition Letters 27(8), 908–917 (2006)
11. Nadeem, M., Zucker, J., Hanczar, B.: Accuracy-rejection curves (arcs) for comparing classification methods with a reject option. Journal of Machine Learning Research - Proceedings Track 8, 65–81 (2010)
12. Pietraszek, T.: Optimizing abstaining classifiers using roc analysis. In: Proceedings of the 22nd International Conference on Machine Learning, ICML 2005, pp. 665–672 (2005)
13. Pietraszek, T.: On the use of roc analysis for the optimization of abstaining classifiers. Machine Learning 68(2), 137–169 (2007)
14. Platt, J.C.: Sequential minimal optimization: A fast algorithm for training support vector machines. Tech. rep., Advances in Kernel Methods - Support Vector Learning (1998)
15. Santos-Pereira, C.M., Pires, A.M.: On optimal reject rules and roc curves. Pattern Recognition Letters 26(7), 943–952 (2005)

16. Schölkopf, B., Platt, J.C., Shawe-Taylor, J.C., Smola, A.J., Williamson, R.C.: Estimating the support of a high-dimensional distribution. Neural Computation 13(7), 1443–1471 (2001)
17. Tohmé, M., Lengelle, R.: Maximum margin one class support vector machines for multiclass problems. Pattern Recognition Letter 32(13), 1652–1658 (2011)
18. Tortorella, F.: A roc-based reject rule for dichotomizers. Pattern Recognition Letters 26(2), 167–180 (2005)
19. Tortorella, F.: An optimal reject rule for binary classifiers. In: Amin, A., et al. (eds.) SPR 2000 and SSPR 2000. LNCS, vol. 1876, p. 611. Springer, Heidelberg (2000)
20. van de Vijver, M.: A gene-expression signature as a predictor of survival in breast cancer. N. Engl. J. Med. 347, 1999–2009 (2002)
21. Weston, J., Watkins, C.: Multi-class support vector machines. Tech. rep., University of London (1998)
22. Yuan, M., Wegkamp, M.: Classification methods with reject option based on convex risk minimization. Journal of Machine Learning Research 11, 111–130 (2010)

Relative Comparison Kernel Learning with Auxiliary Kernels

Eric Heim[1], Hamed Valizadegan[2], and Milos Hauskrecht[1]

[1] University of Pittsburgh, Department of Computer Science, Pittsburgh, PA, U.S.A.
{eth13,milos}@cs.pitt.edu
[2] UARC, University of California at Santa Cruz,
NASA Ames Research Center, Moffett Field, CA, U.S.A.
hamed.valizadegan@nasa.gov

Abstract. In this work we consider the problem of learning a positive semidefinite kernel matrix from relative comparisons of the form: "object A is more similar to object B than it is to C", where comparisons are given by humans. Existing solutions to this problem assume many comparisons are provided to learn a meaningful kernel. However, this can be considered unrealistic for many real-world tasks since a large amount of human input is often costly or difficult to obtain. Because of this, only a limited number of these comparisons may be provided. We propose a new kernel learning approach that supplements the few relative comparisons with "auxiliary" kernels built from more easily extractable features in order to learn a kernel that more completely models the notion of similarity gained from human feedback. Our proposed formulation is a convex optimization problem that adds only minor overhead to methods that use no auxiliary information. Empirical results show that in the presence of few training relative comparisons, our method can learn kernels that generalize to more out-of-sample comparisons than methods that do not utilize auxiliary information, as well as similar metric learning methods.

Keywords: similarity learning, relative comparisons.

1 Introduction

The effectiveness of many kernel methods in unsupervised [24], [5], semi-supervised [29], [28], [26], and supervised [18] learning is highly dependent on how meaningful the input kernel is for modeling similarity among objects for a given task. In practice, kernels are often built by using a standard kernel function on features extracted from data. For example, when building a kernel over clothing items, features can be extracted for each item regarding attributes like size, style, and color. Then, a predefined kernel function (e.g. the Gaussian kernel function) can be applied to these features to build a kernel over clothing. However, for certain tasks, objects may not be represented well by extracted features alone. Consider a product recommendation system for suggesting replacements for out

T. Calders et al. (Eds.): ECML PKDD 2014, Part I, LNCS 8724, pp. 563–578, 2014.

of stock clothing items. Such a system requires a model of similarity based on how humans perceive clothing, which may not be captured entirely by features. For this, it is likely that human input is necessary to construct a meaningful kernel.

In general, obtaining reliable information from humans can be challenging, but retrieving *relative comparisons* of the form "object A is more similar to object B than it is to object C" has several attractive characteristics. First, relative comparison questions are less mentally fatiguing to humans compared to other forms (e.g. questions of the form: "From 1 to 10, how similar are objects A and B?") [10]. Second, there is no need to reconcile individual humans' personal scales of similarity. Finally, relative comparison feedback can be drawn implicitly through certain human-computer interactions, such as mouse clicks. In this work, we consider the specific problem of learning a kernel from relative comparisons; a problem we will refer to as *relative comparison kernel learning* (RCKL).

Current RCKL methods [1], [22], [14] assume that all necessary human feedback to build a useful kernel is provided. This is often not the case in real-world scenarios. A large amount of feedback is needed to build a kernel that represents a meaningful notion of how a human views the relationships among objects, and obtaining feedback from humans is often difficult or costly. Hence, it is a realistic assumption that only a limited amount of feedback can be obtained.

We propose a novel RCKL method that learns a meaningful kernel from a limited number of relative comparisons. Our method learns a kernel similar to traditional RCKL methods, but combines this with a combination of *auxiliary kernels* built from extracted features, similar to Multiple Kernel Learning (MKL) methods. The intuition behind this approach is that while human feedback is necessary to construct an appropriate kernel, some aspects of how humans view similarity among objects are likely captured in easily extractable features. If "auxiliary" kernels are built from these features, then they can be used to reduce the need of many relative comparisons. To learn the aforementioned combination, we formulate a convex optimization that adds a only small amount of computational overhead to traditional RCKL methods. Experimentally, we show that our method can learn a kernel that accurately models the relationships among objects, including relationships not explicitly given. More specifically, when given few relative comparisons, our method is shown to generalize to more held out relative comparisons than traditional RCKL methods, as well as similar state-of-the-art methods in metric learning.

The remainder of the paper is organized as follows. Section 2 provides our formal definition of RCKL. Section 3 motivates our problem. Section 4 introduces a general framework for extending RCKL methods to use auxiliary information. Section 5 overviews related work. Section 6 presents an evaluation of our method. Section 7 concludes with future work.

2   Preliminaries

The RCKL problem considered in this work is defined by a set of n objects, $\mathcal{X} = \{x_1, ..., x_n\} \subseteq \mathbb{X}$, where $\mathbb{X}$ is the set of all possible objects. Similarity information among objects is given in the form of a set $\mathcal{T}$ of triplets:

$$\mathcal{T} = \{(a, b, c) | x_a \text{ is more similar to } x_b \text{ than } x_c\} \tag{1}$$

The goal is to find a positive semidefinite (PSD) kernel matrix $\mathbf{K} \in \mathbb{R}^{n \times n}$ that satisfies the following constraints:

$$\forall_{(a,b,c) \in \mathcal{T}} : d_{\mathbf{K}}(x_a, x_b) < d_{\mathbf{K}}(x_a, x_c)$$
$$\text{Where } d_{\mathbf{K}}(x_a, x_b) = \mathbf{K}_{aa} + \mathbf{K}_{bb} - 2\mathbf{K}_{ab} \tag{2}$$

Here, $\mathbf{K}_{ab}$ is the element in the ath row and bth column of $\mathbf{K}$, representing the similarity between the ath and bth objects. The elements of $\mathbf{K}$ can be interpreted as the inner products of the objects embedded in a Reproducing Kernel Hilbert Space (RKHS), $\mathcal{H}_{\mathbf{K}}$, endowed with a mapping $\mathbf{\Phi}_{\mathbf{K}} : \mathbb{X} \to \mathcal{H}_{\mathbf{K}}$. With this interpretation $\mathbf{K}_{aa} + \mathbf{K}_{bb} - 2\mathbf{K}_{ab} = \|\mathbf{\Phi}_{\mathbf{K}}(x_a) - \mathbf{\Phi}_{\mathbf{K}}(x_b)\|_2^2$. Thus, learning a kernel matrix $\mathbf{K}$ that satisfies the constraints in (2) is equivalent to embedding the objects in a space, such that for all triplets $(a, b, c) \in \mathcal{T}$, x_a is closer to x_b than it is to x_c without explicitly learning the mapping $\mathbf{\Phi}_{\mathbf{K}}$. We say that a triplet (a, b, c) is satisfied if the corresponding constraint in (2) is satisfied.

One interpretation of (2) is that triplets define a "less than" binary relation $(R_{\mathcal{T}})$ over the set of all pairwise distances of the objects in $\mathcal{X}$ $(S_{\mathcal{X}})$. For example, if $\mathcal{X} = \{x_1, x_2, x_3\}$ and $\mathcal{T} = \{(1, 2, 3), (2, 1, 3)\}$, then $S_{\mathcal{X}} = \{(d_{\mathbf{K}}(x_1, x_2), d_{\mathbf{K}}(x_1, x_3), d_{\mathbf{K}}(x_2, x_3))\}$, and $R_{\mathcal{T}} = \{(d_{\mathbf{K}}(x_1, x_2), d_{\mathbf{K}}(x_1, x_3)), (d_{\mathbf{K}}(x_1, x_2), d_{\mathbf{K}}(x_2, x_3))\}$. With this in mind, we continue onto the next section where we discuss the RCKL problem in more depth.

3   The Impact of Few Triplets

To help motivate this work, we provide some insight into why it can be assumed, in practice, that only a limited number of triplets can be obtained from humans, and the potential impact it has on learning an accurate kernel. First, we begin by defining some properties of sets of triplets:

Definition 1. *Given a set of triplets* $\mathcal{T}$, *let* $\mathcal{T}^{\infty}$ *be the* transitive closure *of* $\mathcal{T}$

Definition 2. *Given a set of triplets* $\mathcal{T}$, *let* $\mathcal{T}^{trans} = \mathcal{T}^{\infty} \setminus \mathcal{T}$

Definition 2 simply defines $\mathcal{T}^{trans}$ as the set of triplets that can be inferred by transitivity of triplets in $\mathcal{T}$. For example, if $\mathcal{T} = \{(a, b, c), (c, a, b)\}$ then $\mathcal{T}^{trans} = \{(b, a, c)\}$.

Definition 3. *A set* $\mathcal{T}$ *of triplets is* conflicting *if* $\exists a, b, c : (a, b, c) \in \mathcal{T}^{\infty} \wedge (a, c, b) \in \mathcal{T}^{\infty}$

A set of conflicting triplets given by a source can be seen as inconsistent or contradictory in terms of how the source is comparing objects. In practice, this can be handled by prompting the source of triplets to resolve this conflict or by using simplifying methods such as in [16]. We defer to these methods in terms of how conflicts can be dealt with and consider the *non-conflicting* case. Let $\mathcal{T}^{total}$ be the set of all non-conflicting triplets that would be given by a source, if prompted with every relative comparison question over n objects. We begin by stating the following:

Theorem 1. *For* n *objects,* $|\mathcal{T}^{total}| = \frac{1}{2}(n^3 - 3n^2 + 2n)$

Theorem 1 is proven in the extended version of this work [7]. For even a small number of objects, obtaining most of $\mathcal{T}^{total}$ from humans would be too difficult or costly in many practical scenarios, especially if feedback is gained through natural use of a system, such as an online store, or if feedback requires an expert's opinion, such as in the medical domain. Let $\mathcal{T} \subseteq \mathcal{T}^{total}$ be the set of triplets actually *obtained* from a source. We say that a triplet t is *unobtained* if $t \in \mathcal{T}^{total} \setminus \mathcal{T}$. To build a model that accurately reflects the true notion of similarity given by a source of triplets, an RCKL method should learn a kernel $\mathbf{K}$ that not only satisfies the obtained triplets, but also many of the triplets in $\mathcal{T}^{total}$, including those that were unobtained. This means that given a small number of obtained triplets, an RCKL method should somehow *infer* a portion of the unobtained triplets in order to build an accurate model of similarity. In the remainder of this section we consider two possible scenarios where unobtained triplets could potentially be inferred.

For the following analysis, we assume that triplets are obtained one at a time. Also, we assume that the order in which triplets are obtained is random. This could be a reasonable assumption in applications, such as search engines, where the goal of asking relative comparison questions that are most useful in the learning process comes secondary to providing the best search results, and as such, no assumptions can be made regarding which relative comparison questions are posed to a source. Thus, the worst-case in the following analysis is with adversarial choice of both $\mathcal{T}^{total}$ and the order in which triplets are obtained. Let $\mathcal{T}_i$ be the set of triplets given by an adversary after i triplets are given. Under these assumptions, we state the following theorem:

Theorem 2. *In the worst-case,* $\forall_{i=1,...,|\mathcal{T}^{total}|} : \mathcal{T}_i^{trans} \setminus \mathcal{T}_i = \emptyset$

Theorem 2 is proven in [7]. This states that in the worst case, no unobtained triplet can be inferred by transitive relationship among obtained triplets. As a result, it may fall on the RCKL methods themselves to infer triplets. Many RCKL methods attempt to do this by assuming the learned kernel $\mathbf{K}$ has low rank. By limiting the rank of $\mathbf{K}$ to $r < n$, an RCKL method may effectively infer unobtained triplets by eliminating those that cannot be satisfied by a rank r kernel. For instance, assume an RCKL method attempts to learn a rank r kernel from $\mathcal{T}$, and assume the triplets (a,b,c) and (a,c,b) are not in $\mathcal{T}$. If the set $\mathcal{T} \cup (a,c,b)$ cannot be satisfied by a rank r kernel, but $\mathcal{T} \cup (a,b,c)$ can, then the RCKL method can only learn a kernel in which (a, b, c) is satisfied. Let $\mathcal{T}^{rank-r}$ be the set of all unobtained, not otherwise inferred, triplets that are inferred when an RCKL method enforces $\text{rank}(\mathbf{K}) \leq r$. For adversarial choice of $\mathcal{T}^{total}$ we can state the following theorem:

Theorem 3. *In the worst case,* $\forall_{t \in \mathcal{T}^{rank-r}} : t \notin \mathcal{T}^{total}$

Theorem 3 is proven in [7]. This theorem states that in the worst case, any triplet inferred by limiting the rank of $\mathbf{K}$ is not a triplet a source would give. If a large portion of $\mathcal{T}^{total}$ cannot be obtained or correctly inferred, then much of the information needed for an RCKL method to learn a kernel that reflects how a source views the relationship among objects is simply not available. The goal of this work is to use auxiliary information describing the objects to supplement obtained triplets in order to learn a kernel that can satisfy more unobtained triplets than traditional methods. In the following section we propose a novel RCKL method that extends traditional RCKL methods to use auxiliary information.

4   Learning a Kernel with Auxiliary Information

In this section we introduce a generalized framework for traditional RCKL methods. Then, we expand upon this to create two new frameworks: One that combines auxiliary kernels to satisfy triplets, and another that is a hybrid of the previous two.

4.1   Traditional RCKL

Many RCKL methods can be generalized by the following optimization problem:

$$
\begin{aligned}
\min_{\mathbf{K}} \ & E(\mathbf{K}, \mathcal{T}) + \lambda \text{trace}(\mathbf{K}) \\
\text{s.t. } & \mathbf{K} \succeq 0,
\end{aligned}
\tag{3}
$$

The first term, $E(\mathbf{K}, \mathcal{T})$, is a function of the error that the objective incurs for $\mathbf{K}$ not satisfying triplets in $\mathcal{T}$. The second term regularizes $\mathbf{K}$ by its trace. Here, the trace is used as a convex approximation of the non-convex rank function. The rank of $\mathbf{K}$ directly reflects the dimensionality of the embedding of the objects in $\mathcal{H}_{\mathbf{K}}$. A low setting of the hyperparameter λ favors a more accurate embedding, while a high value prefers a lower rank kernel. The PSD constraint ensures that $\mathbf{K}$ is a valid kernel matrix, and makes (3) a semidefinite program (SDP) over n^2 variables. For the remainder of this paper we will refer to (3) as Traditional Relative Comparison Kernel Learning (RCKL-T).

4.2   RCKL via Conic Combination

In general, if there are few triplets in $\mathcal{T}$ relative to n, there are many different RCKL solutions. Without using information regarding how the objects relate other than $\mathcal{T}$, RCKL-T methods may not generalize well to the many unobtained triplets. However, objects can often be described by features drawn from data. From these features, $A \in \mathbb{Z}^+$ auxiliary kernels $\mathbf{K}_1, ..., \mathbf{K}_A \in \mathbb{R}^{n \times n}$ can be constructed using standard kernel functions to model the relationship among objects. If one or more auxiliary kernels satisfy many triplets in $\mathcal{T}$, they may represent factors that influence how some of the unobtained triplets would have been answered. For instance, if a user considers characteristics such as color and size to be important when comparing clothing items, then kernels built from color and size may model a trend in how the user answers triplets over clothing items. If these kernels do represent a trend, then they could not only satisfy a portion of triplets in $\mathcal{T}$, but also a portion of the unobtained triplets. We wish to identify which of the given auxiliary kernels model trends in given triplets and combine them to satisfy triplets in $\mathcal{T}$. An approach popularized by multiple kernel learning methods is to combine kernels by a weighted sum:

$$\mathbf{K}' = \sum_{a=1}^{A} \mu_a \mathbf{K}_a \quad \boldsymbol{\mu} \in \mathbb{R}_{\geq 0}^{A} \tag{4}$$

$\mathbf{K}'$ is a conic combination of PSD kernels, so itself is a PSD kernel [18]. $\mathbf{K}'$ induces the mapping $\boldsymbol{\Phi}_{\mathbf{K}'} : \mathbb{X} \to \mathbb{R}^D$ [6]:

$$\boldsymbol{\Phi}_{\mathbf{K}'}(x_i) = [\sqrt{\mu_1} \boldsymbol{\Phi}_1(x_i), ..., \sqrt{\mu_A} \boldsymbol{\Phi}_A(x_i)] \tag{5}$$

Here $\boldsymbol{\Phi}_j : \mathbb{X} \to \mathbb{R}^{d_j}$ is a mapping from an object into the RKHS defined by $\mathbf{K}_j \in \mathbb{R}^{n \times n}$, and $D = \sum_{a=1}^{A} d_a$. In short, (4) induces a mapping of the objects into a feature space defined as the weighted concatenation of the individual kernels' feature spaces. Consider, then, the following optimization:

$$\min_{\boldsymbol{\mu}} E(\mathbf{K}', \mathcal{T}) + \lambda \|\boldsymbol{\mu}\|_1$$
$$\text{s.t. } \boldsymbol{\mu} \geq 0 \tag{6}$$

By learning the weight vector $\boldsymbol{\mu}$, (6) scales the individual concatenated feature spaces to emphasize those that reflect $\mathcal{T}$ well, and reduce the influence of those that do not. Because of its relationship to multiple kernel learning, we call this formulation RCKL-MKL.

Since the auxiliary kernels are fixed, regularizing them by their traces has no effect on their rank nor the rank of $\mathbf{K}'$. Instead, we choose to regularize $\boldsymbol{\mu}$ by its ℓ_1-norm, a technique first made popular for its use in the Least Absolute Shrinkage and Selection Operator (LASSO) [23]. For a proper setting of λ, this has the effect of eliminating the contribution of kernels that do not help in reducing the error by forcing their corresponding weights to be exactly zero. Note that RCKL-MKL does not learn the elements of a kernel directly, and as a result is a linear program over A variables.

By limiting the optimization to only a conic combination of the predefined auxiliary kernels, RCKL-MKL does not necessarily produce a kernel that satisfies any triplets in $\mathcal{T}$. To capture the potential generalization power of using auxiliary information while retaining the ability to satisfy triplets in $\mathcal{T}$, we propose to learn a combination of the auxiliary kernels and $\mathbf{K}_0$, a kernel similar to the one in RCKL-T whose elements are learned directly. By doing this, we force RCKL-T to prefer solutions similar to the auxiliary kernels, which could satisfy unobtained triplets. We call this hybrid approach Relative Comparison Kernel Learning with Auxiliary Kernels (RCKL-AK).

4.3 RCKL-AK

RCKL-AK learns the following kernel combination:

$$\mathbf{K}'' = \mathbf{K}_0 + \sum_{a=1}^{A} \mu_a \mathbf{K}_a \quad \boldsymbol{\mu} \in \mathbb{R}_{\geq 0}^{A}, \ \mathbf{K}_0 \succeq 0 \tag{7}$$

(7) is a conic combination of kernel matrices that induces the mapping:

$$\boldsymbol{\Phi}_{\mathbf{K}''}(x_i) = [\boldsymbol{\Phi}_0(x_i), \sqrt{\mu_1}\boldsymbol{\Phi}_1(x_i), ..., \sqrt{\mu_A}\boldsymbol{\Phi}_A(x_i)] \tag{8}$$

The intuition behind this combination is that auxiliary kernels that satisfy many triplets are emphasized by weighing them more, and $\mathbf{K}_0$, which is learned directly, can satisfy the triplets that cannot be satisfied by the conic combination of the auxiliary kernels. Consider, again, the example of a person comparing clothing items from an online store. She may compare clothes by characteristics such as color, size, and material, which are features that can be extracted and used to build the auxiliary kernels. However, other factors may influence how she compares clothes, such as designer or pattern, which may be omitted from the auxiliary kernels. In addition, she may have a personal sense of style that is impossible to be gained from features alone. $\mathbf{K}_0$, and thus features induced by the mapping $\boldsymbol{\Phi}_0$, is learned to model factors she uses to compare clothes that are omitted from the auxiliary kernels or cannot be modeled by extracted features. Using (7), we propose the following optimization:

$$\min_{\mathbf{K}_0, \boldsymbol{\mu}} E(\mathbf{K}'', \mathcal{T}) + \lambda_1 \mathrm{trace}(\mathbf{K}_0) + \lambda_2 \|\boldsymbol{\mu}\|_1$$
$$\text{s.t. } \mathbf{K}_0 \succeq 0, \ \boldsymbol{\mu} \geq 0 \tag{9}$$

This objective has two regularization terms: trace regulation on $\mathbf{K}_0$, and ℓ_1-norm regularization on $\boldsymbol{\mu}$. Increasing λ_1 limits the expressiveness of $\mathbf{K}_0$ by reducing its rank, while increasing λ_2 reduces the influence of the auxiliary kernels by forcing the elements of $\boldsymbol{\mu}$ towards zero. Thus, λ_1 and λ_2 represent a trade-off between finding a kernel that is more influenced by $\mathbf{K}_0$ and one more influenced by the auxiliary kernels. Like RCKL-T, RCKL-AK is an SDP, but with $n^2 + A$ optimization variables. For practical A, RCKL-AK can be solved with minimal additional computational overhead to RCKL-T.

One desirable property of (9) is that under certain conditions, it is a convex optimization problem:

Proposition 1. *If E is a convex function in both $\mathbf{K}_0$ and $\boldsymbol{\mu}$, then (9) is a convex optimization problem.*

Proposition 1 is proven in [7]. While Prop. 1 may seem simple, it allows us to leverage traditional RCKL methods that contain error functions that are convex in $\mathbf{K}_0$ and $\boldsymbol{\mu}$ in order to solve (9) using convex optimization techniques. Two such error functions are discussed in the following subsections.

Algorithm 1. STE-AK Projected Gradient Descent

Input:
$\quad \mathcal{X} = \{x_1, ..., x_n\}$,
$\quad \mathcal{T} = \{(a, b, c) | x_a \text{ is more similar to } x_b \text{ than } x_c\}$,
$\quad \mathbf{K}_1, ..., \mathbf{K}_A \in \mathbb{R}^{n \times n}, \lambda_1 \in \mathbb{R}^+, \lambda_2 \in \mathbb{R}^+, \eta \in \mathbb{R}^+$
Output:
$\quad \mathbf{K}'' \in \mathbb{R}^{n \times n}$

1: $t \leftarrow 0$
2: $\mathbf{K}_0^0 \leftarrow \mathbf{I}^{n \times n}$
3: $\mu_1^0, ..., \mu_A^0 \leftarrow \frac{1}{A}$
4: $\mathbf{K}'' \leftarrow \mathbf{K}_0^0 + \sum_{a=1}^A \mu_a^0 \mathbf{K}_a$
5: **repeat**
6: $\quad \mathbf{K}_0^{t+1} \leftarrow \mathbf{K}^t - \eta * (\nabla_{\mathbf{K}^t} E_{\text{STE}}(\mathbf{K}'', \mathcal{T}) + \lambda_1 * \mathbf{I}^{n \times n})$
7: $\quad \boldsymbol{\mu}^{t+1} \leftarrow \boldsymbol{\mu}^t - \eta * (\nabla_{\boldsymbol{\mu}^t} E_{\text{STE}}(\mathbf{K}'', \mathcal{T}) + \lambda_2)$
8: $\quad \mathbf{K}_0^{t+1} \leftarrow \Pi_{PSD}(\mathbf{K}_0^{t+1})$
9: $\quad \boldsymbol{\mu}^{t+1} \leftarrow \Pi_+(\boldsymbol{\mu}^{t+1})$
10: $\quad \mathbf{K}'' \leftarrow \mathbf{K}_0^{t+1} + \sum_{a=1}^A \mu_a^{t+1} \mathbf{K}_a$
11: $\quad t \leftarrow t + 1$
12: **until** convergence

STE-AK. Stochastic Triplet Embedding (STE) [14] proposes the following probability that a triplet is satisfied:

$$p_{abc}^{\mathbf{K}} = \frac{\exp(-d_{\mathbf{K}}(x_a, x_b))}{\exp(-d_{\mathbf{K}}(x_a, x_b)) + \exp(-d_{\mathbf{K}}(x_a, x_c))}$$

If this probability is high, then x_a is closer to x_b than it is to x_c. As such, we minimize the negative sum of the log-probabilities over all triplets.

$$E_{\text{STE}}(\mathbf{K}'', \mathcal{T}) = -\sum_{(a,b,c) \in \mathcal{T}} \log(p_{abc}^{\mathbf{K}''}) \tag{10}$$

With this error function we call our method STE-AK and can state the following proposition:

Proposition 2. (10) *is convex in both* $\mathbf{K}_0$ *and* $\boldsymbol{\mu}$

Proposition 2 is proven in [7]. By Props. 1 and 2, STE-AK is a convex optimization problem.

GNMDS-AK. Another potential error function is one motivated by Generalized Non-Metric Multidimensional Scaling (GNMDS) [1] which uses hinge loss:

$$E_{\text{GNMDS}}(\mathbf{K''}, \mathcal{T}) = \sum_{(a,b,c)\in\mathcal{T}} \max(0, d_{\mathbf{K''}}(x_a, x_b) - d_{\mathbf{K''}}(x_a, x_c) + 1) \qquad (11)$$

We call our method with this error function GNMDS-AK. The hinge loss ensures that only triplets that are unsatisfied by a margin of one increase the objective. GNMDS-AK is also a convex optimization problem, due to Prop. 1 and the following:

Proposition 3. (11) is convex in both $\mathbf{K}_0$ and $\boldsymbol{\mu}$.

Proposition 3 is proven in [7]. For a more rigorous comparison of RCKL methods see [14]. We propose to solve both STE-AK and GNMDS-AK via projected gradient descent algorithms, one of which (STE-AK) is outlined in the following section. The algorithm to solve GNMDS-AK is very similar to the one below, and can be found in [7].

Projected Gradient Descent for STE-AK Our method for solving STE-AK is outlined in Alg. 1. After initialization, the algorithm repeats the following steps until convergence:

1. **Line 6:** Take a gradient step for $\mathbf{K}_0$ (trace regularization included)
2. **Line 7:** Take a gradient step for $\boldsymbol{\mu}$ (ℓ_1-norm regularization included)
3. **Line 8:** Project $\mathbf{K}_0$ onto the positive semidefinite cone
4. **Line 9:** Project the elements of $\boldsymbol{\mu}$ to be non-negative
5. **Line 10:** Update $\mathbf{K''}$

Projection onto the positive semi-definite cone is done by performing eigendecomposition of the matrix $\mathbf{K}_0$, assigning all negative eigenvalues to zero, and then reassembling $\mathbf{K}_0$ from the original eigenvectors and the new eigenvalues [20]. Projection of the elements of $\boldsymbol{\mu}$ to be non-negative is simply done by assigning all negative elements to be zero. The ℓ_1-norm regularization in Alg. 1 is performed by adding $\boldsymbol{\lambda}_2 = \lambda_2 * \mathbf{1}^A$ to the gradient (Line 7). Since $\boldsymbol{\mu}$ is constrained to the non-negative orthant, the subgradient of the ℓ_1-norm function needs only to be over the non-negative orthant, thus $\boldsymbol{\lambda}_2$ is an acceptable subgradient. Moreover, since we then project the elements of $\boldsymbol{\mu}$ to be non-negative, we get the desired effect of the ℓ_1-norm regularization: the reduction of some elements to be exactly zero.

5   Related Work

RCKL-AK can be viewed as a combination of *multiple kernel learning* (MKL) and *non-metric multidimensional scaling* (NMDS). Learning a non-negative sum of kernels, as in (4), appears often in MKL literature, which is focused on finding efficient methods for learning a combination of predefined kernels for a learning task. The most widely studied problem in MKL has been Support Vector Classification [13], [17], [25], [9]. To our knowledge there has been no application of MKL techniques to the task of learning a kernel from relative comparisons.

The RCKL problem posed in Sec. 2 is a special case of the NMDS problem first formalized in [1], which in turn is a generalization of the Shepard-Kruskal NMDS problem [21]. GNMDS, STE, and Crowd Kernel Learning (CKL) [22] are all methods that can be applied to the RCKL problem. However, none of these methods consider inputs beyond relative comparisons. Our work creates a novel RCKL method that uses ideas popularized in MKL research to incorporate side information into the learning problem.

Relative comparisons have also been considered in *metric learning* [19], [3], [8]. In metric learning, the focus is on learning a distance metric over objects that can be applied to out-of-sample *objects*. This work focuses specifically on finding a kernel over given objects that generalizes well to out-of-sample (unobtained) *triplets*. In this way, the goal of metric learning methods is somewhat different than the one in this work. Two recent works propose methods to learn a Mahalanobis distance metric with multiple kernels: Metric Learning with Multiple Kernels (ML-MKL) [27] and Multiple Kernel Partial Order Embedding (MKPOE) [16]; the latter focusing exclusively on relative distance constraints similar to those in this work. The kernel learned by RCKL-AK induces a mapping that is fundamentally different than those learned by these metric learning techniques. Consider the mapping induced by one of the metric learning methods proposed in both [16] (Section 4.2) and [27] (Equation 6):

$$\mathbf{\Phi}_{\mu,\Omega}(x) = \mathbf{\Omega}\left[\sqrt{\mu_1}\mathbf{\Phi}_1(x), ..., \sqrt{\mu_A}\mathbf{\Phi}_A(x)\right] \tag{12}$$

The derivation of this mapping can be found in [7]. Here $\mathbf{\Omega} \in \mathbb{R}^{m \times D}$ produces a new feature space by transforming the feature spaces induced by the auxiliary kernels. Without $\mathbf{\Omega}$, (12) learns a mapping similar to (5). The matrix $\mathbf{\Omega}$ plays a role similar to the one $\mathbf{K}_0$ plays in RCKL-AK: it is learned to satisfy triplets that the auxiliary kernels alone cannot. Instead of linearly transforming the auxiliary kernel feature spaces, RCKL-AK implicitly learns new features that are concatenated onto the concatenated auxiliary kernel feature spaces (see (8)).

In both works, the authors propose non-convex optimizations to solve for their metrics, and, in addition, different convex relaxations. A critical issue with the convex solutions is that they employ SDPs over $n^2 * A$ (MKPOE-Full) and $n^2 * A^2$ (NR-ML-MKL) optimization variables, respectively. For moderately sized problems these methods are impractical. To resolve this issue, [16] proposes a method that imposes further diagonal structure on the learned metric, reducing the number of optimization variables to $n * A$ (MKPOE-Diag), but in the process, greatly limits the structure of the metric. RCKL-AK is a convex SDP with $n^2 + A$ optimization variables that does not impose strict structure on the learned kernel. Unfortunately, by learning the unique kernel $\mathbf{K}_0$ directly and not the mapping $\mathbf{\Phi}_0$ or a generating function of $\mathbf{K}''$, our method cannot be applied to out-of-sample objects. Data analysis that does not require the addition of out-of-sample objects can be used over that kernel. There are many unsupervised and semi-supervised techniques that fit this use case.

6   Experiments

In order to show that RCKL-AK can learn kernels from few triplets that generalize to unobtained triplets, we perform two experiments: one using synthetic data, and one using real-world data. More specifically, in both experiments, we train Stochastic Triplet Embedding (STE) and Generalized Non-metric Multidimensional Scaling (GNMDS) variants of Traditional RCKL (RCKL-T), Multiple Kernel Learning RCKL (RCKL-MKL), and RCKL with Auxiliary Kernels (RCKL-AK) as well as non-convex and convex variants of Multiple Kernel Partial Order Embedding (MKPOE), on an increasing number of training triplets, and evaluate all models on their ability to satisfy held-out triplets. For the MKPOE methods, we consider a triplet (a, b, c) to be satisfied if $d_\mathbf{M}(x_a, x_b) < d_\mathbf{M}(x_a, x_c)$, where $d_\mathbf{M}$ is the distance function defined by the metric. The STE and GNMDS implementations used are from [14], which are made publicly available on the authors' websites. The MKL and AK versions were extended from these. MKPOE implementations were provided to us by their original authors. All auxiliary kernels are normalized to unit trace, and all hyperparameters were validated via line or grid search using validation sets.

6.1   Synthetic Data

To generate synthetic data we began by randomly generating 100 points in seven, independent, two-dimensional feature spaces where both dimensions were over the interval $[0, 1]$. Then, we created seven linear kernels, $\mathbf{K}_0, ..., \mathbf{K}_6$ from these seven spaces. We combined four of the seven kernels:

$$\mathbf{K}^* = \frac{1}{2}\mathbf{K}_0 + \frac{1}{4}\mathbf{K}_1 + \frac{1}{6}\mathbf{K}_2 + \frac{1}{12}\mathbf{K}_3 \qquad (13)$$

We then used $\mathbf{K}^*$ as the ground truth to answer all possible, non-redundant triplets. Following the experimental setup in [22], we divided these triplets into 100 triplet "rounds". A round is a set of triplets where each object appears once as the head a being compared to randomly chosen objects b and c. From the pool of rounds, 20 were chosen to be the training set, 10 were chosen to be the validation set, and the remaining rounds were the test set. This was repeated ten times to create ten different trials.

Next, each point in all seven feature spaces was perturbed with randomly generated Gaussian noise. From these new spaces, we created seven new linear kernels $\hat{\mathbf{K}}_0, ..., \hat{\mathbf{K}}_6$, of which $\hat{\mathbf{K}}_1, ..., \hat{\mathbf{K}}_6$ were used as the auxiliary kernels in the experiment. Here, $\hat{\mathbf{K}}_1, ..., \hat{\mathbf{K}}_3$ are kernels that represent attributes that influence how the ground truth makes relative comparisons. $\hat{\mathbf{K}}_4, ..., \hat{\mathbf{K}}_6$ contain information that is not considered when making comparisons, and $\mathbf{K}_0$ represents intuition about the objects that was not or cannot be input as an auxiliary kernel.

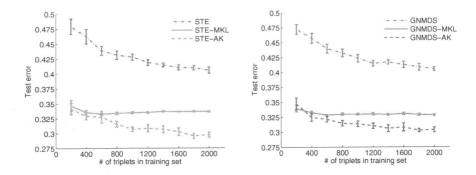

Fig. 1. Mean test error over ten trials of the synthetic data set

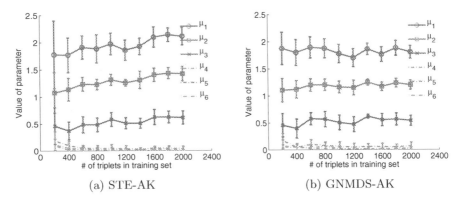

(a) STE-AK   (b) GNMDS-AK

Fig. 2. Mean values of μ on synthetic data

We wish to evaluate the performance of each method as the number of training triplets increases. As more training triplets are added, each method should generalize better to held-out triplets. To show this we performed the following experiment. For each trial, the 20 training rounds and 10 validation rounds are divided into ten subsets, each containing two training rounds and one validation round. Starting with one of the subsets, each model is trained, setting the hyperparameters through cross-validation on the validation set, and evaluated on the test set. Then, another subset is added to the training and validation sets. We repeat this process until all ten subsets are included. We evaluate the methods by the total number of unsatisfied triplets in the test set divided by the total number of triplets in the test set (test error). For all of the following figures, error bars represent a 95% confidence interval.

Discussion: Figure 1 shows the mean test error over the ten trials as a function of the number of triplets in the training set. Both RCKL-MKL methods improve performance initially, but achieve their approximate peak performance at around 600 training triplets and fail to improve as triplets are added. This supports the claim that RCKL-MKL is overly limited by only being able to combine auxiliary kernels. Both RCKL-T methods perform worse than all other methods. Without the side information provided by the auxiliary kernels, RCKL-T cannot generalize to test triplets with few training triplets.

We believe this experiment demonstrates the utility of both $\mathbf{K}_0$ and the auxiliary kernels in RCKL-AK. With very few training triplets, the RCKL-AK methods relied on the auxiliary kernels, thus the performance is similar to the RCKL-MKL methods. As triplets are added, the RCKL-AK methods used $\mathbf{K}_0$ to satisfy the triplets that a conic combination of the auxiliary kernels could not. Further evidence for this is shown by the fact that the rank of $\mathbf{K}_0$ increased as the number of training triplets increased. For example, for STE-AK, the mean rank of $\mathbf{K}_0$ was 85.6, 94.2, and 96.2 for 200, 400, and 600 triplets in the training set, respectively. In other words, the optimal settings of λ_1 and λ_2 made $\mathbf{K}_0$ more expressive as the number of triplets increased.

Ideally, the RCKL-AK methods should eliminate $\hat{\mathbf{K}}_4$, $\hat{\mathbf{K}}_5$, and $\hat{\mathbf{K}}_6$ from the model by reducing their corresponding weights μ_4, μ_5, and μ_6 to exactly zero. Figure 2 shows the values of the $\boldsymbol{\mu}$ parameter for STE-AK and GNMDS-AK as the number of triplets increase. Both RCKL-AK methods correctly identify the three auxiliary kernels from which the ground truth kernel was created by setting their corresponding weight parameters to be non-zero. In addition, they assigned weights to the kernels roughly proportional to the ground truth. The three noise kernels were assigned very low, and often zero weights. The RCKL-MKL methods learned similar values for the elements of $\boldsymbol{\mu}$ than those in Fig. 2. Since RCKL-MKL learned the relative importance of the auxiliary kernels with only few triplets, it had achieved approximately its peak performance and could not improve further with the addition of more triplets.

Figure 3a shows the same STE-AK, GNMDS-AK, and GNMDS-MKL error plots as Fig. 1, but also includes three variations of MKPOE: A non-convex formulation (MKPOE-NC), and two convex formulations (MKPOE-Full and MKPOE-Diag). All metric learning methods perform very similarly, yet worse than RCKL-MKL and RCKL-AK. We believe that the MKPOE methods must transform the auxiliary kernel space drastically to satisfy the few triplets. By doing this they lose much of the information in the auxiliary kernels that allows RCKL-MKL and RCKL-AK methods to form more general solutions.

6.2   Music Artist Data

We also performed an experiment using comparisons among popular music artists. The *aset400* dataset [4] contains 16,385 relative comparisons of 412 music artists gathered from a web survey, and [15] provides five kernels built from various features describing each artist and their music. Two of the kernels were built from text descriptions of the artists, and three were built by extracting acoustic features from songs by each artist.

The *aset400* dataset provides a challenge absent in the synthetic data: not all artists appear in the same number of triplets. In fact, some artists never appear as the head of a triplet at all. As a result, this dataset represents a setting where feedback was gathered non-uniformly amongst the objects. In light of this, instead

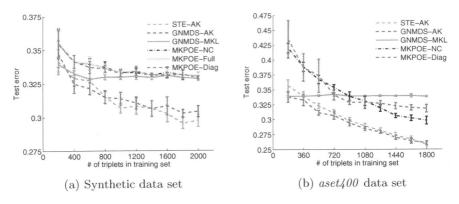

(a) Synthetic data set   (b) *aset400* data set

Fig. 3. Mean test error over ten trials

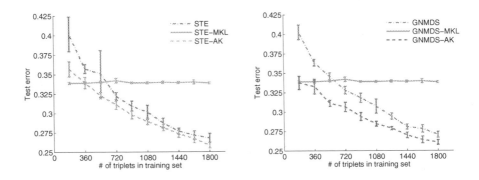

Fig. 4. Mean test error over ten trials of the *aset400* data set

of training the models in rounds of triplets, we randomly chose 2000 triplets as the development set; the rest were used as the test set. As before, we broke the development set into ten subsets. Each subset was progressively added to the working set for training, validation, and testing. Ten percent of the working set was used for validation and 90 percent was used for training. The experiment was performed ten times on different randomly chosen train/validation/test splits.

Discussion: The results, shown in Fig. 4, are similar to those for the synthetic data with a few key differences. First, the RCKL-MKL methods did not perform as well, relative to the RCKL-T methods. This could be attributed to the fact that the auxiliary kernels here did not reflect the triplets as well as those in the synthetic data. Only one kernel was consistently used in every iteration (the kernel built from artist tags). The rest were either given little weight or completely removed from the model. As with the synthetic data, with 200 and 400 training triplets the RCKL-AK methods performed as well as their respective RCKL-MKL counterparts, but as more triplets were added to the training set, the RCKL-AK methods began to perform much better. In this experiment, the RCKL-T methods became more competitive, but were outperformed significantly by RCKL-AK much of the time. This, again, could be because the auxiliary kernels were less useful than with the synthetic data.

Figure 3b compares the performance of MKPOE-NC and MKPOE-Diag on the *aset400* dataset to the RCKL-AK methods as well as GNMDS-MKL. MKPOE-Full could not be included in this experiment due to its impractically long run-time for an experiment of this size. Both MKPOE methods perform similarly; they seem to suffer greatly from the lack of meaningful auxiliary kernels, and over all experiments, have statistically significantly higher test error than both RCKL-AK methods.

7   Conclusions and Future Work

In this work we propose a method for learning a kernel from relative comparisons called Relative Comparison Kernel Learning with Auxiliary Kernels (RCKL-AK) that supplements given relative comparisons with auxiliary information. RCKL-AK is a convex SDP that can be solved by adding slight computational overhead to traditional methods and more efficiently than many metric learning alternatives. Experimentally, we show that RCKL-AK learns kernels that generalize to more out of sample relative comparisons than the aforementioned traditional and metric learning methods.

We believe the results of this work open many avenues for future research. First, while common in solving SDPs, the most time-consuming step in RCKL-AK is projecting the kernel onto the PSD cone after each gradient step. However, Low-rank Kernel Learning (LRKL) [12] is a kernel learning method that can find a solution without this costly projection. We will investigate finding a more efficient method of solving for RCKL-AK using ideas from LRKL, as well as potentially other methods. Second, we will explore methods to extend RCKL-AK to out-of-sample objects much like the work in [2] studied extensions for various popular kernel methods (LLE, Isomap, etc.). Third, the analysis done in Section 3 considered the case where an adversary was providing triplets. We will study the average, or "random" case, which may be more likely in practice. Fourth, there has been recent work that makes the case for non-sparse regularization in MKL problems [11]. While our formulation uses ℓ_1-norm, it would seem possible to generalize RCKL-AK to use ℓ_p-norms for $p > 1$, as well. We will explore the use of these norms in the RCKL-AK framework. Finally, we will explore practical applications of our method, specifically, the use of RCKL-AK for product recommendation.

Acknowledgments. We would like to thank Matthew Berger and Lee Seversky for their guidance and insightful conversations at the onset of this work.

References

1. Agarwal, S., Wills, J., Cayton, L., Lanckriet, G., Kriegman, D., Belongie, S.: Generalized non-metric multidimensional scaling. In: AISTATS (2007)
2. Bengio, Y., Paiement, J., Vincent, P., Delalleau, O., Roux, N.L., Ouimet, M.: Out-of-sample extensions for lle, isomap, mds, eigenmaps, and spectral clustering. In: NIPS 16 (2004)

3. Davis, J., Kulis, B., Jain, P., Sra, S., Dhillon, I.: Information-theoretic metric learning. In: ICML (2007)
4. Ellis, D., Whitman, B., Berenzweig, A., Lawrence, S.: The quest for ground truth in musical artist similarity. In: ISMIR (2002)
5. Filippone, M., Camastra, F., Masulli, F., Rovetta, S.: A survey of kernel and spectral methods for clustering. Pattern Recognition 41(1), 176–190 (2008)
6. Gönen, M., Alpaydın, E.: Multiple kernel learning algorithms. JMLR 12, 2211–2268 (2011)
7. Heim, E., Valizadegan, H., Hauskrecht, M.: Relative comparison kernel learning with auxiliary kernels. arXiv preprint arXiv:1309.0489 (2014)
8. Huang, K., Ying, Y., Campbell, C.: Generalized sparse metric learning with relative comparisons. KAIS 28(1), 25–45 (2011)
9. Jain, A., Vishwanathan, S., Varma, M.: Spg-gmkl: Generalized multiple kernel learning with a million kernels. In: SIGKDD (August 2012)
10. Kendall, M., Gibbons, J.: Rank Correlation Methods, 5th edn. Oxford University Press (1990)
11. Kloft, M., Blanchard, G.: On the convergence rate of lp-norm multiple kernel learning. JMLR 13, 2465–2502 (2012)
12. Kulis, B., Sustik, M., Dhillon, I.: Learning low-rank kernel matrices. In: ICML (2006)
13. Lanckriet, G., et al.: Learning the kernel matrix with semidefinite programming. JMLR 5, 27–72 (2004)
14. van der Maaten, L., Weinberger, K.: Stochastic triplet embedding. In: MLSP (2012)
15. McFee, B., Lanckriet, G.: Heterogeneous embedding for subjective artist similarity. In: ISMIR (2009)
16. McFee, B., Lanckriet, G.: Learning multi-modal similarity. JMLR 12, 491–523 (2011)
17. Rakotomamonjy, A., Bach, F., Canu, S., Grandvalet, Y.: Simplemkl. JMLR 9, 2491–2521 (2008)
18. Schölkopf, B., Smola, A.: Learning with kernels: support vector machines, regularization, optimization and beyond. The MIT Press (2002)
19. Schultz, M., Joachims, T.: Learning a distance metric from relative comparisons. In: NIPS (2004)
20. Schwertman, N., Allen, D.: Smoothing an indefinite variance-covariance matrix. JSCS 9(3), 183–194 (1979)
21. Shepard, R.: The analysis of proximities: Multidimensional scaling with an unknown distance function. i. Psychometrika 27(2), 125–140 (1962)
22. Tamuz, O., Liu, C., Belongie, S., Shamir, O., Kalai, A.: Adaptively learning the crowd kernel. In: ICML (2011)
23. Tibshirani, R.: Regression shrinkage and selection via the lasso. JRSS. Series B, pp. 267–288 (1996)
24. Valizadegan, H., Jin, R.: Generalized maximum margin clustering and unsupervised kernel learning. In: NIPS (2006)
25. Varma, M., Babu, B.: More generality in efficient multiple kernel learning. In: ICML (2009)
26. Wang, F., Zhang, C.: Label propagation through linear neighborhoods. TKDE 20(1), 55–67 (2008)
27. Wang, J., Do, H., Woznica, A., Kalousis, A.: Metric learning with multiple kernels. In: NIPS (2011)
28. Zhou, D., Bousquet, O., Lal, T., Weston, J., Schölkopf, B.: Learning with local and global consistency (2004)
29. Zhu, X., Ghahramani, Z.: Learning from labeled and unlabeled data with label propagation. Tech. rep., CMU-CALD-02-107, Carnegie Mellon University (2002)

Transductive Minimax Probability Machine

Gao Huang[1,2], Shiji Song[1,2], Zhixiang (Eddie) Xu[3], and Kilian Weinberger[3]

[1] Tsinghua National Laboratory for Information Science and Technology (TNList)
[2] Department of Automation, Tsinghua University, Beijing, 100084 China
[3] Department of Computer Science & Engineering, Washington University,
St. Louis, MO 63130 USA
`huang-g09@mails.tsinghua.edu.cn`, `shijis@mail.tsinghua.edu.cn`,
`xuzx@cse.wustl.edu`, `kilian@wustl.edu`

Abstract. The Minimax Probability Machine (MPM) is an elegant machine learning algorithm for inductive learning. It learns a classifier that minimizes an upper bound on its own generalization error. In this paper, we extend its celebrated *inductive* formulation to an equally elegant *transductive* learning algorithm. In the transductive setting, the label assignment of a test set is already optimized during training. This optimization problem is an intractable mixed-integer programming. Thus, we provide an efficient label-switching approach to solve it approximately. The resulting method scales naturally to large data sets and is very efficient to run. In comparison with nine competitive algorithms on eleven data sets, we show that the proposed Transductive MPM (TMPM) almost outperforms all the other algorithms in both accuracy *and* speed.

Keywords: minimax probability machine, transductive learning, semi-supervised learning.

1 Introduction

The Minimax Probability Machine (MPM) was originally introduced by Lanckriet et al. and provides an elegant approach to *inductive* supervised learning. It trains a discriminant classifier that directly minimizes an upper bound on its own generalization error. In particular, it first estimates the first and second moments of the conditional class distributions empirically. Building upon the celebrated work in [8] and [2], it then trains a classifier to minimize the worst case (maximal) probability of a test point falling on "the wrong side" of the decision hyperplane.

In this paper we revisit the MPM and extend it to an equally elegant *transductive* formulation. In transductive learning (TL) [20], the *unlabeled* test data is available during training and the label assignment is optimized directly while the classifier is learned. In classification settings, this results in an integer assignment problem, which is inherently NP-hard [11]. Nevertheless, many approaches have been proposed, typically based on clever heuristics including spectral graph partitioning [9], support vector machines [10], and others [23].

T. Calders et al. (Eds.): ECML PKDD 2014, Part I, LNCS 8724, pp. 579–594, 2014.

The MPM framework can incorporate the transductive label assignment problem much more naturally and efficiently than other learning paradigms, *e.g.*, Support Vector Machines (SVM) [16]. First, MPM has the attractive property that its learning complexity is independent of the size of training set provided the first and second moments of the conditional class distributions are given, enabling it handle large amount of training samples effortlessly. Second, the two steps of MPM, first estimating the conditional data distribution and then optimizing the hyperplane, give rise to an EM-like transductive algorithm [4] that is both highly efficient and accurate. As a first step, the test-label assignments are optimized (by label switching) to give rise to conditional probability distributions that maximize the *worst-case separation probability* with the current hyperplane. As a second step, the hyperplane is retrained based on the updated label assignments.

We first formulate the *Transductive Minimax Probability Machine* (TMPM) as an exact mixed-integer prgramming and then formalize our approximate solution. We show that the proposed algorithm provably increases the problem objective with every update and converges in a finite number of iterations. As both steps of TMPM are highly efficient, the algorithm scales to large data sets effortlessly. Similar to Transductive SVM (TSVM) [9], TMPM is particularly well suited for data sets with inherent cluster structure. In the presence of underlying manifold structure, Laplacian regularization [1] is often used for semi-supervised learning. We show that TMPM can be further extended to also incorporate such manifold smoothing if it is supported by the data set.

Finally, we evaluate the efficacy of TMPM on an extensive set of real world classification tasks. We compare against nine state-of-the-art learning algorithms and show that TMPM clearly outperforms most of them in *speed and accuracy* with an impressive consistency across learning tasks.

2   Related Works

Several extensions to the MPM [13] have been explored before, in particular for handling uncertain or missing data [3,17]. These works can be seen as dealing with missing information in the input space, while our work is dealing with missing information in the label space. The recent work [12] adopted the minimax probability approach for multiple instance learning. Huang et al. [7] proposes a semi-supervised learning method by combining k-nearest neighbors with a robust extension of MPM. Prior work by Nigam el al. [14] utilizes similar structure with the EM algorithm. The low density separation (LDS) semi-supervised algorithm proposed in [6] builds a fully connected graph kernel and trains a transductive SVM [9] to learn a hyperplane that traverses a low density region between clusters.

Perhaps most similar to our work is the Transductive SVM (TSVM) [9], which also iterates between label switching and classifier re-training. In contrast to TSVM, our algorithm is based on MPM, which greatly reduces the computational cost of re-training. Moreover, we further improve the efficiency drastically by adopting the idea of switching multiple class assignments at a time as the large-scale extension of TSVM [18]. Therefore, the proposed algorithm only re-trains MPM very few times. Additionally, TMPM provably optimizes a well-defined global objective function with each iteration, without heuristically adjusting it (gradually up-weight unlabeled data) during training as in TSVM [9].

3 Minimax Probability Machine

Consider the binary classification case, where we are given labeled training inputs $\{\mathbf{x}_1, \ldots, \mathbf{x}_m\} \in \mathcal{R}^d$ and their corresponding labels $\{y_1, \ldots, y_m\} \in \{-1, +1\}$. We are also provided with *unlabeled* test inputs $\{\mathbf{x}_{m+1}, \ldots, \mathbf{x}_n\} \in \mathcal{R}^d$.

Let us denote the two class-conditional data distributions as $\mathbb{P}(\mathbf{x}_+ | y = +1)$ and $\mathbb{P}(\mathbf{x}_- | y = -1)$, respectively. MPM aims to learn a hyperplane $\{\mathbf{w}, b\}$ that separates positive and negative classes with maximum probability. Since the true class-conditional distributions $\mathbb{P}$ are usually unknown, Lanckriet et al. [13] propose to maximize the worst case probability p that the two classes are separated:

$$\max_{p, \mathbf{w} \neq 0, b} \quad p$$

$$\text{s.t.} \quad \inf_{\mathbb{P} \in \mathcal{S}_+} \mathbb{P}(\mathbf{w}^\top \mathbf{x}_+ + b \geq 0| + 1) \geq p, \tag{1}$$

$$\inf_{\mathbb{P} \in \mathcal{S}_-} \mathbb{P}(\mathbf{w}^\top \mathbf{x}_- + b \leq 0| - 1) \geq p.$$

To make this optimization tracktable, the infimums are constrained to sets of distributions $\mathcal{S}_+, \mathcal{S}_-$ that match the empirical first and second order moments of the training data. Let us denote these estimated moments as mean $\hat{\mu}_+$ and covariance $\hat{\Sigma}_+$ for the positive class and $\hat{\mu}_-, \hat{\Sigma}_-$ for the negative class respectively. Then $\mathcal{S}_+$ is defined as:

$$\mathcal{S}_+ = \left\{ \mathbb{P} \ : \ \mathbb{E}[\mathbf{x}] = \hat{\mu}_+ \wedge \mathbb{E}[(\mathbf{x} - \mu_+)(\mathbf{x} - \mu_+)^\top] = \hat{\Sigma}_+ \right\}.$$

Based on the prior work in [8] and [2], Lanckriet et al. [13] show that with this restriction, the separating probability constraints in (1) can be converted into tractable inequality constraints:

$$\mathbf{w}^\top \hat{\mu}_+ + b \geq \kappa \sqrt{\mathbf{w} \hat{\Sigma}_+ \mathbf{w}}, \tag{2}$$

where $\kappa = \sqrt{p/(1-p)}$ and the inequality for the negative class is similarly defined. The above inequality can be proven with the multivariate Chebyshev inequality, and we refer readers to [13] for details.

With inequality (2), the optimization in (1) can be converted into the following unconstrained optimization problem [13], which accesses the data only through the empirical estimates of the first and second order moments,

$$\max_{\mathbf{w}} \ \kappa := \frac{\mathbf{w}^\top (\hat{\mu}_+ - \hat{\mu}_-)}{\sqrt{\mathbf{w}^\top (\hat{\Sigma}_+ + \Sigma_{\delta_+}) \mathbf{w}} + \sqrt{\mathbf{w}^\top (\hat{\Sigma}_- + \Sigma_{\delta_-}) \mathbf{w}}}. \tag{3}$$

Here, $\Sigma_{\delta_+}, \Sigma_{\delta_-}$ are regularization terms, often set to $\sigma^2 \mathbf{I}$ for some small σ, or proportional to the diagonal elements of the covariance matrix of all training inputs.

Let $\mathbf{w}^*$ denote the optimal solution to (3), then the optimal bias term can be computed as $b^* = -\mathbf{w}^{*\top}\hat{\mu}_+ + \kappa^* \sqrt{\mathbf{w}^{*\top}(\hat{\Sigma}_+ + \Sigma_{\delta_+})\mathbf{w}^*}$. The optimal separation probability corresponds to $p^* = \kappa^{*2}/(1 + \kappa^{*2})$. The optimization (3) can be solved by an iterative least-squares method [13] with a worst-case computational complexity of $O(d^3)$. If the cost of estimating μ_+, μ_-, Σ_+ and Σ_- is taken into account, then the total complexity of this approach is $O(d^3 + md^2)$.

4   Transductive Minimax Probability Machine

In this section, we introduce our transductive extension to MPM, which we refer to as TMPM. In transductive learning [20], the unlabeled test data is available during training and it is allowed to assign the labels directly as part of the learning procedure. We first formalize the TMPM optimization problem, which is NP-hard. We then introduce an efficient algorithm to find an approximate solution. Finally, we prove that our algorithm monotonically increases the objective function value and converges in a finite number of steps.

4.1   Setup

Let $\hat{\mathbf{y}} = [\hat{\mathbf{y}}^l; \hat{\mathbf{y}}^u] \in \{-1, +1\}^n$ denote the class assignment vector for both labeled $(\hat{\mathbf{y}}^l)$ and unlabeled $(\hat{\mathbf{y}}^u)$ inputs (the class assignments for labeled inputs are fixed, and thus $\hat{\mathbf{y}}^l = \mathbf{y}$). We also let $\mathcal{D}_+, \mathcal{D}_-$ denote the sets of positive and negative labeled *test* inputs respectively, given the current class assignment vector $\hat{\mathbf{y}}$.

Transductive estimation of μ and Σ.   A key aspect of TMPM is to incorporate test inputs into the empirical estimation of the mean $(\hat{\mu}_+, \hat{\mu}_-)$ and covariance $\hat{\Sigma}_+, \hat{\Sigma}_-$ of the two class distributions. Since they depend on the class assignment, we write them as functions of $\hat{\mathbf{y}}$:

$$\hat{\mu}_+(\hat{\mathbf{y}}) = \frac{1}{|\mathcal{D}_+|} \sum_i \mathbf{x}_i, \forall \mathbf{x}_i \in \mathcal{D}_+$$

$$\hat{\Sigma}_+(\hat{\mathbf{y}}) = \frac{1}{|\mathcal{D}_+|} \sum_i (\mathbf{x}_i - \hat{\mu}_+)(\mathbf{x}_i - \hat{\mu}_+)^\top, \forall \mathbf{x}_i \in \mathcal{D}_+$$

The corresponding $\hat{\mu}_-(\hat{\mathbf{y}})$ and $\hat{\Sigma}_-(\hat{\mathbf{y}})$ can be computed in a similar fashion.

Mixed-Integer Optimization.   Our goal is to find the best class assignment $\hat{\mathbf{y}}^u$ for the test inputs and the corresponding MPM classifier $\mathbf{w}$ simultaneously. The joint search over $\mathbf{w} \in \mathcal{R}^d$ and $\hat{\mathbf{y}}^u \in \{-1, +1\}^{m-n}$ leads to the following mixed-integer optimization problem:

$$\max_{\mathbf{w}, \hat{\mathbf{y}}^u} \kappa := \frac{\mathbf{w}^\top \left(\hat{\mu}_+(\hat{\mathbf{y}}) - \hat{\mu}_-(\hat{\mathbf{y}}) \right)}{\sqrt{\mathbf{w}^\top \left(\hat{\Sigma}_+(\hat{\mathbf{y}}) + \Sigma_{\delta_+} \right) \mathbf{w}} + \sqrt{\mathbf{w}^\top \left(\hat{\Sigma}_-(\hat{\mathbf{y}}) + \Sigma_{\delta_-} \right) \mathbf{w}}},$$

$$\text{s.t.} \quad \frac{1}{n} \sum_{i=1}^{n} \hat{y}_i = 2r - 1 \tag{4}$$

where $\hat{y}_i$ denotes the class assignment for input $\mathbf{x}_i$. The equally constraint enforces the fraction of positive test inputs to match $r(0 < r < 1)$, which can be set according to prior knowledge or estimated from training labels.

Note that in the above formulation, the class conditioned means and convariances are estimated from both *labeled and unlabeled* data. Therefore, the empirical moments are functions of $\hat{\mathbf{y}}^u$, which are also optimization variables.

4.2 The TMPM Algorithm

The optimization problem (4) is computationally intractable to solve globally when the number of input n data is large.

Inspired by Transductive SVM [9], we adopt the strategy of label-switching, and approximate (4) with a iterative greedy procedure. Specifically, we alternately optimize the class assignment $\hat{\mathbf{y}}^u$ and MPM classifier $\mathbf{w}$. First, we keep the MPM classifier $\mathbf{w}$ fixed and optimize the class assignment $\hat{\mathbf{y}}$ through label switching, and then we fix the class assignment, and re-optimize the MPM classifier $\mathbf{w}$.

Initialization. In order to initialize the labels $\hat{\mathbf{y}}^u$, we first train a regular MPM (3) on the labeled training data and then use the resulting classifier to obtain predictions on the test data. To ensure the label assignment is within the feasible set of (4), *i.e.*, its class ratio matches r, we assign the $r(n-m)$ test inputs with highest prediction values to class $+1$, and the rest to class -1.

Classifier Re-optimization. Once the test labels are assigned, we re-train the MPM parameters $\mathbf{w}, b$ on the full (train and test) data set with the actual training labels $\hat{\mathbf{y}}^\ell$ and the (temporarily) assigned labels $\hat{\mathbf{y}}^u$. Note that the re-optimization of $\{\mathbf{w}, b\}$ is actually optimizing κ with fixed $\hat{\mathbf{y}}$. We use the resulting classifier to generate new predictions $t_i = \mathbf{w}^\top \mathbf{x}_i + b$ for all test inputs $\mathbf{x}_i$. These predictions are not immediately used to update the tentative labels of inputs $\mathbf{x}_i$. Instead, it will be used to guide the label switching procedure in the subsequent paragraph.

Label Switching. After the MPM is retrained and the predictions t_i are computed, we re-optimize the assignments of $\hat{\mathbf{y}}^u$ through label switching as in TSVM [9]. However, unlike TSVM in which identifying a candidate pair of labels for switching is straightforward by checking the values of slack variables, TMPM has a more complicated objective function with respect to the label assignments. A naive implementation is to tentatively switch each pair of labels to see if it increases the objective value. But this leads to a worst case complexity of

---

Algorithm 1. The TMPM algorithm

1: **Input:** Labeled training inputs and their corresponding labels $\{\mathbf{x}_i, y_i\}_{i=1}^m$; Unlabeled testing inputs $\{\mathbf{x}_i\}_{i=m+1}^n$.

2: **Parameters:** λ (for regularization)

3: Compute class ratio r on $\mathbf{y}$ or using prior knowledge.

4: Initialize class assignment vector $\hat{\mathbf{y}} = [\mathbf{y}, \hat{\mathbf{y}}^u]$ by training a MPM using labeled inputs, and assign $\lceil r(n - m) \rceil$ unlabeled test inputs with highest predicting value to class $+1$, and the rest to class -1.

5: Compute $\hat{\mu}_+, \hat{\mu}_-, \hat{\Sigma}_+, \hat{\Sigma}_-$.

6: Set $\Sigma_{\delta_+} = \Sigma_{\delta_-} = \lambda diag(\mathbf{v})$, where $\mathbf{v}$ are the diagonal elements of the covariance matrix of all training inputs.

7: **while** *true* **do**

8:     $(\mathbf{w}, b) =$ Train MPM $(\hat{\mu}_+, \hat{\mu}_-, \hat{\Sigma}_+, \hat{\Sigma}_-, \Sigma_{\delta_+}, \Sigma_{\delta_-})$

9:     $(t, \bar{t}_+, \bar{t}_-) =$ Predict MPM $(\mathbf{w}, b, \mathbf{x})$

10:     **if** $\nexists(\mathbf{x}_i, \mathbf{x}_j)$ satisfying conditions in (5) **break**

11:     **while** $\exists(\mathbf{x}_i, \mathbf{x}_j)$ satisfying conditions in (5) **do**

12:         Switch the labels of $\mathbf{x}_i$ and $\mathbf{x}_j$ $(\hat{y}_i \Leftrightarrow \hat{y}_j)$.

13:         Update $\hat{\mu}_+, \hat{\mu}_-, \hat{\Sigma}_+, \hat{\Sigma}_-$.

14:         Update $\bar{t}_+ = \mathbf{w}^\top \hat{\mu}_+ + b$; $\bar{t}_- = \mathbf{w}^\top \hat{\mu}_- + b$;

15:     **end while**

16: **end while**

17: **Output:** Class assignment vector on test inputs $\hat{\mathbf{y}}^u$, MPM classifier $(\mathbf{w}, b)$.

$O(n^4 d^2)$[1], which is computationally inefficient when test data set is large. Here we introduce a method to quickly identify candidate label pairs and update the means and covariances at minimum cost, reducing the complexity of label switching at each iteration to $O(n \log(n) + nd^2)$.

Let us define the average prediction of class $+1$ as $\bar{t}_+ = \mathbf{w}^\top \hat{\mu}_+ + b$ and similarly $\bar{t}_-$. We search for pairs of *test* inputs $(\mathbf{x}_i \in \mathcal{D}_+, \mathbf{x}_j \in \mathcal{D}_-)$ who, if their labels were switched, would improve the objective κ. As we derive in the subsequent section, we can identify such pairs as inputs $\mathbf{x}_i, \mathbf{x}_j$ whose prediction values (t_i, t_j) satisfy the following two conditions:

$$1.\ t_i < t_j$$
$$2.\ \bar{t}_- \le \frac{t_i + t_j}{2} \le \bar{t}_+ \tag{5}$$

Intuitively these two conditions require to check for: 1. the input $\mathbf{x}_i$, which is currently considered *positive*, has a *lower* prediction value than input $\mathbf{x}_j$, which is assumed to be *negative*; 2. The average of the two predictions t_i and t_j lies between the two class averages. We will prove in Theorem 1 that by switching label pairs that meet these conditions, the objective strictly increases.

It is efficient to search for label pairs for switching based on the above conditions. At each iteration, we first find the n_+ positively labeled test data whose

---

[1] First, explicitly computing the objective value requires $O(nd^2)$, and there are $O(n^2)$ candidate pairs for each label switching; Second, the number of label-pairs to be switched at each iteration is proportional to n.

prediction values are smaller than the maximal prediction of $\mathcal{D}_-$, and similarly we find the n_- negatively labeled test inputs whose predictions are above the minimal prediction of $\mathcal{D}_+$. These inputs are the candidates that meet Condition 1. This step requires $O(n)$ time of computation. Second, we sort the prediction values of these candidates, which can be done in $O(n\log(n))$ time in a worst case. Finally, we iteratively match the candidate from $\mathcal{D}_+$ with the lowest prediction with the candidate from $\mathcal{D}_-$ with the highest prediction. If they meet Condition 2, we switch their labels, update the means and covariances, and eliminate both of them from the candidate set; otherwise, we remove the positively (negatively) candidate if the right (left) inequality of Condition 2 is violated. It can be verified that these eliminated instance will never meet the switching conditions in this iteration. This procedure has a worst case complexity of $O(nd^2)$, if we update the empirical moments using the rules in the following lemma.

Lemma 1. *Let* $\{\hat{\mu}'_+, \hat{\Sigma}'_+\}$ *denote the estimated mean and covariance of positive class after switching two instances* $\mathbf{x}_i \in \mathcal{D}_+$ *and* $\mathbf{x}_j \in \mathcal{D}_-$. *Then we have*

$$\hat{\mu}'_+ = \hat{\mu}_+ + \frac{1}{|\mathcal{D}_+|}(\mathbf{x}_j - \mathbf{x}_i), \tag{6}$$

$$\hat{\Sigma}'_+ = \hat{\Sigma}_+ + \frac{|\mathcal{D}_+| - 1}{|\mathcal{D}_+|^2}(\mathbf{x}_j - \mathbf{x}_i)(\mathbf{x}_j - \mathbf{x}_i)^\top + \tag{7}$$

$$\frac{1}{|\mathcal{D}_+|}(\mathbf{x}_i - \mu_+)(\mathbf{x}_j - \mathbf{x}_i)^\top + \frac{1}{|\mathcal{D}_+|}(\mathbf{x}_j - \mathbf{x}_i)(\mathbf{x}_i - \mu_+)^\top,$$

where $\{\hat{\mu}_+, \hat{\Sigma}_+\}$ *are the estimated mean and covariance before label switching.*

Remark 1. Naturally, we can update the mean and covariance of the negative class in a similar fashion. The above expressions enable us to re-estimate the means and covariances after each label switching at a minimum cost, with a complexity of $O(d^2)$.

Termination. We keep iterating between label switching and MPM re-training until no more pairs can be found that satisfy (5). The TMPM algorithm is summarized in pseudo-code in Algorithm 1.

Remark 2. The overall time complexity of Algorithm 1 is $O(L(d^3 + n\log(n) + nd^2))$, where L is the number of outer loop executions. Typically, we have $L \approx 10$. The term $O(d^3)$ results from re-training the MPM, following the method proposed in [13]. The terms $O(n\log(n) + nd^2)$ capture the complexity of finding eligible pairs for switching and updating empirical moments.

4.3   Algorithm Analysis

In this subsection, we show that Algorithm 1 terminates in a finite number of iterations.

Firstly, we prove that each MPM label switching strictly increases κ in (4). We formalize this guarantee as Theorem 1.

Theorem 1. *If two test inputs* $\mathbf{x}_i \in \mathcal{D}_+$, $\mathbf{x}_j \in \mathcal{D}_-$ *and their corresponding predictions* t_i, t_j *satisfy the switching conditions in (5), then by switching the assigned labels of* $\mathbf{x}_i$ *and* $\mathbf{x}_j$ *(*$\hat{y}_i \Leftrightarrow \hat{y}_j$*), the objective function value* κ *in (4) strictly increases.*

Proof: We first show that the numerator of (4) increases after the label switching. This follows from plugging in the mean and covariance updates stated above in Eqs. (6) and (7).

$$
\mathbf{w}^\top (\hat{\mu}'_+ - \hat{\mu}'_-)
$$
$$
= \mathbf{w}^\top (\hat{\mu}_+ - \hat{\mu}_-) + (\frac{1}{|\mathcal{D}_+|} + \frac{1}{|\mathcal{D}_-|})(t_j - t_i)
$$
$$
> \mathbf{w}^\top (\hat{\mu}_+ - \hat{\mu}_-).
$$

The last inequality holds because of the condition $t_i < t_j$.

As a second step, we show that the denominator of (4) decreases after switching the labels. Again, by using the update rules in (6) and (7), we have

$$
\mathbf{w}^\top \hat{\Sigma}'_+ \mathbf{w} = \mathbf{w}^\top \hat{\Sigma}_+ \mathbf{w} + \frac{|\mathcal{D}_+| - 1}{|\mathcal{D}_+|^2} (\mathbf{w}^\top (\mathbf{x}_j - \mathbf{x}_i))^2
$$
$$
+ \frac{2}{|\mathcal{D}_+|} \mathbf{w}^\top (\mathbf{x}_i - \mu_+)(\mathbf{x}_j - \mathbf{x}_i)^\top \mathbf{w}
$$
$$
= \mathbf{w}^\top \hat{\Sigma}_+ \mathbf{w} + \frac{1}{|\mathcal{D}_+|} \mathbf{w}^\top (\mathbf{x}_j - \mathbf{x}_i) \mathbf{w}^\top (\mathbf{x}_j + \mathbf{x}_i - 2\hat{\mu}_+)
$$
$$
- \frac{1}{|\mathcal{D}_+|^2} (\mathbf{w}^\top (\mathbf{x}_j - \mathbf{x}_i))^2
$$
$$
= \mathbf{w}^\top \hat{\Sigma}_+ \mathbf{w} + \frac{(t_j - t_i)(t_i + t_j - 2\bar{t}_+)}{|\mathcal{D}_+|} - \frac{(t_j - t_i)^2}{|\mathcal{D}_+|^2}
$$
$$
< \mathbf{w}^\top \hat{\Sigma}_+ \mathbf{w},
$$

where the last inequality holds because of the switching conditions $(t_i + t_j)/2 \le \bar{t}_+$ and $t_i < t_j$ given in Theorem 1.

Following a similar line of reasoning, we can show that

$$
\mathbf{w}^\top \hat{\Sigma}'_- \mathbf{w} < \mathbf{w}^\top \hat{\Sigma}_- \mathbf{w}. \tag{8}
$$

Further, notice that Σ_{δ_+} and Σ_{δ_-} are independent of class assignment vector $\hat{\mathbf{y}}$, we have the denominator of (4) decreases after label switching.

As the enumerator of κ increases and its denominator decreases, and as the label switching preserves the class ratio of of $\hat{\mathbf{y}}^u$, it follows that the objective κ in (4) strictly increases with each label switching.   ∎

Theorem 2. *Algorithm 1 terminates after a finite number of iterations.*

Proof: Since the label switching strictly increases the objective according to Theorem 1, and the re-training of MPM never decrease the objective, the alternating optimization method in Algorithm 1 strictly improves the value of κ at each iteration. Therefore, the outer loop cannot repeat label assignments (Otherwise, we will solve a same MPM model in two differently iterations, which will yield a same value of κ. This means the objective is not strictly increasing over these two iterations, leading to a contradictory). Since the number of label assignments is finite, the algorithm must terminate after a finite number of iterations. ∎

Note that Theorem 2 is based on a worst case scenario analysis. In practice, each iteration switches many labels and TMPM terminates after a small number(≈ 10) of iterations.

4.4 TMPM with Manifold Regularization

Transductive learning or semi-supervised learning algorithms can often assist classifiers by revealing underlying structure in the data distribution. A successful approach is manifold regularization, introduced by Belkin et al. [1]. Here, a proximity graph is created and the classifier is regularized to make similar predictions on similar inputs. This approach is typically successful if the data distribution obeys some intrinsically low dimensional manifold structure. TMPM can also incorporate manifold regularization naturally. We replace the regularization covariance matrices $\Sigma_{\delta_+}, \Sigma_{\delta_-}$ in the TMPM objective function (4) with a Laplacian regularization term [21]. More formally, we set $\Sigma_{\delta_+} = \Sigma_{\delta_-} = \lambda \mathbf{X}^\top \mathbf{L} \mathbf{X}$, where $\mathbf{X}^\top = [\mathbf{x}_1, \ldots \mathbf{x}_n]$ is a matrix containing both training and test inputs and $\mathbf{L} \in \mathbb{R}^{n \times n}$ denotes the *graph Laplacian* constructed from $\mathbf{x}_1, \ldots, \mathbf{x}_n$. Finally, λ denotes a regularization tradeoff parameter. We refer to this manifold regularization extension as TMPM^{mr}.

5   Results

We evaluate TMPM on a wide variety of synthetic and real world data sets. Our implementation is implemented in MATLAB^{TM}, and is executed on an Intel i7 Quad Core CPU 3.20GHz machine with 32GB RAM.

5.1   Toy Example

We use a toy data set to visualize the transductive learning process of TMPM. The toy data set is a binary classification problem, where each class contains 200 inputs generated from a 2-dimensional Gaussian distribution. We randomly reveal one label from each class to create a training set with two instances. The remaining inputs are unlabeled and constitute the test data, see Figure 1 (upper left panel). On this data, a linear SVM achieves 0.78 test accuracy.

Figure 1 visualizes the decision boundary and label assignments of each iteration of TMPM until termination. Inputs that will be switched in this iteration are highlighted with circles. The inner ellipsoids represent the covariances of the data in $\mathcal{D}_+, \mathcal{D}_-$, and the outer ellipsoids are κ times larger, so that the decision plane is tangent to both of the outer ellipsoids. Since $p = \kappa^2/(1 + \kappa^2)$, the larger κ, the higher the minimax probability p of separating the two subsets. The value

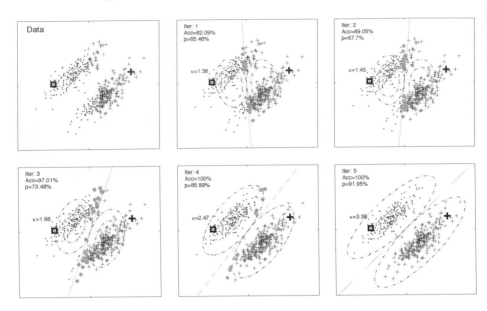

Fig. 1. The TMPM algorithm visualized on a toy data set. Only two inputs are initially labeled (big square and cross, *top left*). Small dots and crosses indicate the labels assigned by TMPM to the test data. Inputs highlighted with circles are those to be switched in a particular current iteration. The inner ellipsoids visualize the covariances of two classes, and the outer ellipsoids represent κ times of the covariances.

of p and the classification accuracy are indicated in the top left of each frame, both of which increase monotonically until the algorithm terminates after the 5th iteration, when no more pairs of inputs satisfy the conditions for switching. As a byproduct, we obtain a lower bound $p = 91.95\%$ probability of separating these two classes for additional unseen data. We also obtain the estimated means and covariances for the two classes:

$$\hat{\mu}_+ = (1.01, -1.00), \quad \hat{\mu}_- = (-1.00, 0.94),$$

$$\hat{\Sigma}_+ = \begin{bmatrix} 0.94 & 0.81 \\ 0.81 & 1.04 \end{bmatrix}, \hat{\Sigma}_- = \begin{bmatrix} 0.89 & 0.72 \\ 0.72 & 0.87 \end{bmatrix},$$

which are very close to the true means and covariances of the two Gaussian distributions that generate the data:

$$\mu_+ = (1, -1), \quad \mu_- = (-1, 1), \quad \Sigma_+ = \Sigma_- = \begin{bmatrix} 1.0 & 0.8 \\ 0.8 & 1.0 \end{bmatrix}.$$

5.2 Transductive Learning Results

We evaluate TMPM on several real-world data sets, and compare against state-of-the-art transductive/semi-supervised learning algorithms. The characteristics of these data sets are summarized in the second and third rows of Table 1.

Experiment Setup. For each data set, we randomly select 10 samples as labeled set, another 10 samples as validation set, and the rest as unlabeled test set. The prediction error rate on the unlabeled test set is used as the evaluation criteria, and we report the average results over 20 runs with randomly selected labeled/validation/unlabeled data. The linear TMPM is used for all the experiments (for high dimensional data where $d > n$, we use *linear kernel* TMPM given in the Appendix). The only hyper-parameter in linear TMPM is the regularization coefficient λ, which is selected from the candidate set $[10^{-4}, 10^{-3}, \ldots, 10^{4}]$ based on the performance on the validation set.

Datasets. The *g50c, g241c, digit1, text* data sets are obtained form Olivier Chapelle's Semi-Supervised Learning benchmark data set collection[2] [5]. The data set *breast, australian* (Australian Credit Approval), *pcmac* (corresponding to two classes of the 20 newsgroups data set), *adults, kddcup,* are taken from the UCI Machine Learning Repository[3].

Baselines. First, the SVM and MPM are trained using only the labeled data, and we report the better results of a linear version and an RBF kernel version of these algorithms. Other baselines include the Transductive SVM (TSVM^{light}) [9], the TSVM with multiple switching strategy (TSVM^{ms}) [18], the semi-supervised EM with Gaussian distribution assumption (for all data sets except *pcmac* and *text*) and with multinomial distribution assumption (for *pcmac* and *text*), the low density separation algorithm (LDS) [6], and the squared-loss mutual information regularization (SMIR) [15]. For TSVM^{light} and TSVM^{ms}, we report both linear and RBF kernel version results. The tradeoff parameter C in these algorithms are selected from the set $[10^{-4}, 10^{-3}, \ldots, 10^{4}]$, and the kernel width is selected from the set $[2^{-5}, 2^{-4}, \ldots, 2^{1}]$ times the average pairwise distance of the training data. The kernel type is indicated in sub-scripts (*e.g.* TSVM$^{light}_{linear}$ and TSVM$^{light}_{rbf}$). For EM^{gauss}, a ridge λI is added to the covariance matrix when computing posterior probability, where λ is selected from the set $[10^{-6}, 10^{-5}, \ldots, 10^{2}]$. For LDS and SMIR, we follow the suggestions in [6] and [15] to create a candidate hyperparameter set and select the best value based on the validation set.

Prediction Accuracy. The experimental results are summarized in Table 1. For each data set, the best performance up to statistical significance is highlighted in **bold**. Standard deviations are provided inside parenthesis. If an algorithm is not able to scale to a particular data set (or fails to converge), it is indicated with N/A.

A few trends can be observed: 1. TMPM obtains the best result (up to statistical significance) on almost all data sets; 2. TMPM's standard deviation of error is always the lowest among all TL/SSL algorithms over all data sets except *pcmac* and *text*, demonstrating that TMPM is insensitive to the initial predictions on unlabeled test data; 3. Generally, non-linear classifiers do not outperform

---

[2] http://olivier.chapelle.cc/ssl-book/benchmarks.html
[3] http://archive.ics.uci.edu/ml/datasets.html

Table 1. Data set statistics and test error rates (in %) on nine benchmark data sets, comparing TMPM against various state-of-the-art algorithms. N/A indicates that an algorithm fails to scale to that specific data set. Best results up to statistical significance are highlighted in **bold**.

Statistics	g50c	g241c	breast	australian	digit1	pcmac	text	adults	kddcup
# features	50	241	10	14	241	3289	11960	123	122
# inputs	550	1500	683	690	1500	1943	1500	32541	500000
Algorithm	Test-error (%)								
SVM	16.7 ± 4.2	36.9 ± 2.4	3.2 ± 0.3	19.3 ± 5.9	20.0 ± 5.0	38.4 ± 4.9	35.5 ± 4.4	26.8 ± 5.6	5.1 ± 1.7
MPM	18.1 ± 3.9	37.4 ± 2.5	3.3 ± 0.2	19.6 ± 3.2	25.1 ± 3.7	39.7 ± 5.6	38.9 ± 4.0	24.2 ± 4.8	1.6 ± 1.0
$\text{TSVM}_{linear}^{light}$	$\mathbf{4.8 \pm 0.7}$	17.5 ± 3.3	$\mathbf{2.8 \pm 0.5}$	16.5 ± 6.7	16.7 ± 2.4	40.9 ± 9.8	$\mathbf{27.0 \pm 4.4}$	N / A	N / A
$\text{TSVM}_{rbf}^{light}$	5.0 ± 0.5	14.3 ± 1.2	3.5 ± 0.7	16.7 ± 4.9	15.6 ± 2.0	41.6 ± 10.4	29.5 ± 6.2	N / A	N / A
$\text{TSVM}_{linear}^{ms}$	6.0 ± 1.3	17.1 ± 4.4	3.2 ± 0.1	16.3 ± 4.1	15.7 ± 4.0	38.8 ± 8.9	$\mathbf{26.3 \pm 3.7}$	22.3 ± 4.0	N / A
TSVM_{rbf}^{ms}	4.9 ± 0.6	15.1 ± 5.5	3.3 ± 0.2	17.0 ± 5.5	15.6 ± 3.3	37.8 ± 4.5	27.4 ± 3.8	N / A	N / A
EM	9.3 ± 4.8	37.7 ± 10.0	8.6 ± 4.3	19.9 ± 5.9	11.3 ± 9.1	$\mathbf{33.1 \pm 9.8}$	33.7 ± 0.3	24.2 ± 8.5	1.0 ± 0.4
LDS	9.2 ± 4.9	15.7 ± 6.7	4.3 ± 0.7	16.8 ± 2.8	13.5 ± 6.9	38.4 ± 7.5	27.4 ± 3.8	N / A	N / A
SMIR	17.1 ± 4.4	36.8 ± 3.3	4.0 ± 1.2	19.9 ± 4.0	12.5 ± 5.0	44.7 ± 5.0	35.2 ± 4.6	N / A	N / A
TMPM	$\mathbf{4.8 \pm 0.4}$	$\mathbf{13.1 \pm 0.4}$	$\mathbf{2.8 \pm 0.1}$	15.2 ± 0.8	9.0 ± 1.8	$\mathbf{30.5 \pm 8.0}$	35.2 ± 5.2	$\mathbf{20.5 \pm 2.8}$	$\mathbf{0.6 \pm 0.1}$

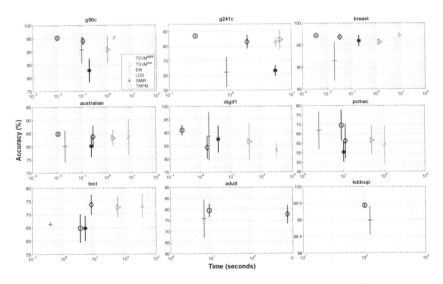

Fig. 2. Classification accuracy (in %) and computational time (in seconds) of different TL algorithms on data sets described in Table 1. Error bar indicates the standard deviation.

linear classifiers, which is not unusual in the TL/SSL setting due to the typically small training set sizes.

Analysis. We explain the strong performance of TMPM in parts on the underlying MPM framework. Compared to TSVM^{light} and TSVM^{ms}, TMPM only yields significantly worse performance on the *text* data, but outperforms or matches both on all other problems.

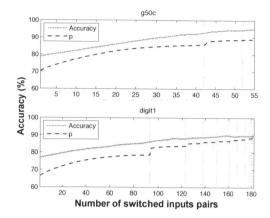

Fig. 3. Transductive learning curve on *g50c* and *digit1*. Each dotted vertical line represents a $(\mathbf{w}, b)$ re-optimization step.

Table 2. Test error rates on data sets with manifold structure. Best results (up to statistical significance) are highlighted in **bold**.

	d	n	SVM	LapSVM	LapMPM	TMPM^{mr}
coil20	1024	576	13.3 ± 4.2	**10.5 + 3.9**	12.9 ± 3.4	**11.7 ± 3.9**
uspst	256	803	16.5 ± 2.3	**14.2 ± 2.4**	16.6 ± 2.4	**13.6 ± 3.4**

Figure 3 shows the transduction accuracy and the value of p with respect to the number of switched label pairs on two representative data sets. Each interval between two vertical lines corresponds to an execution of one inner loop of Algorithm 1. As predicted, p strictly increases after each pair of labels is switched or $\mathbf{w}, b$ are re-optimized. Not surprisingly, the transduction accuracy increases steadily as p increases. We can also observe that the number of switched labels per iteration decrease (roughly) exponentially, indicating TMPM converges quickly in practice.

Speed. In Figure 2 we compare the training time (plotted on a logarithmic scale) and classification accuracy of the above TL/SSL algorithms (we omit TSVM$_{rbf}^{light}$ and TSVM$_{rbf}^{ms}$ here since they are significantly slower than their linear version). The TMPM is the fastest among all the six algorithms on 6/9 of the data sets. It is only slower than the EM^{multi} algorithm on two high dimensional text data (*pcmac*, *text*) sets and *adult* (although the differences are sometimes no more than a few seconds). The TMPM is 1 to 4 orders of magnitude faster than the TSVM^{light}, and is also significantly faster than the TSVM^{ms}. The speed advantage of TMPM over other algorithms is even larger as the unlabeled data increases.

5.3   Performances on Manifold Data Sets

We also evaluate TMPM with manifold regularization (TMPM^{mr}) on two well known data sets considered to have manifold structures (from the UCI data set repository). For each data set, we select 50 inputs as labeled set, 50 as hold-out, and leave the rest as unlabeled test set.

The kernel TMPM with RBF kernel is adopted here, and the kernel width is selected from the set $[2^{-5}, 2^{-4}, \ldots, 2^1]$ times the average pairwise distance of the training data. For comparison, we evaluate two representative manifold-based SSL algorithms, LapSVM [1] and LapMPM [22]. For all algorithms, the same graph *Laplacian* is used, whose parameter setting can be found in [19]. The prediction error rates (in %) on unlabeled test set are summarized in Table 2. The results show that TMPM^{mr} is also competitive with LapSVM and LapMPM on these manifold data sets.

6   Conclusion

In this paper, we propose a novel transductive learning algorithm (TMPM) based on the minimax probability machine. Although TL learning is not new, the TMPM framework provides a fresh and exciting approach to transductive learning. The underlying assumption is that the optimal decision hyperplane should lead to a maximum worst-case separation probability between different data classes. We convert this search problem into a mixed-integer programming, and propose an efficient algorithm to approximate it greedily.

We show that TMPM converges in a finite number of iterations and has a low computational complexity in the number of unlabeled inputs. Experimental results demonstrate that TMPM is promising in generalization performance and scales naturally to large data sets.

Acknowledgments. G. Huang and S. Song were supported by the National Natural Science Foundation of China under Grant 61273233, the Research Fund for the Doctoral Program of Higher Education under Grant 20120002110035 and Grant 20130002130010, the National Key Technology Research and Development Program under Grant 2012BAF01B03, the Project of China Ocean Association under Grant DY125-25-02, and the Tsinghua University Initiative Scientific Research Program under Grant 2011THZ07132. K.Q. Weinberger and Z. Xu were supported by NSF IIS-1149882 and IIS-1137211.

References

1. Belkin, M., Niyogi, P., Sindhwani, V.: Manifold regularization: A geometric framework for learning from labeled and unlabeled examples. The Journal of Machine Learning Research 7, 2399–2434 (2006)
2. Bertsimas, D., Sethuraman, J.: Moment problems and semidefinite optimization. In: Handbook of Semidefinite Programming, pp. 469–509. Springer (2000)

3. Bhattacharyya, C., Pannagadatta, K., Smola, A.J.: A second order cone programming formulation for classifying missing data. In: Neural Information Processing Systems (NIPS), pp. 153–160 (2005)

4. Bishop, C.M., Nasrabadi, N.M.: Pattern recognition and machine learning, vol. 1. Springer, New York (2006)

5. Chapelle, O., Schölkopf, B., Zien, A., et al.: Semi-supervised learning, vol. 2. MIT Press, Cambridge (2006)

6. Chapelle, O., Zien, A.: Semi-supervised classification by low density separation. In: Proceedings of the 10th International Conference on Artificial Intelligence and Statistics, pp. 57–64 (2005)

7. Huang, G., Song, S., Gupta, J.N.D., Wu, C.: A second order cone programming approach for semi-supervised learning. Pattern Recognition 46(12), 3548–3558 (2013)

8. Isii, K.: On sharpness of tchebycheff-type inequalities. Annals of the Institute of Statistical Mathematics 14(1), 185–197 (1962)

9. Joachims, T.: Transductive inference for text classification using support vector machines. In: ICML, vol. 99, pp. 200–209 (1999)

10. Joachims, T., et al.: Transductive learning via spectral graph partitioning. In: ICML, vol. 3, pp. 290–297 (2003)

11. Korte, B.B.H., Vygen, J.: Combinatorial optimization, vol. 21. Springer (2012)

12. Krummenacher, G., Ong, C.S., Buhmann, J.: Ellipsoidal multiple instance learning. In: Proceedings of the 30th International Conference on Machine Learning, pp. 73–81 (2013)

13. Lanckriet, G.R., Ghaoui, L.E., Bhattacharyya, C., Jordan, M.I.: A robust minimax approach to classification. The Journal of Machine Learning Research 3, 555–582 (2003)

14. Nigam, K., McCallum, A.K., Thrun, S., Mitchell, T.: Text classification from labeled and unlabeled documents using em. Machine Learning 39(2-3), 103–134 (2000)

15. Niu, G., Jitkrittum, W., Dai, B., Hachiya, H., Sugiyama, M.: Squared-loss mutual information regularization: A novel information-theoretic approach to semi-supervised learning. In: Proceedings of the 30th International Conference on Machine Learning, pp. 10–18 (2013)

16. Schölkopf, B., Smola, A.J.: Learning with kernels. MIT Press (2002)

17. Shivaswamy, P.K., Bhattacharyya, C., Smola, A.J.: Second order cone programming approaches for handling missing and uncertain data. The Journal of Machine Learning Research 7, 1283–1314 (2006)

18. Sindhwani, V., Keerthi, S.S.: Large scale semi-supervised linear svms. In: Proceedings of the 29th annual international ACM SIGIR Conference on Research and Development in Information Retrieval, pp. 477–484. ACM (2006)

19. Sindhwani, V., Niyogi, P., Belkin, M.: Beyond the point cloud: from transductive to semi-supervised learning. In: ICML 2005, pp. 824–831. ACM (2005)

20. Vapnik, V.N.: Statistical learning theory (1998)

21. Weinberger, K.Q., Sha, F., Zhu, Q., Saul, L.K.: Graph laplacian regularization for large-scale semidefinite programming. In: NIPS, pp. 1489–1496 (2006)

22. Yoshiyama, K., Sakurai, A.: Manifold-regularized minimax probability machine. In: Schwenker, F., Trentin, E. (eds.) PSL 2011. LNCS, vol. 7081, pp. 42–51. Springer, Heidelberg (2012)

23. Zhu, X.: Semi-supervised learning literature survey. Computer Science, University of Wisconsin-Madison 2, 3 (2006)

Appendix: Kernel TMPM

Suppose that the labeled training data are re-arranged so that the first m_+ inputs belong to class $+1$, and the rest m_- inputs belong to class -1. Let $\mathbf{K} \in \mathbb{R}^{m \times m}$ ($m = m_+ + m_-$) denotes the kernel matrix, whose first m_+ rows and last m_- rows are denoted by $\mathbf{K}_+$ and $\mathbf{K}_-$, respectively.

The kernel MPM is formulated as

$$\max_{\boldsymbol{\theta} \in \mathbb{R}^m} \frac{\boldsymbol{\theta}^\top (\boldsymbol{\eta}_+ - \boldsymbol{\eta}_-)}{\sqrt{\boldsymbol{\theta}^\top (\boldsymbol{\Phi}_+^\top \boldsymbol{\Phi}_+ + \lambda_+ \mathbf{K}) \boldsymbol{\theta}} + \sqrt{\boldsymbol{\theta}^\top (\boldsymbol{\Phi}_-^\top \boldsymbol{\Phi}_- + \lambda_- \mathbf{K}) \boldsymbol{\theta}}}$$

where

$$\boldsymbol{\Phi}_+ = (\mathbf{K}_+ - \mathbf{1}_{m_+} \boldsymbol{\eta}_+^\top)/\sqrt{m_+}$$

$$\boldsymbol{\Phi}_y = (\mathbf{K}_- - \mathbf{1}_{m_-} \boldsymbol{\eta}_-^\top)/\sqrt{m_-}$$

$$[\boldsymbol{\eta}_+]_i = \sum_{j=1}^{m_+} \mathbf{K}_{j,i},$$

$$[\boldsymbol{\eta}_-]_i = \sum_{j=m_++1}^{m} \mathbf{K}_{j,i},$$

and $\mathbf{1}_n$ denotes an all one vector of dimension n.

Based on the kernel MPM, we give the kernel TMPM in Algorithm 2.

Algorithm 2. The Kernel TMPM algorithm

1: **Input:** Labeled data $\{\mathbf{x}_i, y_i\}_{i=1}^m$; Unlabeled test inputs $\{\mathbf{x}_i\}_{i=m+1}^n$;
2: **Parameters:** λ (for regularization)
3: Compute class ratio r on $\mathbf{y}$ or using prior knowledge.
4: Initialize class assignment vector $\hat{\mathbf{y}} = [\mathbf{y}, \hat{\mathbf{y}}^u]$ by training a kernel MPM using labeled inputs, and assign $\lceil r(n-m) \rceil$ unlabeled test inputs with highest predicting value to class $+1$, and the rest to class -1.
5: Compute $\boldsymbol{\eta}_+, \boldsymbol{\eta}_-, \mathbf{K}_+$ and $\mathbf{K}_-$.
6: **while** *true* **do**
7:    $(\boldsymbol{\theta}, b) =$ Train kernel MPM $(\boldsymbol{\eta}_+, \boldsymbol{\eta}_-, \mathbf{K}_+, \mathbf{K}_-, \lambda)$
8:    $\bar{t}_+ = \boldsymbol{\theta}^\top \boldsymbol{\eta}_+ + b, \quad \bar{t}_- = \boldsymbol{\theta}^\top \boldsymbol{\eta}_- + b$
9:    $t =$ Predict kernel MPM $(\boldsymbol{\theta}, b, \mathbf{x})$
10:    **if** $\nexists (t_i, t_j)$ satisfying the conditions in (5) **break**
11:    **while** $\exists (t_i, t_j)$ satisfying conditions in (5) **do**
12:       Switch the labels of $\mathbf{x}_i$ and $\mathbf{x}_j$ ($\hat{y}_i \Leftrightarrow \hat{y}_j$).
13:       Update $\boldsymbol{\eta}_+, \boldsymbol{\eta}_-, \mathbf{K}_+$ and $\mathbf{K}_-$.
14:       $\bar{s} = \boldsymbol{\theta}^\top \boldsymbol{\eta}_+ - b; \quad \bar{t} = \boldsymbol{\theta}^\top \boldsymbol{\eta}_- - b;$
15:    **end while**
16: **end while**
17: **Output:** Class assignment vector on test inputs $\hat{\mathbf{y}}^u$, MPM classifier $(\boldsymbol{\theta}, b)$.

Covariate-Correlated Lasso for Feature Selection

Bo Jiang[1], Chris Ding[1,2], and Bin Luo[1]

[1] School of Computer Science and Technology, Anhui University, Hefei, 230601, China
[2] CSE Department, University of Texas at Arlington, Arlington, TX 76019, USA
{jiangbo,luobin}@ahu.edu.cn, chqding@uta.edu

Abstract. Lasso-type variable selection has been increasingly adopted in many applications. In this paper, we propose a covariate-correlated Lasso that selects the covariates correlated more strongly with the response variable. We propose an efficient algorithm to solve this Lasso-type optimization and prove its convergence. Experiments on DNA gene expression data sets show that the selected covariates correlate more strongly with the response variable, and the residual values are decreased, indicating better covariate selection. The selected covariates lead to better classification performance.

1 Introduction

In many regression applications, there are too many unrelated predictors which may hide the relationship between the response and the most related predictors. A common way to resolve this problem is variable selection, that is to select a subset of the most representative or discriminative predictors from the input predictor set. In machine learning and data mining tasks, the main challenge of variable selection is to select a set of predictors, as small as possible, that help the classifier to accurately classify the learning examples. Various kinds of variable selection methods have been proposed to tackle the issue of high dimensionality. One major type of variable selection methods is to use the filter methods, such as: t-test, F-statistic [5], ReliefF [10], mRMR [12] and mutual information [13]. These methods are usually independent of classifiers. Another wrapper-type of variable selection methods is to take classifiers to evaluate subsets of predictors [9]. In addition, some stochastic search techniques have also been used for variable selection [16].

Recently, sparsity regularization receives increasing in variable selection. The well known Lasso (Least Absolute Shrinkage and Selection Operator) is a penalized least square method with ℓ_1-regularization, which is used to shrink/suppress variables to achieve variable selection [3,14,19,17,18]. However, ℓ_1-minimization algorithm is not stable compared with ℓ_2-minimization. Elastic Net added ℓ_2-regularization in Lasso to make the regression coefficients more stable [19]. Group Lasso was proposed where the covariates are assumed to be clustered in groups, and the sum of Euclidean norms of the loadings in each group is used [17]. Supervised Group Lasso performed K-means clustering before Group Lasso [11]. From the covariate point of view, the aim of traditional Lasso-type models is to select a set of covariates from the input covariate set that linearly represent the response approximately. However, they consider data approximation and representation only, without explicitly incorporating the correlation between the response and covariates.

T. Calders et al. (Eds.): ECML PKDD 2014, Part I, LNCS 8724, pp. 595–606, 2014.

In this paper, correlation information is considered into the Lasso-type variable selection, where regression coefficients associated with larger correlations between the response and covariates are penalized less. Therefore, the selected covariates are highly correlated with the response, i.e., the response can be sparsely approximated (represented) by its closer covariates. In the following, we firstly briefly review the normal Lasso and Elastic Net, then present our covariate-correlated Lasso (ccLasso) model. An efficient iterative algorithm, with its proof of convergence, is presented to solve the proposed ccLasso optimization problem. Promising experimental results show the benefits of the proposed ccLasso model.

2 Brief Review of Lasso and Elastic Net

Let $(x_1, y_1), \cdots, (x_n, y_n)$ be the input data, where $x_i \in \mathcal{R}^p$ is a vector of predictors and $y_i \in \mathcal{R}$ is a scalar response for x_i. Formulate them in matrix form $X = (x_1, x_2, \cdots, x_n)^T \in \mathcal{R}^{n \times p}$ and $y = (y_1, y_2, \cdots, y_n) \in \mathcal{R}^n$. Here we adopt the language of LARS (covariate point of view) [6]. The j-th column of X (e.g., j-th dimension or feature throughout the n data points) is the j-th covariate, denoted as a column vector $a_j \in \mathcal{R}^n$. Let $A = (a_1, a_2, \cdots, a_p)$. The goal of Lasso is variable (covariate) selection. It selects a subset of $k < p$ covariates from the p covariates $a_1, \cdots, a_p$ (remember p is the dimension of x_i) that best approximate the response vector y. Lasso minimizes

$$\min_{\beta} \left\| y - \sum_{j=1}^{p} \beta_j a_j \right\|^2 + \lambda \sum_{j=1}^{p} |\beta_j| = \|y - A\beta\|^2 + \lambda \|\beta\|_1, \tag{1}$$

Here ℓ_q-norm of vector v is defined as $\|v\|_q = \left[\sum_{i=1}^{n} |v_i|^q \right]^{1/q}$. For simplicity, we ignore the subscript 2 for the Euclidean distance $q = 2$: $\|v\| = \|v\|_2$. $\lambda \geq 0$ is a penalty parameter. When λ is large, many components of β are zero. The nonzero components give the selection of covariates. This covariate point of view is identical to the compressed sensing of Donoho et al[4].

In general, ℓ_1-minimization is not stable compared with ℓ_2-minimization [15]. To compensate for this, Elastic Net [19] further adds the ridge regression penalty term into Lasso objective function, which can be formulated as

$$\min_{\beta \in \mathcal{R}^p} \|y - A\beta\|^2 + \lambda \|\beta\|_1 + \zeta \|\beta\|^2, \tag{2}$$

where $\lambda, \zeta \geq 0$ are model parameters. Apart the sparsity, Elastic Net usually encourages a grouping effect, i.e., strongly correlated covariates tend to be in or out of the model together.

3 Covariate-Correlated Lasso

In this section, we present our covariate-correlated Lasso (ccLasso).

3.1   Covariate-Response Vector Correlation

First, we rescale the data. Note that in Lasso we can normalize y such that $\|y\| = 1$; This change is absorbed by β through an overall proportional constant. Second, we can also normalize each covariate a_j to $\|a_j\| = 1$. The difference is absorbed into β_j. Our covariate-correlated Lasso (ccLasso) is motivated by the following two observations.

First, since each covariate a_j has the same dimension as y, we may consider the correlation between y and covariate a_j. This is useful information. Intuitively, if we select a few covariates to form a linear combination that best approximates the response vector y, then the covariates correlated more with y would be good choices. In fact, this correlation information has been emphasized and successfully used in analysis of gene expression microarray data of gene selection [8]. To the best of our knowledge, this correlation information has not been explored or emphasized in Lasso-type covariate selection.

Then, we can prove that if we restrict β to have only one nonzero component, the selected covariate must be the covariate which correlates with y the most, i.e., the highest correlation coefficient w.r.t. y.

Lemma 1. *If we select one covariate among the p covariates, the selected one has the highest correlation coefficient with y.*

Proof. Selecting one covariate a_j that minimizes the error most is the following minimization problem,

$$\min_{j,\beta_j} J = \|y - \beta_j a_j\|^2 = y^T y + \beta_j^2 a_j^T a_j - 2\beta_j y^T a_j. \tag{3}$$

Since y and a_j are normalized, i.e., $y^T y = 1$, $a_j^T a_j = 1$ and y and a_j are already centered as in standard regression, the correlation coefficient is

$$\rho(y, a_j) = \frac{y^T a_j}{\|y\|\|a_j\|} = y^T a_j.$$

Thus $J = 1 + \beta_j^2 - 2\beta_j \rho(y, a_j)$. Setting the derivative w.r.t. β_j to zero, we obtain $\beta_j = \rho(y, a_j)$. Thus, $J = 1 - [\rho(y, a_j)]^2$ and the selection problem becomes

$$\min_j 1 - [\rho(y, a_j)]^2. \tag{4}$$

Therefore the selected one must has the highest (absolute value) correlation coefficient with y.   □

The above result is intuitively appealing: if we select one covariate to represent y approximately, the selected covariate must be the one closest (most correlated) to y. If we select two covariates to represent y, the standard LASSO results are not necessarily the two covariates most correlated to y. Our covariate-correlated Lasso (ccLasso) is motivated by the desire to encourage the selected covariates to correlate more with y.

3.2   Covariate-Correlated Lasso

By imposing the correlation information into the variable selection, our covariate-correlated Lasso can be formulated as follows,

$$\min_{\beta} \ \|y - A\beta\|^2 + \lambda \sum_{j=1}^{p} \mu_j |\beta_j|, \tag{5}$$

where
$$\mu_j = (1 - |\rho(y, a_j)|)^2. \tag{6}$$

The intuition is that when a_j correlates strongly (either positively or negatively) with y, μ_j is close to zero, thus a small penalty. As λ increases, β_j with large penalty tend to go to zero. Thus the final selected covariates tend to have larger correlation with y.

We now use α to replace β to denote/emphasize the regression coefficients obtained from ccLasso. Let $D = \mathrm{diag}(\mu_1, \cdots, \mu_p)$, then ccLasso can be written compactly as

$$\min_{\alpha} \ \|y - A\alpha\|^2 + \lambda\|D\alpha\|_1. \tag{7}$$

4   Computational Algorithm

4.1   Update Algorithm

Problem Eq.(7) is a convex formulation and we seek the global optimal solution. In this section, an efficient algorithm is derived to solve this problem. The detailed algorithm is given in Algorithm 1. In Algorithm 1, the initialization is the solution of the ridge regression problem

$$\min_{\alpha} \ \|y - A\alpha\|^2 + \lambda\alpha^T D\alpha. \tag{8}$$

The solution of this ridge regression problem is given by

$$\alpha^{(0)} = (A^T A + \lambda D/2)^{-1} A^T y. \tag{9}$$

4.2   Convergence Analysis

In this section, we provide a convergence analysis for Algorithm 1. Since $L(\alpha)$ is a convex function of α, thus, we only need to prove that the objective function value $L(\alpha)$ is non-increasing in each iteration in Algorithm 1. This is summarized in Theorem 1.

Theorem 1. *The objective function value $L(\alpha)$ of Eq.(7) for ccLasso minimization problem is non-increasing,*

$$L(\alpha^{t+1}) \leq L(\alpha^t), \tag{12}$$

upon the updating formulae Eq.(11) in Algorithm 1.

To prove Theorem 1, we need the help of the following two Lemmas, which are needed to be proved firstly.

Algorithm 1. Algorithm for covariate-correlated Lasso

1: **Input:** Training data $A \in \mathcal{R}^{n \times p}$ and corresponding response $y \in \mathcal{R}^n$, parameters λ, maximum number of iteration t_{max}, and convergence tolerance $\epsilon > 0$;
2: Compute $B = A^T A$, D, μ_j as in Eqs.(6,7)
3: Initialize $t = 0$, $\alpha^{(t)} = (A^T A + \lambda D/2)^{-1} A^T y$.
4: Update diagonal matrix

$$M^{(t)} = \text{diag}\left(\sqrt{|\alpha_1^{(t)}|}, \cdots, \sqrt{|\alpha_p^{(t)}|} \right); \tag{10}$$

5: Update combination coefficients

$$\alpha^{(t+1)} = M^{(t)} \left[M^{(t)} B M^{(t)} + \frac{\lambda}{2} D \right]^{-1} M^{(t)} A^T y; \tag{11}$$

6: If $t > t_{max}$ or $\|\alpha^{(t+1)} - \alpha^{(t)}\| < \epsilon$, go to step 7; otherwise, set $t = t + 1$ and go to step 4;
7: **Output:** The converged regression coefficients $\alpha^* = \alpha^{(t+1)}$.

Lemma 2. *Define an auxiliary function*

$$G(\alpha) = \|y - A\alpha\|^2 + \lambda \sum_{i=1}^{p} \frac{\alpha_i^2}{2|\alpha_i^{(t)}|} d_i. \tag{13}$$

Along with the $\{\alpha^{(t)}, t = 0, 1, 2, \cdots\}$ sequence obtained in Algorithm 1, the following inequality holds,

$$G(\alpha^{(t+1)}) \leq G(\alpha^{(t)}). \tag{14}$$

Proof. Since both two terms in auxiliary function $G(\alpha)$ are semi-definite programming (SDP) problems, we can obtain the global optimal solution of $G(\alpha)$ by taking the derivatives and let them equal to zero.

Making use of $M^{(t)}$ denotation in Eq.(10), the auxiliary function $G(\alpha)$ can be rewritten as

$$G(\alpha) = \|y - A\alpha\|^2 + \frac{\lambda}{2} \alpha^T (M^{(t)})^{-2} D\alpha. \tag{15}$$

Take the derivative of Eq.(15) with respect to α, and we get

$$\frac{\partial G(\alpha)}{\partial \alpha} = 2A^T A\alpha - 2A^T y + \lambda(M^{(t)})^{-2} D\alpha. \tag{16}$$

The second order derivatives are

$$\frac{\partial^2 G(\alpha)}{\partial \alpha_i \partial \alpha_j} = 2A^T A + \lambda(M^{(t)})^{-2} D. \tag{17}$$

This is clearly a positive semi-definite matrix. Thus function $G(\alpha)$ is a convex function and its global optimal solution α^* is unique. By setting $\frac{\partial G(\alpha)}{\partial \alpha} = 0$, we obtain

$$\alpha^* = \left[A^T A + \frac{\lambda}{2}(M^{(t)})^{-2}D\right]^{-1} A^T y \tag{18}$$

$$= M^{(t)}\left[M^{(t)}BM^{(t)} + \frac{\lambda}{2}D\right]^{-1} M^{(t)}A^T y. \tag{19}$$

The solution α^* is the global optima of $G(\alpha)$. Thus $G(\alpha^*) \leq G(\alpha)$ for any α. In particular, $G(\alpha^*) \leq G(\alpha^{(t)})$. Comparing Eq.(11) with Eq.(19), $\alpha^{(t+1)} = \alpha^*$. This completes the proof of Lemma 2.

Remark. It is important to note that we use Eq.(19) instead of the seemingly simpler Eq.(18). This is because as iteration progresses, some elements of $\alpha^{(t)}$ could become zero due to the sparsity of l_1-penalty. This causes the failure of the inverse of $M^{(t)}$ in Eq.(18). Thus Eq.(18) is ill-defined. However, $M^{(t)}$ is well-defined. Thus Eq.(19) is well-defined, which is chosen as the updating rule Eq.(11) in Algorithm 1.

Lemma 3. The $\{\alpha^{(t)}, t = 0, 1, 2, \cdots\}$ sequence obtained by iteratively computing Eqs.(10,11) in Algorithm 1 has the following property

$$L(\alpha^{(t+1)}) - L(\alpha^{(t)}) \leq G(\alpha^{(t+1)}) - G(\alpha^{(t)}). \tag{20}$$

Proof. Setting $\Delta = (L(\alpha^{(t+1)}) - L(\alpha^{(t)})) - (G(\alpha^{(t+1)}) - G(\alpha^{(t)}))$, substitute Eq.(7) and Eq.(13) in it, we obtain

$$\Delta = (\lambda\|D\alpha^{(t+1)}\|_1 - \lambda\|D\alpha^{(t)}\|_1) - \left(\lambda\sum_{i=1}^{p} d_i \frac{(\alpha_i^{(t+1)})^2}{2|\alpha_i^{(t)}|} - \lambda\sum_{i=1}^{p} d_i \frac{(\alpha_i^{(t)})^2}{2|\alpha_i^{(t)}|}\right)$$

$$= -\frac{\lambda}{2}\sum_{i=1}^{p} \frac{d_i}{|\alpha_i^{(t)}|}\left(-2|\alpha_i^{(t+1)}||\alpha_i^{(t)}| + 2|\alpha_i^{(t)}|^2 + (\alpha_i^{(t+1)})^2 - (\alpha_i^{(t)})^2\right)$$

$$= -\frac{\lambda}{2}\sum_{i=1}^{p} \frac{d_i}{|\alpha_i^{(t)}|}\left(|\alpha_i^{(t+1)}| - |\alpha_i^{(t)}|\right)^2 \leq 0. \tag{21}$$

This completes the proof of Lemma 3.

Proof of Theorem 1. From Lemma 2 and Lemma 3, we have,

$$L(\alpha^{(t+1)}) - L(\alpha^{(t)}) \leq G(\alpha^{(t+1)}) - G(\alpha^{(t)}) \leq 0, \tag{22}$$

which is to say

$$L(\alpha^{(t+1)}) \leq L(\alpha^{(t)}). \tag{23}$$

This completes the proof of Theorem 1. Therefore, Algorithm 1 converges to the global optimal solution of ccLasso model starting from any initial coefficient $\alpha^{(0)}$, due to the convexity of optimization problem. Setting $d_i = 1$, the same algorithm can solve the standard Lasso problem.

5   Experiments

We evaluate the effectiveness of the proposed covariate-correlated Lasso (ccLasso) on the two well known data sets: Colon Cancer Data [1] and Leukemia Dataset [8]. The performance in variable selection and classification accuracy of the ccLasso will be compared with other methods. Once the variables are selected by our ccLasso method, the standard regression has been used to achieve classification [3].

5.1   Colon Cancer Data

The data is Affymetrix Oligonucleotide Array measurements of gene expression levels of 40 tumor and 22 normal colon tissues for 6500 human genes [1]. A subset of 2000 genes based on highest minimal intensity across the samples was selected[1]. These data were first preprocessed by taking a base 10 logarithmic of each expression level, and then each sample is centerized and normalized to zero mean and unit variance across the genes [3].

Classification Comparison. Because this dataset does not contain test set, we use the leave-one-out cross validation (LOOCV) method to evaluate the performance of the classification methods on a selected subset of genes [3]. The external LOOCV proce-dure is performed as follows: 1) remove one observation from the training set; 2) Select top 150 genes as ranked in terms of the t statistic; 3) Re-selected the k most important genes from the 150 genes by the proposed ccLasso algorithm; 4) Use these k genes to classify the left out sample. This process was repeated for all observations in the train-ing set, and the average classification performance has been computed. Figure 1 shows the comparison results across different k genes selected out. Here, we can note that (1) the performances of all three methods are better as more genes are picked out for clas-sification. (2) Lasso performs better than Elastic Net in this dataset. (3) The proposed ccLasso shows consistent superiority over the Lasso and Elastic Net. The best classifi-cation accuracy and its corresponding genes are summarized in Table 1. The proposed ccLasso is compared with the following classification methods: SVM [7], MAVE-LD [2], gsg-SSVS [16], Lasso [14] and Elastic Net [19]. It is clear demonstrated that the proposed ccLasso is better than the other popular classification methods using only moderate number of genes.

Table 1. Classification results on Colon Cancer Data

Method	No. of genes	LOOCV accuracy
SVM	1000 or 2000	0.9032
MAVE-LD	50	0.8387
gsg-SSVS	10/14	0.8871
Lasso	18	0.8316
Elastic Net	18	0.9510
ccLasso	18	0.9755

---

[1] http://microarray.princeton.edu/oncology/affydata/

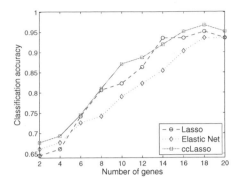

Fig. 1. External LOOCV classification accuracy of Lasso, Elastic Net and ccLasso on Colon Cancer Data

Average Correlation. As discussed in Lemma 1 in Section 3.1, we show that if we select only one covariate using Lasso model, this covariate must be the one with the highest correlation with y. From this, we expect that for small number of selected co-variates, their average correlation with y will be high. But if we select larger number of covariates using Lasso, their average correlation with y will be smaller. In contrast, our ccLasso can select the large number of desired covariates that are highly correlated with the response y. To further illustrate these, we compute the average correlation coefficients between the selected covariates (genes) and y across different number of genes. Figure 2 (a) shows the comparison results. Here we can noted that for small number of selected genes, both Lasso and ccLasso can select the genes that are highly correlated with y. However, if we select large number of genes, the average correlation coefficients for ccLasso are clearly larger than that for Lasso model.

Residual Comparison. Both Lasso and ccLasso are the approximation models for solving the following problem

$$\min_{\beta} \|y - A\beta\|, \quad s.t. \quad \|\beta\|_0 = k. \tag{24}$$

In other words, we select a subset A_S of the covariates A with k entries (training samples) such that we achieve the best representation using A_S. This is a discrete selection problem and is well known to be is NP hard.

When using Lasso and ccLasso, this is done as follows,
(A0) Tuning the model parameter λ such that the optimal solution α contains k nonzero entries. (B0) Select the co-variates corresponding to the nonzero entries of α. This gives the subset A_S. (C0) Compute the optimal representation β by solving the linear regression problem,

$$J_{\text{residual-error}} = \min_{\beta} \|y - A_S\beta\|. \tag{25}$$

We compare the covariate subset A_S selected by ccLasso and Lasso, and then compute the residual errors. The results are shown in Figure 2 (b). It is clear that ccLasso selected

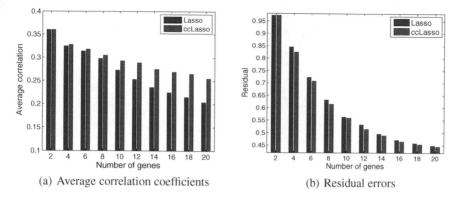

(a) Average correlation coefficients   (b) Residual errors

Fig. 2. Average correlation between selected covariates and y and residual errors for Lasso and ccLasso on Colon dataset

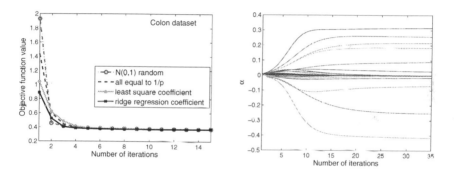

Fig. 3. LEFT: Objective function convergence with different initializations on Colon Dataset; RIGHT: Coefficient vector α during iterations (different colors denote different elements of α)

variables lead to lower residual error than Lasso. Thus, ccLasso model provides better approximate solutions for the discrete selection problem as compared to Lasso model.

Convergence of ccLasso. Figure 3 shows the variation of objective function across the iterations with different initializations in Algorithm 1. We can see that Algorithm 1 converges very quickly and the maximum iteration number is fewer than 30. Regardless of the initializations, the final objective function values are the same and converge almost at the same time, indicting the efficiency and effectiveness of the proposed ccLasso algorithm.

5.2   Leukemia Dataset

The leukaemia data contains DNA gene expressions of 72 tissue samples [8][2]. Following previous work, these tissue samples are divided into the training set of 38 samples

---
[2] http://www.broad.mit.edu/cgi-bin/cancer/datasets.cgi

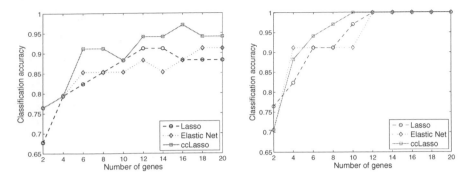

Fig. 4. Classification accuracy of Lasso, Elastic Net and ccLasso on Leukemia testing set (left) and training set (right)

Table 2. Classification results on Leukemia Dataset

Method	No. of genes	Training accuracy	Test accuracy
SVM	25~2000	0.9474	0.8824~0.9412
MAVE-LD	50	0.9737	0.9706
gsg-SSVS	14	0.9737	0.9706
Lasso	20	1.0000	0.8824
Elastic Net	20	1.0000	0.9118
ccLasso	20	1.0000	0.9412

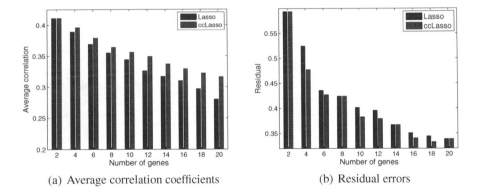

(a) Average correlation coefficients   (b) Residual errors

Fig. 5. Average correlation between selected covariates and y and residual errors for Lasso and ccLasso on Leukemia dataset

and the test set of 34 samples. The data gives expression levels of 7129 genes and DNA products. We use the preprocess method suggested by [3,5].Figure 4 shows the comparison classification accuracy results on training and testing sets across different number of genes, respectively. Here we can note that (1) all of the three methods perform well on training set (all classify correct). (2) On test set, Elastic Net generally performs better than Lasso with the same number of genes selected out. (3) Our ccLasso consistently outperforms the other two methods. Table 2 summarizes the comparison results. Noted that the proposed ccLasso performs better than other methods with moderate number of genes. Figure 5 shows the average correlation coefficients and residual error results, respectively. Noted that as the number of selected genes increases, ccLasso model can return the genes that are more correlated with y (Fig. 5 (a)). Also it returns lower residual errors and thus approximates the discrete variable selection problem more closely than Lasso model (Fig. 5 (b)). Figure 6 shows the variation of objective function across the iterations with different initializations on this dataset. We can see that Algorithm 1 converges very quickly regardless of the different initializations. The above results are general consistent with that on Colon data, and further demonstrates the benefits of the proposed ccLasso.

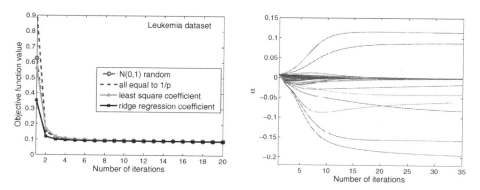

Fig. 6. LEFT: Objective function convergence with different initializations on Leukemia Dataset; RIGHT: Coefficient vector α during iterations (different colors denote different elements of α)

6   Conclusion

Covariate-correlated Lasso (ccLasso) naturally promotes correlation of the selected variable (covariate) with response y; this leads to smaller residual values, indicating a better solution to the discrete variable selection problem. The model achieves this with no extra parameters and same level of computation as standard Lasso. An efficient algorithm has been derived to solve ccLasso. Experiments on two well known gene datasets show that the proposed ccLasso consistently outperforms several state-of-the-art feature selection methods.

Acknowledgments. This work was supported in part by the National High Technology Research and Development Program of China (863 Program) under Grant 2014AA012204, and by the National Natural Science Foundation of China under Grant 61202228.

References

1. Alon, U., Barkai, N., Notterman, D., Gish, K., Ybarra, S., Mack, D., Levine, A.: Broad patterns of gene expression revealed by clustering analysis of tumor and normal colon tissues probed by oligonucleotide arrays. Proceedings of the National Academy of Sciences 96(12), 6745–6750 (1999)
2. Antoniadis, A., Lambert-Lacroix, S., Leblanc, F.: Effective dimension reduction methods for tumor classification using gene expression data. Bioinformatics 19(5), 563–570 (2003)
3. Chen, S., Ding, C., Luo, B., Xie, Y.: Uncorrelated lasso. In: AAAI, pp. 166–172 (2013)
4. Donoho, D.: Compressed sensing. Technical Report, Stanford University (2006)
5. Dudoit, S., Fridlyand, J., Speed, T.P.: Comparison of discrimination methods for the classification of tumors using gene expression data. Journal of the American Statistical Association 97(457), 77–87 (2002)
6. Efron, B., Hastie, T., Johnstone, I., Tibshirani, R.: Least angle regression. Annals of Statistics 32, 407–451 (2004)
7. Furey, T.S., Cristianini, N., Duffy, N., Bednarski, D.W., Schummer, M., Haussler, D.: Support vector machine classification and validation of cancer tissue samples using microarray expression data. BMC Bioinformatics 16(10), 906–914 (2000)
8. Golub, T.R., Slonim, D.K., Tamayo, P., Huard, C., Gaasenbeek, M., Mesirov, J.P., Coller, H., Loh, M.L., Downing, J.R., Caligiuri, M.A., et al.: Molecular classification of cancer: class discovery and class prediction by gene expression monitoring. Science 286(5439), 531–537 (1999)
9. Kohavi, R., John, G.H.: Wrappers for feature subset selection. Artif. Intell. 97(1-2), 273–324 (1997)
10. Kononenko, I.: Estimating attributes: Analysis and extensions of RELIEF. In: Bergadano, F., De Raedt, L. (eds.) ECML 1994. LNCS, vol. 784, Springer, Heidelberg (1994)
11. Ma, S., Song, X., Huang, J.: Supervised group lasso with applications to microarray data analysis. BMC Bioinformatics 8 (2007)
12. Peng, H., Long, F., Ding, C.H.Q.: Feature selection based on mutual information: Criteria of max-dependency, max-relevance, and min-redundancy. IEEE Trans. Pattern Anal. Mach. Intell. 27(8), 1226–1238 (2005)
13. Raileanu, L.E., Stoffel, K.: Theoretical comparison between the gini index and information gain criteria. Ann. Math. Artif. Intell. 41(1), 77–93 (2004)
14. Tibshirani, R.: Regression shrinkage and selection via the Lasso. Journal of the Royal Statistical Society. Series B (Methodological) 58(1), 267–288 (1996)
15. Xu, H., Caramanis, C., Mannor, S.: Sparse algorithms are not stable: A no-free-lunch theorem. IEEE Trans. Pattern Anal. Mach. Intell. 34(1), 187–193 (2012)
16. Yang, A.J., Song, X.Y.: Bayesian variable selection for disease classification using gene expression data. Bioinformatics 26(2), 215–222 (2010)
17. Yuan, M., Lin, Y.: Model selection and estimation in regression with grouped variables. Journal of the Royal Statistical Society: Series B (Statistical Methodology) 68(1), 49–67 (2006)
18. Zou, H.: The adaptive lasso and its oracle properties. Journal of the American Statistical Association 101, 1418–1429 (2006)
19. Zou, H., Hastie, T.: Regularization and variable selection via the elastic net. Journal of the Royal Statistical Society: Series B (Statistical Methodology) 67(2), 301–320 (2005)

Random Forests with Random Projections of the Output Space for High Dimensional Multi-label Classification

Arnaud Joly, Pierre Geurts, and Louis Wehenkel

Dept. of EE & CS & GIGA-R
University of Liège, Belgium

Abstract. We adapt the idea of random projections applied to the output space, so as to enhance tree-based ensemble methods in the context of multi-label classification. We show how learning time complexity can be reduced without affecting computational complexity and accuracy of predictions. We also show that random output space projections may be used in order to reach different bias-variance tradeoffs, over a broad panel of benchmark problems, and that this may lead to improved accuracy while reducing significantly the computational burden of the learning stage.

1 Introduction

Within supervised learning, the goal of multi-label classification is to train models to annotate objects with a subset of labels taken from a set of candidate labels. Typical applications include the determination of topics addressed in a text document, the identification of object categories present within an image, or the prediction of biological properties of a gene. In many applications, the number of candidate labels may be very large, ranging from hundreds to hundreds of thousands [2] and often even exceeding the sample size [9]. The very large scale nature of the output space in such problems poses both statistical and computational challenges that need to be specifically addressed.

A simple approach to solve multi-label classification problems, called binary relevance, is to train independently a binary classifier for each label. Several more complex schemes have however been proposed to take into account the dependencies between the labels (see, e.g. [19,13,7,21,8,23]). In the context of tree-based methods, one way is to train multi-output trees [3,12,15], ie. trees that can predict multiple outputs at once. With respect to single-output trees [5], the score measure used in multi-output trees to choose splits is taken as the sum of the individual scores corresponding to the different labels (e.g., variance reduction) and each leaf is labeled with a vector of values, coding each for the probability of presence of one label. With respect to binary relevance, the multi-output tree approach has the advantage of building a single model for all labels. It can thus potentially take into account label dependencies and

T. Calders et al. (Eds.): ECML PKDD 2014, Part I, LNCS 8724, pp. 607–622, 2014.

reduce memory requirements for the storage of the models. An extensive experimental comparison [17] shows that this approach compares favorably with other approaches, including non tree-based methods, both in terms of accuracy and computing times. In addition, multi-output trees inherit all intrinsic advantages of tree-based methods, such as robustness to irrelevant features, interpretability through feature importance scores, or fast computations of predictions, that make them very attractive to address multi-label problems. The computational complexity of learning multi-output trees is however similar to that of the binary relevance method. Both approaches are indeed $O(pdn \log n)$, where p is the number of input features, d the number of candidate output labels, and n the sample size; this is a limiting factor when dealing with large sets of candidate labels.

One generic approach to reduce computational complexity is to apply some compression technique prior to the training stage to reduce the number of outputs to a number m much smaller than the total number d of labels. A model can then be trained to make predictions in the compressed output space and a prediction in the original label space can be obtained by decoding the compressed prediction. As multi-label vectors are typically very sparse, one can expect a drastic dimensionality reduction by using appropriate compression techniques. This idea has been explored for example in [13] using compressed sensing, and in [8] using bloom filters, in both cases using regularized linear models as base learners. This approach obviously reduces computing times for training the model. At the prediction stage however, the predicted compressed output needs to be decoded, which adds computational cost and can also introduce further decoding errors.

In this paper, we explore the use of random output space projections for large-scale multi-label classification in the context of tree-based ensemble methods. We first explore the idea proposed for linear models in [13] with random forests: a (single) random projection of the multi-label vector to an m-dimensional random subspace is computed and then a multi-output random forest is grown based on score computations using the projected outputs. We exploit however the fact that the approximation provided by a tree ensemble is a weighted average of output vectors from the training sample to avoid the decoding stage: at training time all leaf labels are directly computed in the original multi-label space. We show theoretically and empirically that when m is large enough, ensembles grown on such random output spaces are equivalent to ensembles grown on the original output space. When d is large enough compared to n, this idea hence may reduce computing times at the learning stage without affecting accuracy and computational complexity of predictions.

Next, we propose to exploit the randomization inherent to the projection of the output space as a way to obtain randomized trees in the context of ensemble methods: each tree in the ensemble is thus grown from a different randomly projected subspace of dimension m. As previously, labels at leaf nodes are directly computed in the original output space to avoid the decoding step. We show, theoretically, that this idea can lead to better accuracy than the first idea and, empirically, that best results are obtained on many problems with very low values of m, which leads to significant computing time reductions at the learning stage. In addition, we study the interaction between input randomization (à la Random Forests) and output randomization (through random projections), showing that there is an interest, both in terms of predictive performance and in terms of computing times, to optimally combine these two ways of randomization. All in all, the proposed approach constitutes a very attractive way to address large-scale multi-label problems with tree-based ensemble methods.

The rest of the paper is structured as follows: Section 2 reviews properties of multi-output tree ensembles and of random projections; Section 3 presents the proposed algorithms and their theoretical properties; Section 4 provides the empirical validations, whereas Section 5 discusses our work and provides further research directions.

2   Background

We denote by $\mathcal{X}$ an input space, and by $\mathcal{Y}$ an output space; without loss of generality, we suppose that $\mathcal{X} = \mathbb{R}^p$ (where p denotes the number of input features), and that $\mathcal{Y} = \mathbb{R}^d$ (where d is the dimension of the output space). We denote by $P_{\mathcal{X},\mathcal{Y}}$ the joint (unknown) sampling density over $\mathcal{X} \times \mathcal{Y}$.

Given a learning sample $\left((x^i, y^i) \in (\mathcal{X} \times \mathcal{Y})\right)_{i=1}^{n}$ of n observations in the form of input-output pairs, a supervised learning task is defined as searching for a function $f^* : \mathcal{X} \to \mathcal{Y}$ in a hypothesis space $\mathcal{H} \subset \mathcal{Y}^{\mathcal{X}}$ that minimizes the expectation of some loss function $\ell : \mathcal{Y} \times \mathcal{Y} \to \mathbb{R}$ over the joint distribution of input / output pairs: $f^* \in \arg\min_{f \in \mathcal{H}} E_{P_{\mathcal{X},\mathcal{Y}}} \{\ell(f(x), y)\}$.

NOTATIONS: Superscript indices (x^i, y^i) denote (input, output) vectors of an observation $i \in \{1, \ldots, n\}$. Subscript indices (e.g. x_j, y_k) denote components of vectors.

2.1   Multi-output Tree Ensembles

A classification or a regression tree [5] is built using all the input-output pairs as follows: for each node at which the subsample size is greater or equal to a pre-pruning parameter $n_{\min}$, the best split is chosen among the p input features combined with the selection of an optimal cut point. The best sample split (S_r, S_l) of the local subsample S minimizes the average reduction of impurity

$$\Delta I((y^i)_{i \in S}, (y^i)_{i \in S_l}, (y^i)_{i \in S_r}) = I((y^i)_{i \in S}) - \frac{|S_l|}{|S|} I((y^i)_{i \in S_l}) - \frac{|S_r|}{|S|} I((y^i)_{i \in S_r}). \quad (1)$$

Finally, leaf statistics are obtained by aggregating the outputs of the samples reaching that leaf.

In this paper, for multi-output trees, we use the sum of the variances of the d dimensions of the output vector as an impurity measure. It can be computed by (see Appendix A, in the supplementary material[1])

$$\text{Var}((y^i)_{i \in S}) = \frac{1}{|S|} \sum_{i \in S} ||y^i - \frac{1}{|S|} \sum_{i \in S} y^i||^2, \tag{2}$$

$$= \frac{1}{2|S|^2} \sum_{i \in S} \sum_{j \in S} ||y^i - y^j||^2. \tag{3}$$

Furthermore, we compute the vectors of output statistics by component-wise averaging. Notice that, when the outputs are vectors of binary class-labels (i.e. $y \in \{0,1\}^d$), as in multi-label classification, the variance reduces to the so-called Gini-index, and the leaf statistics then estimate a vector of conditional probabilities $P(y_j = 1 | x \in leaf)$, from which a prediction $\hat{y}$ can be made by thresholding.

Tree-based ensemble methods build an ensemble of t randomized trees. Unseen samples are then predicted by aggregating the predictions of all t trees. Random Forests [4] build each tree on a bootstrap copy of the learning sample [4] and by optimising the split at each node over a locally generated random subset of size k among the p input features. Extra Trees [11] use the complete learning sample and optimize the split over a random subset of size k of the p features combined with a random selection of cut points. Setting the parameter k to the number of input features p allows to filter out irrelevant features; larger $n_{\min}$ yields simpler trees possibly at the price of higher bias, and the higher t the smaller the variance of the resulting predictor.

2.2   Random Projections

In this paper we apply the idea of random projections to samples of vectors of the output space $\mathcal{Y}$. With this in mind, we recall the Johnson-Lindenstrauss lemma (reduced to linear maps), while using our notations.

Lemma 1. *Johnson-Lindenstrauss lemma [14] Given $\epsilon > 0$ and an integer n, let m be a positive integer such that $m \geq 8\epsilon^{-2} \ln n$. For any sample $(y^i)_{i=1}^n$ of n points in $\mathbb{R}^d$ there exists a matrix $\Phi \in \mathbb{R}^{m \times d}$ such that for all $i, j \in \{1, \dots, n\}$*

$$(1 - \epsilon)||y^i - y^j||^2 \leq ||\Phi y^i - \Phi y^j||^2 \leq (1 + \epsilon)||y^i - y^j||^2. \tag{4}$$

Moreover, when d is sufficiently large, several random matrices satisfy (4) with high probability. In particular, we can consider Gaussian matrices which elements are drawn i.i.d. in $\mathcal{N}(0, 1/m)$, as well as (sparse) Rademacher matrices which elements are drawn in $\{-\sqrt{\frac{s}{m}}, 0, \sqrt{\frac{s}{m}}\}$ with probability $\{\frac{1}{2s}, 1 - \frac{1}{s}, \frac{1}{2s}\}$, where $1/s \in (0, 1]$ controls the sparsity of Φ [1,16].

Notice that if some Φ satisfies (4) for the whole learning sample, it obviously satisfies (4) for any subsample that could reach a node during regression tree

---

[1]   static.ajoly.org/files/ecml2014-supplementary.pdf

growing. On the other hand, since we are not concerned in this paper with the 'reconstruction' problem, we do not need to make any sparsity assumption 'à la compressed sensing'.

3 Methods

We first present how we propose to exploit random projections to reduce the computational burden of learning single multi-output trees in very high-dimensional output spaces. Then we present and compare two ways to exploit this idea with ensembles of trees. Subsection   3.3 analyses these two ways from the bias/variance point of view.

3.1 Multi-output Regression Trees in Randomly Projected Output Spaces

The multi-output single tree algorithm described in section 2 requires the computation of the sum of variances in (2) at each tree node and for each candidate split. When $\mathcal{Y}$ is very high-dimensional, this computation constitutes the main computational bottleneck of the algorithm. We thus propose to approximate variance computations by using random projections of the output space. The multi-output regression tree algorithm is modified as follows (denoting by LS the learning sample $((x^i, y^i))_{i=1}^n$):

- First, a projection matrix Φ of dimension $m \times d$ is randomly generated.
- A new dataset $LS_m = ((x^i, \Phi y^i))_{i=1}^n$ is constructed by projecting each learning sample output using the projection matrix Φ.
- A tree (structure) $\mathcal{T}$ is grown using the projected learning sample LS_m.
- Predictions $\hat{y}$ at each leaf of $\mathcal{T}$ are computed using the corresponding outputs in the original output space.

The resulting tree is exploited in the standard way to make predictions: an input vector x is propagated through the tree until it reaches a leaf from which a prediction $\hat{y}$ in the original output space is directly retrieved.

If Φ satisfies (4), the following theorem shows that variance computed in the projected subspace is an ϵ-approximation of the variance computed over the original space.

Theorem 1. *Given $\epsilon > 0$, a sample $(y^i)_{i=1}^n$ of n points $y \in \mathbb{R}^d$, and a projection matrix $\Phi \in \mathbb{R}^{m \times d}$ such that for all $i, j \in \{1, \ldots, n\}$ condition (4) holds, we have also:*

$$(1 - \epsilon)\,\mathrm{Var}((y^i)_{i=1}^n) \leq \mathrm{Var}((\Phi y^i)_{i=1}^n) \leq (1 + \epsilon)\,\mathrm{Var}((y^i)_{i=1}^n). \tag{5}$$

Proof. See Appendix B, supplementary material.

As a consequence, any split score approximated from the randomly projected output space will be ϵ-close to the unprojected scores in any subsample of the complete learning sample. Thus, if condition (4) is satisfied for a sufficiently small ϵ then the tree grown from the projected data will be identical to the tree grown from the original data[2].

For a given size m of the projection subspace, the complexity is reduced from $O(dn)$ to $O(mn)$ for the computation of one split score and thus from $O(dpn \log n)$ to $O(mpn \log n)$ for the construction of one full (balanced) tree, where one can expect m to be much smaller than d and at worst of $O(\epsilon^{-2} \log n)$. The whole procedure requires to generate the projection matrix and to project the training data. These two steps are respectively $O(dm)$ and $O(ndm)$ but they can often be significantly accelerated by exploiting the sparsity of the projection matrix and/or of the original output data, and they are called only once before growing the tree.

All in all, this means that when d is sufficiently large, the random projection approach may allow us to significantly reduce tree building complexity from $O(dtpn \log n)$ to $O(mtpn \log n + tndm)$, without impact on predictive accuracy (see section 4, for empirical results).

3.2   Exploitation in the Context of Tree Ensembles

The idea developed in the previous section can be directly exploited in the context of ensembles of randomized multi-output regression trees. Instead of building a single tree from the projected learning sample LS_m, one can grow a randomized ensemble of them. This "shared subspace" algorithm is described in pseudo-code in Algorithm 1.

Algorithm 1. Tree ensemble on a single shared subspace Φ

Require: t, the ensemble size
Require: $((x^i, y^i) \in (\mathbb{R}^p \times \mathbb{R}^d))_{i=1}^n$, the input-output pairs
Require: A tree building algorithm.
Require: A sub-space generator
    Generate a sub-space $\Phi \in \mathbb{R}^{m \times d}$;
    for $j = 1$ to t **do**
        Build a tree structure $\mathcal{T}_j$ using $((x^i, \Phi y^i))_{i=1}^n$;
        Label the leaves of $\mathcal{T}_j$ using $((x^i, y^i))_{i=1}^n$;
        Add the labelled tree $\mathcal{T}_j$ to the ensemble;
    end for

Another idea is to exploit the random projections used so as to introduce a novel kind of diversity among the different trees of an ensemble. Instead of building all the trees of the ensemble from a same shared output-space projection, one could instead grow each tree in the ensemble from a different output-space projection. Algorithm 2 implements this idea in pseudo-code. The randomization

---

[2] Strictly speaking, this is only the case when the optimum scores of test splits as computed over the original output space are isolated, i.e. when there is only one single best split, no tie.

introduced by the output space projection can of course be combined with any existing randomization scheme to grow ensembles of trees. In this paper, we will consider the combination of random projections with the randomizations already introduced in Random Forests and Extra Trees. The interplay between these different randomizations will be discussed theoretically in the next subsection by a bias/variance analysis and empirically in Section 4. Note that while when looking at single trees or shared ensembles, the size m of the projected subspace should not be too small so that condition (4) is satisfied, the optimal value of m when projections are randomized at each tree is likely to be smaller, as suggested by the bias/variance analysis in the next subsection.

Algorithm 2. Tree ensemble with individual subspaces Φ_j

Require: t, the ensemble size
Require: $((x^i, y^i) \in (\mathbb{R}^p \times \mathbb{R}^d))_{i=1}^n$, the input-output pairs
Require: A tree building algorithm.
Require: A sub-space generator
  for $j = 1$ to t **do**
    Generate a sub-space $\Phi_j \in \mathbb{R}^{m \times d}$;
    Build a tree structure $\mathcal{T}_j$ using $((x^i, \Phi_j y^i))_{i=1}^n$;
    Label the leaves of $\mathcal{T}_j$ using $((x^i, y^i))_{i=1}^n$;
    Add the labelled tree $\mathcal{T}_j$ to the ensemble;
  end for

From the computational point of view, the main difference between these two ways of transposing random-output projections to ensembles of trees is that in the case of Algorithm 2, the generation of the projection matrix Φ and the computation of projected outputs is carried out t times, while it is done only once for the case of Algorithm 1. These aspects will be empirically evaluated in Section 4.

3.3   Bias/Variance Analysis

In this subsection, we adapt the bias/variance analysis carried out in [11] to take into account random output projections. The details of the derivations are reported in Appendix C (supplementary material).

Let us denote by $f(.; ls, \phi, \epsilon) : \mathcal{X} \to \mathbb{R}^d$ a single multi-output tree obtained from a projection matrix ϕ (below we use Φ to denote the corresponding random variable), where ϵ is the value of a random variable ε capturing the random perturbation scheme used to build this tree (e.g., bootstrapping and/or random input space selection). The square error of this model at some point $x \in \mathcal{X}$ is defined by:

$$Err(f(x; ls, \phi, \epsilon)) \stackrel{\text{def}}{=} E_{Y|x}\{||Y - f(x; ls, \phi, \epsilon)||^2\},$$

and its average can decomposed in its residual error, (squared) bias, and variance terms denoted:

$$E_{LS,\Phi,\varepsilon}\{Err(f(x; LS, \Phi, \varepsilon))\} = \sigma_R^2(x) + B^2(x) + V(x)$$

where the variance term $V(x)$ can be further decomposed as the sum of the following three terms:

$$V_{LS}(x) = \text{Var}_{LS}\{E_{\Phi,\varepsilon|LS}\{f(x; LS, \Phi, \varepsilon)\}\}$$
$$V_{Algo}(x) = E_{LS}\{E_{\Phi|LS}\{\text{Var}_{\varepsilon|LS,\Phi}\{f(x; LS, \Phi, \varepsilon)\}\}\},$$
$$V_{Proj}(x) = E_{LS}\{\text{Var}_{\Phi|LS}\{E_{\varepsilon|LS,\Phi}\{f(x; LS, \Phi, \varepsilon)\}\}\},$$

that measure errors due to the randomness of, respectively, the learning sample, the tree algorithm, and the output space projection (Appendix C, supplementary material).

Approximations computed respectively by algorithms 1 and 2 take the following forms:

- $f_1(x; ls, \epsilon^t, \phi) = \frac{1}{t}\sum_{i=1}^{t} f(x; ls, \phi, \epsilon_i)$
- $f_2(x; ls, \epsilon^t, \phi^t) = \frac{1}{t}\sum_{i=1}^{t} f(x; ls, \phi_i, \epsilon_i),$

where $\epsilon^t = (\epsilon_1, \ldots, \epsilon_t)$ and $\phi^t = (\phi_1, \ldots, \phi_t)$ are vectors of i.i.d. values of the random variables ε and Φ respectively.

We are interested in comparing the average errors of these two algorithms, where the average is taken over all random parameters (including the learning sample). We show (Appendix C) that these can be decomposed as follows:

$$E_{LS,\Phi,\varepsilon^t}\{Err(f_1(x; LS, \Phi, \varepsilon^t))\}$$
$$= \sigma_R^2(x) + B^2(x) + V_{LS}(x) + \frac{V_{Algo}(x)}{t} + V_{Proj}(x),$$
$$E_{LS,\Phi^t,\varepsilon^t}\{Err(f_2(x; LS, \Phi^t, \varepsilon^t))\}$$
$$= \sigma_R^2(x) + B^2(x) + V_{LS}(x) + \frac{V_{Algo}(x) + V_{Proj}(x)}{t}.$$

From this result, it is hence clear that Algorithm 2 can not be worse, on the average, than Algorithm 1. If the additional computational burden needed to generate a different random projection for each tree is not problematic, then Algorithm 2 should always be preferred to Algorithm 1.

For a fixed level of tree randomization (ε), whether the additional randomization brought by random projections could be beneficial in terms of predictive performance remains an open question that will be addressed empirically in the next section. Nevertheless, with respect to an ensemble grown from the original output space, one can expect that the output-projections will always increase the bias term, since they disturb the algorithm in its objective of reducing the errors on the learning sample. For small values of m, the average error will therefore decrease (with a sufficiently large number t of trees) only if the increase in bias is compensated by a decrease of variance.

The value of m, the dimension of the projected subspace, that will lead to the best tradeoff between bias and variance will hence depend both on the level of tree randomization and on the learning problem. The more (resp. less) tree randomization, the higher (resp. the lower) could be the optimal value of m, since both randomizations affect bias and variance in the same direction.

4 Experiments

4.1 Accuracy Assessment Protocol

We assess the accuracy of the predictors for multi-label classification on a test sample (TS) by the "Label Ranking Average Precision (LRAP)" [17], expressed by

$$\text{LRAP}(\hat{f}) = \frac{1}{|TS|} \sum_{i \in TS} \frac{1}{|y^i|} \sum_{j \in \{k : y^i_k = 1\}} \frac{|\mathcal{L}^i_j(y^i)|}{|\mathcal{L}^i_j(1_d)|}, \tag{6}$$

where $\hat{f}(x^i)_j$ is the probability (or the score) associated to the label j by the learnt model $\hat{f}$ applied to x^i, 1_d is a d-dimensional row vector of ones, and

$$\mathcal{L}^i_j(q) = \left\{ k : q_k = 1 \text{ and } \hat{f}(x^i)_k \geq \hat{f}(x^i)_j \right\}.$$

Test samples without any relevant labels (i.e. with $|y^i| = 0$) were discarded prior to computing the average precision. The best possible average precision is thus 1. Notice that we use indifferently the notation $|\cdot|$ to express the cardinality of a set or the 1-norm of a vector.

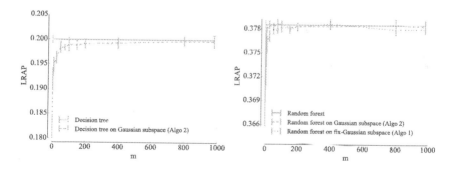

Fig. 1. Models built for the "Delicious" dataset ($d = 983$) for growing numbers m of Gaussian projections. Left: single unpruned CART trees ($n_{\min} = 1$); Right: Random Forests ($k = \sqrt{p}$, $t = 100$, $n_{\min} = 1$). The curves represent average values (and standard deviations) obtained from 10 applications of the randomised algorithms over a same single LS/TS split.

4.2   Effect of the Size m of the Gaussian Output Space

To illustrate the behaviour of our algorithms, we first focus on the "Delicious" dataset [20], which has a large number of labels ($d = 983$), of input features ($p = 500$), and of training ($n_{LS} = 12920$) and testing ($n_{TS} = 3185$) samples.

The left part of figure 1 shows, when Gaussian output-space projections are combined with the standard CART algorithm building a single tree, how the precision converges (cf Theorem 1) when m increases towards d. We observe that in this case, convergence is reached around $m = 200$ at the expense of a slight decrease of accuracy, so that a compression factor of about 5 is possible with respect to the original output dimension $d = 983$.

The right part of figure 1 shows, on the same dataset, how the method behaves when combined with Random Forests. Let us first notice that the Random Forests grown on the original output space (green line) are significantly more accurate than the single trees, their accuracy being almost twice as high. We also observe that Algorithm 2 (orange curve) converges much more rapidly than Algorithm 1 (blue curve) and slightly outperforms the Random Forest grown on the original output space. It needs only about $m = 25$ components to converge, while Algorithm 1 needs about $m = 75$ of them. These results are in accordance with the analysis of Section 3.3, showing that Algorithm 2 can't be inferior to Algorithm 1. In the rest of this paper we will therefore focus on Algorithm 2.

4.3   Systematic Analysis over 24 Datasets

To assess our methods, we have collected 24 different multi-label classification datasets from the literature (see Section D of the supplementary material, for more information and bibliographic references to these datasets) covering a broad spectrum of application domains and ranges of the output dimension ($d \in [6; 3993]$, see Table 1). For 21 of the datasets, we made experiments where the dataset is split randomly into a learning set of size n_{LS}, and a test set of size n_{TS}, and are repeated 10 times (to get average precisions and standard deviations), and for 3 of them we used a ten-fold cross-validation scheme (see Table 1).

Table 1 shows our results on the 24 multi-label datasets, by comparing Random Forests learnt on the original output space with those learnt by Algorithm 2 combined with Gaussian subspaces of size $m \in \{1, d, \ln d\}^3$. In these experiments, the three parameters of Random Forests are set respectively to $k = \sqrt{p}$, $n_{\min} = 1$ (default values, see [11]) and $t = 100$ (reasonable computing budget). Each model is learnt ten times on a different shuffled train/testing split, except for the 3 EUR-lex datasets where we kept the original 10 folds of cross-validation.

We observe that for all datasets (except maybe SCOP-GO), taking $m = d$ leads to a similar average precision to the standard Random Forests, i.e. no difference superior to one standard deviation of the error. On 11 datasets, we see that $m = 1$ already yields a similar average precision (values not underlined in

---

[3] $\ln d$ is rounded to the nearest integer value; in Table 1 the values of $\ln d$ vary between 2 for $d = 6$ and 8 for $d = 3993$.

Table 1. High output space compression ratio is possible, with no or negligible average precision reduction ($t = 100$, $n_{\min} = 1$, $k = \sqrt{p}$). Each dataset has n_{LS} training samples, n_{TS} testing samples, p input features and d labels. Label ranking average precisions are displayed in terms of their mean values and standard deviations over 10 random LS/TS splits, or over the 10 folds of cross-validation. Mean scores in the last three columns are underlined if they show a difference with respect to the standard Random Forests of more than one standard deviation.

| Datasets | | | | | Random Forests | | Random Forests on Gaussian sub-space | | |
Name	n_{LS}	n_{TS}	p	d		$m=1$	$m=\lfloor 0.5+\ln d\rfloor$	$m=d$
emotions	391	202	72	6	0.800 ±0.014	0.800 ±0.010	0.810 ±0.014	0.810 ±0.016
scene	1211	1196	2407	6	0.870 ±0.003	0.875 ±0.007	0.872 ±0.004	0.872 ±0.004
yeast	1500	917	103	14	0.759 ±0.008	0.748 ±0.006	0.755 ±0.004	0.758 ±0.005
tmc2017	21519	7077	49060	22	0.756 ±0.003	0.741 ±0.003	0.748 ±0.003	0.757 ±0.003
genbase	463	199	1186	27	0.992 ±0.004	0.994 ±0.002	0.994 ±0.004	0.993 ±0.004
reuters	2500	5000	19769	34	0.865 ±0.004	0.864 ±0.003	0.863 ±0.004	0.862 ±0.004
medical	333	645	1449	45	0.848 ±0.009	0.836 ±0.011	0.842 ±0.014	0.841 ±0.009
enron	1123	579	1001	53	0.683 ±0.009	0.680 ±0.006	0.685 ±0.009	0.686 ±0.008
mediamill	30993	12914	120	101	0.779 ±0.001	0.772 ±0.001	0.777 ±0.002	0.779 ±0.002
Yeast-GO	2310	1155	5930	132	0.420 ±0.010	0.353 ±0.008	0.381 ±0.005	0.420 ±0.010
bibtex	4880	2515	1836	159	0.566 ±0.004	0.513 ±0.006	0.548 ±0.007	0.564 ±0.008
CAL500	376	126	68	174	0.504 ±0.011	0.504 ±0.004	0.506 ±0.007	0.502 ±0.010
WIPO	1352	358	74435	188	0.490 ±0.010	0.430 ±0.010	0.460 ±0.010	0.480 ±0.010
EUR-Lex (subj.)	19348	10-cv	5000	201	0.840 ±0.005	0.814 ±0.004	0.828 ±0.005	0.840 ±0.004
bookmarks	65892	21964	2150	208	0.453 ±0.001	0.436 ±0.002	0.445 ±0.002	0.453 ±0.002
diatoms	2065	1054	371	359	0.700 ±0.010	0.650 ±0.010	0.670 ±0.010	0.710 ±0.020
corel5k	4500	500	499	374	0.303 ±0.012	0.309 ±0.011	0.307 ±0.011	0.299 ±0.013
EUR-Lex (dir.)	19348	10-cv	5000	412	0.814 ±0.006	0.782 ±0.008	0.796 ±0.009	0.813 ±0.007
SCOP-GO	6507	3336	2003	465	0.811 ±0.004	0.808 ±0.005	0.811 ±0.004	0.806 ±0.004
delicious	12920	3185	500	983	0.384 ±0.004	0.381 ±0.003	0.382 ±0.002	0.383 ±0.004
drug-interaction	1396	466	660	1554	0.379 ±0.014	0.384 ±0.009	0.378 ±0.013	0.367 ±0.016
protein-interaction	1165	389	876	1862	0.330 ±0.015	0.337 ±0.016	0.337 ±0.017	0.335 ±0.014
Expression-GO	2485	551	1288	2717	0.235 ±0.005	0.211 ±0.005	0.219 ±0.005	0.232 ±0.005
EUR-Lex (desc.)	19348	10-cv	5000	3993	0.523 ±0.008	0.485 ±0.008	0.497 ±0.009	0.523 ±0.007

column $m = 1$). For the 13 remaining datasets, increasing m to $\ln d$ significantly decreases the gap with the Random Forest baseline and 3 more datasets reach this baseline. We also observe that on several datasets such as "Drug-interaction" and "SCOP-GO", better performance on the Gaussian subspace is attained with high output randomization ($m = \{1, \ln d\}$) than with $m = d$. We thus conclude that the optimal level of output randomization (i.e. the optimal value of the ratio m/d) which maximizes accuracy performances, is dataset dependent.

While our method is intended for tasks with very high dimensional output spaces, we however notice that even with relatively small numbers of labels, its accuracy remains comparable to the baseline, with suitable m.

To complete the analysis, Appendix F considers the same experiments with a different base-learner (Extra Trees of [11]), showing very similar trends.

4.4   Input vs Output Space Randomization

We study in this section the interaction of the additional randomization of the output space with that concerning the input space already built in the Random Forest method.

To this end, we consider the "Drug-interaction" dataset ($p = 660$ input features and $d = 1554$ output labels [10]), and we study the effect of parameter k controlling the input space randomization of the Random Forest method with the randomization of the output space by Gaussian projections controlled by the parameter m. To this end, Figure 2 shows the evolution of the accuracy for growing values of k (i.e. decreasing strength of the input space randomization), for three different quite low values of m (in this case $m \in \{1, \ln d, 2 \ln d\}$). We observe that Random Forests learned on a very low-dimensional Gaussian subspace (red, blue and pink curves) yield essentially better performances than Random Forests on the original output space, and also that their behaviour with respect to the parameter k is quite different. On this dataset, the output-space randomisation makes the method completely immune to the 'over-fitting' phenomenon observed for high values of k with the baseline method (green curve).

We refer the reader to a similar study on the "Delicious" dataset given in the Appendix E (supplementary material), which shows that the interaction between m and k may be different from one dataset to another. It is thus advisable to jointly optimize the value of m and k, so as to maximise the tradeoff between accuracy and computing times in a problem and algorithm specific way.

4.5   Alternative Output Dimension Reduction Techniques

In this section, we study Algorithm 2 when it is combined with alternative output-space dimensionality reduction techniques. We focus again on the "Delicious" dataset, but similar trends could be observed on other datasets.

Figure 3(a) first compares Gaussian random projections with two other dense projections: Rademacher matrices with $s = 1$ (cf. Section 2.2) and compression matrices obtained by sub-sampling (without replacement) Hadamard matrices [6]. We observe that Rademacher and subsample-Hadamard sub-spaces behave very similarly to Gaussian random projections.

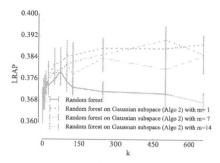

Fig. 2. Output randomization with Gaussian projections yield better average precision than the original output space on the "Drug-Interaction" dataset ($n_{\min} = 1$, $t = 100$)

In a second step, we compare Gaussian random projections with two (very) sparse projections: first, sparse Rademacher sub-spaces obtained by setting the sparsity parameter s to 3 and $\sqrt{d}$, selecting respectively about 33% and 2% of the original outputs to compute each component, and second, sub-sampled identity subspaces, similar to [22], where each of the m selected components corresponds to a randomly chosen original label and also preserve sparsity. Sparse projections are very interesting from a computational point of view as they require much less operations to compute the projections but the number of components required for condition (4) to be satisfied is typically higher than for dense projections [16,6] Figure 3(b) compares these three projection methods with standard Random Forests on the "delicious" dataset. All three projection methods converge to plain Random Forests as the number of components m increases but their behaviour at low m values are very different. Rademacher projections converge faster with $s = 3$ than with $s = 1$ and interestingly, the sparsest variant ($s = \sqrt{d}$) has its optimum at $m = 1$ and improves in this case over the Random Forests baseline. Random output subspaces converge slower but they lead to a notable improvement of the score over baseline Random Forests. This suggests that although their theoretical guarantees are less good, sparse projections actually provide on this problem a better bias/variance tradeoff than dense ones when used in the context of Algorithm 2.

Another popular dimension reduction technique is the principal component analysis (PCA). In Figure 3(c), we repeat the same experiment to compare PCA with Gaussian random projections. Concerning PCA, the curve is generated in decreasing order of eigenvalues, according to their contribution to the explanation of the output-space variance. We observe that this way of doing is far less effective than the random projection techniques studied previously.

4.6   Learning Stage Computing Times

Our implementation of the learning algorithms is based on the *scikit-learn* Python package version 0.14-dev [18]. To fix ideas about computing times, we report these obtained on a Mac Pro 4.1 with a dual Quad-Core Intel Xeon processor at 2.26 GHz, on the "Delicious" dataset. Matrix operation, such as random projections, are performed with the BLAS and the LAPACK from the Mac OS

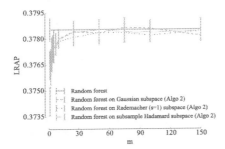

(a) Computing the impurity criterion on a dense Rademacher or on a subsample-Hadamard output sub-space is another efficient way to learn tree ensembles.

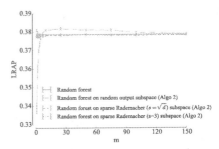

(b) Sparse random projections output sub-space yield better average precision than on the original output space.

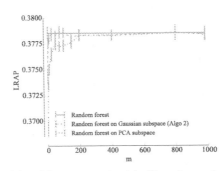

(c) PCA compared with Gaussian sub-spaces.

Fig. 3. "Delicious" dataset, $t = 100$, $k = \sqrt{p}$, $n_{\min} = 1$

X *Accelerate* framework. Reported times are obtained by summing the user and sys time of the *time* UNIX utility.

The reported timings correspond to the following operation: (i) load the dataset in memory, (ii) execute the algorithm. All methods use the same code to build trees. In these conditions, learning a random forest on the original output space ($t = 100$, $n_{\min} = 1$, $k = \sqrt{d}$) takes 3348 s; learning the same model on a Gaussian output space of size $m = 25$ requires 311 s, while $m = 1$ and $m = 250$ take respectively 236 s and 1088 s. Generating a Gaussian sub-space of size $m = 25$ and projecting the output data of the training samples is done in less than 0.25 s, while $m = 1$ and $m = 250$ takes around 0.07 s and 1 s respectively. The time needed to compute the projections is thus negligible with respect to the time needed for the tree construction.

We see that a speed-up of an order of magnitude could be obtained, while at the same time preserving accuracy with respect to the baseline Random Forests method. Equivalently, for a fixed computing time budget, randomly projecting the output space allows to build more trees and thus to improve predictive performances with respect to standard Random Forests.

5   Conclusions

This paper explores the use of random output space projections combined with tree-based ensemble methods to address large-scale multi-label classification problems. We study two algorithmic variants that either build a tree-based ensemble model on a single shared random subspace or build each tree in the ensemble on a newly drawn random subspace. The second approach is shown theoretically and empirically to always outperform the first in terms of accuracy. Experiments on 24 datasets show that on most problems, using gaussian projections allows to reduce very drastically the size of the output space, and therefore computing times, without affecting accuracy. Remarkably, we also show that by adjusting jointly the level of input and output randomizations and choosing appropriately the projection method, one could also improve predictive performance over the standard Random Forests, while still improving very significantly computing times. As future work, it would be very interesting to propose efficient techniques to automatically adjust these parameters, so as to reach the best tradeoff between accuracy and computing times on a given problem.

To best of our knowledge, our work is the first to study random output projections in the context of multi-output tree-based ensemble methods. The possibility with these methods to relabel tree leaves with predictions in the original output space makes this combination very attractive. Indeed, unlike similar works with linear models [13,8], our approach only relies on Johnson-Lindenstrauss lemma, and not on any output sparsity assumption, and also does not require to use any output reconstruction method. Besides multi-label classification, we would like to test our method on other, not necessarily sparse, multi-output prediction problems.

Acknowledgements. Arnaud Joly is research fellow of the FNRS, Belgium. This work is supported by PASCAL2 and the IUAP DYSCO, initiated by the Belgian State, Science Policy Office.

References

1. Achlioptas, D.: Database-friendly random projections: Johnson-lindenstrauss with binary coins. J. Comput. Syst. Sci. 66(4), 671–687 (2003)
2. Agrawal, R., Gupta, A., Prabhu, Y., Varma, M.: Multi-label learning with millions of labels: recommending advertiser bid phrases for web pages. In: Proceedings of the 22nd International Conference on World Wide Web, pp. 13–24. International World Wide Web Conferences Steering Committee (2013)
3. Blockeel, H., De Raedt, L., Ramon, J.: Top-down induction of clustering trees. In: Proceedings of ICML 1998, pp. 55–63 (1998)
4. Breiman, L.: Random forests. Machine Learning 45(1), 5–32 (2001)
5. Breiman, L., Friedman, J.H., Olshen, R.A., Stone, C.J.: Classification and Regression Trees. Wadsworth (1984)
6. Candes, E.J., Plan, Y.: A probabilistic and ripless theory of compressed sensing. IEEE Transactions on Information Theory 57(11), 7235–7254 (2011)

7. Cheng, W., Hüllermeier, E., Dembczynski, K.J.: Bayes optimal multilabel classification via probabilistic classifier chains. In: Proceedings of the 27th International Conference on Machine Learning (ICML 2010), pp. 279–286 (2010)
8. Cisse, M.M., Usunier, N., Artières, T., Gallinari, P.: Robust bloom filters for large multilabel classification tasks. In: Burges, C., Bottou, L., Welling, M., Ghahramani, Z., Weinberger, K. (eds.) Advances in Neural Information Processing Systems 26, pp. 1851–1859 (2013)
9. Dekel, O., Shamir, O.: Multiclass-multilabel classification with more classes than examples. In: International Conference on Artificial Intelligence and Statistics, pp. 137–144 (2010)
10. Faulon, J.L., Misra, M., Martin, S., Sale, K., Sapra, R.: Genome scale enzyme–metabolite and drug–target interaction predictions using the signature molecular descriptor. Bioinformatics 24(2), 225–233 (2008)
11. Geurts, P., Ernst, D., Wehenkel, L.: Extremely randomized trees. Machine Learning 63(1), 3–42 (2006)
12. Geurts, P., Wehenkel, L., d'Alché Buc, F.: Kernelizing the output of tree-based methods. In: Proceedings of the 23rd International Conference on Machine Learning, pp. 345–352. ACM (2006)
13. Hsu, D., Kakade, S., Langford, J., Zhang, T.: Multi-label prediction via compressed sensing. In: Bengio, Y., Schuurmans, D., Lafferty, J., Williams, C.K.I., Culotta, A. (eds.) Advances in Neural Information Processing Systems 22, pp. 772–780 (2009)
14. Johnson, W.B., Lindenstrauss, J.: Extensions of Lipschitz mappings into a Hilbert space. Contemporary mathematics 26(189-206), 1 (1984)
15. Kocev, D., Vens, C., Struyf, J., Dzeroski, S.: Tree ensembles for predicting structured outputs. Pattern Recognition 46(3), 817–833 (2013)
16. Li, P., Hastie, T.J., Church, K.W.: Very sparse random projections. In: Proceedings of the 12th ACM SIGKDD International Conference on Knowledge Discovery and Data Mining, pp. 287–296. ACM (2006)
17. Madjarov, G., Kocev, D., Gjorgjevikj, D., Dzeroski, S.: An extensive experimental comparison of methods for multi-label learning. Pattern Recognition 45(9), 3084–3104 (2012)
18. Pedregosa, F., Varoquaux, G., Gramfort, A., Michel, V., Thirion, B., Grisel, O., Blondel, M., Prettenhofer, P., Weiss, R., Dubourg, V., et al.: Scikit-learn: Machine learning in python. The Journal of Machine Learning Research 12, 2825–2830 (2011)
19. Read, J., Pfahringer, B., Holmes, G., Frank, E.: Classifier chains for multi-label classification. In: Buntine, W., Grobelnik, M., Mladenić, D., Shawe-Taylor, J. (eds.) ECML PKDD 2009, Part II. LNCS (LNAI), vol. 5782, pp. 254–269. Springer, Heidelberg (2009)
20. Tsoumakas, G., Katakis, I., Vlahavas, I.: Effective and efficient multilabel classification in domains with large number of labels. In: Proc. ECML/PKDD 2008 Workshop on Mining Multidimensional Data (MMD 2008), pp. 30–44 (2008)
21. Tsoumakas, G., Katakis, I., Vlahavas, I.P.: Random k-labelsets for multilabel classification. IEEE Trans. Knowl. Data Eng. 23(7), 1079–1089 (2011)
22. Tsoumakas, G., Vlahavas, I.P.: Random k-labelsets: An ensemble method for multilabel classification. In: Kok, J.N., Koronacki, J., Lopez de Mantaras, R., Matwin, S., Mladenič, D., Skowron, A. (eds.) ECML 2007. LNCS (LNAI), vol. 4701, pp. 406–417. Springer, Heidelberg (2007)
23. Zhou, T., Tao, D.: Multi-label subspace ensemble. In: International Conference on Artificial Intelligence and Statistics, pp. 1444–1452 (2012)

Communication-Efficient Distributed Online Prediction by Dynamic Model Synchronization[*]

Michael Kamp[1], Mario Boley[1], Daniel Keren[2],
Assaf Schuster[3], and Izchak Sharfman[3]

[1] Fraunhofer IAIS & University Bonn
{surname.name}@iais.fraunhofer.de
[2] Haifa University
dkeren@cs.haifa.ac.il
[3] Technion, Israel Institute of Technology
{assaf,tsachis}@technion.ac.il

Abstract. We present the first protocol for distributed online prediction that aims to minimize online prediction loss and network communication at the same time. This protocol can be applied wherever a prediction-based service must be provided timely for each data point of a multitude of high frequency data streams, each of which is observed at a local node of some distributed system. Exemplary applications include social content recommendation and algorithmic trading. The challenge is to balance the joint predictive performance of the nodes by exchanging information between them, while not letting communication overhead deteriorate the responsiveness of the service. Technically, the proposed protocol is based on controlling the variance of the local models in a decentralized way. This approach retains the asymptotic optimal regret of previous algorithms. At the same time, it allows to substantially reduce network communication, and, in contrast to previous approaches, it remains applicable when the data is non-stationary and shows rapid concept drift. We demonstrate empirically that the protocol is able to hold up a high predictive performance using only a fraction of the communication required by benchmark methods.

1   Introduction

We consider distributed online prediction problems on multiple connected high-frequency data streams where one is interested in minimizing predictive error and communication at the same time. This situation abounds in a wide range of machine learning applications, in which communication induces a severe cost. Examples are parallel data mining [23, 10] and M2M communication [20] where communication constitutes a performance bottleneck, learning with mobile sensors [16, 18] where communication drains battery power, and, most centrally,

---

[*] A preliminary extended abstract of this paper was presented at the BD3 workshop at VLDB'13. This research has been supported by the EU FP7-ICT-2013-11 under grant 619491 (FERARI).

T. Calders et al. (Eds.): ECML PKDD 2014, Part I, LNCS 8724, pp. 623–639, 2014.

prediction-based real-time services [8] carried out by several servers, e.g., for social content promotion, ad placement, or algorithmic trading. Here, due to network latency, the cost of communication can also be a loss of prediction quality itself, because, in order to avoid inconsistent system states some data points have to be discarded for learning whenever a communication event is triggered. In this paper, we abstract on all these various motivations and provide a protocol that aims to minimize communication as such. In particular, we provide the first protocol that dynamically adapts communication to exploit the communication reduction potential of well-behaved input sequences but at the same time retains the predictive performance of static communication schemes.

In contrast to work on the communication complexity of batch learning [3, 17, 2, 7], we consider the online in-place performance of a streaming distributed prediction system. For this setting, earlier research focused on strategies that communicate periodically after a fixed number of data points have been processed [13, 8]. For these static communication schemes Dekel et al. [8] shows that for smooth loss functions and stationary environments optimal asymptotic regret bounds can be retained by updating a global model only after observing a mini-batch of examples. While such a fixed periodic communication schedule reduces the communication, further reduction is desirable: the above mentioned costs of communication can have a severe impact on the practical performance—even if they are not reflected in asymptotic performance bounds. Moreover, distributed learning systems can experience periodical or singular target drifts. In these settings, a static schedule is bound to either provide only little to none communication reduction or to insufficiently react to changing data distributions.

In this work, we give the first data-dependent distributed prediction protocol that dynamically adjusts the amount of communication performed depending on the hardness of the prediction problem. It aims to provide a high online in-place prediction performance and, at the same time, explicitly tries to minimize communication. The underlying idea is to perform model synchronizations only in system states that show a high variance among the local models, which indicates that a synchronization would be most effective in terms of its correcting effect on future predictions. While the model variance is a non-linear function in the global system, we describe how it can be monitored locally in a communication-efficient way. The resulting protocol allows communicative quiescence in stable phases, while, in hard phases where variance reduction is crucial, the protocol will trigger a lot of model synchronizations. Thus, it remains applicable when the data is non-stationary and shows rapid concept drifts—cases in which a static scheme is doomed to either require a high communication frequency or suffer from low adaption. We show theoretically (Sec. 3.1), that, despite the communication reduction achieved by our dynamic protocol, it retains any shifting regret bounds provided by its static counterpart. We also demonstrate its properties empirically (Sec. 4) with controlled synthetic data and real-world datasets from stock markets and the short-message service Twitter.

2   Preliminaries

In this section we formally introduce the distributed online prediction task. We recall simple sequential learning algorithms and discuss a basic communication scheme to utilize them in the distributed scenario.

2.1   Distributed Online Prediction

Throughout this paper we consider a distributed online prediction system of k **local learners** that maintain individual **linear models** $w_{t,1}, \ldots, w_{t,k} \in \mathbb{R}^n$ of some global environment through discrete time $t \in [T]$ where $T \in \mathbb{N}$ denotes the total time horizon with respect to which we analyze the system's performance. This environment is represented by a **target distribution** $\mathcal{D}_t \colon X \times Y \to [0,1]$ that describes the relation between an input space $X \subseteq \mathbb{R}^n$ and an output space $Y \subseteq \mathbb{R}$. The nature of Y varies with the learning task at hand; $Y = \{-1, 1\}$ is used for binary classification, $Y = \mathbb{R}$ for regression. Generally, we assume that all training examples $x \in X$ are drawn from a ball of **radius** R and also that $x_n = 1$ for all $x \in X$, i.e., $\|x\| \in [1/n, R]$—two common assumptions in online learning (the latter avoids to explicitly fit a bias term of the linear models). All learners sample from $\mathcal{D}_t$ independently in parallel using a constant and uniform sampling frequency, and we denote by $(x_{t,l}, y_{t,l}) \sim \mathcal{D}_t$ the **training example** received at node l at time t. Note that, while the underlying environment can change over time, we assume that at any given moment t there is one fixed distribution governing the points observed at all local nodes.

Conceptually, every learner first observes the input part $x_{t,l}$ and performs a real time service based on the linear **prediction score** $p_{t,l} = \langle w_{t,l}, x_{t,l} \rangle$, i.e., the inner product of $x_{t,l}$ and the learner's current model vector. Only then it receives as feedback the true label $y_{t,l}$, which it can use to locally update its model to $w_{t+1,l} = \varphi(w_{t,l}, x_{t,l}, y_{t,l})$ by some **update rule** $\varphi \colon \mathbb{R}^n \times X \times Y \to \mathbb{R}^n$. Let $W_t \in \mathbb{R}^{k \times n}$ denote the complete **model configuration** of all local models at time t (denoting by $w_{t,l}$ the model at learner l at time t as above). The learners are connected by a communication infrastructure that allows them to jointly perform a **synchronization operation** $\sigma \colon \mathbb{R}^{k \times n} \to \mathbb{R}^{k \times n}$ that resets the whole model configuration to a new state after local updates have been performed. This operator may take into account the information of all local learners simultaneously. The two components (φ, σ) define a **distributed learning protocol** that, given the inputs of the environment, produces a sequence of model configurations $\mathbf{W} = W_1, \ldots, W_T$. Its performance is measured by:

1. the in-place predictive performance $\sum_{t=1}^{T} \sum_{l=1}^{k} f(w_{t,l}, x_{t,l}, y_{t,l})$ measured by a **loss function** $f \colon \mathbb{R}^n \times \mathbb{R}^n \times Y \to \mathbb{R}_+$ that assigns positive penalties to prediction scores based on how (in-)appropriately they describe the true label; and

2. the amount of **communication** within the system that is measured by the number of bits send in-between learners in order to compute the synchronization operation σ.

Regarding the predictive performance, one is typically interested in bounding the average **regret** of the model configurations produced by the protocol with respect to a **reference sequence** $\mathbf{U} = U_1, \ldots, U_T$. For technical reasons, in this paper we focus on the squared regret, i.e.,

$$R(\mathbf{W}, \mathbf{U}) = \frac{1}{T} \sum_{t=1}^{T} \frac{1}{k} \sum_{l=1}^{k} (f(w_{t,l}, x_{t,l}, y_{t,l}) - f(u_{t,l}, x_{t,l}, y_{t,l}))^2 .$$

This type of regret is often referred to as shifting regret (see, Herbster and Warmuth [9]) and typically bounds are given in the total shift per node of the reference sequence $\sum_{t=1}^{T} \sum_{l=1}^{k} \|u_{t,l} - u_{t-1,l}\|^2$. Traditional results often restrict regret analysis to the case of a static reference sequence, i.e., $u_{1,1} = u_{1,2} = \cdots = u_{t,l}$. This is particularly useful if we consider the **stationary scenario** where $\mathcal{D}_1 = \cdots = \mathcal{D}_T$.

2.2   Loss-Proportional Convex Update Rules

Principally, the protocol developed in this paper can be applied to a wide range of update rules for online learning (from, e.g., stochastic gradient descend [24] to regularized dual averaging [21]). For the formal analysis, however, we focus on update rules covered by the following definition.

Definition 1. *We call an update rule φ an f-**proportional convex update** for a loss function f if there are a constant $\gamma > 0$, a closed convex set $\Gamma_{x,y} \subseteq \mathbb{R}^n$, and $\tau_{x,y} \in (0, 1]$ such that for all $w \in \mathbb{R}^n$, $x \in X$, and $y \in Y$ it holds that*

(i) $\|w - \varphi(w, x, y)\| \geq \gamma f(w, x, y)$, *i.e., the update magnitude is a true fraction of the loss incurred, and*

(ii) $\varphi(w, x, y) = w + \tau_{x,y} (P_{x,y}(w) - w)$ *where $P_{x,y}(w)$ denotes the projection of w onto $\Gamma_{x,y}$, i.e., the update direction is identical to the direction of a convex projection that only depends on the training example.*

As a first example for update rules satisfying these conditions, consider the **passive aggressive update rules** [6]. These rules are defined for a variety of learning tasks including classification, regression, and uni-class prediction and can be uniformly described by

$$\varphi(w, x, y) = \arg\min_{w' \in \mathbb{R}^n} \frac{1}{2} \|w - w'\|^2 \ \ s.t. \ \ f(w', x, y) = 0 \tag{1}$$

where for classification f is the **hinge loss**, i.e., $f(w, x, y) = \max(1 - y\langle w, x\rangle, 0)$, for regression the ϵ-**insensitive loss**, i.e., $f(w, x, y) = \max(|\langle w, x\rangle - y| - \epsilon, 0)$, and for uni-class prediction (where no x is observed and $Y = \mathbb{R}^n$) the loss is given by $f(w, y) = \max(|w - y| - \epsilon, 0)$. It can be observed immediately that, in all three cases, these update rules are an actual projection on the convex set $\Gamma_{x,y} = \{w \in \mathbb{R}^n : f(w, x, y) = 0\}$, which corresponds to a half-space, a 2ϵ-strip, and an ϵ-ball, respectively. Hence, Cond. (ii) of the definition follows immediately with $\tau_{x,y} = 1$. Cond. (i) can then be verified from the closed form solution of Eq. 1, which in case of classification is given by

$$\varphi(w, x, y) = w + \frac{f(w, x, y)}{\|x\|^2} yx \ .$$

Using the data radius R, we can easily bound the update magnitude from below as $\|w - \varphi(w, x, y)\| \geqslant R^{-1}f(w, x, y)$, i.e., Cond. (i) holds with $\gamma = R^{-1}$. The other cases follow similarly. Crammer et al. [6] also gives other variants of passive aggressive updates that have a reduced learning rate determined by an aggressiveness parameter $C > 0$. These rules also satisfy the conditions of Def. 1. For example the rule for classification then becomes

$$\varphi(w, x, y) = w_t + \frac{f(w, x, y)}{\|x\|^2 + \frac{1}{2C}} yx \ .$$

Using $\|x\| \in [1/n, R]$, one can show that this variant remains hinge-loss proportional with $\gamma = n^{-1}(R^2 + 1/(2C))^{-1}$, and the update direction is identical to the same convex projection as in the standard case.

Another popular family of update rules for differentiable loss functions is given by **stochastic gradient descent**, i.e., rules of the form

$$\varphi(w, x, y) = w - \eta \nabla_w f(w, x, y)$$

with a positive learning rate $\eta > 0$. If one uses the squared hinge loss, $f(w, x, y) = 1/2 \max(1 - y\langle w, x \rangle, 0)^2$, we have $\nabla_w f(w, x, y) = y(1 - y\langle w, x \rangle)x$. Hence, this update rule is hinge loss proportional with $\gamma = \eta/n$, and the update direction is identical to the passive aggressive update rule for classification— that is, in the direction of a convex projection. The same can be checked for regression using the squared ϵ-insensitive loss and many other variants of gradient descent.

In the following we will define a static averaging protocol that reduces the communication cost in a distributed online learning scenario and serves as baseline to our dynamic synchronization protocol.

2.3   Static Averaging

In terms of cost, every synchronization operator lies between two extreme baselines—constant broadcast of all training examples and quiescence, i.e., no communication at all. The predictive performance of these two extremes in terms of static regret lies between $O(\sqrt{kT})$ for serial learning (which is optimal for the stationary setting, see Cesa-Bianchi and Lugosi [5] and [1]) and $O(k\sqrt{T})$ for no communication, which corresponds to solving k separate online learning problems in parallel.

An intermediate solution is to only reset all local models to their joint average every b rounds where $b \in \mathbb{N}$ is referred to as **batch size** (see Mcdonald et al. [13] and Dekel et al. [8]). Formally, this **static averaging operator** is given by $\sigma(W_t) = (\overline{W}_t, \ldots, \overline{W}_t)$ if $t \mod b = 0$ and $\sigma(W_t) = W_t$, otherwise. Here,

---

Algorithm 1. Static Averaging Protocol

Initialization:
    local models $w_{1,1}, \ldots, w_{1,k} \leftarrow (0, \ldots, 0)$

Round t at node l:
    observe $x_{t,l}$ and provide service based on $p_{t,l}$
    observe $y_{t,l}$ and **update** $w_{t+1,l} \leftarrow \varphi(w_{t,l}, x_t, y_t)$
    if $t \mod b = 0$ **then**
        send $w_{t,l}$ to coordinator

At coordinator every b rounds:
    receive local models $\{w_{t,l} : l \in [k]\}$
    For all $l \in [k]$ **set** $w_{t,l} \leftarrow \sigma(w_{t,1}, \ldots, w_{t,k})_l$

$\overline{W}_t = 1/k \sum_{l=1}^{k} w_{t,l}$ denotes the mean model. This choice of a (uniform) model mixture is often used for combining linear models that have been learned in parallel on independent training data (see also McDonald et al. [14], Zinkevich et al. [24]). The motivation is that the mean of k models provides a variance reduction of $\sqrt{k}$ over an individual random model (recall that all learners sample from the same distribution, hence their models are identically distributed). For certain learning problems in the stationary setting, it can even be shown that this protocol retains the asymptotically optimal regret of $O(\sqrt{kT})$ [8] for small enough batch sizes[1].

For assessing the communication cost of this operation, we use a simplified cost model that only counts the number of model vectors sent between the learners: independently of the exact communication infrastructure, the number of model messages asymptotically determines the true bit-based communication cost. Using a designated coordinator note as in Alg. 1, σ can be applied to a configuration of the distributed prediction system simply by all nodes sending their current model to the coordinator, who in turn computes the mean model and sends it back to all the nodes. Hence, the **communication cost** of static averaging over k nodes with batch size b is $O(kT/b)$.

While this is less than the naive baseline by a factor of b, in many scenarios the achieved reduction might still be insufficient. In particular, for non-stationary settings the batch size has to be chosen small enough for the protocol to remain adaptive to changes in the environment so that the communication reduction effect can be marginal. A big weakness of the scheme is that it is oblivious to the actual model configuration observed, so that it also induces a lot of communication in situations where all models are approximately identical. In the

---

[1] Dekel et al. [8] consider a slightly modified algorithm, which accumulates updates and then only applies them delayed at the end of a batch. However, the expected loss of eager updates (as used in Alg. 1) is bounded by the expected loss of delayed updates in the stationary setting (as used in Dekel et al. [8]) as long as the updates reduce the distance to a loss minimizer on average (which is the case for sufficient regularization; see again Zhang [22, Eq. 5]).

following section, we present a data-dependent dynamic averaging operator that can substantially reduce the communication cost while approximately retaining the performance of static averaging.

3   Dynamic Synchronization

In this section, we develop a dynamic protocol for synchronizations based on quantifying their effect. In order to assess the performance of this protocol from a learning perspective, we compare it to the static protocol as described in Alg. 1. After showing that this approach is sound from a learning perspective, we discuss how it can be implemented in a distributed prediction system in a communication-efficient way.

3.1   Partial Averaging

Intuitively, the communication for performing model averaging is not well invested in situations where all models are already approximately equal. A simple measure to quantify the effect of synchronizations is given by the **variance** of the current local model configuration space, i.e., $\delta(W) = \frac{1}{k} \sum_{l=1}^{k} \|\overline{W} - W_l\|^2$. In the following definition we provide a relaxation of the static averaging operation that allows to omit synchronization in cases where the variance of a model configuration is low.

Definition 2. *A **partial averaging operator** with positive variance threshold $\Delta \in \mathbb{R}$ and batch size $b \in \mathbb{N}$ is a synchronization operator σ_Δ such that $\sigma_\Delta(W_t) = W_t$ if $t \mod b \neq 0$ and otherwise: (i) $\overline{W_t} = \overline{\sigma_\Delta(W_t)}$, i.e., it leaves the mean model invariant, and (ii) $\delta(\sigma_\Delta(W)) \leq \Delta$, i.e., after its application the model variance is bounded by Δ.*

An operator adhering to this definition does not generally put all nodes into sync (albeit the fact that we still refer to it as *synchronization* operator). In particular it allows to leave all models untouched as long as the variance remains below the threshold Δ or to only average a subset of models in order to satisfy the variance constraint. This is the basis for our dynamic averaging protocol. In the following, we analyze the impact on the learning performance of using partial averaging instead of static averaging. We start with showing that, given two model configurations D and S, applying the partial averaging operator σ_Δ to D and the static averaging operator σ to S increases their average squared pairwise model distances by at most Δ.

Lemma 3. *Let $D_t, S_t \in \mathbb{R}^{k \times n}$ be model configurations at time $t \in \mathbb{N}$. Then*

$$\frac{1}{k} \sum_{l=1}^{k} \|\sigma_\Delta(D_t)_l - \sigma(S_t)_l\|^2 \leq \frac{1}{k} \sum_{l=1}^{k} \|d_{t,l} - s_{t,l}\|^2 + \Delta .$$

Proof. We consider the case $t \mod b = 0$ (otherwise the claim follows immediately). Expressing the pairwise squared distances via the difference to $\overline{D_t}$ and using the definitions of σ and σ_Δ we can bound

$$\frac{1}{k}\sum_{l=1}^{k}\|\sigma_\Delta(D_t)_l - \sigma(S_t)_l\|^2 = \frac{1}{k}\sum_{l=1}^{k}\|\sigma_\Delta(D_t)_l - \overline{D_t} + \overline{D_t} - \overline{S_t}\|^2$$

$$= \underbrace{\frac{1}{k}\sum_{l=1}^{k}\|\sigma_\Delta(D_t)_l - \overline{D_t}\|^2}_{\leq \Delta,\ \text{by (ii) of Def. 2}} + \underbrace{2\langle\frac{1}{k}\sum_{l=1}^{k}\sigma_\Delta(D_t)_l - \overline{D_t}, \overline{D_t} - \overline{S_t}\rangle}_{=0,\ \text{by (i) of Def. 2}} + \|\overline{D_t} - \overline{S_t}\|^2$$

$$\leq \Delta + \|\frac{1}{k}\sum_{l=1}^{k}(d_{t,l} - s_{t,l})\|^2 = \Delta + \frac{1}{k}\sum_{l=1}^{k}\|d_{t,l} - s_{t,l}\|^2 \ .$$

$\square$

In order to prove a regret bound of partial over static averaging it remains to show that this increase in distance cannot separate model configurations too far during the learning process. For this we show that f-propotional convex updates on the same training example reduce the distance between a pair of models proportional to their loss difference.

Lemma 4. *Let φ be an f-proportional convex update rule with constant $\gamma > 0$. Then for all models $d, s \in \mathbb{R}^n$ it holds that*

$$\|\varphi(d, x, y) - \varphi(s, x, y)\|^2 \leq \|d - s\|^2 - \gamma^2 \left(f(d, x, y) - f(s, x, y)\right)^2 .$$

Proof. For $w \in \mathbb{R}^n$ we write $P_{x,y}(w) = P(w)$ for the projection of w on $\Gamma_{x,y}$ and $w' = \varphi(w, x, y)$. Since $P(\cdot)$ is a projection on a convex set, it holds for all $v, w \in \mathbb{R}^n$ that

$$\|P(v) - P(w)\|^2 \leq \|v - w\|^2 - \|v - P(v) - w + P(w)\|^2 \tag{2}$$

(e.g., by lemma 3.1.4 in Nesterov [15]). Also since $w' = \tau_{x,y}P(w) + (1 - \tau_{x,y})w$ by (ii) of the definition of f-proportional convex updates, the idempotence of $P(\cdot)$ implies that $P(w) = P(w')$. Applying (2) to the models d, s and to the updated models d', s', respectively, and subtracting the two inequalities gives

$$0 \leq \|d - s\|^2 - \|d' - s'\|^2 - \|d - P(d) - s + P(s)\|^2 + \|d' - P(d) - s' + P(s)\|^2 \ .$$

By inserting $w' = w + \tau_{x,y}(P(w) - w)$ and using $\tau_{x,y} \in (0, 1]$ it follows that

$$\begin{aligned}
\|d' - s'\|^2 &\leq \|d - s\|^2 - \|(d - P(d)) - s + P(s)\|^2 \\
&\quad + (1 - \tau_{x,y})^2\|(d - P(d)) - s + P(s)\|^2 \\
&\leq \|d - s\|^2 - \tau_{x,y}\left(\|d - P(d)\| - \|s - P(s)\|\right)^2 \\
&\leq \|d - s\|^2 - \gamma^2\left(f(d, x, y) - f(s, x, y)\right)^2
\end{aligned} \tag{3}$$

as required, where the last inequality follows from $\tau_{x,y} \in (0,1]$ and (i) of the definition of f-proportionality by noting that

$$\|w - P(w)\| = \frac{1}{\tau_{x,y}}\|w - (w + \tau_{x,y}(P(w) - w))\| = \frac{\|w - w'\|}{\tau_{x,y}} \geq \frac{\gamma}{\tau_{x,y}}f(w,x,y) \ .$$

$\square$

From the two lemmas above we see that, while each synchronization increases the distance between the static and the dynamic model by at most Δ, with each update step, the distance is decreased proportional to the loss difference. In the following theorem, we state that the average squared regret of using a partial averaging operator σ_Δ over a static averaging operator σ with batch size b is bounded by $\Delta/(b\gamma^2)$. We use the notion $\varphi(W_t) = (\varphi(w_{t,1}, x_{t,1}, y_{t,1}), \dots, \varphi(w_{t,k}, x_{t,k}, y_{t,k}))$.

Theorem 5. *Let* $\mathbf{D} = D_0, \dots, D_T$ *and* $\mathbf{S} = S_0, \dots, S_T$ *be two sequences of model configurations such that* $D_0 = S_0$ *and for* $t = 1, \dots, T$ *defined by* $D_{t+1} = \sigma_\Delta(\varphi(D_t))$ *and* $S_{t+1} = \sigma(\varphi(S_t))$, *respectively (with an identical batch size* $b \in N$). *Then it holds that* $R(\mathbf{D}, \mathbf{S}) \leq \Delta/(b\gamma^2)$.

Proof. Let $\beta_t = 1$ if $t \mod b = 0$ and $\beta_t = 0$ otherwise. By combining Lm. 3 and 4 we have for all $t \in [T]$ that

$$\frac{1}{k}\sum_{l=1}^{k}\|d_{t+1,l} - s_{t+1,l}\|^2 \leq \frac{1}{k}\sum_{l=1}^{k}\|d_{t,l} - s_{t,l}\|^2 - \frac{\gamma^2}{k}\sum_{l=1}^{k}(f(d_{t,l}) - f(s_{t,l}))^2 + \beta_t\Delta \ .$$

Applying this inequality recursively for $t = 0, \dots, T$, it follows that

$$\frac{1}{k}\sum_{l=1}^{k}\|d_{T+1,l} - s_{T+1,l}\|^2 \leq \frac{1}{k}\sum_{l=1}^{k}\|d_{0,l} - s_{0,l}\|^2 + \left\lfloor\frac{T}{b}\right\rfloor\Delta$$

$$-\sum_{t=1}^{T}\frac{\gamma^2}{k}\sum_{l=1}^{k}(f(d_{t,l}) - f(s_{t,l}))^2.$$

Using $D_0 = S_0$ we can conclude

$$\sum_{t=1}^{T}\frac{1}{k}\sum_{l=1}^{k}(f(d_{t,l}) - f(s_{t,l}))^2 \leq \frac{1}{\gamma^2}\left(\left\lfloor\frac{T}{b}\right\rfloor\Delta - \frac{1}{k}\sum_{l=1}^{k}\|d_{T+1,l} - s_{T+1,l}\|^2\right) \leq \frac{T}{b\gamma^2}\Delta$$

which yields the result after dividing both sides by T. $\square$

We remark that Thm. 5 implies that partial averaging retains the optimality of the static mini-batch algorithm of Dekel et al. [8] for the case of stationary targets: by using a time-dependent variance threshold based on $\Delta_t \in O(1/\sqrt{t})$ the bound of $O(\sqrt{T})$ follows. From Thm. 5 it follows that if a shifting bound exists for the static protocol then this bound also applies to the dynamic protocol.

Formally, suppose the shifting regret $R(\mathbf{S}, \mathbf{U})$ of using the static averaging operator is bounded by $c_1 \sum_{t=1}^{T} \sum_{l=1}^{k} \|u_{t,l} - u_{t-1,l}\|_2^2 + c_2$, for a reference sequence $\mathbf{U}$ and positive constants $c_1, c_2 \in \mathbb{R}_+$ (as, e.g., in [9]). Then the shifting regret of using dynamic averaging is bounded by

$$R(\mathbf{D}, \mathbf{U}) \leq c_1 \sum_{t=1}^{T} \sum_{l=1}^{k} \|u_{t,l} - u_{t-1,l}\|_2^2 + c_2 + \frac{1}{\gamma^2}\Delta \ ,$$

where $\mathbf{D}$ denotes the sequence of model configurations produced by σ_Δ. For the proof let furthermore $\mathbf{S}$ denote the sequence of model configurations produced by σ. With this we can directly derive the bound by using the definition of shifting regret, i.e.,

$$R(\mathbf{D}, \mathbf{U}) = \frac{1}{T} \sum_{t=1}^{T} \frac{1}{k} \sum_{l=1}^{k} (f(d_{t,l}) - f(u_{t,l}))^2$$

$$= \frac{1}{T} \sum_{t=1}^{T} \frac{1}{k} \sum_{l=1}^{k} ((f(d_{t,l}) - f(s_{t,l})) + (f(s_{t,l}) - f(u_{t,l})))^2$$

$$\overset{Thm.5}{\leq} \frac{1}{\gamma^2}\Delta + \frac{1}{T} \sum_{t=1}^{T} \frac{1}{k} \sum_{l=1}^{k} (f(s_{t,l}) - f(u_{t,l}))^2$$

$$\leq \frac{1}{\gamma^2}\Delta + R(\mathbf{S}, \mathbf{U}) = \frac{1}{\gamma^2}\Delta + c_1 \sum_{t=1}^{T} \sum_{l=1}^{k} \|u_{t,l} - u_{t-1,l}\|_2^2 + c_2 \ .$$

Intuitively, this means that the dynamic protocol only adds a constant to any shifting bound of static averaging.

3.2   Communication-Efficient Protocol

After seeing that partial averaging operators are sound from the learning perspective, we now turn to how they can be implemented in a communication-efficient way. Every distributed learning algorithm that implements a partial averaging operator has to implicitly control the variance of the model configuration. However, we cannot simply compute the variance by centralizing all local models, because this would incur just as much communication as static full synchronization. Our strategy to overcome this problem is to first decompose the global condition $\delta(W) \leq \Delta$ into a set of local conditions that can be monitored at their respective nodes without communication (see, e.g., Sharfman et al. [19]). Secondly, we define a resolution protocol that transfers the system back into a valid state whenever one or more of these local conditions are violated. This includes carrying out a sufficient amount of synchronization to reduce the variance to be less or equal than Δ.

For deriving local conditions we consider the domain of the variance function restricted to an individual model vector. Here, we identify a condition similar

---

Algorithm 2. Dynamic Synchronization Protocol

Initialization:
    local models $w_{1,1}, \ldots, w_{1,k} \leftarrow (0, \ldots, 0)$
    reference vector $r \leftarrow (0, \ldots, 0)$
    violation counter $v \leftarrow 0$

Round t at node l:
    observe $x_{t,l}$ and provide service based on $p_{t,l}$
    observe $y_{t,l}$ and **update** $w_{t+1,l} \leftarrow \varphi(w_{t,l}, x_t, y_t)$
    if $t \mod b = 0$ **and** $\|w_{t,l} - r\|^2 > \Delta$ **then**
        send $w_{t,l}$ to coordinator (violation)

At coordinator on violation:
    let B be the set of nodes with violation
    $v \leftarrow v + |B|$
    if $v = k$ **then** $B \leftarrow [k]$, $v \leftarrow 0$
    while $B \neq [k]$ **and** $\frac{1}{B} \sum_{l \in B} \|w_{t,l} - r\|^2 > \Delta$ **do**
        augment B by augmentation strategy
        receive models from nodes added to B
    send model $\overline{\mathbf{w}} = \frac{1}{B} \sum_{l \in B} w_{t,l}$ to nodes in B
    if $B = [k]$ also set new reference vector $r \leftarrow \overline{\mathbf{w}}$

---

to a safe-zone (see Keren et al. [12]) such that the global variance can not cross the Δ-threshold as long as all local models satisfy that condition.[2]

Theorem 6. *Let $D_t = d_{t,1}, \ldots, d_{t,k} \in \mathbb{R}^n$ be the model configuration at time t and $r \in \mathbb{R}^n$ be some **reference vector**. If for all $l \in [k]$ the **local condition** $\|d_{t,l} - r\|^2 \leq \Delta$ holds, then the global variance is bounded by Δ, i.e.,*

$$\frac{1}{k} \sum_{l=1}^{k} \|d_{t,l} - \overline{D}_t\|^2 \leq \Delta \ .$$

Proof. The theorem follows directly from the fact that the current average vector $\overline{D}_t$ minimizes the squared distances to all $d_{t,i}$, i.e.,

$$\frac{1}{k} \sum_{i=1}^{k} \|d_{t,l} - \overline{D}_t\|^2 \leq \frac{1}{k} \sum_{i=1}^{k} \|d_{t,l} - r\|^2 \leq \Delta$$

$\square$

We now incorporate these local conditions into a distributed prediction algorithm. As a first step, we have to guarantee that at any time all nodes use the same reference vector r, for which a natural choice is the last average model that has been set to all local nodes. If the reference vector is known to all local

---

[2] Note that a direct distribution of the threshold across the local nodes (as in, e.g., Keralapura et al. [11]) is in-feasible, because the variance function is non-linear.

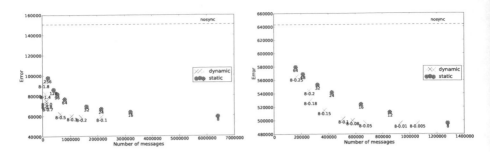

Fig. 1. Performance of static and dynamic model synchronization that track (left) a rapidly drifting disjunction over 100-dim. data with 512 nodes; and (right) a neural network with one hidden layer and 150 output vars. with 1024 nodes.

learners a local learners l can then monitor its local condition $\|d_{t,l} - r\|^2 \leq \Delta$ in a decentralized manner.

It remains to design a resolution protocol that specifies how to react when one or several of the local conditions are violated. A direct solution is to trigger a full synchronization in that case. This approach, however, does not scale well with a high number of nodes in cases where model updates have a non-zero probability even in the asymptotic regime of the learning process. When, e.g., PAC models for the current target distribution are present at all local nodes, the probability of one local violation, albeit very low for an individual node, increases exponentially with the number of nodes. An alternative approach that can keep the amount of communication low relative to the number of nodes is to perform a local balancing procedure: on a violation, the respective node sends his model to a designated node we refer to as coordinator. The coordinator then tries to balance this violation by incrementally querying other nodes for their models. If the mean of all received models lies within the safe zone, it is transferred back as new model to all participating nodes, and the resolution is finished. If all nodes have been queried, the result is equal to a full synchronization and the reference vector can be updated. In both cases, the variance of the model configuration is bounded by Δ at the end of the balancing process, because all local conditions hold. Also, it is easy to check that this protocol leaves the global mean model unchanged. Hence, it is complying to Def. 2.

While balancing can achieve a high communication reduction over direct resolution particularly for a large number of nodes, it potentially degenerates in certain special situations. We can end up in a stable regime in which local violations are likely to be balanced by a subset of the nodes; however a full synchronization would strongly reduce the expected number of violations in future rounds. In other words: balancing can delay crucial reference point updates indefinitely. A simple hedging mechanism for online optimization can be employed in order to avoid this situation: we count the number of local violations using the current reference point and trigger a full synchronization whenever this number exceeds the total number of nodes. This concludes our dynamic protocol for distributed prediction. All components are summarized in Alg. 2

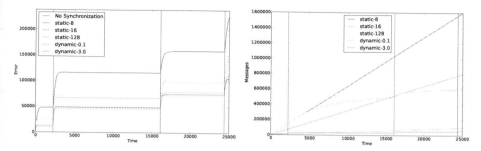

Fig. 2. Cumulative error (left) and communication (right) over time for tracking a rapidly drifting disjunction for different synchronization protocols; vertical lines depict drifts.

4   Empirical Evaluation

In this section we investigate the practical performance of the dynamic learning protocol for settings ranging from clean linearly separable data, over unseparable data with a reasonable linear approximation, up to real-world data without any guarantee. Our main goal is to empirically confirm that the predictive gain of static full synchronizations (using a batch size of 8) over no synchronization can be approximately preserved for small enough thresholds, and to assess the amount of communication reduction achieved by these thresholds.

4.1   Linearly Separable Data

We start with the problem of tracking a rapidly drifting random disjunction. In this case the target distribution produces data that is episode-wise linearly separable. Hence, we can set up the individual learning processes so that they converge to a linear model with zero classification error within each episode. Formally, we identify a target disjunction with a binary vector $z \in \{0,1\}^n$. A data point $x \in X = \{0,1\}^n$ is labeled positively $y = 1$ if $\langle x, z \rangle \geq 1$ and otherwise receives a negative label $y = -1$. The target disjunction is drawn randomly at the beginning of the learning process and is randomly re-set after each round with a fixed drift probability of 0.0001. In order to have balanced classes, the disjunctions as well as the data points are generated such that each coordinate is set independently to 1 with probability $\sqrt{1 - 2^{-1/n}}$. We use the unregularized passive aggressive update rule with hinge loss.

In Fig. 1 (left) we present the result for dimensionality $n = 100$, with $k = 512$ nodes, processing $m = 12.8M$ data points through $T = 100000$ rounds. For divergence thresholds up to 0.3, dynamic synchronization can retain the error number of statically synchronizing every 8 rounds. At the same time the communication is reduced to 9.8% of the original number of messages. An approximately similar amount of communication reduction can also be achieved using static synchronization by increasing the batch size to 96. This approach, however, only retains 61.0% of the accuracy of statically synchronizing every 8 rounds.

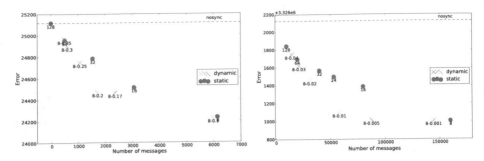

Fig. 3. Performance of static and dynamic synchronization with 256 nodes that predict (left) Twitter retweets over 1000 textual features and (right) stock prices based on 400 prices and sliding averages.

Fig. 2 provides some insight into how the two evaluation metrics develop over time. Target drifts are marked with vertical lines that frame episodes of a stable target disjunction. At the beginning of each episode there is a relatively short phase in which additional errors are accumulated and the communicative protocols acquire an advantage over the baseline of never synchronizing. This is followed by a phase during which no additional error is made. Here, the communication curve of the dynamic protocols remain constant acquiring a gain over the static protocols in terms of communication.

4.2   Non-separable Data with Noise

We now turn to a harder experimental setting, in which the target distribution is given by a rapidly drifting two-layer neural network. For this target even the Bayes optimal classifier per episode has a non-zero error, and, in particular, the generated data is not linearly separable. Intuitively, it is harder in this setting to save communication, because a non-zero residual error can cause the linear models to periodically fluctuate around a local loss minimizer—resulting in crossings of the variance threshold even when the learning processes have reached their asymptotic regime. We choose the network structure and parameter ranges in a way that allow for a relatively good approximation by linear models (see Bshouty and Long [4]). The process for generating a single labeled data point is as follows: First, the label $y \in Y = \{-1, 1\}$ is drawn uniformly from Y. Then, values are determined for hidden variables H_i with $1 \leq i \leq \lceil \log n \rceil$ based on a Bernoulli distribution $P[H_i = \cdot | Y = y] = \text{Ber}(p_{i,y}^h)$. Finally, $x \in X = \{-1, 1\}^n$ is determined by drawing x_i for $1 \leq i \leq n$ according to $P[X_i = x_i, |H_{p(i)} = h] = \text{Ber}(p_{i,h}^o)$ where $p(i)$ denotes the unique hidden layer parent of x_i. In order to ensure linear approximability, the parameters of the output layer are drawn such that $|p_{i,-1}^o - p_{i,1}^o| \geq 0.9$, i.e., their values have a high *relevance* in determining the hidden values. As in the disjunction case all parameters are re-set randomly after each round with a fixed drift probability (here, 0.01). For this non-separable setting we choose again to optimize the hinge loss, this time with regularized passive aggressive updates with $C = 10.0$ and a batch size of $b = 8$.

Fig. 1 (right) contains the results for dimensionality 150, with $k = 1024$ nodes, processing $m = 2.56M$ data points through $T = 10000$ rounds. For variance thresholds up to 0.08, dynamic synchronization can retain the error of the baseline. At the same time, the communication is reduced to 45% of the original number of messages. Moreover, even for thresholds up to 0.2, the dynamic protocol retains more than 90% of the accuracy of static synchronization with only 20% of its communication.

4.3 Real-World Data

We conclude our experimental section with tests on two real-world datasets containing stock prices and Twitter short messages, respectively.

The data from Twitter has been gathered via its streaming API (https://dev.twitter.com/docs/streaming-apis) during a period of 3 weeks (Sep 26 through Oct 15 2012). Inspired by the content recommendation task, we consider the problem of predicting whether a given tweet will be re-tweeted within one hour after its posting—for a number of times that lies below or above the median hourly re-tweet number of the specific Twitter user. The feature space are the top-1000 textual features (stemmed 1-gram, 2-gram) ranked by information gain, i.e., $X = \{0, 1\}^{1000}$. Learning is performed with $C = 0.25$. The stock price data is gathered from Google Finance (http://www.google.com/finance) and contains the daily closing stock prices of the S&P100 stocks between 2004 and 2012. Inspired by algorithmic trading, we consider the problem of predicting tomorrow's closing price, i.e., $Y = \mathbb{R}$, of a single target stock based on all stock prices and their moving averages (11, 50, and 200 days) of today, i.e., $X = \mathbb{R}^{400}$. The target stock is switched with probability 0.001. Here, we use the epsilon insensitive loss, $\epsilon = 0.1$, and a regression parameter of $C = 1.0$ for regularized passive aggressive updates.

The results for 1.28M data points distributed to $k = 256$ nodes are presented in Fig. 3. Again, the gap between no synchronization and the baseline is well preserved by partial synchronizations. For Twitter (left), a threshold of 0.1 performs even better then the static baseline with less communication (0.97%). With a threshold of 0.2 the dynamic protocol still preserves 74% of predictive gain using only 27% communication. For the stock prices (right), a threshold of 0.005 preserves 99% of the predictive gain using 54% of the communication. The trade-off is even more beneficial for threshold 0.01 which preserves 92% of the gain using only 36% communication.

5   Conclusion

We presented a protocol for distributed online prediction that can save communication by dynamically omitting synchronizations in sufficiently stable phases of a modeling task, while at the same time being adaptive in phases of concept drifts. The protocol has a controlled predictive regret over its static counterpart and experiments show that it can indeed reduce the communication substantially— up to 90% in settings where the linear learning processes are suitable to model the data well and converge reasonably fast. Generally, the effectiveness of the approach appears to correspond to the effectivity of linear modeling with f-proportional convex update rules in the given setting.

For future research a theoretical characterization of this behavior is desirable. A practically even more important direction is to extend the approach to other model classes that can tackle a wider range of learning problems. In principle, the approach of controlling model variance remains applicable, as long as the variance is measured with respect to a distance function that induces a useful loss bound between two models. For probabilistic models this can for instance be the KL-divergence. However, more complex distance functions constitute more challenging distributed monitoring tasks, which currently are open problems.

References

[1] Abernethy, J., Agarwal, A., Bartlett, P.L., Rakhlin, A.: A stochastic view of optimal regret through minimax duality. In: 22nd Annual Conference on Learning Theory (2009)

[2] Balcan, M.-F., Blum, A., Fine, S., Mansour, Y.: Distributed learning, communication complexity and privacy. CoRR, abs/1204.3514 (2012)

[3] Bar-Or, A., Wolff, R., Schuster, A., Keren, D.: Decision tree induction in high dimensional, hierarchically distributed databases. In: Proceedings of the SIAM International Conference on Data Mining (2005)

[4] Bshouty, N.H., Long, P.M.: Linear classifiers are nearly optimal when hidden variables have diverse effects. Machine Learning 86(2), 209–231 (2012)

[5] Cesa-Bianchi, N., Lugosi, G.: Prediction, learning, and games. Cambridge University Press (2006) ISBN 978-0-521-84108-5

[6] Crammer, K., Dekel, O., Keshet, J., Shalev-Shwartz, S., Singer, Y.: Online passive-aggressive algorithms. Journal of Machine Learning Research 7, 551–585 (2006)

[7] Daumé III, H., Phillips, J.M., Saha, A., Venkatasubramanian, S.: Efficient protocols for distributed classification and optimization. CoRR, abs/1204.3523 (2012)

[8] Dekel, O., Gilad-Bachrach, R., Shamir, O., Xiao, L.: Optimal distributed online prediction using mini-batches. Journal of Machine Learning Research 13, 165–202 (2012)

[9] Herbster, M., Warmuth, M.K.: Tracking the best linear predictor. Journal of Machine Learning Research 1, 281–309 (2001)

[10] Hsu, D., Karampatziakis, N., Langford, J., Smola, A.J.: Parallel online learning. CoRR, abs/1103.4204 (2011)

[11] Keralapura, R., Cormode, G., Ramamirtham, J.: Communication-efficient distributed monitoring of thresholded counts. In: SIGMOD, pp. 289–300 (2006)

[12] Keren, D., Sharfman, I., Schuster, A., Livne, A.: Shape sensitive geometric monitoring. Transactions on Knowledge and Data Engineering 24(8), 1520–1535 (2012)

[13] Mcdonald, R., Mohri, M., Silberman, N., Walker, D., Mann, G.S.: Efficient large-scale distributed training of conditional maximum entropy models. In: Advances in Neural Information Processing Systems (NIPS), vol. 22, pp. 1231–1239 (2009)

[14] McDonald, R.T., Hall, K., Mann, G.: Distributed training strategies for the structured perceptron. In: HLT-NAACL, pp. 456–464 (2010)

[15] Nesterov, Y.: Introductory Lectures on Convex Optimization: A Basic Course, vol. 87. Kluwer Academic Publisher (2003)

[16] Nguyen, X., Wainwright, M.J., Jordan, M.I.: Decentralized detection and classification using kernel methods. In: ICML, page 80. ACM (2004)
[17] Ouyang, J., Patel, N., Sethi, I.: Induction of multiclass multifeature split decision trees from distributed data. Pattern Recognition 42(9), 1786–1794 (2009)
[18] Predd, J.B., Kulkarni, S., Poor, V.: Distributed learning in wireless sensor networks. Signal Processing Magazine 23(4), 56–69 (2006)
[19] Sharfman, I., Schuster, A., Keren, D.: A geometric approach to monitoring threshold functions over distributed data streams. Transactions on Database Systems (TODS) 32(4), 23 (2007)
[20] Wang, J.-P., Lu, Y.-C., Yeh, M.-Y., Lin, S.-D., Gibbons, P.B.: Communication-efficient distributed multiple reference pattern matching for m2m systems. In: Proceedings of the International Conference on Data Mining (ICDM). IEEE (2013)
[21] Xiao, L.: Dual averaging methods for regularized stochastic learning and online optimization. Journal of Machine Learning Research 11, 2543–2596 (2010)
[22] Zhang, T.: Solving large scale linear prediction problems using stochastic gradient descent algorithms. In: Proceedings of the International Conference on Machine Learning (ICML), page 116. ACM (2004)
[23] Zinkevich, M., Smola, A.J., Langford, J.: Slow learners are fast. In: Advances in Neural Information Processing Systems (NIPS), vol. 22, pp. 2331–2339 (2009)
[24] Zinkevich, M., Weimer, M., Smola, A.J., Li, L.: Parallelized stochastic gradient descent. In: Advances in Neural Information Processing Systems (NIPS), pp. 2595–2603 (2010)

Hetero-Labeled LDA: A Partially Supervised Topic Model with Heterogeneous Labels

Dongyeop Kang[1,*], Youngja Park[2], and Suresh N. Chari[2]

[1] IT Convergence Laboratory, KAIST Institute, Daejeon, South Korea
[2] IBM T.J. Watson Research Center, Yorktown Heights, NY 10598, USA
dykang@itc.kaist.ac.kr, {young_park,schari}@us.ibm.com

Abstract. We propose Hetero-Labeled LDA (hLLDA), a novel semi-supervised topic model, which can learn from multiple types of labels such as document labels and feature labels (i.e., heterogeneous labels), and also accommodate labels for only a subset of classes (i.e., partial labels). This addresses two major limitations in existing semi-supervised learning methods: they can incorporate only one type of domain knowledge (e.g. document labels or feature labels), and they assume that provided labels cover all the classes in the problem space. This limits their applicability in real-life situations where domain knowledge for labeling comes in different forms from different groups of domain experts and some classes may not have labels. hLLDA resolves both the label heterogeneity and label partialness problems in a unified generative process.

hLLDA can leverage different forms of supervision and discover semantically coherent topics by exploiting domain knowledge mutually reinforced by different types of labels. Experiments with three document collections–*Reuters*, *20 Newsgroup* and *Delicious*– validate that our model generates a better set of topics and efficiently discover additional latent topics not covered by the labels resulting in better classification and clustering accuracy than existing supervised or semi-supervised topic models. The empirical results demonstrate that learning from multiple forms of domain knowledge in a unified process creates an enhanced combined effect that is greater than a sum of multiple models learned separately with one type of supervision.

1 Introduction

Motivated by a diverse set of requirements such as information management and data security, there is an increasing need for large scale topic classification in large distributed document repositories. In these environments, documents are generated and managed independently by many different divisions and domain experts in the company. Often, it is prohibitively expensive to perform supervised topic classification at an enterprise scale, because it is very challenging to catalog what topics exist in the company let alone provide labeled samples for all the topics.

In recent years, probabilistic topic modeling, most notably Latent Dirichlet Allocation (LDA) has been widely used for many text mining applications as an alternative to expensive supervised learning approaches. Probabilistic topic modeling approaches can

---

* This work was conducted while the author was an intern at IBM Research.

T. Calders et al. (Eds.): ECML PKDD 2014, Part I, LNCS 8724, pp. 640–655, 2014.
© Springer-Verlag Berlin Heidelberg 2014

discover underlying topics in a collection of data without training a model with labeled samples. However, unsupervised topic modeling relies primarily on feature (word) occurrence statistics in the corpus, and the discovered topics are often determined by dominant collocations and do not match with the true topics in the data.

A more realistic approach would be to use a semi-supervised learning in which the topic discovery process is guided by some form of domain knowledge. In recent years, many extensions to LDA, in both supervised and semi-supervised ways, have been proposed to generate more meaningful topics incorporating various side information such as correlation of words [16], word constraints [2, 12], document labels [20], and document network structure [7, 11]. Typically, these models extend LDA by constraining the model variables with newly observed variables derived from side information.

These methods have shown some success but are constrained by two major limitations: Firstly, they assume labels are present for all latent topics. This assumption can be satisfied in situations where all topics are known in advance and obtaining side information is relatively easy, such as a collection of user generated content and tags as in [20]. However, in a large distributed complex environment, this is not a realistic assumption. Secondly, they support only one type of supervision, e.g., the domain knowledge should be provided as either document labels or feature labels. In a large distributed environment, labeling is typically done by a diverse set of domain experts, and labels can be provided in different forms. For instance, some experts may be willing to label a small set of sample documents; while others can provide some topic-indicative features (i.e. features which are known *a priori* to be good indicators of the topics).

In this paper, we propose a new semi-supervised topic model to address these limitations in a unified generative process. It provides a unified framework that discovers topics from data that is *partially labeled* with *heterogenous labels*:

Heterogeneous Supervision: We assume that multiple types of supervision can exist in the training data. For instance, some training data are provided with document labels, and some others are associated with topic-indicative features. Further, we assume that a topic can receive multiple types of labels, e.g., feature and document labels. A simplistic approach to support multiple label types is to sequentially build topic models, i.e, build a model with one label type and use this model's output to bootstrap the next iteration with another label type. This naive approach is inefficient due to multiple learning steps and fail to capture new information reinforced by different label types. Instead, we develop a unified model to simultaneously learn from different types of domain knowledge.

Partial Supervision: hLLDA also can handle the label partialness problem, where the training data are partially labeled. We allow for two types of partial labels:

- *Partially labeled document*: The labels for a document cover only a subset of all the topics the document belongs to. Our goal is to predict all the topics for the document.
- *Partially labeled corpus*: Only a small number of documents in a corpus are provided with labels. Our goal is to find the labels for all the documents.

We validate our algorithm using *Reuters*, *20 Newsgroup* and *Delicious*, which have been widely used in previous topic models and are adequate for testing the label partialness problem, since the documents contain multiple topics. The experiments for the label heterogeneity shows that hLLDA achieves about 3 percentage points higher classification and clustering accuracy than LLDA by adding feature labels comprising only 10 words for each topic. The experiments for the label partialness shows that hLLDA produces 8.3 percentage points higher clustering accuracy and 34.4% improvement on Variational Information compared with LLDA. The results confirm that hLLDA significantly enhances the applicability of topic modeling for situations where partial, heterogenous labels are provided. Further we show that learning from multiple forms of domain knowledge in a unified process creates an enhanced combined effect that is greater than a sum of multiple models learned separately with one type of supervision.

In summary, the main contributions of the paper include:

- We propose a novel unified generative model that can simultaneously learn from different types of domain knowledge such as document labels and feature labels.
- hLLDA effectively solves the label partialness problem when the document label set is a subset of the topic set and/or the training data contain unlabeled documents.
- hLLDA is simple and practical, and it can be easily reduced to LDA, zLDA and LLDA depending on the availability of domain information.

The remainder of this paper is structured as follows. We first compare hLLDA with existing supervised and semi-supervised topic modeling algorithms in Section 2. Section 3 describes the generative process of hLLDA and the learning and inference algorithm in details. Experimental data and evaluation results are shown in Section 4 and Section 5. Section 6 provides final discussions and future work.

2   Related Work

hLLDA is broadly related to semi-supervised and supervised topic models. Existing (semi-)supervised topic models can be categorized into two groups based on the type of domain knowledge they utilize: *document supervision* and *feature supervision*.

Document Supervision

Existing approaches that utilize document labels fall in supervised learning assuming that all the documents in the training data have document labels. Supervised methods such as sLDA [5], discLDA [15], and medLDA [24] have shown a comparable performance on classification and regression tasks as general discriminative classifiers, but they support only one topic for a document. Labeled LDA (LLDA) [20] extends previous supervised models to allow multiple topics of documents, and Partially labeled LDA (PLDA) [21] further extends LLDA to have latent topics not present in the document labels. PLDA supports one-to-many mapping between labels and topics, but the number of latent topics is fixed constant for all documents. Recently, [14] propose a non-parametric topic model using Dirichlet Process with Mixed Random Measures (DP-MRM) that allows one label to be mapped with multiple topics. [18] propose a Dirichlet-multinomial regression (DMR) topic model that can incorporate arbitrary types of observed document features, such as author and publication venue, by providing a log-linear prior on document-topic distributions. DMR can be viewed as a supervised topic model by treating document labels as document features.

Table 1. Comparison of *h*LLDA with supervised and semi-supervised topic models using document labels

	No. of Topics per Document	Label-Topic Mapping	Label Partialness
sLDA	single	one-to-one	no
LLDA	multiple	one-to-one	no
PLDA	multiple	one-to-many	yes
DP-MRM	multiple	one-to-many	no
*h*LLDA	multiple	one-to-one	yes

Table 2. Comparison of *h*LLDA with supervised and semi-supervised topic models using word labels

	Label Type	Label-Topic Mapping	Label Partialness
zLDA	unlabeled groups of features	one-to-one	no
SeededLDA	unlabeled groups of features	one-to-one	no
*h*LLDA	labeled or unlabeled groups of features	one-to-many	yes

Feature Supervision

A feature label is typically provided as a set of words that are likely to belong to the same topic. Feature labels are helpful for discovering non-dominant or secondary topics by enforcing the words be assigned to the labeled topics, while standard LDA usually ignore them in favor of more prominent topics. Andrzejewski *et al.* proposed three different approaches for incorporating feature labels. In zLDA, they constrain latent topic assignment of each word to a set of seed words [2]. [3] applies Dirichlet Forest which allows must-links and cannot-links on topics, and [4] uses First-Order-Logic to generate human friendly domain knowledge. [12] described Seeded LDA that restricts latent topics to specific interests of a user by providing sets of seed words. To maximize the usage of seed words in learning, they jointly constrain both document-topic and topic-word distributions with the seed word information.

To our knowledge, *h*LLDA is the only semi-supervised topic model that combine heterogeneous side information together in one generative process, and discover the topics of documents using partially labeled documents and/or corpus. Table 1 and Table 2 summarize the differences of *h*LLDA with other existing algorithms that support document supervision and word supervision respectively.

3 Hetero-Labeled LDA

In this section, we describe *h*LLDA in detail and discuss how it handles heterogeneous labels and partially labeled data. We propose a unified framework that can incorporate multiple types of side information in one simple generative process.

Preliminaries

We first introduce some notations that will be used in the paper as shown in Table 3.

Table 3. Notations

$\mathcal{D}$	a document collection, $\{d_1, d_2, \ldots, d_M\}$
M	the number of documents in $\mathcal{D}$
$\mathcal{V}$	the vocabulary of $\mathcal{D}$, $\{w_i, w_2, \ldots, w_N\}$
N	the size of $\mathcal{V}$, i.e., the number of unique words in $\mathcal{D}$
$\mathcal{T}$	the set of topics in $\mathcal{D}$, $\{T_1, T_2, \ldots, T_K\}$
K	the number of topics in $\mathcal{T}$
$\mathcal{L}_W$	the set of topics provided by word labels
K_W	the number of unique topics in $\mathcal{L}_W$
$\mathcal{L}_D$	the set of topics provided by document labels
K_D	the number of unique topics in $\mathcal{L}_D$
$\mathcal{L}$	the label space, i.e., $\mathcal{L} = \mathcal{L}_W \cup \mathcal{L}_D$
$\mathcal{D}_L$	labeled documents
$\mathcal{D}_U$	unlabeled documents, i.e., $\mathcal{D} = \mathcal{D}_L \cup \mathcal{D}_U$

We also define three different levels of side information for both document supervision and feature supervision.

Definition 1 (Side Information) *Any domain knowledge that can constrain the topic distributions of documents or words.* hLLDA *supports the following three different levels of side information.*

- Group Information: *It only specifies that a group of documents or words that belong to a same set of topics (e.g., $L_d = \{d_1, d_2, \ldots, d_c\}$) and $L_w = \{w_1, w_2, \ldots, w_g\}$).*
- Label Information: *This side information provides a group of labels with associated topic labels. For instance, $L_d = \{d_1, d_2, \ldots, d_c; T_1, T_2, \ldots, T_k\}$ specifies that the documents belong to topics $T_1, \ldots, T_k$, where $1 \leq k \leq K$.*
- Topic Distribution: *This information further provides topic distributions of the label information. For instance, $L_d = \{d_1, \ldots, d_c; T_1, \ldots, T_k; p_1, \ldots, p_k\}$ indicates that the documents belonging to the topic T_i with the likelihood of p_i. We note that p_i is a perceived likelihood value by domain experts, and $\sum_i p_i < 1$ in many cases.*

hLLDA Model

The main goals of *h*LLDA are to build a topic model that can incorporate different types of labels in a unified process and to discover all underlying topics when only a small subset of the topics are known in advance. We solve the problems by modifying both the document topic distribution (θ) and word topic assignment (z) with the side information. Figure 1 depicts the graphical representation of *h*LLDA. In *h*LLDA, the global topic distribution θ is generated by both a Dirichlet topic prior α and a label-specific topic mixture ψ obtained from the document labels Λ_d with a Dirichlet prior γ. Then, the word topic assignment z is generated from the global topic mixture θ constrained by word labels Λ_w.

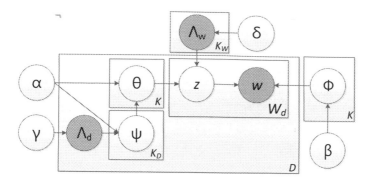

Fig. 1. Graphical representation of *h*LLDA. $|\Lambda_d| = K_D$ and $|\Lambda_w| = K_W$. Note that z is influenced by both the word side information (Λ_w) and the document side information (Λ_d) in *h*LLDA, producing synergistic effect of heterogeneous side information.

Table 4 describes the generative process of *h*LLDA in more detail. In *h*LLDA, the total number of topics (K) is set to the sum of the numbers of unique topics present in the document and word labels (i.e., $|\mathcal{L}_D \cup \mathcal{L}_W|$) and the number of additional latent topics (K_B) the user wants to discover from the corpus. Here, the number of latent topics (K_B) is an input parameter.

We first draw multinomial topic distributions over the words for each topic k, ϕ_k, from a Dirichlet prior β as in the LDA model [6] (line 1–2). However, unlike other LDA models, *h*LLDA has an additional initialization step for word topic assignment z, when word (feature) labels are provided as side information (line 3–5). For each topic appearing in the word labels, k_W, we draw multinomial topic distributions, Λ_{k_W}, over the vocabulary using a smoothed *Bernoulli* distribution, i.e., $\Lambda_{k_W}^{(w)} = (l_1, l_2, ..., l_V)$ where $l_v \in \{\delta, 1 - \delta\}$. The *Bernoulli*$_{smooth}$ distribution generates smoothed values δ ($0 < \delta < 1$) with success probability p or $1 - \delta$ with failure probability $1 - p$, rather than value 1 with probability p and value 0 with probability $1 - p$ as in the *Bernoulli* distribution. We propose the *Bernoulli*$_{smooth}$ distribution to handle the label partialness. Note that the *Bernoulli* distribution does not allow words or documents to be assigned to the topics not provided in the document or feature labels. However, with *Bernoulli*$_{smooth}$, documents and words can be assigned to topics from other latent topics with a low probability $1 - \gamma$ and $1 - \delta$ respectively.

The *Bernoulli*$_{smooth}$ distribution drawn from word label information, Λ_{k_W}, contains a vector of topics for each word and is later used to constrain the global topic mixture θ as described in line 16. We multiply Λ_{k_W} with θ to generate the multinomial distribution z (line 16). The topic assignment z_i for each word i in a document d is chosen from a multinomial distribution $\{\lambda_1^{(d)}, ..., \lambda_K^{(d)}\}$, where $\lambda_i^{(d)}$ denotes the assigned topic for word i in document d and is generated by multiplying the global topic mixture θ and the word label constraint Λ_{k_W}. Applying soft constraints on word topic assignment z using word labels is similar to zLDA [2], but, zLDA puts constraints on word instances, while *h*LLDA puts constraints over the vocabulary elements. Further, by influencing z with the mixture of the word side information and the document side information (see

Table 4. Generative process for hLLDA. $Bernoulli_{smooth}$ distribution generates smoothed values (e.g., value $v, 0 < v < 1$ with success probability p or $1 - v$ with failure probability $1 - p$) rather than value 1 or value 0.

1	For each topic $k \in \{1, ..., K\}$
2	Generate $\phi = (\phi_{k,1}, ..., \phi_{k,V})^T \sim Dir(\cdot \mid \beta)$
3	For each topic $k_W \in \{1, ..., K_w\}$
4	For each word $w \in \{1, ..., N\}$
5	Generate $\Lambda_{k_W}^{(w)} \sim Bernoulli_{smooth}(\cdot \mid \delta)$
6	For each document d:
7	if $d \in \mathcal{D}_U$
8	Generate $\theta^{(d)} = (\theta_1, ..., \theta_K)^T \sim Dir(\cdot \mid \alpha)$
9	if $d \in \mathcal{D}_L$
10	For each topic $k_D \in \{1, ..., K_D\}$
11	Generate $\Lambda_{k_D}^{(d)} \sim Bernoulli_{smooth}(\cdot \mid \gamma)$
12	Generate $\Psi^{(d)} = (\psi_1, ..., \psi_{K_d})^T \sim Dir(\cdot \mid \alpha \cdot \Lambda_{k_D}^{(d)})$
13	Generate $\theta^{(d)} = (\theta_{K_d+1}, ..., \theta_{(K)})^T \sim Dir(\cdot \mid \alpha_{K_d+1:K})$
14	Generate $\theta^{(d)} = (\Psi^{(d)T} \mid \theta^{(d)T})^T$
15	For each i in $\{1, ..., N_d\}$
16	Generate $z_i \in \{\lambda_1^{(d)}, ..., \lambda_K^{(d)}\} \sim Mult(\cdot \mid \Lambda_{k_W}^{(i)} \cdot \theta^{(d)})$
17	Generate $w_i \in \{1, ..., V\} \sim Mult(\cdot \mid \phi_{z_i})$

Figure 1), hLLDA can benefit from the combined effect of multiple heterogeneous side information.

hLLDA generates the document topic distribution θ differently for documents with document side information and for documents without document labels (line 7–14). If the document is unlabeled (i.e., $d \in \mathcal{D}_U$), we generate topics using the Dirichlet prior α in the same way as in LDA (line 8). If the document is labeled (i.e., $d \in \mathcal{D}_L$), we first generate the document labels over topics $\Lambda_{K_D}^{(d)} = (l_1, l_2, ..., l_{K_D})$, where $l_k \in \{\gamma, 1 - \gamma\}$ is drawn from the smoothed $Bernoulli$ distribution, $Bernoulli_{smooth}(\cdot \mid \gamma)$ (line 10-11). The soft constraints on document labels enable hLLDA to discover other latent topics for *partially labeled documents or corpus*, which do not exist in the document labels. We note that this is different from both Labeled LDA (LLDA) [20] and Partially Labeld LDA (PLDA) [21]. In LLDA, a document is strictly constrained to generate topics only from the provided document labels. PLDA relaxes this restriction and allows a document to be assigned a set of latent topics that are unseen in the document labels, but the number of the latent topics is arbitrarily fixed constant for all documents.

Note that, in $Bernoulli_{smooth}(\cdot \mid \delta)$ and $Bernoulli_{smooth}(\cdot \mid \gamma)$, the values for δ and γ are larger than $1 - \delta$ and $1 - \gamma$ respectively, ensuring that the topic distributions from the side information have more weights than the topics not covered by the side information. Further, when the document side information is provided in the form of Topic Distribution as described in Definition 1, the perceived likelihoods, p_i, are used as biased priors.

We generate a document label-topic mixture $\Psi^{(d)}$ of size K_D using the Dirichlet topic prior α and the document label constraints $\Lambda_{K_D}^{(d)}$ (line 12) and then generate a latent topic mixture $\theta^{(d)}$ of size K-K_D using the Dirichlet prior α (line 13). Finally, we concatenate the document label-topic mixture Ψ and the latent topic mixture θ to generate θ with size K (line 14). The concatenation together with the soft constraints on document topics allow the document to generate new topics that are not included in the document labels from *partially labeled documents or corpus*. Even though the concatenation of Dirichlet random variables does not produce a value that is an element of the simplex, our experiments show that it solves the label partialness very well.

The remaining steps (line 15–17) are similar to the processes in LDA. For each word i in document d, we generate topic assignment z_i from multinomial distribution $\theta^{(d)}$ and word label constraint $\Lambda_{k_W}^{(i)}$ and generate the word from multinomial distribution ϕ_{z_i}.

Learning and Inference

We use the Gibbs sampling algorithm [9] to estimate the latent variables θ, ψ, and ϕ. We note that the word and document label priors δ and γ are independent from the rest of model parameters, and, since we simply concatenate ψ into θ (line 14), we can use the same inference as in LDA. Thus, our inference process follows the Gibbs sampling procedure that estimates only θ and ϕ.

At each iteration, the topic of ith document, z_i, is estimated by the conditional probability

$$
\begin{aligned}
&P(z_i = k | \mathbf{z}_{-i}, \mathbf{w}, \Lambda_W, \Lambda_D, \alpha, \eta, \gamma, \delta) \\
&\propto P(z_i = k | \mathbf{z}_{-i}, \mathbf{w}, \Lambda_W, \alpha, \eta, \gamma) \\
&\propto \Lambda_k^{(w_i)} \times \left(\frac{n_{-i,k}^{(w_i)} + \eta}{\sum_{w'}^{W} \left(n_{-i,k}^{(w')} + \eta \right)} \right) \left(\frac{n_{-i,k}^{(d)} + \alpha}{\sum_{k'}^{T} \left(n_{-i,k'}^{(d)} + \alpha \right)} \right)
\end{aligned}
\tag{1}
$$

where $\Lambda_k^{(w_i)}$ is a word label constraint that outputs γ, $0 < \gamma < 1$ when $w_i \in \Lambda_W$, and 1-γ when $w_i \notin \Lambda_W$. The soft constraints on sampling procedure is similar to zLDA [2], except that the topic k can be a new topic not in the word labels. Then, we obtain the estimated probability ϕ_{kw} of word w in topic k and the estimated probability θ_{dk} of topic k in document d using Equation 2 and 3 respectively.

$$
\phi_{kw} = \frac{n_{-i,k}^{(w_i)} + \eta}{\sum_{w'}^{W} \left(n_{-i,k}^{(w')} + \eta \right)}
\tag{2}
$$

$$
\theta_{dk} = \frac{n_{-i,k}^{(d)} + \alpha}{\sum_{k'}^{T} \left(n_{-i,k'}^{(d)} + \alpha \right)}
\tag{3}
$$

When no side information is provided, hLLDA is reduced to LDA. Compared to LLDA, θ in hLLDA is limited by soft constraints drawn from the documents labels, and, thus, becomes the same as LLDA, when only document side information is considered, and the document label prior γ is a binary vector representing the existence of topic labels for each document. Compared to zLDA, z in hLLDA is softly constrained by both the word labels and the document labels in assigning topics for each word in each document. hLLDA can be reduced to zLDA, when the side information contains only word labels, and K_W is equal to K. Based on these observations, hLLDA can be viewed as a generalized version of LDA, LLDA and zLDA. Further, we note that the existence of latent topic mixture θ enables hLLDA to find latent topics not covered by the document or word labels without harming the original distribution of topics from the labels.

4   Experiments

We conduct experiments to answer the following questions:

Q1 How effective is learning from mixture of heterogeneous labels for topic categorization?

Q2 How well does hLLDA discover latent topics from partially labeled documents and corpus?

Q3 How accurate are the generated topics?

Data

All experiments are conducted with three public data sets–*Reuters-21578* [22], *20 Newsgroup* [1], and *Delicious* [8]. The *Reuters-21578* data set contains a collection of news articles in 135 categories, and we chose the 20 most frequent topics for the experiments (hereafter called *Reuters*). For the *20 Newsgroup* dataset, we use all 20 categories in the data set (hereafter called *20News*). For the *Delicious* data set, we first selected the 50 most frequent tags in *Delicious.com*, and then manually chose 20 tags from the 50 tags and 5,000 documents for the selected 20 categories (hereafter called *Delicious*). Table 5 shows the topic categories in the the experiment data sets. We then conducted the following text processing on the documents: First, all stopwords were removed and words were stemmed using Porter's Stemmer [19]. Then, all words occurring in fewer than 5 documents were discarded. After the preprocessing, *Reuters* contains $11,305$ documents and $19,271$ unique words; *20News* has $19,997$ documents with $57,237$ unique words; and *Delicious* contains $5,000$ with $141,787$ unique words.

Domain Knowledge

We use the topic labels in the data sets as document side information. To evaluate the label heterogeneity (**Q1**) and partialness problems (**Q2**), we conduct experiments with varying amount of document side information comprising the first 5, 10, 15 and 20 labels from the topics in Table 5. We treat the documents belonging to the selected categories as labeled and the remaining documents as unlabeled.

For word side information, we extracted top 20 words for each class based on TF-IDF (term frequency-inverse document frequency), manually filtered irrelevant words

Table 5. The 20 topics in *Reuters*, *20News*, and *Delicious* data sets

Reuters	earn, acq, money-fx, crude, grain, trade, interest, wheat, ship, corn, dlr, oilseed, sugar, money-supply, gnp, coffee, veg-oil, gold, nat-gas, soybean
20News	alt.atheism, sci.space, comp.os.ms-windows, rec.sport.baseball, misc.forsale, soc.religion.christian, rec.autos, sci.crypt, talk.religion.misc, sci.med, comp.sys.ibm.pc.hardware, rec.sport.hockey, talk.politics.guns, sci.electronics, comp.graphics, rec.motorcycles, talk.politics.misc, comp.sys.mac.hardware, talk.politics.mideast, comp.windows.x
Delicious	design, web, software, reference, programming, art, education, resources, photography, music, business, technology, research, science, internet, shopping, games, marketing, typography, graphics

out and chose top 10 words as final word labels. When a word appears in multiple classes, we remove the word from all the classes except the class for which the word has the highest TF-IDF value. In real world, word labels are given by domain experts so they have more meaningful information than our artificially generated word labels. Even though we have conducted an experiment with real business data that contains document and word labels with successful experimental results, they are not included in this paper due to confidential information.

Evaluation Methods

We implement two variations of *h*LLDA and compare them with three existing topic modeling algorithms–LDA [6], LLDA [20] and zLDA [2]. (For multi-label classification task such as *Reuters* and *Delicious*, sLDA is not appropriate to compare with [20] so we does not include sLDA in our experiment.) The first version, *h*LLDA (L=T), assumes that all the topics are present in the labels to directly compare it with LLDA. The second version. *h*LLDA (L<T), is for cases where the label set is a subset of the topic set and validate the label partialness problem. For all the models, we use a Collapsed Gibbs sampler [10] for inference with standard hyper-parameter values $\alpha = 1.0$ and $\beta = 0.01$ and run the sampler for 1,000 iterations.

All comparisons are done using 5-fold cross validation over 10 random runs. For question **Q1** and **Q2**, we measure the following three evaluation metrics. For **Q3**, we compare the discovered topics qualitatively by visualizing the topics.

Prediction Accuracy: We predict a label of a new document by choosing the topic with the highest probability from the posterior document-topic distribution θ and check whether the label exists in the topic set of the document.

Clustering F-measure: We simulate clustering by assigning each document to the topic (i.e., cluster) that has the highest probability in θ. If two documents belong to the same topic by both the ground truth and by the simulated clustering, then it is regarded as correct. The F-measure is then calculated for all the pairs of documents. Even though clustering may not be a general metric to evaluate topic modeling algorithms, it can be a good indicator of how topics are coherently grouped together especially when label information is incomplete (i.e., label partialness). Section explains the details.

Variational Information(VI): VI measures the amount of information lost and gained in changing clustering C_1 to clustering C_2 [17]. The VI of two clusters X and Y is calculated as $VI(X, Y) = H(X) + H(Y) - 2 * I(X, Y)$ where $H(X)$ (or $H(Y)$) denotes the entropy of the clustering X (or Y), and $I(X, Y)$ is the mutual information between X and Y. Lower VI values indicate better clustering results.

5   Experimental Results

We measure the performance of hLLDA and the baseline systems for the label heterogeneity the label partialness problems and also visually compare the discovered topics by hLLDA and LLDA.

Label Heterogeneity

We first validate the effectiveness of hLLDA in dealing with heterogenous labels. In this experiment, we used document labels and feature labels as heterogeneous domain knowledge for hLLDA, but we can easily extend to other types of labels such as document structure labels. Further, we assume that all topics appear in the labels, and all training documents are labeled with document labels or feature labels.

Figure 2(a) shows the accuracy of multi-class prediction. As we can see, both versions of hLLDA perform well for all three data sets. The accuracy levels of hLLDA are significantly better than LDA and zLDA and slightly higher than LLDA. This indicates that mixture of two heterogeneous domain information improve the prediction accuracy. Figure 2(b) shows the F-measure of the multi-class clustering task. The F-measure of both hLLDA algorithms show similar performance as LLDA while significantly outperforming LDA and zLDA. We notice that, however, for *Delicious*, hLLDA is better than LLDA confirming that adding feature label information is beneficial. These results indicate that hLLDA can combine different types of supervision successfully, and the combination of heterogeneous label types is beneficial for both classification and clustering tasks.

Label Partialness

For the label partialness problem, we consider two types of label partialness: *partially labeled document* and *partially labeled corpus*.

Partially Labeled Documents: The goal is to predict the full set of topics for a document when only a subset of topics is provided as labels for the document. We conduct experiments for different levels of partialness ranging from 10% to 100% with 10% interval. For p% partialness, we include a topic in the document's label set with probability p. In this experiment, *20News* and *Delicious* were used because most documents in the data sets have multiple topics. As we can see from the results shown in Figure 3, hLLDA, especially hLLDA ($L < T$), outperforms all other algorithms both in terms of clustering F1-measure and VI.

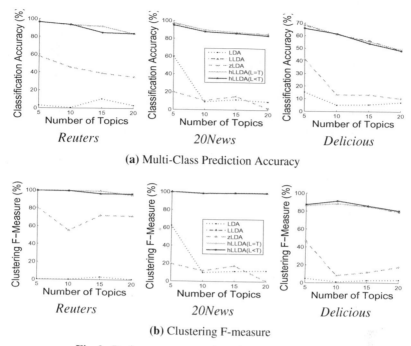

(a) Multi-Class Prediction Accuracy

(b) Clustering F-measure

Fig. 2. Performance comparison for label heterogeneity

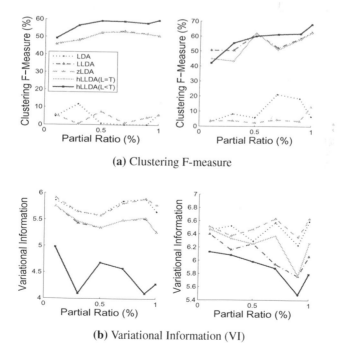

(a) Clustering F-measure

(b) Variational Information (VI)

Fig. 3. Clustering F-measure and VI (the lower the better) for partially labeled documents on *20News* (left) and *Delicious* (right). *PartialRatio* indicates the probability of each topic being included in the labels.

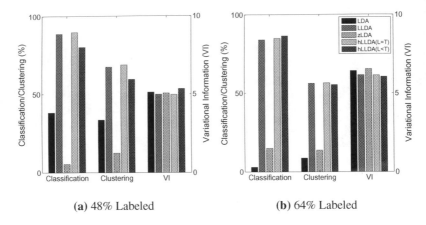

(a) 48% Labeled   (b) 64% Labeled

Fig. 4. Performance comparison for partially labeled corpus on *Delicious*

Table 6. Number of topically irrelevant (Red) and relevant (Blue) words marked by users in Table 7. The more red words are, the lower the topic quality is. Similarly, the more blue words are, the higher the topic quality is.

	LLDA		**hLLDA**	
	#RedWords	#BlueWords	#RedWords	#BlueWords
20News	15	11	2	35
Delicious	17	12	6	30

Partially Labeled Corpus: The goal is to find the labels for all the documents in the corpus when only a subset of the documents are labeled ($|\mathcal{D}_L| \ll |\mathcal{D}|$). We conduct the same experiments as for label heterogeneity using *Delicious*, but introduced unlabeled documents in the training data. Figure 4a and Figure 4b show the results when only the documents belonging to the first 5 topics (48% of the documents) and the first 10 topics (64% of the documents) are considered labeled respectively. As we can see, *h*LLDA outperforms both LDA and zLDA significantly in all cases. Further, the results show that *h*LLDA achieves a comparable performance to LLDA while using less than half of the labels and even better performance only with about 60% of the labels!

Quality of Discovered Topics

We compare the quality of topics discovered by *h*LLDA with partial labels and by LLDA with full labels. We ran *h*LLDA using only 10 topics as the documents labels and discovered 20 topics. To keep the amount of domain information the same, we split the data set into two subsets with 10 topics each and ran LLDA separately for each subset. Table 7 shows the discovered topics for *20News* (top) and *Delicious* (bottom): The first column shows the the true topics, and the second and the third columns show the

Table 7. Comparison of topics generated by LLDA with full labels and *h*LLDA with partial labels. Each row depicts a topic label and top five words for the topic discovered by the two algorithms. Words marked in red or blue show the differences between the two algorithms. The words in red indicate topically irrelevant words, and the words in blue denote relevant words for the topic.

	Labels	LLDA(L=10,T=10) & LLDA(L=10,T=10)	*h*LLDA (L=10,T=20)
20News	atheism	peopl, dont, god, moral, believ	peopl, god, dont, moral, believ
	space	space, launch, orbit, time, system	space, launch, orbit, system, time
	ms-windows	window, file, program, imag, run	window, file, driver, run, program
	baseball	game, team, plai, player, win	game, player, team, dont, hit
	forsale	drive, card, scsi, system, sale	sale, email, price, plea, drive
	christian	god, christian, peopl, believ, church	god, christian, peopl, believ, church
	autos	car, dont, bike, im, time	car, bike, dont, engin, im
	crypt	govern, kei, peopl, gun, encrypt	kei, encrypt, chip, govern, secur
	religion.misc	peopl, armenian, dont, jew, israel	god, peopl, dont, christian, moral
	med	medic, dont, health, peopl, drug	medic, effect, dont, disea, studi
	pc.hardware	drive, scsi, card, id, control	drive, card, scsi, mac, monitor
	hockey	game, team, plai, hockei, player	game, team, plai, hockei, win
	politics.guns	gun, peopl, dont, weapon, fire	gun, law, weapon, peopl, crime
	electronics	wire, ground, dont, circuit, power	power, wire, batteri, circuit, ground
	graphics	imag, file, graphic, program, format	-
	motorcycles	bike, dod, ride, dont, motorcycl	-
	politics.misc	peopl, dont, presid, govern, time	▷ presid, dont, peopl, govern, job
			▷ parti, polit, vote, convent, univ
	mac.hardware	mac, appl, drive, monitor, system	-
	politics.mideast	armenian, peopl, israel, isra, turkish	▷ armenian, turkish, muslim, armenia, turk
			▷ israel, isra, jew, arab, jewish
	windows.x	window, file, program, server, run	ile, imag, program, displai, window
			+ fire, peopl, start, didnt, dont, children
Delicious	design	design, comment, repli, post, thank	design, comment, post, thank, repli
	software	file, softwar, download, support, web	file, download, softwar, window, free
	art	post, art, begin, map, comment	art, post, begin, artist, book
	education	learn, student, educ, talk, world	learn, student, educ, talk, world
	science	scienc, peopl, time, page, link	scienc, peopl, time, page, depress
	photography	photo, am, photographi, comment, jul	photo, am, photographi, post, photograph
	music	music, record, rock, band, de	music, record, rock, band, song
	business	xpng, twitter, busi, search, blog	busi, search, blog, inform, servic
	games	game, element, function, code, html	game, comment, articl, appl, app
	marketing	de, que, la, social, en	twitter, social, post, media, market
	shopping	tshirt, shop, de, product, top	ship, free, price, shop, offer
	typography	font, design, thank, type, comment	-
	graphics	icon, file, free, graphic, brush	-
	programming	code, function, post, file, page	▷ element, function, code, exampl, content
			▷ python, tornado, thread, framework, server
	research	research, start, post, search, comment	-
	web	xpng, web, css, user, site	xpng, scalablesvg, xsvg, flash, arduino
	internet	de, que, le, da, la	-
	technology	comment, googl, technolog, inform, app	-
	reference	element, pdf, html, content, map	pdf, html, sheet, cheat, intel
	resources	repli, design, post, free, thank, web, site	
			+ stack, librari, sentenc, data, scholar
			+ oct, plugin, jul, commentcont, jan
			+ de, le, la, un, et
			+ de, que, la, para, el
			+ die, und, der, map, da

top 5 words discovered by LLDA and *h*LLDA respectively. We marked the topics that *h*LLDA did not find with '-' , and the topics *h*LLDA generated but do not exist in the data set with '+'. The topics with '▷' indicate that multiple topics were generated for one true topic. As we can see, *h*LLDA discovers topics very accurately with the first 10 topics matching very well with the true topics for both *20News* and *Delicious*. Further note that, for both *20News* and *Delicious* data sets, *h*LLDA discovered new latent topics even though no labels were provided for these topics. For example, *h*LLDA discovered 6 out of 10 latent topics for *20News*, such as *pc.hardware, hockey, politics.guns, electronics, politics.misc* and *windows.x*.

We also examine the top 5 words for each topic: The words discovered by both algorithms are marked in black, and words discovered by only one algorithm are marked in red or blue– blue denoting relevant words and red denoting irrelevant words respectively. As we can see, *h*LLDA generates much more relevant (blue) words at the top and also extract more general words than LLDA, even when both cases were judged topically relevant. For instance, LLDA generates "drive", "card", "scsi" for topic *forsale*, while *h*LLDA produces "sale", "price", and "offer". The same trend is seen for *Delicious* data set, especially for topics *business, games* and *marketing*. Table 6 shows the total number of blue and red words generated by LLDA and *h*LLDA. As we can see, *h*LLDA produced much more relevant words and much fewer irrelevant words for both data sets, yielding 87% and 65% reduction in red words and 218% and 150% increase in blue words for *20News* and *Delicious* respectively. The results clearly show the effectiveness of *h*LLDA in handling partial labels.

6 Conclusion

We proposed *h*LLDA, a partially supervised topic model to deal with the heterogeneity and partialness of labels. Our algorithm is simple and flexible and can deal with different label types in a unified framework. Experimental results demonstrate the effectiveness of *h*LLDA for both label heterogeneity and label partialness problems. Experiments also validate that *h*LLDA can discover latent topics for which no label or side information was provided. Further, *h*LLDA produces comparable classification performance and much better clustering performance than existing semi-supervised models while using much smaller amount of labels.

In the future, we plan to incorporate additional type of label information such as partial or full taxonomy of topics [13]. Also, to further improve the performance of label prediction for partially labeled documents, we consider generating topic hierarchies such as Hierarchical Dirichlet Process (HDP) [23].

References

1. 20 Newsgroup, http://qwone.com/~jason/20Newsgroups/
2. Andrzejewski, D., Zhu, X.: Latent dirichlet allocation with topic-in-set knowledge. In: NAACL HLT 2009 Workshop on Semi-Supervised Learning for Natural Language Processing (2009)
3. Andrzejewski, D., Zhu, X., Craven, M.: Incorporating domain knowledge into topic modeling via dirichlet forest priors. In: ICML (2009)

4. Andrzejewski, D., Zhu, X., Craven, M., Recht, B.: A framework for incorporating general domain knowledge into latent dirichlet allocation using first-order logic. In: IJCAI (2011)
5. Blei, D.M., McAuliffe, J.D.: Supervised topic models. In: NIPS (2007)
6. Blei, D.M., Ng, A.Y., Jordan, M.I.: Latent dirichlet allocation. JMLR (2003)
7. Chang, J., Blei, D.M.: Relational topic models for document networks. Journal of Machine Learning Research - Proceedings Track 5 (2009)
8. Delicious, http://arvindn.livejournal.com/116137.html
9. Griffiths, T.L., Steyvers, M.: Proceedings of the National Academy of Sciences (2004)
10. Griffiths, T.L., Steyvers, M.: Finding scientific topics. PNAS (2004)
11. Ho, Q., Eisenstein, J., Xing, E.P.: Document hierarchies from text and links. In: WWW (2012)
12. Jagarlamudi, J., Daumé, H., Udupa, R.: Incorporating lexical priors into topic models. In: EACL, pp. 204–213 (2012)
13. Kang, D., Jiang, D., Pei, J., Liao, Z., Sun, X., Choi, H.J.: Multidimensional mining of large-scale search logs: a topic-concept cube approach. In: WSDM (2011)
14. Kim, D., Kim, S., Oh, A.: Dirichlet process with mixed random measures: a nonparametric topic model for labeled data. In: ICML (2012)
15. Lacoste-Julien, S., Sha, F., Jordan, M.I.: Disclda: Discriminative learning for dimensionality reduction and classification. In: NIPS (2008)
16. Lafferty, J.D., Blei, D.M.: Correlated topic models. In: NIPS (2005)
17. Meilă, M.: Comparing clusterings—an information based distance. J. Multivar. Anal. (2007)
18. Mimno, D.M., McCallum, A.: Topic models conditioned on arbitrary features with dirichlet-multinomial regression. In: UAI, pp. 411–418 (2008)
19. Porter, M.F.: An algorithm for suffix stripping. Program: electronic library and information systems (1980)
20. Ramage, D., Hall, D., Nallapati, R., Manning, C.D.: Labeled lda: A supervised topic model for credit attribution in multi-labeled corpora. In: EMNLP (2009)
21. Ramage, D., Manning, C.D., Dumais, S.T.: Partially labeled topic models for interpretable text mining. In: KDD (2011)
22. Reuters-21578, http://kdd.ics.uci.edu/databases/reuters21578/
23. Teh, Y.W., Jordan, M.I., Beal, M.J., Blei, D.M.: Hierarchical dirichlet processes. Journal of the American Statistical Association (2006)
24. Zhu, J., Ahmed, A., Xing, E.P.: Medlda: maximum margin supervised topic models for regression and classification. In: ICML (2009)

Bayesian Multi-view Tensor Factorization

Suleiman A. Khan[1] and Samuel Kaski[1,2]

Helsinki Institute for Information Technology HIIT,
[1] Department of Information and Computer Science, Aalto University, Finland
[2] Department of Computer Science, University of Helsinki, Finland
{first.last}@aalto.fi

Abstract. We introduce a Bayesian extension of the tensor factorization problem to multiple coupled tensors. For a single tensor it reduces to standard PARAFAC-type Bayesian factorization, and for two tensors it is the first Bayesian Tensor Canonical Correlation Analysis method. It can also be seen to solve a tensorial extension of the recent Group Factor Analysis problem. The method decomposes the set of tensors to factors shared by subsets of the tensors, and factors private to individual tensors, and does not assume orthogonality. For a single tensor, the method empirically outperforms existing methods, and we demonstrate its performance on multiple tensor factorization tasks in toxicogenomics and functional neuroimaging.

1   Introduction

Tensor Factorization methods decompose data into underlying latent factors or components, taking advantage of the natural tensor structure in the data. A wide range of low-dimensional representations of tensors have been proposed earlier [1]. The most well-known models include the CP CANDECOMP/PARAFAC [2,3] and the Tucker 3-mode factor analysis [4]. Tucker is a more generic model for complex interactions, whereas CP as an additive combination of rank-1 contributions is easier interpretable analogously to matrix factorizations. Recently well-regularized probabilistic tensor factorization methods have been introduced for both CP [5] and Tucker [6], though they are limited to single tensors only.

Two-View Tensor Models. In order to discover shared patterns between two co-occuring tensors, joint factorization approaches decompose them into correlated factors [7]. Recently, several non-probabilistic methods for Tensor Canonical Correlation Analysis have been introduced [8,9,10] extending the matrix counterparts. The methods impose different constraints but all aim at finding a common latent representation of two paired tensors.

Two-View Matrix Models. For paired matrices, integration approaches have been thoroughly studied. For an overview on nonlinear Canonical Correlation Analysis (CCA) see [11] and Bayesian CCA see [12].

Multi-view Models. Multi-view modeling integrates information from multiple coupled datasets. For unsupervised multi-view modelling, a method has recently

T. Calders et al. (Eds.): ECML PKDD 2014, Part I, LNCS 8724, pp. 656–671, 2014.

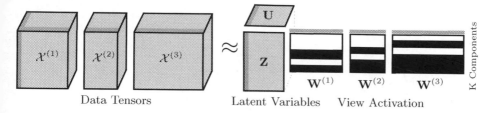

Fig. 1. Multi-view tensor factorization. Datasets $\mathcal{X}^{(1)}, \mathcal{X}^{(2)}, \mathcal{X}^{(3)}$ are simultaneously decomposed into K components. The $\mathbf{Z}$ and $\mathbf{U}$ loadings are common to all tensors, while the view-specific loadings $\mathbf{W}^{(m)}$ show the intrinsic component-view structure in the data. The structure is highlighted in $\mathbf{W}^{(m)}$ with *black* representing a component active in a view (non-zero loadings), while *white* is switched off (zero-loadings).

been proposed for decomposing several coupled matrices, into components shared by subsets of the matrices, and components private to each matrix. The method was called Group Factor Analysis [13]. As far as we know, methods for analysing multiple coupled tensors have not been proposed earlier.

In this paper we formulate and address the novel *multi-view tensor factorization problem*, where the task is to decompose multiple coupled or co-occuring tensors into factors that are shared by subsets of the tensors: one, some or all of them. We formulate a Bayesian model to solve the task, allowing automatic model complexity selection and an intrinsic solution for degeneracies. For two views, our model is the first Bayesian Tensor Canonical Correlation Analysis.

The rest of the paper is structured as follows: In section 2 we formulate the novel multi-view tensor factorization problem. In section 3 we present our Bayesian multi-view tensor factorization model and describe its relationship to existing works. In section 4 we validate the model's performance in various settings and demonstrate its application in a novel toxicogenomics setting and a neuroimaging case. We conclude with discussion in section 5.

Notations: We will denote a tensor as $\mathcal{X}$, a matrix $\mathbf{X}$, vector $\mathbf{x}$ and a scalar x. The Frobenius norm of a tensor is defined as $\|\mathcal{X}\| = \sqrt{\sum_n \sum_d \sum_l \mathcal{X}_{n,d,l}^2}$. The Mode-2 product $\times_2$ between a tensor $\mathcal{A} \in \mathbb{R}^{N \times K \times L}$ and a matrix $\mathbf{B} \in \mathbb{R}^{D \times K}$ is the projected tensor $(\mathcal{A} \times_2 \mathbf{B}) \in \mathbb{R}^{N \times D \times L}$. A reshaped Khatri Rao product $\odot$ of two matrices $\mathbf{A} \in \mathbb{R}^{N \times K}$ and $\mathbf{C} \in \mathbb{R}^{L \times K}$ is the "column-wise matched" outer product of K vector-pairs that results in the tensor $(\mathbf{A} \odot \mathbf{C}) \in \mathbb{R}^{N \times K \times L}$. The outer product of two vectors is denoted $\circ$. The *rank* of a tensor $\mathcal{X}$ is the smallest number of rank-1 tensors that generate $\mathcal{X}$ as their sum. The *order* of a tensor is the number of axes in the tensors, also called ways or modes. For notational simplicity the model is presented for third order tensors, while it is trivially extendable to higher orders.

2   Multi-view Tensor Factorization

We formulate the novel Multi-view Tensor Factorization (MTF) problem for a collection of $m = 1, \ldots, M$ paired tensors (views), $\mathcal{X}^{(1)}, \mathcal{X}^{(2)}, \ldots, \mathcal{X}^{(M)} \in$

$\mathbb{R}^{N \times D_m \times L}$, as the combined factorization that decomposes the tensors into factors shared between all, some, or a single tensor. In MTF, each tensor is factorized into a view-specific matrix of loadings $\mathbf{W}^{(m)} \in \mathbb{R}^{D_m \times K}$ and a low-dimensional tensor $\mathcal{Y} \in \mathbb{R}^{N \times K \times L}$ common for all views:

$$\mathcal{X}^{(m)} = \mathcal{Y} \times_2 \mathbf{W}^{(m)} + \boldsymbol{\epsilon}^{(m)} .$$

Here $\boldsymbol{\epsilon}^{(m)} \in \mathbb{R}^{N \times D_m \times L}$ is the noise tensor.

The view-specific matrix of loadings $\mathbf{W}^{(m)}$ then controls which of the factors k from the common tensor are active in each view. For convenience we assume a fixed number of K factors, with the understanding that for methods capable of choosing the number of factors, K is set large enough, and the loadings of extra components will automatically become set to zero.

The tensor $\mathcal{Y}$ forms the shared latent tensor and can be left unconstrained (equivalent to Tucker1 factorization), or can be further constrained to represent any decomposition including Tucker2, Tucker3 or CP. The CP decomposition factorizes a tensor into a sum of rank-1 tensors, where each rank-1 tensor is the outer product of vector loadings in all modes, whereas in Tucker variants the factor interactions are modelled via a core tensor $\mathcal{G}$. This rank-1 component decomposition of CP and its intrinsic axis property from parallel proportional profiles [14], along with uniqueness of solutions [15], gives it a very strong interpretive power. The Tucker model is more flexible, though, the complex interactions via $\mathcal{G}$ and non-uniqueness of solutions make its interpretation more difficult. Therefore, we adapt an underlying CP decomposition for our model.

Figure 1 illustrates MTF for the joint CP-type factorization. More formally,

$$\mathcal{X}^{(m)} = \sum_{k=1}^{K} \mathbf{Z}_k \circ \mathbf{U}_k \circ \mathbf{W}_k^{(m)} + \boldsymbol{\epsilon}^{(m)} \tag{1}$$
$$= (\mathbf{Z} \odot \mathbf{U}) \times_2 \mathbf{W}^{(m)} + \boldsymbol{\epsilon}^{(m)} .$$

Here $\mathbf{Z} \in \mathbb{R}^{N \times K}$ and $\mathbf{U} \in \mathbb{R}^{L \times K}$ are the common latent variables and the $\mathbf{W}^{(m)}$ are loadings for each view m.

Figure 1 shows the MTF formulation for three tensors, where components (rows) of $\mathbf{W}^{(m)}$ can be active in all, two, or a single view. The loadings $\mathbf{W}_k^{(m)}$ are zero for the components k that are not active in view m. A component active in two or more views has non-zero loadings in the corresponding $\mathbf{W}_k^{(m)}$ and is hence shared between them. This specification comprehensively represents the intrinsic structure of the tensor collection.

3   Bayesian Multi-view Tensor Factorization

We formulate a Bayesian treatment of the MTF problem of Equation 1, by complementing it with priors for model parameters. Figure 2 summarizes the dependencies between the variables in the decomposition of the M observed tensors $\mathcal{X}^{(m)}$ as a graphical model. The main idea is incorporated in plate M,

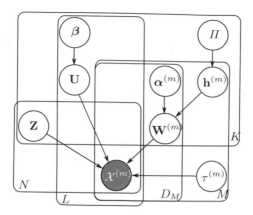

Fig. 2. Plate diagram for Bayesian multi-view tensor factorization

which represents the view-specific loadings $\mathbf{W}^{(m)}$, having two layers of sparsity:
1) *view-wise* sparsity controlled by $\mathbf{h}^{(m)}$ and 2) *feature-wise* sparsity (across the
D_M features) controlled by $\boldsymbol{\alpha}^{(m)}$. The view-wise sparsity acts as an on/off switch
and allows the model to automatically learn which views share each factor, and
also the total number of factors in the data. The plate K represents probabilistic
CP decomposition for each view, where $\mathbf{Z}$ and $\mathbf{U}$ are the latent variables.

The distributional assumptions of our model (explained in detail below) are:

$$\mathcal{X}_{n,l}^{(m)} \sim \mathcal{N}((\mathbf{Z}_n \odot \mathbf{U}_l) \times_2 \mathbf{W}^{(m)}, \mathbf{I}(\tau^{(m)})^{-1})$$
$$\mathbf{Z} \sim \mathcal{N}(0, I)$$
$$\mathbf{U}_{l,k} \sim \mathcal{N}(0, (\boldsymbol{\beta}_{l,k})^{-1})$$
$$\mathbf{W}_{d,k}^{(m)} \sim \mathbf{h}_k^{(m)} \mathcal{N}(0, (\boldsymbol{\alpha}_{d,k}^{(m)})^{-1}) + (1 - \mathbf{h}_k^{(m)})\delta_0$$
$$\mathbf{h}_k^{(m)} \sim Bernoulli(\pi_k)$$
$$\pi_k \sim Beta(a^\pi, b^\pi)$$
$$\boldsymbol{\beta}_{l,k} \sim Gamma(a^\beta, b^\beta)$$
$$\boldsymbol{\alpha}_{d,k}^{(m)} \sim Gamma(a^\alpha, b^\alpha)$$
$$\tau^{(m)} \sim Gamma(a^\tau, b^\tau)$$

where $Gamma(a, b)$ is parameterized by *shape a, rate b*.

The coupled $N \times L$ samples in each tensor $\mathcal{X}^{(m)}$ are modelled via the product
of loadings, with a view-specific observation precision $\tau^{(m)}$. For the latent variables, we assume *a priori* independence, and induce an element-wise automatic
relevance determination ARD prior [16] on $\mathbf{U}_{l,k}$ to encourage sparsity.

To infer the interactions between views and components, we make the model
view-wise sparse via a Spike and Slab prior [17] on the projection weights $\mathbf{W}^{(m)}$.
The spike and slab prior has two parts, one being a delta δ_0 function centered at
zero and the other some continuous distribution (usually Gaussian). We replace

the Gaussian with an element-wise ARD prior to additionally allow feature-level sparsity in our model. The ARD is a Normal-Gamma prior that specifies the precision $\alpha_{d,k}^{(m)}$ controling the scale of each variable. Our element-wise d, k, m formulation of ARD encourages the loadings within a component-view pair to be sparse. In the spike and slab construct, the binary value $\mathbf{h}_k^{(m)}$ drawn from a Bernoulli distribution gives the component-view activation. If $\mathbf{h}_k^{(m)} = 1$, the component k is active in view m and the loadings $\mathbf{W}_k^{(m)}$ are sampled from a corresponding element-wise ARD prior, whereas if $\mathbf{h}_k^{(m)} = 0$, the component-view pair is not active and the loadings $\mathbf{W}_k^{(m)}$ are set to zero via δ_0, inducing view-wise sparsity.

Learning the $\mathbf{h}^{(m)}$ activities allows automatic determination of the number and sharing of factors between the views. This is because if K is set to be large enough, the model will switch off $\mathbf{h}_k^{(m)}$, for all the extra k, m pairs. This yields the underlying sharing pattern of the views, even producing empty components that are not active in any view. The presense of empty components indicates that K was set to a large enough value, and the amount of non-empty components gives the rank of the view collection. In the construct, π_k represents probability of activation of each component.

The joint probability of data and parameters can be factorized as follows, and inference is performed via Gibbs sampling:

$$p(\mathcal{X}^{(1)}, \mathcal{X}^{(2)}, ..., \mathcal{X}^{(M)}, \Theta) = \prod_{m=1}^{M} \prod_{n=1}^{N} \prod_{l=1}^{L} p(\mathcal{X}_{n,l}^{(m)} | \mathbf{Z}_n, \mathbf{U}_l, \mathbf{W}^{(m)}, \tau^{(m)})$$

$$p(\tau^{(m)}) p(\mathbf{Z}_n) \prod_{k=1}^{K} p(\mathbf{U}_{l,k} | \beta_{l,k}) p(\beta_{l,k})$$

$$\prod_{d=1}^{D^{(m)}} p(\mathbf{W}_{d,k}^{(m)} | \alpha_{d,k}^{(m)}, \mathbf{h}_k^{(m)}) p(\alpha_k^{(m)}) . p(\mathbf{h}_k^{(m)} | \pi_k) p(\pi_k)$$

Degeneracies can complicate the practical use of CP when analyzing real data [18]. Most degeneracies occur due to non-trilinear structure in the data and are identified by strong negative correlations between two components. To overcome the problem, researchers have proposed adding orthogonality and non-negativity constraints that address it by hindering correlations [18,19], but may also effect the model's ability to discover PARAFAC's intrinsic axes.

In our Bayesian formulation, we impose an element-wise ARD prior on the component loadings $\mathbf{W}^{(m)}$,$\mathbf{U}$. The element-wise prior regularizes the solution allowing determination of precise factor loadings, and is a construct less strict than orthogonality. Our model should therefore be able to handle weak degeneracies, via a flexible composition that still allows identifying PARAFAC's intrinsic axes.

3.1   Special Cases and Related Problems

We next present special cases of our model and relate them to the existing works.

Sparse Bayesian CP. For $m = 1$ (a single view) our model reduces to sparse Bayesian CP factorization, which can automatically infer the number of components. In this special case our formulation goes very close to the Bayesian CP [20], the main differences being that they use MAP estimation and do not have feature-level sparsity.

Other Bayesian versions of CP include a variant specialized for temporal datasets [5], the fully conjugate model [21], and an exponential family framework [22]. For Tucker factorizations, Chu and Ghahramani [6] formulated Tucker in a probabilistic framework (pTucker) while [23] presented a non-linear variant using Gaussian processes. All of these follow different assumptions; however, unlike our method, none of them automatically learns the rank of the tensors. Instead, repetitive methods of rank identification are used, though they pose serious scalablity issues for large tensors [1].

Bayesian Tensor CCA. For $m = 2$, our model is the first Bayesian Tensor CCA. The model is related to tensor-CCAs in the classical domain, specifically to [8,10]. An additional technical difference, besides our Bayesian treatment, is that the earlier works assume the two tensors to be paired in a single mode (N), while we assume pairing in both N and L. Both settings are sensible and applicability depends on the nature of the data.

There have also been fusion studies on coupled matrix-tensor factorization, where values in a tensor were predicted with side information from a matrix, or vice versa. A gradient-based least squares optimization approach was presented in [24], while [25,26] used generalized linear models in a coupled matrix-tensor factorization framework to solve link prediction and audio processing tasks.

In the matrix domain, a related multi-view problem was recently studied under the name of Group Factor Analysis [13]. The goal there was to perform a joint factor analysis of multiple matrices to find relationships between datasets. Their method also finds components shared between subsets of views but, naturally, works only for matrices.

4   Experiments

We have applied our model on both simulated and real datasets. We will first demonstrate in a simulated example the model's ability to correctly separate shared and view-specific components, as well as precisely identify the factor mode loadings. We next compare our model to the existing state-of-the-art methods on benchmark single-view datasets, to validate that in the single-view special case our algorithms are comparable. We then validate our model's performance on simulated multi-view tensors and compare to the single-view tensor methods and the multi-view matrix methods as the existing baselines, ascertaining the advantage gained by the multi-view tensor decomposition. Finally, we apply our method on multi-view real data tensors on a new problem from toxicogenomics and a functional neuroimaging dataset, demonstrating the interpretative power and diverse applicability of the model.

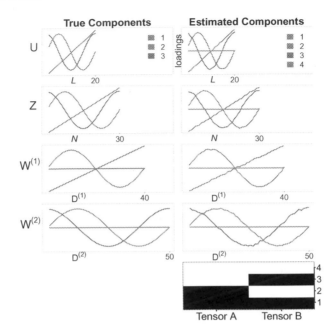

Fig. 3. Demonstration of BMTF decomposing two tensors $\mathcal{A}$ and $\mathcal{B}$ simultaneously, finding the one shared and two view-specific components. **Left:** Loadings are drawn for the three components (1 shared, 2 specific) embedded in the data. **Right-Bottom:** component-view activation $\mathbf{h}_k^{(m)}$ for a $K = 4$ BMTF run. **Right:** Loadings of the four BMTF components reveal the shared and specific components.

Our model detects the number and type of components automatically, as long as it is run with a large enough K, resulting in the extra components getting zero loadings. The practical procedure we followed is to increase K until empty components are found. The experiments were run with the hyperparameters $a^\pi, b^\pi, a^\alpha, b^\alpha, a^\beta, b^\beta, a^\tau, b^\tau$ initialized to 1. To account for high noise in real datasets, the noise hyperparameters a^τ, b^τ were initialized assuming a signal-to-noise ratio of 1. All remaining model parameters were learned using Gibbs sampling while discarding the first 10,000 samples as the burn-in and using the next 10,000 samples for estimating the posterior. Our R implementation of the model is available at http://research.ics.aalto.fi/mi/software/bmtf/.

4.1   Simulated Illustration

We first demonstrate the ability of our BMTF to decompose the data into factors in a two-view setting. For this purpose two tensor datasets $\mathcal{A}$ and $\mathcal{B}$ were created using three underlying components, one of which is shared between both tensors, while one is specific to each. Figure 3-left shows the 3-mode loadings used to create the two tensors, where $\mathbf{Z}$ and $\mathbf{U}$ are the common 1st and 2nd mode loadings between both tensors while $\mathbf{W}^{(1)}$ and $\mathbf{W}^{(2)}$ are the 3rd mode loadings for tensor $\mathcal{A}$ and tensor $\mathcal{B}$, respectively. The shared component (blue) has non-zero loadings in both $\mathbf{W}^{(1)}$ and $\mathbf{W}^{(2)}$ while the specific ones have non-zero $\mathbf{W}^{(m)}$ loadings in only the corresponding view.

BMTF was run with $K = 4$, i.e., larger than the number of embedded components (=3). Figure 3-bottom-right plots the learned $\mathbf{h}_k^{(m)}$ values for the $M = 2$ views and $K = 4$ components. The plot shows that one component is active in both views (black) while one component active in each view, demonstrating that the model correctly separates the shared and view-specific effects. The fourth component was rightly detected as not active in any of the views, as the data come from only three components, indicating that the model identifies the correct number of components by switching off the extra ones. The discovered loadings for the 4 components are plotted in Figure 3-right. The plots show that the loadings are identified correctly in this simulated example.

4.2   Single View

As discussed in Section 3.1, our method also solves the CP problem as a special case when run on a single dataset. We compare our formulation to the existing state-of-the-art single-view methods on benchmark datasets to validate that our performance is at least comparable. These single-view methods have not been generalized to multi-view tensors where our main contribution lies.

Comparison Methods. We compare to the following state-of-the-art approaches.

ARDCP: Mørup *et. al.* [20] formulated CP in a Bayesian framework and automatically learn the number of components, using MAP estimation. In comparison to them, our model is fully Bayesian and additionally element-wise sparse.

pTucker: Chu and Ghahramani [6] presented a probabilistic version of the Tucker model. Tucker is more flexible than CP, though not easy to interpret.

CP: We also compare to the most widely used and updated classical CP implementation from the *N-way Toolbox* (v3.31 of July 2013, http://www.models.life.ku.dk/nwaytoolbox). The implementation solves the factorization using the well established Alternating Least Squares ALS algorithm [27]. On the computational side, per-iteration complexity of BMTF exceeds ARDCP and CP only due to computing $K \times K$ covariance matrices, which is small compared to the rest of the computation. Tucker is costlier than CP as it needs to solve for the core tensor as well, while pTucker reduces its costs with custom solutions.

Datasets. We use the three commonly used benchmark datasets in tensor modeling from http://www.models.life.ku.dk/nwaydata, namely Amino Flow Injection Analysis, and Kojima Girls datasets.

We test our model for both its ability to find the number of components to model the data correctly in a missing value setting. We randomly selec half of the values in the datasets for training the models and predicted t remaining half. The split was repeated independently 100 times. BMTF a ARDCP learned the number of components for each split. CP and pTucker wer

Table 1. Detection of number of factors, and ability to find the intrinsic structure. The table lists the number of factors of the three real datasets determined by *pftest (on full data)* from *N-way Toolbox* and compares the ability of BMTF with other state of the art methods in a) learning the number of factors and b) prediction error, when data contains missing values.

DATA SET	AMINO ACID	FLOW INJECTION	KOJIMA GIRLS
SIZE	5 x 201 x 61	12 x 50 x 45	4 x 153 x 20
		FACTORS	
pftest	3	4	2
BMTF	3.0 ± 0.0	4.5 ± 0.5	2.0 ± 0.1
ARDCP	3.1 ± 0.3	4.0 ± 0.0	1.2 ± 0.4
		PREDICTION RMSE	
BMTF	0.0257±0.0003	**0.045±0.010**	**0.189±0.025**
ARDCP	0.0278±0.0035	0.065±0.001	0.305±0.051
CP	0.0256±0.0003	0.053±0.001	1.643±4.098
PTUCKER	**0.0250±0.0003**	0.049±0.001	0.236±0.055

run with the number of components estimated from the full data using the de-facto standard *pftest* from *N-way toolbox* [27].

Results are presented in Table 1. Both BMTF and ARDCP recovered the number of components well despite 50% missing values, with the mean being close to the number obtained by *pftest* on *full data*. The result clearly shows that automatic component selection works even in the presence of missing values.

Prediction RMSE results for the first two datasets Amino Acids and Flow Injection show that all methods perform almost comparably and none goes exceedingly wrong, confirming that our method compares well with state-of-the-art single-view methods. The third dataset Kojima Girls shows a major difference in the performance of the methods. This dataset is known to have a degeneracy problem, and hence the standard CP fails to model the data correctly. ARDCP seems to perform better in comparison to CP, and close examination reveals that this is because ARDCP tends to skip the degenerate component as can be seen from the mean component number of 1.2. Using fewer components is one way of avoiding the effect of degeneracies. Our method does both, finding the correct number of components and being able to cope with degeneracies as is shown by the best performance. With its flexible parametrization the Tucker is also able to correctly model non-trilinear structure in the data, which is a characteristic of degeneracies [28]; hence does not suffers from the degeneracy problem.

4.3   Multi-view

To validate the performance of our model in multi-view settings, we applied it to simulated data sets that have all types of factors, i.e., factors specific to just one view, factors shared between a small subset of views and factors shared between

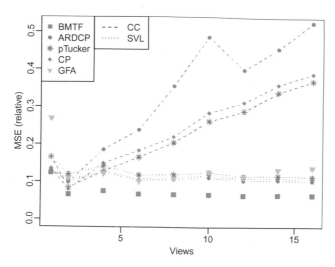

Fig. 4. Performance of Bayesian multi-view tensor factorization compared to single-view tensor methods and multi-view matrix methods (baselines). The number of views increases on the x-axis while the relative mean square error of recovering the underlying data is plotted on the y-axis. The single-view methods were tested in two settings a) CC marked with dashed-lines, where all the tensors are concatenated; b) SVL as dotted-lines, where models are learned for each tensor seperately.

most of the views. We show that the model can correctly discover the structure as the number of views is increased, while the baseline approaches are unable to find the correct result.

We simulated a data set consisting of $M = 16$ views with dimensions $N = 20$, $L = 5$ and D_m randomly sampled between 10 and 100, using a manually constructed set of K=31 factors of the various types. For each component, the loadings $\mathbf{Z}_{:,k}$, $\mathbf{U}_{:,k}$ and $\mathbf{W}_{:,k}^{(m)}$ were randomly sampled from the standard normal distribution for all active m. For the non-active views m in the component k, the $\mathbf{W}_{:,k}^{(m)}$ were set to zero. The views were then created as:

$$\overline{\mathcal{X}}^{(m)} = \sum_k \mathbf{Z}_{:,k} \circ \mathbf{U}_{:,k} \circ \mathbf{W}_{:,k}^{(m)}$$

$$\mathcal{X}^{(m)} = \overline{\mathcal{X}}^{(m)} + \boldsymbol{\epsilon}^{(m)}$$

where $\overline{\mathcal{X}}^{(m)}$ is the true underlying data while $\boldsymbol{\epsilon}^{(m)}$ is a noise tensor sampled from a normal distribution with mean zero and variance equivalent to that of $\overline{\mathcal{X}}^{(m)}$.

We ran BMTF for $M = 1, \ldots, 16$. The single-view tensor methods were run in two settings, a) on a concatenation of all views [CC], b) single view learning [SVL], where a model is learned for each view seperately. BMTF found the correct number of components in all cases while ARDCP[CC] failed to detect the correct number for $M \geq 4$. The other two methods, CP and pTucker, were run with the true number of factors. In single view learning [SVL], the methods were unable

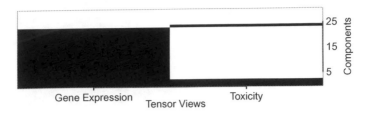

Fig. 5. Component activations in the toxicogenomics dataset indicate 3 shared components between the disease-specific gene expression responses and toxicity measurements of the drugs. The presence of several empty components indicates that $K = 30$ was enough to model the data.

to identify the sharing between components, as they do not solve the multi-view problem addressed by BMTF. For completeness, we also compare our method to multi-view matrix FA (GFA) [13] by matricizing the tensors $\mathcal{X}^{(M)} \in \mathbb{R}^{N \times D_m \times L}$ into matrices $\mathbf{X}^{(M)} \in \mathbb{R}^{(N \times L) \times D_m}$.

We measured the models' performance in terms of the recovery error of the missing data. Defining $\hat{\mathcal{X}}^{(m)}$ as the model's estimate of the data, the recovery error is computed as the relative mean square error $\|\hat{\mathcal{X}}^{(m)} - \overline{\mathcal{X}}^{(m)}\|^2 / \|\mathcal{X}^{(m)} - \overline{\mathcal{X}}^{(m)}\|^2$ averaged over all the views.

Figure 4 plots the recovery error of our method as a function of the number of views. Our model's performance is stable as the number of views increases and outperforms all the baseline tensor and matrix alternatives. Single-view methods, applied to a data set which contains all tensors concatenated, deteriorate rapidly; while by learning each tensor seperately they are unable to discover the shared pattern. The matricized method (GFA) performs comparably to the single-view tensor methods. The experiment confirms that the specific multi-view tensor problem cannot be optimally solved with methods not designed for the purpose, and that our method fulfills its promise.

4.4   Application Scenarios and Interpretation

We next demonstrate the method at work on multi-view tensor datasets in potential use cases of BMTF. The first application represents a new problem at the juncture of toxicity and bioinformatics, while the second is a functional neuroimaging case.

Toxicogenomics. We analyzed a novel drug toxicity response problem, where the tensors arise naturally when gene expression responses of multiple drugs are measured for multiple diseases (different cancers) across the genes. The data contain two views, the measurement of post-treatment gene expression, and sensitivity of the cells to the drug. The key question that BMTF can answer is, which parts of the responses are specific to individual types of cancer and which occur across cancers, and which of them are related to drugs effectiveness. These patterns, if uncovered, can help understand the mechanisms of toxicity [29].

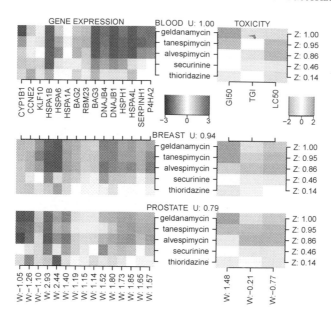

Fig. 6. Component 1 captures the well-known heatshock protein response. The top genes (left) and toxicity indicators (right) from the two views are plotted as columns, and the three different cancers as rows. The component links the strong upregulation of the heatshock protein genes (red) to high toxicity (green) in the top three drugs, all of which are heatshock protein inhibitors.

The dataset contained two views. The first, $m = 1$, contained the post-treatment differential gene expression responses $D_1 = 1106$ of several drugs $N = 78$ as measured over multiple cancer types $L = 3$. The second, $m = 2$, contained the corresponding drug sensitivity measurements $D_2 = 3$. The gene expression data were obtained from the connectivity map [30] that contained response measurements of three different cancers: Blood Cancer, Breast Cancer and Prostate Cancer. The data were processed so that gene expression values represent up (positive) or down (negative) regulation from the untreated (base) level. Strongly regulated genes were selected, resulting in $D_1 = 1106$. The drug screen data for the three cancer types were obtained from the NCI-60 database [31], measuring toxic effects of drug treatments via three different criteria: GI50 (50% growth inhibition), LC50 (50% lethal concentration) and TGI (total growth inhibition). The data were processed to represent the drug concentration used in the connectivity map to be positive when toxic, and negative when non-toxic.

BMTF was run with K=30, resulting in 3 components shared between both the gene expression and toxicity views, revealing that some patterns are indeed shared (Figure 5). These shared components form hypotheses about underlying biological processes that characterize toxic responses of the drugs.

The first component captures the well-known "Heatshock Protein" response. The response is characterized by strong upregulation of heatshock genes in all

Table 2. Prediction RMSE of BMTF in comparison to existing methods on toxicoge-nomics and neuroimaging datasets. The mean prediction performance over 100 runs of independent sets of missing values (50% missing) is given, along with one standard error of the mean. BMTF outperformed all other methods significantly with t-test p-values $< 10^{-6}$ on toxicogenomics data, and p-values $< 10^{-4}$ on neuroimaging data.

				CC		SVL	
		BMTF	GFA	ARDCP	CP	ARDCP	CP
Toxicogenomics	Mean	**0.4811**	0.5223	0.8919	5.3713	0.6438	5.0699
	StdError	**0.0061**	0.0041	0.0027	0.0310	0.0047	0.0282
Neuroimaging	Mean	**0.5105**	0.5144	0.6224	0.5740	0.5725	0.5611
	StdError	**0.0004**	0.0004	0.0003	0.0004	0.0003	0.0010

cancers (Figure 6-left) and corresponding high toxicity indications (Figure 6-right). The response is being activated by the heat shock protein (HSP90) in-hibitor drugs, all of which have the highest loadings in the component (the top three drugs). The HSP inhibition response has been well studied for treatment of cancers [32] evaluating its therapeutic efficacy. Had the biological action not already been discovered, our component could have been a key in revealing it.

Component 2 represents toxic mechanisms via inhibition of protein synthesis (details not shown) and Component 3 via damaging of cell DNA. Both of these components reveal interesting cancer type-specific findings, detailed interpreta-tions of which are under way. The experiment validates that the model is able to find useful factors from multiple-tensor data.

We also evaluated BMTF for predicting missing values on the toxicogenomics data. BMTF outperformed the single-view methods[1] and matrix methods signif-icantly with t-test p-values $< 10^{-6}$, on the prediction RMSE of 100 independent runs (Table 2). Additionally, the tensors of BMTF are easier to interpret than the corresponding $(L \times N) \times D_m$ matrices of matricized GFA, and the reformed tensors of single view CP.

Functional Neuroimaging. As the second demonstration we analysed a multi-view functional neuroimaging dataset, which comes from subjects exposed to multiple audiovisual stimuli. The data contained $M = 7$ views, representing the different audio and audiovisual stimuli, each composed of three songs. The different views are brain recordings made under different "presentations" of the same songs: purely auditory ones including singing (A:Sing), piano (A:Piano) *etc*, and audiovisual speaking with both voice and image of speaker (AV:Speech) *etc*. The views have a natural tensor structure where brain activity was recorded with fMRI from $L = 10$ subjects over the course of the experiment ($N = 162$ time points) in $D_m = 32$ regions of interest (data from [13]).

BMTF was run with $K = 300$ and the $\mathbf{h}^{(m)}$ profile is shown in Figure 7. The plot indicates that there exist several potentially interesting components shared between different subsets of views. The large number of view-specific components model "structured noise", i.e., mostly brain activity not related to the stimuli.

---

[1] pTucker failed to complete even on 50GB of RAM, hence was excluded.

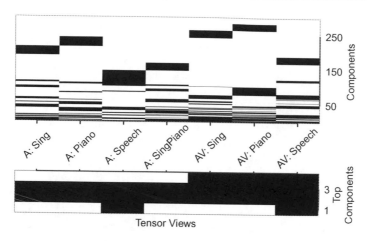

Fig. 7. Top: Component activations in the neuroimaging dataset. The components shared between subsets of views capture potentially interesting variation, separated from the view-specific "structured noise" or non-interesting variation. **Bottom:** Zoomed inset of top components based on subject (**U**) loadings. The first component is active in both speech views.

The goal of the fMRI study was to find responses that generalize across the subjects and describe relationships of the different presentation conditions (views). We selected components generalizing across subjects by sorting them based on the subject (**U**) loadings, and explain the first one here to concretize what the method can produce. The first component is active in the speech-related views, pure audio (A:Speech), and combined audio-visual (AV:Speech) views, indicating that it captures speech-related activity. A closer look at the $\mathbf{W}^{(m)}$ loadings for the views shows activation of the same auditory regions of the brain, demonstrating the signal is neuroscientifically relevant.

Quantitatively, BMTF fits the data better than simpler alternatives as demonstrated by the missing value prediction in Table 2, while in comparison to the analysis of [13], it extracts more components having consistent behaviour over the subjects, indicating that taking the tensorial nature of data into account improves detection of structure.

5   Discussion

We introduced a novel multi-view tensor factorization problem, of collectively decomposing multiple paired tensors into factors. We factorize the tensors into PARAFAC-type (equivalently, CP-type) components, each shared by a subset of the tensors, from one to all. We introduced a Bayesian multi-view tensor factorization (BMTF) model that solves the problem via a joint CP-type decomposition of tensors while learning the precise type and number of factors automatically. In the special case of two tensors, our method is simultaneously also the first Bayesian tensor canonical correlation analysis (CCA) method. The model can also be considered as an extension of the matrix-based Group Factor Analysis method [13] to tensors.

We validated the model's performance in identifying components on simulated data. The model was then demonstrated on a new toxicogenomics problem and a neuroimaging dataset, yielding interpretable findings with detailed interpretations on-going. Initial evidence suggests that taking the tensor nature of data into account makes the results more accurate and precise. In particular, the model is able to handle degenerate solutions well, making the formulation applicable to a wider set of datasets.

Acknowledgments. The work was supported by the Academy of Finland (140057; Finnish Centre of Excellence COIN, 251170) and the FICS doctoral programme. We also acknowledge Aalto Science-IT resources.

References

1. Kolda, T., Bader, B.: Tensor decompositions and applications. SIAM Review 51(3), 455–500 (2009)
2. Carroll, J.D., Chang, J.J.: Analysis of individual differences in multidimensional scaling via an n-way generalization of Eckart-Young decomposition. Psychometrika 35(3), 283–319 (1970)
3. Harshman, R.A.: Foundations of the parafac procedure: models and conditions for an explanatory multimodal factor analysis. UCLA Working Papers in Phonetics 16, 1–84 (1970)
4. Tucker, L.R.: Some mathematical notes on three-mode factor analysis. Psychometrika 31(3), 279–311 (1966)
5. Xiong, L., Chen, X., Huang, T.K., Schneider, J., Carbonell, J.G.: Temporal collaborative filtering with bayesian probabilistic tensor factorization. In: Proceedings of SIAM Data Mining, vol. 10, pp. 211–222 (2010)
6. Chu, W., Ghahramani, Z.: Probabilistic models for incomplete multi-dimensional arrays. In: Proceedings of AISTATS. JMLR W&CP, vol. 5, pp. 89–96 (2009)
7. Lee, S.H., Choi, S.: Two-dimensional canonical correlation analysis. IEEE Sig. Proc. Letters 14(10), 735–738 (2007)
8. Kim, T.K., Cipolla, R.: Canonical correlation analysis of video volume tensors for action categorization and detection. IEEE Transactions on Pattern Analysis and Machine Intelligence 31(8), 1415–1428 (2009)
9. Yan, J., Zheng, W., Zhou, X., Zhao, Z.: Sparse 2-d canonical correlation analysis via low rank matrix approximation for feature extraction. IEEE Sig. Proc. Letters 19(1), 51–54 (2012)
10. Lu, H.: Learning canonical correlations of paired tensor sets via tensor-to-vector projection. In: Proceedings of IJCAI, pp. 1516–1522 (2013)
11. Hardoon, D.R., Szedmak, S., Shawe-Taylor, J.: Canonical correlation analysis: An overview with application to learning methods. Neural Computation 16(12), 2639–2664 (2004)
12. Klami, A., Virtanen, S., Kaski, S.: Bayesian canonical correlation analysis. Journal of Machine Learning Research 14, 965–1003 (2013)
13. Virtanen, S., Klami, A., Khan, S.A., Kaski, S.: Bayesian group factor analysis. In: Proceedings of AISTATS. JMLR W&CP, vol. 22, pp. 1269–1277 (2012)

14. Cattell, R.B.: Parallel proportional profiles and other principles for determining the choice of factors by rotation. Psychometrika 9(4), 267–283 (1944)
15. Kruskal, J.B.: Three-way arrays: rank and uniqueness of trilinear decompositions, with application to arithmetic complexity and statistics. Linear Algebra and its Applications 18(2), 95–138 (1977)
16. Neal, R.M.: Bayesian learning for neural networks. Springer (1996)
17. Mitchell, T.J., Beauchamp, J.J.: Bayesian variable selection in linear regression. Journal of the American Statistical Association 83(404), 1023–1032 (1988)
18. Krijnen, W., Dijkstra, T., Stegeman, A.: On the non-existence of optimal solutions and the occurrence of degeneracy in the candecomp/parafac model. Psychometrika 73(3), 431–439 (2008)
19. Srensen, M., Lathauwer, L., Comon, P., Icart, S., Deneire, L.: Canonical polyadic decomposition with a columnwise orthonormal factor matrix. SIAM Journal on Matrix Analysis and Applications 33(4), 1190–1213 (2012)
20. Mørup, M., Hansen, L.K.: Automatic relevance determination for multiway models. Journal of Chemometrics 23(7-8), 352–363 (2009)
21. Hoff, P.D.: Hierarchical multilinear models for multiway data. Computational Statistics & Data Analysis 55(1), 530–543 (2011)
22. Hayashi, K., Takenouchi, T., Shibata, T., Kamiya, Y., Kato, D., Kunieda, K., Yamada, K., Ikeda, K.: Exponential family tensor factorization: an online extension and applications. Knowledge and Information Systems 33(1), 57–88 (2012)
23. Xu, Z., Yan, F., Qi, A.: Infinite tucker decomposition: Nonparametric bayesian models for multiway data analysis. In: Proceedings of ICML, pp. 1023–1030 (2012)
24. Acar, E., Rasmussen, M.A., Savorani, F., Naes, T., Bro, R.: Understanding data fusion within the framework of coupled matrix and tensor factorizations. Chemometrics and Intelligent Laboratory Systems 129, 53–63 (2013)
25. Ermis, B., Acar, E., Cemgil, A.T.: Link prediction in heterogeneous data via generalized coupled tensor factorization. Data Mining and Knowledge Discovery, 1–34 (2013)
26. Yilmaz, K.Y., Cemgil, A.T., Simsekli, U.: Generalised coupled tensor factorisation. In: Proceedings of NIPS, pp. 2151–2159 (2011)
27. Andersson, C.A., Bro, R.: The N-way toolbox for MATLAB. Chemometrics and Intelligent Laboratory Systems 52(1), 1–4 (2000)
28. Lundy, M.E., Harshman, R.A., Kruskal, J.B.: A two-stage procedure incorporating good features of both trilinear and quadrilinear models. Multiway Data Analysis, 123–130 (1989)
29. Hartung, T., Vliet, E.V., Jaworska, J., Bonilla, L., Skinner, N., Thomas, R.: Food for thought.. systems toxicology. ALTEX 29(2), 119–128 (2012)
30. Lamb, J., et al.: The connectivity map: Using gene-expression signatures to connect small molecules, genes, and disease. Science 313(5795), 1929–1935 (2006)
31. Shoemaker, R.H.: The nci60 human tumour cell line anticancer drug screen. Nature Reviews Cancer 6(10), 813–823 (2006)
32. Kamal, A., et al.: A high-affinity conformation of HSP90 confers tumour selectivity on HSP90 inhibitors. Nature 425(6956), 407–410 (2003)

Conditional Log-linear Models
for Mobile Application Usage Prediction

Jingu Kim and Taneli Mielikäinen

Nokia, Sunnyvale, CA, USA
{jingu.kim,taneli.mielikainen}@nokia.com

Abstract. Over the last decade, mobile device usage has evolved rapidly
from basic calling and texting to primarily using applications. On average, smartphone users have tens of applications installed in their devices.
As the number of installed applications grows, finding a right application at a particular moment is becoming more challenging. To alleviate
the problem, we study the task of predicting applications that a user
is most likely going to use at a given situation. We formulate the prediction task with a conditional log-linear model and present an online
learning scheme suitable for resource-constrained mobile devices. Using
real-world mobile application usage data, we evaluate the performance
and the behavior of the proposed solution against other prediction methods. Based on our experimental evaluation, the proposed approach offers
competitive prediction performance with moderate resource needs.

1 Introduction

The number of applications installed to smartphones is increasing rapidly. In the
U.S., the average number of installed applications on a device increased from 32
in 2011 to almost 41 in 2012 [1]. While installing many applications is an easy
way to extend device functionalities, it makes finding a particular application
more difficult. In mainstream mobile user interfaces, users need to browse a grid
or a list of applications to locate a desired application. This is tedious with a
large number of applications. Many mobile user interfaces offer means to organize
applications into folders, but the fundamental problem of browsing and filtering
through folders remains.

A complementary approach to mitigate the problem is to learn from a user's
behavior and predict the most relevant applications for a given situation. A
general idea is to model the relationship between a user's application use and
context, such as time and location. In addition to building user interfaces that
offer applications the most likely to be used [9], such predictors could be used
to improve user experience by pre-launching applications [14].

Despite the popularity of smartphones, the development and understanding
of machine learning methods for application usage prediction have been limited.
Hand-crafted techniques have been designed to utilize temporal or spatial patterns of application usage [12,14], but they have difficulties in combining various
types of context information. The naïve Bayes and the nearest neighbor methods

T. Calders et al. (Eds.): ECML PKDD 2014, Part I, LNCS 8724, pp. 672–687, 2014.

have been employed in a number of work [7,9,13]. These methods take a variety of context information into account and have advantages in their simplicity. However, they have limitations in prediction accuracy due to strict modeling assumptions or in the use of computation resources.

We propose a prediction method based on a conditional log-linear model. The model describes the conditional probability of application usage given observed context variables. This model is one of discriminative models that include logistic regression and conditional random fields. Unlike the naïve Bayes method where independence between features is assumed, our method makes no assumptions on the distribution of features and does not suffer from inaccurate predictions with correlated features. Our method quickly generates predictions by evaluating a parametric linear model, with no additional cost for an increased size of usage data. We present an online training scheme that can be easily accommodated in smartphones, where computation resources are limited.

We demonstrate the effectiveness of the proposed approach through detailed experimental analysis using real-world mobile application usage data. We define a few evaluation measures and evaluate them to compare our approach with other prediction methods proposed for the task. Our evaluation shows that our method consistently outperforms existing ones in each of evaluation measures. We offer in-depth analysis on the behavior of prediction methods by showing their effects on individual users and learning curves on usage sequences. Our analysis illustrates the advantages of the proposed method.

The rest of this paper is organized as follows. We describe the problem setup and related work in Section 2 and Section 3, respectively. In Section 4, we describe a conditional log-linear model and our prediction method. In Section 5, we explain the setup of our experiments including compared methods, evaluation measures, and a data set. The results of experiments and our interpretation are in Section 6. We conclude this paper with discussion in Section 7.

2   Mobile Application Usage Prediction

Suppose a sequence of previously used applications and associated context information are given. Context information includes time stamps, location, and other sensor readings available at the time of application use. When predicting, we use context information available at that time to find applications to be likely used. Table 1 shows an example case, in which context variables for prediction are shown at the bottom row.

The output of a prediction method is a list of applications ordered from the most likely to the least likely used ones. Our goal is to have the user's selection in the beginning of the list: The earlier the user's selection is in the list, the better. In some cases, one might want the output to be only one application or a set of applications presented without an order. These outputs can be generated by taking one or more items from the beginning of the ordered list.

Training and prediction occur in a consecutive manner. Table 1 shows only one stage of prediction. The output of a prediction method is presented to a

Table 1. An example case of application usage prediction. Our task is to make a prediction for applications to be likely launched in the bottom row.

Application	Time stamp (UTC)	Time zone	Latitude	Longitude	Wi-Fi network
Facebook	2014-02-13 20:13:46	-8	n/a	n/a	WORK
WhatsApp	2014-02-13 20:15:20	-8	37.37	-122.03	WORK
Twitter	2014-02-13 20:19:01	-8	37.37	-122.03	WORK
Email	2014-02-13 21:35:02	-8	n/a	n/a	n/a
Twitter	2014-02-13 21:39:38	-8	n/a	n/a	n/a
Facebook	2014-02-14 01:22:55	-8	37.35	-121.92	HOME
?	2014-02-14 01:23:01	-8	n/a	n/a	HOME

user or used to pre-launch applications. When the user makes a new selection, it serves as an additional training case to be used for the following stages.

In our task, only the usage data of one user are used to make predictions for the same user. In this scenario, all computation occurs within a user's device without having to transmit usage data among users or to cloud servers. See Section 7 for comments on potential extensions.

It is worth distinguishing our prediction task from related ones. Our task is different from recommending new applications. We deal with a situation that users launch applications on a regular basis. They use the Email, Facebook, or Twitter application typically multiple times a day. Our task is to estimate which applications are to be the most likely used under given context. In contrast, an application recommendation task is to discover new applications that have not been used before. News or movie recommendation is similar to application recommendation as users typically do not repeat reading the same article or watching the same movie.

There is a similarity between our task and sequence labeling, but there is a distinction, too. In both cases, unknown variables and observations are organized in a sequence. Sequence labeling (e.g., part-of-speech tagging in natural language processing) involves predicting all unknown labels at the same time. In contrast, in application usage prediction, once a prediction is generated for one stage, the user's selection is observed and used for the prediction of the next stage. In other words, application usage prediction focuses on predicting one stage at a time, while sequence labeling concerns predicting the entire unknowns altogether.

A primary goal of our task is to design a method that offers accurate predictions. In addition, due to restrictions in the mobile environment, resource consumption is an important issue. Users interact with smartphones very often, but CPU and memory resources are limited there. Fast generation of predictions is critical in providing responsive user interface or pre-launching applications seamlessly.

3   Related Work

Methods for mobile application usage prediction have been discussed in a number of publications. Tan et al. interpreted the problem as time-series prediction and proposed methods that take advantage of periodic patterns [12]. Yan et al. proposed spatial and temporal features that are in turn used to determine if an application is relevant at a moment [14]. These work revealed insights on important sources of context information, but their approaches lack flexibility to combine various types of context information. The naïve Bayes method [5,9] and the nearest neighbor method [7,13] have been commonly used due to their simplicity. We include them in our experiments to further understand them through comparisons with our proposed approach.

Researchers have used prediction methods for designing predictive user interface or improving system responsiveness. Shin et al. [9] have developed a predictive home-screen application and conducted a user study to validate its effectiveness. Yan et al. [14] and Xu et al. [13] have used prediction methods to pre-launch applications and showed that the launch delay can be reduced. Our discussion in this paper is focused on the design of algorithms as we aim to understand algorithm choices. Understandings in our paper can of course be used for implementing or improving those predictive systems.

Previous work also addressed the discovery of useful context information. Time stamps have been found useful in multiple literature as application usage tends to vary according to the time of day or the day of week [9,12]. Location information, often measured through GPS readings or the name of Wi-Fi network, has been shown useful [9,14]. Application transitions and temporal usage patterns have been shown useful too [7,14,15]. These information can be incorporated into our method without making changes in its core model. Our focus is understanding the effects of algorithm choices instead of analyzing the effects of particular context information.

Learning methods based on conditional log-linear models include logistic regression, maximum entropy prediction, and conditional random fields [2,11]. Our method is similar to multi-class logistic regression since the target variable is a discrete random variable with more than two possible values. We discuss the use of conditional log-linear models specifically for predicting mobile application usage, present a suitable training scheme, and provide performance analysis.

4   Prediction with Conditional Log-Linear Model

In this section, we describe variables, context features, a log-linear model, and our prediction method based on the model. Our notations are as follows. We use an uppercase letter, such as X, to denote a random variable. We use a lowercase letter, such as x, to denote an instantiation of a random variable. We use bold letters, such as $\mathbf{X}$ or $\mathbf{x}$, to denote a vector of variables or instantiations. Subscripts are used to denote elements within a vector, such as in $\mathbf{X} = (X_1, \cdots, X_q)$. Superscripts are used to denote elements in a sequence, such as in $\mathbf{X}^{(1)}, \cdots, \mathbf{X}^{(k)}$.

4.1   Context Variables and Features

Let Y be a discrete random variable representing an application. Let $\mathbf{X} = (X_1, \cdots, X_q)$ be a vector of random variables representing context. Instances

Table 2. Variables representing target application and context information

Variable	Description	Value for the case in Table 1
Y	Target application	*to be predicted*
X_1	UTC (Coordinated Universal Time)	2014-02-14 01:23:01
X_2	Time zone offset	-8
X_3	Longitude from GPS	n/a
X_4	Latitude from GPS	n/a
X_5	Wi-Fi network	HOME
X_6	The most recently used application	Facebook
X_7	The second recently used application	Twitter

Table 3. Examples of context features defined for variables in Table 2.

Feature	Description	Variables
f_1	Y is the most recently used application.	Y,X_6
f_2	Y is the second recently used application.	Y,X_7
f_3	Y is Email, and the time of day is morning - within [6AM,12PM).	Y,X_1,X_2
f_4	Y is Email, and the day of week is Monday.	Y,X_1,X_2
f_5	Y is Email, and latitude and longitude are within $[37.0, 38.0)$ and $[-122.0, -121.0)$, respectively.	Y,X_3,X_4
f_6	Y is Facebook, and Wi-Fi network is HOME.	Y,X_5
f_7	Y is Facebook, and the most recently used application is Email.	Y,X_6
f_8	Y is Facebook, and the second recently used application is Twitter.	Y,X_7

of $\mathbf{X}$ and Y are observed as a sequence. A prediction task is, given an observed instance of the context variable, $\mathbf{x}$, to predict an application, $\hat{\mathbf{y}}$. Example variables are shown in Table 2 along with values for the case shown in Table 1.

We use context features in order to easily incorporate variables into a probabilistic model. Context features are simply functions defined on context variables and the target variable. See Table 3 for examples of context features, defined with variables in Table 2. Features in Table 3 are shown as statements which can be evaluated to be true or false, which in turn produces a binary output. A feature can also be a real-valued function. Observe that all feature examples in Table 3 involve the target variable, Y. Because our goal is to fit conditional probabilities, features that do not involve the target variable are not used. Although all feature examples in Table 3 have human-understandable meanings, that does not need to be the case in general. As long as a feature can be constructed from data, it can be used in the model.

4.2   Conditional Log-Linear Models

We parameterize conditional probability with a log-linear model:

$$P(Y \mid \boldsymbol{\theta}, \mathbf{X}) = \frac{\exp\left\{\boldsymbol{\theta}^T \mathbf{f}(\mathbf{X}, Y)\right\}}{\sum_{Y'} \exp\left\{\boldsymbol{\theta}^T \mathbf{f}(\mathbf{X}, Y')\right\}}. \tag{1}$$

Here, $\mathbf{f}(\mathbf{X}, Y) = (f_1(\mathbf{X}, Y), \cdots, f_p(\mathbf{X}, Y)) \in \{0, 1\}^p$ is a p-dimensional feature vector, and $\boldsymbol{\theta} = (\theta_1, \cdots, \theta_p) \in \mathbb{R}^p$ is a p-dimensional parameter vector. Each of $\theta_j \in \{\theta_1, \cdots, \theta_p\}$ indicates a weight assigned to a corresponding feature, f_j. The denominator is needed to ensure that the sum of probabilities is one.

Model (1) is a discriminative model, described for the conditional probability, $P(Y \mid \boldsymbol{\theta}, \mathbf{X})$. This is a key difference between our method and the naïve Bayes method, which uses a generative model described for $P(Y, \mathbf{X} \mid \boldsymbol{\theta})$. The main advantage of discriminative modeling is that it is more reliable when features are correlated. Discriminative models make no assumption on the distribution of features and use full expressive power for making predictions. On the other hand, generative models make additional independence assumptions to obtain a tractable method. When some context features are highly correlated, the naïve Bayes method, which assumes the independence between features, becomes less accurate for application usage prediction. For more information on discriminative and generative models, see, e.g., [11].

Given a training set, $\mathcal{D} = \left\{\left(\mathbf{x}^{(i)}, y^{(i)}\right)\right\}_{i=1}^{k}$, it is possible to fit $\boldsymbol{\theta}$ by maximum likelihood:

$$\hat{\boldsymbol{\theta}} \leftarrow \arg\max_{\boldsymbol{\theta}} L(\boldsymbol{\theta} \mid \mathcal{D}), \tag{2}$$

where L is a log-likelihood function. L is expressed as

$$L(\boldsymbol{\theta} \mid \mathcal{D}) = \sum_{i=1}^{k} \log P(y^{(i)} \mid \mathbf{x}^{(i)}, \boldsymbol{\theta}) = \sum_{i=1}^{k} \left(\boldsymbol{\theta}^T \mathbf{f}(\mathbf{x}^{(i)}, y^{(i)}) - \log Z_{\boldsymbol{\theta}, \mathbf{x}^{(i)}}\right) \tag{3}$$

where

$$Z_{\boldsymbol{\theta}, \mathbf{X}} = \sum_{Y} \exp\left\{\boldsymbol{\theta}^T \mathbf{f}(\mathbf{X}, Y)\right\}. \tag{4}$$

A common way to solve optimization problem (2) is using the gradient-descent method. This approach works well for data sets of moderate size in a batch learning setting, where training is completed before the parameters are used for prediction. However, application usage prediction utilizes the model in the online learning setting where the parameters get immediately updated each time a new training example is obtained. For that, an online gradient-descent approach is more suitable.

Online gradient-descent is performed after user's each selection. Suppose $\boldsymbol{\theta}^{(k)}$ represents coefficients before the kth selection. When a user makes a new selection, $\left(\mathbf{x}^{(k)}, y^{(k)}\right)$, coefficients are updated as follows. Likelihood for $\left(\mathbf{x}^{(k)}, y^{(k)}\right)$ and corresponding gradient are written as

$$L\left(\boldsymbol{\theta} \mid \left(\mathbf{x}^{(k)}, y^{(k)}\right)\right) = \boldsymbol{\theta}^T \mathbf{f}(\mathbf{x}^{(k)}, y^{(k)}) - \log Z_{\boldsymbol{\theta}, \mathbf{x}^{(k)}},$$

$$\frac{\partial}{\partial \boldsymbol{\theta}} L\left(\boldsymbol{\theta} \mid \left(\mathbf{x}^{(k)}, y^{(k)}\right)\right) = \mathbf{f}(\mathbf{x}^{(k)}, y^{(k)}) - \mathbb{E}_{Y \mid \mathbf{X}^{(k)}, \boldsymbol{\theta}}\left[\mathbf{f}(\mathbf{x}^{(k)}, Y)\right], \tag{5}$$

where $\frac{\partial}{\partial \boldsymbol{\theta}} \log Z_{\boldsymbol{\theta}, \mathbf{x}^{(k)}} = \mathbb{E}_{Y|\mathbf{X}^{(k)}, \boldsymbol{\theta}} \left[\mathbf{f}(\mathbf{x}^{(k)}, Y) \right]$ can be easily verified using (4). An update scheme for $\boldsymbol{\theta}$ is

$$\boldsymbol{\theta}^{(k+1)} \leftarrow \boldsymbol{\theta}^{(k)} - \alpha \left(-\frac{\partial}{\partial \boldsymbol{\theta}} L \left(\boldsymbol{\theta}^{(k)} \mid \left(\mathbf{x}^{(k)}, y^{(k)} \right) \right) \right)$$
$$= \boldsymbol{\theta}^{(k)} + \alpha \left(\mathbf{f}(\mathbf{x}^{(k)}, y^{(k)}) - \mathbb{E}_{Y|\mathbf{X}^{(k)}, \boldsymbol{\theta}^{(k)}} \left[\mathbf{f}(\mathbf{x}^{(k)}, Y) \right] \right). \tag{6}$$

Parameter α, called the learning rate, represents how much update is taken. Online learning is typically used with decreasing learning rates, such as $\frac{1}{k}$. In application usage prediction, however, algorithms need to adapt to user's usage pattern that might change over time. In this case, decreasing learning rates could prevent an algorithm from adapting to user's recent behavior. We use a constant learning rate as in (6) and demonstrate prediction performance for different values of α.

Update scheme (6) is effective in resource-limited environments, such as in mobile devices. Unlike batch gradient-descent, which needs to use all previous data each time, online gradient-descent in (6) utilizes only one case, $\left(\mathbf{x}^{(k)}, y^{(k)} \right)$. This not only makes the update step faster but also reduces memory requirements since previous data do not need to be held. Online gradient-descent is also effective with sparse features. The feature vector, $\mathbf{f}(\mathbf{x}^{(k)}, y^{(k)})$, is typically very sparse; only a small number of features are nonzero. The expectation in (6), $\mathbb{E}_{Y|\mathbf{X}^{(k)}, \boldsymbol{\theta}^{(k)}} \left[\mathbf{f}(\mathbf{x}^{(k)}, Y) \right]$, is also similarly sparse. As a result, only a small number of coefficients need to be updated at each step, making an update cheaper.

A prediction for given context $\mathbf{x}$ is made as follows. Using trained $\boldsymbol{\theta}$, we evaluate $P(y \mid \boldsymbol{\theta}, \mathbf{x})$ for each mobile application y using (1). We then rank applications in order of decreasing probabilities.

5   Experiment Setup

Using mobile phone applications usage data from Nokia Mobile Data Challenge [6], we have evaluated the proposed approach together with various other methods previously known in the literature or folklore (see Section 5.1) using various accuracy measures (see Section 5.2). With a sequence of application usage, we evaluated each prediction method as follows. For each application selection in the sequence, we first used only the context of the selection to generate a ranking of predicted applications. The ranking was evaluated using the ground-truth selection based on accuracy measures. We then updated the model of the prediction method using the ground-truth selection and moved on to the next application selection.

5.1   Algorithms Compared

We denote our proposed method as the conditional log-linear (CLL) method. We have compared it with the following methods.

- **Most Recently Used (MRU).** MRU suggests recently selected applications from the most recent to the least recent ones. MRU has been used as a baseline method in a few previous work [7,9,13]. In cache algorithms, it is known as Least Recently Used [10], as the focus there is replacing the least likely used items rather than identifying the most likely used ones.
- **Most Frequently Used (MFU).** MFU counts the selections of each application and suggests applications in order of decreasing usage counts [7,9].
- **Weight Decay (WD).** Weight Decay assigns a weight to each item. The weight is increased when a corresponding item is selected, and weights decay otherwise. After the selection of $y^{(k)}$, weights are updated as

$$W(y) \leftarrow \begin{cases} 1 + W(y), & \text{if } y = y^{(k)}, \\ W(y)\exp(-\lambda), & \text{otherwise,} \end{cases}$$

where $W(y)$ represents y's weight, and λ is decay rate. The larger the λ is, the faster weights disappear. Applications are predicted in order of decreasing weights. Weight Decay is in between of MRU and MFU. It is similar to MRU if λ is large, and it is similar to MFU if λ is small. This method is known also as exponentially weighted average and has been explored in, e.g., high-frequency trading [8].
- **Naïve Bayes (NB).** Naïve Bayes is a method based on a generative model with independence assumptions among features [5,9]. Prediction probabilities are computed as

$$P(Y \mid \mathbf{X}) = \frac{P(Y)P(\mathbf{X} \mid Y)}{P(\mathbf{X})} = \frac{P(Y)\prod_{j=1}^{p} P(f_j(\mathbf{X}, Y) \mid Y)}{P(\mathbf{X})}.$$

- **K Nearest Neighbor (KNN).** In KNN [7,13], k previous events in which user's context is the most similar to the current context are searched. Those events, called neighbors, make weighted votes for applications, where the weights are given by the degrees of context similarities. Applications are predicted in order of decreasing votes.

Prediction methods have some parameters such as the decay rate for Weight Decay (λ), the number of neighbors for KNN (k), and the learning rate for CLL (α). For each parameter, we assessed various values and selected the best performing one. See Section 6.

We have used the same features for all algorithms, whenever applicable. MRU, MFU, and WD only utilize usage counts and do not incorporate context features. Whereas context features used for CLL involve both $\mathbf{X}$ and Y, features for NB and KNN involve only $\mathbf{X}$. For NB and KNN, we have used features that are defined with only $\mathbf{X}$ but otherwise equivalent to features used for CLL. See Section 5.3 for the description of features.

5.2   Evaluation Measures

Let $\hat{\mathbf{y}} = (\hat{y}_1, \cdots, \hat{y}_d)$ be a predicted ranking, and let y_g denote the ground-truth selection. Let $\text{Hit}(\hat{\mathbf{y}}, y_g)$ be the hit position: $\text{Hit}(\hat{\mathbf{y}}, y_g) = \min_i\{i : \hat{y}_i = y_g\}$. In

Table 4. Scores from evaluation measures

Measure	Hit position						
	1st	2nd	3rd	4th	5th	6th	7th
Recall@6	1	1	1	1	1	1	0
DCG	1	1	0.63	0.5	0.43	0.38	0.35
MRR	1	1/2	1/3	1/4	1/5	1/6	1/7

general, the smaller $\mathrm{Hit}(\hat{\mathbf{y}}, y_g)$, the better $\hat{\mathbf{y}}$. There are different ways to quantify accuracy, and we consider a few choices as follows.

- **Recall@N.** Since there is only one relevant item taken by a user, Recall@N is obtained by checking whether hit occurs within the top N items:

$$\mathrm{Recall}_N(\hat{\mathbf{y}}, y_g) = \begin{cases} 1, & \text{if } \mathrm{Hit}(\hat{\mathbf{y}}, y_g) \leq N, \\ 0, & \text{otherwise.} \end{cases}$$

- **Discounted Cumulative Gain (DCG).** DCG is commonly used in information retrieval [4]. The relevance of an item is either 0 or 1 in our case, and it is discounted by the hit position:

$$\mathrm{DCG}(\hat{\mathbf{y}}, y_g) = \begin{cases} 1, & \text{if } \mathrm{Hit}(\hat{\mathbf{y}}, y_g) = 1, \\ 1/\log_2 \mathrm{Hit}(\hat{\mathbf{y}}, y_g), & \text{otherwise.} \end{cases}$$

- **Mean Reciprocal Rank (MRR).** Reciprocal rank is another measure that discounts relevance based on the hit position. Its average, called mean reciprocal rank (MRR), is often used to assess the quality of ordered items.

$$\mathrm{RR}(\hat{\mathbf{y}}, y_g) = 1/\mathrm{Hit}(\hat{\mathbf{y}}, y_g).$$

Table 4 shows scores from these evaluation measures for hit positions from 1 to 7. Recall@N only cares whether the true selection is included in the first N items. DCG and MRR give discounted scores if the hit position is away from the beginning, except that DCG does not differentiate the first two positions. Precision@N was not considered because the reciprocal rank provides roughly the same information.

Depending on how a prediction method is used, an appropriate evaluation measure might vary. However, it is great to have an algorithm that performs the best in each of these measures. We demonstrate that this is the case for our proposed method.

5.3   Data Set and Features

Nokia Mobile Data Challenge data set[1] is the result of Lausanne Data Collection Campaign, conducted by Nokia Research Center Lausanne from 2009 to 2011

---

[1] https://research.nokia.com/page/12000

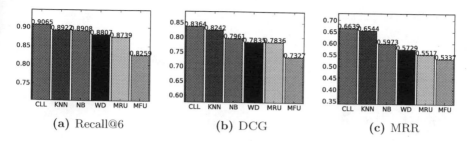

Fig. 1. Average of user scores

in the Lake Geneva region. The data consist of smartphone usage of nearly 200 participants for one year or more. We selected data of 142 users for which at least 2,500 usage events are available. For more information on this data, see [6].

The data contain time stamps, GPS coordinates, the name of the Wi-Fi network, and the identifier of the GSM tower, when available. We have used the following features for prediction methods: discretized GPS coordinates, discretized time of day, day of week, Wi-Fi network name, GSM tower identifier, and the list of recently used applications. These features are used for the CLL, NB, and KNN methods.

6   Experimental Results

We have assessed $\lambda, \alpha \in \{10^{-3}, 10^{-2.5}, 10^{-2}, 10^{-1.5}, \cdots, 1\}$ for WD and CLL and $k \in \{1, 3, 5, 10, 20, 40, \cdots, 2^8 \times 10\}$ for KNN. The best performing values for (λ, k) were $(10^{-1}, 80)$ for Recall@6, $(10^{-0.5}, 40)$ for DCG, and $(10^{-1}, 20)$ for MRR. The best performing value for α was $10^{-1.5}$ in all cases. For each evaluation measure, we present results from these best values in order to compare the best cases of prediction methods. We used $N = 6$ for measuring recall. See more information about α's and N's in Section 6.5.

Experiments were executed on a Linux computer with 16 cores of Intel(R) Xeon processor and 48GB memory. All algorithms and the experimentation software were implemented in Python.

6.1   Average Prediction Accuracy

We first present average prediction accuracy in Fig. 1. We took the averages of accuracy scores from the usage sequence of each user and then took an average from all users' averages. For example, the average Recall@6 score can be interpreted as follows: For users whose data were used for our test, six applications predicted by the CLL method include the correct selection with 0.9065 probability on average.

CLL showed the highest accuracy for each of Recall@6, DCG, and MRR measures, followed by KNN, NB, WD, MRU, and MFU. WD, MRU, and MFU showed relatively low accuracy because they do not utilize context information. Among CLL, KNN, and NB that utilize context information and features, NB was less accurate than CLL and KNN. This is partly due to an assumption in the naïve Bayes model that features are conditionally independent with each other. Only KNN showed performance comparable to that of CLL, but its accuracy scores were consistently smaller than those of CLL.

6.2 Individual User Analysis

While the average scores shown in Fig. 1 summarize the overall prediction performance, further analysis is necessary. For example, the average scores do not tell much whether a method worked perfectly for some users and rather poorly for others, or equally well for all.

In this section we first analyze how each algorithm performed for different users. The left column of Fig. 2 shows the percentiles of user scores. For example, the 5th percentile of Recall@6 for CLL is interpreted as follows: For 95 percent of users, with at least 0.838 probability, six applications predicted by CLL include the true selection. A desirable algorithm should perform well for all users, and it should consistently appear at the top of these graphs, as demonstrated by the CLL method.

Another interesting aspect is a switch from a baseline method. We selected MRU as a baseline as it is lightweight, simple, widely used, and relatively well performing. Our question here is "If MRU is replaced with another algorithm, would the replacement improve the accuracy scores for each user?" In the right column of Fig. 2, we show the offsets of the scores of CLL, KNN, NB, and WD from the scores of MRU. Offsets for MFU are not shown because its average scores shown in Fig. 1 are worse than those of MRU. From Fig. 2, see, e.g., the median of Recall@6 offsets for CLL was 0.03. The interpretation of this observation is that, when a prediction method is changed from MRU to CLL, the Recall@6 scores improve by at least 0.03 for at least half of users.

CLL showed the most desirable behavior in that its score offsets were always positive. In more detail, improvements in prediction scores for using CLL instead of MRU were at least 0.005 for Recall@6, 0.014 for DCG, and 0.024 for MRR. This is in strong contrast with the behavior of KNN, NB and WD, which all performed worse than MRU for some users. As shown in Fig. 2, the score offsets of these methods were substantially negative for some users, suggesting that predictions would become less accurate for the users if MRU is replaced with one of these methods.

6.3 Learning Curves

It might take a while before prediction methods adapt to a user's usage patterns, while the early user experience is often formative for the user's opinion about the usefulness of predictions. In this section, we investigate how quickly the accuracy scores of prediction methods improve from the beginning of the usage.

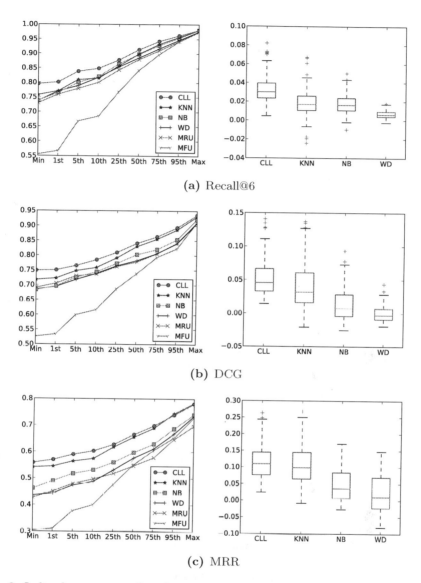

(a) Recall@6

(b) DCG

(c) MRR

Fig. 2. Left column: percentiles of user scores of each algorithm, right column: distribution of the score offsets of each algorithm from MRU

Fig. 3 shows the learning curves of prediction algorithms, where the averages of 100 consecutive values at each point are shown.

With enough usage data, the accuracy scores of CLL dominated those of other methods in each evaluation measure. In early stages, WD, MRU, and NB performed well based on Recall@6, DCG, and MRR, respectively. It took about 100 to 200 application usage events for the CLL method to provide more accurate

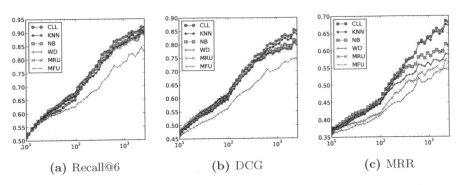

(a) Recall@6   (b) DCG   (c) MRR

Fig. 3. Learning curves of prediction methods. The x-axis represent the number usage events.

predictions than compared ones. It took around 200 usage events for the average Recall@6 score of CLL to reach 0.8. To reach 0.85, it needed 500 to 600 usage counts.

6.4   Efficiency Aspects

In Table 5, we summarize time and space complexity as well as observations on CPU and memory usages from our experiments. In our executions, on average $d \approx 60$, $p \approx 160000$, and $k \approx 10000$, where d, p, and k represent the numbers of applications, features, and usage data, respectively. Since probabilities or weights need to be sorted to determine the order of applications, $\mathcal{O}(d \log d)$ is involved in prediction complexity except in MRU. The time and space complexity of WD, MRU, and MFU only involves d and their average CPU and memory usages are overall very small.

Note that p denotes the number of all features involved, such as those shown in Table 3. The prediction complexity of CLL and NB involves $\mathcal{O}(p)$. This is because roughly $\mathcal{O}(\frac{p}{d})$ features are used for one application, and probabilities need to be evaluated for all applications. For KNN, there are roughly $\mathcal{O}(\frac{p}{d})$ features per one usage case, and features from all usage data need to be accessed for making predictions.

The training complexity of CLL is $\mathcal{O}(p)$ as its training involves gradient-descent. For KNN and NB, it is $\mathcal{O}(\frac{p}{d})$ because their training is no more than generating and counting features. Experimental observations are consistent with complexity: Average CPU time used by CLL for training appears larger than those used by KNN or NB. Albeit more expensive than KNN and NB, the average training time of CLL was kept at a moderate level of 4 milliseconds due to use of online gradient-descent. Training would be more expensive if a standard batch gradient-descent scheme is used.

In accuracy assessments, KNN is closest to CLL, so we make more comments comparing the two. A drawback of KNN is that its prediction time and space requirements increase with the number of training data (k) because KNN needs

Table 5. Complexity and observations for the CPU and memory usage. p: # of features, d: # of applications, k: # of training data.

	Computational complexity			Average resource consumption observed		
	Prediction	Training	Space	Prediction (ms)	Training (ms)	Memory (KB)
CLL	$\mathcal{O}(p + d\log d)$	$\mathcal{O}(p)$	$\mathcal{O}(p)$	1.53	4.05	5 869
KNN	$\mathcal{O}(\frac{p}{d}k + d\log d)$	$\mathcal{O}(\frac{p}{d})$	$\mathcal{O}(\frac{p}{d}k)$	16.95	0.114	16 973
NB	$\mathcal{O}(p + d\log d)$	$\mathcal{O}(\frac{p}{d})$	$\mathcal{O}(p)$	1.93	0.116	1 691
WD	$\mathcal{O}(d\log d)$	$\mathcal{O}(d)$	$\mathcal{O}(d)$	0.0567	0.0444	12
MRU	$\mathcal{O}(1)$	$\mathcal{O}(d)$	$\mathcal{O}(d)$	0.0030	0.0037	6
MFU	$\mathcal{O}(d\log d)$	$\mathcal{O}(1)$	$\mathcal{O}(d)$	0.0451	0.0023	11

to access all usage events in order to find neighbors. This is inevitable for KNN to be comparable with CLL in terms of prediction accuracy. On the other hand, CLL needs to store and process only the coefficients of a log-linear model, so its time and space requirements do not depend on the size of usage history. In Table 5, the observed prediction time and the memory usage of KNN were much larger than those of CLL. The prediction complexity of KNN can be improved toward $\mathcal{O}(\frac{p}{d}\log k + d\log d)$ using a tree-like data structure [3], but it is difficult to do for high-dimensional data, which commonly occur when various context features are used.

When it comes to trade-offs between the prediction and the training time, what is more directly related to a user's experience is the prediction time. For predictive home-screen system, the prediction time determines system responsiveness, and for predictive pre-launching system, it determines the cost of pre-launch attempts. In this respect, CLL appears to be more suitable than KNN.

6.5   Learning Rates and Recall Window Size

We report additional information related to our analysis. Fig. 4(a) shows the accuracy scores of CLL for various learning rates. The best results were found from $\alpha = 10^{-1.5}$ for each of Recall@6, DCG, and MRR. Fig. 4(b) shows the recall scores for various N cases. Overall, the relative performance of algorithms did not vary much according to N, and CLL appears to perform the best overall. We used $N = 6$ for our analysis in this paper.

7   Conclusions and Outlook

We presented a method for predicting mobile application usage based on a conditional log-linear model and an online gradient-descent scheme. Experimental results demonstrate that the proposed method outperforms previous ones consistently for different evaluation measures. Our analysis on the behavior of prediction methods for individual users and on the learning curves illustrates the

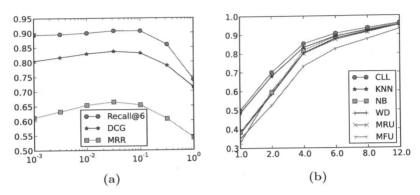

Fig. 4. (a) Accuracy scores of CLL for various learning rates (α in x-axis) (b) Recall@N scores (N in x-axis)

advantages of the proposed method. Our method maintains moderate usage of system resources and offer preferable efficiency for predictions.

In fact, the proposed method is not limited to mobile application usage prediction. It can be applied to other situations where users make repeated selections among a list of available items. As long as context information is available along with users' selections, our method can be considered and used.

While the results are promising, there are a few aspects to investigate further. First, the learning curves in Fig. 3 show that the CLL method was not optimal in the beginning of a usage sequence. A simple approach would be to use another method in the early stages and switch to CLL after a sufficient number of usages. However, it is unclear which one to use because the best choice varies for the Recall@6, DCG, and MRR measures. There are a number of machine learning problems related to combining multiple prediction methods. In addition, devising a scheme that improves the prediction accuracy of CLL in the early stage would be valuable.

Another direction is to investigate this problem in a cloud-assisted setting, where (parts of) model training can be off-loaded to a cloud server [14]. Cloud assistance allows the use of training schemes more expensive than online gradient-descent. Furthermore, cloud-based learning would also allow building methods that use the usage data of many users. For example, the distributed training of models among a group of users could improve each user's personal predictor without completely compromising their privacy. Finally, cloud assistance would make it easy to bring in information sources beyond users' mobile devices, such as the index of the web, movie archives, and so on.

Acknowledgments. We thank Mikko Honkala, Leo Käkkäinen, and Tany-oung Kim for their insightful comments. We thank Matti Kääriäinen and Jarno Seppänen for their contribution to an early version of the CLL algorithm.

References

1. State of the appnation - a year of change and growth in us smartphones (2012),
   http://www.nielsen.com/us/en/insights/news/2012/state-of-the-appnation-a-year-of-change-and-growth-in-u-s-smartphones.html
   (accessed on February 11, 2014)
2. Bishop, C.M.: Pattern Recognition and Machine Learning. Springer (2006)
3. Chávez, E., Navarro, G., Baeza-Yates, R., Marroquín, J.L.: Searching in metric
   spaces. ACM Computing Surveys 33(3), 273–321 (2001)
4. Järvelin, K., Kekäläinen, J.: IR evaluation methods for retrieving highly relevant
   documents. In: Proceedings of the 23rd Annual International ACM SIGIR Conference on Research and Development in Information Retrieval, pp. 41–48 (2000)
5. Kamisaka, D., Muramatsu, S., Yokoyama, H., Iwamoto, T.: Operation prediction
   for context-aware user interfaces of mobile phones. In: Proceedings of the Ninth Annual International Symposium on Applications and the Internet, pp. 16–22 (2009)
6. Laurila, J.K., Gatica-Perez, D., Aad, I., Blom, J., Bornet, O., Do, T.M.T., Dousse,
   O., Eberle, J., Miettinen, M.: The mobile data challenge: Big data for mobile
   computing research. In: Proceedings of the 2012 Pervasive Workshop on Nokia
   Mobile Data Challenge (2012)
7. Liao, Z.X., Li, S.C., Peng, W.C., Yu, P.S., Liu, T.C.: On the feature discovery
   for app usage prediction in smartphones. In: Proceedings of the 2013 13th IEEE
   International Conference on Data Mining, pp. 1127–1132 (2013)
8. Loveless, J., Stoikov, S., Waeber, R.: Online algorithms in high-frequency trading.
   Commun. ACM 56(10), 50–56 (2013)
9. Shin, C., Hong, J.H., Dey, A.K.: Understanding and prediction of mobile application usage for smart phones. In: Proceedings of the 2012 ACM Conference on
   Ubiquitous Computing, pp. 173–182 (2012)
10. Sleator, D.D., Tarjan, R.E.: Amortized efficiency of list update and paging rules.
    Commun. ACM 28(2), 202–208 (1985)
11. Sutton, C., McCallum, A.: An introduction to conditional random fields for re-
    lational learning. In: Introduction to Statistical Relational Learning, pp. 93–128.
    MIT Press (2006)
12. Tan, C., Liu, Q., Chen, E., Xiong, H.: Prediction for mobile application usage
    patterns. In: Proceedings of the 2012 Pervasive Workshop on Nokia Mobile Data
    Challenge (2012)
13. Xu, Y., Lin, M., Lu, H., Cardone, G., Lane, N., Chen, Z., Campbell, A., Choudhury,
    T.: Preference, context and communities: A multi-faceted approach to predicting
    smartphone app usage patterns. In: Proceedings of the 2013 International Sympo-
    sium on Wearable Computers, pp. 69–76 (2013)
14. Yan, T., Chu, D., Ganesan, D., Kansal, A., Liu, J.: Fast app launching for mobile
    devices using predictive user context. In: Proceedings of the 10th International
    Conference on Mobile Systems, Applications, and Services, pp. 113–126 (2012)
15. Zou, X., Zhang, W., Li, S., Pan, G.: Prophet: What app you wish to use next. In:
    Proceedings of the 2013 ACM Conference on Pervasive and Ubiquitous Computing
    Adjunct Publication, pp. 167–170 (2013)

Pushing-Down Tensor Decompositions over Unions to Promote Reuse of Materialized Decompositions*

Mijung Kim and K. Selçuk Candan

Arizona State University, USA
{mijung.kim.1,candan}@asu.edu

Abstract. From data collection to decision making, the *life cycle* of data often involves many steps of integration, manipulation, and analysis. To be able to provide end-to-end support for the full data life cycle, today's data management and decision making systems increasingly combine operations for data manipulation, integration as well as data analysis. Tensor-relational model (TRM) is a framework proposed to support both relational algebraic operations (for data manipulation and integration) and tensor algebraic operations (for data analysis). In this paper, we consider joint processing of relational algebraic and tensor analysis operations. In particular, we focus on data processing workflows that involve data integration from multiple sources (through unions) and tensor decomposition tasks. While, in traditional relational algebra, the costliest operation is known to be the join, in a framework that provides both relational and tensor operations, tensor decomposition tends to be the computationally costliest operation. Therefore, it is most critical to reduce the cost of the tensor decomposition task by manipulating the data processing workflow in a way that reduces the cost of the tensor decomposition step. Therefore, in this paper, we consider data processing workflows involving tensor decomposition and union operations and we propose a novel scheme for pushing down the tensor decompositions over the union operations to reduce the overall data processing times and to promote reuse of materialized tensor decomposition results. Experimental results confirm the efficiency and effectiveness of the proposed scheme.

1 Introduction

As a higher-order generalization of matrices, tensors provide a suitable data representation for multidimensional data sets and tensor decomposition (which is a higher-order generalization of SVD/PCA for multi-aspect data analysis) helps capture the higher-order latent structure of such datasets. Consequently, the tensor data model is increasingly being used by many application domains including scientific data management [6,9,18,25], sensor data management [24], and social network data analysis [15,14,17]. On the other hand, from data collection to decision making, the *life cycle* of data often involves many steps of integration, manipulation, and analysis. Therefore, to be able to provide end-to-end support for the full data life cycle, today's data management and decision making systems increasingly need to combine different types of operations for data manipulation, integration, and analysis.

---

* This work is partially funded by NSF grants #116394, RanKloud: Data Partitioning and Resource Allocation Strategies for Scalable Multimedia and Social Media Analysis and #1016921, One Size Does Not Fit All: Empowering the User with User-Driven Integration.

T. Calders et al. (Eds.): ECML PKDD 2014, Part I, LNCS 8724, pp. 688–704, 2014.
© Springer-Verlag Berlin Heidelberg 2014

Relational operation	Tensor manipulation
Select	Slicing of a tensor (or taking a single or subset of elements across a given mode)
Project	Creating a sub-cube with a smaller set of modes
Cartesian-Product and Equi-Join	Composition of multiple tensors through outer-product
Union	Cell-wise OR (and row/slice insertion)
Intersection	Cell-wise AND (and row/slice elimination)

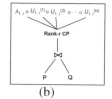

(a)   (b)

Fig. 1. (a) Implementation of relational operations through tensor manipulation and (b) a query plan with a join operation of two tensors, $\mathcal{P}$ and $\mathcal{Q}$, preceding a tensor decomposition operation

We are currently building TensorDB, which extends a native array database, SciDB [5], with operations needed to support the full *life cycle* of data. TensorDB is based on a tensor-relational model (TRM) [11], which brings together relational algebraic operations (for data manipulation and integration) and tensor algebraic operations (for data analysis) and supports complex data processing plans where multiple relational algebraic and tensor algebraic operations are composed with each other (Figure 1(b)).

1.1 Tensor-Based Relational Model (TRM)

Let $A_1, \ldots, A_n$ be a set of attributes in the schema of a relation, R, and $D_1, \ldots, D_n$ be the attribute domains. Let the relation instance $\mathcal{R}$ be a finite multi-set of tuples, where each tuple $t \in D_1 \times \ldots \times D_n$. [11] defines various types of tensors representing relations, including occurrence tensors and value tensor. For example, an *occurrence tensor* $\mathcal{R}_o$ corresponding to the relation instance $\mathcal{R}$ as an n-mode tensor, where each attribute $A_1, \ldots, A_n$ is represented by a mode. For the ith mode, which corresponds to A_i, let $D'_i \subseteq D_i$ be the (finite) subset of the elements such that $\forall v \in D'_i \exists t \in \mathcal{R}$ s.t. $t.A_i = v$ and let $idx(v)$ denote the rank of v among the values in D'_i relative to an (arbitrary) total order, $<_i$, defined over the elements of the domain, D_i. The cells of the *occurrence tensor* $\mathcal{R}_o$ are such that $\mathcal{R}_o[u_1, \ldots, u_n] = 1$ iff $\exists t \in \mathcal{R}$ s.t. $\forall_{1 \leq j \leq n} idx(t.A_j) = u_j$ and 0 otherwise. Intuitively, each cell indicates whether the corresponding tuple exists in the multi-set corresponding to the relation or not.

[11] also discusses the implementation of various relational algebraic operations to manipulate relations represented as tensors in TRM (Figure 1(a)) as well as other tensor manipulation operations, such as tensor decomposition.

1.2 Tensors Decomposition

The two most popular tensor decompositions are the CP [6,9] and Tucker [25] decompositions. CP decomposes the input tensor into a sum of component rank-one tensors; i.e., the rank-r CP Decomposition, $CP(\mathcal{P}_{I_1 \times I_2 \times \cdots \times I_N})$, of the tensor $\mathcal{P}_{I_1 \times I_2 \times \cdots \times I_N}$ is defined as $\mathbf{P}^{(1)}, \ldots, \mathbf{P}^{(N)}$ such that

$$\mathcal{P}_{I_1 \times I_2 \times \cdots \times I_N} \approx \sum_{k=1}^{r} P_k^{(1)} \circ P_k^{(2)} \circ \cdots \circ P_k^{(N)}. \tag{1}$$

We also use the formulation where the column vectors of each factor are normalized to the unit length with the weights absorbed into a vector λ; i.e., $CP(\mathcal{P}_{I_1 \times I_2 \times \cdots \times I_N}) =$

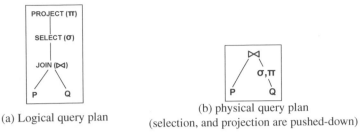

(a) Logical query plan

(b) physical query plan
(selection, and projection are pushed-down)

Fig. 2. Query optimization in relational algebra: (a) A logical query plan involving selection, projection, and join operations: (b) an equivalent physical plan where the selection and projection operations are *pushed-down* to minimize the amount of data fed into the join operator

$\langle \lambda, \mathbf{P}^{(1)}, \ldots, \mathbf{P}^{(N)} \rangle$, such that

$$\mathcal{P}_{I_1 \times I_2 \times \cdots \times I_N} \approx \sum_{k=1}^{r} \lambda_k \circ P_k^{(1)} \circ P_k^{(2)} \circ \cdots \circ P_k^{(N)}, \tag{2}$$

where λ_i is the ith element of vector λ of size r and $U_i^{(n)}$ is the ith unit-length column vector of the matrix $\mathbf{P}^{(n)}$ of size $I_n \times r$, for $n = 1, \cdots, N$.

While, as described above, CP decomposition can be represented in the form of a diagonal core tensor and one factor matrix (also called a factor) per mode, the Tucker decomposition results in a dense core tensor multiplied by a matrix along each mode. Many of the algorithms for decomposing tensors are based on an iterative process, such as alternating least squares (ALS), that approximates the solution though iterations until a convergence condition is reached [6,9].

1.3   Decomposition Push-Down Strategy for Optimizating TRM Workflows

One key goal of TensorDB is to deploy optimization strategies for complex queries involving both tensor decomposition and tensor manipulation operations, such as join and union operations that integrate data from multiple sources.

In relational algebra, the costliest operation is the join operation. Consequently, given a complex query plan, the relational optimizers *push-down* data reduction operations, such as selections (which reduce the number of tuples) and projections (which reduce the number of data attributes) over join-operations to reduce the amount of data fed into the join operators (Figure 2). In TensorDB, based on TRM, however, tensor decomposition operation tends to be the computationally costliest operation: for dense tensors, the cost is exponential in the number of modes of the data. While the operation is relatively cheaper for sparse tensors, the cost and memory requirement still outweigh other more traditional relational operators.

Therefore, *a key criterion for optimizing query workplans in TensorDB is to reduce the number of data modes and non-zero data entries in the tensors that need to be decomposed.* In [11], for example, we considered query plans that involve join operations and tensor decompositions (Figures 1(b) and 3(a)) and proposed a *decomposition push-down* strategy that reduces the number of modes of the data tensors being decomposed. This *join-by-decomposition* (JBD) strategy *pushes-down* the tensor-decomposition operation so that the input tensors (which have smaller number of modes than the join

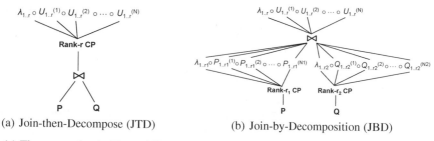

(a) Join-then-Decompose (JTD)   (b) Join-by-Decomposition (JBD)

Fig. 3. (a) The query plan in Figure 1(b) and (b) an alternative query plan where the tensor decomposition operation is pushed-down [11]

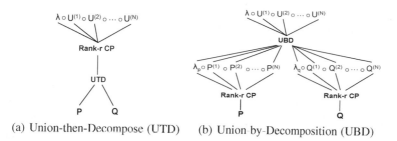

(a) Union-then-Decompose (UTD)   (b) Union-by-Decomposition (UBD)

Fig. 4. (a) A query plan with an union of two tensors, $\mathcal{P}$ and $\mathcal{Q}$ preceding tensor decomposition and (b) an alternative query plan where the decomposition is pushed-down over union

tensor) are decomposed into their spectral components and then these decompositions are combined to obtain the final decomposition as shown in Figure 3(b).

In this paper, *we focus on query plans that involve tensor decomposition and union operations (as in Figure 4(a)) and propose novel* decomposition push-down *strategies (as in Figure 4(b)) that help reduce the overall cost of the query plan.* We refer to the query plan that first performs the union operation on the data and then applies the tensor decomposition on the union of the data as *union-then-decompose* (UTD) plan. The query plan with decomposition push-down, which first performs the tensor decompositions on each input data source and then combines these decomposed tensors as the *union-by-decomposition* (UBD) plan.

1.4 Contributions of This Paper: Union-by-Decomposition (UBD)

A *union-by-decomposition* (UBD) plan, with decomposition push-down, has various advantages over the conventional *union-then-decompose* (UTD) plan:

- Firstly, especially when the overlaps between the input data sources are small, the union operation can combine relatively small and sparse tensors into a larger and denser tensor. Consequently, the decomposition over the union data can be much more expensive than the decompositions over the input data sources. Moreover multiple tensor decompositions on input tensors can run in parallel, which will further reduce the cost.

– Secondly, a *union-by-decomposition* (UBD) based plan provides opportunities for materializing decomposition of data tensors and re-using these materialized decompositions in more complex queries requiring integration of data.

Despite these advantages, however, implementing the UBD strategy requires us to address a number of key challenges:

– **Challenge 1: How can we combine the factor matrices of tensor decompositions with their own eigen basis into the eigen basis of the union tensor?** If tensor decomposition is thought of as a group of clusters, combining different groups of clusters for different tensors into another group of clusters for the union of the tensors is not straightforward.
– **Challenge 2: For the common data elements at the intersection of multiple data sources, which factors (clusters when the clustering analogy is used) among the different tensor decompositions should we choose?** This is critical as the choice can impact the final accuracy of the UBD based plan.

In this paper, we present algorithms and techniques to address these questions. We first review the related work in Section 2. In Section 3, we extend TRM with the proposed *union-by-decomposition* operation: we discuss strategies for combining the tensor decompositions for the union of the tensors from different sources and consider alternative selection measures to choose a group of factors for data entries common to input data sources. We also consider query plans that include both join and union operations along with tensor decomposition. We, then, experimentally evaluate the proposed scheme in Section 5 and conclude the paper in Section 6.

2   Related Work

2.1   Tensors and Tensor Decomposition

The two most popular tensor decompositions are the CANDECOMP/PARAFAC (CP [6,9]) and Tucker [25] decompositions. CANDECOMP [6] and PARAFAC [9] decompositions (together known as the CP decomposition) decompose the input tensor into a sum of component rank-one tensors. While CP decomposition can be represented in the form of a diagonal core tensor and one factor matrix (also called a factor) per mode, the Tucker decomposition results in a dense core tensor. Many of the algorithms for decomposing tensors are based on an iterative alternating least squares (ALS) process that approximates the solution by iteratively improving the decomposition until a convergence condition is reached [6,9]. In [21], the complexity of ALS schemes has been discussed. Non-iterative approaches to tensor decomposition include closed form solutions, such as generalized rank annihilation method (GRAM) [19] and direct trilinear decomposition (DTLD) [20], which fit the model by solving a generalized eigenvalue problem. [13] provides an overview of the tensor decomposition algorithms.

Tensor decomposition is a costly process. In dense tensor representation, the cost increases exponentially with the number of modes of the tensor. While decomposition cost increases more slowly (linearly with the number of nonzero entries in the tensor) for sparse tensors, the operation can still be very expensive for large data sets. [24] uses

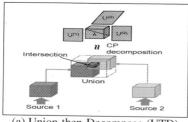
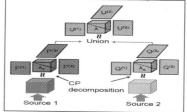

(a) Union-then-Decompose (UTD)   (b) Union-by-Decomposition (UBD)

Fig. 5. (a) Tensor decomposition on the union of the two relations and (b) the union operation on the two tensor decompositions of the input relations

randomized sampling to approximate the tensor decomposition where the tensor does not fit in the available memory. A modified ALS algorithm proposed in [18] computes Hadamard products instead of Khatri-Rao products for efficient PARAFAC for large-scale tensors. [15] developed a greedy PARAFAC algorithm for large-scale, sparse tensors in MATLAB. [14] proposed a memory-efficient Tucker (MET) decomposition to address the intermediate blowup problem in Tucker decomposition. A parallelization strategy of tensor decomposition on MapReduce has been proposed in [10]. ParCube proposed in [17] is a parallelizable tensor decomposition algorithm, which produces sparse approximation of tensor decompositions. In [11,12], we proposed parallelized tensor decompositions within a tensor relational algebraic framework.

2.2 Array Databases

There are several in-database data models for modeling tensor data. Column-oriented organizations [22] are efficient when many or all rows are accessed, such as during an aggregate computation. Row-oriented organizations, on the other hand, are efficient when many or all of the columns on a single row are accessed or written on a single disk seek. Key-value organizations [1] are useful when working with less structured data, such as documents, which tend not to be relational. The array model [4,5,8,26] is a natural representation to store multidimensional data and facilitate multidimensional data analysis. Approaches to represent array based data can be broadly categorized into four types. (a) The first approach is to represent the array in the form of a table [7,26]. (b) A second approach is to use blob type in a relational database as a storage layer for array data [4,8]. (c) Sparse matrices can also be represented using a graph-based abstraction [16]. For example, in [16], ALS (alternating least squares) is solved using a graph algorithm that represents a sparse matrix as a bipartite graph. (d) The last approach is to consider a native array model and an array-based storage scheme, such as a chunk-store, as in [5].

3 Union-by-Decomposition (UBD) and Decomposition Push-Down

In this section, we describe our proposed union-by-decomposition (UBD) approach that pushes down tensor decompositions over union operators: Unlike the more conventional

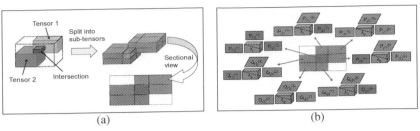

Fig. 6. Naive grid-based UBD: (a) Input tensors are partitioned into an intersecting sub-tensor and non-intersecting sub-tensors; (b) intermediary decompositions of grid-based UBD

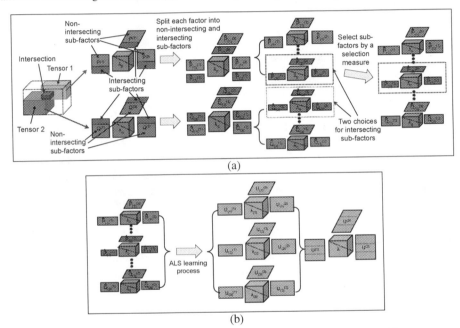

Fig. 7. UBD: (a) first the inputs tensors are decomposed and (b) these decompositions are recombined by considering the common and non-intersection parts of the factor matrices.

union-then-decompose (UTD) scheme, which applies decomposition on the union of the two relations (Figure 5(a)), UBD first performs the tensor decomposition on the input tensors then these decompositions are combined into the final result (Figure 5(b)).

3.1 Challenge 1: Implementing UBD through Partition-Based ALS

Naive Grid-Based UBD. One way to implement the UBD operation is to divide the input tensors into *common (or intersection)* and ($2^N - 1$ many when the number of modes is N) *uncommon* sub-tensors as shown in Figure 6(a) and then considering each partition as a cell of a larger tensor partitioned into a grid as shown in Figure 6(b) and applying the grid-based tensor decomposition strategy proposed in [18] to combine these into a single decomposition.

Algorithm 1. Union-By-Decomposition (UBD) (input: two tensors $\mathcal{P}_{I_1 \times I_2 \times \cdots \times I_N}$ and $\mathcal{Q}_{J_1 \times J_2 \times \cdots \times J_N}$, optional input: CP decompositions of $\mathcal{P}$ and $\mathcal{Q}$, $\langle \mathbf{P}^{(1)}, \ldots, \mathbf{P}^{(N)} \rangle$ and $\langle \mathbf{Q}^{(1)}, \ldots, \mathbf{Q}^{(N)} \rangle$, respectively, output: factors $\mathbf{U}^{(1)}, \ldots, \mathbf{U}^{(N)}$ for $\mathcal{P} \cup \mathcal{Q}$)

1: **if** no existing decompositions given **then**
2:     Run any available CP algorithm on $\mathcal{P}$ and $\mathcal{Q}$ *in parallel* to get factors $\mathbf{P}^{(1)}, \ldots, \mathbf{P}^{(N)}$ and $\mathbf{Q}^{(1)}, \ldots, \mathbf{Q}^{(N)}$
3: **end if**
4: **for** each mode n **do**
5:     create sub-factors $\hat{\mathbf{P}}_{(1)}^{(n)}$ and $\hat{\mathbf{P}}_{(2)}^{(n)}$, and $\hat{\mathbf{Q}}_{(2)}^{(n)}$ and $\hat{\mathbf{Q}}_{(3)}^{(n)}$ with non-intersecting and intersecting sub-factors of $\mathbf{P}^{(n)}$ and $\mathbf{Q}^{(n)}$, respectively (see Figure 7(a))
6: **end for**
7: select either $\hat{\mathbf{P}}_{(2)}^{(n)}$ and $\hat{\mathbf{Q}}_{(2)}^{(n)}$ for factors $\mathbf{T}^{(n)}$ for intersection $\mathcal{P} \cap \mathcal{Q}$ by a selection measure (see Section 3.2)
8: repeat the update process for sub-factors $\mathbf{U}_{(1)}^{(n)}$, $\mathbf{U}_{(2)}^{(n)}$, and $\mathbf{U}_{(3)}^{(n)}$ using Equation 9 until a stopping condition is satisfied, which are combined to $\mathbf{U}^{(n)}$ by Equation 7 (see Figure 7(b))

Proposed Implementation of UBD. An obvious shortcoming of the naive grid-based UBD discussed above is that it leads to a very large number of intermediary decompositions and this number increases quickly with the number of modes of the input tensors. To tackle this challenge, we propose to decompose input tensors directly (through decomposition push-down) and recombine the resulting factor matrices in a way that reflects the common and non-intersecting sub-factors of these decompositions as shown in Figure 7. The high-level pseudocode of this partition-based UBD scheme is shown in Algorithm 1. We next present the details of the proposed UBD process:

Let us assume that we are given two tensors $\mathcal{P}_{I_1 \times I_2 \times \cdots \times I_N}$ and $\mathcal{Q}_{J_1 \times J_2 \times \cdots \times J_N}$ and let us assume we have already computed their CP decompositions

$$CP(\mathcal{P}) = \hat{\mathcal{P}} = \langle \mathbf{P}^{(1)}, \ldots, \mathbf{P}^{(N)} \rangle \quad \text{and} \quad CP(\mathcal{Q}) = \hat{\mathcal{Q}} = \langle \mathbf{Q}^{(1)}, \ldots, \mathbf{Q}^{(N)} \rangle. \quad (3)$$

Our goal is to estimate $CP(\mathcal{P} \cup \mathcal{Q}) = \langle \mathbf{U}^{(1)}, \ldots, \mathbf{U}^{(N)} \rangle$ efficiently using these decompositions. To achieve this, we solve the ALS problem

$$\min \| (\mathcal{P} \cup \mathcal{Q}) - \langle \mathbf{U}^{(1)}, \ldots, \mathbf{U}^{(N)} \rangle \| \quad (4)$$

by appropriately combining sub-factors of the input tensors. More specifically, each factor of $\mathcal{P}$ and $\mathcal{Q}$ are split into two: a non-intersecting ($\mathbf{P}_{(1)}^{(n)}$ and $\mathbf{Q}_{(3)}^{(n)}$) and intersecting ($\mathbf{P}_{(2)}^{(n)}$ and $\mathbf{Q}_{(2)}^{(n)}$) partitions. Given these, the CP decompositions of $[k_1, k_2, \ldots, k_N]$-th sub-tensor of $\mathcal{P}$ and $\mathcal{Q}$ are

$$CP(\mathcal{P}^{(\bar{\mathbf{k}})}) = \langle \mathbf{P}_{(k_1)}^{(1)}, \ldots, \mathbf{P}_{(k_N)}^{(N)} \rangle \quad \text{and} \quad CP(\mathcal{Q}^{(\bar{\mathbf{k}})}) = \langle \mathbf{Q}_{(k_1)}^{(1)}, \ldots, \mathbf{Q}_{(k_N)}^{(N)} \rangle, \quad (5)$$

respectively, where $\bar{\mathbf{k}} = [k_1, k_2, \ldots, k_N]$ for $k_n \in \{1, 2\}$ for $\mathcal{P}^{(\bar{\mathbf{k}})}$ and $k_n \in \{2, 3\}$ for $\mathcal{Q}^{(\bar{\mathbf{k}})}$. Given these, we can approximate the decompositions of each sub-tensor of $\mathcal{P}$ and $\mathcal{Q}$ with the CP decompositions of $\mathcal{P}$ and $\mathcal{Q}$, respectively (see Figure 7(a)):

$$CP(\mathcal{P}^{(\bar{\mathbf{k}})}) \approx \langle \hat{\mathbf{P}}_{(k_1)}^{(1)}, \ldots, \hat{\mathbf{P}}_{(k_N)}^{(N)} \rangle \quad \text{and} \quad CP(\mathcal{Q}^{(\bar{\mathbf{k}})}) \approx \langle \hat{\mathbf{Q}}_{(k_1)}^{(1)}, \ldots, \hat{\mathbf{Q}}_{(k_N)}^{(N)} \rangle. \quad (6)$$

Let us denote the CP decomposition of $[k_1, k_2, \ldots, k_N]$-th sub-tensor of $\mathcal{P} \cup \mathcal{Q}$ as

$$CP((\mathcal{P} \cup \mathcal{Q})^{(\bar{k})}) = CP(\mathcal{Y}^{(\bar{k})}) = \langle \mathbf{U}_{(k_1)}^{(1)}, \ldots, \mathbf{U}_{(k_N)}^{(N)} \rangle,$$

where $\bar{k} = [k_1, k_2, \ldots, k_N]$ for $k_n \in \{1, 2, 3\}$. Note that each factor of $CP(\mathcal{P} \cup \mathcal{Q})$ can be split into three partitions

$$\mathbf{U}^{(n)} = [\mathbf{U}_{(1)}^{(n)T} \mathbf{U}_{(2)}^{(n)T} \mathbf{U}_{(3)}^{(n)T}]^T, \tag{7}$$

one corresponding to a non-intersecting sub-factor from one input matrix, the other corresponding to a common sub-factor, and the last corresponding to a non-intersecting sub-factor from the second input matrix. Given these, we can re-formulate the minimization problem in Equation (4) for each sub-tensor $\mathcal{Y}^{(\bar{k})}$ of $\mathcal{P} \cup \mathcal{Q}$ as minimizing D, where

$$D = \frac{1}{2} \sum_{k_1=1}^{3} \cdots \sum_{k_N=1}^{3} \|\mathcal{Y}^{(\bar{k})} - \langle \mathbf{U}_{(k_1)}^{(1)}, \ldots, \mathbf{U}_{(k_N)}^{(N)} \rangle \|,$$

or, considering the n-mode matricized tensor $\mathbf{Y}_{(n)}^{(\bar{k})}$ of $\mathcal{Y}^{(\bar{k})}$, as minimizing

$$D = \frac{1}{2} \sum_{\bar{k}} \|\mathbf{Y}_{(n)}^{(\bar{k})} - \mathbf{U}_{(k_n)}^{(n)} \{\mathbf{U}_{(k_1)}^{(1)} \odot \mathbf{U}_{(k_2)}^{(2)} \odot \cdots \odot \mathbf{U}_{(k_{n-1})}^{(n-1)} \odot \mathbf{U}_{(k_{n+1})}^{(n+1)} \odot \cdots \odot \mathbf{U}_{(k_N)}^{(N)} \} \|,$$

where $\odot$ is the Khatri-Rao product.

This minimization problem can be solved using an ALS problem by identifying gradient components with respect to sub-factors as in [18]. More specifically, the gradient component with respect to sub-factor $\mathbf{U}_{(k_n)}^{(n)}$ is

$$\Delta_{\mathbf{U}_{(k_n)}^{(n)}} D = \sum_{\bar{k}_n = k_n} \left(-\mathbf{Y}_{(n)}^{(\bar{k})} \mathbf{U}_{(\bar{k})}^{\odot -n} + \mathbf{U}_{(k_n)}^{(n)} \mathbf{U}_{(\bar{k})}^{\odot -n T} \mathbf{U}_{(\bar{k})}^{\odot -n} \right)$$
$$= \sum_{\bar{k}_n = k_n} \left(-\mathbf{Y}_{(n)}^{(\bar{k})} \mathbf{U}_{(\bar{k})}^{\odot -n} + \mathbf{U}_{(k_n)}^{(n)} \{\mathbf{U}_{(\bar{k})}^T \mathbf{U}_{(\bar{k})} \}^{\circledast -n} \right), \tag{8}$$

where $\circledast$ is the Hadamard (element-wise) product. Given this, each sub-factor $\mathbf{U}_{(k_n)}^{(n)}$ can be updated using the update rule

$$\mathbf{U}_{(k_n)}^{(n)} \leftarrow \left(\sum_{\bar{k}_n = k_n} \mathbf{Y}_{(n)}^{(\bar{k})} \mathbf{U}_{(\bar{k})}^{\odot -n} \right) \left(\sum_{\bar{k}_n = k_n} (\mathbf{U}_{(\bar{k})}^T \mathbf{U}_{(\bar{k})})^{\circledast -n} \right)^{-1}. \tag{9}$$

Note that, from Equation 6, for each sub-tensor $\mathcal{Y}^{(\bar{k})} = \mathcal{P}^{(\bar{k})}$, considering to the first input matrix we have

$$\mathbf{Y}_{(n)}^{(\bar{k})} \mathbf{U}_{(\bar{k})}^{\odot -n} \approx \hat{\mathbf{P}}_{(k_n)}^{(n)} \hat{\mathbf{P}}_{(\bar{k})}^{\odot -n T} \mathbf{U}_{(\bar{k})}^{\odot (-n)}. \tag{10}$$

Similarly, for each sub-tensor $\mathcal{Y}^{(\bar{k})} = \mathcal{Q}^{(\bar{k})}$, considering to the second input matrix, we have

$$\mathbf{Y}_{(n)}^{(\bar{k})} \mathbf{U}_{(\bar{k})}^{\odot -n} \approx \hat{\mathbf{Q}}_{(k_n)}^{(n)} \hat{\mathbf{Q}}_{(\bar{k})}^{\odot -n T} \mathbf{U}_{(\bar{k})}^{\odot (-n)}. \tag{11}$$

Finally, for each sub-tensor $\mathcal{Y}^{(\bar{k})}$ such that $\mathcal{Y}^{(\bar{k})} = \mathcal{P} \cap \mathcal{Q}$,

$$\mathbf{Y}_{(n)}^{(\bar{k})} \mathbf{U}_{(\bar{k})}^{\odot -n} \approx \mathbf{T}^{(n)} \mathbf{T}^{\odot -n T} \mathbf{U}_{(\bar{k})}^{\odot(-n)}, \tag{12}$$

where $\mathbf{T}^{(n)}$ are the factors of $CP(\mathcal{P} \cap \mathcal{Q})$. Note that $\mathbf{T}^{(n)}$ can be estimated from either the CP decomposition of $\mathcal{P}^{(\bar{2})}$

$$CP(\mathcal{P} \cap \mathcal{Q}) = CP(\mathcal{P}^{(\bar{2})}) \approx \langle \hat{\mathbf{P}}_{(2)}^{(1)}, \ldots, \hat{\mathbf{P}}_{(2)}^{(N)} \rangle,$$

where $\bar{2} = [k_1, k_2, \ldots, k_N]$ for all $k_n = 2$, or the CP decomposition of $\mathcal{Q}^{(\bar{2})}$

$$CP(\mathcal{P} \cap \mathcal{Q}) = CP(\mathcal{Q}^{(\bar{2})}) \approx \langle \hat{\mathbf{Q}}_{(2)}^{(1)}, \ldots, \hat{\mathbf{Q}}_{(2)}^{(N)} \rangle.$$

The choice is critical and can impact significantly on the accuracy of the overall process. Therefore, we next discuss how to select whether to use $\hat{\mathbf{P}}_{(2)}^{(n)}$ or $\hat{\mathbf{Q}}_{(2)}^{(n)}$ to estimate $\mathbf{T}^{(n)}$.

3.2 Challenge 2: Selection of Sub-factors for the Overlapping Sub-tensor

As described above, the factors $\mathbf{T}^{(n)}$ of the overlapping sub-tensor, $\mathcal{P} \cap \mathcal{Q}$ (used in the computation of $CP(\mathcal{P} \cup \mathcal{Q})$) can be selected from either $\hat{\mathbf{P}}_{(2)}^{(n)}$ or $\hat{\mathbf{Q}}_{(2)}^{(n)}$. As also explained before, the choice is critical as it may impact the accuracy of the final decomposition, $CP(\mathcal{P} \cup \mathcal{Q})$. Therefore, in this subsection, we explore alternative ways for choosing the sub-factors, $\mathbf{T}^{(n)}$, of $CP(\mathcal{P} \cap \mathcal{Q})$.

Intersection-Based Selection Criteria. When we are choosing between $\hat{\mathbf{P}}_{(2)}^{(n)}$ and $\hat{\mathbf{Q}}_{(2)}^{(n)}$ to use as $\mathbf{T}^{(n)}$, one criteria would be to consider how well $\hat{\mathcal{P}}^{(\bar{2})} = \langle \hat{\mathbf{P}}_{(2)}^{(1)} \cdots \hat{\mathbf{P}}_{(2)}^{(N)} \rangle$ and $\hat{\mathcal{Q}}^{(\bar{2})} = \langle \hat{\mathbf{Q}}_{(2)}^{(1)} \cdots \hat{\mathbf{Q}}_{(2)}^{(N)} \rangle$ fit $\mathcal{P} \cap \mathcal{Q}$:

$$IC_1(\hat{\mathcal{P}}^{(\bar{2})}) = 1 - \frac{\|(\mathcal{P} \cap \mathcal{Q}) - \hat{\mathcal{P}}^{(\bar{2})}\|}{\|\mathcal{P} \cap \mathcal{Q}\|} \quad \text{and} \quad IC_1(\hat{\mathcal{Q}}^{(\bar{2})}) = 1 - \frac{\|(\mathcal{P} \cap \mathcal{Q}) - \hat{\mathcal{Q}}^{(\bar{2})}\|}{\|\mathcal{P} \cap \mathcal{Q}\|}.$$

One obvious difficulty with this fit-based intersection criterion, IC_1, is that the fit computations can be very costly. Alternatively, if we consider the two tensor decompositions, $\hat{\mathcal{P}}^{(\bar{2})}$ and $\hat{\mathcal{Q}}^{(\bar{2})}$ as two groups of clusters, then we need to choose the group of clusters on which the membership of the shared elements (the overlapping part) is more tight and we can use the norms of the sub-factors to quantify how strongly elements belongs to the corresponding clusters. Intuitively, norms of the sub-factors corresponding to the overlapping region

$$IC_2(\hat{\mathcal{P}}^{(\bar{2})}) = \|\langle \hat{\mathbf{P}}_{(2)}^{(1)}, \ldots, \hat{\mathbf{P}}_{(2)}^{(N)} \rangle\|, \quad IC_2(\hat{\mathcal{Q}}^{(\bar{2})}) = \|\langle \hat{\mathbf{Q}}_{(2)}^{(1)}, \ldots, \hat{\mathbf{Q}}_{(2)}^{(N)} \rangle\|,$$

explain the contribution of each element to these clusters and the one with the larger intersection criterion measure, IC_2, can be used to $\mathbf{T}^{(n)}$.

Note that the norm of the sub-factors of the overlapping region excludes any knowledge about how the groups fit with the rest of the tensors. Alternatively, we can account for the strengths of the groups in the whole tensor by also considering the core tensor

$$IC_3(\hat{\boldsymbol{\mathcal{P}}}^{(\bar{2})}) = \|\langle \lambda_p, \hat{\mathbf{P}}_{(2)}^{(1)}, \ldots, \hat{\mathbf{P}}_{(2)}^{(N)}\rangle\|, \quad IC_3(\hat{\boldsymbol{\mathcal{Q}}}^{(\bar{2})}) = \|\langle \lambda_q, \hat{\mathbf{Q}}_{(2)}^{(1)}, \ldots, \hat{\mathbf{Q}}_{(2)}^{(N)}\rangle\|,$$

and select the tensor which leads to the larger intersection criterion, IC_3, measure. Here, λ_p and λ_q are core vectors of $\hat{\boldsymbol{\mathcal{P}}}^{(\bar{2})}$ and $\hat{\boldsymbol{\mathcal{Q}}}^{(\bar{2})}$, respectively.

Note that for IC_2 and IC_3, the columns of $\hat{\mathbf{P}}_{(2)}^{(n)}$ and $\hat{\mathbf{Q}}_{(2)}^{(n)}$ are normalized to length one with the weights absorbed into the vector λ_p and λ_q, respectively.

Union-Based Selection Criteria. The aforementioned intersection-based selection criteria have a potential weakness: as we see later in Section 5, the selection measures based on intersection fit and norm work well when the two input tensors are balanced in size. If the two tensors are unbalanced in size (i.e. one of the tensors is much larger than the other) the non-overlapping region of the larger tensor is likely to have a large impact on the final accuracy and the intersection-based selection criteria which primarily focus on the overlapping region of the tensors may fail to capture this. To address this limit of intersection-based selection criteria, we also consider *union-based* selection criteria that take into account both non-overlapping and overlapping parts of the tensors.

Firstly, we consider the fit of the union of the decomposed tensors to the union of the two original tensors

$$UC_1(\langle \mathbf{U}^{(1)}, \ldots, \mathbf{U}^{(N)}\rangle) = 1 - \frac{\|(\boldsymbol{\mathcal{P}} \cup \boldsymbol{\mathcal{Q}}) - \langle \mathbf{U}^{(1)}, \ldots, \mathbf{U}^{(N)}\rangle\|}{\|\boldsymbol{\mathcal{P}} \cup \boldsymbol{\mathcal{Q}}\|},$$

and we choose between the two alternatives by setting the initial $\mathbf{U}^{(n)}$ to $[\hat{\mathbf{P}}_{(1)}^{(n)T} \hat{\mathbf{P}}_{(2)}^{(n)T} \hat{\mathbf{Q}}_{(3)}^{(n)T}]^T$ and to $[\hat{\mathbf{P}}_{(1)}^{(n)T} \hat{\mathbf{Q}}_{(2)}^{(n)T} \hat{\mathbf{Q}}_{(3)}^{(n)T}]^T$ and observing which one leads to a better fit. UC_1 is the *initial* fit of the union of the decomposed tensors to the union of the two original tensors in the beginning of the update process of $\mathbf{U}_{(k_n)}^{(n)}$ for $k_n = 1, 2, 3$ (see Equation 9). Intuitively, this initial fit can be thought of as a rough indicator of whether the final fit of the union of the decomposed tensors solved by the learning process will be close to the decomposition on the union of two tensors or not.

As a second criterion, we consider the density of the input tensors, $\boldsymbol{\mathcal{P}}_{I_1 \times I_2 \times \cdots \times I_N}$ and $\boldsymbol{\mathcal{Q}}_{J_1 \times J_2 \times \cdots \times J_N}$,

$$UC_2(\boldsymbol{\mathcal{P}}) = \frac{|\boldsymbol{\mathcal{P}}|}{\prod_{i=1}^{N} I_i}, \quad UC_2(\boldsymbol{\mathcal{Q}}) = \frac{|\boldsymbol{\mathcal{Q}}|}{\prod_{i=1}^{N} J_i},$$

where $|\boldsymbol{\mathcal{X}}|$ is the number of non-zeros of $\boldsymbol{\mathcal{X}}$. Given this, we set the initial $\mathbf{U}^{(n)}$,

$$\mathbf{U}^{(n)} = [\hat{\mathbf{P}}_{(1)}^{(n)T} \hat{\mathbf{P}}_{(2)}^{(n)T} \hat{\mathbf{Q}}_{(3)}^{(n)T}]^T, \text{ if } \boldsymbol{\mathcal{P}} \text{ has a larger density, or}$$

$$\mathbf{U}^{(n)} = [\hat{\mathbf{P}}_{(1)}^{(n)T} \hat{\mathbf{Q}}_{(2)}^{(n)T} \hat{\mathbf{Q}}_{(3)}^{(n)T}]^T, \text{ if } \boldsymbol{\mathcal{Q}} \text{ has a larger density.}$$

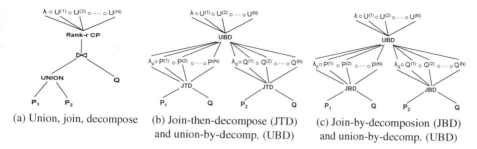

(a) Union, join, decompose   (b) Join-then-decompose (JTD)   (c) Join-by-decomposion (JBD)
and union-by-decomp. (UBD)   and union-by-decomp. (UBD)

Fig. 8. Three alternative query plans for implementing a complex query plan with union, join, and decompose operations

Intuitively, the overlapping part will be more tightly connected with the non-overlapping part in the input tensor with the larger density – simply because there are less chances that an entry will be seen only in the overlapping part. Thus, given the choice between using the decompositions (for the overlapping part) of the input tensor with the larger density and of the tensor with the smaller density, the former is likely to lead to lesser errors.

4   Parallelization, Materialization, and Further Optimizations

The proposed union-by-decomposition (UBD) scheme leads to various optimization opportunities. First of all, assuming the availability of multiple computation units, the individual data sources can be decomposed in parallel. Moreover, each individual decomposition of the sub-tensors can also be obtained in parallel, leading to highly parallelizable execution plans. Secondly, as we see in Section 5, in situations where the same data source is integrated (unioned) with different data sources over time, we can decompose this data source once and materialize the decomposition for later reuse within a UBD process, thereby avoiding significant amount of runtime work.

In addition, the proposed union-by-decomposition (UBD) operator is compatible with other novel (decomposition push-down based) operators that are part of TensorDB, including the *join-by-decomposition* (JBD) operator, discussed in Section 1.3, and can be used as part of a general optimization framework. Figure 8 provides an example: in Figure 8(b) first the join is pushed down over union and then the decomposition is pushed down over union, whereas in Figure 8(c) the decomposition is pushed down also over the join operator leading to (as we see in Section 5) a highly efficient query plan.

5   Experimental Evaluation

In this section, we present experimental results assessing the efficiency and effectiveness of the proposed union-by-decomposition (UBD) scheme and the selection criteria.

5.1   Experimental Setup

For these experiments, we used real data tensors (Table 1): (a) MovieLens 1M data set [2] with a 3-mode tensor (user, movie, rating) and (b) a 4-mode tensor

Table 1. Tensor data sets

Data set	Attributes	Size	Density (%)
3-mode MovieLens 1M	(user, movie, rating)	$6000 \times 3400 \times 5$	0.8451
3-mode book rating	(user, book, rating)	$105283 \times 340556 \times 11$	0.0003
4-mode Epinions	(user, product, category, rating)	$22111 \times 296000 \times 26 \times 5$	0.000007
4-mode MovieLens 1M	(user, movie, genre, rating)	$6000 \times 3400 \times 18 \times 5$	0.0994

(user movie, genre, rating), (c) a book rating data set [27] with a 3-mode tensor (user, book, rating), and (d) Epinions data set [23] with a 4-mode tensor (user, product, category, rating). From each data tensor, we created pairs of sub-tensors (chosen randomly) with different degrees of intersection (10%, 20%, 40%, 60%). The target rank that we consider for the CP decomposition is 10. The default selection measure is the density-based selection measure, UC_2.

For evaluation, we consider both *execution time* and *degree of fit* defined as

$$\text{fit}(\mathcal{X}, \mathcal{P} \,\hat{\cup}\, \mathcal{Q}) = 1 - \frac{\|\mathcal{X} - (\mathcal{P} \,\hat{\cup}\, \mathcal{Q})\|}{\|\mathcal{X}\|}, \tag{13}$$

where $\mathcal{X}$ is the union of $\mathcal{P}$ and $\mathcal{Q}$ and $\mathcal{P} \,\hat{\cup}\, \mathcal{Q}$ is the tensor obtained by re-composing the decomposition of $\mathcal{P} \cup \mathcal{Q}$ in the considered scheme. Comparing the fit with respect to $\mathcal{X}$ enables us not only to measure how well $\mathcal{P} \cup \mathcal{Q}$ approximates the entries in $\mathcal{P} \cup \mathcal{Q}$, but also whether $\mathcal{P} \,\hat{\cup}\, \mathcal{Q}$ includes any spurious entries that are not originally in $\mathcal{P} \cup \mathcal{Q}$.

We ran all the experiments on a machine with Intel Core i5-2400 CPU @ 3.10GHz ×4 with 7.7 GB RAM. We used MATLAB Version 7.13.0.564 (R2011b) 64-bit for the general implementation and MATLAB Parallel Computing Toolbox for the parallel implementations. We used the MATLAB Tensor Toolbox [3] to represent relational tensors as sparse tensors.

5.2   Results #1: UBD vs. UTD (with and without Materialization)

We first compare the proposed UBD against the more conventional UTD scheme. As a second competitor, we also consider the naive grid-based UBD discussed in Section 3.

Firstly, as we see in Figure 9(a), when there are opportunities for reusing existing materialized decompositions of the input tensors, as expected, UBD is much faster than the UTD as well as the naive grid-based UBD.

Secondly, in Figure 9(c), we consider the case where there are no opportunities for reusing existing decompositions. As we see in this figure, as expected, when the input tensors have to be decomposed as part of the UBD process, whether UBD outperforms UTD depends on the characteristics of the input tensors: in particular, as expected, UBD is faster than UTD when (a) *the degree of intersection is low* ($\leq 20\%$) and (b) *the input tensors are not extremely sparse*: if these conditions are not satisfied, the size of the union result is close to the sizes of input tensors and, if the result is also sparse, there is no gain in pushing down the decompositions.

Note that, when materialized decompositions of the input tensors do not exist, grid-based UBD can out-pace the proposed UBD and UTD in many configurations. However, as we see in Figure 9(b), this comes at the cost of a significant drop in accuracy: the proposed UBD scheme achieves fits close to the fit of UTD, whereas the accuracy of the grid-based UBD is much lower. Note also that the accuracy of UBD is especially good in data sets that are not *extremely sparse*.

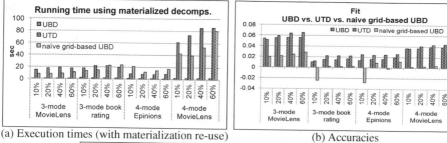

(a) Execution times (with materialization re-use)   (b) Accuracies

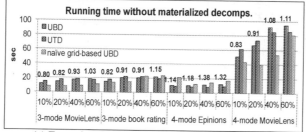

(c) Execution times (without materialization re-use)
Running time ratio (UBD/UTD) is shown on each bar
(the smaller the ratio, the better is the performance of UBD)

Fig. 9. UBD vs. UTD vs. naive grid-based UBD on pairs of tensors with different intersection sizes (10%, 20%, 40%, 60%)

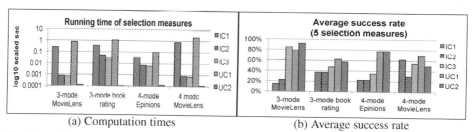

(a) Computation times   (b) Average success rate

Fig. 10. Efficiency and accuracy of the different selection measures in average of different intersection sizes (10%, 20%, 40%, and 60%)

5.3 Results #2: Evaluation of the Alternative Selection Measures

In Section 3.2, we considered various approaches (IC_1, IC_2, IC_3, UC_1, and UC_2) for choosing the sub-factors for the overlapping parts of the input tensors. Figure 10(a) shows that fit-based measures (intersection fit, IC_1 and union fit, UC_1) are more expensive than norm-based measures (IC_2, IC_3). The *density*-based approach (UC_2) has an almost 0 execution cost. Note that, when we compare the computation times of these selection measures to the execution times of the UBD operators (Figure 9), we see that even the most expensive selection strategy is, in practice, affordable. Therefore, the major criterion for selecting among these measures should be accuracy.

For measuring the accuracy of different selection measures, we considered the percentage of the cases where each selection measure returned the best alternative. As

Table 2. Average fit of the different selection measures (The highest fits for each data set are highlighted in bold)

Data set	IC_1	IC_2	IC_3	UC_1	UC_2
3-mode MovieLens 1M	0.0538	0.0539	0.0551	0.0551	**0.0553**
3-mode book rating	0.0127	0.0127	0.0134	**0.0141**	0.0138
4-mode Epinions	0.0133	0.0133	0.0144	**0.0164**	**0.0164**
4-mode MovieLens 1M	**0.0380**	0.0376	0.0378	0.0377	**0.0380**

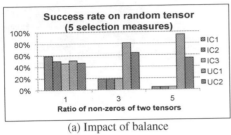

(a) Impact of balance

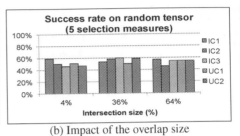

(b) Impact of the overlap size

Fig. 11. Success rate in predicting the best fit of UBD using the 5 selection measures compared among different (a) ratios of non-zeros of two tensors and (b) intersection sizes

shown in Figure 10(b), the union-based fit (UC_1) measure works best overall. The density measure (UC_2) also works well. The figure also shows that the intersection-based measures (IC_1, IC_2, IC_3) are not good indicators, even behave negatively in some cases: among them the IC_3 works the best since it also accounts for the non-overlapping regions through the cluster strength indicated by the core. Table 2 further studies the average degree of fits returned by the different strategies. The table confirms that the average fits obtained by the union-based selection measures are overall better than the intersection-based selection measures. While the numbers vary, the degrees of fit based on the union-based selection measures are up to 20% better than IC_1 and IC_2.

To further study the impacts of various parameters on the selection accuracy, we also created random tensors with different configurations, varying the balance (ratio of densities) of the input tensors and intersection sizes. For each experiment, we created 10 different random tensors of size $5000 \times 5000 \times 10$ and measured the percentage cases in which each measure selected the better fitting tensor. As the default configuration, we set the ratio of non-zeros to 1 (most balanced), intersection size to 4%, and the density of the union tensor to 0.01%.

In Figure 11(a), we first study the impact of balance. Here, the configuration with $ratio = 1$ corresponds to the most balanced configuration. As we expected, when the tensors are balanced, all measures work similarly (with a slight edge to the intersection-based measures); however, as the imbalance among tensors increases, intersection-based measures get worse, while the union based measures, especially UC_1, improve.

Unlike balance, the size of the intersection has no significant impact on the selection accuracy (Figure 11(b)), indicating that all measures are robust in this respect.

5.4 Results #3: Impact of Composition of UBD with Other Operators

As we discussed in Section 4, the proposed union-by-decomposition (UBD) operator is compatible with other operators that are part of TensorDB and can be used as part

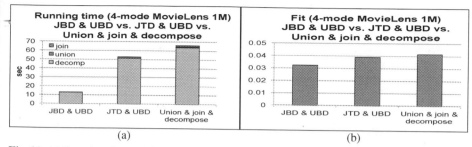

Fig. 12. (a) Running times and (b) fits of three alternative query plans "JBD and UBD" vs. "JTD and UBD" vs. "union, join, and decompose" (see Figure 8) on 4-mode MovieLens 1M

of a general optimization framework. In Figure 12 for a sample data, we study the alternative query plans considered in Figure 8. As expected, the figure shows that pushing decompositions down the join and union operations (i.e., using UBD, proposed in this paper, and/or JBD, proposed in [11]) provides a much faster execution times than the union operation and join operation followed by a final CP decomposition step. As shown in Figure 12(a), among these three alternative query plans, the query plan using JBD and UBD is the fastest (faster than $5\times$ of the union, join, and decompose strategy) but comes with $\sim 20\%$ drop in accuracy (Figure 12(b)). On the other hand, using UBD proposed in this paper along with the conventional join-then-decompose (JTD) strategy instead of JBD reduces the execution time relative to "union, join, and decompose" by $\sim 20\%$ (Figure 12(a)), with a negligible impact on accuracy (Figure 12(b)).

6   Conclusion

TensorDB, which extends array databases with a tensor-relational model (TRM), supports both relational algebraic operations (for data manipulation and integration) and tensor algebraic operations (for data analysis) for the complete life cycle of data that involves consecutive steps of integration, manipulation, and analysis. In TensorDB, we focused on data processing workflows involving both tensor decomposition and data integration (union) operations and proposed a novel scheme for pushing down the tensor decompositions over the union operations to reduce the overall data processing times and to promote reuse of materialized tensor decomposition results. Experimental results confirmed the efficiency and effectiveness of the proposed decomposition push-down strategy and the corresponding union-by-decomposition (UBD) operator.

References

1. Hash tables and associative arrays. In: Algorithms and Data Structures, pp. 81–98. Springer, Heidelberg (2008)
2. Movielens dataset from grouplens research group, http://www.grouplens.org
3. Bader, B.W., Kolda, T.G.: Matlab tensor toolbox version 2.2 (January 2007), http://csmr.ca.sandia.gov/~tgkolda/TensorToolbox/
4. Baumann, P., et al.: The multidimensional database system rasdaman. In: SIGMOD (1998)

5. Brown, P.G.: Overview of scidb: large scale array storage, processing and analysis. In: SIGMOD, pp. 963–968 (2010)
6. Carroll, J., Chang, J.-J.: Analysis of individual differences in multidimensional scaling via an n-way generalization of "eckart-young" decomposition. Psychometrika (1970)
7. Cohen, J., et al.: Mad skills: new analysis practices for big data. In: VLDB (2009)
8. Dobos, L., et al.: Array requirements for scientific applications and an implementation for microsoft sql server. In: AD, pp. 13–19 (2011)
9. Harshman, R.A.: Foundations of the PARAFAC procedure: Models and conditions for an "explanatory" multi-modal factor analysis. In: UCLA Working Papers in Phonetics (1970)
10. Kang, U., et al.: Gigatensor: scaling tensor analysis up by 100 times - algorithms and discoveries. In: KDD, pp. 316–324 (2012)
11. Kim, M., Candan, K.S.: Approximate tensor decomposition within a tensor-relational algebraic framework. In: CIKM, pp. 1737–1742 (2011)
12. Kim, M., Candan, K.S.: Decomposition-by-normalization (dbn): Leveraging approximate functional dependencies for efficient tensor decomposition. In: CIKM, pp. 355–364 (2012)
13. Kolda, T., Bader, B.: Tensor decompositions and applications. In SIAM Review 51(3), 455–500 (2009)
14. Kolda, T., Sun, J.: Scalable tensor decompositions for multi-aspect data mining. In: ICDM, pp. 363–372 (December 2008)
15. Kolda, T.G., et al.: Higher-order web link analysis using multilinear algebra. In: ICDM (2005)
16. Low, Y., et al.: Distributed graphlab: a framework for machine learning and data mining in the cloud. VLDB 5(8), 716–727 (2012)
17. Papalexakis, E.E., Faloutsos, C., Sidiropoulos, N.D.: Parcube: Sparse parallelizable tensor decompositions. In: Flach, P.A., De Bie, T., Cristianini, N. (eds.) ECML PKDD 2012, Part I. LNCS, vol. 7523, pp. 521–536. Springer, Heidelberg (2012)
18. Phan, A.H., Cichocki, A.: Parafac algorithms for large-scale problems. Neurocomputing 74(11), 1970–1984 (2011)
19. Sanchez, E., Kowalski, B.R.: Generalized rank annihilation factor analysis. Analytical Chemistry 58(2), 496–499 (1986)
20. Sanchez, E., Kowalski, B.R.: Tensorial resolution: A direct trilinear decomposition. Journal of Chemometrics 4(1), 29–45 (1990)
21. Sorber, L., et al.: Optimization-based algorithms for tensor decompositions: canonical polyadic decomposition, decomposition in rank-(L_r,L_r,1) terms, and a new generalization. SIAM Journal on Optimization 23(2), 695–720 (2013)
22. Stonebraker, M., et al.: C-store: a column-oriented dbms. In: VLDB, pp. 553–564 (2005)
23. Tang, J., et al.: eTrust: Understanding trust evolution in an online world. In: KDD (2012)
24. Tsourakakis, C.E.: Mach: Fast randomized tensor decompositions. In: SDM, pp. 689–700 (2010)
25. Tucker, L.: Some mathematical notes on three-mode factor analysis. Psychometrika 31(3), 279–311 (1966)
26. van Ballegooij, A.R., Cornacchia, R., de Vries, A.P., Kersten, M.L.: Distribution rules for array database queries. In: Andersen, K.V., Debenham, J., Wagner, R. (eds.) DEXA 2005. LNCS, vol. 3588, pp. 55–64. Springer, Heidelberg (2005)
27. Ziegler, C.-N., et al.: Improving recommendation lists through topic diversification. In: WWW (2005)

Author Index